Water Treatment Plant Operation

Volume 1

Seventh Edition

Pearson Education, Inc., 330 Hudson Street, New York, New York 10013
A Pearson Education Company
www.pearsoned.com

Printed in the United States of America

21 2022

000200010271997877

EEB/AD

PEARSON

ISBN 10: 1-323-41617-X
ISBN 13: 978-1-323-41617-4

About the Office of Water Programs

The Office of Water Programs (OWP) is a nonprofit, self-supporting unit of University Enterprises, Inc., an auxiliary of California State University, Sacramento. We are here to deliver cost-effective solutions for protecting and enhancing water resources, public health, and the environment through training, scientific research, and public eduction.

The distance learning training programs we offer for individuals interested in the operation and maintenance of drinking water and wastewater facilities are developed in collaboration with experienced industry professionals who have extensive knowledge of all types of facilities. The university, fully accredited by the Western Association of Schools and Colleges, administers and monitors our training programs, under the direction of Dr. Ramzi J. Mahmood.

As the leading provider of drinking water and wastewater operator training for over 40 years, OWP offers nearly 50 courses via print, online, and DVD that provide critical training and development resources. Operators, managers, and administrators rely on OWP training materials and courses to help them prepare for certification and to earn academic credit, continuing education units, and contact hours in their chosen career fields.

Our training materials and programs meet the needs of operators of water treatment plants, water distribution systems, wastewater collection systems, and wastewater treatment and reclamation facilities. We also offer programs and materials for pretreatment facility inspectors, environmental compliance inspectors, and utility managers.

Please visit us online at www.owp.csus.edu to see our full catalog of training programs and materials. Other ways to contact us include:

Phone: (916) 278-6142
Fax: (916) 278-5959

Office of Water Programs
California State University, Sacramento
6000 J Street, Modoc Hall, Suite 1001
Sacramento, CA 95819-6025

About This Training Manual

Objective

The basic objective in the operation of a water treatment plant is to produce a drinking water that is safe and aesthetically pleasing at a reasonable cost. This manual is designed to provide water treatment plant operators with the knowledge and skills needed to effectively operate their systems, by covering these topics:

1. Introduction to Water Treatment
2. Source Water, Reservoir Management, and Intake Structures
3. Coagulation and Flocculation
4. Sedimentation
5. Filtration
6. Disinfection
7. Corrosion Control
8. Taste and Odor Control
9. Laboratory Procedures

Scope

Material in this manual covers situations encountered by water treatment plant operators in most locations. The goal is to provide you with an understanding of basic operational concepts for water treatment plants and with the ability to recognize, analyze, and solve problems when they occur. Operational programs for water treatment plants vary with the age of the system, the extent and effectiveness of existing programs, and local conditions. You can adapt the information and procedures presented in our training materials to fit your needs. The table of contents seen later in this section provides a detailed list of topics discussed.

Technology is advancing rapidly in the field of water treatment plant operation. To keep pace with scientific and industry advances in addition to regulatory requirements, the material in this manual is periodically revised and updated. You, as the water treatment professional, also need to be aware of industry advances and recognize the need for continuing your professional training and development beyond what is offered in our training programs. We suggest that you seek out opportunities in your daily work life to increase your knowledge and skills through contact with colleagues and supervisors and from attending professional meetings, workshops, conferences, and classes.

Uses

Our training manuals are a sought-after source for operators preparing for state certification. This training manual was developed to serve the needs of operators in a variety of ways, including: as a self-paced, independent-study distance learning course, as a textbook in traditional college and university classes (often taught by experienced operators), and as the basis of a course offered by their employer. Some cities or utility agencies use this manual as part of on-the-job training programs for their operators. In this scenario, a manual is purchased for each operator and a senior operator or a group of operators are designated as instructors who help answer questions and provide assistance to operators-in-training.

We encourage you to use this training manual in the manner that best fits your training needs.

Course Enrollment

Operators have the option to enroll in distance learning courses directly with OWP, which can be used to earn CEUs or contact hours. In order to certify that a person has successfully completed a training course, OWP administers course exams and issues grade reports and certificates to enrolled students. For more information about our course enrollments, visit www.owp.csus.edu.

Acknowledgments

The Office of Water Programs is indebted to the many industry professionals who contributed to this manual—which represents the collaboration and commitment of many. We extend our thanks and acknowledgment to the following individuals (in alphabetical order) for their assistance with this edition: Camille Campeon, Fabien Espinasse, John Goodrich, John Johnston, Mary Krizanosky, Rich Parkhurst, Thomas Perry, Kirk Van Rooyan.

We also extend our thanks to the following individuals who contributed to previous editions (in alphabetical order): Leonard Ainsworth, E.E. "Skeet" Arasmith, Ted Bakker, Jo Boyd, John Brady, Dean Chausee, Walter Cockrell, Gerald Davidson, Terry Engelhardt, Fred Fahlen, David Fitch, Richard Haberman, Dempsey Hall, Larry Hannah, Lee Harry, Jerry Hayes, Ed Henley, Jerry Higgins, Andrew Holtan, Deborah Horton, Charles Jeffs, George Kinias, Kirk Laflin, Chet Latif, Frank Lewis, Perry Libby, D. Mackay, William Maguire, Nancy McTigue, Rich Metcalf, Joe Monscvitz, Angela Moore, Harold Mowry, Theron Palmer, Eugene Parham, Catherine Perman, William Redman, David Rexing, Jack Rossum, William Ruff, Gerald Samuel, Carl Schwing, Jim Sequeira, David Sorenson, Russell Sutphen, R. Rhodes Trussell, Kenneth Walimaa, Robert Wentzel, James Wright, Mike Yee, Clarence Young, Mike Young, Anthony Zigment.

Related Training Materials

Meet your professional development needs with these related training materials:

- Water Treatment Plant Operation, volume two in the series—offered in print with a course enrollment option
- Water Distribution System Operation and Maintenance—offered in print, CD, and five online courses with a course enrollment option
- Small Water System Operation and Maintenance—offered in print, CD, and five online courses with a course enrollment option
- Small Water Systems Video Information Series—offered in DVD with a supplemental learning booklet and a course enrollment option
- Water Systems Operation and Maintenance Video Training Series—offered in DVD with a supplemental learning booklet and a course enrollment option
- Utility Management—offered in print with a course enrollment option
- Manage for Success—offered in print with a course enrollment option

Instructions to Distance Learning Course Participants

To progress steadily through this training course, we suggest you establish a regular study schedule. Some of the chapters are longer and more complex than others, which may require more time and effort depending on your industry knowledge and experience.

Each chapter is arranged for you to read a section and answer questions in the Check Your Understanding section. Based on how well you answer the questions, you can decide if you understand the material well enough to continue or whether you need to review the section before starting the next one. The important thing is that you understand the material presented in each chapter before starting the next one. You may find this procedure slower than how you might read a typical textbook, but you will probably remember much more after you finished this course.

A chapter review is provided at the end of each chapter to help you review the important points covered in the material. Writing your answers to these questions and reviewing the related chapter content for questions you cannot answer can help you better retain the information you covered in each chapter.

Appendix A, "Introduction to Basic Math for Operators," contains a discussion of basic math concepts for you to review or and practice as needed before getting into more complex concepts and calculations that are directly related to water treatment plant operation.

As a distance learning course participant, you get to set the pace and determine the method of learning that best meets your educational goals, which may include:

- Starting a career in the field of water treatment
- Improving your job performance
- Preparing for a job promotion
- Earning CEUs or contact hours
- Preparing for a certification exam

Making the time to review the material, taking notes or highlighting the text, and answering review questions can increase your understanding of the material presented and assist you in achieving your career goals and improving your job performance as a water treatment industry professional.

Preface

Thank you for purchasing this title in our water treatment operator training series. This two-volume series focuses on the knowledge and skills needed by operators of water treatment systems. This first volume covers topics of importance to operators of all types of water treatment systems, including an introduction to plant management, water sources and treatment, reservoir management and intake structures, coagulation and flocculation, sedimentation, filtration, disinfection, corrosion control, taste and odor control, and laboratory procedures. The second volume focuses on controlling of iron and manganese, water softening, arsenic removal, and controlling trihalomethanes.

As you may know, our training program started over four decades ago and was established by the late Dr. Ken Kerri. Our training materials represent the culmination of knowledge gathered from many operators and experts in this field. The primary purpose of our training materials continues to be reaching operators and giving them the opportunity to do their jobs better. With our single-minded focus on this objective, we embarked on updating our training manuals with a new design and layout to enhance the operator learning experience—as you can see in this new edition. Features of the new design include:

- Two-third/one-third column layout, allowing for better readability. The one-third column is flexible in that we use the space to expand on topics with key term definitions, figures, tables, and photos. Operators can also use the column to make notes.
- Key terms are shown in close proximity to where they are mentioned in text. This new feature provides context for operators to better understand terms and leads to more effective learning.
- Printing in color enhances the clarity and detail of graphics and photos, improves navigation of the materials, and uses a color scheme to connect similar elements.
- A new math appendix, "Introduction to Basic Math for Operators," is presented to make this sometimes challenging subject more accessible and relevant to operators. We use a more standardized approach to present units, apply conversion, and show equations. The appendix is divided into four sections: basic math, intermediate math, advanced math, and applications in water treatment. The first three sections move operators gradually through mathematical concepts and calculations from simple arithmetic to the more advanced. Our goal was to improve your ability to grasp math concepts and confidently apply them in daily plant operations.

The Office of Water Programs extends thanks to the reviewers and contributors listed in the "Acknowledgments" section on page viii for generously sharing their expertise for the benefit of the industry professionals served by our training programs.

I wish you success in your chosen program of study.

Dr. Ramzi J. Mahmood, Director
Office of Water Programs

Table of Contents

Chapter 3
Coagulation and Flocculation

Chapter 4
Sedimentation

Chapter 5
Filtration

Chapter 6
Disinfection

Chapter 7
Corrosion Control

Chapter 8
Taste and Odor Control

CHAPTER 1

Introduction to Water Treatment

CHAPTER OUTLINE

LEARNING OBJECTIVES

1. Discuss the need for, uses of, and regulations governing the production of safe, pleasant drinking water.
2. Explain the flow pattern through conventional surface water treatment plants.
3. Use daily operating procedures to safely operate and maintain a water treatment plant by regulating flows, applying chemicals and adjusting doses, maintaining equipment and facilities, and responding to emergency conditions.
4. Communicate operational information using verbal and written reports and records.
5. Implement energy conservation measures.

KEY TERMS

acid rain	coagulation	jar test	sludge
alkalinity	coliforms	pathogenic organisms	toxic
BOD (biochemical oxygen demand)	composite sample	pH	turbidity
chlorine residual	corrosion	power factor	weir
clear well	floc	raw water	
coagulant	flocculation	safety data sheet (SDS)	
	grab sample	sedimentation	

1.1   Water as a Limited Resource

For decades, Americans have used water as though their supply would never run out. In recent years, drought conditions have forcibly brought the need to conserve and properly budget our water resources to the minds of water supply managers. Even in the driest years, though, rain across the country enormously exceeds water use. The trouble is that the nation's water resources are unevenly distributed. The Pacific Northwest has a big surplus. The agricultural states of the Southwest fight for the last salty drop of river water from the lower Colorado River. The federal government has spent billions of dollars building and operating facilities to divert water for use in arid areas and those without ready access to water. Contamination is a problem, too. Mineral residues from irrigation have damaged once fertile soil. **Acid rain** is killing the fish in mountain lakes. America's drinking water has been tainted with substances as unfamiliar as trichloroethylene (TCE) and as commonplace as highway salt. Vast underground basins of water, deposited over many years, have been seriously depleted in a matter of decades.

All water comes as rain or precipitation from the sky, but 92 percent of the water either evaporates immediately or runs off eventually into the oceans. One-quarter of the water that irrigates, powers, and bathes America is taken from an ancient network of underground aquifers. According to US Geological Survey (USGS) data*, in 1950 the United States took some 12 trillion gallons (45 billion cubic meters) of water out of the ground; by 1980 the figure had more than doubled to 30.7 trillion gallons. In 2010 (the last year for which data is available), the rate had dropped to 28.8 trillion gallons. This decrease can be attributed to technological innovations that allow more efficient use of water and to conservation efforts in response to water shortages caused by droughts and other climate factors.

Water shortages directly influence energy consumption. As groundwater levels fall, more energy is required to pump water from deeper levels in the basin. In several areas, vast water projects use large amounts of electricity to pump water many miles along the project. As energy becomes more expensive, the users of the water will see the increased cost reflected in their water rates. This link between water demands and pumping costs drives water conservation programs, even when water is abundant.

Water is regarded as commonplace because it is the most plentiful liquid on earth and because of our familiarity with it. Water is present in almost all natural objects and in almost every part of the earth that people can reach. There is water vapor in the air, and liquid water in rocks and soil. In addition to the water that wets them, clay and certain kinds of rocks contain water in chemical combination with other substances. Both plant life and animal life depend upon water for survival. A plant receives the greater part of its food from the soil in water solutions and manufactures the rest of its food in the presence of water. Like other animals, all of the tissues of human bodies are bathed in water. Our foods must be suspended or dissolved in water solutions to be carried to the different parts of the body. Also, most waste products are eliminated from the body as water-soluble substances.

(i) *"Estimated Use of Water in the United States in 2010," US Geological Survey Circular 1405. pubs.usgs.gov/circ/1405

acid rain
Precipitation that has been rendered (made) acidic by airborne pollutants.

Water may be commonplace, but useful water is not always readily available. Even before the discovery of America, one of the common causes of war between Native American tribes was water rights. Among the first considerations of any new land development is water. Useful water is only rarely free, and it is not very abundant in many parts of the United States. There are not many places left where a person can feel safe in drinking water from a spring, stream, or pond. Even the groundwater produced by wells must be tested regularly. In some areas, human activities have made it difficult to locate a safe water supply of any sort. In certain coastal areas, for example, overpumping from the ground has depleted the groundwater basins. As a result, the intrusion of seawater is ruining the basin for most useful purposes. Other sources of groundwater contamination include seepage from septic tank leaching systems and agricultural drainage systems, the improper disposal of hazardous wastes in sanitary landfills and dumps, and the entry of surface runoff into poorly constructed wells. Some human activities clearly pose a serious threat to life on this planet.

Check Your Understanding

1. Why has it become necessary to conserve and properly budget our water resources?

2. How are water shortages and energy consumption linked together?

3. Why do many sources of water need treatment?

4. Name three ways groundwater may become contaminated.

1.2   Water Treatment

The basic objective in the operation of water treatment plants is to produce a drinking water that is safe and aesthetically pleasing at a reasonable cost with respect to capital as well as operation and maintenance.

From a public health perspective, production of a safe drinking water, one that is free of harmful bacteria and toxic materials, is the first priority. It is also important to produce a high-quality water that appeals to the consumer. Generally, this means that the water must be clear (free of turbidity), colorless, and free of objectionable tastes and odors. Consumers also show a preference for water supplies that do not stain plumbing fixtures and clothes, do not corrode plumbing fixtures and piping, and do not leave scale deposits or spot glassware.

Consumer sensitivity to the environment (air quality, water quality, noise) has significantly increased in recent years. With regard to water quality, consumer demands have never been greater. In some instances, consumers have substituted bottled water to meet specific needs, namely, for drinking water and cooking purposes.

Design engineers select water treatment processes on the basis of the type of water source, source water quality, and desired finished water quality established by drinking water regulations and consumer desires. Table 1.1 is a summary of typical water treatment processes as they relate to the source and quality of the **raw water**.

raw water
(1) Water in its natural state, before any treatment. (2) Water entering the first treatment process of a water treatment plant.

Table 1.1   Water sources, quality problems, and treatment

Source	Water Quality Problems	Treatment
Groundwater Only	Nitrate	a. Anion Exchange b. Reverse Osmosis
	Sand	Sand Separators
	Sulfide Odors	a. Aeration b. Oxidation (Chlorination) c. Desulfuration (Sulfur Dioxide)
	Corrosivity	Carbon Dioxide Stripping by Aeration
Both Groundwater and Surface Water	Coliforms or Microbial Contamination	Disinfection (Chlorination)
	Excessive Hardness	a. Ion Exchange Softening b. Lime (and Soda) Softening
	Iron or Manganese	a. Sequestration (Polyphosphates) b. Removal by Special Ion Exchange c. Permanganate and Greensand d. Oxidation by Aeration Followed by Filtration e. Oxidation with Chlorine Followed by Filtration f. Oxidation with Permanganate Followed by Filtration
	Dissolved Minerals (High Total Dissolved Solids)	a. Ion Exchange b. Reverse Osmosis
	Corrosivity (Low pH)	a. pH Adjustment with Chemicals b. Corrosion Inhibitor Addition (Zinc Phosphate, Silicate)
	Fluoridation (Preventive)	Add Fluoride Chemicals
Surface Water Only	Coliforms or Microbial Contamination	a. Disinfection (Other Oxidants–Ozone, Chlorine Dioxide, Chloramination) b. Coagulation, Flocculation, Sedimentation, Filtration, and Disinfection
	Turbidity, Color	Coagulation, Flocculation, Sedimentation, and Filtration
	Odors (Organic Materials)	a. Clarification (Coagulation, Flocculation, Sedimentation, and Filtration) b. Oxidation (Chlorination or Permanganate) c. Special Oxidation (Chlorine Dioxide) d. Adsorption (Granular Activated Carbon)
	Trihalomethanes (THMs)	a. Disinfection with Ozone, Chlorine Dioxide, or Chloramination (Do Not Prechlorinate) b. Remove THM precursors c. Remove THMs After Formed

Water has many important uses and each requires a specific level of water quality. The major concern of the operators of water treatment plants and water distribution systems is to produce and deliver to consumers water that is safe and pleasant to drink. The water should be acceptable to domestic and commercial water users and many industries. Some industries, such as food and drug processors and the electronics industry, require higher quality water. Many industries will locate where the local water supply meets their specific needs, while other industries have their own water treatment facilities to produce water suitable for their needs.

Check Your Understanding

1. What is the first priority for operating a water treatment plant?

2. What type of water is appealing to consumers?

3. Which industries require extremely high-quality water?

1.3 The Safe Drinking Water Act

Most of the regulations concerning public water systems are focused on the production of a water that is safe for people to drink. These regulations continue to change over time to address new information about the health effects of different substances that may be in drinking water. Operators are urged to develop close working relationships with their local regulatory agencies to keep themselves informed of the frequent changes in regulations and requirements. An excellent source of up-to-the-minute information about drinking water regulations is the US Environmental Protection Agency's (EPA's) toll-free Safe Drinking Water Hotline at (800) 426-4791. Also see the chapter on drinking water regulations in *Water Treatment Plant Operation*, Volume II, for more information about primary and secondary contaminants; and monitoring, sampling, and reporting requirements.

In 1974, the Safe Drinking Water Act (SDWA) was signed into law, giving the federal government, through the EPA, the authority to:

- Set national standards regulating the levels of contaminants in drinking water
- Require public water systems to monitor and report their levels of identified contaminants
- Establish uniform guidelines specifying the acceptable treatment technologies for cleansing drinking water of unsafe levels of pollutants

The EPA sets two kinds of drinking water standards. Primary regulations (Table 1.2) establish maximum contaminant levels (MCLs) based on the health significance of the contaminants

Table 1.2 Primary drinking water regulations

Constituent	Maximum Contaminant Level
Inorganic Chemicals	
Antimony	0.006 mg/L
Arsenic	0.05 mg/L
Asbestos	7 million fibers/L
Barium	2 mg/L
Beryllium	0.004 mg/L
Cadmium	0.005 mg/L
Chromium	0.1 mg/L
Copper	1.3 mg/L[a] (at tap)
Cyanide	0.2 mg/L
Fluoride	4 mg/L
Lead	0.015 mg/L[b] (at tap)
Mercury	0.002 mg/L
Nitrate (as N)	10 mg/L
Nitrite (as N)[b]	1 mg/L
Selenium	0.05 mg/L
Thallium	0.002 mg/L
Organic Chemicals	
Volatile Organics	
Benzene	0.005 mg/L
Carbon Tetrachloride	0.005 mg/L
o-Dichlorobenzene	0.6 mg/L
p-Dichlorobenzene	0.075 mg/L
1,2-Dichloroethane	0.005 mg/L

Table 1.2 Primary drinking water regulations *(continued)*

Constituent	Maximum Contaminant Level
Organic Chemicals *(continued)*	
1,1-Dichloroethylene	0.007 mg/L
cis-1,2-Dichloroethylene	0.07 mg/L
trans-1,2-Dichloroethylene	0.1 mg/L
Dichloromethane	0.005 mg/L
1,2-Dichloropropane	0.005 mg/L
Ethylbenzene	0.7 mg/L
Monochlorobenzene	0.1 mg/L
Styrene	0.1 mg/L
Tetrachloroethylene	0.005 mg/L
Toluene	1 mg/L
1,2,4-Trichlorobenzene	0.07 mg/L
1,1,1-Trichloroethane	0.2 mg/L
1,1,2-Trichloroethane	0.005 mg/L
Trichloroethylene (TCE)	0.005 mg/L
Vinyl Chloride	0.002 mg/L
Xylenes (total)	10 mg/L
Pesticides and Synthetic Organics	
Acrylamide	treatment technique
Alachlor	0.002 mg/L
Atrazine	0.003 mg/L
Benzo(a)pyrene	0.0002 mg/L
Carbofuran	0.04 mg/L
Chlordane	0.002 mg/L
Dalapon	0.2 mg/L
Dibromochloropropane (DBCP)	0.0002 mg/L
Di(2-ethylhexyl)adipate	0.4 mg/L
Di(2-ethylhexyl)phthalate	0.006 mg/L
Dinoseb	0.007 mg/L
Diquat	0.02 mg/L
Endothall	0.1 mg/L
Endrin	0.002 mg/L
Epichlorhydrin	treatment technique
Ethylene Dibromide (EDB)	0.00005 mg/L
Glyphosate	0.7 mg/L
Heptachlor	0.0004 mg/L
Heptachlor Epoxide	0.0002 mg/L
Hexachlorobenzene	0.001 mg/L
Hexachlorocyclopentadiene	0.05 mg/L
Lindane	0.0002 mg/L
Methoxychlor	0.04 mg/L
Oxamyl (Vydate)	0.2 mg/L
PCBs	0.0005 mg/L
Pentachlorophenol	0.001 mg/L

while the secondary standards (Table 1.3) are established based on aesthetic considerations and are a state option.

While the SDWA gave EPA responsibility for developing drinking water regulations, it gave state regulatory agencies the opportunity to assume primary responsibility for enforcing those regulations.

Implementation of the SDWA has greatly improved basic drinking water purity across the nation. However, EPA surveys of surface water and groundwater indicate the presence of synthetic organic chemicals in the nation's water sources, with a small percentage at levels of concern. In addition, research studies suggest that some naturally occurring contaminants may pose even greater risks to human health than the synthetic contaminants. Further, there is growing concern about microbial and radon contamination.

In the years following passage of the SDWA, Congress felt that EPA was slow to regulate contaminants and states were lax in enforcing the law. Consequently, in 1986 and again in 1996 Congress enacted amendments designed to strengthen the 1974 SDWA. These amendments set deadlines for the establishment of maximum contaminant levels, placed greater emphasis on enforcement, authorized penalties for tampering with drinking water supplies, mandated the complete elimination of lead from drinking water, and placed considerable emphasis on the protection of underground drinking water sources.

All public water systems must comply with the SDWA regulations. A public water system is any publicly or privately owned water supply system that:

1. Has at least 15 service connections, or
2. Regularly serves an average of at least 25 individuals daily at least 60 days out of the year.

Drinking water regulations also take into account the type of population served by the system and classify water systems as community or noncommunity systems. Therefore, to understand what requirements apply to any specific system, it is first necessary to determine whether the system is considered a community system or a noncommunity system. A community water system is defined as one that:

1. Has at least 15 service connections used by all-year residents, or
2. Regularly serves at least 25 all-year residents.

Any public water system that is not a community water system is classified as a noncommunity water system. Restaurants, campgrounds, and hotels could be considered noncommunity systems for purposes of drinking water regulations.

In addition to distinguishing between community and noncommunity water systems, EPA identifies some small systems as nontransient noncommunity systems if they regularly serve at least 25 of the same persons over 6 months per year. This classification applies to water systems for facilities such as schools or factories where the consumers served are nearly the same every day but do not actually live at the facility. In general, nontransient noncommunity systems must meet the same requirements as community systems.

A transient noncommunity water system is a system that does not regularly serve drinking water to at least 25 of the same persons over 6 months per year. This classification is used by EPA only in regulating nitrate levels and total coliform. Examples of a transient noncommunity system might be campgrounds or service stations if those facilities do not meet the definition of a community, noncommunity, or nontransient noncommunity system.

In 1996, the president signed new Safe Drinking Water Act (SDWA) amendments into law as Public Law 104-182. These amendments changed the existing SDWA, created new programs, and included more than $12 billion in federal funds for drinking water programs and activities from fiscal year (FY) 1997 through FY 2003.

Topics covered in the amendments include arsenic research, assistance for water infrastructure and watersheds, assistance to colonias (low-income communities located along the US–Mexico border), backwash water recycling, bottled water, capacity development (technical, financial, and managerial), consumer awareness, contaminant selection and standardsetting authority, definitions (public water system, community water system and noncommunity water system), disinfectants and disinfection byproducts, drinking water studies and research, effective date of regulations, enforcement, environmental finance centers and capacity clearinghouse, estrogenic substances screening program, conditions that could qualify a water system for an exemption, groundwater disinfection, groundwater protection programs, lead plumbing and pipes, monitoring and information gathering, monitoring relief, monitoring for unregulated

Table 1.2 Primary drinking water regulations (continued)

Constituent	Maximum Contaminant Level
Pesticides and Synthetic Organics (continued)	
Picloram	0.5 mg/L
Simazine	0.004 mg/L
Toxaphene	0.003 mg/L
2,4-D	0.07 mg/L
2,3,7,8-TCDD (Dioxin)	0.00000003 mg/L
2,4,5-TP (Silvex)	0.05 mg/L
Microbial	
Total Coliform	1 per 100 mL
	<40 samples/mo—no more than 1 positive
	>40 samples/mo—no more than 5% positive
Giardia lamblia	3-log (99.9%) removal[c]
Legionella	treatment technique[c]
Enteric viruses	4-log (99.99%) removal[c]
Heterotrophic bacteria	treatment technique[c]
Physical	
Turbidity[c]	0.5 to 5 NTU
Radionuclides	
Gross alpha particles	15 pCi/L
Gross beta particles[d]	4 mrem/yr
Radium 226 & 228	5 pCi/L
Uranium	30 µg/L
Disinfection Byproducts	
TTHMs[e]	0.08 mg/L

[a]Action level for treatment.
[b]Applies to community, nontransient noncommunity, and transient noncommunity water systems.
[c]Applies to systems using surface water or groundwater under the influence of surface water.
[d]Applies to surface water systems serving more than 100,000 persons and any system determined by the state to be vulnerable.
[e]Applies to systems serving more than 10,000 persons and to all surface water systems that meet the criteria for avoiding filtration.

Table 1.3 Secondary drinking water regulations

Constituent	Maximum Contaminant Level[a]
Aluminum	0.05–0.2
Chloride	250
Color	15 Color Units
Copper	1.0
Corrosivity	Noncorrosive
Fluoride	2
Foaming Agents (MBAS)	0.5
Iron	0.3
Manganese	0.05
Odor	3 (Threshold Odor Number)
pH	6.5–8.5
Silver	0.10
Sulfate	250
Total Dissolved Solids (TDS)	500
Zinc	5

[a]mg/L unless noted.

(i) "Overview of the Safe Drinking Water Act Amendments of 1996," by Frederick Pontius, *Journal American Water Works Association*, October 1996, pages 22–33.

Water Treatment Plant Operation, Volume II, chapter on drinking water regulations.

contaminants, occurrence of contaminants in drinking water database, operator certification, primacy, public notification, drinking water regulations for radon, review of National Primary Drinking Water Regulations (NPDWRs), risk assessment application to establishing NPDWRs, small systems (technical assistance, treatment technology, variances), source water quality assessment and petition programs, state revolving loan fund, authorization to promulgate an NPDWR for sulfate, Surface Water Treatment Rule (SWTR) compliance, variance treatment technologies, water conservation, and waterborne disease study and training.

Because not all contaminants can be controlled through treatment processes, source protection is an important component of producing a high-quality drinking water. For more information on source water protection, see Chapter 2, "Source Water, Reservoir Management, and Intake Structures."

Check Your Understanding

1. Who must comply with the Safe Drinking Water Act (SDWA) regulations?
2. What is the definition of a public water system?
3. Define community water system.

1.4   Water Treatment Plants

Most communities have established water treatment plants to produce safe and pleasant drinking water that meets SWDA requirements. This water must be free of disease-causing organisms and toxic substances. Also, the water should not have a disagreeable taste, odor, or appearance.

A water treatment plant takes raw water from a source, such as a stream or lake, and passes the water through a series of interrelated treatment processes; if the quality of the raw water changes, or any process fails to do its intended job, all of the downstream processes will be affected. The raw water flows through tanks or basins where chemicals are added and mixed with it. Then the water slowly flows through larger tanks that allow the heavier suspended solids to settle out. Any remaining solids are removed by filtration and the water is disinfected. The size of a water treatment plant as well as the number and specific types of processes it uses will depend on several factors: (1) the impurities in the raw water, (2) water quality (purity) standards, (3) the demand for water by the population being served, (4) fire protection, and (5) cost considerations.

1.4.1   Conventional Surface Water Treatment

To describe a water treatment plant, the following paragraphs discuss the flow of water through a typical or conventional surface water treatment plant. Most surface waters receive this type of treatment. Figure 1.1 shows a flow diagram of water treatment processes and the purpose or function of each process.

Raw water usually enters a water treatment plant through some type of intake structure. The main purpose of the intake structure is to draw in water while preventing leaves and other debris from clogging or damaging pumps, pipes, and other pieces of equipment in the treatment plant.

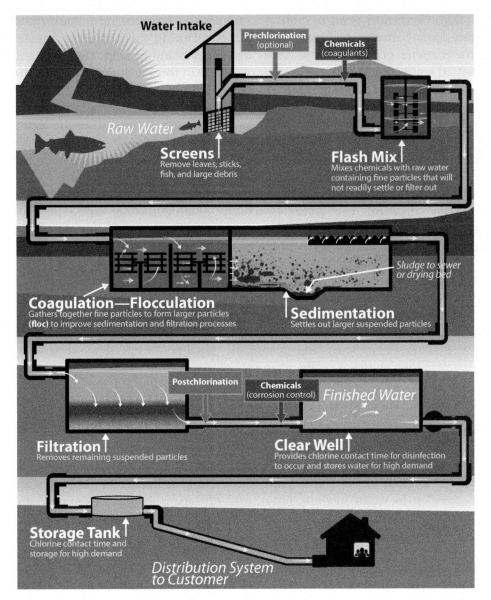

Figure 1.1   Flow diagram of conventional surface water treatment plant processes

Various types of screens are often found in intake structures or in the suction line to raw water pumps.

Chlorine is added to water to kill **pathogenic organisms**. Chlorination at the beginning of a water treatment plant (prechlorination) can help control tastes and odors and prevent the growth of algae and slimes in other treatment processes. Also, the use of prechlorination often reduces chlorine requirements for postchlorination. Some waters should not be prechlorinated because they contain substances that will react with chlorine and form cancer-causing compounds (trihalomethanes).

Coagulant chemicals, such as alum, are added to help remove light, fine particles and other materials suspended in the water. Coagulants cause these very fine particles to clump together into larger particles. A flash mixer is used to thoroughly mix the coagulating chemicals with the water being treated. **Flocculation** is the name of the treatment process in which paddles gently mix the water. The clumps of particles formed by coagulation come together and form larger and larger floc particles. These larger **floc** particles are easier to remove by sedimentation and filtration.

pathogenic (path-o-JEN-ick) organisms
Organisms, including bacteria, viruses, protozoa, or internal parasites, capable of causing disease (such as giardiasis, cryptosporidiosis, typhoid fever, cholera, or infectious hepatitis) in a host (such as a person). There are many types of organisms that do not cause disease. These organisms are called nonpathogenic.

coagulant (ko-AGG-yoo-lent)
A chemical that causes very fine particles to clump (floc) together into larger particles. This makes it easier to separate the solids from the liquids by settling, skimming, draining, or filtering.

flocculation (flock-yoo-LAY-shun)
The gathering together of fine particles after coagulation to form larger particles by a process of gentle mixing. This clumping together makes it easier to separate the solids from the water by settling, skimming, draining, or filtering.

floc
Clumps of bacteria and particles or coagulants and impurities that have come together and formed a cluster. Found in flocculation tanks and settling or sedimentation basins.

Sedimentation is an operation in which the water being treated flows very slowly through a large tank or basin. During this time, the heavier floc particles gradually settle out of the water being treated. The flocs and settled solids that reach the bottom of the basin form a sludge that must be removed and either discharged to a sewer or disposed of in a landfill after being dried in drying beds. Treated water leaves the sedimentation basin by flowing over **weirs** (flow control devices) at the outlet end of the basin.

After sedimentation, the water passes through some type of filter to remove the remaining suspended impurities and flocs. The filter may be made of sand, anthracite coal, or some other type of granular material or a combination of these materials.

After filtration, the water is usually disinfected by some type of chlorination process. The purpose of disinfection is to kill the remaining disease-causing organisms in the water.

If the treated water is **corrosive** (capable of deteriorating metal pipe), chemicals should be added to reduce the corrosivity of the water or to prevent scale (rust) formation.

Treated water is stored in a large tank or basin (**clear well**) until it is pumped into the distribution system for use or to serve as storage during low demand periods for later use during periods of high demand. Storage also provides chlorine contact time for disinfection.

1.4.2 Additional Treatment

Some water treatment plants also use other treatment processes to address particular source water characteristics. For example, waters are softened to remove excess hardness caused by calcium and magnesium. Extra soap is needed to clean or wash with hard water. Also, hard waters will cause scale to develop in water heaters, pipes, and fittings. Iron and manganese are undesirable because they will cause undesirable color in water and stain clothes and plumbing fixtures. Iron and manganese also can promote the growth of iron bacteria, which can cause tastes and odors. Flow diagrams and treatment processes for removing hardness as well as iron and manganese are presented in Volume II of this series of training manuals.

Check Your Understanding

1. What is the purpose of a water treatment plant?
2. Why do intake structures at water treatment plants have screens?
3. How is the sludge disposed of after it is removed from a sedimentation basin?
4. Why is excessive hardness removed from drinking water?

sedimentation (SED-uh-men-TAY-shun)
The process of settling and depositing of suspended matter carried by water or wastewater. Sedimentation usually occurs by gravity when the velocity of the liquid is reduced below the point at which it can transport the suspended material.

weir (WEER)
(1) A wall or plate placed in an open channel and used to measure the flow of water. The depth of the flow over the weir can be used to calculate the flow rate, or a chart or conversion table may be used to convert depth to flow. (2) A wall or obstruction used to control flow (from settling tanks and clarifiers) to ensure a uniform flow rate and avoid short-circuiting.

corrosion
The gradual decomposition or destruction of a material by chemical action, often due to an electrochemical reaction. Corrosion may be caused by (1) stray current electrolysis, (2) galvanic corrosion caused by dissimilar metals, or (3) differential-concentration cells. Corrosion starts at the surface of a material and moves inward.

clear well
A reservoir for the storage of filtered water of sufficient capacity to prevent the need to vary the filtration rate with variations in demand. Also used to provide chlorine contact time for disinfection.

1.5   Plant Operation

Simply described, water treatment plant operators keep a treatment plant operating to produce a safe, pleasant, and adequate supply of water. They monitor the raw water entering the plant and keep an eye on the water as it flows through all of the various treatment processes. Flows into the plant are adjusted according to conditions of the raw water and system demands for water. Equipment and facilities are maintained and repaired as necessary to keep the water flowing and the plant working today and

Table 1.4 Typical duties of a water treatment plant operator

1. Start up, shut down, and make periodic operating checks of plant equipment such as pumping systems, chemical feeders, auxiliary equipment (compressors), and measuring and control systems.

2. Perform routine preventive maintenance such as lubrication, operating adjustments, cleaning, and painting equipment.

3. Load and unload chemicals, such as chlorine cylinders, bulk liquids, powdered chemicals, and bagged chemicals, either by hand or using chemical-handling equipment such as forklifts and hoists.

4. Perform minor corrective maintenance on plant mechanical equipment, for example, chemical feed pumps and small units.

5. Maintain plant records, including operating logs, daily diaries, chemical inventories, and data logs.

6. Monitor the status of plant operating guidelines, such as flows, pressures, chemical feeds, levels, and water quality indicators, by reference to measuring systems.

7. Collect representative water samples and perform laboratory tests on samples for turbidity, color, odor, coliforms, chlorine residual, and other tests as required.

8. Order chemicals, repair parts, and use tools.

9. Estimate and justify budget needs for equipment and supplies.

10. Conduct safety inspections, follow safety rules for plant operations, and develop and conduct tailgate safety meetings.

11. Discuss water quality with the public, conduct tours of your plant (especially for school children), and participate in your employer's public relations program.

12. Communicate effectively with other operators and supervisors on the technical level expected for your position.

13. Make math calculations to determine chemical feed rates, flow quantities, detention and contact times, and hydraulic loadings as required for plant operations.

into the future. Typical duties performed by water treatment plant operators are summarized in Table 1.4.

Operators who are familiar with all aspects of the treatment plants they operate often have better records of producing safe, pleasant, and adequate supplies of water. The easiest way to become familiar with a plant is to be involved in the plant from the beginning. Once the need for a new or improved water treatment plant has been recognized by the community, the community votes to issue the necessary bonds to finance the project and requests that consulting engineers submit plans and specifications. In the best interests of the community and the consulting engineer, you should be present (or at least available) during both the design and construction periods to be completely familiar with the entire plant layout, including the piping, equipment, and machinery and their intended operation. This will provide you with the opportunity to relate your plant drawings to actual facilities. At this time, you should gather together all the data and literature for the equipment to prepare a regular maintenance schedule. You and the engineer should discuss how the water treatment plant should best be run and the means of operation the designer had in mind when the plant was designed. Records of these activities and subsequent changes should be available to operators who join the staff at an existing plant.

If the plant is an existing one that is being remodeled, you may be in a position to offer excellent advice to the consulting engineer. Your experience provides valuable technical knowledge concerning the characteristics of the raw water and the limitations of the present facilities. Together with the consultant, you can be a member of an expert team able to advise your water utility.

1.5.1 Daily Operating Procedures

The following chapters of this manual explain the step-by-step procedures to effectively operate a water treatment plant. This section outlines a checklist of daily activities for the operators of a typical plant. You should prepare a similar list for your plant. Consider this list and all of the information in this manual. Take into account the size of your plant, type and condition of your plant, raw water characteristics, your distribution system, and the number of operators available to help and their skills. A sample checklist is shown in Figure 1.2. By using this example and the information contained in this manual, you should be able to prepare a checklist for your plant that reflects your specific needs.

1.5.1.1 At the Start of the Shift

1. Review what has happened during the last shift or since you left the plant and respond to any problems.

    a. An important part of reviewing plant operations is exchanging information with the operator on duty. Shift assignments should be arranged to provide an overlap at shift changes to permit a brief period (perhaps 15 to 30 minutes) for operators to exchange important information on the status of the plant and special problems that need attention or monitoring.

    b. Your review should include a check of raw and finished water quality for any changes. Especially important are the turbidity and chlorine residual of the finished water in the clear well.

    c. Review system pressures.

    d. Check clear well storage and levels of other reservoirs in the distribution system.

    e. Examine records of raw water and distribution system pumping. Inspect records or charts of flowmeters or total hours run by each pump.

    f. Check the status of each filter. Do any of them need to be backwashed?

2. Prepare for your shift. You have two major concerns: to provide your consumers with drinking water of suitable quality and sufficient quantity.

    a. Quality
    Water quality is controlled mainly by the proper application of chemicals. Be sure you have sufficient chemicals to meet today's demands and future demands.

        (1) Alum. Check daily use and amount in storage.
        (2) Polymers. Record daily use, level of day tank, and amount in storage.
        (3) Chlorine. Record daily use, weigh containers, and determine amount in storage.

    b. Quantity
    The ideal influent pumping rate will allow the plant to operate at a constant rate for the next 24 hours. This rate should allow the plant to meet all demands and maintain adequate reserves in the clear well and service storage reservoirs. The items listed below should be considered when selecting an influent pumping rate.

        (1) Examine the previous day's circumstances.
            (a) How much water was treated and pumped to the distribution system?
            (b) What were the weather conditions? Was it hot? Overcast? Lawns being watered? Freezing conditions?

DATE_____   OPERATOR_____

| | 1 | | 2 | | 3 |
SHIFT   ☐ Midnight-8 a.m.   ☒ 8 a.m.-4 p.m.   ☐ 4 p.m.-Midnight

1. *INTAKE STRUCTURE*
   a. Status of Bar Screens   1 [N] 2 [A]
   b. Status of Intake Pumps   1 [N] 2 [N] 3 [O]
   COMMENTS: Bar screen in Bay No. 2 partially clogged with debris. Will clean this shift.

2. *CHLORINATION SYSTEM*
   a. Status of Evaporators   1 [N] 2 [N]
   b. Status of Chlorinators   1 [N] 2 [N]
   c. Status of Chlorine Residual Analyzer   [N]
   d. Status of Booster Pumps   1 [O] 2 [O]
   e. Chlorine Feed Points   Pre- [N]   Intermediate [N]   Post [N]
   COMMENTS:_____

3. *COAGULANT FEED SYSTEMS*
   a. Status of Alum Feed Pumps   1 [N] 2 [O]
   b. Status of Polymer Feed Pumps   1 [N] 2 [O]
   COMMENTS:_____

4. *FLASH MIXER/FLOCCULATION SYSTEMS*
   a. Status of Flash Mixer   [N]
   b. Status of Flocculators-Basin No. 1   1 [N] 2 [N] 3 [A] 4 [N]
                  -Basin No. 2   1 [O] 2 [O] 3 [O] 4 [O]
   COMMENTS: Flocculator No. 3 in Basin No. 1 starting to develop a whine. Will monitor and advise maintenance.

5. *SEDIMENTATION BASINS*
   a. Basins in Service   1 [N] 2 [O]
   b. Status of Sludge Pumps   1 [N] 2 [O]
   c. Sludge Lagoons in Service   1 [O] 2 [N] 3 [O]
   COMMENTS:_____

6. *FILTER STATUS*
   a. Filters in Service   1 [N] 2 [N] 3 [N] 4 [O]
   b. Backwash Cycle: Auto [N]   Manual [O]
   c. Filter Run Termination: Loss of Head [N]   Time [O]
   d. Status of Filter Aid System [O]
   COMMENTS:_____

7. *FINISHED WATER CLEAR WELL AND PUMP STATION*
   a. Clear Well Level   [A]
   b. Status of Pumps   1 [N] 2 [N] 3 [O]
   c. Status of NaOH Feed System   [N]
   COMMENTS: Reservoir level at 9:30 a.m. is low. Will evaluate system demand and adjust plant flow rate.

USE OF CHECKLIST

1. An entry in a box indicates that the unit is in service or out of service and its status (condition).

2. The status of an item is indicated by inserting an "N" (meaning normal), an "A" (meaning abnormal), or an "O" (meaning out of service) in the proper box.

3. Any "A" entry should be briefly explained in the comments section for that system.

4. This checklist is designed to supplement other plant records such as the daily operations log, which requires more detailed information and entries. The checklist should not be used in lieu of these other records.

Figure 1.2   Routine plant checklist

(2) Consider the previous day's average flow based on raw water pumping and records of water pumped to the distribution system.

(3) Consider storage in clear well. Ideally, the clear well will be nearly full at the start of each day.

(4) After considering items (1), (2), and (3), select a raw water pumping rate that will allow the plant to operate at a constant rate during the next 24 hours. The clear well will be drawn down during the high demand period during the day and filled at night. See Section 1.5.2, "Regulation of Flows," for additional details.

3. Walk through your plant. Start at the intake or headworks and follow the flow of water through your plant to the clear well.

   a. Look for anything unusual or different in the appearance of the water as it goes through each process. Inspect each piece of equipment.

   b. Listen for any unusual or different noises from the equipment.

   c. Feel the equipment for excessive temperature and vibrations.

   d. Smell for any signs of developing odors in your finished water. Also, be aware of any signs of equipment burning or overheating.

   e. Taste and smell your finished water for any changes or undesirable characteristics.

   f. Monitor your plant for safety hazards. If you observe anyone using unsafe procedures or if you observe an unsafe condition, correct the situation immediately.

4. Respond to minor problems not taken care of previously.

5. Collect samples for quality control and analyze samples.

   a. Check turbidity, chemical doses (jar tests), chlorine residual, pH, alkalinity, and coliforms.

   b. Check calibrations on finished water recording turbidimeter and chlorine residual analyzer.

6. Perform all scheduled preventive maintenance.

7. Record all necessary data and be sure all records are up to date.

8. Order supplies, including chemicals.

9. Review safety program.

1.5.1.2 During the Shift

1. Backwash filters. (Check automatic systems for proper operation.)

2. Observe system pressures regularly. If system pressures are monitored and transmitted to a control room, then the operator on duty will be responsible for constant surveillance. If system pressures fluctuate beyond established ranges, adjust pumping rates or number of pumps in service as necessary. However, always check your instrumentation to be sure the system pressure readings are accurate and not the result of an instrumentation malfunction.

3. Monitor the main control board. Charts provide important information as to present plant status. Alarm panel lights also provide important status information. Do not ignore lit indicator or warning lights, particularly red lights as these usually require immediate attention.

jar test
A laboratory procedure that simulates coagulation/flocculation with differing chemical doses. The purpose of the procedure is to estimate the minimum coagulant dose required to achieve certain water quality goals. Samples of water to be treated are placed in six jars. Various amounts of chemicals are added to each jar, stirred, and the settling of solids is observed. The lowest dose of chemicals that provides satisfactory settling is the dose used to treat the water.

chlorine residual
The concentration of chlorine present in water after the chlorine demand has been satisfied. The concentration is expressed in terms of the total chlorine residual, which includes both the free and combined or chemically bound chlorine residuals. Also called residual chlorine.

4. Check the storage level in the clear well. If the clear well level is monitored in a control room, then the operator on duty will be responsible for surveillance and any necessary adjustments of influent pumping rates.

5. Adjust pumping rates.

   a. Pumping rates into the distribution system should be adjusted to maintain system pressures and demands. Pressures should be monitored throughout the system and whenever the pressures start to drop, increase the pump speed or place an additional pump in service.
   b. Raw water pumping rates should be adjusted as necessary to maintain desired levels in the clear well throughout the day.

6. Quality control checks.

   a. Collect samples for quality control and analyze.
   b. Check turbidity, chemical doses (jar tests), chlorine residual, pH, and alkalinity.
   c. Inspect chlorination system. Is it working properly? Are there any leaks?

7. Repeat these tasks as often as necessary.

1.5.1.3 At the End of the Shift

Perform the following tasks at the end of your shift or before leaving your plant at night:

1. Repeat the tasks listed under Section 1.5.1.2, "During the Shift."

2. Report plant operations to the operator coming on duty. If the next shift is a few hours away:

   a. Anticipate raw water and finished water pumping requirements until the next operator comes on duty.
   b. Be sure all chemical dosage facilities are prepared to operate until the next operator comes on duty.
   c. Secure the plant for the night by checking outside lighting and security systems and locking the plant.

Check Your Understanding

1. Why should a water treatment plant operator be present when a new plant is being constructed?

2. What should be an operator's first task upon arrival at a water treatment plant?

3. How does an operator control finished water quality?

4. What tasks should an operator do during the shift?

5. What tasks should an operator do at the end of the shift?

1.5.2 Regulation of Flows

Water treatment plant flow rates fluctuate with total system demand. Thus, distribution system or consumer demands generally control water treatment plant operations. Demands for water vary depending on the following considerations:

1. Time of day

2. Day of week

3. Season of the year
4. Prevailing weather conditions
5. Manufacturing demands
6. Unusual events (fire, main breaks)

To meet these variable system demands, adequate source capacity and treated water storage volume are essential. Usually, a minimum distribution system operating pressure is established for all water service connections, and this value establishes the basic guideline for system operation. When the pressure falls below a predetermined value, additional flow must be provided. This demand establishes the finished water demand. Thus, most water treatment plants are operated on a demand-feedback basis, whereby finished water flow requirements (consumer demands) establish raw water and in-process flow requirements. In small water treatment systems, the plant may only need to be operated for a portion of the day (one 8-hour shift) in order to produce sufficient water for reservoir storage to meet distribution system demands during a 24-hour period.

Clear wells (or plant storage reservoirs) provide the necessary operational storage to average out high and low flow demands. The reservoir is filled when demands are low to compensate for peak periods, which draw the level down (see Figure 1.3). This reservoir also acts as a buffer that prevents frequent ON/OFF cycling of finished water pumps and permits planned changes in treatment plant operation.

To estimate how much water needs to be treated each day, analyze current storage levels in clear wells (or plant storage reservoirs) and in distribution system service storage reservoirs. Also, the expected consumer demand for the day must be estimated. This is where historical flow records can be informative. Important items to consider include trends during the past few days. Is the weather changing and causing an increase or decrease in demand for water? In the late spring or early summer, increasing temperatures can cause an increase in demand for water to irrigate gardens and lawns. High winds and freezing conditions can also create shifts in demand. In many communities, the lifestyle of the people can create predictable demands for water on certain days of the week, for example, when people wash their clothes or cars or irrigate their yards. Another factor that may influence demand for water is whether or not children are in school or at home playing with water. In summary, to estimate the flow rate or amount of water to be treated each day by the water treatment plant, review:

1. Clear well and distribution storage needs
2. Yesterday's and historical consumer demands
3. Weather forecasts

Actual regulation of the flow to water treatment plants depends on the method used to deliver water to the plant. If the raw water is transmitted by gravity from a reservoir through a pipe under pressure, then the flow can be changed by adjusting a butterfly or plug-type valve. If pumping is involved, then a change in the pumping rate or the number of pumps in service will be required.

Raw water pumps are either constant speed or variable speed units. In constant rate pumping units using multiple pumps, adjustment of flow is accomplished by adding or removing pumps in service to produce

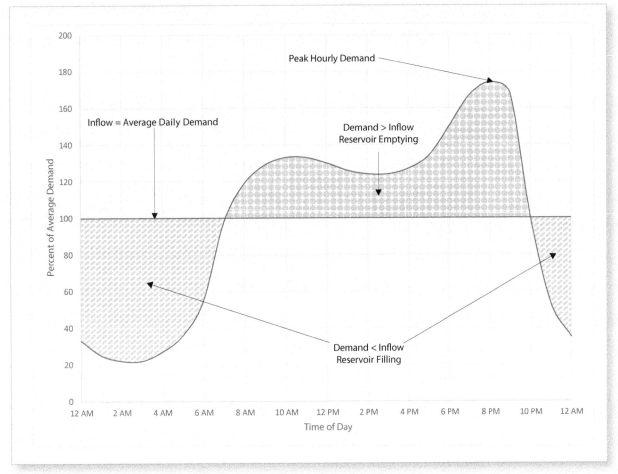

Figure 1.3 Typical daily variation of system demand (Lauer, W.C., 2013. *Water Distribution Operator Training Handbook,* 4th ed. Copyright © American Water Works Association. Republished with permission.)

the desired flow. Variable speed pumps can be adjusted by changing the drive speed of the pumps to produce the desired flow. Electricity costs for variable speed pumps are much higher than for constant speed pumping units. Pumps used to pump finished water into the distribution system operate in a similar manner. Many pumps are designed to operate most efficiently in a fairly narrow range of flow and pressure conditions. Therefore, the individual operating characteristics of each pump should be considered so that the most efficient pumping mode can be selected.

The following example shows how to adjust the raw water flow rate to meet system demands. Let us assume that a clear well has a maximum storage capacity of 5 million gallons (MG), with 3 million gallons currently in storage, and the plant demand flow rate is 1.5 million gallons per day (MGD).

If the plant is being operated on a continuous basis, the raw water flow rate should be adjusted to approximately three MGD depending on weather conditions and the day of the week.

Example 1

The 3 MGD will meet the demand of 1.5 MGD and allow us to add 1.5 MG during the 24-hour period to storage. We did not try to make up the entire 2 million gallon deficit in storage because this could possibly cause overfilling of the reservoir if distribution system demands suddenly drop.

In this same example, if the plant only operated for 8 hours per day, then the flow rate would have to be increased by a factor of 3 [(3)(3 MGD) = 9 MGD] to achieve the same results.

Let's further assume that the plant has an operating capacity of 5 MGD and the raw water pumping station has one 1,500 gallons per minute (GPM) constant speed pump and two 1,000 GPM constant speed pumps. (A flow of 1 MGD is approximately equal to a flow of 700 GPM for 24 hours.) At the initial flow rate of 1.5 MGD, only one of the 1,000 GPM pumps would be needed to provide this flow (1,000 GPM for 24 hours is almost equal to 1.5 MGD). Both 1,000 GPM units in service would almost provide the desired 3 MGD flow (actual flow would be 2.9 MGD). Constant rate distribution system pumps would be adjusted in a manner similar to that just described.

After you have made the flow rate change, you should verify the actual flow rate by reading the raw water flow measuring device. You should periodically check the storage levels (elevations) in the clear well and service storage reservoirs to determine if they are maintaining the desired storage volumes. Plant flows should be adjusted whenever major changes in consumer demand occur.

1.5.2.1   Treatment Process Changes

To maintain adequate clear well and distribution system water storage levels, raw or source water flow changes may be required (either increases or decreases). Raw water flow changes should take into account the travel or detention time between the source of supply (river or lake) and the treatment plant.

When storage demands change and require adjustments in the flow of water through a plant, you may also be required to perform the following functions:

1. Adjust chemical feed rates
2. Change filtration rates
3. Perform jar tests
4. Observe floc formation and floc settling characteristics
5. Monitor process performance
6. Collect process water quality samples
7. Visually inspect overall process conditions

Some of these changes may occur automatically if your plant has flow-paced chemical feeders. Changes in chemical feed rates (coagulants and chlorine) are required when using manually operated chemical feeders because they are generally set to feed a specific amount of chemical, and this amount is dependent on the rate of flow. Adjustment and calibration of chemical feeders is discussed in Chapter 3, "Coagulation and Flocculation."

Filters are usually operated at a constant production rate, as described in Chapter 5, "Filtration." Automatic control systems typically maintain uniform flow rates, but the number of filters in service can be changed by starting or stopping individual filter units to meet changing needs.

In addition to the above considerations, each of the other unit treatment processes (coagulation, flocculation, and sedimentation) is designed to operate over a broad range of flow rates. However, in some instances, major flow changes may require either adding or removing facilities from service.

coagulation(ko-agg-yoo-LAY-shun)
The clumping together of very fine particles into larger particles (floc) caused by the use of chemicals (coagulants). The chemicals neutralize the electrical charges of the fine particles, allowing them to come closer and form larger clumps. This clumping together makes it easier to separate the solids from the water by settling, skimming, draining, or filtering.

1.5.3   Chemical Use and Handling

A wide variety of chemicals are used in the water treatment plant in the production of a safe and palatable drinking water supply. They play a crucial role in controlling process performance and producing a high-quality water. Chemicals are used in the following aspects of water treatment:

1. Clarification (turbidity reduction)
2. Disinfection
3. Taste and odor control
4. Algae control
5. Corrosion/scaling control
6. Water softening
7. Fluoridation

The most commonly used chemicals for water treatment are described in Table 1.5. The American Water Works Association has developed standards for many of these chemicals, which help to ensure that only quality chemicals are used in water treatment. These standards should be referred to when ordering treatment chemicals (see AWWA Standards, Section B—Treatment, www.awwa.org).

The choice of specific chemicals to use in a given water treatment plant will vary depending on source water quality, type of treatment to be performed, availability of chemicals, and, to some degree, economic considerations.

The initial selection of specific chemicals and anticipated feed rates for a given application are frequently determined by pilot-plant testing of the specific source water. Pilot-plant tests are usually performed by the water treatment plant designer by constructing a smallscale treatment plant at or near the source of supply. Pilot tests provide the designer, as well as the treatment plant operator, with valuable information on the treatability of a given source of supply over a broad range of water quality conditions.

Operators should be thoroughly familiar with the types of chemicals used in water treatment, specific chemical selection and applications, evaluation methods for determining performance, and safe storage and handling techniques. All chemicals are potentially dangerous and all necessary precautions must be taken before handling any chemical. All containers, no matter what the use, should bear some form of precautionary labeling.

A **safety data sheet (SDS)** is your best source of information about dangerous chemicals. Ask your chemical supplier to furnish you with the SDS for each chemical you purchase. No chemical should be received,

safety data sheet (SDS)
Safety data sheets (SDSs) are an essential component of the Globally Harmonized System of Classification and Labeling of Chemicals (GHS) and are intended to provide comprehensive information about a substance or mixture for use in workplace chemical management. They are used as a source of information about hazards, including environmental hazards, and to obtain advice on safety precautions. In the GHS, they serve the same function that the material safety data sheet (MSDS) does in OSHA's Hazard Communication Standard. The SDS is normally product related and not specific to the workplace; nevertheless, the information on an SDS enables the employer to develop an active program of worker protection measures, including training, which is specific to the workplace, and to consider measures necessary to protect the environment.

Table 1.5   Chemical types and characteristics[a]

Chemical Name	Chemical Formula	Commercial Concentration	Comments
Coagulants			
Aluminum Sulfate (Alum, granular)	$Al_2(SO_4)_3 \bullet 14\ H_2O$	47–50% $(Al_2(SO_4)_3)$	Acidic
Ferric Chloride	$FeCl_3 \bullet 6\ H_2O$	47–50% $(Al_2(SO_4)_3)$	Acidic
Ferric Sulfate	$Fe_2(SO_4)_3 \bullet 9\ H_2O$	59–61% $FeCl_3$	Acidic, stinging
Ferrous Sulfate	$FeSO_4 \bullet 7\ H_2O$	90–94% $Fe_2(SO_4)_3$	Cakes Dry
Cationic Polymer	—	55% $(FeSO_4)$	Positively Charged
Anionic Polymer	—	Varies	Negatively Charged
Nonionic Polymer	—	Varies	
Disinfection			
Sodium Hypochlorite	$NaOCl$	12–15% (Cl_2)	Solution
Calcium Hypochlorite (HTH)	$Ca(OCl)_2 \bullet 4\ H_2O$	65–70% (Cl_2)	Powder
Chlorine	Cl_2	99.8% (Cl_2)	Gas/Liquid
Anhydrous Ammonia	NH_3	99–100% (NH_3)	Gas/Liquid
Ammonium Hydroxide	NH_4OH	29.4% (NH_3)	Solution
Ammonium Sulfate	$(NH_4)_2SO_4$	6.3% (NH_3)	Cakes Dry
Chlorine Dioxide	ClO_2	26.3% (Cl_2)	Generated On Site
Ozone	O_3	—	Generated On Site
Taste and Odor			
Activated Carbon	C	—	Insoluble
Potassium Permanganate	$KMnO_4$	100%	Very Soluble
Algae Control			
Copper Sulfate	$CuSO_4 \bullet 5\ H_2O$	99% $(CuSO_4)$	
Corrosion Control			
Calcium Hydroxide (Hydrated Lime)	$Ca(OH)_2$	75–99% (CaO)	Basic
Sodium Hydroxide (Caustic Soda)	$NaOH$	98.9% $(NaOH)$	Very Basic
Softening			
Calcium Oxide (Quicklime)	CaO	75–99% (CaO)	
Sodium Carbonate (Soda Ash)	Na_2CO_3	99.4% (Na_2CO_3)	
Fluoridation			
Sodium Silicofluoride	Na_2SiF_6	59.8% (F)	Powder
Sodium Fluoride	NaF	43.6% (F)	Powder or Crystal
Fluosilicic Acid	H_2SiF_6	23.8% (F)	Solution

[a]Adapted from *Standard Methods for the Examination of Water and Wastewater* (see Section 1.13, "Additional Resources").

stored, or handled until essential safety information is provided to those who come into contact with the substance. Among other information, the SDS will provide:

1. Identification of the product and the manufacturer or distributor
2. Identification of all hazards regarding the chemical
3. Composition/information on ingredients
4. First-aid measures including important symptoms/effects and required treatment

5. Fire-fighting measures including suitable extinguishing techniques and equipment and chemical hazards from fire

6. Accidental release measures listing emergency procedures, protective equipment, and proper methods of containment and cleanup

7. Handling and storage precautions, including incompatibilities

8. Exposure controls/personal protection including exposure and threshold limits and personal protective equipment

9. Physical and chemical properties

10. Stability and reactivity, including possible hazardous reactions

11. Toxic effects on living organisms including routes of exposure as well as related symptoms, acute and chronic effects, and numerical measures of toxicity

Remember, do not work with a chemical unless you understand the hazards involved and are using the protective equipment necessary to protect yourself. Contact your local safety regulatory agency about specific chemicals you may deal with if there is any doubt about safe procedures.

1.5.3.1   Storage and Supply

Water treatment chemicals can be stored in a number of ways, including:

1. Solid (dry) form (bags, cartons, drums)

2. Liquid form (drums, tanks, cylinders)

3. Gaseous form (cylinders)

Chemicals should be stored in accordance with the manufacturer's written recommendations and requirements established by regulatory agencies. Regardless of the storage method, always anticipate future chemical requirements so that an ample supply will be on hand when needed. A good practice is to maintain at least a 30-day supply of all commonly used treatment chemicals. Minimum storage quantities should be established for each type of chemical, and this information will indicate when chemicals should be ordered. For example, many plants do not allow the supply of chlorine to drop below a 15-day supply at the plant site. Keep a running account of chemical use and storage inventory on a daily basis. Use these records to calculate how many days' supply are on hand.

Example 2

Shown below is the amount of chlorine used by a small water treatment plant during one week.

Day of the Week	Mon	Tues	Wed	Thurs	Fri	Sat	Sun
Chlorine Used, lb	43	39	34	38	39	37	29

What was the daily average use of chlorine in pounds per day?

toxic
A substance that is poisonous to a living organism. Toxic substances may be classified in terms of their physiological action, such as irritants, asphyxiants, systemic poisons, and anesthetics and narcotics. Irritants are corrosive substances that attack the mucous membrane surfaces of the body. Asphyxiants interfere with breathing. Systemic poisons are hazardous substances that injure or destroy internal organs of the body. Anesthetics and narcotics are hazardous substances that depress the central nervous system and lead to unconsciousness.

	Known		Unknown
Chlorine Used Each Day of Week, lb/d		F_A =	Daily Average Use of Chlorine, lb/d

Calculate the daily average use of chlorine in pounds per day.

$$F_A\left[\frac{lb}{d}\right] = \frac{\text{Sum of Chlorine Used Each Day}}{\text{Total Number of Days}}$$

$$= \frac{43[lb] + 39[lb] + 34[lb] + 38[lb] + 39[lb] + 37[lb] + 29[lb]}{7[d]}$$

$$= \frac{259[lb]}{7[d]}$$

$$= 37\frac{lb}{d}$$

Note: We used a 7-day average so each day of the week would be considered.

Example 3

The chlorine cylinder in service has less than one day's chlorine supply remaining. Four 150-pound (lb) chlorine cylinders are in storage. The plant uses an average of 37 pounds of chlorine per day (lb/d). How many days' (d) supply of chlorine is available?

	Known		Unknown
n	= Number of Chlorine Cylinders	= 4 cyl	Supply of Chlorine, d
M_{cyl}	= Cylinder Weight	= 150 lb	
F	= Average Use	= 37 lb/d	

Calculate the available supply of chlorine in days.

$$\text{Supply of Chlorine}[d] = \frac{M_{cyl} \times n}{F}$$

$$= \frac{150\left[\dfrac{lb}{cyl}\right] \times 4[cyl]}{37\left[\dfrac{lb}{d}\right]}$$

$$= 16\ d$$

1.5.3.2 Safe Handling

In the routine operation of a water treatment plant, you will come in contact with a variety of potentially dangerous chemicals. While some chemicals are inactive (inert), it is good practice to consider all chemicals as a potential hazard.

When unloading or transferring chemicals, be especially careful. Know the locations of all safety showers and eye wash fountains. Be familiar with their use and test them periodically to be sure that they function properly.

Wear protective clothing when working with chemicals. Goggles and face shields will protect your eyes and face. Protect other exposed portions of the body by wearing rubber or neoprene gloves, aprons, or other protective clothing. Chemical dust can irritate the eyes and respiratory system. Use respirators, when appropriate, and always use dust collectors when such equipment is provided. Promptly wash down or clean up all chemical spills to prevent falls and physical contact with the chemical.

A few treatment chemicals, such as caustic soda and chlorine, can be hazardous to the operator, and extreme care should be taken in the handling of these chemicals. Caustic soda is one of the most dangerous of the common alkalies, and direct contact will cause severe burns. Caustic can quickly and permanently cloud vision if not immediately flushed out of the eyes. Determine the location of safety showers and eye wash stations before starting to work with caustic soda. Wash down caustic soda spills immediately. When handling caustic soda, wear safety goggles and a face shield; cover your head with a wide brim hat; wear rubber or neoprene gloves, apron, and boots (a full-body protective suit is preferable); and do not tuck pant legs inside boots or shoes. If dust or mist is encountered, use a respirator.

Chlorine is a strong respiratory irritant, and either prolonged exposure to chlorine gas or high concentrations of chlorine gas could be fatal. Wherever chlorine liquid or gas is stored or used, the following safety equipment should be provided:

1. Shower and eye wash facility.
2. Emergency breathing apparatus (air pack).
3. Chlorine gas detector.
4. Floor-level vents.
5. Fans that maintain a positive air pressure in the storage facility.

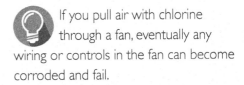

If you pull air with chlorine through a fan, eventually any wiring or controls in the fan can become corroded and fail.

All water treatment plant operators should be fully trained in chlorine safety and leak detection procedures. Whenever you enter a chlorine facility, make sure that the fan is operating. If a chlorine leak is suspected or the chlorine gas concentration in the room is not known, wear a self-contained air pack and use the buddy system. Rubber- or plastic-coated gloves should be worn when handling chlorine containers. When in doubt—always use the buddy system and have another operator standing by with an air pack. (See Chapter 6, Section 6.9, "Chlorine Safety Program," for additional information about chlorine safety.)

1.5.3.3 First-Aid Procedures

Every operator should be familiar with the following first-aid procedures:

EYE BURNS (GENERAL)

1. Call 911 immediately.
2. Apply a steady flow of water to eyes for at least 15 minutes.
3. Do not remove burned tissue from the eyes or eyelids.
4. Do not apply medication (except as directed by a physician).
5. Do not use compresses.

SKIN BURNS (GENERAL)

1. Call 911 immediately.
2. Remove contaminated clothing immediately (preferably in a shower).
3. Flush affected areas with generous amounts of water.
4. Do not apply medication (except as directed by a physician).

SWALLOWING OR INHALATION (GENERAL)

1. Call 911 immediately.
2. Read antidote on label of any chemical swallowed. For some chemicals, vomiting should be induced, while for other chemicals, vomiting should not be induced.

CHLORINE GAS CONTACT

1. Call 911 immediately.
2. Check for breathing. Tilt the victim's head back (tilting the head back opens the airway and may restore breathing), put your ear over the victim's mouth and nose, and listen and feel for air. Look at the victim's chest to see if it is rising and falling. Watch for breathing for 3 to 5 seconds.

   a. If victim is breathing, place person face up with head and back in a slightly elevated position. Keep victim warm and comfortable.

   b. If there is no breathing, perform hands-only cardio-pulmonary resuscitation (CPR). With the victim face up, push hard and fast in the center of the chest. Continue hands-only CPR until breathing starts or help arrives.

3. Eye irritation caused by chlorine gas should be treated by flushing the eyes with generous amounts of water for not less than 15 minutes. Hold eyelids apart to ensure maximum flushing of exposed areas. Do not attempt to neutralize with chemicals. Do not apply any medication (except as directed by a physician).
4. Minor throat irritation can be relieved by drinking milk. Do not give the victim any drugs (except as directed by a physician).

LIQUID CHLORINE CONTACT

1. *Skin contact.* Call 911 immediately. Flush the affected area with water. Remove contaminated clothing while flushing (preferably in a shower). Wash affected skin surfaces with soap and water while continuing to flush. Do not attempt to neutralize with chemicals. Do not apply medication (except as directed by a physician).
2. *Ingestion.* Call 911 immediately. If liquid chlorine has been swallowed, immediately give victim large amounts of water or milk, followed with milk of magnesia, vegetable oil, or beaten eggs. Do not give sodium bicarbonate. Never give anything by mouth to an unconscious victim.

If you are trained in conventional CPR using chest compressions and mouth-to-mouth breathing, use that method instead of hands-only CPR. Training in both hands-only and conventional CPR is available from local Red Cross and American Heart Association chapters as well as many public health agencies.

Video training for hands-only CPR and other information about CPR is available from the American Heart Association at cpr.heart.org.

Check Your Understanding

1. What is a pilot plant?
2. How are water treatment chemicals stored?
3. What protective clothing should be worn when working with chemicals?
4. What safety equipment should be provided wherever chlorine liquid or gas is stored or used?

1.5.4 Water Quality Monitoring

Operators perform a variety of laboratory tests on source water samples, process water samples, and finished water samples to monitor overall water quality and to evaluate process performance. The sampling location and type of water sample used for a particular analysis will vary depending on the purpose of the analysis. Compliance with the MCLs is usually measured at the point where water enters the distribution system. However, water can degrade in the distribution or delivery system. In all cases, it is important to stress that the water sample must be representative of actual conditions of the entire flow being sampled.

Grab samples, single samples collected at particular places and times, are usually adequate for making periodic measurements of water quality indicators. They are especially useful when measuring indicators that may change quickly after collection such as coliforms, pH, and temperature. When variations in water quality are expected, composite samples may be more appropriate. Ideally, process control measurements should be made on a continuous basis by special instrumentation.

Sample points for common water quality indicators in a conventional treatment plant are shown in Figure 1.4. The frequency of sampling for individual process control water quality indicators will vary from hourly to once per day, depending on the quality of the source water and the importance of the indicator being evaluated. Thus, certain water quality indicators such as turbidity may be routinely monitored (every four hours or continuously), while others such as alkalinity are sampled less frequently (once per shift).

1.5.4.1 Turbidity Removal

Most municipal plants built to treat surface water are designed to remove turbidity, and turbidity is the single water quality indicator over which water treatment plant designers and operators have the greatest control.

In a well-designed and operated treatment plant, high turbidity removals can be achieved under optimum conditions, while poorly operated treatment plants (inadequate pretreatment and filtration) may only achieve relatively low turbidity removals (filtered water turbidities greater than 1.0 turbidity units or TU). This range of turbidity removal effectiveness is important because of the relationship between turbidity and

grab sample
A single sample of water collected at a particular time and place that represents the composition of the water only at that time and place.

coliform (KOAL-i-form)
A group of bacteria found in the intestines of warm-blooded animals (including humans) and in plants, soil, air, and water. The presence of coliform bacteria is an indication that the water is polluted and may contain pathogenic (disease-causing) organisms. Fecal coliforms are those coliforms found in the feces of various warmblooded animals, whereas the term coliform also includes other environmental sources.

pH (pronounce as separate letters)
pH is an expression of the intensity of the basic or acidic condition of a liquid. Mathematically, this is equivalent to the negative of the base 10 logarithm.
If $\{H^+\} = 10^{-6.5}$, then pH = 6.5. The pH may range from 0 to 14, where 0 is most acidic, 14 most basic, and 7 neutral.

$$pH = Log\frac{1}{\{H^+\}} \ or \ -Log(\{H^+\})$$

composite (come-PAH-zit) (proportional) sample
A composite sample is a collection of individual samples obtained at regular intervals, usually every one or two hours during a 24-hour time span. Each individual sample is combined with the others in proportion to the rate of flow when the sample was collected. The resulting mixture (composite sample) forms a representative sample and is analyzed to determine the average conditions during the sampling period.

alkalinity (AL-kuh-LIN-it-tee)
The capacity of water or wastewater to neutralize acids. This capacity is caused by the water's content of carbonate, bicarbonate, hydroxide, and occasionally borate, silicate, and phosphate. Alkalinity is expressed in milligrams per liter of equivalent calcium carbonate. Alkalinity is not the same as pH because water does not have to be strongly basic (high pH) to have a high alkalinity. Alkalinity is a measure of how much acid must be added to a liquid to lower the pH to 4.5.

Water Quality Complaint Investigator's Guide, 2nd Edition. William C. Lauer. American Water Works Association (AWWA). 2014. ISBN: 9781583219928. www.awwa.org. www.epa.gov/waterresilience

bacteria and other pathogenic organisms. There is considerable evidence that shows that filtered waters with high levels of turbidity cannot be effectively disinfected. Turbidity must be removed for disinfection to be effective in killing or inactivating disease-causing organisms.

Thus, the goal of all water treatment plants should be to produce a filtered water with the lowest practical level of turbidity. Many water treatment plants in the United States treating surface water routinely produce filtered water turbidities in the range of 0.05 to 0.3 TU. A recommended target turbidity level is 0.1 TU.

1.5.4.2   Water Quality Complaints

Consumers are another source of information about water quality. Consumer complaints or concerns may indicate problems with the water delivered to their homes and businesses. Often consumer complaints stem from problems in the water distribution system, not the treatment plant. However, treatment plant operations such as changes in supply and pH, chlorine, and copper residual levels may prompt some consumers to contact their water supply agency to report problems.

The basic process for dealing with a consumer complaint is:

1. Collect as much information from the consumer as possible, including the consumer's contact information.
2. Direct the complaint to the proper department or person for evaluation, if necessary.
3. Initiate an investigation.
4. Consider different potential causes of the problem and methods of testing.
5. Gather data from sources such as operations records of the plant or distributions system, laboratory or field testing results on water samples, and distribution system inspection results (cross connections, proper valve operation).
6. Determine the cause of the problem. Possible causes include private plumbing problems (customer's responsibility), transient events (service work on pipes that is now complete or temporary changes in chemistry), or treatment plant problems.
7. If the problem is caused by the public water system, initiate necessary corrective action.
8. Document the problem, cause, and solution (if any) for future reference.
9. Inform the customer of your conclusion and any necessary corrective action taken or underway or future plans for corrective action.

1.5.5   Sludge Handling and Disposal

In water treatment, the major source of **sludge** solids to be disposed of is the suspended solids in the source water. The treatment chemicals themselves, especially alum, constitute a secondary source of sludge solids. Another source of sludge is the precipitate from the lime–soda ash softening process.

turbidity (ter-BID-it-tee)
The cloudy appearance of water caused by the presence of suspended and colloidal matter. In the waterworks field, a turbidity measurement is used to indicate the clarity of water. Technically, turbidity is an optical property of the water based on the amount of light reflected by suspended particles. Turbidity cannot be directly equated to suspended solids because white particles reflect more light than dark-colored particles and many small particles will reflect more light than an equivalent large particle.

sludge (SLUJ)
The settleable solids separated from liquids during processing.

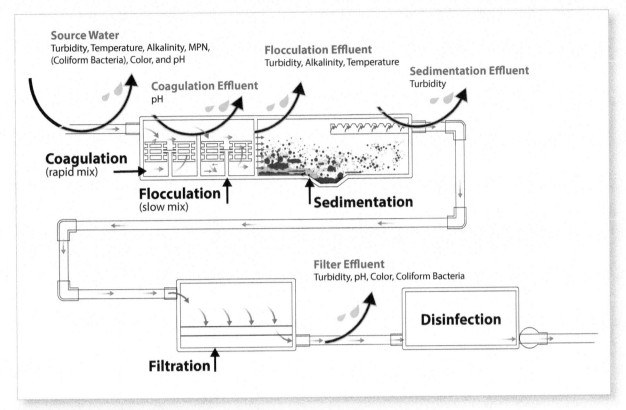

Figure 1.4 Sample points for common water quality indicators

In most water treatment plants, over 99 percent of the suspended solids in the source water are removed by the sedimentation and filtration processes. These processes concentrate the source water solids and treatment chemicals, which are then collected and processed to reduce their volume before final disposal.

Federal laws include sludge from a water treatment plant as an industrial waste that requires proper handling and disposal. Under the National Pollutant Discharge Elimination System (NPDES) provision of the federal laws, a permit must be obtained for wastewater discharges (process sludge) from a water treatment plant into a surface water or groundwater source. This permit sets discharge limits on water quality characteristics such as pH, total suspended solids, settleable solids, flow, and **BOD**.

For procedures on how to process and dispose of sludge, see the chapter on handling and disposal of process wastes in Volume II of this series of training manuals. Another helpful reference in this series of training manuals is *Advanced Waste Treatment*, Chapter 3, "Residual Solids Management."

Check Your Understanding

1. Grab samples are used when measuring what types of water quality indicators?

2. Why are high levels of turbidity removal important?

3. How are most suspended solids in the source water removed at water treatment plants?

4. What do the letters NPDES stand for?

BOD
(pronounce as separate letters) Biochemical Oxygen Demand. The rate at which organisms use the oxygen in water while stabilizing decomposable organic matter under aerobic conditions. In decomposition, organic matter serves as food for the bacteria and energy results from its oxidation. BOD measurements are used as a measure of the organic strength of wastes in water.

1.5.6   Process Instrumentation and Controls

To assist operators in providing consumers with a safe and palatable supply of drinking water, instruments and controls are used to indicate, measure, monitor, record, control, and signal the breakdown of many of the process functions on a continuous or intermittent basis. Regardless of how simple or complex the instrumentation and control systems are for a plant, operators are responsible for controlling plant operation. Table 1.6 lists typically monitored functions. In smaller water treatment plants, the number of functions monitored may be limited to the essential functions only. The most common methods of sensing these process functions are also listed in Table 1.6.

Table 1.6   Commonly monitored functions and sensing methods[a]

Function	Monitored Locations	Sensing Methods
Flow	• Raw Water • Service Water • Chemical Solution • Filters • Wash Water • Sludge • Finished Water	• Venturi Meters (Differential Pressure) • Propeller Meter • Magnetic Meter • Sonic Meter • Rotameter
Level	• Chemical Tanks and Hoppers • Filters • Wash Water Tank • Clear Well • Recovery Basins	• Float • Bubbler • Probe • Pressure Cell • Sonic
Chlorine Residual	• Each Unit Process • Finished Water	• Amperometric (Electrode) • Colorimetric
Turbidity	• Raw Water • Each Unit Process • Individual Filters • Finished Water	• Surface Scatter (High Turbidity) • Nephelometric (Low Turbidity)
pH	• Raw Water • Each Unit Process • Finished Water	• Amperometric (Electrode)
Weight	• Chlorine Cylinders • Chemical Feeders (Loss-in-Weight Type)	• Scales (Mechanical) • Load Cells (Electronic or Pressure)
Pressure	• Service Water • Plant Air Supply • Water Level (Bubblers) • Effluent Pump Station • General Piping	• Bourdon Tube (Mechanical) • Electronic • Differential Pressure (DP Cell)
Loss-in-Head	• Individual Filters	• Differential Pressure
Sludge Density	• Sedimentation Basin • Solids Disposal Piping	• Ultrasonic • Infrared
Conductivity	• Raw Water • Finished Water	• Electronic
Temperature	• Raw Water	• Electronic

[a]Based on B.G. Stone, Notes from "Design of Water Treatment Systems." See Section 1.13, "Additional Resources."

Instrumentation and controls are communication devices that transmit information (data) from measuring locations in the water treatment plant to a central data collection point (usually a control room). In most modern water treatment plants, this central data collection point is the main control panel (see Figure 1.5). This is the nerve center of the plant and is located in the operations control room. From this single location, the operator can monitor and control most of the major process functions. In many modern water treatment plants, sophisticated electronic methods provide virtually automatic control over major process functions. However, manual controls must be available to back up critical functions in water treatment plants so operators can override the automatic system when necessary.

Figure 1.5   Control station

1.5.6.1   Signal Transmission Methods

There are numerous methods of transmitting data from sensing or measuring locations in the water treatment plant to a central control point. Methods used to transmit data include mechanical, pneumatic, hydraulic, electronic, and electrical. However, electronic systems (millivolt or milliamp) have become the popular choice. When certain data must be transmitted over long distances (greater than about 1,500 feet or 450 m), other systems such as ethernet network, telephone tone, microwave, wireless technology, or radio transmission may be used.

MECHANICAL

The most basic type of instrumentation is purely mechanical. These devices are easily understood, reliable, and generally the most economical. Examples of mechanical devices include:

1. Float valves
2. Pressure relief valves
3. Pressure gauges
4. Indicators (valve position and fluid level)
5. Switches (high/low level, pressure)
6. Scales (chlorine and other chemical weighing devices)

These mechanical devices have been used for many years and are still in use at many of the older plants. Many newer types of sensing and transmitting devices are based on the principles of these mechanical devices.

PNEUMATIC

Pneumatic (using air pressure or vacuum) measuring and transmission devices are safe and reliable, and service can be maintained during short-term power outages. In these systems, data are usually transmitted through small-diameter (⅛- to ¼-inch or 3 to 6 mm) copper tubing. Usually, the air supply for pneumatic instrumentation must be dried to prevent condensation in equipment signal lines, and it must be filtered to remove particulate contaminants that could clog orifices. Most pneumatic transmitters produce a signal in the range of 3 to 15 psi (0.2 to 1.0 kg/cm² or 1.4 kPa to 6.9 kPa), and the signal is usually linear within that range (that is, 3 psi = 0%, 9 psi = 50%, and 15 psi = 100% of scale).

One drawback of pneumatic systems is that the air volume in the signal lines causes some dampening of the signal. This means that on longer transmission lines (say 300 to 500 feet or 90 to 150 m), delays of 5 to 30 seconds may be experienced. Normally, the maximum transmission distance for pneumatic signals is about 1,000 feet (300 m). Therefore, pneumatic devices frequently provide a direct readout at the measuring location. For remote readings, the pneumatic signal is converted to an electronic signal and then transmitted to a receiver in a control room.

HYDRAULIC

Hydraulic (using liquid pressure) instrumentation, unlike pneumatic signals, operates over a broad range of pressures using water, water and oil, or glycerol. The fluid is normally conveyed through small-diameter pipe (¼ to 1 inch or 6 to 25 mm) constructed of copper or steel.

Hydraulic signals are not dampened like pneumatic signals so signal transmission is virtually instantaneous. These systems can be used to transmit weight (load cells) and pressure data, and are commonly used to operate (power) mechanical equipment such as valves.

Hydraulic systems are probably the least used signal transmission method in water treatment plants.

ELECTRONIC

Electronic signal transmission has become the most commonly used data gathering and transmission method.

Water treatment plant process data are usually transmitted within a range of 4- to 20-millivolt signals (often expressed as milliamp signals). A major advantage of electronic signal transmission is that extremely low potential (voltage) electronic signals from field devices (such as electrodes in water) can be amplified to millivolt signals. Another advantage of electronics is that converting signals (transducing) from one form to another is quite easy. For example, solid-state circuitry can be used to convert variable sensing data to linear signals.

Electronic signals are used to activate common electrical power and control circuitry (120 and 240 volt) to start and stop or otherwise control equipment, and to activate audible and visual alarms. The following list summarizes the functions commonly performed using electronic instrumentation:

1. Indication (gauges, digital indicators, and cathode ray tube displays)
2. Recording (strip or circular charts)
3. Data logging (magnetic tape or disks)
4. Alarm (audible or visual)
5. Control (computer, solid-state circuitry, or relay logic systems)

ELECTRICAL

Electrical signals (other than electronic) are also commonly used to perform a variety of signaling, telemetry, and control functions. Ordinary alternating current (AC), either 120 volt or 240 volt, is frequently used. However, other voltages, such as 6, 12, and 24 volts (AC or DC) are commonly used for control when a shock hazard may present a problem.

Common functions performed with electrical signals include the following:

1. ON/OFF control (make or break by means of switches)
2. Time-impulse control (signals are transmitted for only a portion of a cycle, for example, 15 seconds)
3. Pulse rate control (pulses or short transmissions of power)

1.5.6.2 Control Methods

Many of the instrumentation systems used in water treatment plants are combinations of the previously described systems, and are commonly referred to as hybrid systems.

Perhaps the most widely used method of automatically controlling pumps, valves, chemical feeders, and other devices is known as relay logic. Relay logic is a method of switching electrical power on and off in accordance with a predetermined sequence (logic) by means of relays, process switches and contacts, timers, and manual switches. Virtually any sequence of operational control can be achieved using relay logic. This flexible approach allows the system to be operated by manual controls if the need arises.

1.5.6.3 Computers

In many of the newer water treatment plants, computers (microprocessors) are being used to monitor and record data on process functions and status. However, critical functions are wired for manual operation if the computer fails. A cathode ray tube (CRT) is commonly used to visually display selected data when requested by an operator.

Increasing use of computers will minimize the number of visual indicators (such as stripchart recorders) required to monitor, record, and display water treatment process data.

Computer control systems are computer-monitored alarm, response, control, and data acquisition systems used by operators to monitor and adjust their treatment processes and facilities. The computer control system collects, stores, and analyzes information about all aspects of operation and maintenance; transmits alarm signals, when necessary; and allows fingertip control of alarms, equipment, and processes. The computer control system provides the information that operators need to solve minor problems before they become major incidents. As the nerve center at the treatment plant, the system allows operators to enhance the efficiency of their facility by keeping them fully informed and fully in control. Some systems allow remote monitoring and control through portable devices such as laptops, tablets, and smartphones.

Plant instrumentation does not relieve you of your responsibility to make operational decisions and to exercise operational control. Automation provides operators with accurate, timely information so that they can determine the necessary process changes for optimal plant operation.

Check Your Understanding

1. How are data transmitted over distances greater than about 1,500 feet?

2. What precautions must be exercised when using air for pneumatic data transmission?

3. What types of transmission fluids are commonly used in hydraulic systems?

4. List the functions commonly performed by electronic signals.

5. What is a hybrid instrumentation system?

6. What is relay logic?

1.5.7 Emergency Conditions and Procedures

In the operation of any water treatment plant, abnormal or emergency conditions will occasionally arise that require calm, quick action on the part of the operator. Emergency conditions you may encounter include:

1. Treatment process failures
2. Process equipment failures
3. Power failures
4. Fires
5. Floods, earthquakes, or other natural disasters

You must be able to distinguish between an abnormal condition and a red-alert emergency condition. A red alert means that you must immediately seek outside help. Typical red-alert emergencies include sabotage, raw water contamination, chemical spills, fires, serious injuries, and chlorine leaks.

1.5.7.1 Treatment Process Failures

Treatment process failures generally result from an abrupt or unexpected change in source water quality. A typical example of this condition occurs when the source water suspended solids concentration abruptly increases (high turbidity) as a result of precipitation and runoff into the source water supply. Other, less common, examples are accidental waste-water contamination or chemical spills in the source water system.

Operators should anticipate turbidity fluctuations resulting from precipitation and runoff, and they should obtain samples of the water for jar testing as soon as possible. This will allow operators to make planned adjustments to the treatment process and avoid major process upsets.

Contamination of the source water system by wastewater or chemical spills is nearly impossible to anticipate, so planned adjustments to the treatment process to correct for these problems is unlikely. However, an early warning of wastewater contamination of the source water may be a sudden drop in chlorine residual in the treatment process or a sudden increase in the chlorine demand of the water being treated. Immediate adjustment of the chlorine dosage should be made and additional bacteriological tests should be performed to define the extent of the problem. Do not wait for the results from bacteriological tests to tell you that you have a problem because by then it is too late.

Accidental chemical spills are perhaps the most hazardous situation to deal with since normal treatment process monitoring techniques may not detect the problem. In most cases, the operator must rely on outside notification of this event. Special sampling may be required to define the extent of the problem.

In cases where a treatment process upset results in the failure to meet a specific water quality standard, the operator must promptly notify supervisory personnel and the appropriate local health authorities. In extreme cases, complete process shutdown and public notification may be required. During these periods, the operator must work closely with health authorities.

Another cause of treatment process failures is operator error. Occasionally, all of us make mistakes. If we are on top of everything and operating our plant as intended, an error in one process may be eliminated or reduced by another process.

Proper chemical doses can be difficult to maintain. If you discover the chemical dose is too high or too low, immediately make the proper adjustment. Try to monitor the doses more frequently.

If you discover insufficient or no chlorine residual in your clear well, immediately increase the chlorine dosage to the finished water. Review your records. If everything is working properly, all of your quality control tests are looking good, and the turbidity level in the clear well is low, you probably will experience no serious problems.

However, if you discover no chlorine residual in your clear well and your plant is having operating difficulties, you are in trouble. If the turbidity is high in the clear well, try to add chlorine to the clear well or to the finished water pump discharge to achieve the desired chlorine residuals in the clear well and through the entire distribution system. Review your records and the operation of your chlorination system. Determine why there is no chlorine residual in the clear well and correct the situation.

Unfortunately, errors can and do happen. After an error has occurred, try to develop procedures that will prevent the error from occurring again. Share your experience with other operators so they will not make the same error. Working together is important and can help everyone.

1.5.7.2 Process Equipment Failures

Process equipment failures may also result in treatment process upsets in the event that chemical feeders, chlorinators, or other primary process equipment items fail to operate satisfactorily. The best safeguard against premature process equipment failures is a good preventive maintenance program. The operator plays a vital role in the preventive maintenance program by performing daily inspections of process equipment and making minor adjustments and repairs when necessary.

In certain essential processes, such as in chlorination systems, extra equipment is usually built into the system to provide backup in case an individual system part fails. This feature should also be included in other primary process systems such as chemical feed systems.

In the event of a process failure that results in the failure to meet a water quality standard, promptly notify supervisory personnel and the appropriate local health authorities. In extreme cases, process shutdown and public notification may be necessary. Let us examine possible equipment failures and how you might respond to them. We are assuming that you do not have standby facilities or they have failed, too. If you consider what would happen at your plant if these failures occur, you may be able to justify the installation of essential standby equipment.

1. Intake screens

   If intake screens become plugged or broken, shut the plant down and unplug or repair the screens. Standby or alternate screens obviously are essential. A bypass system may allow continuous operation and avoid the need to shut down the plant.

2. Grit basin

   If the mechanical collector fails and cannot be corrected or adjusted by above-water repair procedures, the facilities must be dewatered for repairs. Whenever facilities must be dewatered for emergency repairs, try to fill up all water storage facilities by early evening. Dewater the facility and make repairs at night when demands are low.

3. Prechlorination facility

   a. Shut down facilities and repair immediately. Try to avoid allow-ing unchlorinated water to pass through your plant and having to rely solely on postchlorination.

   or

   b. If postchlorination facilities are adequate, you may wish to rely strictly on postchlorination. Under these conditions, increase sur-veillance of chlorine residuals.

4. Alum or polymer feeder

   Shut down influent pumps. Repair the chemical feeder. Do not allow water to flow past the point of chemical application without alum or polymer. Otherwise, turbidity will pass through the filters and may exceed EPA Primary Drinking Water Standards.

5. Rapid mix or flash mix

   Consider moving the point of chemical application to a location where water turbulence can help to achieve hydraulic mixing.

6. Flocculators

   a. Underwater units can be repaired during scheduled dewatering of the facility.

   b. Mechanical units should be repaired as soon as possible.

7. Sedimentation tank

   If the mechanical sludge collector fails and cannot be repaired or adjusted by above-water procedures, dewater the facility and repair it at night when demands are low.

8. Filters

   If the valve or backwash system fails, take the failed portion (bank) out of service and repair.

9. Postchlorination facility

   a. Increase prechlorination doses, if possible.

   or

   b. If you are using postchlorination only, shut down and repair immediately. Notify supervisors and proper authorities.

10. Corrosion-control chemical feeder

    Repair as soon as possible.

1.5.7.3 Power Failures

A backup electrical power source is usually provided at water treatment plants for use in the event of commercial power failure. Engine-generator sets powered by diesel fuel, natural gas, or liquid petroleum gas provide the standby capability to furnish a limited amount of electrical power to keep the water treatment plant in service during periods of commercial power failure. In most cases, it is not practical to provide emergency power to meet all treatment plant demands. Therefore, only critical process func-tions (such as chemical feeders, mixers, flocculators, and process pumps) are included on the emergency power bus (a metallic strip or bar that conducts electricity in a distribution board). This power bus is usually con-nected to primary process equipment items by a transfer switch that auto-matically transfers power to the backup or standby source during failures.

At the onset of a commercial power failure, take the following actions:

1. Notify the commercial power supplier of the outage.

2. If the power failure originated at the treatment plant, notify electrical maintenance personnel immediately.

3. Restart process equipment that shut off during the power failure. (Prepare a sequence for your plant so that only one piece of equipment at a time is restarted to avoid overload.)

4. Check chlorination equipment and safety devices for proper operation.

5. Check the engine-generator set for proper operation.

6. Notify supervisory personnel of the condition.

7. Visually inspect all process equipment and check the performance of unit treatment processes.

During brief periods of power outage, most primary plant process functions can continue to operate. However, for extended periods of power outage, it may be necessary to reduce plant production because filter backwashing systems are usually not connected to the emergency power bus due to the high energy demand. As filters stop working because of head loss buildup or turbidity breakthrough, they should simply be removed from service. If insufficient clean filters are available to replace them, the plant flow rate will have to be reduced.

When commercial power is restored, take the following actions:

1. Restart process equipment that shut off during the transfer or was off line during the outage (one at a time).

2. Backwash dirty filters and return them to service.

3. Increase plant flow rate as appropriate.

4. Visually inspect all process equipment and performance of unit treatment processes.

5. Verify process and treated water quality.

6. Notify supervisory personnel of conditions.

1.5.7.4   Fires

If a fire occurs at the water treatment plant, immediately notify the local fire department and then determine the source and severity of the fire. Depending on the type of fire (structure, chemical, electrical), use the appropriate fire safety equipment at the plant in an attempt to extinguish the fire. Do not try to be a hero. If the fire is too involved, wait for the fire department to arrive (response time is usually short).

After calling the fire department, notify plant supervisory personnel promptly of the emergency condition at the plant.

If you have not already done so, make yourself thoroughly familiar with the care and use of fire safety equipment and learn the special procedures to be observed in dealing with chemical and electrical fires. You will not have time to study the equipment after a fire starts.

1.5.7.5   Natural Disasters

Fortunately, natural disasters such as floods and earthquakes are relatively rare events. Most water treatment plants are designed with these events in mind and adequate safety features are usually built into the plant to minimize damage caused by floods or earthquakes. Water treatment plants are normally located on sites that are above the standard flood plain, or special measures are taken to prevent facilities from flooding during a heavy rainstorm.

The design of new structures should include provisions for emergency preparedness and earthquake safety. Since there is not much that operators, or anyone else, can do during catastrophic events, only additional planning and emergency preparation will help protect water supplies.

Following any major flood, earthquake, or other natural disaster, take the following actions:

1. Inspect accessibility of all facilities.
2. Check condition and function of all process equipment.
3. Check structures and chemical storage tanks for structural or other damage.
4. Check the plant piping system for leaks or other visible signs of damage.
5. Prepare a preliminary damage report.
6. Report conditions to plant supervisory personnel.

1.5.7.6 Communications

In the event of an emergency, you will be required to advise other plant personnel of the existing conditions or events that have occurred.

An emergency response procedure should be developed for every water treatment plant so that notification of the proper personnel can be readily accomplished and the emergency resolved. Emergency response procedures should list the names and telephone numbers of persons to be notified under specified conditions, including health department authorities. Guidelines should be developed to assist the operator in determining when to implement these procedures. Alternate communication methods must be considered because telephone service may be lost during an emergency. Be sure to review the emergency response procedures at least once a year and verify the accuracy of all names and telephone numbers.

Check Your Understanding

1. What are three common causes of treatment process failures?
2. How would you safeguard against premature process equipment failures?
3. What would you do if the prechlorination facility in your plant failed?
4. What happens at most water treatment plants when commercial power fails?
5. What would you do if a fire occurred at your water treatment plant?
6. After a major flood, what action should be taken by a water treatment plant operator?

1.5.8 Operating Records and Reports

Records and reports are important aspects of both operation and maintenance of water treatment plants. Tracking processes and operator activities to ensure the plant produces safe water not only helps all operators work together and adjust processes as needed, but also ensures compliance with regulating agencies. In addition, good records contribute to an effective maintenance program that helps avoid emergency situations (see Section 1.6.2, "Records Management").

Effective operators use both verbal and written communications to present information about plant operations to other operators, supervisors, and regulatory agencies. The most common oral reports are those given by the operator whose shift is ending to the operator who is coming on duty as part of the daily tasks (see Section 1.5.1, "Daily Operating Procedures"). The most basic written records are entries in a daily diary or pocket notebook used by each operator to record unique or unusual events. A typical entry in an operator's diary or daily operating log may include:

- *8:30 am*
- *Raw water pump No. 1 starting to develop a whine during low flows.*
- *JQO* (operator should initial entry)

Many plant records are documented and stored for later reference on paper forms or using desktop, laptop, or handheld computers. These operating records can be separated into two major categories: physical records and performance records.

Physical records describe the water treatment plant physical facilities and equipment. These records include:

1. Plant design criteria
2. Construction plans and contract specifications
3. As-built (record) drawings
4. Equipment fabrication drawings and specifications
5. Manufacturers' operation and repair manuals for all equipment items
6. Detailed piping plans and electrical wiring diagrams
7. Equipment records, including manufacturer's name, model number, rated capacity, and date of purchase
8. Maintenance records on each equipment item
9. Hydraulic profiles showing pertinent operating water surface elevations throughout the water treatment plant
10. Cost records for all major equipment item purchases and repairs

Performance records describe the operation of the water treatment plant and provide the operator, as well as others, with a running account of plant operations (historical records). Figures 1.6–1.8 are performance records that show the kinds of information included in daily operation records, chemical inventory records, and water quality records. These records are a valuable resource for the operator trying to solve current process problems and anticipate future needs. Performance records also provide a factual account of the operation that is required to meet legal and regulatory agency requirements.

Typical performance records include the following:

1. Daily operation records (process production inventory, process changes, process equipment performance—see Figure 1.6)
2. Water quality records (source water, process water, finished water—see Figures 1.7 and 1.8)
3. Equipment failure records
4. Accident records
5. Consumer complaint records (include follow-up investigations and corrective actions taken)

6. Chemical inventory records (include storage amounts, safe storage levels, procurement records—see Figure 1.6)
7. Charts produced by process records (strip-chart recorders)
8. Visitor information

Check Your Understanding

1. What is the reason for keeping adequate, reliable records?

2. What are performance records?

1.6 Plant Maintenance

An effective water treatment plant maintenance program will ensure continuing satisfactory operation of the treatment plant facilities under a variety of different operating conditions. Such a program would include routine and preventive maintenance (PM) as well as provisions for effectively handling emergency breakdowns.

All plants should have written instructions on how to operate and maintain the equipment. These materials will serve as reference for all operators and training instructions for new operators. If your plant does not have written instructions, prepare them now or budget the necessary funds to have someone prepare them.

The major elements of a good PM program include the following:

1. Planning and scheduling
2. Records management
3. Spare parts management
4. Cost and budget control
5. Emergency repair procedures
6. Training program

1.6.1 Planning and Scheduling

Planning and scheduling are the foundation of the maintenance program. An important source of information for planning routine maintenance is equipment manufacturers' operating and maintenance instructions. These instructions are usually furnished with the equipment at the time of purchase, or can be obtained directly from the manufacturer. Another important reference source is the plant operation and maintenance manual.

The planning and scheduling effort should define the specific maintenance tasks to be done and the time intervals between tasks. The following items are important considerations in developing a good maintenance plan:

1. Routine procedures
2. Special procedures (equipment overhaul)
3. Skills needed
4. Special tools and equipment requirements
5. Parts availability

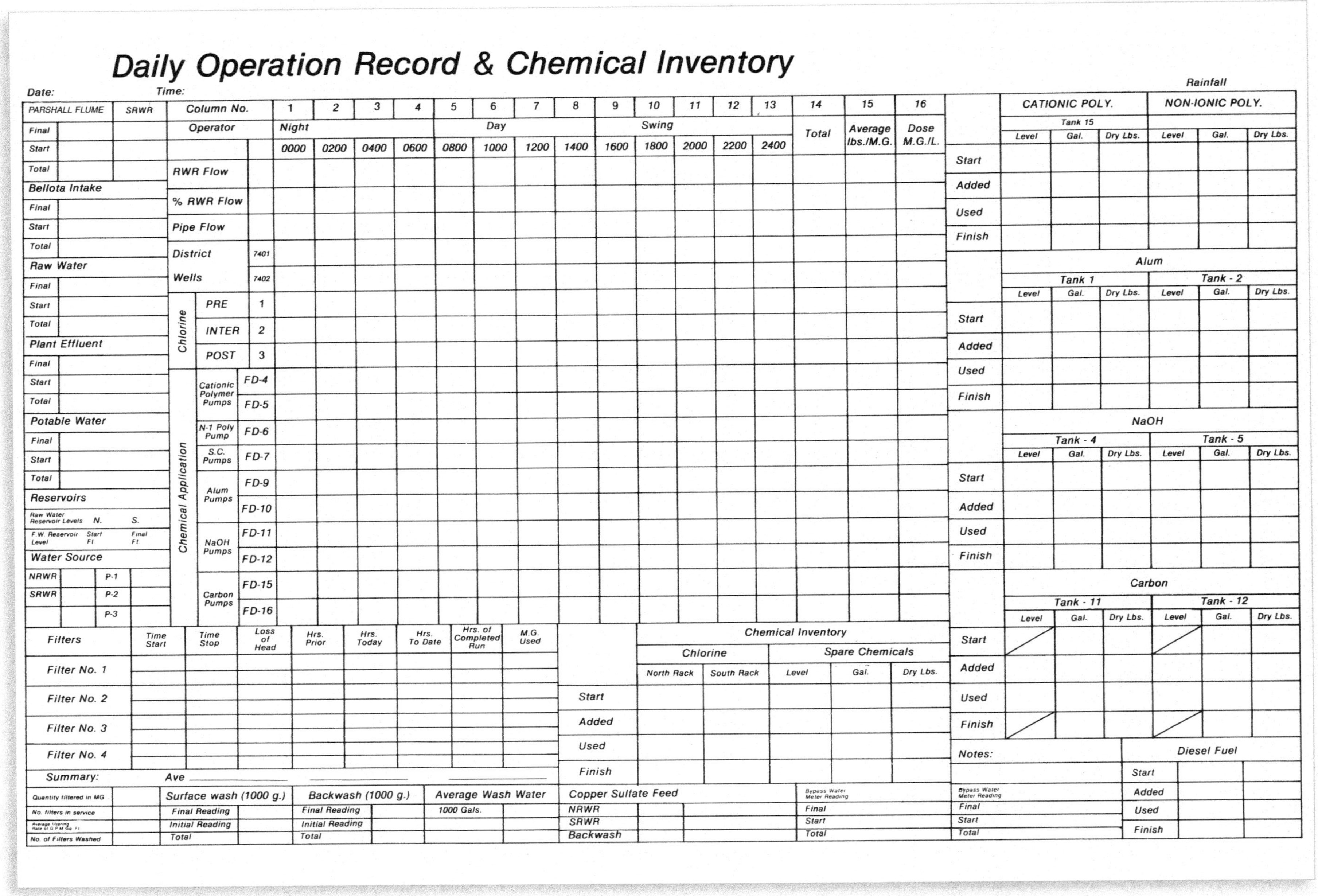

Daily Operation Record & Chemical Inventory

Date: Time:

PARSHALL FLUME	SRWR	Column No.	1	2	3	4	5	6	7	8	9	10	11	12	13	14	15	16
		Operator	Night					Day				Swing				Total	Average lbs./M.G.	Dose M.G./L.
			0000	0200	0400	0600	0800	1000	1200	1400	1600	1800	2000	2200	2400			

Final / Start / Total

RWR Flow

Bellota Intake — % RWR Flow — Pipe Flow — Final / Start / Total

Raw Water — District 7401 — Wells 7402 — Final / Start / Total

Chlorine — PRE 1 — INTER 2 — POST 3

Plant Effluent — Final / Start / Total

Chemical Application — FD-4, FD-5 (Cationic Polymer Pumps); FD-6, FD-7 (N-1 Poly Pump, S.C. Pumps); FD-9, FD-10 (Alum Pumps); FD-11, FD-12 (NaOH Pumps); FD-15, FD-16 (Carbon Pumps)

Potable Water — Final / Start / Total

Reservoirs

Raw Water Reservoir Levels: N. S.   Start Fi Final Fl
F.W. Reservoir Level: Start Fi Final Fl

Water Source — NRWR (P-1), SRWR (P-2), (P-3)

Summary: Ave

Quantity filtered in MG / No. filters in service / Average flow Raw 0.5 and 0.4 / No. of Filters Washed

Filters	Time Stop	Time Start	Loss of Head	Hrs. Prior	Hrs. Today	Hrs. To Date	Hrs. of Completed Run	M.G. Used
Filter No. 1								
Filter No. 2								
Filter No. 3								
Filter No. 4								

Surface wash (1000 g.) — Final Reading / Initial Reading / Total
Backwash (1000 g.) — Final Reading / Initial Reading / Total
Average Wash Water — 1000 Gals.

Copper Sulfate Feed — Start / Added / Used / Finish — NRWR / SRWR / Backwash

Chemical Inventory

Chlorine — North Rack / South Rack

Spare Chemicals — Level / Gal. / Dry Lbs.

CATIONIC POLY. — Tank 15 — Level / Gal. / Dry Lbs.
NON-IONIC POLY. — Level / Gal. / Dry Lbs. — Rainfall
Start / Added / Used / Finish

Alum — Tank 1 / Tank - 2 — Level / Gal. / Dry Lbs.
Start / Added / Used / Finish

NaOH — Tank - 4 / Tank - 5 — Level / Gal. / Dry Lbs.
Start / Added / Used / Finish

Carbon — Tank - 11 / Tank - 12 — Level / Gal. / Dry Lbs.
Start / Added / Used / Finish

Diesel Fuel — Start / Added / Used / Finish

Notes:

Bypass Water Meter Reading — Final / Start / Total

Figure 1.6 Combined operation record and chemical inventory

DAILY LABORATORY RECORD

Operator:			Shift One				Shift Two				Shift Three				Date		
Sample Site	Test	Units	Time												Daily Mean	Range	
			0000	0200	0400	0600	0800	1000	1200	1400	1600	1800	2000	2200		Low	High
Plant Influent (SA-1)	Turb.	NTU															
	R.Cl	mg/l															
	pH																
	Temp.	°C															
	Color	C.U.															
	Odor	TON															
	Alk.	mg/l															
	Hard	mg/l															
Settled Water (SA-3)	Turb.	NTU															
	R.Cl	mg/l															
	pH																
	Temp.	°C															
Filtered Water (SA-4)	Turb.	NTU															
	R.Cl	mg/l															
	pH																
	Temp.	°C															
Plant Effluent (SA-5)	Turb.	NTU															
	R.CL	mg/l															
	pH																
	Temp.	°C															
	Color	CU.															
	Odor	TON															
	Alk.	mg/l															
	Hard.	mg/l															
	Flow	MGD															

Plant Flow			Polymer		Alum		NaOH		Chlorine						PAC		Sp. Chem	
									Pre		Inter		Post					
Time	MGD	Source	Time	mg/l	Time	mg/l	Time	mg/l	Time	lb/D	Time	lb/D	Time	lb/D	Time	mg/l	Time	mg/l

Remarks:

Figure 1.7 Water quality record showing laboratory tests

DAILY BACTERIOLOGY RECORD

DATE:			Routine Samples							Raw Water			Special	
Sample Location			SA-5	SA-5	SA-5				Blank					
Sample Date														
Sample Time														
Residual Cl_2														
Collected By														
Total Coliform														
Presumptive MPN		Vol.												
Medium:	Positives	24h												
Date:														
Control:		48h												
Confirmed MPN		Vol.												
Medium:	Positives	24h												
Date:														
Control:		48h												
MPN Index/100 ml														
Membrane Filter		Vol.												
Medium:		Total Count												
Date:														
Control		Coliform Count												
Coliforms/100 ml														
Fecal Coliform														
MPN		Vol.												
Medium:	Positives	24h												
Date:														
Control:		48h												
Fecal MPN Index/100 ml														
Membrane Filter		Vol.												
Medium:		Total Count												
Date:														
Control		Fecal Count												
Fecal Coliforms/100 ml														
Std. Plate Count		Vol.												
Medium:		Count												
Date:		SPC												
Control:		MI												

Set up by:	Temp (°C)	AM	PM	Notes:
Time:	Incubator			
Read D-1:	Water Bath			
Day 2	Laboratory			
Day 3	Time			
Day 4	By			

Figure 1.8 Water quality record showing coliform tests

ROUTINE PROCEDURES

In the routine operation of the water treatment plant, the operator inspects various mechanical equipment items (valves, pumps) and electrical equipment items (motors) to check for proper operation, and performs maintenance functions including:

1. Keep motors free of dirt and moisture.
2. Ensure good ventilation (air circulation) in equipment work areas.
3. Check motors and pumps for leaks, unusual noise, vibrations, or overheating.
4. Maintain proper lubrication and oil levels.
5. Check for alignment of shafts and couplings.
6. Check bearings for overheating and proper lubrication.
7. Check for proper valve and pump operation.
8. Check calibration of chemical feeders.

These routine tasks should generally be performed on a daily basis. For additional details on how to develop a routine maintenance program, see the chapter on maintenance in Volume II of this series of training manuals.

SPECIAL PROCEDURES

Effective maintenance programs rely on proper scheduling and performance of these procedures:

1. Plan equipment shutdowns to minimize adverse impacts on plant operations.
2. Use equipment repair records to plan and schedule maintenance.
3. Prepare step-by-step procedures or refer to manufacturer's instructions for performing equipment overhauls or other special maintenance tasks.

TOOLS

To be effective in the routine maintenance of plant equipment, you will need to know how to properly use common hand tools to protect equipment and for your own safety. Remember that you cannot perform maintenance procedures safely and properly if you do not have the proper tools.

The exact tools each operator should be familiar with is difficult to specify because the maintenance performed by a plant operator will vary considerably depending on the size of the treatment plant and the number of operators. In larger plants, maintenance personnel will perform virtually all of the routine as well as specialized maintenance functions. In smaller plants, the operator will be expected to perform most of the routine maintenance functions. Operators should at least be familiar with the following types of common hand tools:

1. Screwdrivers (slotted and Phillips)
2. Pipe wrenches
3. Crescent wrenches
4. Socket wrenches
5. Allen wrenches
6. Open-end and box wrenches

7. Hammers (claw, sledge, mallet)

8. Pliers (vise-grips)

9. Files

10. Wire brushes

11. Putty knives

Tool loss and replacement can be a problem at any water treatment plant unless procedures are implemented to effectively manage the inventory of tools and equipment. A commonly used procedure is the checkout list. An individual desiring to use a special tool writes the type of tool, date, and name or initials on the list. When the tool is returned, the name is struck off the list. Index cards can be used for this purpose. A color-coded tag system or computer programs may also be used to help inventory and keep track of tools and equipment.

In the event that a tool is damaged or lost, the description of the tool and any special circumstances or information should be noted so that it can be repaired or replaced.

For an additional list of tools used by operators, see *Small Water System Operation and Maintenance,* Chapter 4, "Small Water Treatment Plants," in this series of operator training manuals.

1.6.2 Records Management

Records management is an important part of the overall recordkeeping process. Records should be filed and cataloged or indexed for future reference. Regulatory agencies may require you to keep certain water quality analyses (bacteriological test results) and customer complaint records on file for specified time periods (10 years for chemical analyses and bacteriological tests). Other records that may have historical value to the operator (source water quality changes and resultant process changes) should be kept as long as they are useful. These records should be maintained in appropriate files properly labeled for easy reference. A comprehensive records management system provides the basis for daily task assignments, provides a permanent record of work performed, and becomes a historical reference source for reviewing equipment performance. A good records management system includes equipment inventory cards, preventive maintenance schedules, spare parts lists and reorder information, and records of work performed as well as operating records and reports.

1.6.3 Spare Parts Management

Certain parts of mechanical equipment items, such as shaft bearings, require periodic replacement because they have a useful life that is considerably shorter than the predicted overall equipment life. This requires that an adequate stock of spare parts be kept on hand at the treatment plant to facilitate planned replacement. Spare parts should be stocked on the basis of:

1. The importance of the part to operation

2. Availability

3. The effect on operation if the part is defective

4. Storage space

If a part is readily available from a supplier, let the supplier stock it for you. It costs money to stock and warehouse unnecessary spare parts. The actual performance history of a particular piece of equipment may indicate the type and number of spare parts that must be stocked. Spare parts should be promptly reordered whenever they are used.

1.6.4   Cost and Budget Control

Accurate records of labor and equipment expenditures are an important part of the overall budget and cost control program. Operation and maintenance budgets are usually prepared on an annual basis. A thorough and up-to-date written performance history of equipment operations, repair, and replacement costs will significantly improve the budget planning process. The work order system is a way of keeping track of how much time and money is spent doing various types of work. This information becomes a planning tool and a reporting system to indicate what is being done and how much it costs. This procedure records the parts used in the repairs and the amount of labor required to perform the repair or other maintenance procedures, as well as other kinds of work. These records will also provide valuable information for deciding when to replace a given piece of equipment due to excessive repair costs.

1.6.5   Emergency Repair Procedures

Identify those pieces of equipment that are critical for your facility to meet the demands for safe drinking water from your consumers. Critical pieces of equipment include raw water pumps, chlorinators, and pumps that deliver finished water to the distribution system. All of these items must have standby or backup equipment. Also, your plant should have standby generators in case of a power outage. If any of these facilities fail, you must have emergency repair procedures to follow that will enable you to put your facility back in service as soon as possible.

1.6.6   Training Program

Perhaps as important as any other single element of the operation and maintenance program is training. Training should be an ongoing feature of the operation program and operators should be encouraged to participate. Such training can increase the expertise of maintenance and operations personnel in the general repair of equipment, in specialized procedures required to calibrate and repair selected equipment items, and in their ability to quickly and properly respond to changes in raw and finished water.

Major equipment manufacturers periodically conduct training programs designed to provide operations and maintenance personnel with a hands-on familiarity with common mechanical and electrical equipment items.

The certification requirements in most states require successful completion of some form of education or training to qualify for taking a certification examination. In many states, some type of education requirement must be met before a certificate can be renewed. In order to do a good job, people need an opportunity to improve their knowledge and skills. The best way for people to improve themselves is with a well-planned training program.

1.6.7 Security

Many public water treatment and delivery systems are required to conduct vulnerability assessments and develop emergency response plans by the Bioterrorism Act of 2002. EPA provides resources to improve drinking water and wastewater system resiliency to disasters, and to quickly recover from contamination involving chemical, biological, and radiological (CBR) agents. EPA also helps water utilities to enhance the cybersecurity of their water systems.

One way to protect facilities from threats is to limit access to the plant and grounds. Fences and gated accesses help to discourage trespassers and other unwanted visitors from entering treatment plant grounds and facilities. Gates should be securely locked during nonroutine working hours and, in some instances, automatic remotely controlled gates may be required to limit access during all hours of operation. Routinely inspect the plant facilities (at least once per shift) and report any unauthorized persons or unusual events to the proper authorities.

Check Your Understanding

1. List the major elements of a good preventive maintenance program.
2. What items should be included in a cost and budget control program?
3. How often should an operator routinely inspect plant facilities for any evidence of unauthorized persons or unusual events?

1.7 Energy Conservation

In the operation of a water treatment plant, a considerable amount of energy may be consumed for lighting, heating and air conditioning, and powering numerous electric motors located throughout the plant. Operators can have a positive impact on overall treatment plant operating costs if energy conservation procedures are followed on a routine basis. Some energy conservation measures at water treatment plants are similar to those for residential properties.

Even though lighting usually represents less than 5 percent of the electric energy use in a water treatment plants, costs can be reduced by:

1. Turning lights off when leaving a room or work area
2. Turning lights off in unoccupied areas
3. Limiting yard lighting to essential areas
4. Replacing existing lamps with high-efficiency lamps as the old ones burn out or require replacing
5. Converting mercury vapor lamps to more efficient high-pressure sodium vapor lamps

Heating, ventilating, and air conditioning equipment consume substantial amounts of fuel and electric energy. Energy savings can be gained here by adjusting thermostats to more efficient settings. Thermostats should be set at 78°F (26°C) for cooling, and at 68°F (20°C) for heating.

Electric motors consume the greatest amount of electricity in most water treatment plants, often over 90 percent of total electric energy consumed. Considerable savings can be achieved by replacing old electric motors with high-efficiency motors as the old ones burn out.

Consideration should also be given to installing capacitors at the treatment plant to correct a low plant power factor. Capacitors offset the reaction power used by inductive devices (electric motors) and improve the overall **power factor** of the plant. This can result in considerable energy savings.

1.7.1   Power Management

The most sophisticated approach to energy conservation, and perhaps the most beneficial, is through a power management program. The first step in developing a power management program to help conserve energy at a water treatment plant is to identify each source of energy use in the plant. These sources are then tabulated, ranked by size of load, and prioritized to define the importance of each load source in the overall program. This basic evaluation will frequently point to areas where immediate savings can be achieved through simple changes in routine operations (for example, avoid backwashing a filter when another high-load source is in operation to reduce peak power charges).

Some utility suppliers have a time-of-use billing schedule, which provides the user with significant price breaks during off-peak demand periods. If possible, filter backwashing or other discretionary functions can be performed during off-peak periods to take advantage of these lower rates.

In its most sophisticated form, power management can be used to control all of the electrical loads in the water treatment plant. With the aid of a computer, each load can be monitored and controlled to provide the most cost-effective operating mode. In some instances, noncritical loads can be turned off (shed) for short time periods while higher priority loads are on line. This can result in significant operational cost savings without compromising the safety and reliability of plant operations.

1.7.2   Power Cost Analysis

A monthly power cost analysis (Figure 1.9) will help assess the impact of power costs on your overall plant operating budget.

The first step of a power cost analysis is reading and recording monthly electric energy use. This is accomplished by reading the plant utility meter or meters. Using the proper meter multiplier factor (1,000 in this example, Row C), you can then determine the total amount of electric energy used in kilowatt hours (kWh) during the month (shown in Row E).

The kilowatt (kW) demand can also be read on the utility meter. This value represents the greatest single energy demand during the month (or prior months), and is generally billed as a separate cost component. Overall energy costs can generally be reduced by keeping demand to

power factor
The ratio of the true power passing through an electric circuit to the product of the voltage and amperage in the circuit. This is a measure of the lag or lead of the current with respect to the voltage. In alternating current the voltage and amperes are not always in phase; therefore, the true power may be slightly less than that determined by the direct product.

a minimum. Notice in Row D that the demand normally varies with plant flow. This results from the higher demand for electrical equipment (pumps and motors) to produce the higher flow rates and operating pressures required during the warmer months. If you are careful, you can avoid creating high peak demands during the warmer months by shifting nonessential operations requiring electrical energy to off-peak hours. Some examples, such as backwashing filters during non-peak hours, have been discussed in the preceding sections.

Total power cost, as shown in Row G, can be obtained from the monthly energy bill. Power costs can be calculated in convenient units such as $/MG as shown in Row J. In this example, Row J was obtained by dividing the total power cost shown in Row G ($) by the total gallons of water pumped as shown in Row I (MG). Monthly power costs are useful for checking the current budget allocation for energy and for preparing the following year's budget.

Keeping track of other treatment costs such as chemical costs (shown in Rows K and L) and water cost (shown in Row M) can also be useful to the operator as well as to supervisory personnel. If large variations in treatment costs appear in any given month, this analysis form will provide you with important clues to help explain or solve the problem.

Row P is provided for making comments on any unusual conditions that caused higher energy or chemical demands. These remarks will be helpful in budgeting for the next year, as well as reporting on the current year's performance (annual operating report).

Check Your Understanding

1. List the major sources of energy consumption in the operation of a water treatment plant.

2. How can the consumption of electric energy at a water treatment plant be reduced?

1.8 Supervision and Administration

In addition to operation and maintenance duties for your water treatment plant, you may also be responsible for supervision of personnel. Chief operators frequently have the responsibility of training new operators and should encourage all operators to strive for higher levels of certification.

As a plant administrator, you may be in charge of recordkeeping. In this case, you will be responsible for operating and maintaining the facilities as efficiently as possible, keeping in mind that the primary objective is to produce safe and pleasant drinking water from your plant. Without adequate, reliable records of the important phases of operation and maintenance, the effectiveness of your operation will not be properly documented (recorded). Also, accurate records are required by regulatory agencies for compliance with the National Primary Drinking Water Regulations of the Safe Drinking Water Act.

Records are an excellent operating tool. Reference to past records can be helpful in adjusting treatment processes for various changes in raw water.

Power Cost Analysis Fiscal Year								
(A) Reading Date	Jun-30	Jul-31	Aug-31	Sep-30	Oct-31	Nov-30	Dec-31	Jan-31
(B) Reading	425.1	969.9	1,525.9	2,089.9	2,518.7	2,909.9	3,210.7	3,522.7
(C) Multiplier	1,000.0	1,000.0	1,000.0	1,000.0	1,000.0	1,000.0	1,000.0	1,000.0
(D) KW Demand	N/A	1,072	1,064	1,072	1,080	1,016	904	920
(E) KWH	N/A	544,800	556,000	504,000	428,800	391,200	300,800	312,000
(F) KWH per KW Demand	N/A	508.2	522.6	526.1	397.0	385.0	332.7	339.1
(G) Total Power Cost	N/A	32,661.82	33,283.36	33,572.11	24,179.31	22,191.25	17,435.19	17,801.25
(H) Water Meter Reading (x 1,000)	7,402,853	8,167,061	8,988,688	9,793,313	10,552,931	11,204,146	11,735,615	12,163,703
(I) Water Pumped (MG)	N/A	764.2	821.6	804.6	759.6	651.2	531.5	428.1
(J) KWH per MG	N/A	713	677	701	565	601	566	729
(K) Power Cost per MG	N/A	42.74	40.51	41.73	31.83	34.08	32.80	41.58
(L) Cost of Chemicals	N/A	27,079.38	25,290.26	27,457.87	20,801.23	19,004.29	7,190.50	8,571.69
(M) Chemical Cost per MG	N/A	35.43	30.78	34.13	27.38	29.18	17.29	20.02
(N) Water Cost per MG	N/A	11.80	11.80	11.80	11.80	11.80	11.80	14.50
(O) Total Power, Chemical & Water Cost per MG*	N/A	89.97	83.09	87.66	71.01	75.06	61.89	76.10
(P) Remarks		Fed activated carbon for taste & odor control	Fed activated carbon for taste & odor control	Fed activated carbon for taste & odor control	Fed activated carbon for taste & odor control	Fed activated carbon for taste & odor control		

Power Cost Analysis Fiscal Year *(continued)*							Annual Total	Annual Average
(A) Reading Date	Feb-28	Mar-31	Apr-30	May-31	Jun-30		Annual Total	Annual Average
(B) Reading	3,839.5	4,157.9	4,543.5	5,021.9	5,649.1		N/A	N/A
(C) Multiplier	1,000.0	1,000.0	1,000.0	1,000.0	1,000.0		N/A	N/A
(D) KW Demand	945	952	1,120	1,300	1,215		12,660	1,055
(E) KWH	316,800	318,400	385,600	478,400	627,200		5,164,000	430,333
(F) KWH per KW Demand	335.2	334.5	344.3	368.0	516.2		4,908.9	409.1
(G) Total Power Cost	18,069.16	18,034.57	21,570.10	26,560.71	34,246.07		299,604.90	24,967.08
(H) Water Meter Reading (x 1,000)	12,588,965	12,975,671	13,393,901	14,049,165	14,810,375		N/A	N/A
(I) Water Pumped (MG)	425.3	386.7	418.2	655.3	761.2		7,407.5	617.3
(J) KWH per MG	745	823	922	730	824		8,596	716
(K) Power Cost per MG	42.49	46.64	51.58	40.53	44.99		491.50	40.96
(L) Cost of Chemicals	9,601.35	22,305.77	11,395.31	10,689.01	10,052.36		199,439.02	16,619.92
(M) Chemical Cost per MG	22.58	57.68	27.25	16.31	13.21		331.24	27.60
(N) Water Cost per MG	14.50	14.50	14.50	14.50	14.50		157.80	13.15
(O) Total Power, Chemical & Water Cost per MG*	79.57	118.82	93.33	71.34	72.70		980.54	81.71
(P) Remarks		High coagulant dose due to high turbidity						

*For comparison purposes only. Labor and fixed charges not included.

Figure 1.9 Power cost analysis

You may also be the budget administrator. Here you will be in the best position to give advice on budget requirements, management problems, and utility planning. You should be aware of the necessity for additional expenditures, including funds for plant maintenance and enlargement, equipment replacement, laboratory requirements, and personnel needs. You should recognize and define such needs in sufficient time to inform the proper officials to enable them to accomplish early planning and budgeting.

1.9   Public Relations

As an operator, you are in the field of public relations and must be able to explain the purpose and operation of your water treatment plant to visitors, civic organizations, school classes, representatives of the news media, and even to city council members or directors of your district. A well-guided tour for officials of regulatory agencies or other operators may provide these people with sufficient understanding of your plant to allow them to suggest helpful solutions to operational problems. One of the best results from a well-guided tour is gaining support from your city council and the public to obtain the funds necessary to run a good operation.

The overall appearance of your water treatment plant indicates to visitors the type of operation you maintain. If the plant looks dirty and run-down, you will be unable to convince your visitors that you are doing a good job. Your records showing that you are producing a safe drinking water will mean nothing to visitors unless your plant appears clean and well maintained.

Another aspect of your job may be handling customer inquiries or complaints. Whether the customer is asking questions or making a complaint, be courteous and respond as thoroughly as possible. It may take more than one conversation to completely answer a question or to resolve a complaint, so ask for contact information. For more on dealing with water quality complaints, see Section 1.5.4.2, "Water Quality Complaints."

1.10   Safety

Safety is an important operator responsibility and not to be taken for granted. Every operator has the responsibility to be sure that the water treatment plant is a safe place to work and visit. Everyone must follow safe procedures and understand why they must be followed. Operators must be aware of the safety hazards in and around treatment plants. Most accidents result from carelessness or negligence. You should plan or be a part of an active safety program. Chief operators frequently have the responsibility of training new operators and ensuring that safe procedures are employed in all tasks.

In the routine operation of the water treatment plant, the operator will be exposed to many potential hazards, including:

1. Electrical equipment (shocks)
2. Rotating mechanical equipment
3. Open-surface, water-filled structures (drowning)
4. Underground structures (toxic and explosive gases, lack of oxygen, or too much oxygen)

5. Water treatment chemicals (acids, alkalies, chlorine gas)
6. Laboratory reagents (chemicals)
7. Pump stations (high noise levels)

Ample safety devices are generally provided at each water treatment plant to protect the operator, as well as others, from accidents and exposure to chemicals, dust, and other hazardous environments. However, these safety devices are of limited value unless you practice safe procedures. For instance, if an object appears too heavy to lift, do not try to lift it. Get help or use a lifting device such as a forklift.

When working around mechanical and electrical equipment, plant structures, or chemicals, follow the safety procedures listed below to avoid accidents or injury.

ELECTRICAL EQUIPMENT

1. Avoid electric shock by using protective gloves.
2. Avoid grounding yourself in water or on pipes.
3. Ground all electric tools.
4. Lock out electric circuits and tag out remote controls when working on equipment.
5. Always assume all electrical wires are live.
6. Never use metal ladders around electrical equipment.
7. When in doubt about a procedure or repair, ask for help.
8. Use the buddy system and be sure your buddy knows how to rescue you when you need help.

MECHANICAL EQUIPMENT

1. Do not remove protective guards on rotating equipment.
2. Do not wear loose clothing around rotating equipment.
3. Secure and lock out drive motors before working on equipment. Tag out remote controls.
4. Clean up all lubricant spills (oil and grease).

OPEN-SURFACE, WATER-FILLED STRUCTURES

1. Do not avoid or defeat the purpose of protective devices such as handrails by removing them when they are in the way.
2. Close all openings when finished working.
3. Wear Coast Guard-approved personal flotation devices (PFDs) when working around basins or tanks, standing in moving water, or sampling from watercraft. Know the location of all PFDs.

UNDERGROUND STRUCTURES AND CONFINED SPACES

1. Use all confined spaces procedures before entering and while working in confined spaces. (See the section on confined spaces in the chapter on safety in *Water Treatment Plant Operation*, Volume II.) Know the condition of the environment before entering. Determine if there are any toxic gases present, explosive conditions, or an excess or lack of oxygen. Use detection devices that are capable of monitoring the atmosphere continuously.

2. Use portable ventilation fans to ensure good air circulation.

3. Use the buddy system. Also, be certain your buddy is trained and knows what to do in the event that you get into trouble.

CHEMICALS

1. Wear protective clothing when handling or unloading chemicals.

2. Wear goggles and face shields around all potentially hazardous chemicals.

3. Know the location of all safety showers and eye wash facilities and be sure they work.

4. Be familiar with the care and use of air packs.

5. Use chlorine leak detection and safe handling procedures.

6. Promptly clean up all chemical spills.

PUMP STATIONS

1. Use hearing protection devices.

2. Observe the precautions previously listed for working around electrical and mechanical equipment.

If a hazardous situation exists or if a particular procedure is unsafe, do not proceed—call for help.

Always be sure to report any injury, no matter how slight, to your immediate supervisor. This procedure protects you as well as your employer.

For details on how to develop a safety program and safety procedures, see the chapter on safety in Volume II of this series of training manuals.

Check Your Understanding

1. Why are well-guided tours for officials of regulatory agencies or other operators important?

2. Who is often responsible for ensuring that safe procedures are used in all tasks?

3. List the potential hazards an operator could be exposed to during the routine operation of a water treatment plant.

1.11   Water Treatment Plant Operators

Qualified, capable, and dedicated operators perform the day-to-day work of safely operating, maintaining, repairing, and managing water treatment plants, as well as interacting with the public they serve. These operators have the responsibility of producing safe and pleasant drinking water from their plants. With that responsibility comes the additional responsibility for the health and well-being of the community served by the treatment plant. If operators fail to do their jobs, they could be responsible for an outbreak of a waterborne disease, which could even result in death.

The water treatment field is changing rapidly. New treatment plants are being constructed and old plants are being modified and enlarged to meet the water demands of our growing population and industries. In addition to operators, maintenance personnel, supervisors, managers, instrumentation experts, and laboratory technicians are needed now and will be into the future. The Bureau of Labor Statistics (BLS) projects that the number of jobs in the water and wastewater treatment industry will continue to increase as the population grows, current operators retire, regulatory requirements increase, and treatment processes become more sophisticated.

The knowledge and skills required of an operator depend to a large degree on the size and type of water treatment plant where the operator works. Most water treatment jobs require at least a high school education. While some jobs will always exist for manual labor, the real and expanding need is for qualified operators. Most US states and other local governments require certification to work in water treatment plants. Generally, certified operators have demonstrated some combination of experience and education through reports from supervisors, continuing education or contact hours, and examinations. Certification examinations are usually administered by state regulatory agencies or professional associations. The specifics vary among states and provinces, but most certifying agencies have four levels of certification that require increasing experience and knowledge to pass the certification examination. Even operators who choose to maintain a certain level of certification may be required to participate in continuing education to do so, especially as new techniques, advanced equipment, and increasing use of complex instrumentation become industry norms.

Jobs are available for water treatment plant operators wherever people live and need someone to treat water for their homes, offices, or industrial processes. The different types and locations of water treatment plants offer a wide range of working conditions. From a single process operator or a computer control center operator at a complex municipal treatment plant serving several hundred thousand persons and employing 15 to 25 operators to a one-person operation at a small town water treatment plant where the operator must be a jack-of-all-trades because of the diversity of tasks, water treatment plant operators can choose positions that suit their individual preferences and ambitions.

The salary, benefits, and opportunities vary as much as the locations of water treatment plants. At a large municipality, the pay is good and advancement prospects are tops. In a small town, the pay may not be as good, but job satisfaction, freedom from time-clock hours, community service, and prestige may well add up to a more desirable personal achievement. Many of these positions are represented by an employee organization that will try to obtain higher pay and other benefits for operators.

Operators' paychecks usually come from a city, water agency or district, or a private utility company. In addition, operators may be employed by one of the many large industries that operate their own water treatment facilities. Operators are always responsible to their employers for operating and maintaining an economical and efficient water treatment plant. An even greater obligation rests with operator because of the great number of people who drink the water from water treatment plants. In the final analysis, operators are really working for the people who depend on them to produce safe and pleasant drinking water from the treatment plant.

For additional information about the types of jobs a water treatment plant operator performs, working conditions, wages, and the job outlook, refer to the Bureau of Labor Statistics (BLS) website: www.bls.gov.
Visit your local water agency's website for more information about jobs in your area.

Check Your Understanding

1. Why is there a need for water treatment plant operators?
2. What influences the knowledge and skills operators are required to have for their jobs?
3. Who hires water treatment plant operators?

1.12   Math Assignment

A good way to learn how to solve math problems is to work on them a little bit at a time. In this training manual, a short math assignment is included at the end of the chapters. Working through each math assignment can increase your ability to solve math problems common to water treatment plant operation. Some of you may need to spend more time on the basics depending on your background and how long it has been since you used math concepts.

Turn to Appendix A, "Introduction to Basic Math for Operators," and read sections:

- "Learning Objectives"
- "Introduction"
- A.1, "Numbers and Operations"

Read and work through the example problems in sections:

- A.1.1, "Addition"
- A.1.2, "Subtraction"
- A.1.3, "Multiplication"
- A.1.4, "Division"
- A.9.1, "Plant Operation" (English System)
- A.10.1, "Plant Operation" (Metric System)

1.13   Additional Resources

Your local library may be a good source for obtaining these additional resources related to the field of water treatment:

Basic Science Concepts and Applications. American Water Works Association (AWWA), www.awwa.org, ISBN: 1-58321-233-7.

Chemistry for Environmental Engineering and Science, 5th Edition. 2003. Sawyer, Clair N., Perry L. McCarty, and Gene F. Parkin. McGraw-Hill Companies. www.mheducation.com. ISBN: 0072480661.

Manual of Water Utilities Operations (Texas Manual). Texas Water Utilities Association, www.twua.org. ASIN: B000KT9EIU. Chapters on emergency operation, effective public relations in water works operations, and the storage of potable water.

Operation and Maintenance Manual for Stockton East Water Treatment Plant, prepared by James M. Montgomery, Consulting Engineers, Inc., Walnut Creek, CA. 1979.

Standard Methods for the Examination of Water and Wastewater, 22nd Edition. American Public Health Association (APHA), American Water Works Association (AWWA), and the Water Environment Federation (WEF). www.awwa.org. ISBN: 9780875530130.

Stone, B. G. Notes from "Design of Water Treatment Systems," CE-610. Loyola Marymount University, Los Angeles, CA. 1977.

Water Quality. American Water Works Association (AWWA), www.awwa.org. ISBN: 1-58321-232-9.

Water Quality and Treatment: A Handbook of Public Water Supplies, 6th Edition. 2011. Edzwald, James, ed. American Water Works Association (AWWA). www.awwa.org. ISBN: 9780071630115.

Water Treatment. American Water Works Association (AWWA). www.awwa.org. ISBN: 1-58321-230-2.

Water Treatment Plant Design, 5th Edition. 2012. American Water Works Association (AWWA) and American Society of Civil Engineers (ASCE). www.awwa.org. ISBN: 9780071745727.

1.1 Water as a Limited Resource

1. Intrusion of seawater into groundwater basins results from _____.

1.2 Water Treatment

2. What is the basic objective of the operation of water treatment plants?
   1. Ensure that people drink plenty of water each day to support healthy lifestyles
   2. Increase water use and encourage water-intensive landscaping
   3. Produce safe, aesthetically pleasing drinking water at a reasonable cost
   4. Provide opportunities for professional advancement and salary increases

1.3 The Safe Drinking Water Act

3. What is one characteristic of a community water system?
   1. Regularly serves at least 25 full-time residents
   2. Regularly serves at least 25 people who are not residents
   3. Regularly serves fewer than 25 full-time residents
   4. Regularly serves fewer than 25 of the same individuals

1.4 Water Treatment Plants

4. Which of the following influence the size of a water treatment plant as well as the number and types of processes it uses?
   1. Expected water rates and projected profits
   2. Impurities in raw water and cost considerations
   3. National average rainfall and snowpack levels
   4. Local landscaping styles and native plant programs

1.5 Plant Operation

5. Which plant operating guidelines are routinely monitored by water treatment plant operators?
   1. Equipment cleaning and painting schedules
   2. Flows and water quality indicators
   3. Number and frequency of public tours
   4. Vehicle mileage and fuel usage

6. What is one purpose of clear wells or plant storage reservoirs?
   1. Act as a trigger for ON/OFF cycling of finished water pumps
   2. Allow clarification of water by providing time for particulates to settle out
   3. Allow drawdown when demands are low to perform maintenance
   4. Provide operational storage to average out high and low flow demands

7. How would you verify the actual raw water flow rate after adjusting the pumping rate to a water treatment plant?
   1. Check reservoir levels
   2. Count the number of pumps in service
   3. Read the raw water flow measuring device
   4. Rework the flow calculations

8. How are the initial selection of chemical types and anticipated feed rates for a given application in a water treatment plant usually determined?
   1. Computer modeling
   2. Jar tests
   3. Marble tests
   4. Pilot-plant tests

9. What is the best source of information about dangerous chemicals?
   1. Chemical analysis from a certified laboratory
   2. Safety data sheets (SDS)
   3. Trial and error tests
   4. Verbal reports from other operators

10. The chlorine cylinder in service has less than 1 day's chlorine supply remaining. Three 150 lb chlorine cylinders are in storage. The plant uses an average of 28 pounds of chlorine per day (lb/d). How many days' supply of chlorine is available?

    1. 13 days
    2. 14 days
    3. 15 days
    4. 16 days

11. Which of these chemicals is one of the most dangerous common alkalies?

    1. Calcium hydroxide
    2. Caustic soda
    3. Chlorine
    4. Sodium bicarbonate

12. What should you do if an operator is not breathing due to chlorine gas exposure?

    1. Cover victim with a blanket
    2. Flush victim's eyes with water
    3. Perform hands-only or conventional CPR
    4. Provide victim with milk

13. Water samples taken to measure compliance with water quality standards are usually collected at _____.

14. In a water treatment plant, turbidity is commonly monitored at _____, _____, _____, _____, and _____.

15. The range of turbidity removal effectiveness is important because of the relationship between turbidity and _____.

16. Federal laws include sludge from a water treatment plant as a _____ waste that requires proper handling and disposal.

17. What should an operator do promptly when a treatment process upset results in the failure to meet a specific drinking water quality standard?

    1. Apply to EPA for a waiver
    2. Notify appropriate local health authorities
    3. Stop operating process
    4. Switch sources of raw water

18. If a water treatment plant must be dewatered for emergency repairs, how would you prepare for this event?

    1. Empty all storage reservoirs before repairs
    2. Fill all storage reservoirs by early evening
    3. Rewrite preventive maintenance program documents
    4. Take photos of failed equipment

1.6 Plant Maintenance

19. What kinds of records are operators required to keep on file for specified time periods for regulatory compliance?

    1. Annual vehicle mileage and maintenance
    2. Breakroom and restroom cleaning schedules
    3. Timesheets and retirement contribution records
    4. Water quality analysis and customer complaints

20. What is one way of keeping track of how much time and money is spent doing various types of work?

    1. Electronic timesheets
    2. Operator diaries
    3. Preventive maintenance program
    4. Work order system

1.7 Energy Conservation

21. How can energy requirements for lighting be reduced in a water treatment plant?

    1. Keep lights on in unoccupied spaces
    2. Light all areas of the yard
    3. Turn off lights in unoccupied spaces
    4. Use mercury vapor lamps

1.8 Supervision and Administration

22. What regulations require that accurate records of water treatment plant operations be maintained?
    1. Americans with Disabilities Act
    2. Freedom of Information Act
    3. Occupational Safety and Health Act
    4. Safe Drinking Water Act

1.10 Safety

23. What contributes to most treatment plant accidents?
    1. Application and knowledge of safe procedures
    2. Availability and convenience of safety equipment
    3. Carelessness and negligence
    4. Plant design and construction

1.11 Water Treatment Plant Operators

24. What is the responsibility of water treatment plant operators?
    1. Avoid boil-water orders
    2. Minimize costs of producing drinking water
    3. Produce safe and pleasant drinking water
    4. Successfully pass operator certification examinations

25. What do most states require for someone to work as a water treatment plant operator?
    1. College diploma
    2. Driver's license
    3. First-aid training
    4. Operator certificate

CHAPTER 2

Source Water, Reservoir Management, and Intake Structures

CHAPTER OUTLINE

LEARNING OBJECTIVES

1. Use a sanitary survey to evaluate the suitability of an identified water source for drinking purposes and as a general water supply.

2. Develop and implement a reservoir management program using identified water quality problems and data from laboratory and monitoring programs.

3. Identify various types of intake structures, gates, and screens and describe their purposes.

4. Safely operate, maintain, and troubleshoot intake facilities.

5. Record necessary information on the operation and maintenance of reservoir water quality management programs and intake structures.

KEY TERMS

adsorption	capillary fringe	detention time	flushing
aeration	cathodic protection	dimictic	geological log
algae	chelation	direct filtration	head
algal bloom	coliform	direct runoff	hypolimnion
aliphatic hydroxy acids	complete treatment	epidemiology	impermeable
anaerobic	conductivity	epilimnion	littoral zone
aquifer	cross-connection	eutrophic	mesotrophic
BOD	density	eutrophication	metalimnion
capillary action	destratification	evaportransporation	methyl orange alkalinity

molecular weight	pathogenic organisms	reduction	topography
monomictic	pH	sanitary survey	toxic
nutrient	photosynthesis	Secchi disc	transpiration
oligotrophic	phytoplankton	short-circuiting	trihalomethane
overturn	potable water	soft water	turbidity
oxidation	precipitate	stratification	water cycle
oxidation-reduction	precursor, THM	subsidence	water table
potential (ORP)	reaeration	threshold odor	zooplankton

2.1   Sources of Water

All water on earth is a potential source of drinking water, though few water sources are completely safe to drink without treatment to remove harmful components. Many water sources are unsuitable for treatment because of distance to points of use, expense of required treatment, and lack of appropriate treatment methods. All this makes treated drinking water a valuable resource.

2.1.1   Ocean

At some time in its history, virtually all water resided in the oceans. Water in the oceans is too salty for drinking and irrigation uses. Fresh water can be produced from the ocean through the **water cycle** (see Figure 2.1). Water evaporates from the ocean by the energy of the sun at an overall rate of about 6 feet (1.8 m) of water annually. The water that is evaporated is salt-free water because the heavier mineral salts are left behind. This water vapor rises, is carried along by winds, and eventually condenses into clouds. When these clouds become chilled, the small particles of water collect into larger droplets, which may precipitate over land or water. As the water falls in the form of rain, snow, sleet, or hail, it clings to and carries with it all the dust and dirt in the air. The first water that falls during a storm picks up the greatest concentration of contamination. After a short period of fall, the precipitation is relatively free of pollutants. A large part of the evaporated water is carried over land masses by the winds and the droplets that fall there make up our supply of fresh water. These droplets may soak into the ground, fall as snow on the mountain tops, collect in lakes, or evaporate once more and return immediately to the atmosphere, but in one form or another, the droplets eventually return to the ocean.

Another way to make ocean water suitable for beneficial uses is desalination. A common desalination process uses distillation to collect fresh water that has evaporated through heat. Another process is reverse osmosis, which uses pressure to push the water through a semi-permeable membrane to remove unwanted particles from the water. These processes are not widely used because the amount of energy required makes them prohibitively expensive.

2.1.2   Surface Water

Surface water accumulates mainly as precipitation (rain or snow) that does not enter the ground through infiltration or is not returned to the atmosphere by evaporation flows over the ground surface. This **direct runoff**

water cycle
The process of evaporation of water into the air and its return to earth by precipitation (rain or snow). This process also includes transpiration from plants, groundwater movement, and runoff into rivers, streams, and the ocean. Also called the hydrologic cycle.

direct runoff
Water that flows over the ground surface directly into streams, rivers, or lakes. Also called storm runoff.

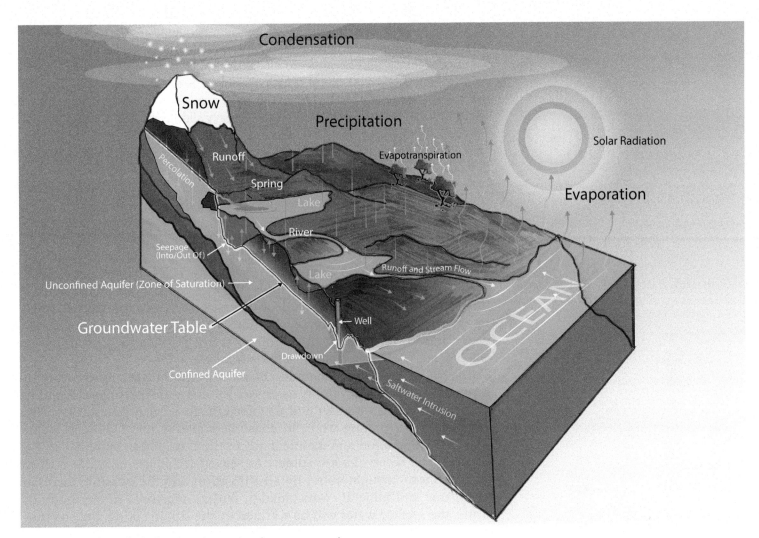

Figure 2.1   Water (hydrologic) cycle as related to water supply

drains from saturated or **impermeable** surfaces, into stream channels, and then into natural or artificial storage sites (or into the ocean in coastal areas).

The amount of available surface water depends largely upon rainfall. When rainfall is limited, the supply of surface water will vary considerably between wet and dry years. In areas of scant rainfall, people build individual cisterns to store rain that drains from the catchment areas of roofs. This type of water supply is used extensively in areas such as the Bermuda Islands, where groundwater is virtually nonexistent and there are no streams.

Surface water supplies may be further divided into river, lake, and reservoir supplies. In general, they are characterized by low calcium and magnesium content (**soft water**), **turbidity**, suspended solids, some color, and microbial contamination.

2.1.2.1   Rivers and Streams

Many of the largest cities in the world depend entirely upon large rivers for their water supplies. Upstream conditions are always a concern when using a river or stream supply. Some cities draw drinking water from a stream or river into which the treated wastewater (sewage) from upstream cities has been discharged. This can present serious problems in water treatment. Because

impermeable
(im-PURR-me-uh-BULL) Not easily penetrated. The property of a material or soil that does not allow, or allows only with great difficulty, the movement or passage of water.

soft water
Water having a low concentration of calcium and magnesium ions. According to US Geological Survey guidelines, soft water is water having a hardness of 60 milligrams per liter or less.

turbidity (ter-BID-it-tee)
The cloudy appearance of water caused by the presence of suspended and colloidal matter. In the waterworks field, a turbidity measurement is used to indicate the clarity of water. Technically, turbidity is an optical property of the water based on the amount of light reflected by suspended particles. Turbidity cannot be directly equated to suspended solids because white particles reflect more light than dark-colored particles and many small particles will reflect more light than an equivalent large particle.

of upstream pollution (municipal wastewater, agricultural drainage, or industrial waste), the proper treatment of river and stream supplies is extremely important. Rivers and streams are also susceptible to scouring of the bottom, changing channels, and silting, which should all be studied before locating the intake for a water supply. Provisions must be made in the design of the intake to make sure that it can withstand floods, heavy silting, ice, and adverse runoff. Because of variations in the quality of water supplied by a river or stream, purification effectiveness must be continually checked. This is especially true if there are industries upstream from the intake that may dump undesirable wastes into the supply. Sudden pollutant loads might not be discovered unless the treatment plant operator constantly monitors the raw water.

2.1.2.2   Lakes and Reservoirs

The selection and use of water from any surface storage source requires considerable study and thought. When ponds, lakes, or open reservoirs are used as sources of water supply, there is the danger of contamination and the consequent spread of diseases, such as typhoid, hepatitis, dysentery, giardiasis, and cryptosporidiosis. Clear water is not always safe water and the old saying that running water purifies itself to drinking water quality within a stated distance is false.

The potential for contamination of surface water makes it necessary to regard such sources of supply as unsafe for domestic use unless properly treated, including filtration and disinfection. To ensure the delivery of a constant, safe drinking water to consumers also requires diligent attention to the operation and maintenance of the distribution system.

Lakes and reservoirs are subject to seasonal changes in water quality such as those brought about by **stratification** and the possible increase of organic and mineral contamination during **overturn**. In any body of water, the surface water will be warmed by the sun in spring and summer causing higher temperatures on the surface. Then in the fall, the cooler air temperatures cool the surface water until it reaches the same temperature as the subsurface waters. At this point, the water temperature is fairly uniform (the same) throughout the entire depth of the lake or reservoir. A breeze will start the surface water circulating and cause the lake to turn over, thus bringing poor quality deeper water to the surface.

Lakes and reservoirs are susceptible to **algal blooms**, especially after fall or spring overturns. The rapid growth of **algae** (blooms) will occur when the temperature is right and the water contains enough nutrients to support rapid algal growth. In any given body of water, blooms of various types of algae can occur several times during a season, depending on what algae are present and whether the conditions are right for algal growth.

Water supplies drawn from large lakes and reservoirs through multilevel intake facilities (openings at several depths) are generally of good quality because the water can be drawn from a depth where algal growths are not prevalent. A large lake or reservoir also dilutes any contamination that may have been discharged into it or one of its tributaries.

Large bodies of water are generally attractive recreational areas. If the water is also used for domestic supplies, however, it must be protected from contamination. This will require proper construction and location of recreational facilities such as boat launching ramps, boat harbors, picnic and camping areas, fishing, and open beach areas away from the intake area. The location and construction of wastewater collection, treatment, and disposal facilities must also be carefully studied to protect domestic water supplies from contamination.

stratification (STRAT-uh-fuh-KAY-shun)
The formation of separate layers (of temperature, plant, or animal life) in a lake or reservoir. Each layer has similar characteristics such as all water in the layer has the same temperature.

overturn
The almost spontaneous mixing of all layers of water in a reservoir or lake when the water temperature becomes similar from top to bottom. This may occur in the fall/winter when the surface waters cool to the same temperature as the bottom waters and also in the spring when the surface waters warm after the ice melts. Also called turnover.

algal (AL-gull) bloom
Sudden, massive growths of microscopic and macroscopic plant life, such as green or bluegreen algae, which can, under the proper conditions, develop in lakes, reservoirs, and lagoons.

algae (AL-jee)
Microscopic plants containing chlorophyll that live floating or suspended in water. They also may be attached to structures, rocks, or other submerged surfaces. Excess algal growths can impart tastes and odors to potable water. Algae produce oxygen during sunlight hours and use oxygen during the night hours. Their biological activities appreciably affect the pH, alkalinity, and dissolved oxygen of the water.

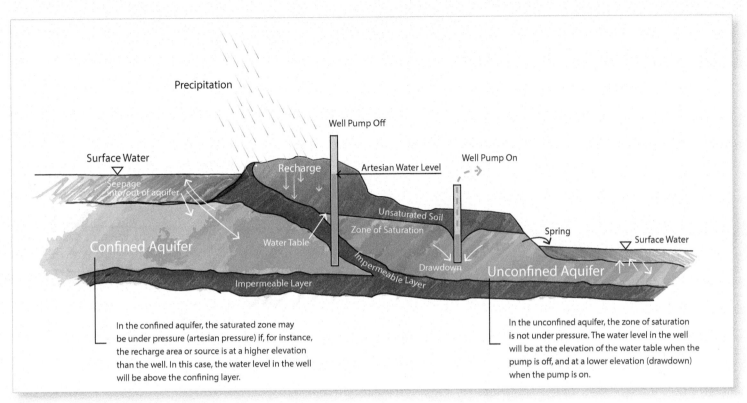

In the confined aquifer, the saturated zone may be under pressure (artesian pressure) if, for instance, the recharge area or source is at a higher elevation than the well. In this case, the water level in the well will be above the confining layer.

In the unconfined aquifer, the zone of saturation is not under pressure. The water level in the well will be at the elevation of the water table when the pump is off, and at a lower elevation (drawdown) when the pump is on.

Figure 2.2   Aquifers and groundwater

2.1.3   Groundwater

As shown in Figure 2.2, part of the precipitation that falls infiltrates the soil. This water replenishes the soil moisture, or is used by growing plants and returned to the atmosphere by **transpiration** (see Figure 2.1). Water that drains downward (percolates) below the root zone can be stored in an **aquifer**. (An aquifer, or water-bearing formation, is an underground layer of rock or soil that stores and permits the passage of water.) If all of the openings or voids in the earth's materials are filled with water, this is known as the zone of saturation. Water in the zone of saturation is referred to as groundwater. Groundwater, unlike surface water, is characterized by higher concentrations of dissolved solids, dissolved gases, and a relatively high hardness. On the other hand, groundwater is free of suspended solids and usually free of microbial contamination.

The upper surface of the zone of saturation, if not confined by impermeable material, is called the **water table**. When an overlying, impermeable formation confines the water in the zone of saturation under pressure, the groundwater is said to be under artesian pressure. The name artesian comes from the ancient province of Artesium in France where, in the days of the Romans, water flowed to the ground surface from a well. However, not all water from wells that penetrate artesian formations flows to ground level. For a well to be artesian, the water in the well must rise above the top of the confining layer.

The porous material just above the water table may contain water by **capillary action**, the movement of water through very small spaces due to molecular forces. This zone is referred to as the **capillary fringe**. Because the water held in the capillary fringe will not drain freely by gravity, this zone is not considered a true source of supply.

Because of the irregularities in underground deposits or layers and in surface **topography**, the water table occasionally intersects (meets) the surface

transpiration (TRAN-spur-RAY-shun)
The process by which water vapor is released to the atmosphere by living plants. This process is similar to people sweating.

aquifer (ACK-wi-fer)
A natural, underground layer of porous, water-bearing materials (sand, gravel) usually capable of yielding a large amount or supply of water.

water table
The upper surface of the zone of saturation of groundwater in an unconfined aquifer.

capillary (KAP-uh-larry) action
The movement of water through very small spaces due to molecular forces.

capillary (KAP-uh-larry) fringe
The porous material just above the water table that may hold water by capillarity (a property of surface tension that draws water upward) in the smaller void spaces.

topography (toe-PAH-gruh-fee)
The arrangement of hills and valleys in a geographic area.

of the ground at a spring or in the bed of a stream, lake, or the ocean. As a result, groundwater moves to these locations as seepage out of the aquifer (see Figure 2.1). Thus, groundwater is continually moving within aquifers even though the movement may be very slow. This movement is caused by a difference in **head** (energy) between two locations (for example, a higher area of recharge and a lower area of discharge). The difference in heads over a distance causes a downward slope in the water table (or artesian pressure surface) in the direction of the water flow. Seasonal variations in the supply of water to the underground reservoir cause considerable changes in the elevation and slope of the water table and the artesian pressure level.

2.1.3.1 Wells

A well that penetrates the water table can be used to extract water from the groundwater basin (see Figures 2.1 and 2.2). The removal of water by pumping will naturally cause a lowering of the water table near the well (drawdown). If pumping continues at a rate that exceeds the rate of replacement by the water-bearing formations, the sustained yield of the well or group of wells will be exceeded. Pumping at that rate over a period of time will deplete the aquifer and bring about other undesired results (such as seawater intrusion and land **subsidence**). Though a poor practice, it occurs frequently in many areas of the United States.

2.1.3.2 Springs

Groundwater that flows naturally from the ground is called a spring. Depending upon whether the discharge is from a water table or an artesian aquifer, springs may flow by gravity or by artesian pressure. The flow from a spring may vary considerably; when the water table or artesian pressure fluctuates, so does the flow from the spring.

2.1.4   Reclaimed Water

The use of treated wastewater as a source of water for non-food crop irrigation is an established practice in many regions of the world. The type of crop that can be safely irrigated depends somewhat on the quality of wastewater and method of irrigation. More than 20,000 acres (8,000 hectares) of agricultural lands in California are irrigated, all or in part, with reclaimed water. Two of the largest operations are at Bakersfield and at Fresno, California. The City of Bakersfield has used wastewater effluent for irrigation since 1912. Most growing seasons, over 2,000 acres (over 800 hectares) of alfalfa, cotton, barley, sugar beets, and pasture are irrigated. Fresno irrigates over 3,000 acres (over 1,200 hectares) of the same types of crops. These two operations use almost 30,000 acre-feet (37 million cubic meters) of reclaimed water per year. The Irvine Ranch Water District in California used over 22,000 acre-feet of reclaimed water in a year, mainly for landscape irrigation. Almost 90 percent of the 2,000,000 acre-feet (2.5 billion cubic meters) of reclaimed wastewater in California is used for crop irrigation. Another state with high reclaimed wastewater use is Florida, which used over 690,000 acre-feet of reclaimed wastewater for beneficial uses in a year, mainly for landscape irrigation.

Reclaimed wastewater is used for:

1. Greenbelt (parks) areas
2. Golf course irrigation
3. Landscape irrigation

head
The vertical distance, height, or energy of water above a reference point. A head of water may be measured in either height (feet or meters) or pressure (pounds per square inch or kilograms per square centimeter).

subsidence (sub-SIDE-ence)
The dropping or lowering of the ground surface as a result of removing excess water (overdraft or overpumping) from an aquifer. After excess water has been removed, the soil will settle, become compacted, and the ground surface will drop, which can cause the settling of underground utilities.

4. Industrial reuse
5. Groundwater recharge
6. Landscape impoundments
7. Wetlands/marsh enhancement

Reclaimed water can be used safely for any of these purposes, with the possible exception of groundwater recharge. Health experts have serious questions regarding organic compounds that are present in wastewater and about our ability to reduce them to safe levels. These doubts increase as new and potentially toxic chemicals are identified each year. Laboratories are unable to adequately detect all of these chemicals without expensive monitoring programs. To protect groundwater resources, regulations generally require that reclaimed water used for groundwater recharge of domestic water supply aquifers by surface spreading is of a quality that fully protects public health. Proposed groundwater recharge projects must be investigated on an individual basis where the use of reclaimed water involves a potential risk to public health.

The uses for reclaimed water are expanding as treatment technologies improve and health studies show the safety of the reclaimed water. While few utilities are currently using reclaimed water for direct potable use, population growth and resource depletion point to a need for making the most of the available water supply.

Treatment of reclaimed water should be appropriate for the intended use. The greater the potential exposure to the public, the more extensive the treatment needs to be. State regulations often specify not only the degree of treatment for the usage of water, but also the reliability features that must be incorporated into the treatment processes to ensure a continuous high degree of finished water quality. Reclamation plants must achieve the quality of treatment expected on a continuous basis. This is of concern to the water supplier and the regulatory agencies that are charged with the responsibility of ensuring that the health of the public is protected.

Check Your Understanding

1. What is the water cycle?
2. What are the general water quality characteristics of surface water supplies?
3. What are the general water quality characteristics of groundwater supplies?
4. What causes the flow of groundwater within an aquifer?
5. How can the sustained yield of an aquifer be exceeded?
6. How much treatment should reclaimed water receive before use?

2.2 Selection of a Water Source

Safe drinking water is essential for individual survival as well as for economic growth in communities. As the populations and industries of cities grow, so does the demand for high-quality drinking water. Many cities in the US have established sources of drinking water, but they are evaluating other sources to supply water to a growing population.

2.2.1   Water Rights

Many potential drinking water sources are unavailable because the rights to their use are already allocated. Water rights holders possess permission to use water but they do not own the water itself. The rights of an individual to use water for domestic, irrigation, or other purposes varies in different states. Some water rights stem from ownership of the land bordering or overlying the source, while others are acquired by a performance of certain acts required by law. Depending on state laws, water rights may be sold, leased, or transferred like other property. Before developing a new water source, the rights to use that water must be available to the organization.

There are three basic types of water rights:

1. Riparian—rights that are acquired with title to the land bordering a source of surface water.
2. Appropriative—rights that are acquired for the beneficial use of water by following a specific legal procedure.
3. Prescriptive—rights that are acquired by diverting water, to which other parties may or may not have prior claims, and putting it to use for a period of time specified by statute.

When there is any question regarding the right to the use of water, a property owner should consult with the appropriate authority and clearly establish rights to its use.

2.2.2   Sanitary Survey

Even if the water rights are available, the source may not be appropriate for drinking water uses. **Sanitary surveys** are used to determine if a new source is appropriate or an existing supply continues to offer a safe and high-quality water. With a new supply, the sanitary survey should be made during the collection of initial engineering data covering the development of a given source and its capacity to meet existing and future needs. The sanitary survey should include the location of all potential and existing health hazards and the determination of their present and future importance. People trained in public health engineering and the **epidemiology** of waterborne diseases should conduct the sanitary survey. In the case of an existing supply, sanitary surveys should be made frequently enough to control health hazards and to maintain high water quality.

The information furnished by a sanitary survey is essential to evaluating the bacteriological and chemical water quality data. The following outline lists the essential factors that should be investigated or considered in a sanitary survey. These items are essential to identify potential hazards, determine factors that affect water quality, and select treatment requirements. Not all of the items are important to any one supply and, in some cases, items not in the list could be found to be significant during the field investigation.

GROUNDWATER SUPPLIES

- Character of local geology; slope (topography) of ground surface.
- Nature of soil and underlying porous material, whether clay, sand, gravel, rock (especially porous limestone); coarseness of sand or gravel; thickness of water-bearing stratum; depth of water table; location and **geological log** (well log) of nearby wells.

sanitary survey
A detailed evaluation and/or inspection of a source of water supply and all conveyances, storage, treatment and distribution facilities to ensure protection of the water supply from all pollution sources.

epidemiology
(EP-uh-DE-me-ALL-o-gee)
A branch of medicine which studies epidemics (diseases which affect significant numbers of people during the same time period in the same locality). The objective of epidemiology is to determine the factors that cause epidemic diseases and how to prevent them.

geological log
A detailed description of all underground features discovered during the drilling of a well (depth, thickness and type of formations).

- Slope of water table, preferably as determined from observation wells or as indicated by slope of the ground surface.
- Extent of the drainage area likely to contribute water to the supply.
- Nature, distance, and direction of local sources of pollution.
- Possibility of surface-drainage water entering the supply and of wells becoming flooded.
- Methods used for protecting the supply against contamination from wastewater collection and treatment facilities and industrial waste disposal sites.
- Well construction: materials, diameter, depth of casing and concrete collar; depth to well screens or perforations; length of well screens or perforations.
- Protection of wellhead at the top and on the sides.
- Pumping station construction (floors, drains); capacity of pumps; storage or direct to distribution system.
- Drawdown when pumps are in operation; recovery rate when pumps are off.
- Presence of an unsafe supply nearby, and the possibility of **cross-connections** causing a danger to the public health.
- Disinfection: equipment, supervision, test kits, or other types of laboratory control.

SURFACE WATER SUPPLIES

- Nature of surface geology; character of soils and rocks.
- Character of vegetation; forests; cultivated and irrigated land.
- Population and wastewater collection, treatment, and disposal on the watershed.
- Methods of wastewater disposal, whether by diversion from watershed or by reclamation treatment.
- Distance to sources of fecal pollution (especially birds) to intake of water supply.
- Proximity to watershed and character of sources of contamination, including industrial wastes, oil field brines, acid waters from mines, sanitary landfills, and agricultural drain waters.
- Adequacy of supply as to quantity (safe yield).
- For lake or reservoir supplies: wind direction and velocity data; drift of pollution; and algal growth potential.
- Character and quality of raw water: typical coliform counts (MPN or membrane filter), algae, turbidity, color, and objectionable mineral constituents.
- Normal period of **detention time**.
- Probable minimum time required for water to flow from sources of pollution to reservoir and through the reservoir to the intake tower.
- The possible currents of water within the reservoir (induced by wind or reservoir discharge) that could cause **short-circuiting** to occur.
- Protective measures in connection with the use of the watershed to control fishing, boating, landing of airplanes, swimming, wading, ice cutting, and permitting animals on shoreline areas.
- Efficiency and constancy of policing activities on the watershed and around the lake.
- Treatment of water: kind and adequacy of equipment; duplication of parts for reliable treatment; effectiveness of treatment; numbers and competency of supervising and operating personnel; contact

cross-connection
A connection between a drinking (potable) water system and an unapproved water supply. For example, if you have a pump moving non-potable water and hook into the drinking water system to supply water for the pump seal, a cross-connection or mixing between the two water systems can occur. This mixing may lead to contamination of the drinking water.

detention time
(1) The theoretical (calculated) time required for a small amount of water to pass through a tank at a given rate of flow. (2) The actual time in hours, minutes or seconds that a small amount of water is in a settling basin, flocculating basin or rapid-mix chamber. In storage reservoirs, detention time is the length of time entering water will be held before being drafted for use (several weeks to years, several months being typical).

$$\text{Detention Time [h]} = \frac{\text{Basin Volume [gal]}}{\text{Flow Rate}\left[\frac{\text{gal}}{\text{d}}\right] \times \frac{\text{d}}{24\,\text{h}}}$$

short-circuiting
A condition that occurs in tanks or basins when some of the flowing water entering a tank or basin flows along a nearly direct pathway from the inlet to the outlet. This is usually undesirable since it may result in shorter contact, reaction, or settling times in comparison with the theoretical (calculated) or presumed detention times.

period after disinfection; free chlorine residuals and monitoring of the water supply both during treatment and following treatment.

- Pumping facilities: pump station design, pump capacity, and standby unit(s).
- Presence of an unsafe supply nearby, and the possibility of cross-connections causing a danger to the public health.

2.2.3   Contamination

Precipitation in the form of rain, snow, hail, or sleet contains very few impurities. (However, there are exceptions such as acid rain and dust from dust bowl areas.) Trace amounts of mineral matter, gases, and other substances may be picked up by precipitation as it forms and falls through the earth's atmosphere. Precipitation, however, has virtually no microbial content.

Once precipitation reaches the earth's surface, many opportunities are presented for the introduction of mineral and organic substances, microorganisms, and other forms of contamination. When water runs over or through the ground surface it may pick up particles of soil. This is noticeable in the water as cloudiness or turbidity. This water also picks up particles of organic matter and microorganisms. As surface water seeps downward into the soil and through the underlying material to the water table, most of the suspended particles are filtered out. This natural filtration may be partially effective in removing microorganisms and other particulate materials; however, the chemical characteristics of the water usually change considerably when it comes in contact with underground mineral deposits.

The widespread use of synthetically produced chemical compounds, especially pesticides, has raised concern for their potential to contaminate water. Many of these materials are known to be **toxic** (a substance that is poisonous to a living organism), some cause cancer, and others have certain undesirable characteristics even when present in a relatively small concentration.

Agents that alter the quality of water as it moves over or below the surface of the earth may be classified under four major headings:

1. Physical—Physical characteristics relate to the sensory qualities of water for domestic use; for example, the water's observed color, turbidity, temperature, taste, and odor.
2. Chemical—Chemical differences between waters include mineral content and the presence or absence of constituents such as fluoride, sulfide, and acids. The comparative performance of hard and soft waters in laundering is one visible effect.
3. Biological—The presence of organisms (viruses, bacteria, algae, insect larvae), alive or dead, and their metabolic products determine the biological character of water. These may also be significant in modifying the physical and chemical characteristics of water.
4. Radiological—Radiological factors must be considered because there is a possibility that the water may have come in contact with radioactive substances.

Consequently, in the development of water supply systems, it is necessary to examine carefully all the factors that might adversely affect the water supply.

2.2.3.1 Physical Characteristics

To be suitable for human use, water should be free from all impurities that are offensive to the senses of sight, taste, and smell. The physical characteristics that might be offensive include turbidity, color, temperature, taste, and odor.

Turbidity: The presence of suspended material in water causes cloudiness, which is known as turbidity. Clay, silt, finely divided organic material, plankton, and other inorganic materials give water this appearance. Turbidities in excess of 5 turbidity units are easily visible in a glass of water, and this level is usually objectionable for aesthetic reasons. Turbidity's major danger in drinking water is that it can harbor bacteria. It may also exert a high demand on chlorine. Water that has been filtered to remove the turbidity should have considerably less than 1 turbidity unit. Good treatment plants consistently obtain finished water turbidity levels of 0.05 to 0.3 units.

Color: Dissolved organic material from decaying vegetation and certain inorganic matter cause color in water. Occasionally, excessive blooms of algae or the growth of other aquatic microorganisms may also impart color. Iron and manganese may be the cause of consumer complaints (red or black water). While the color itself is not objectionable from the standpoint of health, its presence is aesthetically objectionable and suggests that the water needs better treatment. In some instances, however, a color in the water indicates more than an aesthetic problem. For example, an amber color in the water could indicate the presence of humic substances, which could later be formed into trihalomethanes, or it could indicate acid waters from mine drainage.

Temperature: The most desirable drinking waters are consistently cool and do not have temperature fluctuations of more than a few degrees. Groundwater and surface water from mountainous areas generally meet these requirements. Most individuals find that water having a temperature between 50° and 60°F (10° and 15°C) is most pleasing while water over 86°F (30°C) is not acceptable. The temperature of groundwaters varies with the depth of the aquifer. Water from very deep wells (more than 1,000 feet or 300 meters) may be quite warm. Temperature also affects sensory perception of tastes and odors.

Tastes: Each area's natural waters have a distinctive taste related to the dissolved mineral characteristics of local geology. Occasionally, algal growths also impart a distinctive taste. However, taste is rarely measured since most water treatment plants cannot alter a water's mineral characteristics.

Odors: Growths of algae in a water supply can give the water an unpleasant odor. Some groundwaters may contain hydrogen sulfide, which will produce a disagreeable rotten egg odor.

2.2.3.2 Chemical Characteristics

The nature of the materials that form the earth's crust affects not only the quantity of water that may be recovered, but also its chemical makeup. As surface water infiltrates and percolates downward to the water table, it dissolves some of the minerals contained in soils and rocks. Groundwater, therefore, sometimes contains more dissolved minerals than surface water. The use and disposal of chemicals by society can also affect water quality.

Chemical analysis of a domestic water supply is broken down into three areas:

1. Inorganic chemicals, which include the toxic metals (arsenic, barium, cadmium, chromium, lead, mercury, selenium, and silver) and the nonmetals (fluoride and nitrate)

2. Organic chemicals, which include the volatile organics such as benzene, methylene chloride, and trichloroethylene and other organic chemicals such as the pesticides (chlorinated hydrocarbons), pentachlorophenol, aldicarb, dibromodichloropropane (DBCP), polychlorinated biphenyls (PCBs), and simazine

3. The general mineral constituents, which include alkalinity, calcium, chloride, copper, foaming agents (methylene blue active substances or MBAS), iron, magnesium, manganese, pH, sodium, sulfate, zinc, specific conductance, total dissolved solids, and hardness (calcium and magnesium)

Upper limits for the concentrations of the chemicals listed in this section have been established by the Safe Drinking Water Act. See the chapter on drinking water regulations in *Water Treatment Plant Operation*, Volume II for a detailed discussion of the Safe Drinking Water Act.

2.2.3.3   Biological Factors

Water for domestic uses must be made free from **pathogenic organisms**. These organisms include bacteria, protozoa, spores, viruses, cysts, and helminths (parasitic worms).

Many organisms that cause disease in humans originate with the fecal discharges of infected individuals. To monitor and control the activities of human disease carriers is seldom practical. For this reason, it is necessary to take precautions to prevent contamination of a normally safe water source or to institute treatment methods that will produce a safe water.

Unfortunately, the specific disease-producing organisms present in water are not easily isolated and identified. The techniques for comprehensive bacteriological examination are complex and time-consuming. Therefore, it has been necessary to develop tests that indicate the relative degree of contamination in terms of an easily defined quality. The most widely used test involves estimation of the number of bacteria of the coliform group, which are always present in fecal wastes and vastly outnumber specific disease-producing organisms. Coliform bacteria normally inhabit the intestinal tract of humans, but are also found in most animals and birds, as well as in the soil. The Drinking Water Standards in the Safe Drinking Water Act have established upper limits for the concentration of coliform bacteria in a series of water samples with a goal of zero coliforms in all samples. The Maximum Contaminant Level (MCL) for coliforms for systems analyzing fewer than 40 samples per month is no more than one sample per month may be total coliform positive.

To further ensure protection against the spread of waterborne diseases, the Surface Water Treatment Rule (SWTR), which took effect December 31, 1990, requires that all public water systems disinfect their water. If the

pathogenic (path-o-JEN-ick) organisms Organisms, including bacteria, viruses, protozoa, or internal parasites, capable of causing disease (such as giardiasis, cryptosporidiosis, typhoid fever, cholera, or infectious hepatitis) in a host (such as a person). There are many types of organisms that do not cause disease. These organisms are called nonpathogenic.

source of supply (the raw water) is surface water or groundwater under the influence of surface water, the SWTR also requires water suppliers to install filtration equipment unless the source water meets very high standards for purity.

2.2.3.4 Radiological Factors

Most drinking water sources have very low levels of radioactive contaminants (radionuclides), which are not considered to be a public health concern. Of the small percentage of drinking water systems with radioactive contaminant levels high enough to be of concern, most of the radioactivity is naturally occurring. These radioactive contaminants, depending on their chemical properties, may accumulate in drinking water sources at levels of concern.

A very small percentage of drinking water systems are located in areas that have potential sources of radioactive contamination from facilities that use, manufacture, or dispose of radioactive substances. Drinking water contamination may occur through accidental releases of radioactivity or through improper disposal practices. Water systems that are vulnerable to this type of contamination are required to perform extensive monitoring for radioactive contamination to ensure that their drinking water is safe.

The 2000 Radionuclides Rule set new monitoring requirements only for community water systems, which are water systems with at least 15 service connections or that serve 25 or more persons year-round. EPA may consider a future proposal to regulate radionuclides levels in drinking water served by nontransient noncommunity water systems. These are water systems that serve at least 25 of the same people more than 6 months per year. Examples include schools, churches, nursing homes, and factories that supply their own drinking water.

Exposure to uranium in drinking water may cause toxic effects to the kidneys. Also, some people who drink water with excessive levels of radionuclides may have an increased risk of cancer.

Check Your Understanding

1. List the three basic types of water rights.

2. What is the purpose of a sanitary survey?

3. How frequently should a sanitary survey be conducted for an existing water supply?

4. When conducting a sanitary survey, what protective measures should be investigated regarding use of the watershed?

5. List the common physical characteristics of water.

6. What causes turbidity in water?

7. Chemical analysis of a domestic water supply measures what three general types of chemical concentrations?

8. Why are coliform bacteria used to measure the bacteriological quality of water?

9. How can drinking water become contaminated by radionuclides?

 Water Sources, Third Edition. American Water Works Association. www.awwa.org. ISBN: 1-58321-229-9.

Manual of Water Utilities Operations (Texas Manual). Texas Water Utilities Association, www.twua.org. ASIN: B000KT9EIU. Chapters on groundwater supplies, surface water supplies, and raw water quality management.

2.3   Surface Reservoirs as Domestic Water Supplies

During the past few decades, more and more people in both cities and rural areas have become either partially or wholly dependent on surface reservoirs (also called impoundments) and lakes as a source for their water supplies. As populations have increased, domestic, municipal, industrial, recreational, and agricultural water usage has also increased, creating demands on water supplies that cannot be met directly by groundwater or surface water diversions from streams and rivers.

These increased demands have been met for the most part by constructing dams and reservoirs, which provide carryover storage for excess runoff and provide a dependable water supply during the dry season of the year and during periods of prolonged drought. Particularly in the western United States, a majority of the major cities receive domestic water from surface lakes and reservoirs. In most cases, the water is stored in one or more major reservoirs before it is delivered to the consumers. In those areas that do depend directly on local water supplies, there may be several large reservoirs that capture runoff from local watersheds and store it for future use.

The capacities of reservoirs used as domestic water supplies range from less than 100 acre-feet to several million acre-feet. The time water may be stored ranges from weeks or months to several years.

Methods of managing lakes and reservoirs used for domestic water supplies vary widely depending on local situations. In addition to serving domestic water needs, a reservoir may be used for flood control purposes; hydroelectric power generation; regulating downstream releases; recreational purposes; or providing water for agricultural, municipal, and industrial uses. The amount and type of public use allowed on reservoirs also varies widely according to individual situations. Some allow motorboats, some allow only boats without motors; most do not allow any body-contact water sports but some allow complete body-contact sports such as swimming and water skiing.

Small lakes in remote areas may be open for public use only a few days each year while large lakes and reservoirs located near metropolitan areas may accommodate several million visitors annually. The methods of treating water supplies from reservoirs range from disinfection only, to **direct filtration**, to **complete treatment**, which may even include softening and activated carbon filtration. Each reservoir should have a water quality management program that is designed to meet the reservoir's individual requirements.

2.3.1   Factors Affecting Water Quality

Water quality within lakes and reservoirs is influenced and controlled by many factors. Many of the conditions that adversely affect water quality in domestic water supply reservoirs result from our use of the environment, so to control and maintain water quality, our activities must be controlled. The occurrence of acid rainfall is of concern in many areas of the United States and Europe. Pollution from both motor vehicles and industrial plants has increased the acidity of rain in some areas to the point that when runoff reaches lakes and reservoirs, the biological balance is severely affected. Fish die-offs are obvious and easily detected but other biological upsets or trends may not be so obvious.

direct filtration
A method of treating water which consists of the addition of coagulant chemicals, flash mixing, coagulation, minimal flocculation, and filtration. The flocculation facilities may be omitted, but the physical–chemical reactions will occur to some extent. The sedimentation process is omitted.

complete treatment
A method of treating water which consists of the addition of coagulant chemicals, flash mixing, coagulation-flocculation, sedimentation and filtration. Also called conventional filtration.

The impacts of our activities within a given reservoir's drainage area are also a major concern. Wastewater, agricultural runoff, livestock grazing, drainage from mining areas, runoff from urban areas, and industrial discharges may all lead to deterioration in physical, chemical, or biological water quality within a reservoir. Increased turbidity and siltation may result from farming practices, fires, construction (soil grading), and logging operations. If not properly controlled, public use of a reservoir and its drainage may result in reduced water quality.

Natural factors that may affect the quality of water in a given lake or reservoir include the following:

1. Climate: temperature; intensity and direction of wind movements; and type, pattern, intensity, and duration of precipitation
2. Watershed and drainage areas: geology, topography, type and extent of vegetation, and use by native animals
3. Wildfires
4. Reservoir area: geology; land form, including depth, area, and bottom topography; and surface vegetation at the time the reservoir is filled

In addition to these natural factors, specific characteristics of the watershed that feeds the reservoir and the water in the reservoir can cause water quality problems.

2.3.1.1 Watershed Conditions

Domestic water supply reservoirs may at times experience major problems with water quality as a result of conditions within the drainage area or watershed in combination with climatic conditions. In many areas of the United States, the major portion of surface runoff into a reservoir occurs during a very short period of time. In semiarid areas such as the Southwest, 75 percent or more of the annual runoff may occur as a result of only three or four major storms. Most runoff from these storms occurs within a few days during and following the storm. Likewise, in mountainous areas where snowmelt is the major source of runoff, most of the inflow occurs within a relatively short period of time during and following the spring snowmelt. Reservoirs subject to these conditions may experience sudden and dramatic increases in turbidity, nutrient loading, and organic loading depending on geological, topographical, and vegetative conditions within the watershed.

Turbidity problems, which occur during and following periods of major runoff, are reflected in reduced rates of flow through the filters and shortened filter runs at water treatment plants. At reservoirs that do not filter water before service to consumers, federal and state maximum allowable turbidity levels may be exceeded. The length of time that turbidity will affect water quality and water treatment practices depends on the extent of turbidity loading that occurs, mixing in the reservoir as a result of wind and other currents, and the nature of the particles causing turbidity. Larger suspended particles, such as sand and silt, may settle out within a few days or weeks; colloids, such as fine clays, may cause problems for an extended period of time. When stormwaters are colder than reservoir waters, the high-turbidity stormwater sometimes flows into the reservoir underneath the warmer reservoir waters. This can cause the greatest turbidity increases to occur within the deeper zones. Within a few days, mixing takes place and turbidity becomes fairly uniform throughout the reservoir. Later, suspended particles begin to settle out and the waters within the upper zones show the least turbidity while the deepest waters have the greatest.

Increased levels of turbidity are a serious concern for water treatment plant operators because increased turbidity has a high chlorine demand. This could result in a decreased chlorine residual and an increasing possibility of bacterial contamination if the operator is not alert. An outbreak of giardiasis *(Giardia lamblia)* occurred in Pennsylvania when a small reservoir turned over and high-turbidity conditions developed.

Nutrient loading of a reservoir from its drainage area may result in increased productivity (algal blooms). This tends to occur during wet years when increased runoff raises the nutrient loading in the reservoir.

In watersheds containing large quantities of vegetation such as chaparral, there is a noticeable increase in organic loading and in associated THM precursors immediately following periods of major runoff. This condition appears to be related to the large quantities of organic material that are associated with the vegetation. Figure 2.3 illustrates total

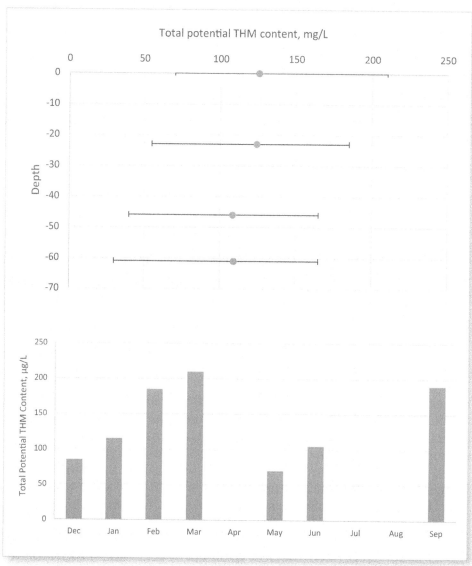

Figure 2.3 Total potential trihalomethane (THM) content at various reservoir depths

potential trihalomethane content in a California reservoir at various depths during a 10-month testing period. (Total potential trihalomethane content was determined by attempting to duplicate actual reservoir conditions in the laboratory. Lake samples were collected at various depths, a 3 mg/L chlorine dose was added to each sample, the sample was stored for 100 hours, and then the sample was analyzed for trihalomethanes.) The dramatic increases in total potential trihalomethane content during the period of January through April are associated with an unusually wet winter. Following the end of the runoff season, the total potential trihalomethane content dropped substantially. The summertime increases in total potential trihalomethane levels are most likely a result of increased organic loading during algal blooms.

2.3.1.2 Thermal Stratification

Water quality problems related to anaerobic conditions often occur in reservoirs and lakes that are thermally stratified. Thermal stratification develops in lakes and reservoirs in the spring when the surface waters begin to warm. As summer approaches, the weather warms, the longer days mean more time for the sun to heat the water, and the spring winds subside. Under these conditions, the surface waters warm rapidly, expand, and become lighter than the lower waters. Although the wind may continue to blow, its contribution to mixing of the lake waters diminishes because of the resistance to mixing resulting from different water densities caused by the increased water temperatures. The greater the difference in water temperatures, the greater the difference in water densities that create the resistance to mixing. When layers of different temperatures occur within a lake, the lake is considered thermally stratified.

Reservoirs and lakes within the United States generally fall into one of two classifications relative to annual thermal stratification cycles. Those relatively deep-water lakes and reservoirs that do not freeze over during winter months undergo a single stratification and destratification (mixing) cycle and are classed as monomictic. In some areas, lakes have one winter overturn that may last from September to mid-May (whenever the wind blows the lake turns over). Lakes and reservoirs that freeze over normally go through two stratification and destratification (mixing) cycles and are classed as dimictic.

In a monomictic lake, the water temperature during winter months is uniform (the same) from top to bottom; the water density throughout the lake is uniform; and the water is mixed only by wind currents. With continued cooling, the surface water becomes more dense and sinks to the bottom. As the season progresses into spring, the sun's rays warm the upper portion of the lake or reservoir faster than the deeper portion. The decrease in density of the warmer water on top slows the vertical mixing action within the lake and a barrier is formed between the upper and lower layers. The upper layer, which continues mixing, is known as the epilimnion (Figure 2.4). The middle layer is the zone of rapid temperature decrease with depth and is called the metalimnion or thermocline. The lowest layer of colder, denser water is the hypolimnion. When these conditions exist, the lake is said to be thermally stratified. The lake remains in this stratified or layered condition through the summer and into the fall or early winter when the surface waters become as cool as deeper waters, the density barrier is broken, and destratification

monomictic (mah-no-MICK-tick)
Lakes and reservoirs that are relatively deep, do not freeze over during the winter months, and undergo a single stratification and mixing cycle during the year. These lakes and reservoirs usually become destratified during the mixing cycle, usually in the fall of the year.

dimictic (dye-MICK-tick)
Lakes and reservoirs that freeze over and normally go through two stratification and two mixing cycles within a year.

epilimnion (EP-uh-LIM-nee-on)
The upper layer of water in a thermally stratified lake or reservoir. This layer consists of the warmest water and has a fairly uniform (constant) temperature. The layer is readily mixed by wind action.

metalimnion (met-uh-LIM-nee-on)
The middle layer in a thermally stratified lake or reservoir. In this layer there is a rapid decrease in temperature with depth. Also called thermocline.

hypolimnion (HI-poe-LIM-nee-on)
The lowest layer in a thermally stratified lake or reservoir. This layer consists of colder, denser water, has a constant temperature, and no mixing occurs.

or mixing (overturn) takes place. Figure 2.5 illustrates the thermal stratification cycle in a monomictic lake by showing average temperature profiles in a Southern California reservoir on January 1, April 25, August 15, and October 25 for a 3-year period.

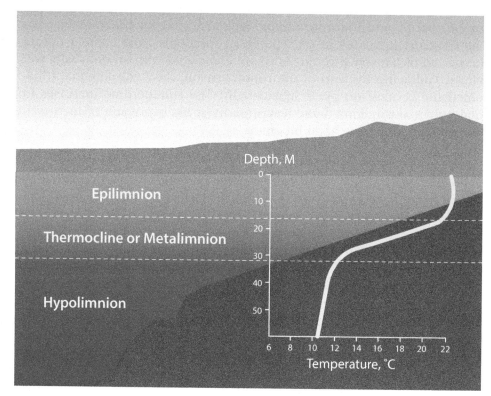

Figure 2.4 Thermally stratified lake or reservoir

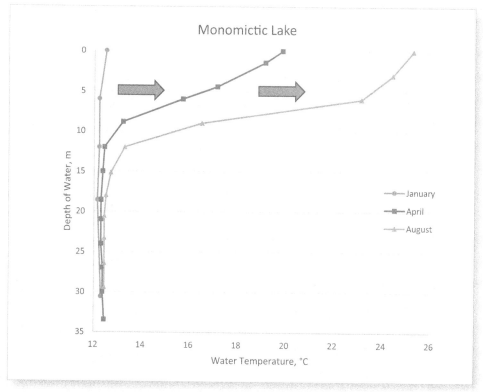

Figure 2.5 Average reservoir temperature profiles on January 1, April 25, August 15, and October 25 for a 3-year period

Once thermal stratification occurs and natural mixing ceases in the metalimnion and hypolimnion zones of a productive lake, major changes in water quality begin to take place. **BOD** (biochemical oxygen demand) within the metalimnion and hypolimnion may lead to total dissolved oxygen (DO) depletion, resulting in **anaerobic** conditions within these zones. Major contributors to the biochemical oxygen demand are the organisms that decompose dead algal cells as they fall into these zones. Depending upon specific conditions, oxygen depletion may be completed in a few weeks or any time up to several months after thermal stratification begins. When the hypolimnion becomes anaerobic, it usually remains so until the reservoir overturn (destratification) occurs in late fall or early winter. Figure 2.6 illustrates the progress of DO stratification in a monomictic lake in California over 1 year. Dissolved oxygen profiles are shown for April, June, and September.

An easy way to measure the temperature profile in a thermally stratified lake or reservoir during the summer is to use a maximum-minimum thermometer. Lower the thermometer to various depths from top to bottom. The minimum temperature is the water temperature at each depth.

2.3.1.3 Nutrients

Many water quality problems in domestic water supply reservoirs occur in reservoirs containing moderate or large quantities of **nutrients** such as phosphate, nitrate, and organic nitrogen compounds. These nutrients may act as a fertilizer in a lake to stimulate the growth of algae just as they stimulate growth on a lawn, garden, or orchard. Reservoirs and lakes that are rich in nutrients and thus very productive in terms of aquatic animal and plant life are commonly referred to as **eutrophic**. Reservoirs that are nutrient-poor and contain little plant or animal life are classed as **oligotrophic**. Between these two types of reservoirs are

BOD (pronounce as separate letters)
Biochemical oxygen demand. The rate at which organisms use the oxygen in water while stabilizing decomposable organic matter under aerobic conditions. In decomposition, organic matter serves as food for the bacteria and energy results from its oxidation. BOD measurements are used as a measure of the organic strength of wastes.

anaerobic (AN-air-O-bick)
A condition in which atmospheric or dissolved molecular oxygen is not present in the aquatic (water) environment.

nutrient
Any substance that is assimilated (taken in) by organisms and promotes growth. Nitrogen and phosphorus are nutrients which promote the growth of algae. There are other essential and trace elements which are also considered nutrients.

eutrophic (yoo-TRO-fick)
Reservoirs and lakes that are rich in nutrients and very productive in terms of aquatic animal and plant life.

oligotrophic (ah-lig-o-TRO-fick)
Reservoirs and lakes that are nutrient poor and contain little aquatic plant or animal life.

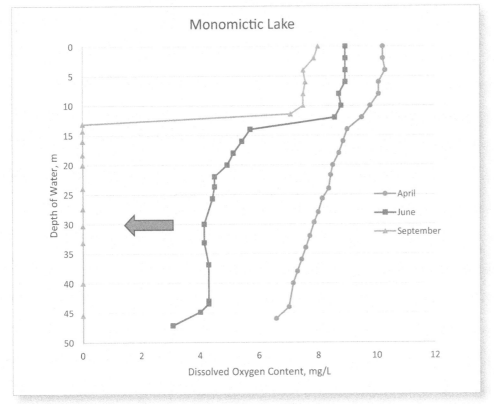

Figure 2.6   Reservoir dissolved oxygen profiles over one summer

mesotrophic reservoirs containing moderate amounts of nutrients able to support moderate levels of plant and animal life.

In productive reservoirs, aquatic plants such as pond weeds, water hyacinths, tules, and sedges may be abundant in the shallow water or **littoral zones** (Figure 2.7). Productive lakes usually support large populations of **phytoplankton** (very small plants) or **zooplankton** (very small animals) at various times during the year. A sudden large increase in plankton populations is commonly referred to as a bloom. Phytoplankton blooms in particular are referred to as algal blooms. The duration and the amount of population growth of an individual algal bloom depends upon various environmental factors, including light, temperature, and nutrient conditions. An individual bloom may contain from one to several types of algae and may last from a few days to several weeks or even months.

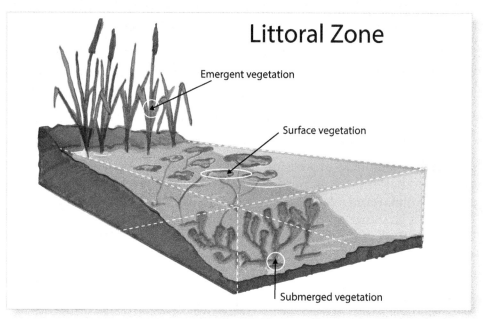

Figure 2.7   Reservoir zones

mesotrophic (MESS-o-TRO-fick)
Reservoirs and lakes that contain moderate quantities of nutrients and are moderately productive in terms of aquatic animal and plant life.

littoral (LIT-or-al) zone
(1) That portion of a body of fresh water extending from the shoreline lakeward to the limit of occupancy of rooted plants. (2) The strip of land along the shoreline between the high and low water levels.

phytoplankton (FIE-tow-plank-ton)
Small, usually microscopic plants (such as algae), found in lakes, reservoirs, and other bodies of water.

zooplankton (ZOE-uh-PLANK-ton)
Small, usually microscopic animals (such as protozoans), found in lakes and reservoirs.

Check Your Understanding

1. List two common sources of water other than lakes or reservoirs.

2. What methods are used to treat domestic water delivered from water supply reservoirs?

3. How do our activities cause deterioration of water quality in reservoirs?

4. What problems can occur in reservoirs that experience large inflows during short periods?

5. When a lake warms in the spring or summer, how does the decrease in density of the warmer surface water influence mixing action within the lake?

6. How soon may oxygen depletion be completed after thermal stratification occurs?

7. Large quantities of what nutrients are undesirable in a water supply reservoir?

8. List the three classes of reservoirs based on nutrient content and productivity in terms of aquatic animal and plant life.

9. What is an algal bloom?

2.3.1.4 Algal Blooms

Several common water quality problems in domestic water supply reservoirs may be related to algal blooms. These problems will be discussed individually but are summarized as follows:

1. Taste and odor problems
2. Shortened filter runs
3. Increased pH (which reduces chlorination efficiency)
4. Organic loading
5. Dissolved oxygen depletion

TASTES AND ODORS

Objectionable tastes and odors in the domestic water supply are often related to the occurrence of algal blooms. The nature of these tastes and odors is influenced by the particular type of algae but may change as the intensity of the algal bloom changes. For example, some algae produce a grassy odor when populations are moderate, but a much more intense odor, described as a septic or pigpen odor, when populations are large or algae are dying and decaying. Among the more common types of tastes and odors produced by algae are the following: fishy, aromatic, grassy, septic, musty, and earthy. Approximately forty types of algae have been identified as taste and odor producers. The extent of taste and odor problems caused by algal blooms ranges from slight consumer objection to total rejection of the supply for domestic uses. Taste complaints often arise when the water is used for drinking or for making coffee or tea because objectionable tastes seem to be more noticeable when the water is at room temperature and above than when the water is cold. Odors are frequently most noticeable when the hot water supply is in use, particularly when it is used for showers, cooking, and dishwashing.

In many situations, chlorination of the water supply reduces the level of tastes and odors; however, there are some instances in which the tastes and odors appear to be stronger following chlorination. Water treatment costs increase significantly when tastes and odors must be removed. Fortunately, many conventional plants are capable of reducing or eliminating tastes and odors when properly operated.

In connection with algae, however, tastes and odors are usually strongest in the thermal layer of the reservoir where the bloom occurs. In most cases, this takes place within the upper layer of water. Figure 2.8 illustrates this fact by showing **threshold odor** profiles during three separate algal blooms in a reservoir in California. These blooms occurred when thermal stratification existed. On the three dates shown, samples were collected from various depths and then analyzed to determine the threshold odor number (TON). The depth and TON were plotted and these points were connected to produce the odor profiles. The odors (represented by higher TON readings) are concentrated within the upper, warmer layer of water and they have not been mixed into the deeper colder layers to any significant extent. See Section 2.3.1.2, "Thermal Stratification."

When the fall overturn occurs in several Southern California domestic water supply reservoirs, obnoxious tastes and odors are brought downward from the upper portion of the reservoir and mixed throughout the entire body of water. Even when multilevel intakes (water inlets) are

threshold odor
The minimum odor of a water sample that can just be detected after successive dilutions with odorless water. Also called odor threshold.

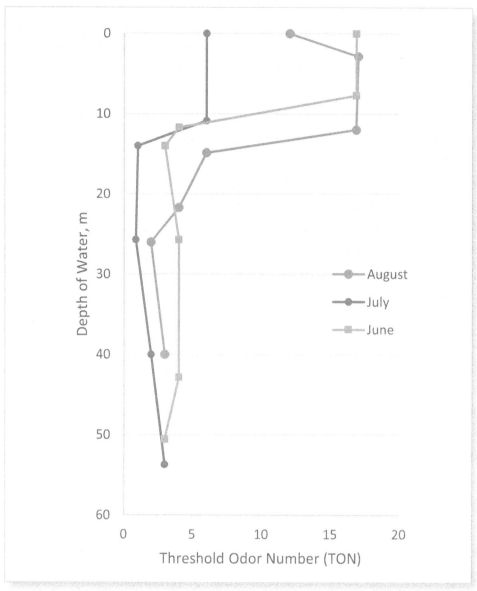

Figure 2.8   Reservoir odor profiles

available, it is not possible to select a depth where taste and odor problems are minimized. In lakes and reservoirs that freeze over during winter months, algal blooms (late algae) have been known to occur underneath the ice, causing taste and odor problems in the deeper waters.

Considerable research has been devoted to identifying taste and odor problems that result from blooms of free-floating (planktonic) algae. One species of blue-green algae in particular, *oscillatoria curviceps,* has been identified as being responsible for certain earthy, musty tastes and odors. To identify algae species and estimate population levels, water samples are collected either with or without the aid of various types of plankton nets (Figure 2.9). Samples are then examined in the laboratory using standard procedures (see Chapter 9, "Laboratory Procedures"). Algae that are growing attached (periphyton) to bottom sediments and structures or submerged plants have been shown to be a contributor to taste and odor problems. These algal growths cannot be sampled and evaluated by conventional methods, so trained scuba divers collect samples and map the extent of algal growths.

Figure 2.9   Plankton sampling net (*Courtesy of Bioquip Products, Inc.*)

Throughout the United States researchers are identifying and quantifying organic compounds that cause taste and odor and linking the particular compounds to the species of algae that produce them. Two compounds, geosmin and methylisoborneol (MIB), which have been linked to earthy, musty tastes and odors, can generate consumer complaints when they are present in the water supply in concentrations as low as a few parts per trillion (nanograms per liter).

SHORTENED FILTER RUNS

A second major problem associated with algal blooms is that certain species of algae, specifically diatoms, tend to clog filters at water treatment plants and thereby reduce both filtration rates and the duration of filter runs. Zooplankton, microscopic, planktonic animals can also clog filters when found in large numbers. Filter runs also may be shortened due to gases released by algae. Normal filter runs commonly extend from 30 to 100 hours before cleaning is required, while short filter runs caused by the presence of algae may be less than 10 hours in length. In extreme cases, the clogging may occur so frequently that the amount of water needed to backwash filters for cleaning is greater than the amount of water treated and sent to the distribution system.

Reduced filtration rates and increased frequency of filter backwashing are reflected in an inability to meet system water demands and increased water treatment costs.

INCREASED pH

Algal blooms are often associated with marked fluctuations in **pH** of the water in the upper layer of the reservoir where blooms occur. The pH is frequently raised from a level near 7 to near 9 or above as a result of these blooms. Chlorination efficiency and coagulation are greatly reduced at the higher pH levels. This increases water treatment costs because pH must be adjusted down, or more chlorine must be added. Since algae remove carbon dioxide from solution and convert it into cellular material as they grow, the carbonate equilibrium (balance) is affected. pH values as high as 9.8 can be reached where algae are in high concentrations and under favorable light conditions.

Increases and decreases in pH are caused by **photosynthesis** during daylight hours and respiration by algae during darkness. During the daylight hours, the pH will increase, while the pH is lowered at night. Respiration by algae results in an increase in carbon dioxide in water

pH (pronounce as separate letters)
pH is an expression of the intensity of the basic or acidic condition of a liquid. Mathematically, pH is the logarithm (base 10) of the reciprocal of the hydrogen ion activity. If $\{H^+\} = 10^{-6.5}$, then pH = 6.5. The pH may range from 0 to 14, where 0 is most acidic, 14 most basic, and 7 neutral.

$$pH = Log\frac{1}{\{H^+\}} \quad \text{or} \quad -Log(\{H^+\})$$

photosynthesis (foe-toe-SIN-thuh-sis)
A process in which organisms, with the aid of chlorophyll (green plant enzyme), convert carbon dioxide and inorganic substances into oxygen and additional plant material, using sunlight for energy. All green plants grow by this process.

(lowering pH), while photosynthesis results in a decrease in carbon dioxide in the water (increasing pH). These fluctuations in pH can adversely affect both the coagulation and disinfection treatment processes.

ORGANIC LOADING

As a result of algal blooms, major increases in organic matter naturally occur within water supply reservoirs. The most notable impacts of this increased organic loading are increased color in the water supply and a major increase in chlorine demand. Solutions to both of these problems contribute to increased water treatment costs.

The organic loading resulting from algal blooms is often associated with high **trihalomethane** levels following free residual chlorination (a water treatment disinfection process). The organic matter contains **THM precursors**, which react with the chlorine to form trihalomethanes. The US Environmental Protection Agency (EPA) has adopted a maximum contaminant level (MCL) of 80 micrograms per liter (μg/L) for trihalomethanes in domestic water supplies, based upon the average of quarterly samplings within the water distribution system. The EPA trihalomethane standards apply to water systems serving more than 10,000 people and to all surface water systems that meet the criteria to avoid filtration.

Increased trihalomethane (THM) levels resulting from algal blooms may exceed the maximum contaminant levels (MCLs). Either a change in disinfection methods or use of activated carbon filtration will normally be required to lower THM levels to within acceptable limits. If either of these procedures is required, it will result in substantial increases in water treatment costs.

DISSOLVED OXYGEN DEPLETION

As algal blooms progress, the DO content at depths where the bloom occurs normally increases markedly as a result of photosynthesis. Supersaturation of DO occurs when the DO exceeds the saturation value for the existing water temperature. This type of supersaturation in surface waters is common during major algal blooms.

When algal cells die, however, this abundant oxygen is used up by the bacteria that feed upon (metabolize) the algal cells. Following severe algal blooms, DO in both surface and deeper waters may be reduced to the point that fish kills occur. Fish kills due to low oxygen levels may also occur when there is a combination of an extremely heavy algal bloom and a sudden reduction in the amount of sunlight available. Under these conditions, the photosynthetic activity of algae slows down. With less oxygen being produced, the algae use oxygen stored in the water for respiration. If this condition persists for a considerable length of time, the water may lose most of its oxygen, causing both the algae and the fish to die of oxygen starvation. Frequently, fish kills from dead algae are the result of the algae clogging the gills of the fish.

Major fish die-offs in domestic water supply reservoirs nearly always generate complaints from the general public, particularly from those who use the water as a drinking supply. The cleanup and disposal of dead fish from a large reservoir can place a large financial and staffing burden on the water utility.

In stratified lakes, oxygen depletion in the colder, deeper waters often occurs following algal blooms. See Section 2.3.1.2, "Thermal Stratification."

Standard Methods for the Examination of Water and Wastewater, 22nd Edition. American Water Works Association. www.awwa.org. ISBN: 9780875530130.

Identification of Algae in Water Supplies, 2001 Edition. CD-ROM. American Water Works Association. www.awwa.org. ISBN: 1-58321-161-6.

trihalomethanes (THMs) (tri-HAL-o-METH-hanes)
Derivatives of methane, CH_4, in which three halogen atoms (chlorine or bromine) are substituted for three of the hydrogen atoms. Often formed during chlorination by reactions with natural organic materials in the water. The resulting compounds (THMs) are suspected of causing cancer.

precursor, THM (pre-CURSE-or)
Natural organic compounds found in all surface and groundwaters. These compounds MAY react with halogens (such as chlorine) to form trihalomethanes (tri-HAL-o-METH-hanes) (THMs); they must be present in order for THMs to form.

2.3.1.5   Anaerobic Conditions

When anaerobic conditions exist in either the metalimnion or hypolimnion of a stratified lake or reservoir, water quality problems may make the water unappealing for domestic use without costly water treatment procedures. Most of these problems are associated with **reduction** in the stratified waters. The first notable problem may be the presence of a strong rotten egg odor in waters drawn from the anaerobic zone. This odor usually indicates the presence of hydrogen sulfide (H_2S) that occurs as sulfate and is reduced to sulfide within the anaerobic bottom sediments. The reduction to H_2S is brought about by certain anaerobic bacteria. Another group of sulfate-reducing bacteria are capable of producing H_2S by attacking organic matter and liberating (freeing) H_2S from protein material. The presence of hydrogen sulfide in domestic water supplies is generally unacceptable to consumers.

A second major problem in anaerobic water occurs when iron or manganese exist in bottom sediments in the reduced state and pass into solution. The presence of either iron or manganese in appreciable quantities within the domestic supply can lead to dirty water problems. In addition to appearing reddish, brown, or just plain dirty, the water may stain clothes during washing and stain porcelain fixtures such as sinks, bathtubs, and toilet bowls. The color or staining occurs as a result of the iron or manganese being changed into the oxidized state after it enters the distribution system. **Oxidation** may occur during disinfection (chlorine is a strong oxidant) or within tanks and reservoirs when the water becomes **aerated**. Water containing either iron or manganese in quantities exceeding the MCLs listed in Federal Secondary Drinking Water Standards (iron, 0.3 mg/L, and manganese, 0.05 mg/L) is generally unacceptable to consumers. If iron or manganese accumulates in the distribution system, extensive **flushing** is required to clean out the system.

When significant levels of DO are present, iron and manganese exist in an oxidized state and normally **precipitate** into the reservoir bottom sediments. Under reducing conditions, however, iron is changed from the oxidized ferric state into the soluble ferrous state, and manganese changes from the oxidized manganic state into the soluble manganous state. Once either or both of these metals pass into solution within the anaerobic zones, they remain there until reservoir overturn occurs and they are oxidized and again precipitate into the bottom sediments. When a lake or reservoir becomes anaerobic, manganese normally passes into solution earlier than iron. Following overturn, manganese usually precipitates out after the iron has precipitated.

In summary, it is often difficult if not impossible to select a level from which to draw acceptable domestic water in a monomictic, productive lake during summer and fall months when thermal stratification exists. Water in the upper levels may contain high quantities of taste- and odor-causing compounds, unacceptably warm water, high organic content, and a pH much higher than desired. The deeper, cooler water may not be acceptable due to the presence of hydrogen sulfide gas, iron and manganese, or other problems related to anaerobic conditions.

Water quality problems within dimictic lakes and reservoirs may be similar to those that occur in monomictic lakes and reservoirs, except that instead of occurring primarily during summer and fall months, they may also exist during winter months. Water normally reaches its greatest

reduction (re-DUCK-shun)
Reduction is the addition of hydrogen, removal of oxygen, or the addition of electrons to an element or compound. Under anaerobic conditions (no dissolved oxygen present), sulfur compounds are reduced to odor-producing hydrogen sulfide (H_2S) and other compounds. The opposite of *oxidation*.

oxidation (ox-uh-DAY-shun)
Oxidation is the addition of oxygen, removal of hydrogen, or the removal of electrons from an element or compound. In the environment, organic matter is oxidized to more stable substances. The opposite of reduction.

aeration (air-A-shun)
The process of adding air to water. Air can be added to water by either passing air through water or passing water through air.

flushing
A method used to clean water distribution lines. Hydrants are opened and water with a high velocity flows through the pipes, removes deposits from the pipes, and flows out the hydrants.

precipitate (pre-SIP-uh-TATE)
(1) An insoluble, finely divided substance which is a product of a chemical reaction within a liquid. (2) The separation from solution of an insoluble substance.

density at 39°F (4°C). As the freezing point of water is near 32°F (0°C), the icecovered upper, colder water is less dense than deeper, slightly warmer waters. This condition results in thermal stratification during periods when surface water temperatures fall below 39°F (4°C). In addition to the normal fall overturn that occurs in monomictic lakes, a second overturn takes place following the spring thaw. When the surface waters warm up to 39°F (4°C), they are denser than the deeper, warmer waters and over-turn takes place once again. Anaerobic conditions and their related water quality problems may exist in these lakes and reservoirs during most of the winter months as well as during summer and fall months.

Water quality improvement programs should be designed to evaluate water quality problems and their causes within a given lake or reservoir. Since no two reservoirs or lakes are exactly alike, each one will require a water quality improvement program prepared specifically for its individual situation.

Check Your Understanding

1. What types of tastes and odors are produced by algae?

2. How can chlorine affect tastes and odors?

3. Where are tastes and odors found in reservoirs?

4. What problems do algae cause on filters?

5. What is the influence of algal blooms on pH?

6. Increased organic loadings from algal blooms can cause what kinds of water quality problems?

7. What is the influence of algal blooms on dissolved oxygen (DO)?

8. What problems are caused by anaerobic conditions in reservoirs?

2.4 Reservoir Management Programs

Reservoir management programs differ based on the uses of the reservoir. As previously mentioned, other uses for reservoirs serving domestic water needs include agricultural, municipal, and industrial purposes; flood control; hydroelectric power generation; fisheries or other environmental uses; and recreation. Water treatment operators usually focus on managing reservoirs for water quality; reducing treatment costs; and fishery, recreational, and property values.

Reservoirs are affected by many factors because they are parts of larger ecological systems that include both natural and manmade impacts. Because of these factors, a variety of tools are necessary to meet management objectives.

2.4.1 Improvement and Maintenance of Water Quality

Water quality management programs can be a very effective tool for controlling water quality problems in domestic water supply lakes and reservoirs. By evaluating the water quality problems that occur in a given reservoir and the various water quality management and control alternatives available, it may be possible to design a program to eliminate or at least control most problems within acceptable limits. Progress has been made toward managing water quality within productive and eutrophic lakes and reservoirs. In many cases, the frequency and intensity of algal blooms and the water quality effects associated with these blooms have been controlled. Frequently, DO depletion can be controlled or eliminated within the metalimnion and hypolimnion, thereby eliminating or controlling iron, manganese, and hydrogen sulfide problems.

Properly designed water quality management programs can also be effective in controlling silt loading, turbidity levels, nutrient loading, and organic loading in many water supply reservoirs. The actual procedures for preparing water quality management programs are discussed in the remainder of this lesson. To be feasible, water quality management programs must be specific to each reservoir and economically as well as technically justifiable.

2.4.2   Reduction of Water Treatment Costs

Operation, maintenance, and capital costs of water treatment facilities have all increased substantially during recent years. Proper management of water quality within the reservoir can prove to be an effective tool in controlling these costs. Cost savings may be realized through increased length of filter runs, reduced chemical costs, and alternative treatment methods. For example, if taste and odor problems and organic loading are controlled within the reservoir, the need for activated carbon treatment at the plant may be reduced or eliminated. By controlling algal blooms and silt loading, it may be possible to consider using direct filtration only instead of complete treatment. Iron and manganese control or removal within the treatment plant may not be necessary if levels are controlled within the reservoir. Cost savings must be evaluated on an individual basis, but may range from a few dollars to tens of dollars per million gallons treated. Major savings of hundreds of thousands to millions of dollars may be realized if capital costs for facilities such as sedimentation basins and activated carbon filters can be eliminated.

2.4.3   Improvement and Maintenance of Fishery, Recreational, and Property Values

In productive reservoirs and lakes where DO depletion occurs within the metalimnion and hypolimnion during summer and fall months, fish are forced to move into the warmer waters of the epilimnion. This not only reduces the size of habitat available to fish, but it may also limit the number of species that exist within a given lake or reservoir. In many areas of the United States, summer temperatures within the epilimnion reach maximums of 75 to 80°F (24 to 27°C). Cold water species of fish, such as trout and salmon, may not survive at these temperatures. If the deeper, colder waters can be prevented from becoming oxygen depleted, both warm and cold water species of fish can be maintained throughout the year.

Proper reservoir management techniques can prevent or minimize fish kills. Reduction in the intensity and severity of algal blooms reduces the hazard of fish kills as a result of rapid oxygen depletion or clogging of their gills with algae. Fish kills that occur in frozen lakes as a result of DO depletion in waters beneath the ice may also be reduced or eliminated by proper management techniques.

Proper reservoir management naturally results in increased appeal of the reservoir for recreational purposes. Recreational values are also increased when algal problems are reduced. Large mats and scums of algae are unappealing to swimmers, bathers, and water skiers. Objectionable odors associated with these blooms reduce the reservoir area's appeal as a site for camping and picnicking. Property values around the edge of a lake or reservoir may increase significantly when major algal problems are reduced or eliminated.

2.4.4   Removal of Trees and Brush from Areas to Be Flooded

Some reservoir management tools must be employed before a reservoir holds water. In areas where new reservoirs are to be formed for the purpose of providing domestic water supplies by construction of dams or other means, it is often advisable to remove trees and brush from the areas to be flooded. The purpose of vegetation removal is to reduce the organic and nutrient loading the reservoir will receive as it fills.

If a large quantity of trees, brush, and other vegetation is left within the reservoir site, organisms will decompose and thus recycle the material after the area is flooded. This will release nutrients and organic matter into reservoir waters. Organisms that decompose the vegetative material will consume DO, thereby increasing the rate of oxygen depletion when thermal stratification of the reservoir takes place. Nutrients that are released during decomposition may lead to more and bigger algal blooms. Organics released during decomposition may contribute to increased color, chlorine demand, and trihalomethane levels following chlorination.

Removal of major vegetation from the reservoir area is best accomplished by mechanical means. Little is accomplished if the material is cut and left on site so that it can decompose by natural processes. When practical, trees can be cut for firewood or timber and the stumps removed mechanically. In some cases, wood is burned on the site and the ashes are removed from the area. In other cases, trees, including stumps, are removed by mechanical means.

Brush can be removed by a single tractor or by two or more tractors operating parallel to each other with a large chain stretched between them.

Once vegetation removal is accomplished, regrowth must be controlled by mechanical means until the reservoir is filled. When the filling cycle is completed, reservoir levels will seldom be lowered long enough for major vegetation regrowth to occur.

2.4.5   Watershed Management

The primary purpose of watershed management should be to control, minimize, or eliminate any practices within the watershed area that are harmful to water quality within the domestic water supply reservoir. Effective watershed management programs are both technically and economically feasible. The general public must be convinced that they are getting their money's worth when investing in watershed management programs. If it is much less costly to cure a water quality problem with specific treatment procedures than it is to prevent the problem from occurring through watershed management, it may be difficult, and even unwise, to try to convince the public that watershed controls should be implemented. If factors such as aesthetic or recreational values of the reservoir are also affected by the reduced water quality, the public should be informed so that the value of the water resource from an aesthetic or recreational viewpoint can be evaluated. While the public is very dollar conscious, it is also sensitive to environmental issues and will support well-designed programs for protecting the environment.

Because of the tremendous differences in size, topography, vegetative conditions, and state of development in watersheds, it is impossible to discuss methods of watershed management that would apply in all areas. As with reservoir management programs, each watershed management program should be designed on the basis of potential water quality problems that must be solved or prevented and the options available for solving or preventing them. In most cases, the best tool for managing watersheds is probably the regulatory process. Regulations and ordinances that control or eliminate practices within a watershed that are detrimental to water quality can be adopted as needed on a local, county, state, or even federal level. All types of practices that may have to be controlled within a given watershed are too numerous to discuss in detail, but the most common practices causing water quality problems and means of controlling them are discussed in the following sections.

2.4.5.1 Wastewater

Contamination of the domestic water reservoir by wastewater (sewage) can lead to water quality problems ranging from significant to severe. Two major types of problems may result from raw wastewater contamination: nutrient loading of the lake or reservoir and microbial contamination. In some cases, nutrient loading due to prolonged wastewater contamination may be great enough to convert a previously unproductive or moderately productive body of water into a very productive, highly eutrophic lake. Examples of lakes that have experienced water quality deterioration due to nutrient loading from wastewater are numerous.

Microbial contamination of a lake or reservoir by wastewater may pose a hazard not only to the domestic water supply but also to people using the reservoir for recreational purposes, such as swimming and water skiing. Diseases caused by protozoans, bacteria, and viruses may all result if the microbial contamination is severe. While conventional water treatment practices do provide protection from microbial contamination to domestic consumers, the source of supply should still be protected. Some states have regulations relating to the maximum allowable **coliform** bacteria content in domestic water supply reservoirs.

The major source of wastewater contamination in domestic water supply reservoirs is usually wastewater disposal systems such as septic tank leaching systems. There are two major methods by which contamination from septic tank systems can be controlled. The first and most dependable solution is to replace all septic tank leaching systems with well-designed sewer systems that collect all wastewater and transport it to centralized treatment facilities. Wastewater treatment facilities should be located outside of the reservoir watershed or at least located so that the treatment plant effluent does not enter the domestic water supply. Unless

coliform (COAL-i-form)
A group of bacteria found in the intestines of warmblooded animals (including humans) and also in plants, soil, air and water. Fecal coliforms are a specific class of bacteria which only inhabit the intestines of warm-blooded animals. The presence of coliform bacteria is an indication that the water is polluted and may contain pathogenic (disease-causing) organisms.

very expensive nutrient removal treatment processes are used, the nutrient loading in a reservoir from a treatment plant effluent may have a significant impact on algae productivity.

The second major solution is to adopt ordinances or conditions that regulate the design and installation of septic tank leaching systems to ensure that they function properly and do not contaminate the domestic water supply. Provisions should be made for all systems in the watershed to be brought up to established standards. Periodic inspections should verify that the septic tank leaching systems continue to function properly.

2.4.5.2   Fertilization

Fertilization of crops and landscaping with materials containing high concentrations of nitrogen compounds can result in these nutrients being carried away in surface runoff and contributing to productivity in the water supply reservoir if fertilization practices are not controlled. Large quantities of fertilizers containing nitrogen compounds are often used in commercial agricultural production; on golf courses and other park-like areas; and for home lawns, orchards, and gardens. Nitrogen compounds in excess of those required for plant growth may be leached downward into the groundwater following irrigation or precipitation. In many instances, the groundwater eventually ends up in the surface water supply reservoir through underflow or surface runoff after it is pumped and used for irrigation. Phosphate-base fertilizers usually do not present the problems that nitrogen fertilizers do because of the soil binding characteristics of phosphate compounds. Once phosphate compounds have entered the soil, they tend to remain there unless taken up by plants.

Fertilization practices can be partially controlled by prohibiting use of fertilizers on nonessential crops, but are best controlled through long-term public education programs. People who need to use large quantities of fertilizers can be properly trained to apply only enough for adequate plant growth and to apply fertilizers at times when there is minimal chance of excess fertilizer being washed into the water supply. In some major agricultural areas, it is becoming common practice to have a leaf analysis done on crop plants to determine fertilizer needs and application rates. This practice not only protects the water supply from overfertilization, it is good business practice for the farmer.

2.4.5.3   Industrial Discharges

Industrial discharges are usually best monitored and controlled by the discharger and a regulatory agency other than the local agency that manages and operates the water supply. In most areas, this responsibility falls under the regulation of a state water pollution control agency. Local water supply agencies can work closely with the pollution control agency in identifying sources of industrial discharges and regulating them so that pollution problems are minimized. Water pollution control programs have been established in each state and are administered under federal legislation, which sets controls on industrial discharges.

If mining occurs within the watershed and materials mined present a hazard to the water supply, strict regulations should be adopted to prevent water supply contamination. Cyanide contamination may occur in the runoff from gold mining operations. Erosion from mine tailings and exposed areas should be strictly controlled. Field investigations have shown very high counts of asbestos fibers in portions of some water supplies. A major source of these asbestos fibers is abandoned mining areas through which surface runoff drains. Asbestos fiber concentrations in domestic water supplies present serious concern regarding links between asbestos and some types of cancer. A limit of 7 million fibers per liter of water has been established.

Oil and gas exploration and drilling must also be regulated so that major contamination of the water supply does not occur. Every agency that operates a surface water supply should prepare and have available an emergency plan, which can be implemented immediately if a major spill of hazardous material occurs. County and state health departments, the EPA, and state pollution control agencies may provide assistance in preparing and implementing hazardous spill emergency plans.

2.4.5.4   Soil Grading and Farming Practices

Some watersheds contain soils which, when eroded, may significantly contribute to turbidity in the surface water supply reservoir. It may be necessary to control soil grading and farming practices in sensitive areas. Of particular concern are soils that contain colloidal clays and similar particles that do not settle out in a reasonable period of time within the reservoir. Soil grading and farming practices can be controlled through regulations and ordinances. The major thrust of these rules is not to prohibit soil disturbances (such as those that occur when farming, logging, or constructing housing subdivisions and industrial parks), but to limit disturbances to those times of the year when danger of erosion from surface runoff is at a minimum. Ordinances can also be used to limit the length of time that soil is left exposed and to specify methods of replanting and replacing ground cover.

2.4.5.5   Livestock Grazing

Livestock grazing can be an invaluable tool for controlling vegetative growth and the potential for wildfires in certain watersheds. However, grazing can also contribute to significant deterioration in water quality within surface water supply reservoirs. Overgrazing can expose soils and lead to increased erosion and turbidity problems. Also, nutrients concentrated in cattle manure may be washed into the reservoir during periods of high runoff and cause eutrophication. In addition, accumulations of manure may contribute to microbial contamination of the domestic water reservoir during major runoff periods. The most notable microbial pathogen associated with cattle feces is *Cryptosporidium*, a protozoan parasite that causes gastroenteritis in humans. Some of the diseases contracted by humans are carried by animals. In most watersheds, animal grazing can be held within acceptable limits through the implementation of controls and regulations. Grazing can be eliminated or controlled for a limited period of time before and during periods of high runoff when fresh manure accumulations would create the greatest hazard to the water supply.

2.4.5.6   Pesticides and Herbicides

Unregulated general use of pesticides and herbicides can cause serious problems of contamination in domestic water supply reservoirs. This problem has been minimized on a national level through federal and local regulations. The federal government monitors and regulates the use of pesticides and herbicides on watershed lands it owns or controls. Certain materials are prohibited from use on these lands, some are restricted in

eutrophication (you-TRO-fi-KAY-shun)
The increase in the nutrient levels of a lake or other body of water; this usually causes an increase in the growth of aquatic animal and plant life.

their application, and some are approved for general use. All uses are reported on at least an annual basis. In some areas, agricultural commissioners regulate the use of pesticides and herbicides and will usually cooperate with local agencies in setting up controls for specific problem areas such as watersheds. Operators should be notified when pesticides and herbicides that may pose a threat to the water supply are being applied.

2.4.5.7   Wildfires

Wildfires in certain watersheds can create more problems for a domestic water supply reservoir than any other source of contamination or pollution. During the runoff period following the fire, large amounts of debris, nutrients, silt, and other pollutants may enter the reservoir. Turbidity will usually increase adversely effecting water treatment plants. Depending on individual situations and the extent of burning within the watershed, water quality may be degraded (reduced) for a relatively short period of time or for several years or more. Reservoir storage space lost to debris and silt accumulations may never be recovered without expensive silt removal.

Fire prevention and control programs are an absolute must in watersheds where major fire hazards exist. Public information programs can often be helpful in preventing fires from occurring. Federal and local fire control agencies can often cooperate in firebreak, vegetation control, and other fire control programs. Controlled burning is a valuable tool in many areas. One fire control measure that is being researched and developed in the western United States is the conversion of chaparral and brushlands to grasslands. This conversion reduces the amount of fuel available to burn. In some instances, the total available water supply can be increased because of increased runoff and reduced **evapotranspiration** from the grasses.

2.4.5.8   Land Use Control

The ownership and management of lands surrounding a reservoir and in its watershed influences the quality of water in the reservoir. Limiting land use can reduce contamination from sources such as erosion, pesticides and herbicides, and biological agents. Many drinking water suppliers, in cooperation with government agencies and other interested organizations, use a variety of strategies to limit land use, including zoning and other local ordinances, state or federal regulation, conservation easements, use permits, and cooperative agreements among land owners and users.

For example, over time the land use around one reservoir in California changed to protect the water quality in the reservoir. Most of the watershed lies within a National Forest and presents little hazard to the water quality of the reservoir except from wildfires and those mining and oil and gas operations permitted. When it was demonstrated that mining operations could significantly reduce the water quality, the Secretary of Interior withdrew permission to mine any US forest lands within the watershed. In this instance, the mining permits were under federal control, but state or local governments may be able to similarly reduce the hazards from mining operations.

Approximately 3,000 acres of the watershed around the reservoir were privately owned and considered prime land for subdivision and other types of development. Studies indicated that any major development would increase the nutrient loading of the lake and lead to serious deterioration of water quality. Because the reservoir is located near a major metropolitan area, development was deemed very likely. As a result of the

evapotranspiration (ee-VAP-o-TRANS-purr-A-shun)
(1) The process by which water vapor passes into the atmosphere from living plants. Also called transpiration. (2) The total water removed from an area by transpiration (plants) and by evaporation from soil, snow and water surfaces.

studies and efforts by local citizens and governmental agencies and representatives, the US Congress approved funds to buy the privately owned lands and preserve them as open space. A local ordinance was passed that limits public use of the open space lands. Residents in the watershed area who wished to remain were granted 25-year or life estates. The remaining residents must comply with conditions of an agreement controlling activities that would significantly deteriorate water quality. In this case, watershed management proved to be a major factor in the overall program to control and maintain water quality within the reservoir. As in this case, single agency ownership of all portions of a watershed may be the only politically and economically feasible way to ensure control of the land in a watershed, but many watersheds are managed through cooperative agreements among a variety of agencies, governments, residents, and land users.

2.4.5.9   Highway Stormwater Runoff

Stormwater runoff from highways has the potential to cause adverse impacts on reservoir water quality. Automobiles, trucks, and buses can leave pollutants on roadway surfaces. The motoring public can discard trash and litter. Winds can blow pollutants on roadways from adjacent lands, and precipitation can deposit pollutants on highways from great distances. Another source of pollutants is highway maintenance activities.

Pollutants of concern in highway stormwater runoff include toxic metals, nutrients, bacteriological constituents, oil and grease, floating materials, trash and litter, pesticides, herbicides, and deicing salts. Pollutants that might enter waterways as a result of accidents or spills are also a concern.

Check Your Understanding

1. What reservoir water quality problems can be controlled or eliminated by reservoir management programs?

2. How can reservoir management programs reduce water treatment costs?

3. What happens to trout and salmon when dissolved oxygen (DO) depletion occurs within the metalimnion and hypolimnion during summer and fall months?

4. Why should trees and brush be removed from areas to be flooded by reservoirs?

5. What water quality problems may be caused by organics released during the decomposition of vegetation covered by water in a reservoir?

6. How can vegetation be removed from a reservoir site?

7. What should be the primary purpose of a watershed management program?

8. What problems can be caused in reservoirs from raw wastewater contamination?

9. How can problems caused by fertilizers be controlled?

10. How can the adverse impacts of soil disturbances from farming, logging, and construction be minimized?

11. How can the use of pesticides and herbicides be controlled in a watershed?

12. What problems can be created as a result of a wildfire?

13. Why might a drinking water supplier acquire title to land in a watershed?

2.4.6   Algae Control by Chemical Methods

Chemical control of both planktonic (free-floating) and attached aquatic growths (periphyton) in domestic water supply reservoirs is primarily used to prevent or control taste and odor problems resulting from algal blooms. Many of these algae control programs, however, are expensive and only minimally effective. In some cases, it may be much more feasible and economical to limit nutrient availability (preventing major algal blooms) than to correct the problems these blooms cause by the application of chemicals.

A second major purpose of controlling algal blooms by chemical means is to reduce the overall biological productivity. This will reduce the rate of oxygen depletion in the lower parts of the lake. Chemical control programs must be carried out in such a manner that the algal bloom is prevented from becoming intense if the rate of oxygen depletion following die-off is to be reduced. The rate of oxygen depletion in the deeper waters will probably be increased if the bloom is allowed to become intense before chemicals are applied. If algal populations are high when chemical control is started, a large die-off may occur over a very short period of time. This will produce a rapid decomposition of the algal bodies that may cause depletion of DO as well as an increase in tastes and odors. Fish die-offs could result from dead algal bodies clogging fish gills or from low DO levels.

A third reason for using chemicals for control of algae is to maintain acceptable aesthetic conditions in the lake or reservoir. Unsightly algal scums, odors, and lack of water clarity may all be controlled by proper application of chemicals for algae control.

2.4.6.1   Chemicals Available

There are very few chemicals that can be economically used for controlling algal growths in domestic water supply lakes and reservoirs. Extreme caution must be exercised in selecting chemicals. Many of the most effective materials are not approved for use in domestic waters due to potential hazards for the health of people, fish, and crops. State and local health agencies should be consulted before any chemical algae control program is started. Some states or local areas may require permits from, or consultation with, agencies such as the Department of Fish and Game and Department of Agriculture.

Chlorine, used as a bactericide or oxidizing agent, may also produce the effects of an algicide. Several commercially prepared chelated copper compounds for use in alkaline waters are presently available and may prove to be economical for use in many situations. However, copper sulfate pentahydrate $(CuSO_4 \cdot 5 H_2O)$ (also called bluestone), either by itself or in conjunction with certain other chemicals, is the primary algicide in common use in domestic water reservoirs. Copper sulfate is relatively inexpensive when compared with other chemicals. This copper compound is toxic to many species of algae at relatively low concentrations, but does not present a health hazard to either the workers applying it or to domestic water users if proper application and safety procedures are followed. However, copper sulfate may be a hazard to trout at levels below those necessary to control some algae and may cause toxins to be released by algae.

Research by the Metropolitan Water District of Southern California and others, carried out in conjunction with the EPA, has pinpointed

chelation (key-LAY-shun)
A chemical complexing (forming or joining together) of metallic cations (such as copper) with certain organic compounds, such as EDTA (ethylene diamine tetracetic acid). Chelation is used to prevent the precipitation of metals (copper).

one major area of concern when copper compounds are regularly added to the water supply source. Water that contains even very low concentrations of copper has been found to cause significant corrosion problems in water distribution systems, particularly those with galvanized piping. The problem appears to be most severe in newer systems where no buildup of calcium or other compounds (which serve to protect the inside of the pipe) has occurred. This problem is more severe in systems supplied with highly mineralized water. Copper residuals in water entering the distribution system must be monitored closely following copper sulfate treatments to compile a record of copper concentrations for future reference.

The action level (EPA-designated concentration requiring treatment) for copper in drinking water is 1.3 mg/L at the consumer's tap, but the acceptable level of copper in some receiving waters is much lower due to the potential toxicity of copper to fish and other aquatic organisms. Due to these concerns, some reservoir managers are using hydrogen-peroxide-based herbicides to control algae in drinking water supply reservoirs.

2.4.6.2   Methods of Copper Sulfate Application

Depending on the method of application, copper sulfate may be purchased in dry form with crystal size ranging from snowflake to large diamond of up to 1 inch (25 mm) or more in diameter. Methods of using copper sulfate compounds range from very simple to very elaborate, depending on the size of the reservoir to be treated, frequency of treatment, and rates of application.

1. The simplest method, and one that may be most applicable in very small lakes and reservoirs, is to drag burlap bags containing the copper material through the water using a boat. The reservoir surface is normally crisscrossed in a zigzag fashion by the boat so as to cover all of the water surface. Mixing of the material within the water is accomplished by wind, diffusion, and gravity. Boat speed, number of bags used, and the size of the crystals can all be regulated to produce the desired rate of application.

2. Another method is to load dry copper sulfate crystals into a hopper mounted on a boat in such a way that they can be fed into a broadcaster (such as the type used for spreading grass seed), which distributes them onto the lake surface. If the entire surface area of the reservoir is to be treated, the material can be applied in a crisscross pattern. Holes can be drilled or cut in the bottom of the hopper to control how much material is released onto the broadcaster. Boat speed, width of area covered by the broadcaster, and rate of copper sulfate fed through the holes can be calculated and adjusted to produce the desired application rate.

3. Perhaps the most efficient and safest method of applying copper sulfate is to mix it into solution and spray it onto the reservoir surface or pump it into the reservoir through a length of pipe (preferably plastic) that contains a number of holes. The pipe may be 15 to 20 feet (4.5 to 6 m) long with the holes about 2 feet (0.5 m) apart. The pipe can be mounted behind the boat below the surface of the water, and perpendicular to the direction of travel. The primary advantage of this submerged application instead of spraying

is that the application of the solution is unaffected by wind. The disadvantages of the submerged application are that it is difficult to tell when the holes are plugged and the delivery system may be damaged or caught in submerged trees or brush, especially in shallow water. Figure 2.10 shows the equipment used for the spraying method of applying copper sulfate from a boat.

Mixing algicide chemicals into solution is a good method when citric acid is combined with the copper sulfate to prevent the precipitation of copper in alkaline waters. Copper sulfate in snowflake-size crystals works best for this method of application. Copper sulfate is commonly supplied in 80-pound (36 kg) bags, which are easy to move and store. Citric acid, which is supplied in granular form in bags, is also easy to use.

To get the materials into solution, they are loaded on the boat or barge in bags and fed into a hopper designed so that different sized holes can be fitted into the bottom outlet. The various hole sizes allow adjustment of the rate at which material is released from the hopper. If two chemicals are to be mixed, a divided hopper can be used with different sized holes so that the feed rate of each chemical can be adjusted independently. If the hopper is made of steel, it can be coated with an epoxy material to limit corrosion.

The materials are fed from the hopper into a corrosion-resistant tank where they are mixed into solution using water pumped into the tank from the lake or reservoir. Once in solution, the materials are applied to the lake or reservoir by pumping through the spray nozzles or through submerged outlets. By proper regulation of flow through the pumps, the solution level within the tank can be regulated. The desired rate of application of chemicals can be obtained by considering the speed of the boat, the rate at which material is fed from the hopper and pumped into the lake, and the area covered by the spray apparatus or submerged outlet system.

2.4.6.3 Copper Sulfate Doses

Alkalinity, pH, suspended matter, water temperature, and species of algae affect the efficiency of copper sulfate as an algicide. Alkalinity of the water is the principal factor that reduces the effectiveness of copper sulfate. In alkaline waters, the copper ions react with bicarbonate and carbonate ions to form insoluble complexes (joined together) that precipitate from solution and reduce the amount of biologically active copper. Once the copper is removed from the ionized form, it is no longer effective as an algicide. Botanist Alfred F. Bartsch found that the copper sulfate dosage should be dependent upon the alkalinity of the water and reported that this combination has resulted in successful treatment in various lakes in the midwestern United States. If the methyl orange alkalinity of the water is less than 50 mg/L, the copper sulfate is effective at the rate of 0.9 pounds of copper sulfate per acre-foot (volume of water) (0.00033 kg/m^3 or 0.33 mg/L). If the methyl orange alkalinity is greater than 50 mg/L, the rate should be 5.4 pounds of copper sulfate per acre (surface area of water) (6.06 kg/ha or 0.000605 kg/m^2). In waters with a high alkalinity, the dosage is not dependent upon depth because precipitation of copper would make it ineffective very far below the surface (copper crystals do not dissolve as they fall through the water).

Experience by various researchers and agencies confirms Bartsch's findings that, in most cases, copper sulfate is fully effective as an algicide when the alkalinity is 0 to 50 mg/L. These experiences further reveal that at alkalinity concentrations from 50 to 150 mg/L, copper sulfate dosages

methyl orange alkalinity
A measure of the total alkalinity in a water sample. The alkalinity is measured by the amount of standard sulfuric acid required to lower the pH of the water to a pH level of 4.5, as indicated by the change in color of methyl orange from orange to pink. Methyl orange alkalinity is expressed as milligrams per liter equivalent calcium carbonate.

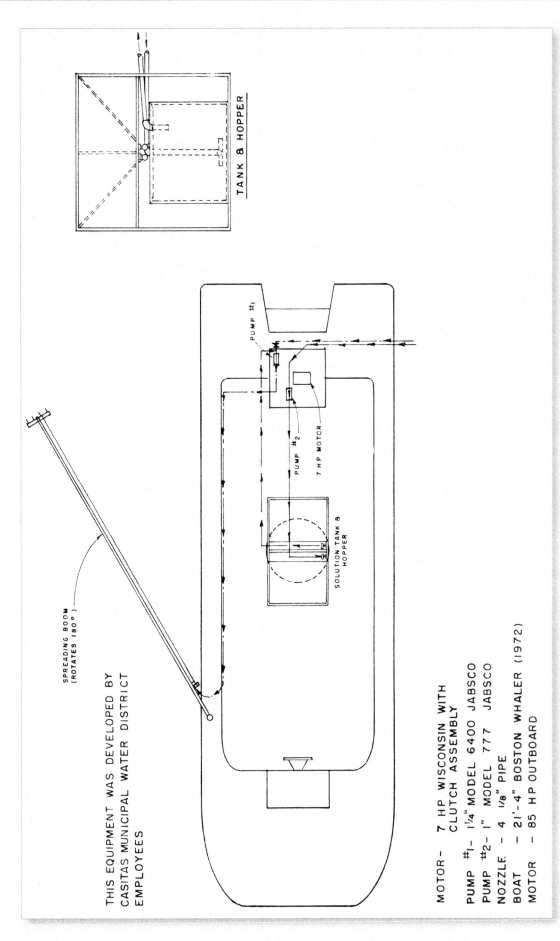

THIS EQUIPMENT WAS DEVELOPED BY
CASITAS MUNICIPAL WATER DISTRICT
EMPLOYEES

MOTOR — 7 HP WISCONSIN WITH
             CLUTCH ASSEMBLY

PUMP #1 — 1¼" MODEL 6400 JABSCO
PUMP #2 — 1" MODEL 777 JABSCO
NOZZLE — 4 ⅛" PIPE
BOAT — 21'-4" BOSTON WHALER (1972)
MOTOR — 85 HP OUTBOARD

Figure 2.10 Copper sulfate spraying equipment

must be increased as the alkalinity increases. When the alkalinity exceeds 150 mg/L, the use of copper sulfate by itself as an algicide would not normally be recommended because of its very low effectiveness.

The effectiveness of copper sulfate as an algicide also depends on the pH. The lower the pH, the more copper ion is present and the more effective copper is as an algicide. Also, the pH level influences the precipitation of copper whose presence is essential to control algae. The higher the pH, the more copper is likely to precipitate and not be available as an algicide.

Suspended matter in the reservoir or lake being treated with copper sulfate can reduce the effectiveness of copper as an algicide. Such suspended matter provides sites or masses other than algal bodies where the copper is **adsorbed**. Organic material, both living and dead, adsorbs the copper. Suspended inorganic sediment is also a significant factor influencing the loss of copper available to kill the algae.

Water temperature plays a major role in how well copper sulfate kills algae. When the water temperature drops to 50°F (10°C) or lower, copper sulfate is not as effective at controlling algae as it is at higher temperatures. So, higher rates of copper sulfate application will generally be required when the water drops below this temperature. In many reservoirs, major blooms of algae do not occur in colder waters with temperatures below 50°F (10°C). In the majority of cases, the problem blooms take place after surface water temperatures have warmed during spring and summer months.

In addition to water quality indicators, the amount of copper sulfate required for effective control of algae is influenced by the species to be treated. Not all algae are alike in their reaction to copper sulfate. Several tiny planktonic green algae, some of the green flagellates, and filamentous blue-green algae are somewhat resistant to the toxic effects of copper sulfate. Most diatoms are quite susceptible to treatment, though they often bloom in large numbers following copper sulfate treatment for other algae. Many of the major taste-and odor-producing algae and filter-clogging algae are controlled effectively with low rates of application.

Copper sulfate products are currently registered through the EPA for use in controlling algae. To comply with these federal requirements, copper sulfate products properly labeled for the intended use must be selected. The maximum rate of application of copper sulfate in domestic water supply sources is currently influenced by regulations limiting the concentration of copper in **potable water**. As stated previously, EPA has established an action level of 1.3 mg/L for copper at the consumer's tap. Another important consideration in the rate of application is the tolerance of fish and other aquatic organisms to copper.

Investigation and field experiments by various individuals and agencies have indicated that effective use of copper sulfate as an algicide can be accomplished in highly alkaline water when the copper sulfate is combined with **aliphatic hydroxy acids**. These acids have proven effective for delaying the chemical reaction of copper with the bicarbonate and carbonate ions of the water and thereby preventing the immediate precipitation of copper in alkaline waters. The most commonly used of these acids is citric acid, as discussed in the previous section. Citric acid is normally mixed with the copper sulfate in a ratio of approximately two parts copper sulfate to one part citric acid regardless of the alkalinity. One water agency used a copper sulfate plus citric acid mixture with satisfactory results for over 10 years in a reservoir in Southern California. That reservoir had an alkalinity of approximately 150 mg/L.

adsorption (add-SORP-shun)
The gathering of a gas, liquid, or dissolved substance on the surface or interface zone of another material.

potable (POE-tuh-bull) water
Water that does not contain objectionable pollution, contamination, minerals, or infective agents and is considered satisfactory for drinking.

aliphatic (AL-uh-FAT-ick) hydroxy acids
Organic acids with carbon atoms arranged in branched or unbranched open chains rather than in rings.

The estimated amount of available copper is an important part of determining the proper chemical dose of copper sulfate. Because of the composition of $CuSO_4 \cdot 5\,H_2O$, the compound contains 25 percent available copper. This percentage is calculated using the **molecular weight** of copper and copper sulfate pentahydrate. The first step of this calculation uses the molecular weight of the constituents of copper sulfate pentahydrate to determine the compound's molecular weight.

$$MW_{CuSO_4 \cdot 5H_2O} = MW_{Cu} + MW_S + 4 \times MW_O + 5 \times MW_{H_2O}$$
$$= 63.5 + 32 + 4 \times 16 + 5 \times 18$$
$$= 249.5 \text{ g}$$

Next, determine the percent of copper in copper sulfate.

$$\frac{MW_{Cu}}{MW_{CuSO_4 \cdot 5H_2O}} \times 100 = \frac{63.5}{249.5} \times 100 = 25\%$$

Now the dose of copper sulfate per million gallons of water can be determined using the following equation:

$$= \frac{1\left[\dfrac{mg}{L}\right] \times \dfrac{3.785\ L}{gal} \times \dfrac{kg}{10^6\ mg} \times \dfrac{2.205\ lb}{kg} \times \dfrac{10^6}{M} \times 100\%}{25\%\ \text{Available Copper}}$$

$$= \frac{33.4 \text{ lb Copper Sulfate}}{\text{Million Gal Water}}$$

Because 8.34 pounds of a substance in a million gallons of water is equal to 1 mg/L, 33.4 pounds of copper sulfate should be added for a 1 mg/L dose of copper to be obtained in 1 million gallons of water.

By knowing the quantity of water to be treated in millions of gallons and the desired dose, it is relatively simple to compute the amount of copper sulfate required per application.

FORMULAS

The dose of copper sulfate to be applied to a reservoir can be calculated using (1) the volume of the reservoir, and (2) the desired dosage in terms of either the reservoir surface area or volume.

1. Reservoir information is usually stated in terms of surface area and volume on the basis of water surface level or elevation. Sometimes, these data are presented as depth vs. surface area or depth vs. volume curves (Figure 2.11). If these curves are available, simply observe the depth of water or surface elevation and obtain the reservoir surface area and volume from the curves. The information used to plot these curves may be available in a table also.

   If the reservoir volume is given in acre-feet, convert this number to a volume in million gallons.

$$\text{Volume}\,[\text{gal}] = \text{Volume}\,[\text{ac-ft}] \times \frac{43{,}560 \text{ ft}^2}{ac} \times \frac{7.48 \text{ gal}}{ft^3}$$

Bartsch, A. F., *Practical Methods for Control of Algae and Water Weeds*, Public Health Reports, 69:749–757, 1954.

molecular weight
The molecular weight of a compound in grams per mole is the sum of the atomic weights of the elements in the compound.

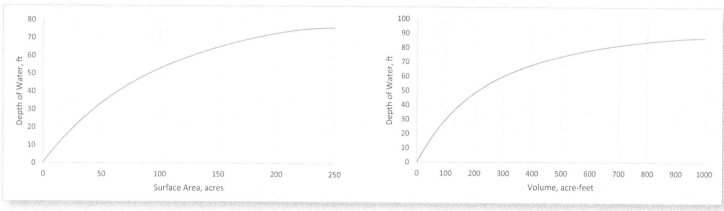

Figure 2.11   Depth vs. surface area and volume curves

The reservoir volume in acre-feet is multiplied by 43,560 square feet per acre to produce a volume in cubic feet. Multiply this number by 7.48 gallons per cubic foot to obtain a volume in gallons.

Frequently, formulas call for the volume in millions of gallons instead of gallons.

$$\text{Volume}[\text{Mgal}] = \text{Volume}[\text{gal}] \times \frac{M}{10^6}$$

$$= 5,000,000[\text{gal}] \times \frac{M}{10^6}$$

$$= 5.0\ \text{Mgal}$$

When a volume in gallons is multiplied by 1 M/1,000,000, only the units change. In the above example, 5,000,000 gallons is the same as 5 million gallons.

2. The copper sulfate required in pounds may be determined on the basis of reservoir volume or reservoir surface area. Another important factor is whether the desired dose is given in terms of a concentration of copper in milligrams of copper per liter of water or as an application of copper sulfate in pounds of copper sulfate per acre of water surface area.

If the desired dose is given as mg/L of copper, calculate the amount of copper needed in pounds as follows:

$$\text{Copper}[\text{lb}] = \text{Volume}[\text{Mgal}] \times \text{Dose}\left[\frac{\text{mg}}{\text{L}}\right] \times \frac{3.785\ \text{L}}{\text{gal}} \times \frac{10^6}{M}$$

$$\times \frac{\text{kg}}{10^6\ \text{mg}} \times \frac{2.205\ \text{lb}}{\text{kg}}$$

Because 25 percent of the copper sulfate is copper, for every 4 pounds of copper sulfate added to the reservoir, 1 pound of copper is added.

$$\text{Copper Sulfate}[\text{lb}] = \frac{\text{Copper}[\text{lb}] \times 100\%}{25\%}$$

The formula for calculating the pounds of copper sulfate needed using the reservoir volume in million gallons (Mgal) and the copper dose in mg/L is a combination of these two formulas.

Copper Sulfate $[lb]$

$$= \frac{Volume[Mgal] \times Dose\left[\dfrac{mg}{L}\right] \times \dfrac{3.785\ L}{gal} \times \dfrac{10^6}{M} \times \dfrac{kg}{10^6\ mg} \times \dfrac{2.205\ lb}{kg} \times 100\%}{25\%}$$

If the copper sulfate dose is given in pounds of copper sulfate per acre of water surface, use the following formula:

$$Copper\ Sulfate\ [lb] = Surface\ Area[ac] \times Dose\left[\frac{lb}{ac}\right]$$

If the alkalinity of the water is greater than 150 mg/L, add 1 pound of citric acid for every 2 pounds of copper sulfate.

$$Citric\ Acid[lb] = Copper\ Sulfate[lb] \times \frac{1\ lb\ Citric\ Acid}{2\ lb\ Copper\ Sulfate}$$

Actual desired copper concentrations, copper sulfate doses, and citric acid doses may vary with each reservoir. Only by monitoring doses and by observing and analyzing results can operators determine the most cost-effective copper sulfate program for their reservoirs.

Example I

A small storage reservoir has a surface area of 5 acres and contains 80 acre-feet (ac-ft) of water. How many pounds (lb) of copper sulfate pentahydrate are needed for a 0.5 mg/L dose of copper? Copper sulfate pentahydrate contains 25 percent copper. Assume the alkalinity is 40 mg/L.

Known			Unknown
A	= Surface Area	= 5 ac	$CuSO_4$ = Copper Sulfate, lb
V	= Volume	= 80 ac-ft	
D_{Cu}	= Copper Dose	= 0.5 mg/L	
Copper Content		= 25%	

$$CuSO_4[lb] = \frac{V[gal] \times D_{Cu}\left[\dfrac{mg}{L}\right] \times \dfrac{3.785\ L}{gal} \times \dfrac{kg}{10^6\ mg} \times \dfrac{2.205\ lb}{kg} \times 100\%}{Copper[\%]}$$

1. Convert acre-feet (ac-ft) to gallons (gal).

$$V[gal] = V[ac\text{-}ft] \times \frac{43,560\ ft^2}{ac} \times \frac{7.48\ gal}{ft^3}$$

$$= 80[ac\text{-}ft] \times \frac{43,560\ ft^2}{ac} \times \frac{7.48\ gal}{ft^3}$$

$$= 26,066,304\ gal$$

2. Calculate the pounds (lb) of copper sulfate needed.

$$CuSO_4[lb] = \frac{V \times D_{Cu} \times 100\%}{Copper[\%]}$$

$$= \frac{26,066,304\,[\text{gal}] \times 0.5 \left[\dfrac{\text{mg}}{\text{L}}\right] \times \dfrac{3.785\,\text{L}}{\text{gal}} \times \dfrac{\text{kg}}{10^6\,\text{mg}} \times \dfrac{2.205\,\text{lb}}{\text{kg}} \times 100\%}{25\%}$$

$$= 435 \text{ lb Copper Sulfate}$$

In large reservoirs, the total volume of water may not need to be treated. From experience, the operator may decide to treat only the top 20 feet (6 meters) or down to the thermocline.

Example 2

How many pounds (lb) of copper sulfate pentahydrate would be required for the reservoir in Example 1 if the alkalinity of the water was 175 mg/L? Because the alkalinity is greater than 150 mg/L, the recommended copper dose is 5.4 lb of copper sulfate per acre (ac) of water surface area.

Known				Unknown
A	=	Surface Area	= 5ac	$CuSO_4$ = Copper Sulfate, lb
V	=	Volume	= 80 ac-ft	
D_{CuSO_4}	=	Copper Dose	= 5.4 lb/ac	
Alkalinity			= 175 mg/L	

Calculate the pounds (lb) of copper sulfate pentahydrate needed.

$$CuSO_4\,[\text{lb}] = A\,[\text{ac}] \times D_{CuSO_4}\left[\frac{\text{lb}}{\text{ac}}\right]$$

$$= 5\,[\text{ac}] \times 5.4\left[\frac{\text{lb}}{\text{ac}}\right]$$

$$= 27 \text{ lb Copper Sulfate}$$

If the alkalinity is greater than 150 mg/L, citric acid may have to be mixed with the copper sulfate and the copper sulfate may have to be added to the reservoir more frequently. When citric acid is added, copper will remain in solution and the dosage should be based on volume of water, not surface area. The dosage of citric acid is usually 1 pound of citric acid for every 2 pounds of copper sulfate, regardless of how much the alkalinity is over 150 mg/L.

Example 3

How many pounds of citric acid would be required for the reservoir in Example 2 if the recommended dose of citric acid is 1 pound of citric acid for every 2 pounds of copper sulfate applied? The copper sulfate dose in Example 2 was 27 pounds (lb).

Known				Unknown
$CuSO_4$	=	Copper Sulfate	= 27 lb	Citric Acid, lb
$D_{Citric Acid}$	=	Citric Acid Dose	= 1 lb Acid/2 lb Copper Sulfate	

Calculate the pounds of citric acid needed.

$$Citric\ Acid\,[\text{lb}] = CuSO_4\,[\text{lb}] \times D_{Citric\ Acid}$$

$$= 27 \text{ lb } CuSO_4 \times \frac{1 \text{ lb Citric Acid}}{2 \text{ lb } CuSO_4}$$

$$= 13.5 \text{ lb Citric Acid}$$

When applying copper sulfate or any other chemical, experiment to find the best dosage for your situation. Whether your doses are based on the entire reservoir volume, or only on the surface area, the amount of chemical required will vary with location, time of year, and water quality. Another important variable is the frequency of application of chemicals. To determine the best dosage and frequency of dosing, develop and analyze results of a reservoir monitoring program.

2.4.6.4   Monitoring

In reservoirs where algae are a potential problem, the operator must have a monitoring program capable of anticipating a possible algal bloom. When the data reveal that a bloom is likely, the operator must take the necessary treatment action to prevent the bloom. After a bloom occurs, it is more difficult, if not almost impossible, to control the bloom and correct the bad effects on water quality.

Whenever a chemical algae control program is started, monitoring should be carried out before, during, and after use of chemicals. Before, and for several days after the chemical application, data on type, amount, and location of algae should be collected to evaluate the effectiveness of treatment at the dosage applied. Careful evaluation must be made to determine if algae die-offs actually occur as a result of the chemical application or if they simply die off due to natural circumstances. This can best be accomplished by monitoring bloom/die-off cycles under natural conditions when no chemical treatment is carried out. Careful monitoring of algicide residual concentrations should be practiced during and following treatment to determine if the desired dose is obtained and the extent of the algicide distribution. For example, if the water body is to be treated to a depth of 15 feet (4.5 m), it may be necessary to adjust application methods to obtain an effective residual to this depth. Accurate data should be kept on the actual algicide concentration (copper, for example) in the reservoir or water supply in case legal questions regarding causes of fish die-off or system corrosion arise. Monitoring levels of copper in the water being released into state-owned streams or rivers should also be conducted and may be required by regulation in some states.

2.4.6.5   Recordkeeping

Full and accurate recordkeeping is an important part of any chemical algae control program. These records are valuable when evaluating current and historical treatment programs, for designing new or revising existing programs, and for showing compliance with federal, state, and local regulations.

2.4.6.6   Safety

Employee and public safety is an important concern in chemical algae control programs. Proper procedures for handling and applying chemicals must be strictly followed. Besides following precautions listed by the chemical manufacturer, anyone applying chemicals should refer to the regulations of federal and state agencies, such as state Occupational Safety and Health Act (OSHA) programs. Particular caution must be exercised when applying copper sulfate in dry form to protect employees and the public from the dust. Special clothing, gloves, and breathing apparatus should be required. In some cases, it may be necessary to close the lake or reservoir to public use during periods of chemical application to protect the public from dust or spray.

Identification of Algae in Water Supplies, 2001 Edition. CD-ROM. American Water Works Association. www.awwa.org. ISBN: 1-58321-161-6.

The Use of Copper Sulfate in Control of Microscopic Organisms, by Frank E. Hale, Ph.D. Presented by Phelps Dodge Refining Corporation.

Investigations of Copper Sulfate for Aquatic Weed Control. A Water Resources Technical Publication, Research Report No. 27, US Department of Interior, Bureau of Reclamation. National Technical Information Service (NTIS). Order No. PB95-145918.

How to Identify and Control Water Weeds and Algae. Created and produced by Applied Biochemists, Inc. Cygnet Enterprises West, Inc.

If chemicals are to be applied from a boat, operators must observe water safety procedures. Coast Guard-approved personal flotation devices (PFDs) should be worn while working on or near bodies of water. Employee training in water safety and first-aid procedures is always a good idea and may be legally required.

Check Your Understanding

1. Why are chemicals used in domestic water supply reservoirs to prevent or control attached and floating aquatic growths?

2. What chemical other than copper sulfate ($CuSO_4 \bullet 5\ H_2O$) may be used as an algicide?

3. List three methods of applying copper sulfate compounds to a reservoir.

4. How does suspended particulate matter in a reservoir reduce the effectiveness of copper as an algicide?

5. What is the major factor limiting the maximum rate of application of copper sulfate in the sources of a domestic water supply?

6. How is the effectiveness of a chemical algae control program evaluated?

7. What safety precautions should be taken by a person applying copper sulfate in the dry form?

2.4.7 Reaeration and Artificial Destratification

Another common way to improve water quality in reservoirs is reaeration-destratification. **Reaeration** is the introduction of air through forced air diffusers into the lower layers of the reservoir. As the air bubbles form and rise through the water, oxygen from the air dissolves into the water and replenishes the DO. The rising bubbles also cause the lower waters to rise to the surface where oxygen from the atmosphere is transferred to the water. This is sometimes called surface reaeration. **Destratification** is the development of vertical mixing within a lake or reservoir to eliminate (either totally or partially) separate layers of temperature, plant, or animal life. This vertical mixing can be caused by mechanical means (pumps) or through the use of forced air diffusers that release air into the lower layers of the reservoir.

The term reaeration-destratification means using air to destratify the reservoir. A relatively small amount of replenishment of the DO in the water actually occurs far below the saturation point. Also, the reservoir may be only partially destratified as a result of this procedure. The entire reservoir may not be destratified, but mixing will occur between the upper and lower layers of water.

The primary purpose of reaeration-destratification programs in domestic water supply reservoirs is usually to eliminate, control, or minimize the negative effects on domestic water quality that occur during periods of thermal stratification and DO depletion. A secondary purpose may be to increase recreational values of the reservoir through expanded and improved fisheries and improved aesthetic conditions. In reservoirs that freeze over during the winter, reaeration-destratification equipment may be used to reduce winter fish kills in waters that normally become anaerobic, and to prevent portions of the lake or pond from freezing over. Prevention of freezing can be accomplished by circulating water

reaeration (RE-air-A-shun)
The introduction of air through forced air diffusers into the lower layers of the reservoir. As the air bubbles form and rise through the water, oxygen from the air dissolves into the water and replenishes the dissolved oxygen. The rising bubbles also cause the lower waters to rise to the surface where oxygen from the atmosphere is transferred to the water. This is sometimes called surface reaeration.

destratification (DEE-strat-uh-fuh-KAYshun)
The development of vertical mixing within a lake or reservoir to eliminate (either totally or partially) separate layers of temperature, plant life, or animal life. This vertical mixing can be caused by mechanical means (pumps) or through the use of forced air diffusers that release air into the lower layers of the reservoir.

vertically within a lake or reservoir. This practice is particularly helpful in marina areas so that open water can be maintained around boat docks and related facilities.

Water quality improvement is obtained by the addition of DO to zones within the lake that would normally become anaerobic during periods of thermal stratification. In reservoirs that have a single outlet gate located at a depth where anaerobic conditions exist during portions of the year, reaeration-destratification systems offer a method of improving water quality for delivery to water treatment and distribution facilities. In reservoirs with multilevel outlets that permit selection of the depth from which water is withdrawn, it is possible to select the depth at which the best water quality exists.

Under certain conditions, reaeration-destratification programs may not solve all the problems that they are intended to solve. In large lakes and reservoirs, it is very difficult to design and operate the system so that an adequate amount of oxygen is added to the water at a reasonable cost. A monitoring program must be in use before and along with reaeration-destratification programs to evaluate both positive and negative impacts on water quality.

Publications that contain valuable references and information relative to design, operation, advantages, disadvantages, and economies of reaeration-destratification systems should be closely reviewed before deciding to install a system in a particular reservoir.

Researchers and reservoir operators who have used and evaluated reaeration-destratification programs are still debating the impact of these programs on algal blooms. Some investigators have reported significant reductions in the intensity of algal blooms and a shift in the species of algae responsible for the blooms. In some cases, taste- and odor-producing bluegreen algae have been replaced by certain species of green algae, which do not cause taste and odor problems. Other investigators have noted increases in intensity of algal blooms and taste and odor problems following startup of reaeration-destratification systems.

Reaeration-destratification effects on algal blooms appear to be related to what happens to nutrient conditions within the reservoir as a result of reaeration-destratification. In lakes and reservoirs that are anaerobic on the bottom and stratified, nutrients (particularly phosphate and nitrate compounds) may be released in large quantities from the bottom sediments into the hypolimnion. Following overturn, these nutrients are mixed into the upper waters where they are available for promoting algal growth under certain environmental conditions. By eliminating anaerobic zones, reaeration-destratification systems may control or eliminate this release of nutrients from the bottom sediments, thereby reducing algal blooms. In some instances, however, reaeration-destratification systems themselves may cause the nutrients within bottom sediments or deeper portions of the reservoir to be mixed upward into the surface waters. When this happens, nutrients become available to algae, thereby increasing algal blooms during certain periods of the year. The number of reservoirs where algal blooms have been decreased by reaeration-destratification appears to be much larger than reservoirs where algal blooms have been increased.

Reservoir operators should collect enough water quality data (temperature, DO, pH, nutrients, alkalinity, suspended matter, turbidity, **Secchi disc** transparency) to make their own analysis of the effects of reaeration-destratification programs upon algal and other environmental conditions.

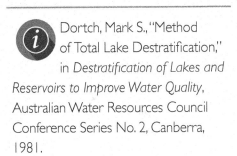

Dortch, Mark S., "Method of Total Lake Destratification," in *Destratification of Lakes and Reservoirs to Improve Water Quality*, Australian Water Resources Council Conference Series No. 2, Canberra, 1981.

A Guide to Aeration/Circulation Techniques for Lake Management, October 1976. Environmental Research Laboratory, US Environmental Protection Agency. National Technical Information Service (NTIS). Order No. PB-264126/4.

Secchi (SECK-key) disc
A flat, white disc lowered into the water by a rope until it is just barely visible. At this point, the depth of the disc from the water surface is the recorded Secchi disc transparency.

2.4.7.1   Methods of Reaeration

There are two basic methods of maintaining or increasing DO concentrations within zones of reservoirs that are partially or fully oxygen-depleted when thermal stratification exists. The first method, commonly referred to as destratification, accomplishes aeration by either altering or totally eliminating thermal stratification. The second method adds DO directly to the hypolimnion without significantly altering the pattern of thermal stratification and is referred to as hypolimnetic aeration or oxygenation. There are many ways (some much more economical than others) in which each of these methods can be used. The application used depends mostly upon the goals of the project, causes of environmental problems being addressed, size and depth of the reservoir, and available budget. A thorough understanding of the reservoir's dynamics is necessary before deciding on the appropriate method.

2.4.7.2   Destratification

Destratification oxygenates the entire reservoir. Destratification is accomplished by inducing vertical mixing using one of two main methods.

The first method uses compressed air, which releases air bubbles through diffusers into the hypolimnion of the reservoir, usually the deepest portion. Commonly, destratification systems using air compressors are located either onshore or on a floating barge or platform as near as possible to the point where the air is to be released into the reservoir. Air is delivered from the compressors into an air supply line, which is connected to a system of diffusers (Figure 2.12). The diffusers release the air near the bottom of the lake or reservoir. As the air bubbles rise toward the surface, they act like a pump, carrying the colder, denser water upward. Dissolved oxygen is added to deeper waters as they mix with upper waters and make contact with the atmosphere at the surface. Some oxygen enters the water through transfer from the bubbles as they rise toward the surface. Because it is more dense, the colder water tends to eventually sink downward, causing vertical circulation. With complete destratification, surface waters are cooled and deeper waters are warmed until an equilibrium is reached and temperatures are nearly equal from top to bottom. One major disadvantage of complete destratification is that deeper waters may become warmer than desired for domestic water use and for certain

Figure 2.12   Compressed air diffuser *(Courtesy of Casitas Municipal Water District)*

species of fish. However, cooler surface temperatures reduce evaporation losses. Through proper design and operation, destratification systems can be used to adjust temperatures as well as DO levels. If airflow is sufficient, this method can reduce algae growth by limiting available light as the algae is circulated into the deeper portions of the lake.

In any given reservoir, the rate of oxygen depletion within the metalimnion and hypolimnion zones may vary considerably from one year to the next depending on algal blooms, die off, and other biological factors. During years of heavy runoff, higher nutrient inflow may increase biological productivity and the rate of oxygen depletion. In high runoff years, it may be necessary to circulate water from the deeper zones through more complete mixing.

The second method of destratification uses either mechanical or hydraulic mixing. Mechanical mixing involves pumping hypolimnetic waters to the surface or pumping surface waters downward using pumps or jets. Hydraulic systems pump water from one level of the reservoir and jet it into another area of different **density**. The pumped water stream induces circulation and mixing. With many of these systems, power requirements are very high when compared to similar diffused-air mixing systems, making them less efficient and more costly to operate.

2.4.7.3 Hypolimnetic Aeration or Oxygenation

Hypolimnetic aeration or oxygenation systems all have one thing in common: they increase the DO content of the hypolimnetic waters while at the same time maintaining thermal stratification. This is accomplished by injecting small air bubbles or pure oxygen into the hypolimnion of a lake or reservoir (Figure 2.13). Hypolimnetic oxygenation has the advantage of oxygenating the sediment/water interface in the lake. This may reduce the

Figure 2.13   Hypolimnetic oxygenation system *(Courtesy of Casitas Municipal Water District)*

density (DEN-sit-tee)
A measure of how heavy a substance (solid, liquid or gas) is for its size. Density is expressed in terms of weight per unit volume, that is, grams per cubic centimeter or pounds per cubic foot. The density of water (at 4°C or 39°F) is 1.0 gram per cubic centimeter or about 62.4 pounds per cubic foot.

cycling of phosphorus, and result in a reduction of algal growth. Additionally, phosphorus may stay trapped in the hypolimnion so that it is unavailable at the surface where sunlight is available for algae growth. Hypolimnetic oxygenation may reduce the formation of other undesirable products (such as metals) that are formed under anaerobic conditions. As a side benefit, hypolimnetic oxygenation helps maintain habitat for cold-water fish.

There are several ways air or oxygen can be added to the hypolimnetic zone. Line diffusers introduce oxygen by releasing bubbles from porous hoses. This type of system has the advantage of applying oxygen close to the sediment-water interface where the chemical/biological reactions producing undesirable byproducts occur. Airlift systems work by pumping hypolimnetic water up where it entrains air and then pumping it back into the hypolimnion, or by pumping naturally oxygenated epilimnetic water down to the hypolimnion. Deep oxygen injection systems inject pure oxygen at the desired depth with a pump. Downflow bubble contact systems pump water through a downflow bubble contactor with a diffuser releasing the oxygenated water.

Layer aeration is a hybrid of destratification and hypolimnetic aeration. This method artificially recirculates a mid-depth layer in the lake and typically uses less energy.

2.4.8   Managing Frozen Reservoirs

In cold climates, impoundment facilities such as open reservoirs and constructed or natural lakes can have ice formations to varying degrees of severity during winter months. To guarantee trouble-free use and operation of these facilities, certain steps should be taken in preparation for cold winters to minimize the harmful effects ice formations have on waterworks operations.

2.4.8.1   Physical Effects of Ice Formation

WATER LEVEL

In anticipation of ice formation on lakes and reservoirs, operators should regulate the water level of the reservoir and maintain an optimum level until the reservoir freezes over. Then, the water level should be lowered, sagging the ice cover and reducing the ice pressure on structures and embankments, thus minimizing damages due to ice formation.

LAKE LEVEL MEASUREMENT

To manage available water resources efficiently, operators have to balance withdrawal from and inflow to the reservoir and determine a withdrawal rate that will guarantee an uninterrupted minimum water supply to the community for domestic and firefighting purposes. This can only be accomplished by continuous monitoring of the water level under the ice.

A float arrangement, though simple and reliable, is impractical when ice formation is encountered. An alternative measuring system should be used, such as a manometer or a bubbler tube. The bubbler tube should be equipped with a heating tape (Pyrotenax Cable) to ensure that no ice will form along the bubbler tube pipe.

INTAKE SCREENS

Inspections of intake screens should be carried out frequently to ensure that debris, ice (sheet) buildup, or frazzle (granular) ice will not obstruct the water flow. In case ice or frazzle ice is encountered, a portable steam generator with sufficient length of hose should be available to counter any serious buildup.

INTAKES

If the intake design allows for withdrawal of water from different depths of the lake, the top intakes should be closed during winter months and water should be withdrawn from lower elevations. Depending on the depth of the lake, the temperature of water withdrawn from lower elevations will be a few degrees warmer than water withdrawn from close to the surface due to density stratification. Under freezing conditions, the heaviest (most dense) water is on the bottom and is warmer than surface water.

Caution should be exercised when withdrawing water from near the bottom of the lake or below the thermocline (transition zone) unless the treatment plant is equipped to routinely handle water from below the thermocline and the increased treatment costs can be justified. Prolonged observation over a period of several years and a thorough understanding of local conditions are mandatory for making ongoing decisions concerning depth of withdrawal. (See Section 2.4.8.2, "Effects on Raw Water Quality," for more details.)

SILT SURVEY

Silt surveys of most reservoirs should be conducted periodically to get an updated measure of silting as it affects available water storage. These surveys can be done conveniently and with great accuracy by putting a grid on the ice surface. At selected points, the water depth can then be measured using sounding equipment or by coring holes through the ice and using a plummet (plumb bob).

2.4.8.2   Effects on Raw Water Quality

Lakes may be classified according to depth as follows:

1. First order, >200 feet (50 meters)
2. Second order, 25 to 200 feet (7.5 to 60 meters)
3. Third order, <25 feet (7.5 meters)

Lakes of the first order exhibit relatively little circulation because the large volume of deep water maintains a stable temperature. Third-order lakes have circulation primarily controlled by wind and wave action and thus are usually well circulated in open water times, thus reducing stagnation effects. These lakes are highly susceptible to stagnation once they are covered by ice. Lakes of the second order may have two circulation or overturn periods, one in the spring and one in the fall. Generally speaking, due to their prevalence, the second-order lakes are more frequently encountered by operators.

In second-order lakes, the temperature at the bottom during winter, when the surface is frozen, is not far from that of maximum density (39.2°F or 4°C). The heaviest water is at the bottom, the lightest is at the top, with the intermediate layers arranged in the order of their density. Under these conditions, the water is in comparatively stable equilibrium, but is inversely stratified. This is the period of winter stagnation.

The degree of water quality problems is highly affected by the depth of the water inlet. Water from the lowest layer of the lake can become stagnant and the reduction of sulfate to sulfide can occur when the DO is depleted. Such conditions are responsible for odor problems in drinking water. The water from the stagnant zone may also become acidified due to higher concentrations of carbon dioxide, which will combine with water to form carbonic acid. If the water becomes sufficiently acidic, reduced forms of iron and manganese may be dissolved from lake bottom materials, resulting in taste and staining problems. After the lake surface is frozen, lake algae might bloom causing subsequent severe odor problems. Prudent treatment with copper sulfate in late autumn, when plankton counts are low, may prevent such occurrence of an algal bloom under the ice.

During periods of stagnation, deposits of organic matter accumulate at the lake bottom; ammonia levels increase; decomposition of organic matter takes place; DO disappears; nitrate, sulfate, and iron compounds become reduced; free ammonia and nitrite increase; and the content of free carbonic acid increases. A monitoring program that collects samples at various depths to determine DO, metals, and nitrogen values can define the extent of this problem.

To prevent stagnant water at lower elevations from rising to the top during spring circulation (lake overturn), the stagnant water should be discarded whenever possible. This can be accomplished by wasting it through flood gates if the design allows for bottom withdrawal to the flood gates.

2.4.8.3   Recreational Use of Reservoir Ice Surfaces

Ideally, reservoir ice surfaces should not be made accessible for recreational use of any kind, including speed skating, skating, ice sailing, and ice fishing. However, if local authorities approve certain activities on the ice, operators should insist on provision of certain safeguards against possible pollution. These provisions include time limits put on activities on the ice, proper and close supervision, leakproof waste containers, adequate toilet facilities, and proper care of these facilities.

2.4.9   Dam and Reservoir Maintenance

The type and frequency of dam and reservoir maintenance will depend on the size and type of dam. Sometimes, these activities are the responsibility of people other than the waterworks operator. The topics listed are presented to give you an idea of some of the items that might be your responsibility. You will have to prepare a program for your facilities depending on what needs to be done.

2.4.9.1   Dam Inspection and Maintenance

Dams must be inspected regularly to avoid a catastrophic disaster. Some of the most serious dam failures have been small water storage reservoirs located above residential subdivisions. Dams should be inspected after heavy rains. Look for evidence of sinkholes (holes in the ground), weep holes (water coming out of holes below the dam), and evidence of burrowing animals.

2.4.9.2   Reservoir Maintenance

Before draining a reservoir for maintenance, determine when, where, and how you will discharge the water in the reservoir. Lower the water level fairly slowly. If the water level drops too quickly, the embankments may slip out and be damaged. If the reservoir is lined and the groundwater is high, the lining could be damaged. Lined reservoirs may have to be drained only during periods of low groundwater levels.

Shoreline vegetation, such as weeds and cattails, should be controlled. Cattails and other weeds can serve as a breeding area for mosquitoes. Mechanical or manual techniques can be used to remove and control vegetation.

Check Your Understanding

1. What is the primary purpose of reaeration-destratification programs in domestic water supply reservoirs?

2. How can water quality be improved by a reaeration-destratification program?

3. List the two basic methods of maintaining or increasing dissolved oxygen (DO) concentrations in reservoirs when thermal stratification exists.

4. How can destratification be accomplished?

5. What factors could cause the rate of dissolved oxygen (DO) depletion in a reservoir to vary considerably from one year to the next?

6. How should the water level in a reservoir be regulated after a reservoir freezes over?

7. How can ice (sheet) buildup or frazzle (granular) ice be prevented from obstructing water flow through intake screens?

8. What water quality problems could develop in a frozen reservoir?

2.5   Laboratory and Monitoring Programs

Using the water quality laboratory and related monitoring programs, information can be collected that is essential in developing and evaluating methods of managing water quality in domestic water supply reservoirs. Laboratory and monitoring programs are essential from both an operations and a legal standpoint to determine whether physical, chemical, and biological water quality indicators are in compliance with federal, state, and local water quality standards. Water quality data are also an essential tool in optimizing operations of the water treatment plant in relation to treatment costs and techniques, and operation and maintenance programs.

Each agency that operates a surface water supply reservoir must design and operate laboratory and monitoring programs that reflect its own financial and technical resources and need for information. Large municipalities or federal, state, or county agencies sometimes provide water to hundreds of thousands of people. Generally, these agencies have much greater analytical and technical capability than the laboratory operated by a local agency that serves small populations. The difference in laboratory capability is often due primarily to the difference in financial resources of the two agencies. The large agency's laboratory may serve several functions such as performing analyses for wastewater treatment facilities, wastewater discharge control programs, and domestic water supply treatment plants, as well as collecting data on sources of water supply. The small agency may operate its laboratory and monitoring program primarily to manage the surface water supply reservoir and related facilities. There are many more small water supply systems in the United States than there are large municipal or regional systems. The small system operator must be able to manage a water supply as well as large system operators. With proper planning, training, and forethought, the small agency can do a thorough and efficient job of conducting laboratory and monitoring programs within its own financial capabilities.

2.5.1 Procedures

In many instances, a small reservoir management agency does not have a large enough volume of samples to justify purchasing equipment and training personnel for certain types of analyses. In these cases, it is often much more practical and economical to contract for the work to be performed by a commercial water laboratory or a consulting firm. Arrangements can sometimes be made with county, state, or federal agencies to have specialized samples collected and analyzed. County, state, and federal agencies can be particularly helpful when a new or unusual problem develops in the water supply. Examples of types of analyses that may be best performed by an outside or commercial laboratory include trihalomethanes and other organics, *Giardia* and *Cryptosporidium*, general and toxic minerals, radioactivity, pesticides, certain toxic heavy metals (such as lead and mercury), and nutrients.

The agency operating a surface water reservoir should be able to perform analyses that produce data needed on a routine basis for conducting day-to-day operations. Much of the data needed to evaluate physical and chemical conditions in a water supply reservoir can be collected using a single multi-probe instrument (Figure 2.14). The cost for this instrument varies depending on the number of probes necessary to measure the

Figure 2.14   Water quality monitoring *(Courtesy of Casitas Municipal Water District)*

water quality indicators and whether the instrument has recording or data transmitting capabilities. Types of data that can be measured at any location and depth within a lake or reservoir with such an instrument include temperature, DO content, conductivity, pH, and oxidation-reduction potential. A thorough survey of all these water quality indicators can be made in a single working day in most reservoirs. For example, at one Southern California reservoir, the above data can be collected at depth intervals of 5 feet (1.5 meters) to 20 feet (6 meters) at seven separate stations in approximately 6 hours. The frequency of the data survey on any given lake is usually related to how rapidly water quality is changing within the lake.

If algal blooms are a problem in a particular reservoir, the operating agency should develop laboratory and personnel capabilities for monitoring and identifying algae. The laboratory should be able to develop information on the intensity and extent of algal blooms, the major species of algae involved in any given bloom, and the water quality problems that develop as a result of the bloom. The local agency should be able to collect data on taste and odor conditions related to algal blooms. If chemical methods for controlling algal blooms are used, the local agency should be capable of monitoring chemical dosage levels and residuals.

On domestic water supply reservoirs that develop anaerobic zones and their related problems, the local agency may need to monitor iron, manganese, and hydrogen sulfide concentrations. Iron and manganese analyses can be contracted out to a commercial laboratory, but are usually performed in house due to the immediate need for the results. The best device for evaluating hydrogen sulfide conditions is often the human nose. If hydrogen sulfide can be smelled in freshly collected unaerated samples, it is present in concentrations objectionable to consumers. However, it is still necessary to measure hydrogen sulfide concentrations to determine improvements. Also, some forms of sulfide may cause odors only after heating.

Information on specific laboratory procedures and training in carrying out these procedures can often be obtained from state health departments. Details on sampling and laboratory procedures are also contained in Chapter 9, "Laboratory Procedures," and also in the chapter on advanced laboratory procedures in Volume II of this series of training manuals.

2.5.2   Recordkeeping

One of the most important functions of a well-designed laboratory and monitoring program is recordkeeping. This is true whether analyses are performed by the local agency or by some outside laboratory. Records provide the basic foundation upon which management programs are designed, implemented, and evaluated. Records can be used to evaluate rates of water quality deterioration or improvement and have value as a predictive tool in determining when water quality problems will occur or cease.

In many agencies, complete and accurate records relating to water quality problems and management programs will aid decision making even if there are personnel changes. Records are of little value if they are only compiled by laboratory personnel and filed away. They must be regularly reviewed and evaluated by people responsible for making decisions

conductivity
A measure of the ability of a solution (water) to carry an electric current.

oxidation-reduction potential (ORP)
The electrical potential required to transfer electrons from one compound or element (the oxidant) to another compound or element (the reductant); used as a qualitative measure of the state of oxidation in water treatment systems. ORP is measured in millivolts, with negative values indicating a tendency to reduce compounds or elements and positive values indicating a tendency to oxidize compounds or elements.

relative to water quality management. Records pertaining to monitoring and laboratory analyses, which are necessary to indicate compliance with federal and state Primary and Secondary Drinking Water Standards, are legally required. State and federal regulations specify how these records should be kept and for how long. Recordkeeping is also a specific requirement of federal and state laboratory certification programs.

2.5.3   Safety

Emphasis on knowing and implementing proper safety procedures should be an important part of any laboratory and monitoring program. Specific information on laboratory safety requirements and procedures can be obtained from either state or federal OSHA offices. For additional information on laboratory safety, see Chapter 9, "Laboratory Procedures," and the chapter on safety in Volume II of this series of training manuals.

Anyone responsible for laboratory or monitoring program safety should try to comply with all state and federal safety requirements at all times. If this attitude is emphasized in day-to-day operations, then inspections by safety enforcement officers will never be a cause for concern.

Safety hazards involved in water quality sampling may be even greater than those encountered within the laboratory. The main safety hazard encountered during reservoir sampling is drowning. Most reservoir sampling programs involve the use of a boat, dock, barge, or some similar piece of equipment. When conducting sampling, proper flotation equipment (such as Coast Guard-approved PFDs) should be worn. People involved in reservoir and stream sampling should at least know how to swim. In many cases, it is advisable to have two people involved in sampling at a specific location. If sampling is to be carried out on streams and rivers that are a source of supply to the reservoir, additional safety measures may be required.

Check Your Understanding

1. What are the purposes of water quality laboratory and monitoring programs?
2. What types of laboratory analyses should be performed by operating agencies?
3. If algal blooms are a problem, what type of laboratory capability should be available to the operating agency?
4. What is the main safety hazard encountered during reservoir sampling?

2.6   Intake Structures

Once good quality water is obtained in the lake, stream, or reservoir, the intake system delivers water free of debris and trash to treatment or distribution facilities efficiently and satisfactorily. Intake structures and related facilities at water supply reservoirs may be more appropriately referred to as intake-outlet facilities, as they take in water from the reservoir for outlet downstream. Intake structure and outlet structure are terms that are often used interchangeably to describe the same facility.

In domestic water supply lakes and reservoirs, these facilities may be used to deliver water to water treatment plants, directly to the distribution system, or for returning water to the river or stream downstream of the reservoir. In some cases, a single intake-outlet system is used to provide for downstream releases to the stream or river and for delivery to the treatment plant or distribution system. In other instances, the facilities that provide for release to the stream or river are entirely separate from those providing service to the domestic water system. River or stream intake structures simply serve to provide raw water for the treatment plant.

Intake facilities should always be constructed on the basis of the specific function that they must serve at a given lake, reservoir, stream, or river. They must be capable of supplying the maximum rate of flow required for the water treatment plant. Water supply lakes and reservoirs must also release adequate flow for downstream uses. In situations where intake facilities provide service to pressurized systems, they must be designed so that minimum operating pressures within the system are maintained when the reservoir is drawn down to its minimum operating level. Newer intake structures are constructed to permit selection of the depth at which water is drawn from the reservoir, lake, stream, or river.

Intake facilities should be constructed to prevent algal scums, trash, logs, and fish from entering the system. To reduce the risk of drawing silt into the intake system, the water inlet should not be located at low points where silt buildup is anticipated. Always be very careful when operating the lowest level valves if they have not been in use for some time in case silt buildup has occurred.

One of the more important considerations in the construction of intake facilities is the ease of operation and maintenance over the expected lifetime of the facility. **Cathodic protection** systems, which minimize the rate of corrosion of metal parts, are a vital part of many intake systems. Every intake structure must be constructed with consideration for operator safety. See Chapter 7, "Corrosion Control," for more information on cathodic protection.

2.6.1 Types of Intake Structures

1. Single-Level Intakes (Figure 2.15). Single- or fixed-level intake systems are commonly used in the distribution systems of domestic water supply streams and reservoirs. The single-level intake is usually located in the deepest portion of the stream or reservoir so that water service can still be provided even when the body of water is down to its minimum operating level, such as during drought conditions. In cases of reservoirs where one intake structure supplies both the domestic water system and releases to streams and rivers, the inlet may be located very close to the bottom so that the reservoir can be drained. Single-inlet intake structures are most suitable in relatively shallow lakes and reservoirs that do not stratify significantly and that exhibit fairly uniform water quality from top to bottom throughout the entire year. These structures may also function well in deeper lakes that are relatively nonproductive and do not experience water quality problems as a result of stratification.

cathodic (ca-THOD-ick) protection
An electrical system for prevention of rust, corrosion, and pitting of metal surfaces which are in contact with water or soil. A low-voltage current is made to flow through a liquid (water) or a soil in contact with the metal in such a manner that the external electromotive force renders the metal structure cathodic. This concentrates corrosion on auxiliary anodic parts which are deliberately allowed to corrode instead of letting the structure corrode.

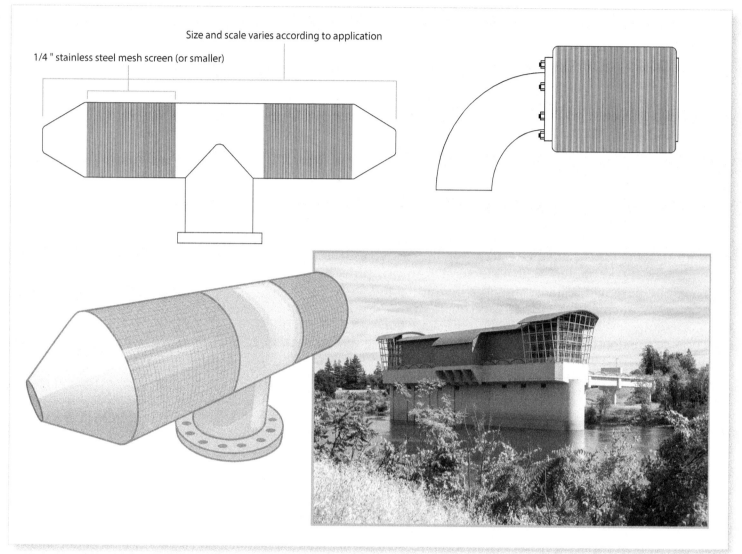

Size and scale varies according to application

1/4 " stainless steel mesh screen (or smaller)

Figure 2.15   River intake facility and single-level intakes

The advantage of single-inlet intake structures is that they are usually much less complicated and, therefore, much less costly to construct than multilevel structures. Because of their simplicity, they are both easier and less costly to operate and maintain than multilevel intake structures. In deep reservoirs that often remain nearly full, however, it may be difficult to inspect these facilities and perform necessary repairs and maintenance.

Major disadvantages of single fixed-level inlet facilities become apparent when they are used in deeper, productive (eutrophic) lakes and reservoirs. If the inlet is located within the hypolimnion, below the depth at which the thermocline forms, major water quality problems may affect water delivered to the treatment plant or distribution system. Water entering the inlet during spring, summer, and fall months may be anaerobic, may contain high concentrations of iron or manganese, and may contain the rotten egg smell caused by the presence of hydrogen sulfide.

Gates or valves that allow water to be taken through the single-level intake may be located either at the point of inlet from the reservoir or stream, in the delivery system at some point downstream, or at both locations.

2. Multilevel Intakes (Figure 2.16). The most satisfactory intake structures in water supply reservoirs are usually those that have inlets to the system at depths ranging from near the surface to the deeper zones. The major advantage of multilevel intake systems in domestic water reservoirs is that they make it possible to serve water from the depth where the best quality of water is located. To obtain good water quality, it may be necessary to draw water from different levels during different seasons of the year. When downstream releases to streams and rivers are required, multilevel structures allow for releases from a depth where temperature and DO conditions are

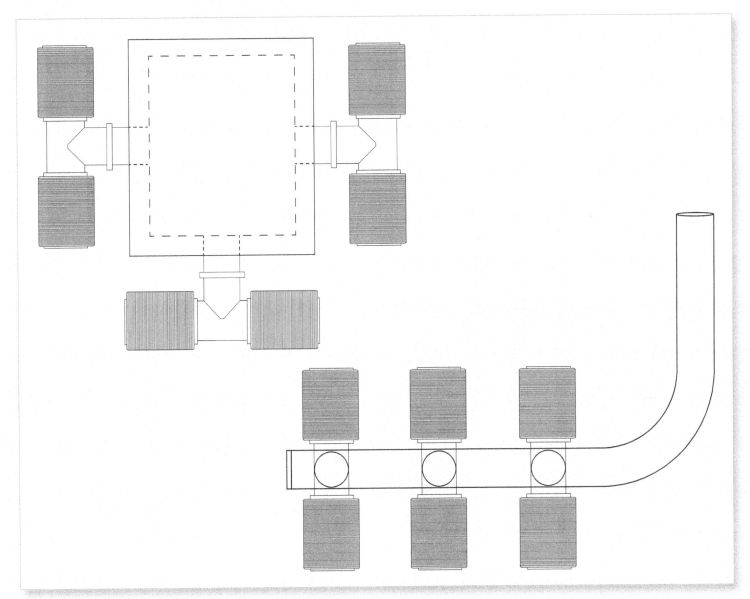

Figure 2.16   Reservoir multilevel intakes

acceptable for protecting downstream fish. Fish kills may result from anaerobic water being released downstream from a reservoir.

Multilevel intake structures are most commonly found in a vertical tower located in the deeper portion of the lake and extending above the water surface. Access to the facility may be by bridge, pier, or boat. Inlet gates may be operated from a deck on top of the tower, from some remote control house, or from another location, depending upon equipment design. Inlet gates are commonly located at specific vertical intervals along the face or faces of the tower. Usually, all inlets feed into the pipeline system, which extends from the bottom of the tower to the treatment plant, distribution system, or point of release to a river. Each inlet is equipped with an individually operated gate or valve at the point of inlet. An additional gate or valve is usually located in the pipeline at some point downstream of the intake structure. This arrangement allows for dual control over waters entering the system. If valves or gates at either the inlet or within the pipeline fail, the system can still be shut down.

Some reservoirs contain multilevel intake structures that are inclined rather than vertical. These facilities are commonly located on the inclined face of an earth-fill dam or some similar slope and extend from the maximum water surface level to deeper portions of the reservoir. Inlets with individually operated intake gates or valves are located at intervals along the inclined structure. As with vertical structures, an additional gate or valve is normally located in the pipeline at some point downstream. The inclined intake structure is often a concrete conduit or tunnel with the pipeline located inside and extending from the bottom of the structure downstream to the transmission pipeline. The intake ports extend from the pipeline through the concrete conduit or tunnel and into the reservoir.

For example, the inclined intake structure shown in Figure 2.17 contains nine intake gates located within the reservoir at depth intervals of 24 feet (7.2 m). When the reservoir is full, water may be drawn from depths ranging from approximately 25 feet (7.5 m) to 217 feet (65 m). Intake gates and related facilities are remotely operated from a control house located on top of the dam at the entrance to the intake structure.

Economy, topography, and ease of access are major considerations in determining whether vertical or inclined intake structures should be installed at a given reservoir or stream. In reservoirs or streams that freeze, the effects of ice on intake structures must be considered. Structures may be endangered not only from ice pressures from the side, but also from uplift if a reservoir is filling and the ice mass lifts vertically. In some cases, reservoir reaeration may be used to prevent ice from forming around the intake. See Section 2.4.8, "Managing Frozen Reservoirs," for additional details.

Selection of the level to withdraw water from a lake or reservoir depends on the water quality in the various layers of water. Once the layer has been selected,

Figure 2.17 Inclined multilevel intake tower (*Courtesy of Casitas Municipal Water District*)

regular monitoring is required to continually withdraw water with a good quality. Winds and uniform water temperatures from top to bottom can cause water quality at various depths to change very quickly. To be prepared for changing conditions, the water treatment plant operator should:

1. Maintain a log of wind direction and velocity, at least during the summer months; in warm climates this should be done all year.

2. Be alert for onshore winds.

3. Maintain close surveillance of threshold odor test results.

4. Be prepared to make necessary changes in treatment plant operation to combat any sudden increases in taste and odor problems.

5. Do everything possible to keep any water with a bad taste or odor that gets through the treatment plant from getting out into the distribution system.

2.6.2   Types of Intake Gates

Any one of several types of intake gates or valves may be used at the inlet to the intake structure or as a control valve within the pipeline system downstream of the intake. The most commonly used types include slide gates (steel or cast iron), gate valves, and butterfly valves. Some gates and valves operate only in the fully open or fully closed positions. Others operate as flow-control or regulating valves and gates at any position from fully open to fully closed. Some gates and valves will perform satisfactorily with head or pressure on only one side while others are designed to perform best with fairly constant head or pressure on both the upstream and downstream sides.

Gates and valves that regulate releases from small impoundments into small distribution facilities are often designed to be operated manually while those in larger installations frequently use electrical power. Both mechanical gates and hydraulically operated gates and valves are in common use in intake facilities.

2.6.3   Intake Screens and Trash Racks

Most intake structures in domestic reservoirs, lakes, streams, and rivers are installed in such a manner that inlet gates, fish screens, and trash racks are all combined into a single structure. The primary purpose of fish screens, trash racks, log stops, and other protective facilities is to prevent or minimize the entry of foreign material and fish into the intake system.

Intake screens are installed in panels or in cylindrical forms. Usually, they are made of stainless steel. Various types of screens include vee wire, traveling screens, woven wire, and slotted plates. See Figures 2.18 and 2.19 for intake screen examples. These facilities must be designed so that they are easy to service and maintain, have a relatively long life, and do an effective job of protecting the water delivery system. In cases where screens, racks, and related structures are constructed of steel or metal,

Figure 2.18 Vee wire passive screen panel

Figure 2.19   Traveling screen

they should be coated with corrosion-resistant material. Sometimes, it is necessary to install a cathodic protection system to extend the service life of these facilities.

The type of screen, trash rack, or log stop used in a given intake structure depends on a number of factors, including depth or depths at which the inlets are located; location of the intake structure in relation to where debris accumulates in the reservoir or stream; frequency and intensity of algal scum or algal mass accumulations; quantity and type of debris encountered; and the size, depth of distribution, and number of forms of aquatic life (fish, crayfish, insects). Nonproductive lakes and rivers that contain little aquatic life and receive little or no debris load may only need bar screens. Lakes and rivers with large algal populations might require fine-mesh screens. In lakes, streams, and reservoirs that contain large quantities of debris and trash, log booms, hanging screen curtains, and similar facilities may be effective in protecting the area surrounding the intake structure. Stream intake structures are usually designed to deflect floating debris away from the inlet ports. This procedure minimizes the time required to clean the screens.

2.6.4   Operation and Maintenance Procedures

Well-designed operation and maintenance programs are necessary if intake structures and related facilities are to perform as intended. Operating criteria, equipment manufacturers' operating instructions, and standard operating procedures should be bound into a manual and used for reference by operators responsible for operating and maintaining intake facilities. If written references containing standard operating and maintenance procedures are not available for a particular facility, they should be prepared with the assistance of knowledgeable operators, design engineers, and equipment manufacturers' representatives.

Intake structures and related facilities should be inspected, operated, and tested periodically, preferably at regular intervals. If the reservoir is not drawn down to the level of the deepest intake at periodic intervals, inspections may be conducted by divers (either scuba or hard hat). In recent years, some agencies have used remote controlled video (TV) units to inspect deep-water facilities. The submarine-type unit containing the camera is operated from above the water and can be maneuvered into any position along the intake structure. A video receiver (TV screen) located on a barge, the shoreline, or other above-water structure allows operators to complete a detailed inspection of gates, screens, and other structures. Broken, worn, and corroded facilities are easily identified. One major advantage of the remotely operated video unit is that it allows a number of persons with different areas of expertise to participate in the inspection. When only divers are used, others must depend entirely on what the divers see or photograph to evaluate conditions.

Proper service and lubrication of intake facilities is particularly important. The following factors are common causes of faulty operation of gates and valves:

1. Settlement or shifting of support structure, which could cause binding of gates
2. Worn, corroded, loose, or broken parts
3. Lack of use
4. Lack of lubrication
5. Vibration
6. Improper operating procedures
7. Design errors or deficiencies
8. Failure of power source or circuit failure
9. Vandalism

Intake screens may be cleaned manually by operators or automatically by mechanical means. Screens and trash racks are often designed so that they can be removed for inspection, maintenance, and cleaning. Screens can be drawn to the surface, worked on, and replaced by one person in a short period of time. Screens and trash racks that are not removable should have some provision for cleaning them in place. Mechanical or hydraulic jet cleaning devices may be used, or divers may be employed to clean the screens. Mechanical cleaning devices require a regular maintenance schedule, including lubrication.

If screens are located out in rivers or lakes, the operator may reverse the flow to clean the screen. Some screens have devices that measure the difference in head or head loss between the water surface upstream and downstream from the screen. When a specified head loss is exceeded, a cleaning cycle is started. The cleaning cycle could consist of high-pressure water sprays, which clean the screen in place, or the screen could be lifted out of the water for cleaning.

Manual methods of cleaning screens include the use of rakes, brooms, bristle brushes, and water sprays. Bristle brushes are made of nylon or polypropylene strips of bristles with lengths from 1.5 to 2.5 inches (38 to 64 mm). Screens may be cleaned manually either under water or removed from the flow stream. Woven wire and slotted plate screens require operators to develop a cleaning program and schedule.

Undersized or improperly designed screens require operators to spend considerable time cleaning them. When this happens, the operator should budget funds to improve the screening facilities.

If frazzle (granular) ice plugs a screen, see Section 2.4.8, "Managing Frozen Reservoirs," for cleaning procedures.

2.6.5   Records

Records containing a history of operations and maintenance performed on intake facilities are vitally important. By keeping a record of when and under what conditions failures or malfunctions occur, it may be possible to take preventive action. Operators come and go, but if adequate records are maintained, new operators are in a better position to perform their jobs properly.

2.6.6   Safety

When working around intake structures, proper safety procedures involving use of electrical and mechanical equipment and water safety should always be observed. Proper safety procedures should be documented and included in the manual containing the standard operating procedures. When working in boats or around open bodies of water, two-person crews are recommended.

Check Your Understanding

1. What is the purpose of intake-outlet facilities in domestic water supply lakes and reservoirs?

2. Why do some intake systems require cathodic protection?

3. What may happen to fish if anaerobic water is released downstream from a reservoir?

4. What are the most common types of intake gates?

5. List the factors that influence the type of screen needed in a specific reservoir.

6. What should be done if written standard operating and maintenance procedures are not available for a particular intake structure?

7. How can inlets that are always under water be inspected?

8. List the major causes of faulty operation of gates and valves.

9. How can screens and trash racks that are not removable be cleaned?

2.7 Math Assignment

Turn to Appendix A, "Introduction to Basic Math for Operators." Read and work through the example problems in sections:

- "Learning Objectives"
- A.5.2, "Area" (A.5.2.1, A.5.2.2, A.5.2.3, A.5.2.4, A.5.2.5, A.5.2.6, and A.5.2.7)
- A.5.3, "Volume" (A.5.3.1, A.5.3.2, A.5.3.3, A.5.3.4, A.5.3.5, and A.5.3.6)
- A.9.4, "Reservoir Management and intake structures" (English System)
- A.10.4, "Reservoir Management and intake structures" (Metric System)

Check the math using a calculator. You should be able to get the same answers.

2.8 Additional Resources

Your local library may be a good source for obtaining these additional resources related to the field of water treatment:

Water Sources, 3rd Edition. American Water Works Association (AWWA). Order No. 1955. ISBN: 1-58321-229-9.

Water Transmission and Distribution, 3rd Edition. American Water Works Association (AWWA). Order No. 1957. ISBN: 1-58321-231-0.

Manual of Water Utilities Operations (Texas Manual). Texas Water Utilities Association, www.twua.org. ASIN: B000KT9EIU. Chapters on raw water quality management and pretreatment of surface water supplies.

Chapter Review

Key Term Matching

_____ 1. Short-Circuiting

_____ 2. Head

_____ 3. Adsorption

_____ 4. Coliform

_____ 5. Cross-Connection

_____ 6. Toxic

_____ 7. Turbidity

_____ 8. Anaerobic

_____ 9. Eutrophic

_____ 10. Potable Water

A. A condition in which atmospheric or dissolved oxygen (DO) is not present in the aquatic (water) environment.

B. A condition that occurs in tanks or basins when some of the flowing water entering a tank or basin flows along a nearly direct pathway from the inlet to the outlet.

C. A connection between a drinking (potable) water system and an unapproved water supply.

D. A substance that is poisonous to a living organism.

E. A group of bacteria found in the intestines of warm-blooded animals (including humans) and in plants, soil, air, and water that indicate the possibility of pathogenic organisms in water.

F. Reservoirs and lakes that are rich in nutrients and are very productive in terms of aquatic animal and plant life.

G. The cloudy appearance of water caused by the presence of suspended and colloidal matter.

H. The gathering of a gas, liquid, or dissolved substance on the surface or interface zone of another material.

I. The vertical distance, height, or energy of water above a reference point.

J. Water that does not contain objectionable pollution, contamination, minerals, or infective agents and is considered satisfactory for drinking.

2.1 Sources of Water

11. What is stratification in lakes and reservoirs?
    1. Breezes starting the circulation of surface water
    2. Formation of separate layers of temperature, plant life, or animal life
    3. Lake turnover due to temperature changes
    4. Uniform water temperature profile from surface to bottom

12. What causes the elevation and slope of water tables and artesian pressure levels to change?
    1. Collection of water into a reservoir
    2. Pollution from agricultural activities
    3. Seasonal variations in the supply of water
    4. Selection of intake structures

13. How extensive should the treatment be for reclaimed water?
    1. Depends on climate conditions at point of use
    2. Depends on potential exposure to the public
    3. Depends on the funds available for treatment
    4. Depends on the NPDES permit

2.2 Selection of a Water Source

14. What are water rights?
    1. Correct methods of treatment
    2. Ownership of water
    3. Permission to discharge water
    4. Permission to use water

15. Who should conduct a sanitary survey?
    1. People trained in phone and in-person survey techniques
    2. People trained in public health engineering and the epidemiology of waterborne diseases
    3. People trained in wastewater collection and treatment
    4. People trained in water treatment plant design and construction

2.3 Surface Reservoirs as Domestic Water Supplies

16. Which of the following water quality problems are not usually related to algal blooms?
    1. Dissolved oxygen depletion
    2. Reduced chlorination efficiency
    3. Reduced water temperature
    4. Taste and odor problems

17. Under what circumstances are tastes and odors in a domestic water supply most noticeable?
    1. At or above room temperature
    2. In glass containers
    3. In plastic containers
    4. Near freezing

18. What is the impact on water treatment plants from increased organic loadings caused by algal blooms?
    1. High trihalomethane levels
    2. Increased chlorine demand and color
    3. Increased public confidence in water supply
    4. Increased water treatment costs

19. How can a water agency meet acceptable trihalomethane levels?
    1. Change disinfection methods
    2. Decrease detention time
    3. Increase disinfection chemicals
    4. Prechlorinate the influent

2.4 Reservoir Management Programs

20. Reservoir water quality management programs can control or eliminate what types of water quality problems?
    1. Excessive turbidity
    2. Frequency and intensity of algal blooms
    3. Insufficient disinfection
    4. Precipitation of iron and manganese

21. How can iron, manganese, and hydrogen sulfide problems be controlled in a reservoir?
    1. By controlling algal productivity
    2. By improving the fishery habitat
    3. By preventing dissolved oxygen depletion
    4. By regulating recreational activities

22. How can the recreational values of a reservoir be improved by proper reservoir management?
    1. Increase in agricultural use
    2. Increase in chemical residuals
    3. Reduction of access points
    4. Reduction of algal problems

23. What is the best tool for managing watersheds?
    1. Political process
    2. Public hearings
    3. Regulatory process
    4. Voluntary compliance

24. How can water quality problems caused by septic tank leaching systems be solved?
    1. Increase nutrient loads to reservoirs
    2. Install more septic tanks
    3. Locate leaching fields within watersheds
    4. Replace septic tanks with sewer systems

25. The volume of a reservoir is estimated to be 9.8 acre-feet. The desired dose of copper is 0.5 mg/L and the copper content of the copper sulfate to be used is 25 percent. How many pounds of copper sulfate will be needed?
    1. 53 lb
    2. 54 lb
    3. 55 lb
    4. 56 lb

26. What is destratification of a lake or reservoir?
    1. Elimination of separate layers by algal blooms
    2. Natural turnover during seasonal changes
    3. Vertical mixing to eliminate separate layers of temperature, plant life, or animal life
    4. Wind blowing across the water surface causing mixing

27. Why might complete destratification of a reservoir be undesirable?
    1. Water may become too cold for fish
    2. Water may become too cold for recreation
    3. Water may become too warm for algal control
    4. Water may become too warm for domestic use

2.5 Laboratory and Monitoring Programs

28. What water quality indicators should be monitored in domestic water supply reservoirs that develop anaerobic zones?
    1. Biochemical oxygen demand (BOD) and turbidity
    2. Hardness, pH, and color
    3. Hydrogen sulfide, iron, and manganese
    4. Total dissolved solids (TDS) and dissolved oxygen (DO)

2.6 Intake Structures

29. How can the entrance of silt into the intake system be minimized?
    1. Locate intake at low point
    2. Locate intake away from low point
    3. Locate intake in shallow water
    4. Locate intake near wetlands

30. What is the major advantage of multilevel intake systems in domestic water reservoirs?
    1. Access to water with different characteristics at different depths
    2. Decreased cost and maintenance requirements
    3. Limited access to water during drought conditions
    4. Little control over quality of water released downstream

CHAPTER 3

Coagulation and Flocculation

CHAPTER OUTLINE

LEARNING OBJECTIVES

1. Safely start up, operate, maintain, and shut down coagulation and flocculation processes, including ballasted flocculation.

2. Select appropriate chemicals and dosages and adjust chemical feed rates using data from jar tests and other laboratory tests performed on water samples collected from coagulation and flocculation basins.

3. Use process data and observations to adjust process variables, including the speed of flash mixers and flocculators, to obtain optimal process performance.

KEY TERMS

alkalinity

anionic polymer

apparent color

baffle

buffer

cationic polymer

coagulation

colloids

color

detention time

disinfection byproduct (DBP)

diversion

floc

flocculation

grab sample

head loss

jar test

laundering weir

molecular weight

nonionic polymer

orifice

particulate

platinum-cobalt unit (pcu)

polyanionic

polyelectrolyte

polymer

reagent

representative sample

safety data sheet (SDS)

short-circuiting

sludge

specific gravity

total organic carbon (TOC)

trihalomethanes (THMs)

true color

turbidity meter

weir

wet chemistry

3.1   Removing Particulates from Water

particulate (par-TICK-you-let)
A very small solid suspended in water which can vary widely in size, shape, density, and electrical charge. Colloidal and dispersed particulates are artificially gathered together by the processes of coagulation and flocculation.

color
The substances in water that impart a yellowish-brown color to the water. These substances are the result of iron and manganese ions, humus and peat materials, plankton, aquatic weeds, and industrial waste present in the water.

colloids (CALL-loids)
Very small, finely divided solids (particles that do not dissolve) that remain dispersed in a liquid for a long time due to their small size and electrical charge. When most of the particles in water have a negative electrical charge, they tend to repel each other. This repulsion prevents the particles from clumping together, becoming heavier, and settling out.

floc
Clumps of bacteria and particles or coagulants and impurities that have come together and formed a cluster. Found in flocculation tanks and settling or sedimentation basins.

coagulation (co-AGG-you-LAY-shun)
The clumping together of very fine particles into larger particles (floc) caused by the use of chemicals (coagulants). The chemicals neutralize the electrical charges of the fine particles, allowing them to come closer and form larger clumps. This clumping together makes it easier to separate the solids from the water by settling, skimming, draining or filtering.

flocculation (FLOCK-you-LAY-shun)
The gathering together of fine particles after coagulation to form larger particles by a process of gentle mixing.

Particulate impurities in water result from land erosion, dissolved minerals, and the decay of plant material. Additional impurities are added by airborne contamination, industrial discharges, and by animal wastes. Thus, surface water sources, polluted by people and nature, are likely to contain suspended and dissolved organic (plant or animal origin) and inorganic (mineral) material, and biological forms such as bacteria and plankton. With few exceptions, surface waters require treatment to remove particulate impurities and **color** before it is distributed to the consumer.

These particulates (commonly called suspended solids) cover a broad size range. Larger sized particles, such as sand and heavy silts, can be removed from water by slowing down the flow to allow for simple gravity settling. These particles are often called settleable solids. Settling of larger sized particles occurs naturally when surface water is stored for a sufficient period of time in a reservoir or a lake. Smaller sized particles, such as bacteria and fine clays and silts, do not readily settle and treatment is required to form larger particles that are settleable. These smaller particles are often called nonsettleable solids or **colloidal** matter.

The purpose of coagulation and flocculation is to remove particulate impurities, especially nonsettleable solids, and color from the water being treated. Nonsettleable particles in water are removed by the use of coagulating chemicals. These chemicals cause the particles to clump together forming **floc**. When pieces of floc clump together, they form larger, heavier floc, which will settle out.

In the **coagulation** process, chemicals are added that will initially cause the particles to become destabilized and clump together. The particles gather together to form larger particles in the **flocculation** process (see Figure 3.1).

Figure 3.2 shows the overall plan view (a diagram or photo showing a facility as it would appear when looking down on top of it) of the coagulation-flocculation processes in a typical plant. Chemicals are added to the source water in the flash-mix chamber and mixed by the flash mixer. The water flows from the flash mixer through a distribution channel to the flocculation basins. From the flocculation basins, the water flows to the sedimentation basins (see Chapter 4, "Sedimentation") and then to the filters (see Chapter 5, "Filtration"). From the treatment plant design drawings (blueprints), the dimensions (length, width, and depth) of the facilities can be determined.

3.2   Coagulation

The term coagulation describes the effect produced when certain chemicals are added to raw water containing slowly settling or nonsettleable particles. The small particles begin to form larger or heavier floc, which will be removed by sedimentation and filtration.

The mixing of the coagulant chemical and the raw water to be treated is commonly referred to as flash mixing. The primary purpose of the

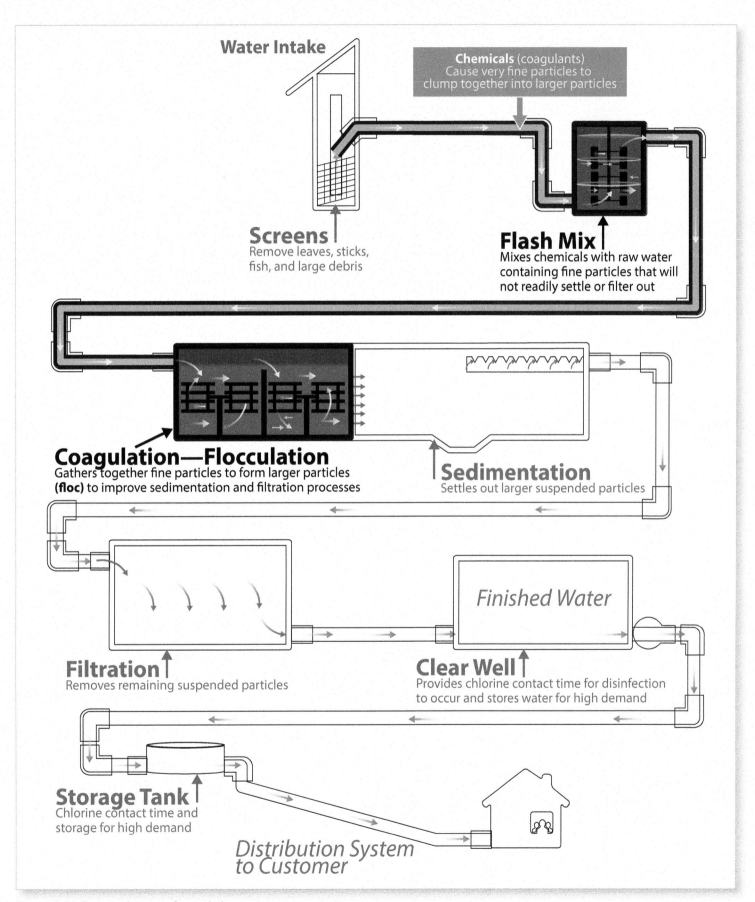

Water Intake

Chemicals (coagulants)
Cause very fine particles to
clump together into larger particles

Screens
Remove leaves, sticks,
fish, and large debris

Flash Mix
Mixes chemicals with raw water
containing fine particles that will
not readily settle or filter out

Coagulation—Flocculation
Gathers together fine particles to form larger particles
(floc) to improve sedimentation and filtration processes

Sedimentation
Settles out larger suspended particles

Finished Water

Filtration
Removes remaining suspended particles

Clear Well
Provides chlorine contact time for disinfection
to occur and stores water for high demand

Storage Tank
Chlorine contact time and
storage for high demand

*Distribution System
to Customer*

Figure 3.1   Process diagram of typical plant

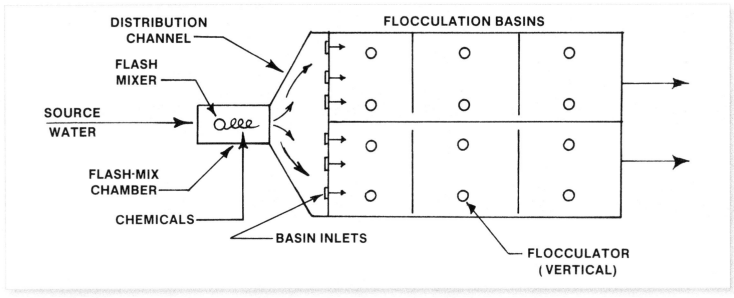

Figure 3.2   Overall plan view of a typical coagulation-flocculation process

flash-mix process is to rapidly mix and equally distribute the coagulant chemical throughout the water. The entire process occurs in a very short time (several seconds), and the first results are the formation of very small particles.

3.2.1   Coagulants

In practice, chemical coagulants are referred to either as primary coagulants or as coagulant aids. Primary coagulants neutralize the electrical charges of the particles, which causes them to begin to clump together. The purpose of coagulant aids is to add density to slow-settling flocs and add toughness so the floc will not break up in the following processes. In view of this definition, coagulant aids could be called flocculation or sedimentation aids. A list of primary coagulants and coagulant aids is shown in Table 3.1.

Metallic salts—aluminum sulfate (commonly called alum), ferric sulfate, ferrous sulfate—and synthetic (manmade) organic **polymers** are commonly used as coagulation chemicals in water treatment because they are effective, relatively low in cost, available, and easy to handle, store, and apply.

When metallic salts, such as aluminum sulfate or ferric sulfate, are added to water, a series of reactions occur with the water and with other

polymer (POLY-mer)
A long chain molecule formed by the union of many monomers (molecules of lower molecular weight). Polymers are used with other chemical coagulants to aid in binding small suspended particles to larger chemical flocs for their removal from water.

Table 3.1   Chemical coagulants used in water treatment

Primary Coagulants	Coagulant Aids	Both
Aluminum Sulfate–$Al_2(SO_4)3 \cdot 14H_2O$	Bentonite (Clay)	Cationic polymers
Ferrous Sulfate–$FeSO_4 \cdot 7H_2O$	Calcium Carbonate–$CaCO_3$	Calcium Hydroxide–$Ca(OH)_2$*
Ferric Sulfate–$Fe_2(SO_4) \cdot 9H_2O$	Sodium Silicate–$NaSiO_3$	Calcium Oxide–CaO*
Ferric Chloride–$FeCl_3 \cdot 6H_2O$	Anionic Polymers & Nonionic Polymers	Sodium Aluminate–$Na_2Al_2O_4$*

*Used as primary coagulants only in water softening processes.

ions in the water. Sufficient chemical quantities must be added to the water to exceed the solubility limit of the metal hydroxide, resulting in the formation of a precipitate (floc). The resulting floc formed will then adsorb on particles (turbidity) in the water.

The synthetic organic polymers used in water treatment consist of a long chain of small subunits or "monomers." The polymer chain can have a linear or branched structure, ranging in length from a fraction of a micron (1 micron = 0.001 millimeter) to as much as 10 microns. The total number of monomers in a synthetic polymer can be varied to produce materials of different molecular weights, which vary from about 100 to 10,000,000. Cationic polymers have a positive electrical charge, anionic polymers have a negative charge, and nonionic polymers have no electrical charge.

The polymers normally used in water treatment contain ionizable groups on the monomeric units (carboxyl, amino, sulfonic groups), and are commonly referred to as polyelectrolytes. Polymers with positively charged groups on the monomeric units are referred to as cationic polyelectrolytes, while polymers with negatively charged groups are called anionic polyelectrolytes. Polymers without ionizable groups are referred to as nonionic polymers.

Cationic polymers have the ability to adsorb on negatively charged particles (turbidity) and neutralize their charge. They can also form an interparticle bridge that collects (entraps) the particles. Anionic and nonionic polymers also form interparticle bridges, which aid in collecting and removing particles from water.

While alum is perhaps the most commonly used coagulant chemical, cationic polymers are used in the water treatment field as both a primary coagulant (in place of alum or other metallic salts) and as a coagulant aid (used in conjunction with alum and other metallic salts). Anionic and nonionic polymers have also proven to be effective in certain applications as coagulant aids and filter aids.

One of the problems that will confront the water treatment plant operator in the selection of an appropriate polymer is that there are a tremendous number of polymers available in the marketplace, and no universal evaluation method has been generally adopted for polymer selection. Thus, the operator should use caution in the selection and use of polymers and should take note of the following considerations regarding polymer use:

1. Polymer overdosing will adversely affect coagulation efficiency and when used as a filter aid, overdosing can result in accelerated head loss buildup.
2. Not all water supplies can be treated with equal success.
3. Some polymers lose their effectiveness when used in the presence of a chlorine residual.
4. Some polymers are dosage limited. The operator should obtain the maximum safe dosage that can be applied from the specific chemical manufacturer.

Because universal standards do not exist for the selection and use of organic polymers, the operator should be careful to select only those products that have been approved by state and federal regulatory agencies for use in potable water treatment. The chemical supplier

molecular weight
The molecular weight of a compound in grams per mole is the sum of the atomic weights of the elements in the compound.

polyelectrolyte (POLY-ee-LECK-tro-lite)
A high-molecular-weight substance, having points of positive or negative electrical charges, that is formed by either natural or synthetic processes. Natural polyelectrolytes may be of biological origin or obtained from starch products or cellulose derivatives. Synthetic polyelectrolytes consist of simple substances that have been made into complex, high-molecular-weight substances. Used with other chemical coagulants to aid in binding small suspended particles to larger chemical flocs for their removal from water. Often called a polymer.

should be required to provide written evidence of this approval. Many chemical suppliers have considerable experience in dealing with many types of water and may be able to recommend the best polymer for your plant.

3.2.2 Basic Coagulant Chemistry

The theory of coagulation is complex, so this discussion of coagulation chemistry is presented to help you understand the coagulation process. Coagulation is a physical and chemical reaction occurring between the **alkalinity** of the water and the coagulant added to the water, which results in the formation of insoluble flocs (floc that will not dissolve).

For a specific coagulant (such as aluminum sulfate or alum), the pH of the water determines which hydrolysis species (chemical compounds) predominate. Lower pH values tend to favor positively charged species, which are desirable for reacting with negatively charged colloids and particulates, forming insoluble flocs, and removing impurities from the water.

The best pH for coagulation usually falls in the range of pH 5 to 7. The proper pH range must be maintained because coagulants generally react with the alkalinity in water. Residual alkalinity in the water serves as a **buffer** to the system (prevent pH from changing) and aids in the complete precipitation of the coagulant chemicals. The amount of alkalinity in the source (raw) water is generally not a problem unless the alkalinity is very low. Alkalinity may be increased by the addition of lime or soda ash.

Polymers are generally added in the coagulation process to stimulate or improve the formation of insoluble flocs.

Generally, the operator has no control over the pH and alkalinity of the source water. Hence, evaluation of these water quality indicators may play a major role in selecting the type of chemical coagulants to be used at a particular water treatment plant, or in changing the type of coagulant normally used if significant changes in pH and alkalinity occur in the raw water.

In some instances, the natural alkalinity in the raw water may be too low to produce complete precipitation of alum. In these cases, lime is often added to ensure complete precipitation. Care must be used to keep the pH within the desired range.

Overdosing as well as underdosing of coagulants may lead to reduced solids removal efficiency. This condition can be corrected by carefully performing **jar tests** and verifying process performance after making any changes in the operation of the coagulation process.

3.2.3 Effective Mixing

In modern water treatment plants, it is desirable to complete the coagulation reaction (mixing of chemicals into the water) in as short a time as possible—preferably within a period of several seconds because the reaction time is short. For complete coagulation and flocculation to take

---

NSF International maintains a searchable database that includes polymers approved for potable water usage. Available information about polymers includes the name of the manufacturer and the maximum concentration recommended. www.nsf.org.

alkalinity (AL-kuh-LIN-it-tee)
The capacity of water or wastewater to neutralize acids. This capacity is caused by the water's content of carbonate, bicarbonate, hydroxide, and occasionally borate, silicate, and phosphate. Alkalinity is expressed in milligrams per liter of equivalent calcium carbonate. Alkalinity is not the same as pH because water does not have to be strongly basic (high pH) to have a high alkalinity. Alkalinity is a measure of how much acid must be added to a liquid to lower the pH to 4.5.

buffer
A solution or liquid whose chemical makeup neutralizes acids or bases, thereby maintaining a relatively constant pH.

jar test
A laboratory procedure that simulates coagulation/flocculation with differing chemical doses. The purpose of the procedure is to estimate the minimum coagulant dose required to achieve certain water quality goals. Samples of water to be treated are placed in six jars. Various amounts of chemicals are added to each jar, stirred, and the settling of solids is observed. The lowest dose of chemicals that provides satisfactory settling is the dose used to treat the water.

place, the coagulant must make contact with all of the suspended particles. This is accomplished by flash mixing.

Several methods can be used to mix the chemicals with the water to be
treated (see Figure 3.3):

1. Hydraulic mixing using flow energy in the system
2. Mechanical mixing
3. Diffusers and grid systems
4. Pumped blenders

Hydraulic mixing with baffles or throttling valves works well in systems that have sufficient water velocity (speed) to cause turbulence in
the water being treated. The turbulence in the flowing water mixes the
chemicals with the water.

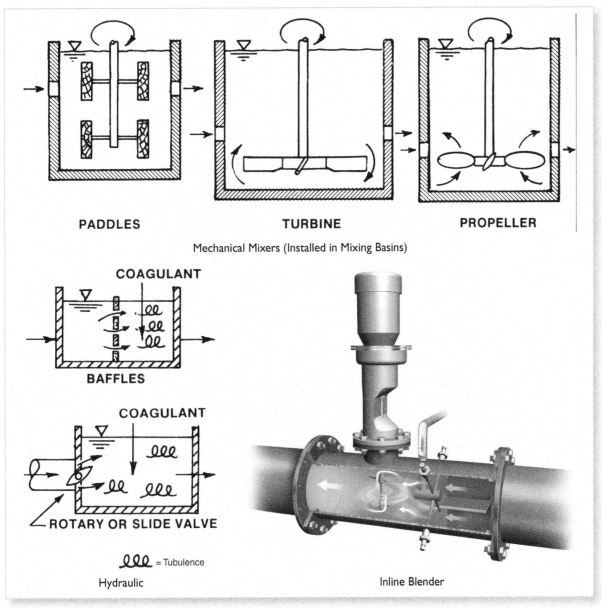

PADDLES   **TURBINE**   **PROPELLER**

Mechanical Mixers (Installed in Mixing Basins)

COAGULANT

BAFFLES

COAGULANT

ROTARY OR SLIDE VALVE

= Tubulence

Hydraulic   Inline Blender

Figure 3.3   Methods of flash mixing *(Courtesy of Walker Process Equipment, division of McNish Corporation)*

Mechanical mixers (paddles, turbines, and propellers) are frequently used in coagulation facilities. Mechanical mixers are versatile and reliable; however, they generally use the greatest amount of electric energy for mixing the coagulant with the water being treated.

Diffusers and grid systems consisting of perforated tubes or nozzles can be used to disperse the coagulant into the water being treated. These systems can provide uniform (equal) distribution of the coagulant over the entire coagulation basin. However, they are generally sensitive to flow changes and may require frequent adjustments to produce the proper amount of mixing.

Pumped blenders (Figure 3.4) have also been used for mixing in coagulation facilities. In this system, the coagulant is added directly to the water being treated through a diffuser in a pipe. This system can provide rapid dispersion of the coagulant and does not create any significant **head loss** in the system. Electric energy consumption is considerably less than that of a comparable mechanical mixer.

Mixing of the chemical coagulant can be satisfactorily accomplished in a coagulation basin, a special rectangular tank with mixing devices. Mixing may also occur in the influent channel or a pipeline to the flocculation basin if the flow velocity is high enough to produce the necessary turbulence. The shape of the basin is part of the flash-mix system design.

head loss
The head, pressure or energy (they are the same) lost by water flowing in a pipe or channel as a result of turbulence caused by the velocity of the flowing water and the roughness of the pipe, channel walls, or restrictions caused by fittings. Water flowing in a pipe loses head, pressure, or energy as a result of friction. The head loss through a filter is due to friction losses caused by material building up on the surface or by the water flowing through the filter media. Also called friction loss.

Figure 3.4 Pumps used in pumped blender system

3.3   Flocculation

Flocculation is a slow stirring process that causes the gathering together of small, coagulated particles into larger, settleable particles. The flocculation process provides contact between particles to promote their gathering together into floc for ease of removal by sedimentation and filtration. Generally, these contacts or collisions between particles result from gentle stirring created by a mechanical or hydraulic means of mixing.

Floc formation is controlled by the rate at which collisions occur between particles and by the effectiveness of these collisions in promoting attachment between particles. The purpose of flocculation is to create a floc of a good size, density, and toughness for later removal in the sedimentation and filtration processes. The best floc size ranges from 0.1 mm to about 3 mm, depending on the type of removal processes used (direct filtration vs. conventional filtration, which is discussed in Chapter 5, "Filtration").

3.3.1   Process Performance

An efficient flocculation process involves the selection of the right stirring time (detention time), the proper stirring intensity, a properly shaped basin for uniform mixing, and mechanical equipment or other means of creating the stirring action. Insufficient mixing will result in ineffective collisions and poor floc formation. Excessive mixing may tear apart the flocculated particles after they have clumped together.

3.3.1.1   Detention Time

Detention time is usually not a critical factor in the coagulation or flash-mixing process if the chemical coagulants are satisfactorily dispersed into the water being treated and mixed for at least several seconds. Detention time is required for the necessary chemical reactions to take place. Some operators have been able to reduce coagulant dosages by increasing the amount of detention time between the point of addition of the coagulant and the flocculation basins. In the flocculation process, however, stirring (detention) time is quite important. The minimum detention time recommended for flocculation ranges from about 5 to 20 minutes for direct filtration systems and up to 30 minutes for conventional filtration. (The size and shape of the flocculation facility also influence the detention time needed for optimum floc development.)

3.3.1.2   Types of Flocculators (Stirrers)

Two types of mechanical flocculators are commonly installed, horizontal paddle wheels and vertical flocculators (see Figures 3.5 and 3.6). Both types can provide satisfactory performance; however, the vertical flocculators usually require less maintenance because they eliminate submerged bearings and packings. Vertical flocculators can be of the propeller, paddle, or turbine types.

Some flocculation can also be accomplished by the turbulence resulting from the roughness in conduits or channels, or by the dissipated energy of head losses associated with **weirs**, **baffles**, and **orifices** used to create turbulence, which will mix chemicals with the water. Weirs and baffles are boards or plates that water flows over while orifices are holes in walls that water flows through. Generally, these methods find only limited use

---

detention time
(1) The theoretical (calculated) time required for a small amount of water to pass through a tank at a given rate of flow. (2) The actual time in hours, minutes or seconds that a small amount of water is in a settling basin, flocculating basin, or rapid-mix chamber. In storage reservoirs, detention time is the length of time entering water will be held before being drafted for use (several weeks to years, several months being typical).

$$\frac{\text{Detention}}{\text{Time [h]}} = \frac{\text{Basin Volume [gal]}}{\text{Flow Rate}\left[\frac{\text{gal}}{\text{d}}\right] \times \frac{\text{d}}{24\text{h}}}$$

weir (WEER)
(1) A wall or plate placed in an open channel and used to measure the flow of water. The depth of the flow over the weir can be used to calculate the flow rate, or a chart or conversion table may be used to convert depth to flow. (2) A wall or obstruction used to control flow (from settling tanks and clarifiers) to ensure a uniform flow rate and avoid short-circuiting.

baffle
A flat board or plate, deflector, guide, or similar device constructed or placed in flowing water, wastewater, or slurry systems to cause more uniform flow velocities, to absorb energy, and to divert, guide, or agitate liquids (water, chemical solutions, slurry).

orifice (OR-uh-fiss)
An opening (hole) in a plate, wall, or partition. An orifice flange or plate placed in a pipe consists of a slot or a calibrated circular hole smaller than the pipe diameter. The difference in pressure in the pipe above and at the orifice may be used to determine the flow in the pipe. In a trickling filter distributor, the wastewater passes through an orifice to the surface of the filter media.

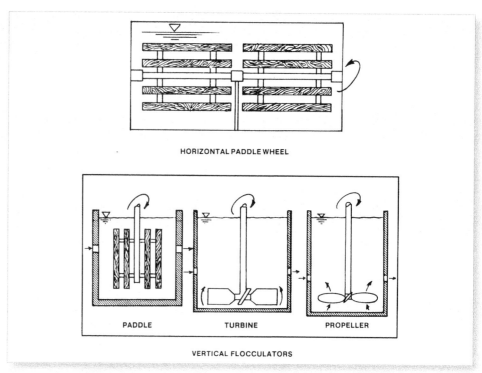

Figure 3.5   Mechanical flocculators

Figure 3.6   Horizontal paddle wheel flocculator

owing to disadvantages such as very localized distribution of turbulence, inadequate detention time, and widely variable turbulence resulting from flow fluctuations.

3.3.1.3   Flocculation Basins

The shape of flocculation basins is determined partially by the flocculator selected, but also for compatibility with adjoining structures (sedimentation basins). Flocculation basins for horizontal flocculators are generally rectangular in shape, while basins for vertical flocculators are nearly square. The depth of flocculation basins is usually about the same as the sedimentation basins.

The best flocculation is usually achieved in a compartmentalized basin. The compartments (most often three) are separated by baffles to prevent **short-circuiting** of the water being treated. The turbulence can be reduced gradually by reducing the speed of the mixers in each succeeding

short-circuiting
A condition that occurs in tanks or basins when some of the flowing water entering a tank or basin flows along a nearly direct pathway from the inlet to the outlet. This is usually undesirable since it may result in shorter contact, reaction, or settling times in comparison with the theoretical (calculated) or presumed detention times.

tank (see Figure 3.7) or by reducing the surface area of the paddles. This is called tapered-energy mixing. The reason for reducing the speed of the stirrers is to prevent breaking apart the large floc particles that have already formed. If you break up the floc you have not accomplished anything and will overload the filters.

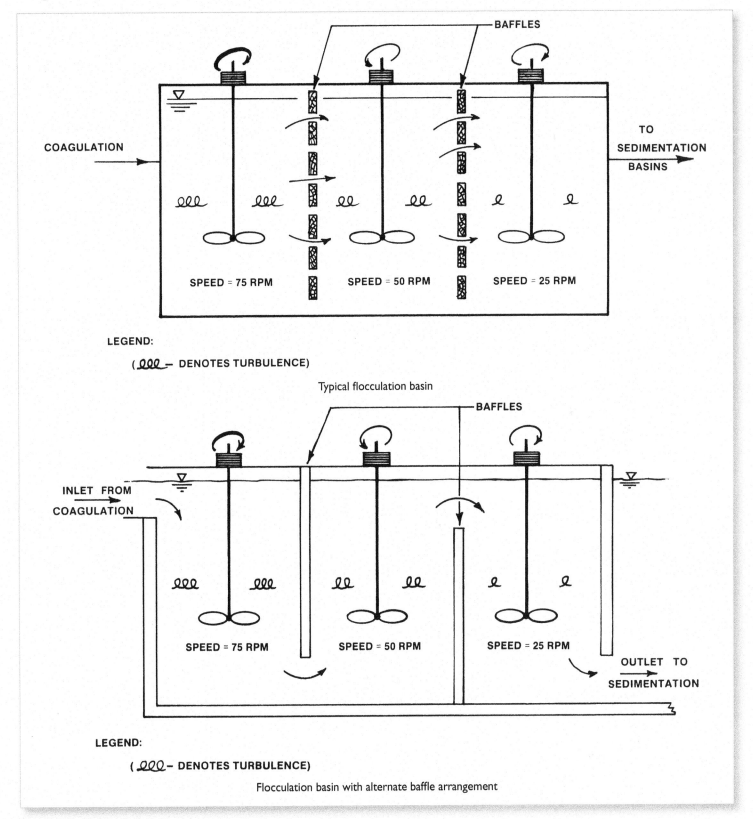

Figure 3.7 Typical flocculation basins

The solids-contact process (upflow clarifiers) is used in some water treatment plants to improve the overall solids removal process. These units combine the coagulation, flocculation, and sedimentation processes into a single basin. A detailed discussion of solids-contact units is given in Chapter 4, "Sedimentation," Section 4.3.1.5, "Solids-Contact Units."

Check Your Understanding

1. What are nonsettleable solids?
2. What happens in the coagulation and flocculation processes?
3. What is the primary purpose of flash mixing?
4. Why are both primary coagulants and coagulant aids used in the coagulation process?
5. What pH range is usually best for coagulation?
6. List four methods of mixing coagulant chemicals into the plant flow.
7. Describe the results of inefficient mixing during flocculation.
8. What is the recommended minimum detention time for flocculation?
9. What is an advantage of vertical flocculators over horizontal flocculators?
10. Why are the compartments in flocculation basins separated by baffles?

3.4 Interaction with Other Treatment Processes

Coagulation and flocculation processes are required to precondition or prepare nonsettleable particles present in the raw water for removal by sedimentation and filtration. Small particles, without proper coagulation and flocculation, are too light to settle out and will not be large enough to be trapped during filtration. In this regard, it is convenient to consider coagulation-flocculation as one treatment process.

Because the purpose of coagulation-flocculation is to promote particulate removal, the effectiveness of the sedimentation and filtration processes, as well as overall plant performance, depends upon successful coagulation-flocculation. Disinfection of the water can also be affected by poor coagulation-flocculation performance. Bacteria and other disease-causing organisms can be bound up in suspended particles and thereby shielded from disinfection if the solids removal processes before final disinfection, especially filtration, are ineffective. Effective coagulation-flocculation promotes the removal of natural organic compounds. Removal of these compounds will reduce the formation of trihalomethanes following the use of chlorine for disinfection.

3.5 Process Control

The theory behind the chemical reactions and the formation of floc associated with the coagulation-flocculation process is rather complex. Yet from a practical viewpoint, the operator of a water treatment plant must be able to measure and control the performance of these processes on a day-to-day basis.

The most important consideration in coagulation-flocculation process control is the selection of the proper type and amount of coagulant chemicals to be added to the water being treated. This determination is commonly made in the laboratory with the aid of a jar testing apparatus (Figure 3.8). When selecting a particular type of coagulant chemical, consideration must be given to the quantity and solids content of the sludge created and the means of ultimate disposal. Jar tests should be run at least daily and more often when the quality of the raw water changes. Changes in the raw water may require changes in the amount of chemical or type of chemical.

Procedures for determining proper coagulant dosage, as well as means for measuring and controlling coagulation-flocculation process performance, are described in the following sections.

Figure 3.8   Jar test apparatus with mechanical stirrers *(Courtesy of Phipps & Bird, Inc.)*

3.6   Normal Operations

Coagulation-flocculation is a pretreatment process for the sedimentation and filtration processes. Most of the suspended solids are removed in the sedimentation basins and filtration is the final step in the solids removal process. Thus, the coagulation-flocculation process should be operated and controlled to improve filtration and thus produce a filtered water that is low in turbidity.

The measurement of filtered water turbidity periodically using **grab samples** or continuously using a **turbidity meter** will give the operator a good indication of overall process performance. However, the operator cannot rely solely on filtered water turbidity for complete process control. The difficulty in relying on a single water quality indicator, such as

grab sample
A single sample of water collected at a particular time and place that represents the composition of the water only at that time and place.

turbidity meter
An instrument for measuring and comparing the turbidity of liquids by passing light through them and determining how much light is reflected by the particles in the liquid. The normal measuring range is 0 to 100 and is expressed as nephelometric turbidity units (NTUs). Also called a turbidimeter.

filtered water turbidity, is that it takes a considerable amount of time to transport the water through the various treatment processes. Depending on the amount of water being processed, the total transit time through the treatment plant can vary from as few as 2 hours to more than 6 hours. This means that a change in coagulant dosage at the front end of the plant will not be noticed in the final finished water quality for at least 2 hours, depending on flow conditions. Thus, turbidity, as well as other water quality indicators, such as pH, temperature, chlorine demand, and floc quality, must be monitored throughout the water treatment process so that poor process performance can be spotted early and corrective measures can be taken.

Process control guidelines for a specific plant are often developed to assist the operator in making these determinations. These guidelines incorporate theory, operator experience, practical knowledge of the source water conditions, and known performance characteristics of the treatment facilities.

3.6.1   Process Actions

In the normal operation of the coagulation-flocculation process, the operator performs a variety of jobs within the water treatment plant. The number and type of functions that each operator will perform vary considerably depending on the size and type of plant and the number of people working in the plant. In smaller plants, the operator is required to control almost all process actions as well as perform most routine maintenance activities. Regardless of the plant size, all operators should be thoroughly familiar with the routine and special operations and maintenance procedures associated with each treatment process.

Typical jobs performed by an operator in the normal operation of the coagulation-flocculation process include the following:

1. Monitor process performance.
2. Evaluate water quality conditions (raw and treated water).
3. Check and adjust process controls and equipment.
4. Visually inspect facilities.

Monitoring process performance is an ongoing activity. As previously discussed, filtered water turbidity levels are controlled to a great extent by the efficiency of the coagulation-flocculation process. Early detection of a failure is extremely important because considerable time elapses while the water flows through the coagulation, flocculation, sedimentation, and filtration processes.

Process performance can be monitored with the aid of continuous water quality analyzers, which automatically measure a specific water quality indicator such as turbidity. However, reliable and accurate water quality analyzers are expensive and, in certain cases, automated equipment is not readily available for measuring all water quality indicators of concern to the operator. Thus, a combination of techniques must be used by the operator to evaluate process performance including visual observations and periodic laboratory tests to supplement any continuous water quality analyzers.

Visual observations and laboratory tests of coagulation-flocculation process performance should be performed on a routine basis. The most common laboratory tests are turbidity, alkalinity, pH, color, temperature, and chlorine demand. The frequency of these observations and tests depends on how much the quality of the source water supply can or does change. In treatment plants where the source water is stored in a large upstream lake or reservoir, the water quality is generally more stable or constant than water taken directly from rivers or streams.

In the case of direct **diversions** from a stream or river, water quality conditions will vary seasonally as well as daily. In extreme cases (during heavy runoff periods), even hourly changes in source water quality can be expected. Thus, the appropriate frequency of performing certain tests may be as often as hourly, or perhaps only once per 8-hour shift.

Visual checks of the coagulation-flocculation process generally include an observation of the turbulence of the water in the flash-mixing channel or chamber (you can see improper flow patterns), and close observation of the size and distribution of floc in the flocculation basins. An uneven distribution of floc could be an indication of short-circuiting in the flocculation basin. Floc particles that are too small or too large may not settle properly and could cause trouble during removal in the sedimentation and filtration processes. These observations are frequently supplemented by laboratory evaluations, which are necessary to provide better data. For example, floc settling characteristics require laboratory evaluation based on trial and experience methods using the jar test.

Speed adjustment of flocculators, if such a feature is provided, should take into account the following items:

1. Volume of floc to be formed. If source water turbidity is low, a small pinpoint floc may be best suited for removal on the filters (direct filtration). Lower flocculator speeds are appropriate here. On the other hand, high turbidity source waters generally require near-maximum flocculator speed to produce a readily settleable floc.

2. Visual observations. Short-circuiting may indicate flocculator mixing intensity is not sufficient, while floc breakup is an indication that the mixing turbulence (speed) may be too high for the type of floc formed (large alum floc).

3. Water temperature. The rate of all chemical reactions decreases with temperature. For example, 50°F (10°C) water slows down chemical reaction times to one-half their speed at 70°F (21°C). So, increase the speed to account for lower water temperature that requires higher mixing turbulence.

Unfortunately, these concepts are not easily measured. Experience and judgment are needed. One of the real limitations in process control is too much reliance on the settled turbidity value. While turbidity gives an indirect measurement of suspended solids concentration, it does not describe particle size, density, volume, nor the ability of a particular filter to handle the applied waters.

diversion
Use of part of a stream flow as a water supply.

Typical coagulation-flocculation monitoring locations are shown in Figure 3.9. Water quality indicators used to evaluate coagulant dosage and process performance include turbidity, temperature, alkalinity, pH, color, and chlorine demand. These water quality indicators are further discussed in Section 3.9, "Laboratory Tests."

Based on an overall evaluation of process performance, the operator may need to make minor changes in chemical feed dosages, or adjust the speed of the flash mixer or flocculators, if variable-speed units are provided. These are normal actions associated with minor changes in source water quality such as turbidity or temperature fluctuations.

Flash mixers are generally less sensitive to speed adjustments than flocculators because their primary purpose is to disperse the chemicals rapidly into the water being treated. This reaction is almost instantaneous in such small quantities.

Process equipment, such as chemical feeders, should be checked regularly to ensure that they are accurately feeding the desired amount (feed rate) of chemical. Operation and maintenance (O&M) of process equipment is discussed in more detail in Section 3.10, "Equipment Operation and Maintenance."

The operator should routinely perform a visual inspection of the overall coagulation-flocculation physical facilities. This is a part of good housekeeping practice. Leaves, twigs, and other debris can easily build up in the influent channel or in the flocculation basins. If ignored, this material may get into other processes where it can foul

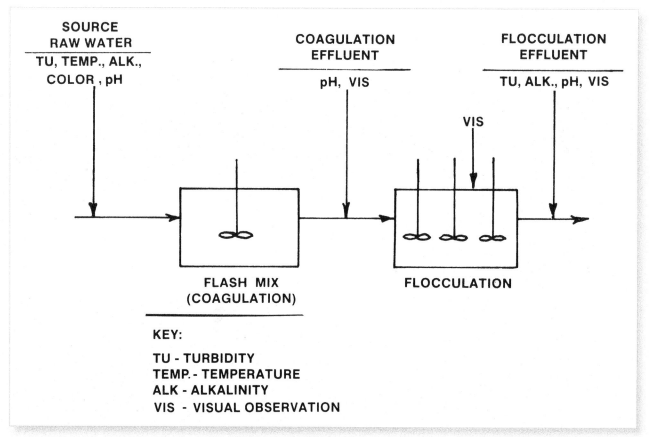

Figure 3.9 Coagulation-flocculation process monitoring guidelines and sample points

meters, water quality monitors, pumps, or other mechanical equipment. In some cases, taste and odor problems can develop from microorganisms that can grow in debris and sediment that accumulate in plant facilities. See Chapter 8, "Taste and Odor Control," for additional information.

Check Your Understanding

1. Why is coagulation-flocculation important to other treatment processes?
2. What is the most important consideration in coagulation-flocculation process control?
3. Which processes remove suspended solids after the coagulation-flocculation process?
4. List the typical functions performed by an operator in the normal operation of the coagulation-flocculation process.
5. How is the effectiveness of the solids removal processes commonly monitored?
6. Which laboratory tests would you use to monitor the coagulation-flocculation process?
7. What would you look for when visually observing the performance of a coagulation-flocculation process?

3.6.2   Process Operation

To illustrate how to operate the coagulation-flocculation process, the procedures used in a typical plant will be discussed on a step-by-step basis. Many plants are different from the example plant and many waters behave differently than the water in this discussion. Operators experiment to find the most effective procedures to produce a high-quality drinking water.

3.6.2.1   Detention Times

To calculate the expected detention times in the flash-mix chamber, distribution channel, and flocculation basins, use the dimensions (length, width, and depth) of the facilities shown in the treatment plant design drawings. These times are important when determining the optimum (best minimum) chemical dosage for the water you are treating by using the jar test. Also, these times are necessary for the desired chemical reactions to occur.

FORMULAS

To calculate the detention times of tanks, basins, or clarifiers, first determine the volume of the container.

1. To calculate the volume (V) of a rectangular tank or basin in cubic feet,

$$V[ft^3] = Length[ft] \times Width[ft] \times Depth[ft]$$

2. To calculate the volume of a circular tank or clarifier in cubic feet,

$$V[ft^3] = \frac{\pi}{4} \times Diameter^2[ft^2] \times Depth[ft]$$

The term $\pi/4$ is equal to 0.785, so this term is commonly used.

$$V[ft^3] = 0.785 \times Diameter^2[ft^2] \times Depth[ft]$$

3. Frequently, we need the volume in gallons, rather than cubic feet.

$$V[gal] = V[ft^3] \times \frac{7.48\,gal}{ft^3}$$

4. To calculate the detention time (t) of any chamber, tank, basin, or clarifier,

$$t[min] = \frac{V[gal]}{Flow\,Rate\left[\dfrac{gal}{d}\right] \times \dfrac{d}{24\,h} \times \dfrac{h}{60\,min}}$$

or

$$t[h] = \frac{V[gal]}{Flow\,Rate\left[\dfrac{gal}{d}\right] \times \dfrac{d}{24\,h}}$$

Detention times are calculated by dividing the volume in gallons by the flow in gallons per day. This produces the detention time in days. Multiply by 24 hours per day to obtain the detention time in hours. To convert the detention time from hours to minutes, multiply by 60 minutes per hour. To convert the detention time from hours to minutes, multiple by 60 minutes per hour and 60 seconds per minute.

Example 1

A water treatment plant treats a flow of 2.4 Mgal/d or 2,400,000 gallons per day (gpd). The flash-mix chamber is 2.5 feet (ft) square (length and width are both 2.5 feet) and the depth of water is 3 feet. Calculate the detention time in seconds (s).

Known			Unknown
Q	= Flow	= 2,400,000 gpd	t = Detention Time, s
L	= Length	= 2.5 ft	
W	= Width	= 2.5 ft	
H	= Depth	= 3.0 ft	

1. Determine the volume (V) of the flash-mix chamber in cubic feet (ft³).

$$V[ft^3] = L[ft] \times W[ft] \times H[ft]$$
$$= 2.5[ft] \times 2.5[ft] \times 3.0[ft]$$
$$= 18.75\ ft^3$$

2. Convert the volume of the flash-mix chamber from cubic feet (ft³) to gallons (gal).

$$V[gal] = V[ft^3] \times \frac{7.48\,gal}{ft^3}$$

$$= 18.75[ft^3] \times \frac{7.48\,gal}{ft^3}$$

$$= 140.25\,gal$$

3. Calculate the detention time of the flash-mix chamber in seconds (s).

$$t[s] = \frac{V[gal]}{Q\left[\dfrac{gal}{d}\right] \times \dfrac{d}{24\,h} \times \dfrac{h}{60\,min} \times \dfrac{min}{60\,s}}$$

$$= \frac{140.25[gal]}{2,400,000\left[\dfrac{gal}{d}\right] \times \dfrac{d}{24\,h} \times \dfrac{h}{60\,min} \times \dfrac{min}{60\,s}}$$

$$= 5.0\,s$$

Example 2

A water treatment plant treats a flow of 2.4 Mgal/d or 2,400,000 gallons per day (gpd). The flocculation basin is 8 feet deep, 15 feet wide, and 45 feet long. Calculate the detention time in minutes (min).

	Known		**Unknown**
Q	= Flow = 2,400,000 gpd	t	= Detention Time, min
H	= Depth = 8 ft		
W	= Width = 15 ft		
L	= Length = 45 ft		

1. Determine the volume of the flocculation basin in cubic feet (ft³).

$$V[ft^3] = L[ft] \times W[ft] \times H[ft]$$

$$= 45[ft] \times 15[ft] \times 8[ft]$$

$$= 5,400\,ft^3$$

2. Convert the volume of the flocculation basin from cubic feet (ft³) to gallons (gal).

$$V[gal] = V[ft^3] \times \frac{7.48\,gal}{ft^3}$$

$$= 5,400[ft^3] \times \frac{7.48\,gal}{ft^3}$$

$$= 40,392\,gal$$

3. Calculate the detention time of the flocculation basin in minutes (min).

$$t[min] = \frac{V[gal]}{Q\left[\frac{gal}{d}\right] \times \frac{d}{24\,h} \times \frac{h}{60\,min}}$$

$$= \frac{40,392[gal]}{2,400,000\left[\frac{gal}{d}\right] \times \frac{d}{24\,h} \times \frac{h}{60\,min}}$$

$$= 24\ min$$

Many operators prepare curves of flow vs. detention time for the basins in their plant. These curves allow for easy selection of stirring times when performing jar tests.

PREPARATION OF DETENTION TIME CURVES

Example 2 was a water treatment plant with a flow of 2.4 Mgal/d. The flocculation basin was 8 feet deep, 15 feet wide, and 45 feet long. We calculated a basin volume of 40,392 gallons (gal) and a detention time of 24 minutes (min). The typical flows for this plant range from 0.8 Mgal/d to 2.4 Mgal/d. To plot a detention time curve, we should have at least four plotting points. This means that we must calculate the detention times for at least four flows. For this example, we will calculate the detention times for flows of 0.8, 1.2, 1.6, 2.0, and 2.4 Mgal/d. This will provide us with five plotting points.

Known	Unknown
V = Basin Volume = 40,392 gal	t = Detention Time, min, for each flow
Q = Flow Rate = 0.8, 1.2, 1.6, 2.0, and 2.4 Mgal/d	

The table below shows the detention time (t) for each of the given flow values:

Flow Rate (Q), (Mgal/d)	Detention Time (t), (min)
0.8	73
1.2	48
1.6	36
2.0	29
2.4	24

Here is how the detention time is calculated:

$$t[min] = \frac{V[gal]}{Q\left[\frac{gal}{d}\right] \times \frac{d}{24\,h} \times \frac{h}{60\,min}}$$

When the flow rate Q is 0.8 Mgal/d,

$$t[min] = \frac{V[gal]}{Q\left[\dfrac{gal}{d}\right] \times \dfrac{d}{24\,h} \times \dfrac{h}{60\,min}} = \frac{40,392[gal]}{800,000\left[\dfrac{gal}{d}\right] \times \dfrac{d}{24\,h} \times \dfrac{h}{60\,min}} = 73\,min$$

Follow the same method to complete the table shown above, and then plot the flow vs. detention time curve as shown in Figure 3.10. This curve can be used to determine the detention time for any flow. If you know the flow you are treating, find the flow on Figure 3.10, move up to the curve, and then across to the detention time scale.

3.6.2.2   The Jar Test

The purpose of jar testing is for operators to observe on a small scale how changes in chemicals, chemical doses, dosages, mixing times, and other parameters will affect the plant's performance. For a step-by-step procedure on how to run a typical jar test for evaluating coagulants doses, see Chapter 9, "Laboratory Procedures." In short, a jar test consists of collecting a sample to be tested, adding the coagulant or other chemical with rapid mixing, followed by slow-mix flocculation, and then settling with the mixer turned off. After settling, a sample of the supernatant (the clear water on top) is taken to determine water quality parameters such as pH, turbidity, and alkalinity. Because the idea is to test variations, particularly in chemicals or doses, multiple jars are usually run in parallel. A typical jar test mixing apparatus is shown in Figure 3.8.

JAR TEST PLANNING

Although the goal is to simulate the full-scale plant, it is almost impossible to exactly duplicate in a static laboratory experiment the flow-through conditions that are occurring in the treatment plant. Nevertheless, the goal is

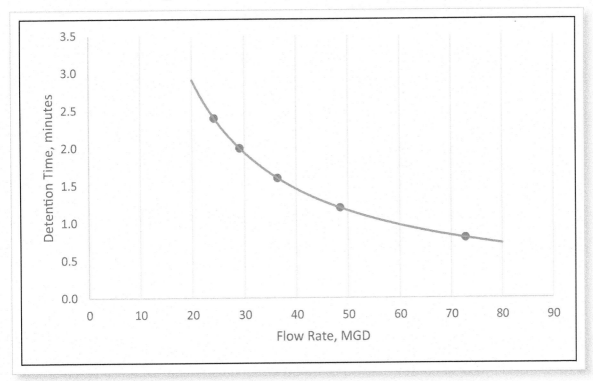

Figure 3.10   Development of flow vs. detention time curve

for the jar test results to accurately reflect how the plant will respond to the chemicals being tested. For instance, if the coagulant dose increases or the pH changes in the lab, the jar test results should be generally the same as the plant water quality analysis under those same changes. For this reason, it is advisable to spend some time fine-tuning the jar test procedure. The mixing time should be the same as the hydraulic detention time in the plant's flocculation basins. On the other hand, the settling time should not be the same as that in the plant because the particles are settling only a few inches in the jar and many feet in the plant. Choose a settling time in which the jar turbidity is approximately the same as the settling tank effluent. Similarly, the rotations per minute (rpm) set on the laboratory mixer should not be the same as those set in the plant. The intensity of the mixing should be the same in both cases, but mixing intensity depends on the amount of mixing power delivered to the water per unit volume. Different kinds of mixers have different relationships between the mixing speed (rpm) and the power delivered to the water. The relationship for the jar test mixer will be different from the plant mixer, so the rpm will be different as well. Adjust the laboratory mixer speed until the jar test results approximate plant performance. Finally, because temperature can affect floc formation, run your jar tests at about the same temperature as the plant. If this is inconvenient to do regularly, try to do a one-time experiment in which you repeat the same jar test at different temperatures to get an idea of how big an impact this variable has on test results.

There are a variety of ways of conducting a jar test, depending on what is being tested. Use procedures that best suit the requirement of the test. Most importantly, change only one operational variable at a time. Although up to six tests may be run simultaneously, remember that comparisons between jars that differ in only one variable (alum dosage, for example) are the most useful. When only one variable is changed and everything else remains the same, any changes in the final outcome will be due to that single changed variable. Analyze the results of one test before changing another variable. Keep good records.

CHEMICAL DOSE. Most often, laboratory jar tests are performed to evaluate how to make better use of coagulants. In these cases, the test parameter is the chemical dose and the most common indicator variable is turbidity. Choose a range of doses that brackets the expected best result. When two chemicals are involved, hold one concentration constant in the test while changing the other. This strategy is illustrated in the example presented later in this section. Compare starting and final concentrations of aluminum (if using alum) or iron (if using ferric chloride or ferrous sulfate), if desired. The difference in concentrations is an indication of the amount of chemicals removed in the process. Large residuals (concentrations) at the ends of tests indicate chemical overdoses or poor sedimentation or filtration efficiencies. These residuals may also cause floc to form in the water distribution system.

OPTIMUM pH. Because some coagulants (especially alum) have an optimum pH range, record the pH during the jar test. The pH in the jar after the flash mix that produces the best results is the target pH. You can run jar tests at different pH levels to determine the optimum pH for a given chemical dose. With this information you can adjust your plant chemical feeders to produce the target pH after flash mixing in their treatment plant.

POST-JAR PROCESSING. What counts most is the turbidity of the effluent from your plant filters and the performance of those filters. To simulate these factors in the laboratory, some operators pass the supernatant from the jars after settling through a qualitative filter paper (8 µm nominal pore size), such as Whatman No. 40 filter paper, and then measure the turbidity of the filtered water. Tests with such filter papers have sometimes shown that the dosage of coagulant needed to achieve a high-quality filtered water may be far below that required to produce a floc that settles well. The lower the chemical dose, the less sludge will be produced. Reducing the chemical dose reduces costs as well.

Another measure of filterability is the time required for the 100 mL supernatant to pass through the filter.

These tests have also shown that in some cases when a non-settling floc forms, the flocculators can be shut off without any effect on filtered water turbidity. This could occur when the raw water has a very low turbidity.

EXAMPLE JAR TEST

This example of the use of jar testing omits some of the details of the procedures. See Chapter 9, "Laboratory Procedures," for the complete procedure.

In a typical plant, and thus in the jar test, liquid alum is used as the primary coagulant and a cationic polymer as a coagulant aid. This example will use 2-liter water samples to fit the jar test apparatus (Figure 3.8).

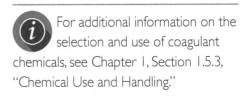

For additional information on the selection and use of coagulant chemicals, see Chapter 1, Section 1.5.3, "Chemical Use and Handling."

Jar test **reagents** can be prepared at several concentrations depending on the desired dosage, as described in Chapter 9.

For the example plant, assume that records indicate the proper range of dosages for current water quality conditions are as follows:

Alum, mg/L = 10 to 20 mg/L

Cationic Polymer, mg/L = 0.5 to 3 mg/L

(For these doses, we will prepare a 2,000 mg/L alum stock solution and 1,000 mg/L polymer stock solution.)

For this example, in test sequence No. 1, evaluate the effect of increasing the polymer dosage while holding the alum dosage constant at 10 mg/L. Increase the polymer dosage from 0.5 mg/L to 3 mg/L.

Test Sequence No. 1						
Jar No.	1	2	3	4	5	6
Alum, mg/L	10	10	10	10	10	10
Cationic Polymer, mg/L	0.5	1.0	1.5	2.0	2.5	3.0

For test sequence No. 2, evaluate the effects of varying the polymer dosage at a higher alum dosage (15 mg/L).

Test Sequence No. 2						
Jar No.	1	2	3	4	5	6
Alum, mg/L	15	15	15	15	15	15
Cationic Polymer, mg/L	0.5	1.0	1.5	2.0	2.5	3.0

reagent (re-A-gent)
A pure chemical substance or solution of that substance that is used to make new products or to measure, detect, or examine other substances in chemical tests.

Other combinations of alum and cationic polymer may be evaluated by trial and experience to determine the optimum (best possible) dosages. Vary the levels of alum doses in a set of tests and try combinations with polymers in a separate test.

Use the following jar test procedure (see Chapter 9 for details):

1. Collect the sample and fill the beakers to the 2-liter mark with the water to be tested.

2. Measure out the required volumes of stock solution separately into small beakers, two for each jar (one for alum and one for polymer).

In test sequence No. 1, add the following reagent volumes:

Test Sequence No. 1						
Jar No.	1	2	3	4	5	6
Alum Dosage, mg/L	10	10	10	10	10	10
Reagent Volume, mL (1%)	10	10	10	10	10	10
Cat. Polymer Dosage, mg/L	0.5	1.0	1.5	2.0	2.5	3.0
Reagent Volume, mL (0.1%)	1.0	2.0	3.0	4.0	5.0	6.0

3. Lower the stirring paddles into the beakers and start them at 80 rpm. Add the stock volumes as simultaneously as possible. Mix for 1 minute.

   Note: The 1-minute mix time was based on the detention times in the flash-mix and distribution channels with a flow of 3.0 Mgal/d. If the flows are less than 3.0 Mgal/d, use more than 1 minute. The 80 rpm speed was chosen based on past experiments. If a higher mixing speed (100 rpm) produces conditions more like hydraulic conditions in the plant, use a higher mixing speed. Table 3.2 shows the detention times in the plant and how they change as the flows change. See "Preparation of Detention Time Curves" on page 133, for more information on determining detention times.

4. Reduce the mixer speed to 20 rpm for 15 minutes to simulate flocculation basin conditions. Record the time required for visible floc to form and describe the floc characteristics (very small, small, medium, large, or very large as well as dense, very dense, or sparse) during mixing. Figure 3.11 shows the data recorded during this jar test.

5. Stop the stirrers. Allow floc to settle for 30 minutes. Observe and note how quickly the floc settles and the floc's appearance. A hazy settled water indicates poor coagulation. Properly coagulated water contains floc particles that are well formed and the water between the particles is clear. Describe the results as poor, fair, good, or excellent.

   Note: The 15 rpm mixing speed and the 30-minute settling time were chosen based on past trials and experience in simulating the plant process with the jar test.

6. Without disturbing the settled floc, take a sample of the supernatant from each jar. Measure the turbidity of each sample.

Table 3.2   Example jar test mixing times

Plant Flow Rate	Flash Mix	Slow Mix
(Mgal/d)	(minutes)	(minutes)
2	1.2	29
2.2	1.0	26
2.4	1.0	24
2.6	0.9	22
2.8	0.8	21
3.0	0.8	19
3.2	0.7	18
3.4	0.7	17
3.6	0.6	16
3.8	0.6	15
4.0	0.6	15

Rapid-Mix Basin[a] — 0.0016 Mgal (Total)
Slow-Mix Basin[a] — 0.0404 Mgal (Total)

[a]Flocculation Basins.
Note: The volumes in million gallons (Mgal) and plant flow rates (Mgal/d) are used to calculate the mixing or detention times.

Results of Test Sequence No. 1

Jar No.	1	2	3	4	5	6
Alum, mg/L	10	10	10	10	10	10
Cationic Polymer, mg/L	0.5	1.0	1.5	2.0	2.5	3.0
Settled Water Turbidity, TU	0.8	0.4	0.2	0.3	0.5	0.9

Date: 1/1/16	Time: 8:00 am	Operator: JJ

Raw Water Data		
Source: Intake Pump #2		
Temperature:	21°C	pH:	7.8
Turbidity:	16 NTU	Color:	240 CU
Alkalinity:	62 mg/L	Hardness:	73 mg/L

Mixing Sequence	
RPM	Time
120	10s
80	10m
20	20m
Settling:	15m

		Jars					
		1	2	3	4	5	6
Chemicals (mg/L)	Alum Liquid/Dry	10	10	10	10	10	10
	Polymer	0.5	1.0	1.5	2.0	2.5	3.0
	Other: _____						
Floc Characteristics	Time After Rapid Mix						
	0 m	Small Sparse	Medium Dense	Medium Sparse	Large Sparse	Large Sparse	Medium Sparse
	10 m	Small Dense	Large Dense	Large Sparse	Large Sparse	Large Dense	Medium Dense
	30 m	Medium Dense	Large Dense	Very Large Dense	Very Large Dense	Large Dense	Medium Dense
Floc Settling	0 m	Fair	Good	Good	Good	Good	Fair
	10 m	Fair	Good	Good	Excellent	Good	Good
	30 m	Good	Good	Excellent	Excellent	Good	Good
Settled Water Quality	Turbidity	0.8	0.4	0.2	0.3	0.5	0.9
	pH	7.1	7.2	7.2	7.2	7.2	7.2
	Color	19	18.5	16.5	17	18	0

Figure 3.11 Jar test data

Plot the settled water turbidity, in nephelometric turbidity units (TU), against the cationic polymer dosage, in mg/L, as shown in Figure 3.12. The results indicate that the lowest turbidity was produced by an alum dose of 10 mg/L and a cationic polymer dose of 1.5 mg/L. The turbidity increases when too much polymer is added, as shown in jars 4, 5, and 6.

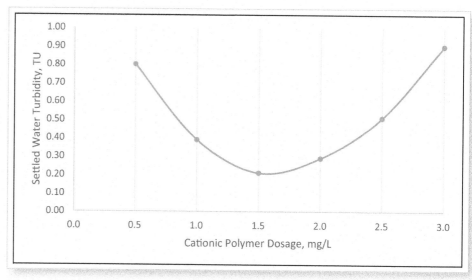

Figure 3.12   Plot of settled water turbidity vs. cationic polymer dosage

EVALUATION OF TEST RESULTS

Several factors are important in evaluating jar test results. These factors include:

1. Rate of floc formation
2. Type of floc particles
3. Clarity of water between floc particles
4. Size of floc
5. Amount of floc formed
6. Floc settling rate
7. Clarity of water (supernatant) above settled floc

Visible floc formation should begin shortly after the flash-mix portion of the jar test. During flocculation mixing, a number of small particles will gradually clump together to form larger particles. Floc particles that are discrete (separate) and fairly dense in appearance are usually better than floc particles that have a light, fluffy appearance. Large floc is impressive but it is neither necessary nor always desirable. Large, light floc does not settle as well as smaller, denser floc, and it is more subject to shearing (breaking up by paddles and water turbulence).

The quantity of floc formed is not as critical as floc quality or clarity of the settled water produced. Ideally, the water between the floc particles should be clear and not hazy or milky in appearance. The best chemical dosage is one that produces a finished water that meets drinking water standards at the lowest cost. Another important consideration is the amount of **sludge** produced. Smaller amounts of sludge are desirable to reduce sludge handling and disposal requirements. Most of the sludge volume consists of precipitates of the added chemicals rather than suspended solids (turbidity) removed.

The rate at which the floc settles after mixing has stopped is another important consideration. The floc should start to settle as soon as the mixer is turned off, and should be almost completely settled (80 to 90 percent) after about 15 minutes. Floc that remains suspended longer than 15 to 20 minutes in the jar test is not likely to settle out in the sedimentation basin, and will increase the load on the filter.

sludge (sluj)
(1) The settleable solids separated from liquids during processing. (2) The deposits of foreign materials on the bottoms of streams or other bodies of water or on the bottoms and edges of wastewater collection lines and appurtenances. Also referred to as biosolids. However, biosolids typically refers to treated waste.

If the floc starts to settle before mixing is completed, or more than about 80 percent of the floc has settled within 1 or 2 minutes after mixing has stopped, the floc is too heavy. In your water treatment plant, this can result in the floc settling out in the flocculation basins rather than in the sedimentation basins. (Fortunately, this is a rather rare occurrence. It indicates that too much chemical has been added.)

FREQUENCY OF PERFORMING TESTS

There is no substitute for experience in evaluating jar test data. Therefore, we recommend that jar tests be performed regularly during periods of high raw water turbidity, even if the plant is producing good quality finished water at the time. This will provide a basis for comparing coagulation-flocculation effectiveness under different conditions and allow fine-tuning of the chemical treatment to achieve the best efficiency. Jar tests of flash-mixer water samples (see "Applying Jar Test Results," page 145) should be performed at the start of every shift. Testing should be done more frequently during periods of high raw water turbidity or rapidly changing turbidity (such as after a storm). The results of these tests may give an early warning of impending (coming) treatment process problems.

Always verify the effectiveness of a change in treatment made on the basis of jar test results. To verify jar test results with treatment plant performance, obtain a water sample just downstream from the flash mixer. Collect the sample after sufficient time has passed for the treatment change to take effect. This sample should be mixed by the jar test apparatus and sampled under the same conditions as the original raw water sample.

"Conduct and Uses of Jar Tests." Herbert E. Hudson, Jr., and E. G. Wagner. *Journal of American Water Works Association,* Volume 73, No. 4, pages 218–222. April 1981.

APPLYING JAR TEST RESULTS

After evaluating the jar test results, apply the dosage used to achieve the best jar test results to your water treatment plant operation.

FORMULAS

The settings on dry chemical feeders differ from liquid chemical feeder settings.

1. Dry chemical feeders are often set on the basis of pounds of chemical fed per day.

$$\text{Feeder Setting}\left[\frac{\text{lb}}{\text{d}}\right] = \text{Flow Rate}\,(\text{Q})\left[\frac{\text{Mgal}}{\text{d}}\right]$$

$$\times \text{Alum}\left[\frac{\text{mg}}{\text{L}}\right] \times \frac{3.785\,\text{L}}{\text{gal}} \times \frac{10^6}{\text{M}} \times \frac{\text{kg}}{10^6\,\text{mg}} \times \frac{2.205\,\text{lb}}{\text{kg}}$$

This formula can also be used if the alum dose is in pounds of alum per million pounds of water because 1 liter weighs 1 million milligrams. Use the following conversion:

$$\text{Feeder Setting}\left[\frac{\text{lb}}{\text{d}}\right]$$

$$= \text{Q}\left[\frac{\text{Mgal}}{\text{d}}\right] \times \text{Alum}\left[\frac{\text{mg}}{\text{L}}\right] \times \frac{3.785\,\text{L}}{\text{gal}} \times \frac{10^6}{\text{M}} \times \frac{\text{kg}}{10^6\,\text{mg}} \times \frac{2.205\,\text{lb}}{\text{kg}}$$

$$\text{Feeder Setting}\left[\frac{\text{lb}}{\text{d}}\right] = Q\left[\frac{\text{Mgal}}{\text{d}}\right] \times \text{Alum}\left[\frac{\text{mg}}{\text{Mmg}}\right]$$

$$\times \frac{3.785\,\text{L}}{\text{gal}} \times \frac{10^6}{\text{M}} \times \frac{1\,\text{Million mg}}{\text{L}} \times \frac{\text{kg}}{10^6\,\text{mg}} \times \frac{2.205\,\text{lb}}{\text{kg}}$$

2. Liquid chemical feeders are often set on the basis of milliliters of chemical solution delivered per minute.

$$\text{Chemical Feeder Setting}\left[\frac{\text{mL}}{\text{min}}\right]$$

$$= \frac{Q\left[\dfrac{\text{Mgal}}{\text{d}}\right] \times \text{Alum}\left[\dfrac{\text{mg}}{\text{L}}\right] \times \dfrac{3.785\,\text{L}}{\text{gal}} \times \dfrac{10^6}{\text{M}} \times \dfrac{\text{d}}{24\,\text{h}} \times \dfrac{\text{h}}{60\,\text{min}}}{\text{Liquid Alum}\left[\dfrac{\text{mg}}{\text{mL}}\right]}$$

Multiply flow by dosage to get the amount of chemical needed per unit of time. The gallons in the flow and the liters in the dose were canceled out by multiplying by 3.785 L/gal. The concentration of liquid alum in mg/mL converted the amount of chemical (alum) needed from one amount in milligrams to a volume in milliliters. The flow in Mgal/d is the amount of chemical needed on a daily basis, so divide that by 24 h/d and 60 min/h to convert to the amount needed per minute. The 1,000,000 over 1 million takes care of the million in Mgal/d.

Liquid alum may be delivered to plants as 8.0 to 8.5 percent Al_2O_3 and contains about 5.36 pounds of dry aluminum sulfate ($Al_2(SO_4)_3 \bullet$ 14.3 H_2O; 48.5 percent dry) per gallon (**specific gravity** 1.325). This converts to 643,000 mg/L, 643 g/L, or 643 mg/mL.

3. To determine a liquid chemical feeder setting in gallons per day (gpd) use the following formula:

$$\text{Chemical Feeder Setting}\left[\frac{\text{gal}}{\text{d}}\right]$$

$$= \frac{Q\left[\dfrac{\text{Mgal}}{\text{d}}\right] \times \text{Alum}\left[\dfrac{\text{mg}}{\text{L}}\right] \times \dfrac{3.785\,\text{L}}{\text{gal}} \times \dfrac{10^6}{\text{M}} \times \dfrac{\text{kg}}{10^6\,\text{mg}} \times \dfrac{2.205\,\text{lb}}{\text{kg}}}{\text{Liquid Alum}\left[\dfrac{\text{lb}}{\text{gal}}\right]}$$

specific gravity
(1) Weight of a particle, substance, or chemical solution in relation to the weight of an equal volume of water. Water has a specific gravity of 1.000 at 39°F (4°C). Particulates with specific gravity less than 1.0 float to the surface and particulates with specific gravity greater than 1.0 sink. (2) Weight of a particular gas in relation to the weight of an equal volume of air at the same temperature and pressure (air has a specific gravity of 1.0). Chlorine has a specific gravity of 2.5 as a gas.

As the first formula in this section shows, when flow in Mgal/d is multiplied by dose in mg/L and the conversion equation, the result is pounds per day. The necessary gallons of liquid alum per day is calculated by dividing the pounds of alum in a gallon of liquid alum into this number.

Example 3

The best dry alum dose from the jar test is 10 mg/L. Determine the setting on the alum feeder in pounds per day (lb/d) when the flow is 3.0 Mgal/d.

Known

D_{alum} = Alum Dose = 10 mg/L

Q = Flow Rate = 3.0 Mgal/d

Unknown

Feeder Setting, lb/d

Calculate the alum feeder setting in pounds per day (lb/d).

$$\text{Feeder Setting}\left[\frac{\text{lb}}{\text{d}}\right]$$

$$= Q\left[\frac{\text{Mgal}}{\text{d}}\right] \times D_{alum}\left[\frac{\text{mg}}{\text{L}}\right] \times \frac{3.785\,\text{L}}{\text{gal}} \times \frac{10^6}{\text{M}} \times \frac{\text{kg}}{10^6\,\text{mg}} \times \frac{2.205\,\text{lb}}{\text{kg}}$$

$$= 3.0\left[\frac{\text{Mgal}}{\text{d}}\right] \times 10\left[\frac{\text{mg}}{\text{L}}\right] \times \frac{3.785\,\text{L}}{\text{gal}} \times \frac{10^6}{\text{M}} \times \frac{\text{kg}}{10^6\,\text{mg}} \times \frac{2.205\,\text{lb}}{\text{kg}}$$

$$= 250\,\frac{\text{lb}}{\text{d}}$$

Set the alum feeder to dose alum at a rate of 250 pounds of alum per day.* Also set the chemical feeder for the cationic polymer level obtained from the jar test results.

*From the 250 lb/d figure, you may easily calculate the lb/hr equivalent or whatever other weight/time feed rate for which your equipment is calibrated.

$$\text{Feeder setting, lb/hr} = \frac{250\left[\frac{\text{lb}}{\text{d}}\right]}{24\left[\frac{\text{h}}{\text{d}}\right]} = 10.4\,\frac{\text{lb}}{\text{h}}$$

Example 4

The optimum liquid alum dose from the jar tests is 10 mg/L. Determine the setting on the liquid alum chemical feeder in milliliters per minute (mL/min) when the flow is 3.0 Mgal/d. The liquid alum delivered to the plant contains 643 milligrams (mg) of alum per milliliter (mL) of liquid solution.

Known

D_{alum} = Alum Dose = 10 mg/L

Q = Flow Rate = 3.0 Mgal/d

Liquid Alum Content = 643 mg/mL

Unknown

Chemical Feeder Setting, mL/min

Calculate the liquid alum chemical feeder setting in milliliters per minute (mL/min).

$$\text{Chemical Feeder Setting}\left[\frac{\text{mL}}{\text{min}}\right]$$

$$= \frac{Q\left[\frac{\text{Mgal}}{\text{d}}\right] \times D_{alum}\left[\frac{\text{mg}}{\text{L}}\right] \times \frac{3.785\,\text{L}}{\text{gal}} \times \frac{10^6}{\text{M}} \times \frac{\text{d}}{24\,\text{h}} \times \frac{\text{h}}{60\,\text{min}}}{\text{Liquid Alum}\left[\frac{\text{mg}}{\text{mL}}\right]}$$

$$= \frac{3.0\left[\frac{\text{Mgal}}{\text{d}}\right] \times 10\left[\frac{\text{mg}}{\text{L}}\right] \times \frac{3.785\,\text{L}}{\text{gal}} \times \frac{10^6}{\text{M}} \times \frac{\text{d}}{24\,\text{h}} \times \frac{\text{h}}{60\,\text{min}}}{643\left[\frac{\text{mg}}{\text{mL}}\right]}$$

$$= 123\,\frac{\text{mL}}{\text{min}}$$

Example 5

The optimum liquid alum dose from the jar tests is 10 mg/L. Determine the setting on the liquid alum chemical feeder in gallons per day (gpd) when the flow is 3.0 Mgal/d. The liquid alum delivered to the plant contains 5.36 pounds (lb) of alum per gallon (gal) of liquid solution.

		Known		**Unknown**
D_{alum}	=	Alum Dose	= 10 mg/L	Chemical Feeder Setting, gpd
Q	=	Flow Rate	= 3.0 Mgal/d	
		Liquid Alum Content	= 5.36 lb/gal	

Calculate the liquid alum chemical feeder setting in gallons per day (gpd).

$$\text{Chemical Feeder Setting} \left[\frac{\text{gal}}{\text{d}} \right]$$

$$= \frac{Q \left[\frac{\text{Mgal}}{\text{d}} \right] \times D_{alum} \left[\frac{\text{mg}}{\text{L}} \right] \times \frac{3.785\,\text{L}}{\text{gal}} \times \frac{10^6}{\text{M}} \times \frac{\text{kg}}{10^6\,\text{mg}} \times \frac{2.205\,\text{lb}}{\text{kg}}}{\text{Liquid Alum} \left[\frac{\text{lb}}{\text{gal}} \right]}$$

$$= \frac{3.0 \left[\frac{\text{Mgal}}{\text{d}} \right] \times 10 \left[\frac{\text{mg}}{\text{L}} \right] \times \frac{3.785\,\text{L}}{\text{gal}} \times \frac{10^6}{\text{M}} \times \frac{\text{kg}}{10^6\,\text{mg}} \times \frac{2.205\,\text{lb}}{\text{kg}}}{5.36 \left[\frac{\text{lb}}{\text{gal}} \right]}$$

$$= 47 \text{ gpd}$$

CHECKING PLANT AGAINST JAR TEST

As noted above, it is always wise to check the effectiveness of process changes based on jar test results. Here is an example procedure.

When the alum and cationic polymer feeders are working properly, collect a sample of well-mixed water from the effluent of the flash-mix chamber to fine-tune process performance. Take the sample to the lab and perform another jar test. First, test the alum dosage to determine whether the alum dosage is too high or too low.

1. Fill two jars with the flash-mixed sample. (The second jar is just a check on the first in case the sample is not representative for some reason.) Because the chemicals have already been mixed in the sample water, floc will start to form. Be sure the sample stays mixed while pouring off the volumes into the six jars.

2. Mix the samples at the speed and time used in the original test.

3. Stop the stirrers. Allow floc to settle for 30 minutes or the time used in the original test. When the stirrers are stopped, immediately collect a sample from the flocculation basin effluent. Fill an experimental jar for comparison with other jars.

4. In all the jars, observe how quickly the floc settles, floc appearance, and turbidity of settled water above the floc. They should be similar.

5. Evaluate these jar test results (see Section 3.6.2.2, "Evaluation of Test Results"), and make further process adjustments as appropriate.

3.6.2.3 Streaming Current Meters

Streaming current meters are devices used by operators to optimize coagulant doses. The streaming current meter is a continuous in-process measuring instrument. Properly used, the streaming current meter can function as an in-process jar test.

Most particles in water are anions (negative charge) and most coagulants are cations (positive charge). The streaming current meter presumes that by bringing the total charge of the water being treated to neutral (zero, 0), the coagulation process has been optimized. Most operators run the charge of their water slightly negative by adjusting the coagulant dose.

3.6.2.4 Evaluation of Plant Performance

One of the best ways to evaluate the performance of your coagulation-flocculation process is to observe the process in the plant. When you walk through the treatment plant, take some clear plastic beakers. Dip some water out of each stage of the treatment process. Hold the sample up to a light and look at the clarity of the water between the floc and study the shape and size of the floc. Study the development of the floc from one flocculation chamber to the next and into the sedimentation basin.

1. Observe the floc as it enters the flocculation basins. The floc should be small and well-dispersed throughout the flow. If not, the flash mixer may not be providing effective mixing or the chemical dose or feed rate may be too low.

2. Tiny alum floc may be an indication that the chemical coagulant dose is too low. A "popcorn flake" is a desirable floc appearance. If the water has a milky appearance or a bluish tint, the alum dose is probably too high.

3. As it moves through the flocculation basins, the size of the floc should be increasing. If the floc size increases and then later starts to break up, the mixing intensity of the downstream flocculators may be too high. Try reducing the speed of these flocculators, or increasing the polymer dosage.

4. Look for the floc settling out in the sedimentation basin. If a lot of floc is observed flowing over the **laundering weirs**, the floc is too light to settle during the detention time produced by that flow rate. By increasing the chemical coagulant dose or adding a coagulant aid such as a polymer, a heavier, larger floc may be produced. The appearance of fine floc particles washing over the effluent weirs could be an indication of too much alum and the dose should be reduced. Regardless of the problem, make only one change at a time and evaluate the results. This topic will be discussed more in Chapter 4, "Sedimentation."

5. Bring some beakers with samples from various locations back to the laboratory, let them sit for awhile, and then observe the floc settling.

6. How are the filters performing? This topic will be discussed more in Chapter 5, "Filtration."

laundering weir (LAWN-der-ing weer)
Sedimentation basin overflow weir. A plate with V-notches along the top to ensure a uniform flow rate and avoid short-circuiting.

Check Your Understanding

1. Estimate the detention time (seconds) in a flash-mix chamber 6 feet long, 4 feet wide, and 5 feet deep when the rate of flow is 10 Mgal/d.

2. Why do many operators prepare flow vs. detention time curves for basins in their plant?

3. What is the goal in setting up a jar test?

4. Why should samples for jar tests be at about the same temperature as water in the plant?

5. Why should only one variable at a time be changed when running a jar test?

6. What factors should be considered when evaluating the results of jar tests?

3.6.3   Chemical Usage for Small Plants

This section contains math problems that the operator of a small treatment plant might be expected to solve. Regardless of the size of your water treatment plant, the procedures for solving these problems are similar.

3.6.3.1   Calculating Amount of Chemical Required

Operators must be able to estimate the amounts of chemicals used on a monthly basis. Always order enough chemicals in advance so your plant will not run out.

Example 6

A water treatment plant treats an average daily flow of 300,000 gpd. Results from a jar test indicate that the desired polymer dosage for coagulation is 2 mg/L. How many pounds (lb) of polymer will be used in 30 days (d)?

		Known		**Unknown**
Q	=	Flow Rate =	300,000 gpd	Polymer Used, lb/d
		=	0.3 Mgal/d	Polymer Used, lb
$D_{polymer}$	=	Polymer Dose = 2 mg/L		
t	=	Time =	30 d	

1. Calculate the polymer used in pounds per day (lb/d).

$$\text{Polymer Used}\left[\frac{lb}{d}\right]$$

$$= Q\left[\frac{Mgal}{d}\right] \times D_{polymer}\left[\frac{mg}{L}\right] \times \frac{3.785\,L}{gal} \times \frac{10^6}{M} \times \frac{kg}{10^6\,mg} \times \frac{2.205\,lb}{kg}$$

$$= 0.3\left[\frac{Mgal}{d}\right] \times 2\left[\frac{mg}{L}\right] \times \frac{3.785\,L}{gal} \times \frac{10^6}{M} \times \frac{kg}{10^6\,mg} \times \frac{2.205\,lb}{kg}$$

$$= 5\frac{lb}{d}$$

2. Estimate the polymer used in 30 days (d).

$$\text{Polymer Used in 30 Days [lb]} = \text{Polymer Used}\left[\frac{lb}{d}\right] \times t[d]$$

$$= 5\left[\frac{lb}{d}\right] \times 30[d]$$

$$= 150\ lb$$

3.6.3.2 Chemical Feeding

The chemical feed rate delivered by chemical feeders must be checked regularly. Jar tests will show the best dosages of chemicals in mg/L. To check on the feed rate delivered by a chemical feeder, measure the volume (in gallons for a liquid chemical feeder) or the weight (in pounds for a dry chemical feeder) delivered during a 24-hour period. The flow during this time period also needs to be known and is recorded in gallons per day (gpd) or million gallons per day (Mgal/d).

Example 7

During a 24-hour period, a plant treated a flow of 1.5 million gallons (Mgal). Ten pounds (lb) of cationic polymer were used in the coagulation process. What is the polymer dosage in mg/L?

Known	Unknown
Q = Flow Rate = 1.5 Mgal/d	$D_{polymer}$ = Polymer Dose, mg/L
Polymer Used = 10 lb/d	

Calculate the polymer dose in mg/L. The basic equation is:

$$Polymer\ Used\left[\frac{lb}{d}\right] = Q\left[\frac{Mgal}{d}\right] \times D_{polymer}\left[\frac{mg}{L}\right] \times \frac{3.785\,L}{gal} \times \frac{10^6}{M} \times \frac{kg}{10^6\ mg} \times \frac{2.205\,lb}{kg}$$

Rearrange the equation to:

$$D_{polymer}\left[\frac{mg}{L}\right] = \frac{Polymer\ Used\left[\frac{lb}{d}\right]}{Q\left[\frac{Mgal}{d}\right] \times \frac{3.785\,L}{gal} \times \frac{10^6}{M} \times \frac{kg}{10^6\ mg} \times \frac{2.205\,lb}{kg}}$$

$$= \frac{10\left[\frac{lb}{d}\right]}{1.5\left[\frac{Mgal}{d}\right] \times \frac{3.785\,L}{gal} \times \frac{10^6}{M} \times \frac{kg}{10^6\ mg} \times \frac{2.205\,lb}{kg}}$$

$$= 0.8\,\frac{mg}{L}$$

This polymer dose of 0.8 mg/L should be very close to the best dose obtained from the jar test. Also, walk through your plant and observe how the water you are treating responds to a polymer dose of 0.8 mg/L. If the floc is not settling properly, the polymer dose may have to be increased or decreased.

Example 8

What is the alum dosage in mg/L if a water treatment plant treats a source water with an average flow of 700 gpm and an average turbidity of 10 TU? A total weight of 150 pounds (lb) of alum is used during a 24-hour (h) period.

	Known		Unknown
Q = Flow Rate	=	700 gpm	D_{alum} = Alum Dose, mg/L
Alum Used	=	150 lb/d	

1. Convert the flow of 700 gallons per minute (gpm) to million gallons per day (Mgal/d).

$$Q\left[\frac{Mgal}{d}\right] = Q\left[\frac{gal}{min}\right] \times \frac{60\,min}{h} \times \frac{24\,h}{d} \times \frac{M}{10^6}$$

$$= 700\left[\frac{gal}{min}\right] \times \frac{60\,min}{h} \times \frac{24\,h}{d} \times \frac{M}{10^6}$$

$$= 1.0\ Mgal/d$$

2. Calculate the alum dose in mg/L.

$$D_{alum}\left[\frac{mg}{L}\right] = \frac{Alum\,Used\left[\dfrac{lb}{d}\right]}{Q\left[\dfrac{Mgal}{d}\right] \times \dfrac{3.785\,L}{gal} \times \dfrac{10^6}{M} \times \dfrac{kg}{10^6\,mg} \times \dfrac{2.205\,lb}{kg}}$$

$$= \frac{150\left[\dfrac{lb}{d}\right]}{1.0\left[\dfrac{Mgal}{d}\right] \times \dfrac{3.785\,L}{gal} \times \dfrac{10^6}{M} \times \dfrac{kg}{10^6\,mg} \times \dfrac{2.205\,lb}{kg}}$$

$$= 18.0\,\frac{mg}{L}$$

Note: For an example of how to calculate the lime dosage to provide the alkalinity when using alum, see Chapter 4, Section 4.5.3, "Calculations."

3.6.3.3   Preparation of Chemical Solutions

Polymers are frequently used as coagulant aids. Polymers are often supplied as dry chemicals, but operators usually need a specific solution concentration (as mg/L or as a percent solution). The solution concentration depends on the type of polymer (cationic, nonionic, or anionic) and the polymer's molecular weight (this weight may vary from 100 to 10 million). The higher the molecular weight of the polymer, the more difficult it is to mix the polymer with dilution water and to feed the resulting solution to the water being treated; the solution becomes very viscous (thick). Therefore, anionic and nonionic dry polymers used as coagulant aids are often prepared as very dilute (weak) solutions (0.25 to 1.0 percent). Cationic polymers in the dry form can be prepared at higher solution concentrations (say 5 to 10 percent) because their molecular weights are typically small.

High molecular weight polymers are very difficult to prepare. To be effective, polymer solutions must be the same throughout (homogeneous). They must be thin enough to be accurately measured and pumped to the flash-mix chamber.

When mixing a dry polymer with water, sift or spread the polymer evenly over the surface of the water in the mixing chamber. The polymer should be sucked evenly into the hole (vortex) of the stirred water. This will ensure that each particle of polymer is wet individually. This will also ensure even dispersion and prevent the formation of large, sticky balls of polymer that have dry polymer in the middle.

Excessive mixing speeds, mixing time, and the buildup of heat can break down the polymer chain and reduce its effectiveness.

High concentrations of polymer result in very thick, sticky solutions. Prepare and use concentrations of polymers that can be metered (measured) easily and pumped accurately to the flash mixer.

Adding dry polymer to water must be done in a closed system or under an efficient dust collector. Polymer powders on floors and walkways become extremely slippery when wet and are very difficult to remove. In the interest of safety, keep polymers off the floors. Use an inert, absorbent material, such as sand or earth, to clean up spills.

FORMULAS

When mixing either dry or liquid polymers, always follow the directions of the polymer supplier. Polymer solutions are usually prepared in batches or as a batch mixture. Often they are stored after mixing in a day tank or an aging tank to allow time for all the powder to dissolve or the solution to become completely mixed.

DRY POLYMERS. To prepare a specific percent polymer solution, use the following formula:

$$\text{Polymer}[\%] = \frac{\text{Dry Polymer}[\text{lb}] \times 100\%}{\text{Dry Polymer}[\text{lb}] + \text{Water}[\text{lb}]}$$

Examining the above formula reveals three possible unknowns. In addition to wanting to know the percent polymer solution, the operator might wish to know the pounds of dry polymer to use or the amount of water in either pounds or probably gallons to mix with the polymer. In any case, two of the three unknowns must be known to solve the problem. The above basic polymer formula to find the polymer concentration as a percent can be rearranged to solve for the other two unknowns.

$$\text{Dry Polymer}[\text{lb}] = \frac{\text{Water}[\text{lb}]}{\dfrac{100\%}{\text{Polymer}[\%]} - 1.0}$$

$$\text{Water}[\text{lb}] = \frac{\text{Dry Polymer}[\text{lb}] \times 100\%}{\text{Polymer}[\%]} - \text{Dry Polymer}[\text{lb}]$$

$$\text{Water}[\text{gal}] = \text{Water}[\text{lb}] \times \frac{\text{gal}}{8.34\,\text{lb}}$$

LIQUID POLYMERS. When working with liquid polymers, usually the supplier provides the percent polymer in the liquid polymer. The problem is to determine how much of the supplier's polymer should be mixed with water to produce a tank or barrel with a diluted or lower percent polymer.

The basic liquid polymer formula is as follows:

$$P_{polymer} \times V_{polymer} = P_{solution} \times V_{solution}$$

Where,

$P_{polymer}$ = Percent Polymer Provided by the Supplier
$V_{polymer}$ = Volume of the Polymer from the Supplier
$P_{solution}$ = Percent Polymer Needed
$V_{solution}$ = Volume of the Polymer After Adding Water to the Solution from the Supplier

$$P_{solution} = \frac{P_{polymer} \times V_{polymer}}{V_{solution}}$$

or

$$V_{solution} = \frac{P_{polymer} \times V_{polymer}}{P_{solution}}$$

Example 9

At the beginning of each shift, or each day, the operator of a small plant mixes a batch of nonionic polymer solution. What is the percent (by weight) polymer solution if the operator adds 0.132 pounds (lb) (60 grams) of dry polymer to 5 gallons (gal) of water?

Known	Unknown
$M_{polymer}$ = Mass of Dry Polymer = 0.132 lb	Polymer Solution in percent (%)
V_{water} = Volume of Water = 5 gal	

1. Convert the 5 gallons (gal) of water to pounds (lb).

$$V_{water}\,[lb] = V_{water}\,[gal] \times \frac{8.34\,lb}{gal}$$

$$= 5\,[gal] \times \frac{8.34\,lb}{gal}$$

$$= 41.7\,lb$$

2. Calculate the polymer solution as a percent (%).

$$Polymer\,[\%] = \frac{M_{polymer}\,[lb] \times 100\%}{M_{polymer}\,[lb] + Water\,[lb]}$$

$$= \frac{0.132\,[lb] \times 100\%}{0.132\,[lb] + 41.7\,[lb]}$$

$$= 0.32\%$$

Example 10

How many pounds (lb) of dry polymer must be added to 50 gallons (417 pounds) of water to produce a 0.5 percent (%) polymer solution?

Known

V_{water} = Water Volume = 50 gal

M_{water} = Water Weight = 417 lb

Polymer Content = 0.5%

Unknown

$M_{polymer}$ = Mass of Dry Polymer, lb

Calculate the pounds of dry polymer that must be added to 50 gallons of water.

$$M_{polymer}[lb] = \dfrac{M_{water}[lb]}{\dfrac{100\%}{Polymer[\%]} - 1.0}$$

$$= \dfrac{417[lb]}{\dfrac{100\%}{0.5\%} - 1.0}$$

$$= 2.1 \text{ lb*}$$

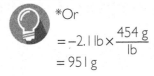

*Or

$$= -2.1\,lb \times \dfrac{454\,g}{lb}$$

$$= 951\,g$$

Example 11

How many gallons (gal) of water should be mixed with 1.5 pounds (lb) of dry polymer to produce a 0.5 percent polymer solution?

Known

Polymer Content = 0.5%

$M_{polymer}$ = Mass of Dry Polymer = 1.5 lb

Unknown

V_{water} = Volume of Water, gal

1. Calculate the pounds (lb) of water needed.

$$Water[lb] = \dfrac{M_{polymer}[lb] \times 100\%}{Polymer[\%]} - M_{polymer}[lb]$$

$$= \dfrac{1.5[lb] \times 100\%}{0.5\%} - 1.5[lb]$$

$$= 300\,[lb] - 1.5\,[lb]$$

$$= 298.5\,lb$$

2. Calculate the volume of water in gallons (gal) to be mixed with the polymer.

$$V_{water}[gal] = Water[lb] \times \dfrac{gal}{8.34\,lb}$$

$$= 298.5[lb] \times \dfrac{gal}{8.34\,lb}$$

$$= 35.8\,gal$$

Example 12

Liquid polymer is supplied to a water treatment plant as a 10 percent solution. How many gallons (gal) of liquid polymer should be mixed in a barrel with water to produce 50 gallons (gal) of 0.5 percent polymer solution?

	Known		**Unknown**
$P_{polymer}$	= Percent Polymer Provided = 0.5% by the Supplier	$V_{solution}$	= Volume of the Polymer After
$V_{polymer}$	= Volume of the Polymer = 50 gal from the Supplier		Adding Water to the Solution from
$P_{solution}$	= Percent Polymer Needed = 10%		the Supplier

Calculate the volume of liquid polymer in gallons (gal).

$$V_{solution} = \frac{P_{polymer} \times V_{polymer}}{P_{solution}} = \frac{0.5\% \times 50\,[gal]}{10\%} = 2.5\,gal$$

The volume of the polymer solution is 50 gallons. This means that we have 2.5 gallons of liquid polymer and 47.5 gallons (50 − 2.5 = 47.5) of water mixed together.

3.6.4   Recordkeeping

One of the most important administrative functions of the water treatment plant operator is the preparation and maintenance of accurate plant operation records.

In the routine operation of the coagulation-flocculation process, the operator usually makes entries in and maintains records such as daily operations logs and diaries. These records should provide an accurate day-to-day account of operating experience.

Accurate records provide the operator, as well as others, with a running account of operations (historical records), and are a great help to the operator in solving current or future process problems. They also provide a factual account of the operation, which is required by law and by regulatory agencies.

In the coagulation-flocculation process, you should keep records of the following items:

1. Source water quality (pH, turbidity, temperature, alkalinity, chlorine demand, and color)
2. Process water quality (pH, turbidity, and alkalinity)
3. Process production inventories (chemicals used, chemical feed rates, amount of water processed, and amount of chemicals in storage)
4. Process equipment performance (types of equipment in operation, maintenance procedures performed, equipment calibration and adjustments)

Record entries should be neat, legible, and easily found. Keep in mind that many records will be kept and may be used for long periods of time. Entries should reflect the date and time of an event and should be initialed by the operator making the entry for future identification.

Operators should maintain a plot of key process variables. A plot of source water turbidity vs. coagulant dosage should be maintained. If other process variables, such as alkalinity or pH, vary significantly, these should also be plotted.

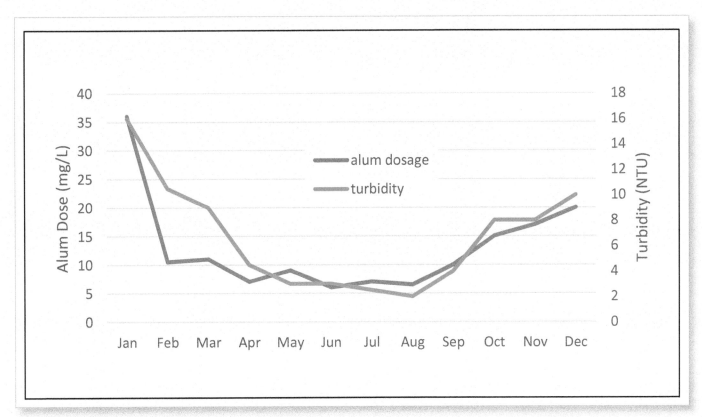

Figure 3.13   Monthly variations in turbidity and alum dosage

These graphs should be designed so that the operator can see one year of data at one time. This will give the operator a better understanding of the seasonal variation in these water quality indicators. A sample graph showing a plot of source water turbidity vs. alum dosage is given in Figure 3.13.

3.6.5   Safety

In the routine operation of the coagulation-flocculation process, the operator will be exposed to a number of potential hazards such as:

1. Electrical equipment
2. Rotating mechanical equipment
3. Water treatment chemicals
4. Laboratory reagents (chemicals)
5. Slippery surfaces caused by wet polymers
6. Open-surface, water-filled structures (drowning)
7. Confined spaces and underground structures, such as valve or pump vaults (toxic and explosive gases, insufficient oxygen)

Accidents do not just happen, they are caused. Therefore, strict and constant attention to safety procedures cannot be overemphasized.

The operator should be familiar with general first-aid practices such as hands-only cardiopulmonary resuscitation (CPR), treatment of common physical injuries, and first aid for chemical exposure (chlorine). See the chapter on safety in Volume II of this series of training manuals.

3.6.6 Communications

Good communications are an essential part of the operator's job. Clear and concise written or oral communications are necessary to advise other operators and support personnel of current process conditions and unique or unusual events. In this regard, good recordkeeping is an essential element of the communication process.

Check Your Understanding

1. Why are anionic and nonionic dry polymers used as coagulant aids often prepared as dilute solutions?

2. Why should operators keep accurate records?

3. What information should be recorded in plant operation records?

4. List the safety hazards an operator may encounter when operating a coagulation-flocculation process.

5. Why are good communications such an essential part of an operator's job?

3.7 Abnormal Conditions

Sudden changes in the source water or filtered water turbidity, pH, alkalinity, temperature, or chlorine demand are signals that the operator should immediately review the performance of the coagulation-flocculation process.

3.7.1 Process Actions

Changes in source water turbidity levels, either increases or decreases, generally require that the operator verify the effectiveness of the coagulant chemicals and dosages being applied at the flash mixer. This is best accomplished by performing a series of jar tests in the laboratory as discussed previously. Remember that decreasing raw water turbidity levels can be just as upsetting to the process and difficult to treat as increasing levels.

Visual observations of flash-mixing intensity as well as the condition of the floc in the flocculation basins may also indicate the need for process changes such as adjustment of mixer speed or coagulant dosage.

Alkalinity, pH, and temperature changes in the source water may have an impact on the clumping together of floc during the coagulation-flocculation process. In addition, water temperature changes may require an adjustment in the level of mixing intensity in flash mixers or flocculators. Temperature changes are usually gradual over time, thus large process adjustments are seldom necessary.

Sudden increases in filtered water turbidity could be caused by poor filter performance (need for backwashing or replacing filter media). However, poor coagulation-flocculation performance is usually the culprit, and the operator must take immediate action to correct the problem, remembering that several hours may pass before changes in the operation of the coagulation-flocculation process are seen in the filter effluent. One quick remedy may be to feed a filter-aid chemical, such as a nonionic polymer, directly to the filter influent. While this may solve the short-term problem, only changes in the coagulation-flocculation process will enhance long-term plant performance. Again, the results of laboratory jar tests should be used as the basis for making process changes. Filter-aid chemicals are discussed further in Chapter 6, "Filtration."

Table 3.3 is a summary of coagulation-flocculation process problems, how to identify the causes of these problems, and how to go about trying to correct the problems.

Table 3.3 Coagulation-flocculation process troubleshooting

Source Water Quality Changes	Operator Actions	Possible Process Changes
• Turbidity	1. Perform necessary analyses to determine extent of change	1. Change coagulant(s)
• Temperature	2. Evaluate overall process performance	2. Adjust coagulant dosage
• Alkalinity	3. Perform jar tests if indicated	3. Adjust flash-mixer/flocculator mixing intensity[a]
• pH	4. Make appropriate process changes (see righthand column, "Possible Process Changes")	4. Add coagulant aid or filter aid
• Color	5. Increase frequency of process monitoring	5. Adjust alkalinity or pH
	6. Verify response to process changes at appropriate time (be sure to allow sufficient time for change to take effect)	

Coagulation Process Effluent Quality Changes	Operator Actions	Possible Process Changes
• Turbidity	1. Evaluate source water quality	1. Change coagulant(s)
• Alkalinity	2. Perform jar tests if indicated	2. Adjust coagulant dosage
• pH	3. Verify process performance:	3. Adjust flash-mixer intensity (if possible)
	a. Coagulant feed rate(s)	4. Adjust alkalinity or pH
	b. Flash-mixer operation	
	4. Make appropriate process changes	
	5. Verify response to process changes at appropriate time	

Flocculation Basin Floc Quality Changes	Operator Actions	Possible Process Changes
• Floc Formation	1. Observe floc condition in basin:	1. Change coagulant
	a. Dispersion	2. Adjust coagulant dosage
	b. Size	3. Adjust flash-mixer/flocculator mixing intensity
	c. Floc strength (breakup)	4. Add coagulant aid
	2. Evaluate overall process performance	5. Adjust alkalinity or pH
	3. Perform jar tests if indicated:	
	a. Evaluate floc size, settling rate, and strength	
	b. Evaluate quality of supernatant: clarity (turbidity), pH, and color	
	4. Make appropriate process changes	
	5. Verify response to process changes at appropriate time	

[a]Note: Very few plants have provisions for adjusting the flash mixer. However, many plants have variable-speed drives on flocculators to allow for adjustment of mixing intensity.

3.7.2 Recordkeeping

During times of abnormal plant operating conditions, good recordkeeping takes on an added importance, that of documenting the unique conditions or special events and the actions taken by the operator to solve the problem. In addition to the normal logbook entry requirements, the operator must keep accurate notes describing unusual conditions and any steps taken to prevent or resolve problems. Often, a "non-result" following an attempted corrective action is as important as a correct solution to the problem. A non-result helps us and others learn from our experience and next time we can try something different. Thus, all actions should be carefully documented by the operator.

Detailed notes may be useful in the design of future water treatment plant facilities, or in the modification of existing facilities. In addition, records such as these provide a historical account of actions taken by the operator, and may be helpful from a legal or regulatory agency perspective. Special reports may need to be prepared describing the events and actions taken. The operator's notes provide the facts for these reports.

3.7.3 Communications

During abnormal periods of operation, you may be required to advise other plant personnel, such as the Senior Operator, Plant Superintendent, Chemist, or Maintenance Mechanic, of the conditions that exist or events that have occurred.

An emergency response procedure should be developed for every water treatment plant so that the proper personnel can be notified quickly and the emergency resolved. Emergency response procedures should list the names and telephone numbers (including off-hours, nights, and weekends) of persons to be notified under specific conditions, including health department authorities. Guidelines should be developed to assist the operator in determining when to use these procedures.

Coagulation-flocculation process changes or other conditions that may require notification of others include:

1. Contamination of source water with a chemical spill
2. Major changes in source or treated water quality (pH, turbidity, alkalinity, and bacteriological quality)
3. Lack of response to changes in coagulant dosage
4. Equipment failure (chlorinators, chemical feeders, mixers, and pumps)
5. Power outages

3.8 Startup and Shutdown Procedures

Startup or shutdown of the coagulation-flocculation process is not a routine operating procedure in most water treatment plants. These procedures generally happen when the plant is shut down for maintenance.

In some rare instances, a shutdown may be required due to a major equipment failure.

Typical actions performed by the operator in the startup or shutdown of the coagulation-flocculation process are outlined below. These procedures may have to be altered depending on the type of equipment in your plant and on the recommendations of the manufacturers of your equipment.

3.8.1   Startup Procedure

1. Check the condition of all mechanical equipment for proper lubrication and operational status.
2. Make sure all chemical feeders are ready. There should be plenty of chemicals available in the tanks and hoppers and ready to be fed to the raw water.
3. Collect a sample of raw water and immediately run a jar test using fresh chemicals from the supply of chemicals to the feeders.
4. Determine the settings for the chemical feeders and set the feed rates on the equipment.
5. Open the inlet gate or valve to start the raw water flowing.
6. Immediately start the selected chemical feed systems.
   a. Open valves to start feeding coagulant chemicals and dilution makeup water.
   b. Start chemical feeders.
   c. Adjust chemical feeders as necessary.
7. Turn on the flash mixer at the appropriate time. You may have to wait until the tank or channel is full before turning on the flash mixer. Follow the manufacturer's instructions.
8. Start the sample pumps as soon as there is water at each sampling location. Allow sufficient flushing time before collecting any samples.
9. Start the flocculators as soon as the first basin is full of water. Be sure to follow manufacturer's recommendations. If possible and appropriate, make necessary adjustments in the speed.
10. Inspect mixing chamber and flocculation basins. Observe formation of floc and make necessary changes.
11. Remove debris floating on the water surface.
12. Perform water quality analyses and make process adjustments as necessary.
13. Calibrate chemical feeders.

Do not allow any untreated water to flow through your plant. All raw water must be treated with alum or other appropriate coagulant. Water that has not been treated with a coagulant could flow through your filter without proper treatment (removal of color and particulates) and into your distribution system.

3.8.2   Shutdown Procedure

1. Close raw water inlet gate or valve to flash-mix chamber or channel.

2. Shut down the chemical feed systems.

   a. Turn off chemical feeders.
   b. Shut off appropriate valves.
   c. Flush or clean chemical feed lines if necessary.

3. Shut down flash mixer and flocculators as water leaves each process. Follow manufacturer's recommendations.

4. Shut down sample pumps before water leaves each sampling location.

5. Waste any water that has not been properly treated.

6. Lock out and tag appropriate electric switches.

7. Dewater basins if necessary. Waste any water that has not been properly treated. Do not dewater below-ground basins without checking groundwater level.

   a. Close basin isolation gates or install stop-logs.
   b. Open basin drain valve(s).
   c. Be careful; basin may float or collapse depending on groundwater, soil, or other conditions.

3.8.3 Recordkeeping

Good records of actions taken during startup/shutdown operations will assist the operator, as well as other plant operations and maintenance personnel, in conducting future start-ups/shutdowns. The results of all inspections, equipment adjustments, and any unusual events should be accurately recorded.

3.8.4 Safety

Safety procedures are extremely important during startup and shutdown operations. In the coagulation-flocculation process, you will be exposed to a variety of potentially hazardous situations that require the use of extreme caution. The operator may be exposed to electrical hazards; rotating and mechanical equipment; open-surface, water-filled basins (drowning); and empty basins. Also, the bottoms of empty basins are very slippery; a fall could be extremely painful and cause a serious injury. Always use safety devices such as handrails. For example, do not remove handrails to make your job easier.

In some instances, you may be required to make repairs in underground structures, such as valve or pump vaults, where you may be exposed to toxic and explosive gases or insufficient oxygen. Always make sure that these areas are properly ventilated. Use proper safety equipment, such as hard hats, goggles, rubber boots, gas detectors, and life jackets, when necessary. Never enter a confined area or space alone. Be sure the person watching you has additional help standing by and knows how to evacuate you if you are injured or lose consciousness. Do not take any chances.

Check Your Understanding

1. What kinds of sudden changes in either raw or filtered water quality are signals that you should immediately review the performance of the coagulation-flocculation process?

2. What should operators do if a significant change in source water turbidity levels is observed?

3. If source water temperature changes suddenly, what changes should an operator consider?

4. Why is good recordkeeping especially important during abnormal plant operating conditions?

5. Why should an emergency response procedure be developed for every water treatment plant?

6. Under what conditions are coagulation-flocculation processes normally shut down?

7. What should be done with any untreated or partially treated water in the plant during shutdown procedures?

8. What safety hazards could be encountered during the startup or shutdown of a coagulation-flocculation process?

3.9   Laboratory Tests

In the operation of the coagulation-flocculation process, the operator will perform a variety of laboratory tests to monitor source water quality and to evaluate process performance. Process control water quality indicators of importance in the operation of the coagulation-flocculation process include turbidity, alkalinity, chlorine demand, color, pH, temperature, odor, and appearance. See Chapter 9, "Laboratory Procedures."

3.9.1   Sampling Procedures

Process water samples will be either grab samples obtained directly from a specific process monitoring location, or continuous samples that are pumped to the laboratory from various locations in the process (raw water, flash-mixer effluent, flocculation basin effluent), as shown in Figure 3.9 on page 140. In either case, it is important to emphasize that process samples must be **representative samples** of actual conditions in the treatment plant. The accuracy and usefulness of laboratory analyses depend on the representative nature of the water samples.

The frequency of sampling for individual process control water quality indicators will vary, depending on the quality of the source water. Certain water quality indicators, such as turbidity, will be routinely monitored, while others, such as alkalinity, are sampled less frequently.

Process grab samples should be collected in clean plastic or glass containers and care should be used to avoid contaminating the sample, especially turbidity and odor samples.

representative sample
A sample portion of material, water, or wastestream that is as nearly identical in content and consistency as possible to that in the larger body.

Samples should be analyzed as soon as possible after the sample is collected. Important water quality indicators, such as turbidity, temperature, chlorine demand, color, odor, pH, and alkalinity, can all change while waiting to be analyzed.

3.9.2   Sample Analysis

Analysis of certain process control water quality indicators, such as turbidity and pH, can be readily performed in the laboratory with the aid of automated analytical instruments, such as turbidity meters and pH meters. Analysis of other process control water quality indicators, such as alkalinity and chlorine demand, may require **wet chemistry** procedures, which are often performed by a chemist or laboratory technician. A more detailed discussion of sampling and laboratory analytical procedures is contained in Volume II, Chapter 21, "Advanced Laboratory Procedures."

One of the most important laboratory procedures in the operation of the coagulation-flocculation process is the jar test procedure. As discussed earlier, this procedure is performed to establish the proper type and amount of chemical(s) to be used in the coagulation or flash-mixing process.

3.9.3   Safety

Laboratory work may expose the operator to a number of different safety hazards. Care should be exercised in the handling of reagents and glassware. Use protective clothing (safety glasses and aprons) while performing wet chemical analyses, especially when handling dangerous chemicals such as acid or caustic solutions. Always perform lab tests in a well-ventilated space, and be familiar with the location and use of safety showers and eye wash facilities.

3.9.4   Recordkeeping

wet chemistry
Laboratory procedures used to analyze a sample of water using liquid chemical solutions (wet) instead of, or in addition to, laboratory instruments.

Record all laboratory test results on appropriate data sheets, and document in detail any unusual results. Upon verification of abnormal test results, be sure to notify the proper personnel of your findings.

3.10   Equipment Operation and Maintenance

In the operation of the coagulation-flocculation process, the operator will be exposed to a variety of mechanical, electrical, and instrumentation equipment, including:

1. Mixers and flocculators
2. Chemical feeders
3. Water quality monitors
4. Pumps
5. Valves
6. Flowmeters and gauges
7. Control systems

Before starting a piece of mechanical equipment, such as a mixer or chemical feeder, be sure that the unit is properly lubricated and its operational status is known. Also, be certain that no one is working on the equipment. Be sure all valves are in the proper position before starting chemical feeders.

After startup, always check for excessive noise and vibration, overheating, and leakage (water, lubricants, and chemicals). When in doubt about the performance of a piece of equipment, always refer to O&M instructions or the manufacturer's technical manual.

Many equipment items, such as valves and mixers, are simple ON/OFF devices with some provision for either speed or position adjustment. Other equipment items, such as pumps and chemical feeders, may require the use of special procedures for priming and calibration. Detailed operating and repair procedures are usually given in the plant operations and maintenance instructions for specific pieces of equipment.

3.10.1   Chemical Feeders

In the coagulation-flocculation process itself, chemical feeders (Figure 3.14) are of particular importance. Chemicals are normally fed at a fixed rate. This can be accomplished by liquid feed (solution) or by dry feed (volumetric or gravimetric). In liquid feed, a diluted solution of known concentration is prepared and fed directly into the water being treated. Liquid chemicals are fed through metering pumps and rotameters. Dry feeders deliver a measured quantity of dry chemical during a specified time interval. Volumetric feeders deliver a specific volume of chemical during a given time interval, while gravimetric feeders deliver a predetermined weight of chemical in a specific unit of time. Generally,

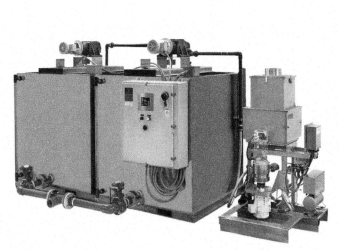

Dry chemical makeup and feed system (polymer)

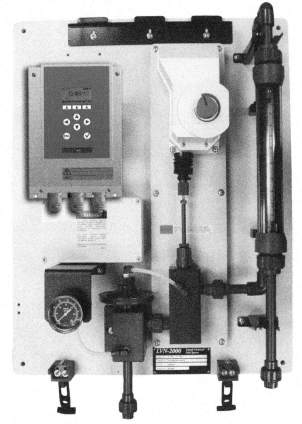

Solution chemical feed system

Figure 3.14   Chemical feeders *(Courtesy of UGSI Chemical Feed, Inc.)*

volumetric feeders can deliver smaller daily quantities of chemicals than gravimetric feeders, but the performance variables are:

1. Volumetric feeders are simpler and of less expensive construction.
2. Gravimetric feeders are usually more easily adapted for recording the quantities of chemicals fed and for automatic control. For this reason, gravimetric feeders are generally used in large treatment plants.

The ultimate decision of which chemical feeder to use for a given application depends on the type of chemical compound, availability of chemical, chemical form (dry or liquid), and the amount to be fed daily.

The capacity rating of solution chemical feeders is usually given in units of gallons per minute (gpm) or gallons per hour (gph), while dry feeders are often rated by the maximum amount of chemical that can be fed in a 24-hour period (pounds per day).

Adjusting or changing the amount of chemical to be fed is generally accomplished by manually changing the feed rate setting on the chemical feeder. Adjustment is physically performed by turning a knob, adjusting a wheel, or by rotating a hand-crank.

Typically, a feed-rate scale is provided on the chemical feeder, which is calibrated over a range from 0 to 100 percent of maximum feed rate, for both solution and dry chemical feeders. In a solution feed system, if the desired feed rate is 3 gph and the chemical feeder has a maximum feed rate of 15 gph, then the feeder would be set at:

$$\text{Scale Setting}[\%] = \frac{\text{Desired Feed Rate}\left[\dfrac{\text{gal}}{\text{h}}\right] \times 100\%}{\text{Maximum Feed Rate}\left[\dfrac{\text{gal}}{\text{h}}\right]}$$

$$= \frac{3\left[\dfrac{\text{gal}}{\text{h}}\right] \times 100\%}{15\left[\dfrac{\text{gal}}{\text{h}}\right]}$$

$$= 20\% \text{ of Full Setting}$$

Likewise, in a dry feed system, if the desired feed rate is 150 pounds per day (lb/d) and the chemical feeder has a maximum feed rate of 300 pounds per day (lb/d), then the feeder would be set at:

$$\text{Feeder Setting}[\%] = \frac{\text{Desired Feed Rate}\left[\dfrac{\text{lb}}{\text{d}}\right] \times 100\%}{\text{Maximum Feed Rate}\left[\dfrac{\text{lb}}{\text{d}}\right]}$$

$$= \frac{150\left[\dfrac{\text{lb}}{\text{d}}\right] \times 100\%}{300\left[\dfrac{\text{lb}}{\text{d}}\right]}$$

$$= 50\% \text{ of Full Setting}$$

3.10.2   Calibration Method

Chemical feed systems should be calibrated at least once per shift to verify proper chemical feed rate. Always calibrate chemical feeders against working system pressures to avoid errors. In liquid chemical feed systems, the volumetric method is probably the most accurate calibration technique. This method involves the use of a calibrated container (usually a graduated cylinder as shown in Figure 3.15) and a stopwatch to determine the volume of chemical fed during a given time period. Ideally, the cylinder and timer are part of the chemical feeder piping system. This procedure can, of course, be used to calibrate the chemical feed pump over the full range of feed rates.

To apply the procedure, select an appropriate time period such as 30 to 90 seconds. The time period should be increased when measuring dilute chemical solutions to ensure accurate results. Fill the graduated container with a convenient amount of the chemical solution. Insert one end of a tube into the container with the chemical and attach the other end to the feeder inlet on the suction side of the feeder. Start the feeder and the stopwatch. After a minimum time period (for example, 30 seconds), read the graduation mark on the container that corresponds to the liquid level drawdown. Record the total elapsed time. The following example uses a 1-liter graduated cylinder.

Figure 3.15   Graduated cylinders
(Courtesy of DesignPrax/Shutterstock)

Example 13

A chemical feeder draws a liquid chemical from a 1-liter (1,000 mL) graduated cylinder for 30 seconds (s). At the end of 30 seconds, the graduated cylinder has 400 mL remaining. What is the chemical feed rate in milliliters per minute (mL/min) and in gallons per minute (gpm)?

Known	Unknown
One Liter (L) of Chemical	Chemical Feed Rate in mL/min and gpm

h_i = Starting Level = 1,000 mL

h_f = Final Level = 400 mL

t = Feed Time = 30 s

1. Determine volume of chemical fed in milliliters.

$$\text{Chemical Fed}[mL] = h_i[mL] - h_f[mL]$$
$$= 1,000[mL] - 400[mL]$$
$$= 600\,mL$$

2. Determine chemical feed rate, mL/min.

$$\text{Chemical Feed Rate}\left[\frac{\text{mL}}{\text{min}}\right] = \frac{\text{Chemical Fed}\,[\text{mL}]}{t\,[\text{min}]}$$

$$= \frac{600\,[\text{mL}]}{30\,[\text{s}] \times \dfrac{\text{min}}{60\,\text{s}}}$$

$$= 1{,}200\,\frac{\text{mL}}{\text{min}}$$

3. Calculate chemical feed rate in GPM.

$$\text{Chemical Feed Rate}\left[\frac{\text{gal}}{\text{min}}\right] = \text{Chemical Feed Rate}\left[\frac{\text{mL}}{\text{min}}\right] \times \frac{\text{gal}}{3{,}785\,\text{mL}}$$

$$= 1{,}200\left[\frac{\text{mL}}{\text{min}}\right] \times \frac{\text{gal}}{3{,}785\,\text{mL}}$$

$$= 0.32\;\text{gpm}$$

Compare your calculated value with the setting on the chemical feeder and any calibration tables you may have. If the values do not agree, recheck your work and then adjust the feeder to deliver the correct amount of chemical. If the chemical feeder has a setting (say, 1, 2, 3), record the chemical feed rate as 0.32 GPM for the setting. Adjust the feeder to another setting (higher or lower) and repeat the test to determine the feed rate at the new setting.

Most chemical feed systems are not furnished with volumetric calibration accessories. However, this feature can be readily added to existing systems by acquiring a standard la-boratory-grade graduated cylinder (preferably plastic) and a stopwatch and installing the necessary piping and valving on the pump suction. Figure 3.16 is a sketch of a typical installation. Another approach is to open a sample tap and pump directly into a graduated cylinder.

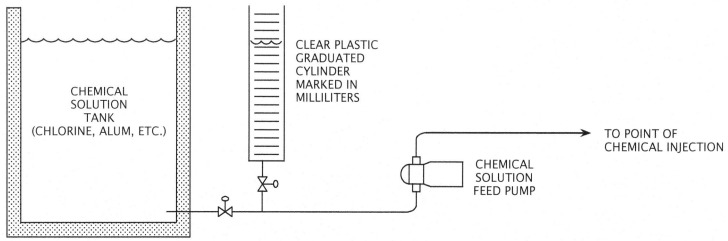

CHEMICAL SOLUTION TANK (CHLORINE, ALUM, ETC.)

CLEAR PLASTIC GRADUATED CYLINDER MARKED IN MILLILITERS

CHEMICAL SOLUTION FEED PUMP

TO POINT OF CHEMICAL INJECTION

Figure 3.16 Calibration system

Procedures similar to those described above can be used for dry feed applications by measuring the dry weight of chemical fed during a given time period.

Example 14

An operator is interested in measuring the dry chemical feed rate in pounds per day (lb/d), so he places an empty bucket weighing 0.8 pounds (lb) under the dry chemical feeder. After 24 minutes (min), the bucket weighs 5.6 pounds (lb). Calculate the feed rate based on the information provided.

Known	Unknown
M_{empty} = Weight of Empty Bucket = 0.8 lb	Dry Chemical Feed Rate, lb/d
M_{full} = Weight of Bucket and Chemical = 5.6 lb	
t = Feed Time = 24 min	

1. Determine the amount of chemical fed in pounds.

$$\text{Chemical Fed}\,[\text{lb}] = M_{full}\,[\text{lb}] - M_{empty}\,[\text{lb}]$$
$$= 5.6\,[\text{lb}] - 0.8\,[\text{lb}]$$
$$= 4.8\,\text{lb}$$

2. Calculate the dry chemical feed rate in pounds per minute (lb/min).

$$\text{Chemical Feed Rate}\left[\frac{\text{lb}}{\text{min}}\right] = \frac{\text{Chemical Fed}\,[\text{lb}]}{t\,[\text{min}]}$$
$$= \frac{4.8\,[\text{lb}]}{24\,[\text{min}]}$$
$$= 0.2\,\frac{\text{lb}}{\text{min}}$$

3. Calculate the dry chemical feed rate in pounds per day (lb/d).

$$\text{Chemical Feed Rate}\left[\frac{\text{lb}}{\text{d}}\right] = \text{Chemical Feed Rate}\left[\frac{\text{lb}}{\text{min}}\right] \times \frac{60\,\text{min}}{\text{h}} \times \frac{24\,\text{h}}{\text{d}}$$
$$= 0.2\left[\frac{\text{lb}}{\text{min}}\right] \times \frac{60\,\text{min}}{\text{h}} \times \frac{24\,\text{h}}{\text{d}}$$
$$= 288\,\frac{\text{lb}}{\text{d}}$$

If you cannot get a bucket this big in your chemical feeder to collect a sample or leave it in the feeder this long, use a pie tin. Some operators will collect three samples of 3 minutes or 5 minutes each and use a total amount of chemical fed for 9 or 15 minutes.

Example 15

An operator is interested in measuring the dry chemical feed rate in grams per minute (g/min), so she places an empty pie tin weighing 30 grams (g) under the dry chemical feeder. After 5 minutes (min), the pie tin weighs 450 grams (g). Calculate the feed rate based on the information provided.

Known	Unknown
M_{empty} = Weight of Empty Tin = 30 g	Dry Chemical Feed Rate, g/min
M_{full} = Weight of Tin and Chemical = 450 g	
t = Feed Time = 5 min	

1. Determine the amount of chemical fed in grams (g).

$$\text{Chemical Fed}[g] = M_{full}[g] - M_{empty}[g]$$
$$= 450[g] - 30[g]$$
$$= 420\,g$$

2. Calculate the dry chemical feed rate in grams per minute (g/min).

$$\text{Chemical Feed Rate}\left[\frac{g}{min}\right] = \frac{\text{Chemical Fed}[g]}{t[min]}$$
$$= \frac{420[g]}{5[min]}$$
$$= 84\frac{g}{min}$$

A check on the amount of liquid chemical used in a given time period, say 24 hours, can be made by measuring the difference (drawdown) in chemical storage tank levels (Figure 3.17). In this case, the operator must compute the volume of chemical used based on the dimensions of the storage tank. Some small chemical storage tanks are mounted on a scale, thus allowing a direct reading of the amount of chemical remaining at the end of each day. The difference between these values is the amount used per day.

Example 16

An alum storage tank has an inside diameter (I.D.) of 10 feet (ft) and a height of 25 feet (ft). During a 24-hour (h) time period (say 8:00 am Monday to 8:00 am Tuesday), the tank level dropped 3 inches (3 in/12 in/ft = 0.25 ft). How many gallons (gal) of chemical were used? If the chemical feed was constant, what was the chemical feed rate in gallons per minute (gpm)?

		Known			Unknown
D	=	Tank Diameter	=	10 ft	Chemical Used, gal
H	=	Tank Height	=	25 ft	Chemical Feed Rate, gpm
h	=	Tank Drop	=	0.25 ft	
t	=	Time	=	24 h	

1. Calculate the volume of chemical used in cubic feet (ft³).

$$\text{Volume}\left[\text{ft}^3\right] = \frac{\pi}{4} \times D^2\left[\text{ft}^2\right] \times h\left[\text{ft}\right]$$

$$= 0.785 \times 10^2 \ [\text{ft}^2] \times 0.25 \ [\text{ft}]$$

$$= 19.63 \ \text{ft}^3$$

2. Determine the volume of chemical used in gallons (gal).

$$\text{Chemical Used}\left[\text{gal}\right] = \text{Volume}\left[\text{ft}^3\right] \times \frac{7.48 \ \text{gal}}{\text{ft}^3}$$

$$= 19.63\left[\text{ft}^3\right] \times \frac{7.48 \ \text{gal}}{\text{ft}^3}$$

$$= 147 \ \text{gal}$$

3. Calculate the chemical feed rate in gallons per minute (gpm).

$$\text{Chemical Feed Rate}\left[\frac{\text{gal}}{\text{min}}\right] = \frac{\text{Chemical Used}\left[\text{gal}\right]}{t[h] \times \dfrac{60 \ \text{min}}{h}}$$

$$= \frac{147\left[\text{gal}\right]}{24\left[h\right] \times \dfrac{60 \ \text{min}}{h}}$$

$$= 0.10 \ \text{gpm}$$

Figure 3.17   Chemical storage tanks

Water treatment plants should have duplicate chemical feeders. This will permit the operator to maintain full service while a chemical feeder is out of service for routine maintenance or major repair. If this feature is not included in your plant, then consider requesting backup equipment as part of the annual operation, maintenance, and repair budget request process.

3.10.3   Preventive Maintenance Procedures

Preventive maintenance programs are designed to ensure the satisfactory long-term operation of treatment plant facilities under a variety of different operating conditions. Scheduled or routine maintenance of valves, mixers, pumps, and chemical feeders is an important part of the preventive maintenance program. The operator will be expected to perform routine maintenance functions as part of an overall preventive maintenance program. Typical functions include:

1. Keeping motors free of dirt and moisture
2. Ensuring good ventilation (air circulation) in equipment work areas
3. Checking pumps for leaks, unusual noise, vibrations, or overheating

4. Maintaining proper lubrication and oil levels
5. Inspecting for alignment of shafts and couplings
6. Checking bearings for wear, overheating, and proper lubrication
7. Exercising infrequently used valves on a regular schedule and checking all valves for proper operation
8. Calibrating flowmeters and chemical feeders (see previous section)

Good recordkeeping is the key to a successful preventive maintenance program. These records provide maintenance and operations personnel with clues for determining the causes of equipment breakdowns, and will often show weaknesses in a particular piece of equipment, which can be corrected before failure.

For more details on preventive maintenance, see the chapter on maintenance in Volume II of this series of training manuals.

3.10.4 Safety

When working around equipment, such as motors, mixers, or flocculators, follow the safety procedures listed below to avoid accidents or injury.

ELECTRICAL EQUIPMENT

1. Always shut off power and lock out and attach a safety tag before working on electrical equipment, instruments, controls, wiring, and all mechanical equipment driven by electric motors.
2. Avoid electric shock by using protective gloves.
3. Use a multimeter to test for live wires and equipment.
4. Check grounds and avoid grounding yourself in water or on pipes.
5. Ground all electric tools.
6. Use the buddy system.

MECHANICAL EQUIPMENT

1. Use protective guards on rotating equipment.
2. Do not wear loose clothing, worn gloves, or long hair around rotating equipment.
3. Clean up all lubricant spills (oil and grease).

OPEN WATER SURFACE STRUCTURES

1. Do not avoid or defeat protective devices, such as handrails, by removing them when they are in the way.
2. Close all openings when finished.
3. Know the location of all personal flotation devices (PFDs) and wear one when necessary. PFDs should be Coast Guard-approved.

VALVE AND PUMP VAULTS

1. Be sure all underground structures are free of hazardous atmospheres (toxic and explosive gases or insufficient oxygen) by using gas detectors.
2. Only work in well-ventilated structures.
3. Use the buddy system.

HANDLING ALUM

Normal precautions should be used to prevent spraying or splashing of liquid alum, especially when the liquid is hot. Face shields can be worn to protect the eyes. A rubber apron and waterproof sleeves may be used to protect clothing.

Both liquid alum and dry alum dust must be flushed from the eyes immediately. Use great amounts of warm water. Alum also must be washed from the skin because prolonged contact can be irritating.

For more details on safety, see the chapter on safety in Volume II of this series of training manuals.

Check Your Understanding

1. List the process control water quality indicators of importance in the operation of the coagulation-flocculation process.

2. When should sample analysis be performed to control the coagulation-flocculation process?

3. What safety hazards may be encountered when working in a laboratory?

4. What types of equipment are used in connection with the coagulation-flocculation process?

5. How do chemical liquid feeders work in the coagulation process?

6. What factors influence the selection of a chemical feeder for a given application?

7. What equipment is used in the volumetric calibration method?

8. What equipment should be part of a preventive maintenance program for a coagulation-flocculation process?

9. What types of hazards could be present in underground structures?

3.11   Enhanced Coagulation

Enhanced coagulation is a process designed to remove natural organic matter (NOM) from water by adjusting both the coagulant dosage and pH to produce the greatest possible reduction of dissolved or suspended organic carbon *(color)*, **total organic carbon (TOC)**, **trihalomethanes (THMs)**, and **disinfection byproducts (DBPs)**. Unlike the sweep treatment, where the pH range is achieved by overdosing the coagulant, the optimum pH is arrived at by adding acid or alkali. The amount of acid will depend on the amount of color and alkalinity present in the raw water. If the color is high and the alkalinity is low, the acid dosage alone may be enough to depress the pH to optimum.

The colloidal and dissolved organics found in some natural waters are the end products of decayed vegetable matter. Generally, natural waters with high organic compounds are low in turbidity (less than 10 NTU). The main constituents of these naturally occurring organic compounds are organic acids called humic substances; they are composed of humic and fulvic acids. Humic acid is a high-molecular-weight, complex macro-molecule. Because of its size, the humic acid molecule is less soluble than the smaller fulvic acid molecule. Both acids exhibit **polyanionic** characteristics and impart a yellowish-brown color to the water.

total organic carbon (TOC)
A measure of the amount of organic carbon in water.

trihalomethanes (THMs) (tri-HAL -o-METH-hanes)
Derivatives of methane, CH_4, in which three halogen atoms (chlorine or bromine) are substituted for three of the hydrogen atoms. Often formed during chlorination by reactions with natural organic materials in the water. The resulting compounds (THMs) are suspected of causing cancer.

disinfection byproduct (DBP)
A contaminant formed by the reaction of disinfection chemicals (such as chlorine) with other substances in the water being disinfected.

polyanionic (poly-AN-eye-ON-ick)
Characterized by many active negative charges.

3.11.1   Chemical Reactions

The fulvic and humic substances found in water are negatively charged anionic polyelectrolytes that owe their stability to their negative charge. To remove these organic acids from water, the first step is to destabilize the molecules. This is accomplished during coagulation, when the organic acids react chemically with inorganic coagulant salts. The negative charge of the organic acids is neutralized and destabilized by the positively charged coagulants. In the flocculation process, which follows coagulation, the destabilized particles come together and form larger floc particles that are subsequently removed by settling.

The chemistry that deals with this particular coagulation process is known as charge chemistry. This is a branch of chemistry in which destabilization and neutralization reactions occur between stable negatively charged particles and stable positively charged particles. One of the major factors affecting charge chemistry is pH. The pH range for color removal with aluminum sulfate is 5.5 to 7.0, with the optimum pH often at 5.8. For ferric sulfate, the pH range is 4.0 to 6.2, with the optimum pH often at 4.5. Optimum pH is a function of the specific raw water to be treated; therefore, optimum pH may vary. At the lower (optimum) pH, four effects take place that enhance coagulation:

1. The humic and fulvic molecules dissociate (separate) to a lesser degree at lower pH.
2. The coagulant demand decreases correspondingly to the degree of molecular dissociation.
3. Flocculation is improved at lower pH.
4. Sulfuric acid addition before coagulant feed preconditions the organic compounds.

3.11.2   Process Control

Color determinations are extremely pH dependent and will always increase as the pH of the water increases. Due to this pH dependency, color determinations should be specified with pH values. Color may be determined by visual comparison or by using a spectrophotometer. Regardless of the method used, the sample must be filtered to determine **true color**. Turbidity in any amount will cause the color to be noticeably higher. Color in a sample that has not been filtered is called **apparent color**.

The process actions described in Section 3.6.1, "Process Actions," for turbidity removal are also appropriate for color removal. In operating an enhanced coagulation process, the operator is attempting to achieve two interrelated objectives at the same time. One objective is to feed the optimal dosage of coagulant to remove the organic color, and the second objective is to adjust pH to the optimal zone for coagulation. Due to the fact that enhanced coagulation is pH dependent, automated pH process control is essential for plant operation. Since acid or alkalinity addition will most likely be required to optimize coagulation, a pH backfeed process control loop is required to maintain the proper coagulation pH. The design of the system uses a monitoring pH meter located after the rapid mix zone and feed valve controllers for acid and alkalinity chemical addition. The pH meter sends a constant

true color
Color of the water from which turbidity has been removed.

apparent color
Color of the water that includes not only the color due to substances in the water but suspended matter as well.

signal to the chemical feed controllers and will communicate to open or close the feed valve, which will increase or decrease chemical feeds, respectively.

The cost-effectiveness of enhanced coagulation is directly related to the process control loop system. Generally, acid or alkali is less expensive than the coagulant, so overdosing the coagulant for pH depression (as in sweep-type treatment) is wasteful and costly. In addition to other process problems, overdosing will generate more treatment residues and will increase the concentration of the coagulant residuals in the finished water.

Optimum dosages for acid, alkalinity, and coagulant are determined by performing a series of jar tests as described in Section 3.6.2.2, "The Jar Test." The procedures listed there are used to determine the coagulant dosage. However, to predetermine the optimum acid/alkalinity dosages for pH adjustment and coagulation, you must first conduct a series of jar tests following the procedures listed below. The following two jar tests illustrate typical results to determine optimum coagulant dosage and pH using a low turbidity, high organic source water, and ferric sulfate as the coagulant. The source water characteristics are: color, 150 **platinum cobalt units (pcu)**; alkalinity, 90 mg/L; pH, 7.3; and turbidity, 1.2 NTU.

1. Fill six 1-liter beakers with raw source water and add varying coagulant dosages based on an estimated treatment range. Mix at medium speed until coagulant is completely dispersed.

2. Adjust pH for optimal coagulation (pH of 4.5 in Test Example No. 1 below) and record milliliters of acid or alkali solution required. To depress the pH, it is best to use sulfuric acid (93 to 96 percent). To increase the pH, the most common chemicals used are lime and sodium hydroxide. A stock solution should be prepared so that the dose in mg/L may be calculated.

Test Example No. 1

Jar Number	1	2	3	4	5	6
$Fe_2(SO_4)_3$, mg/L (Dry Basis)	40	60	80	100	120	140
Acid, mg/L	60	55	50	45	40	35
Alkali, mg/L	0	0	0	0	0	0
Coagulation pH	4.5	4.5	4.5	4.5	4.5	4.5
Settled/Filtered Color	44	30	20	12	12	11

Jar No. 4 shows the optimal color removal without overdosing the coagulant. This coagulant dose should remain constant for the next jar test to determine the optimal coagulation pH. (Test Example No. 1 is an attempt to try to identify both optimum pH and coagulant ranges. Some operators prefer to change only one variable at a time, and would therefore conduct separate tests for coagulant dosage and acid/alkali dosage.)

3. Once the ranges for acid or alkali dosages are determined, proceed with the jar testing by filling six 1-liter beakers with raw source water.

4. Add the predetermined milliliters of acid or alkali to all six beakers and mix well for precoagulation conditioning.

pcu (platinum cobalt units)
A measure of color using platinum cobalt standards by visual comparison.

5. Add coagulant and follow procedure as directed in Section 3.6.2.2, "The Jar Test."

Test Example No. 2						
Jar Number	1	2	3	4	5	6
$Fe_2(SO_4)_3$, mg/L (Dry Basis)	100	100	100	100	100	100
Acid, mg/L	54	48	44	40	35	30
Alkali, mg/L	0	0	0	0	0	0
Coagulation pH	4.0	4.4	4.8	5.2	5.6	6.0
Settled/Filtered Color	16	12	12	14	15	16

Taken together, the results of the two example tests indicate that a coagulant dose of 100 mg/L and an acid dose of 44 to 48 mg/L provide an optimum coagulation pH of 4.4 to 4.8.

6. To pinpoint the optimal pH, run another jar test varying the coagulation pH within the coagulation pH range (pH 4.4 to 4.8) while keeping the coagulant dose constant. The coagulant dose selected should be taken from the prior jar test with the dose that had the best color removal without overdosing.

To further fine-tune the coagulant dose, you could perform another jar test and try variations of coagulant dosage at the optimum pH. In summary, you have to conduct a series of tests to determine both optimum pH and coagulant dose.

Once the optimum pH is determined, the operator can use this information to make process changes to fit source water changes. Changes to the source water that will require adjustments are the increase or decrease in color or alkalinity. Considering that the alkalinity demand for coagulation is greater for aluminum salts than for iron salts, total source water alkalinity may need to be increased to meet the coagulant alkalinity demand before coagulant addition optimization.

Table 3.4 summarizes enhanced coagulation process problems and suggests possible corrective actions.

Table 3.4   Enhanced coagulation process troubleshooting

Treatment Condition Flocculator Effluent	Corrective Action
High coagulation pH with optimum color removal	1. Increase acid feed 2. Decrease alkalinity adjustment in raw water source
High coagulation pH without optimum color removal	1. Increase coagulant 2. Decrease acid feed to maintain optimum pH
Low coagulation pH with optimum color removal	1. Decrease acid feed 2. Increase alkalinity adjustment in raw water source
Low coagulation without optimum color removal	1. Decrease acid if below optimal pH zone 2. Increase coagulant and alkalinity
Loss of acid feed	1. Increase coagulant to achieve optimal pH
Optimal pH without optimized color removal	1. Increase coagulant, decrease acid, or increase alkalinity
Optimal pH and color removal with floc carryover	1. Decrease coagulant 2. Increase polymer 3. Increase removal of settled floc 4. Decrease flow-through velocities in treatment unit
High turbidities and coagulant residuals in settled water	1. Check for floc carryover 2. Adjust polymer feed to enhance settling 3. Jar test to determine optimum acid and coagulant dosage

3.12 Ballasted Flocculation

Ballasted flocculation is a physical-chemical water treatment process in which microsand is added to the flocculation stage to encourage floc settling (ballast the floc). This efficient coagulation and flocculation treatment process is followed by a sedimentation stage that typically consists of a horizontal-flow rectangular sedimentation basin with inclined tube settlers (Figure 3.18).

As with conventional flocculation, ballasted flocculation depends on proper coagulant and polymer dosing and correctly adjusted and calibrated chemical feeders. See Section 3.6.2.2, "The Jar Test," and Section 3.10.1, "Chemical Feeders," for more information on these topics. Laboratory tests are also important for process control. See Section 3.9, "Laboratory Tests," and Chapter 9, "Laboratory Procedures," for more information. Operators will be exposed to the same potential hazardous situations as in conventional coagulation-flocculation processes. See Section 3.6.5, "Safety," for more information.

Proper operation of a ballasted flocculation system requires greater operator expertise than operation of conventional coagulant systems because the addition of ballast requires close monitoring of the recycled microsand concentration. The short retention time also requires prompt operator response to maintain design conditions and to provide optimum coagulant dosages.

The ballasted flocculation process discussed in this section is called ACTIFLO® and was designed by OTV (Veolia), a French company. A similar process called STERAU® was developed by SAUR. Both processes use the same principle.

Typical ACTIFLO® treatment removal efficiencies for raw water are:

Major portions of the material in this section are based on copyrighted information provided by I. KRÜGER, Inc.

- Suspended solids, >85 percent
- Turbidity, 90 to 99 percent
- Green algae, 90 to 98 percent
- Blue-green algae, 70 to 95 percent

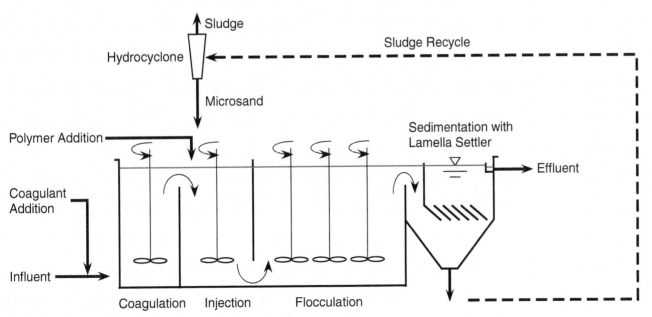

Figure 3.18 Flow diagram of the Actiflo® process

The ACTIFLO® process consists of adding an inorganic coagulant (alum or ferric) to the raw water and allowing floc to form in the first stage of flocculation. Next, a high-molecular-weight polymer and microsand particles (90 to 200 μm) are added to the second stage, and the microsand particles flocculate with the preformed chemical floc particles in the second and third stages. After flocculation, the ballasted floc is settled. The sludge containing the microsand is sent through a hydrocyclone where the microsand is recovered and reused and the sludge is sent for additional treatment and disposal. The much higher density microsand is discharged from the bottom of the hydrocyclone and reinjected into the process for reuse. The lighter density sludge is discharged from the top of the hydrocyclone to the sludge handling equipment for thickening and dewatering before disposal. The chemically inert microsand does not react with the process chemistry, allowing it to be effectively removed from the chemical sludge and reused in the process.

3.12.1   Process Stages

The ballasted flocculation process has four distinct stages as shown in Figure 3.18.

- Rapid mix tank where the coagulant is added and mixed
- Injection tank where polymer and microsand are added and mixed
- Flocculation basin (maturation) where the floc develops and grows
- Sedimentation basin with tube settler where the floc is settled and removed as sludge

The water treated at a typical ACTIFLO® plant first passes through a fine screen and goes to the rapid mix tank. The chemical coagulant—alum, ferric, or polyaluminum chloride (PACL)—is mixed with the raw water and destabilizes the suspended solids by neutralizing surface charges. Optimizing the coagulant dose is essential in achieving an effective clarification.

Next, the mixture flows to the injection tank (not used for the ACTIFLO® Turbo design) where a polymer and microsand are introduced to the water being treated. The polymer helps the microsand to become attached to the flocs. An excess of microsand (greater than 7 g/L) would cause erosion of the flocs, while low microsand dosages (less than 1 g/L) would result in a poor ballasting effect. The efficiency is optimal when the size of microsand particles is less than 160 μm. The contact time is about 2 minutes, but because the kinetic reaction decreases when temperature drops, a longer contact time will increase floc growth in cold water.

It is the responsibility of the operator to ensure that the proper microsand is being used and stored appropriately. Order and use only microsand that meets the specifications of the particular system, and verify when accepting delivery of the microsand that it meets the required specifications.

The rapid mix (coagulation) and the injection tanks are equipped with mixers that provide the appropriate mixing intensity for floc formation and maturation (enlargement).

The flocs that formed in the mixing tank grow and mature in the flocculation basin. The contact time is approximately 6 minutes or

Throughout this section, there will be references to both the classic ACTIFLO® and the newer ACTIFLO® Turbo designs. The differences are minor (and will be indicated), but the concepts are the same for both processes.

3 minutes for ACTIFLO® Turbo. Ballasted flocculation supports higher mixing intensities than conventional flocculation without floc breakage by up to a factor of 10.

The water next flows into a sedimentation basin with a tube settler where the microsand-weighted flocs and microsand ballast are separated from the treated water. A mechanical scraper with rake arms collects the microsand/sludge into a central suction pit, where it is pumped by rubber-lined slurry pumps to the hydrocyclones. The desired surface loading in the settling basin is approximately 131 to 241 ft³/hr/ft² (40 to 74 m³/hr/m²) for drinking water applications.

The ballasted flocs are heavier than floc in conventional systems, which provides for an increased settling velocity. Plant observations indicate that 15 seconds are sufficient for settling, but a 3-minute settling time is typical for practical purposes.

The sludge drawn from the settling basin is pumped by a rubber-lined slurry pump to a set of hydrocyclones (see Section 3.12.4, "Hydrocyclone Performance"). The pumps are V-belt driven with constant-speed motors and operate at a constant discharge pressure, typically 35 to 40 pounds per square inch (psi). The microsand is separated from the sludge by the intense vortex created in the hydrocyclone, and the cleaned microsand is then recycled to the injection tank (or the maturation tank for ACTIFLO® Turbo) to maintain a typical concentration between 2 and 6 g/L. The sludge is pumped to appropriate processing facilities (sewer or sludge treatment units). The sludge recycled flow typically makes up 2 to 3 percent of the total water flow treated.

3.12.2 Startup

Before starting the treatment process, perform the tasks described in this section. Prepare a checklist of items to inspect or check, including:

1. Be sure all piping, tanks, and basins are clean. Remove all dirt, debris, and construction material left by contractors.
2. Fill all tanks and basins with water.
3. Open the manual valves at the suction and discharge of the microsand/slurry recirculation pumps.
4. Check to be sure seal water for the microsand/slurry recirculation pumps is operational.
5. Verify that the coagulant and polymer dosing equipment (pumps and valves) is working properly.
6. Inspect coagulant storage tank to be sure it is full.
7. Check to be sure polymer solution is well mixed and aged properly.
8. Determine that the coagulant and polymer metering pumps are calibrated and are adjusted (or can be adjusted) to provide the flow rates required to obtain the proper dosage in the system.

The removal of all debris from the tanks and basins before startup must be emphasized. Any debris in the system after startup can enter the hydrocyclones and clog the hydrocyclone apex. If the apex becomes clogged, higher than normal levels of microsand can be forced out of the system.

The following procedure is used to load the microsand into the system. For a typical system, a microsand concentration of 2 to 6 grams of microsand per liter of water in the injection tank and flocculation basin is suggested. Based on experience, it is recommended that an initial loading of 800 pounds of microsand per million gallons (96 kg/million liters or ML) of capacity be added to charge the ACTIFLO® process.

After 20 to 30 minutes of operation, the concentration of microsand should be measured at least three times over a 30-minute period. The results should be averaged to estimate the actual concentration. If additional microsand is needed to increase the microsand concentration, add microsand in increments of 500 to 1,000 pounds (230 to 460 kg) to increase the microsand concentration to the desired level. After each microsand addition, wait 30 minutes and then recheck the microsand concentration.

3.12.3   Microsand Management

Microsand is essential for the satisfactory operation of the process, but an accurate dose is not required for the process to operate properly. The microsand concentration can increase or decrease by as much as 30 percent without impeding the process performance. During normal operating conditions, if there is an increase in turbidity or in the color of raw water, be sure that the microsand concentration is at the high end of the desired operating range. During normal operation microsand losses occur because the hydrocyclone microsand recovery is not 100 percent efficient. To determine how often to replace microsand lost during normal operation, estimate the amount of micosand lost per day.

Example 17

Let's convert this example to metric units.

Flow = 27 ML/d

$$M\left[\frac{kg}{d \cdot train}\right] = Flow\left[\frac{ML}{d \cdot train}\right] \times M\left[\frac{mg}{L}\right] \times \frac{kg}{Mmg}$$

$$= 22.7\left[\frac{ML}{d \cdot train}\right] \times 2\left[\frac{mg}{L}\right] \times \frac{kg}{Mmg}$$

$$= 45\frac{kg}{d \cdot train}$$

Assume a typical plant has an average microsand loss of 2 mg microsand/liter of water per day per train. (One hydrocyclone treating 6 Mgal/d or 23 ML/day of water is considered one train of the ballasted flocculation treatment process.) Estimate the amount of microsand loss, M, in pounds per day (using the following formula:

$$M\left[\frac{lb}{d \cdot train}\right] = Flow\left[\frac{Mgal}{d \cdot train}\right] \times M\left[\frac{mg}{L}\right] \times 8.34\frac{lb}{gal}$$

$$= 6\left[\frac{Mgal}{d \cdot train}\right] \times 2\left[\frac{mg}{L}\right] \times 8.34\frac{lb}{gal}$$

$$= 100\frac{lb}{d \cdot train}$$

On a weekly basis, this is a loss of about 700 pounds (315 kg) of microsand per train when operated at an average flow of 6 Mgal/d (23ML/day). Supplemental microsand does not need to be added every day. Weekly or biweekly microsand additions are most common.

Operators should keep accurate records of the microsand concentration in the recirculation pipe in addition to the operations records kept for conventional coagulation-flocculation processes. (See Section 3.6.4, "Recordkeeping.")

3.12.3.1   Microsand Concentration Monitoring

A minor amount of microsand is constantly lost from the process. A microsand concentration as low as 2 g/L is sufficient for satisfactory performance of the ballasted flocculation process. However, the microsand concentration should always be maintained within the recommended range of 2 to 6 g/L. Operators should monitor the microsand concentration two to three times per day for each train in operation.

To determine the system concentration of microsand, collect samples from the vertical microsand recirculation pipe. The concentration of microsand can be estimated using the following procedure.

Always wear protective eyewear when collecting samples to measure the microsand concentration because splashing could injure the eyes.

Verify that each operating hydrocyclone has an uninterrupted conical discharge.

1. Use either a 1,000 mL or 2,000 mL graduated cylinder or Imhoff Cone to take samples. Fill the container as close to the top mark as possible without overspilling. *Caution: A 1,000 mL container may fill quickly and with force.*

2. Allow the sample to settle for 3 minutes.

3. Record the volume of the settled microsand (V_s) in mL and the total sample volume (V) in mL. For example, V = 2,000 mL and V_s = 30 mL. Figure 3.19 shows an operator measuring V_s in a graduated cylinder. Repeat steps 1 through 3 on the same recirculation line two more times.

4. Use the same sampling procedure as you did on the first hydrocyclone and repeat steps 1 through 3 for all hydrocyclones in operation on the system. For example, if you have two hydrocyclones operating on a given system, you should have six values for V_s and six values for V.

To determine the concentration of microsand in the entire system, you must next average all collected values of V_s. Also, average all collected values of V.

For the most accurate determination of microsand in any given system, this sampling procedure should be performed three times throughout a 24-hour period. Three separate values of C_m should be averaged to

Figure 3.19   Operator measuring V_s using a graduated cylinder

determine one final microsand concentration for the day. Record and trend the final daily value of C_m to indicate microsand loss over an extended period of time.

To calculate the concentration of the microsand in the tanks, use the following formula:

$$C_m = \frac{(V_s)(Q_r)(N)(1,700)}{(V)(Q_{train})}$$

Where,

C_m	=	Microsand concentration in tanks, g/L
Q_r	=	Microsand slurry recirculation rate, gpm
Q_{train}	=	Influent flow rate, gpm
N	=	Number of hydrocyclones in operation
V	=	Volume of sample collected in cone, mL
V_s	=	Volume of settled microsand, mL
1,700	=	Conversion factor/Bulk density $(\delta = (1.7\ g/mL)(1,000\ mL/L))$

Example 18

An ACTIFLO® process has one hydrocyclone in operation (one train) with an influent flow of 4,000 gpm and a microsand slurry recirculation rate of 400 gpm. Six V_s samples from the train produced the following volumes of settled microsand: 30 mL, 25 mL, 25 mL, 40 mL, 30 mL, and 30 mL. The six volumes of samples collected in the cone averaged 2,000 mL. Calculate C_m, the microsand concentration in the tanks in grams of microsand per liter (g/L).

Known

Q_r	=	Microsand Slurry Recirculation Rate	= 400 gpm
Q_{train}	=	Influent Flow Rate	= 4,000 gpm
N	=	Number of Hydrocyclones in Operation	= 1
V	=	Volume of Sample Collected in Cone	= 2,000 mL
1,700	=	Conversion Factor/Bulk Density $(\delta = (1.7\ g/mL)(1,000\ mL/L))$	
V_s Values	=	Volumes of Setled Microsand	= 30 mL, 25 mL, 25 mL, 40 mL, 30 mL, and 30 mL

Unknown

V_s	=	Volume of settled microsand, mL
C_m	=	Microsand concentration in tanks, g/L

1. Calculate the average V_s value, the volume of settled microsand, mL, for the six samples collected.

$$V_s[mL] = \frac{Sum\ of\ V_s\ Values\ Collected}{Number\ of\ V_s\ Values\ Collected}$$

$$= \frac{30[\text{mL}] + 25[\text{mL}] + 25[\text{mL}] + 40[\text{mL}] + 30[\text{mL}] + 30[\text{mL}]}{6}$$

$$= \frac{180[\text{mL}]}{6}$$

$$= 30 \text{ mL}$$

2. Calculate the microsand concentration of the microsand in the tanks.

$$C_m = \frac{(V_s)(Q_r)(N)(1,700)}{(V)(Q_{train})}$$

$$= \frac{30.0[\text{mL}] \times 400\left[\dfrac{\text{gal}}{\text{min}}\right] \times 1 \text{ Train} \times 1,700\left[\dfrac{\text{g}}{\text{L}}\right]}{2,000[\text{mL}] \times 4,000\left[\dfrac{\text{gal}}{\text{min}}\right]}$$

$$= 2.6 \frac{\text{g}}{\text{L}}$$

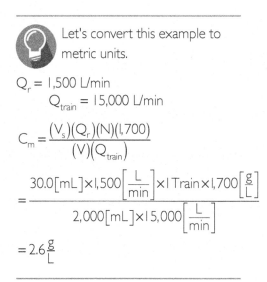

Let's convert this example to metric units.

$Q_r = 1,500$ L/min

$\quad Q_{train} = 15,000$ L/min

$$C_m = \frac{(V_s)(Q_r)(N)(1,700)}{(V)(Q_{train})}$$

$$= \frac{30.0[\text{mL}] \times 1,500\left[\dfrac{\text{L}}{\text{min}}\right] \times 1 \text{ Train} \times 1,700\left[\dfrac{\text{g}}{\text{L}}\right]}{2,000[\text{mL}] \times 15,000\left[\dfrac{\text{L}}{\text{min}}\right]}$$

$$= 2.6 \frac{\text{g}}{\text{L}}$$

3. Record and examine the results of three C_m values calculated during a 24-hour period:

Sample	Time	Average, C_m
1	8:00 am	2.6 g/L
2	Noon	2.4 g/L
3	4:00 pm	2.1 g/L

An examination of the results reveals a downward trend in the microsand concentration, which would indicate increasing losses of microsand. As long as the ACTIFLO® effluent quality is acceptable, microsand does not need to be added until the system concentration drops below 2 g/L.

3.12.3.2   Microsand Addition

Microsand is added to the system at the injection basin or maturation basin of each train wherever possible. Avoid adding microsand immediately near the basin mixer. Large quantities of microsand could adversely impact the mixer impeller, causing dangerous deflections and stresses to the shaft, which may result in damage to the entire mixer assembly. Fine particles are always present in raw microsand. Expect the turbidity of the clarified water to increase slightly after the addition of new microsand.

The microsand concentration in the ballasted flocculation process should be maintained within the recommended operating range of 2 to 6 g/L. Never allow the microsand concentration to exceed 20 g/L or the scraper could become overloaded, resulting in possible damage to the motor assembly or seals. Microsand concentrations greater than 6 g/L can increase the microsand lost and removed with the sludge by overloading the hydrocyclone, but will not harm the process performance. High microsand concentrations will decrease over time until the 2 to 6 g/L range is achieved again.

Microsand may be delivered in sacks or in bulk, depending on the capacity of the treatment system. Operators must follow procedures regarding the delivery and handling of sack or bulk material. Microsand should be stored properly until use to prevent damage or contamination.

When working with microsand, always refer to the product's **safety data sheet (SDS)** for appropriate ventilation and breathing information.

3.12.4 Hydrocyclone Performance

When operating the ballasted flocculation process, several factors can influence the performance of the hydrocyclone (see Figures 3.20, 3.21, and 3.22). Operators typically adjust and control these factors while the cyclone is operating.

Figure 3.20   Hydrocyclones

Figure 3.21   Hydrocyclone, with microsand/sludge feed stream entering and overflow drain leaving

safety data sheet (SDS)
Safety data sheets (SDSs) are an essential component of the Globally Harmonized System of Classification and Labeling of Chemicals (GHS) and are intended to provide comprehensive information about a substance or mixture for use in workplace chemical management. They are used as a source of information about hazards, including environmental hazards, and to obtain advice on safety precautions. In the GHS, they serve the same function that the material safety data sheet (MSDS) does in OSHA's Hazard Communication Standard. The SDS is normally product related and not specific to the workplace; nevertheless, the information on an SDS enables the employer to develop an active program of worker protection measures, including training, which is specific to the workplace, and to consider measures necessary to protect the environment.

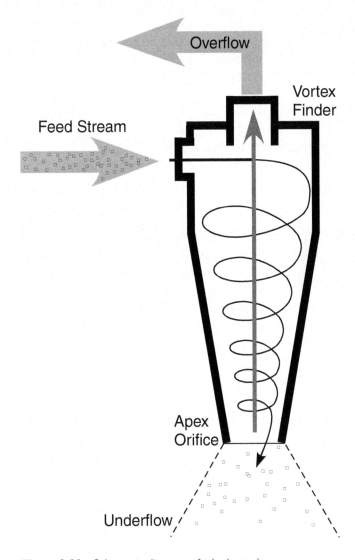

Figure 3.22 Schematic diagram of a hydrocyclone

1. Feed Solution

   The concentration of the hydrocyclone feed solution influences the microsand recovery of the hydrocyclone. The more diluted (less microsand) the feed solution, the better the microsand recovery through the hydrocyclone. Therefore, even though the system is capable of maintaining high microsand concentrations and performing satisfactorily, to minimize microsand losses from the hydrocyclone, do not overload the system with microsand.

2. Underflow

   Typically, the apex produces a 20 to 30 degree cone discharge. Under the condition of extremely high underflow solid concentration, the apex will create a "roping" discharge. A rope is an indication that the apex is not allowing all of the coarse solids out and consequently some are being forced out the cyclone overflow. The operator must keep the apex of the hydrocyclone free from obstructions. If the apex becomes clogged, microsand will be forced out the system.

3. Inlet Pressure

   To ensure efficient microsand/sludge separation, the pressure at the inlet to the hydrocyclone should be approximately 20 to 30 psi (140 to 210 g/cm^2).

4. Overflow

   There should be no detectable presence of microsand in the overflow of the hydrocyclone. Verify this periodically (weekly or when excessive levels of sand loss are encountered) by taking a sample of the overflow with an Imhoff Cone or a graduated cylinder and measuring the microsand concentration after allowing a settling time of 1 minute.

5. Clogging

   If higher than expected microsand usage rates are observed, the problem could be caused by debris being held in one or all of four areas of the hydrocyclone. Always inspect all four areas because debris could be causing obstructions in more than one area. To inspect any part of the hydrocyclone, always take the hydrocyclone out of service by locking out and tagging out the microsand recirculation pump that supplies the slurry.

The tip of the apex is the easiest area to check for debris that is causing an obstruction. The underflow from the hydrocyclone should be a uniform shape throughout the entire spray area. When a portion or even the entire underflow spray pattern is irregular, this is a good indication of debris lodged in the apex tip. To locate debris, take the hydrocyclone out of service and disassemble the lowest section of the hydrocyclone body. Remove debris and reassemble the hydrocyclone.

Depending on the size of the obstruction, it could be lodged in any portion of the hydrocyclone body without showing any irregular underflow patterns. To inspect the body of the hydrocyclone, take the hydrocyclone out of service. Then remove the lower portion of the hydrocyclone body. Inspect and clean out any debris in the removed lower portion and the upper portion of the body still attached. Any debris found should be identified and noted in case similar material is found during other inspections. If the same type of debris is responsible for recurring obstructions, steps should be taken to minimize or eliminate the entrance of this debris to the process trains. Once the hydrocyclone has been inspected and all obstructions removed, reassemble the hydrocyclone body.

Obstructions in the area of the hydrocyclone overflow/vortex finder do not create irregular underflow spray patterns, but higher than normal microsand usage is a good indication of an obstruction. To inspect this area, take the hydrocyclone out of service. Then remove the overflow pipe. After the overflow pipe has been removed and there is a clear view into the vortex finder, inspect for and clean out any obstructions.

Obstructions in the area of the hydrocyclone inlet will not create irregular underflow spray patterns, so higher than normal microsand usage can be an indication of obstructions. To inspect the hydrocyclone inlet area, take the hydrocyclone out of service. Disconnect the influent piping to the hydrocyclone from the inlet. Remove the inlet head to inspect the inlet area and inlet chamber of the hydrocyclone for debris.

3.12.5   Process Performance Monitoring

Procedures should be developed to monitor the performance of the ballasted flocculation process. The procedures depend mainly on the influent raw water, the product effluent or treated water, and the equipment and processes in the water treatment plant. The frequency of monitoring the influent and effluent water quality and equipment performance can depend on influent characteristics and changes and the status of the processes and equipment. Typically, performance monitoring and analysis may be scheduled on an hourly basis, every 4 hours, or on a daily basis for the following factors:

- Influent or Raw Water: flow rate, turbidity, color, alkalinity, pH, temperature
- Effluent or Product Water: daily total flow rate, turbidity, color, alkalinity, pH
- Hydrocyclone: inlet pressure (25 to 30 psi), underflow flow rate, underflow spray pattern
- Chemical Dosing: alum feed rate and dosage, polymer feed rate and dosage, acid feed rate and dosage, base feed rate and dosage
- System Microsand Concentration: 2 to 6 g/L

3.12.6   Process Optimization

This section outlines the information needed and the procedures to follow to optimize the ballasted flocculation process. This process to identify and select the process variable and operating guidelines that can give the best performance should be developed at the plant on the basis of raw water treated, required product water quality, and the equipment at the plant. All important data must be collected, analyzed, stored, and available for future comparisons and decision-making processes. Water quality indicators that should be monitored for optimization include:

- Turbidity
- Total organic carbon (TOC)
- Color
- Algae
- Total suspended solids (TSSs)
- Trihalomethanes (THMs)
- Haloacetic acids (HAAs)

To optimize the process, first determine the optimum pH for coagulation. Then, optimize the coagulant dose at the optimum pH, and finally optimize the polymer dose at the optimum coagulant dose and pH. In addition to following this sequence, operators can ensure proper optimization by verifying proper operation of all process-related equipment, instruments, and chemical feeders before optimization; changing only one variable at a time; waiting a minimum of two hydraulic detention times before recording data or making another change; and recording influent water data and performance for reference when similar conditions exist.

3.12.7 Troubleshooting Procedures

Possible problems operators may encounter when operating a ballasted flocculation process include elevated clarified water turbidity, scraper high torque alarm, low-pressure alarm for microsand recirculation pump pressure switch, microsand leakage from recirculation pump, and microsand loss from the system. Table 3.5 lists potential problems, possible

Table 3.5 Troubleshooting guide for operating a ballasted flocculation process

Problem	Possible Cause	Necessary Check	Remedies
Elevated Clarified Water Turbidity	Change in raw water quality	Check raw water quality	1. Perform jar testing to predetermine the optimum coagulant dosage and polymer dosage 2. Adjust coagulant dosage and polymer dosage
	Loss of coagulant	1. Check coagulant dosage 2. Conduct pump drawdown	Resume normal coagulant dosing
	Loss of polymer	1. Check polymer dosage and pump drawdown 2. Check polymer diffuser to verify that the polymer solution is delivering	Resume normal polymer dosing

	Low microsand concentration	1. Check system microsand concentration
		2. If the microsand concentration is significantly lower than the normal range, perform the following microsand loss checking procedures:
		  a. Check hydrocyclone apex to see if there is any obstruction forcing microsand out of system
		  b. Check hydrocyclone overflow to verify no detectable presence of microsand
		3. Check to ensure that the maturation tank mixer and the sludge scraper are operating correctly
			1. If microsand concentration is less than 2 g/L and there is no noticeable microsand loss, increase microsand concentration to within 2 to 6 g/L range
			2. If microsand loss is occurring, remove the hydrocyclone apex and clean out the obstruction. Add microsand to increase microsand concentration to 2 to 6 g/L range
			3. a. Restart stopped equipment
			  b. After 30 minutes, check microsand concentration and adjust accordingly

Scraper High Torque Alarm	Mechanical condition	Check general mechanical condition of scraper and the lubricating oil level of scraper	Refer to scraper supplier's O&M manual for remedy information
	Foreign object	Check if foreign object has fallen into the tank	It will be necessary to drain the tank if the item cannot be retrieved
	Overloaded	Check if the system is overloaded with microsand	If the system is overloaded with microsand, contact manufacturer
	Unknown	If the cause cannot be determined, do not attempt to restart it until after tank is drained and inspected	
Low-Pressure Alarm for Microsand Recirculation Pump Pressure Switch	Pump inlet pipe clogging	Check pump inlet pipe to make sure there are no obstructions	Perform suction pipe flushing procedures a. Close the discharge side plug valve b. Connect the flush line c. Turn on water for continuous 5 minutes d. Put the equipment back into normal operation e. Restart the pump
Microsand Leakage from Microsand Recirculation Pump	Mechanical seal failure	Check general mechanical condition of microsand recirculation pump	Refer to pump supplier's O&M manual for remedy information

Table 3.5   Troubleshooting guide for operating a ballasted flocculation process *(continued)*

Problem	Possible Cause	Necessary Check	Remedies
Microsand Loss from the System	Loss of Treatment	Check that chemical dosages are correct and that chemicals are being injected into system	Resume normal polymer and coagulant dosing
	Clog in hydrocyclone underflow	Check for underflow blockage	If obstruction is present, remove, clean, and replace apex tip and restart system
	Loss of maturation mixer or settling tank scraper	Make sure maturation mixer and settling tank scraper is operating (shafts are rotating clockwise)	Correct if failure exists and resume normal operation

causes, necessary checks, and appropriate remedies to correct potential problems.

3.12.8   Intermittent Use

Occasionally, a ballasted flocculation process will be operated intermittently. This section provides the information on the proper procedures to prepare for intermittent operation. The ballasted flocculation process can operate effectively under repeated start and stop operations without affecting process performance or equipment. The process has the ability to perform successfully within 10 to 15 minutes of startup. See Section 3.8.2, "Shutdown Procedure," page 167, for additional information regarding procedures to follow when shutting down the system.

3.12.8.1   Short-Term Standby Mode (Wet Storage)

The short-term standby mode is a shutdown condition in which microsand and water are retained in the ballasted flocculation system to keep the system active and ready to be restarted. Most plants use this wet storage standby mode during offline conditions. This method is recommended instead of dry storage because it allows for fast and easy system startup, allows for all mechanical equipment to be exercised under full load every other week (biweekly), and does not require microsand removal from the system.

1. Process Shutdown Sequence

   Ensure that each process train goes through the correct process shutdown sequence. If done through the control system, this should be done automatically. If done manually, refer to the O&M manual for proper sequence. If freezing conditions are not expected, the system can be drained down to a water level just above the tube settler. When the system is drained, minimal microsand will be lost, primarily in the areas around the drains.

2. Exercise the Mechanical Equipment

   The equipment should be exercised every 2 weeks for 20 minutes. This keeps the gears and bearings freshly lubricated with oil and grease and provides for longer equipment life. The microsand that is in the system will be unaffected and will remain in the bottom of the process tanks and basins. During equipment exercising and system restart, the microsand will be readily resuspended.

3.12.8.2 Long-Term Shutdown and Freeze Protection

If the ballasted flocculation system will be out of service for periods longer than 2 months, either of the following methods is recommended.

1. Continuous Cycle Through the System
   Potable water, treated secondary effluent, or nonpotable water should be continuously cycled through the system. This will allow for periodic exercising of the equipment and will control any odor or growth conditions that may arise.
2. Complete Draining of the Tanks and Basins
   If desired, all microsand can be removed from the system by bypassing the hydrocyclones and sending the microsand recirculation pump discharge to a drain or to a hopper. The cost is minimal on a per ton basis.

Some additional maintenance procedures are required for long-term shutdowns, including immersed instrumentation maintenance and chemical feed maintenance.

Steps should be taken to ensure that the pH probes are not allowed to dry out. The pH meters used are commonly the immersion or insertion type, and as long as the water surface elevation remains above the pH probes, drying out will not be a problem. If the pH probes are likely to be exposed above the water surface for any period of time, they should be placed in a container of distilled water. For recommendations on proper long- and short-term storage of pH probes, refer to the manufacturer's instrument manual.

Most turbidity meters are also of the immersion or insertion type. It is recommended that each turbidity meter (both influent and effluent) by cleaned and wiped down at the beginning of each shutdown period. This will help to prevent any type of biological growth or buildup on the sensor. Follow the manufacturer's recommendations for proper long-term storage.

All chemical feed equipment should be flushed with water if it is not anticipated the system will be used again within 14 days. This is easily accomplished when the chemical feed pumps are supplied with a flush water connection and each pump has a dedicated calibration column.

It is good practice to operate all chemical feed equipment during the biweekly exercise. If a dry polymer system is used, it is better to run water rather than polymer through the system and sumps to prevent the need for flushing and cleaning the polymer out of the system.

Freeze protection requirements vary by location. Various methods are used at existing plants. Local site conditions, operation frequency, and economics should be analyzed to determine the best method of freeze protection.

One method of freeze protection that may be used in extremely cold climates (especially during the time period when the system is not expected to run) is to completely drain the system and store the equipment according to long-term equipment storage instructions from the equipment manufacturer. Before draining the system, the microsand can be easily removed from the system by bypassing the hyrdocyclones and sending it to the existing sewer system or trash hopper.

3.12.9 Equipment Maintenance

Because ballasted flocculation uses some of the same equipment as conventional flocculation, the maintenance activities are similar for those items (see Section 3.10, "Equipment Operation and Maintenance"). However, some additional activities are required for maintenance of equipment specific to ballasted flocculation.

3.12.9.1   Tube Settler Modules

Properly cleaned tube settler modules are essential for the proper operation of the ballasted flocculation process. To properly maintain the modules, two types of cleaning procedures need to be performed. The frequency and duration of each cleaning procedure will vary depending on the raw water quality and the amount and types of chemicals used in the process. Table 3.6 outlines the two types of cleaning procedures.

Table 3.6   Cleaning procedures for tube settler modules

Cleaning Type	Drain Water Level to	Procedure	After Procedure
Type 1 1–2 hour duration 3–4 week frequency	At least 1 foot (0.3 m) below bottom of modules	Spray off the tops and all module sections Rinse thoroughly	Place train back in service
Type 2 3–6 hour duration 6–12 month frequency	Low enough for operator to walk on bottom of tank	Spray off material collected on bottom of modules Rinse thoroughly	Inspect underside of modules for proper placement, correct support, and damage Place train back in service

Type 1 cleaning is done on an as needed basis depending on the visual appearance of the tube settler modules. However, Type 2 cleaning should be done regularly within the estimated time periods regardless of the appearance of the tube settler modules because material caught on the bottoms of the modules may not be readily visible. Also, regularly performing Type 1 cleaning may reduce the duration of Type 2 cleaning.

For all cleaning procedures, stop the influent flow to the train and drain the water in the settling basin down to the appropriate level. Use a water hose with no more than 40 to 60 psi (2.8 to 4.2 kilograms/square centimeter) of water pressure to spray the tube settler modules. Remember that the tubes are fragile and will not support your weight. Use appropriate safety precautions.

3.12.9.2   Mixers

Never operate the ballasted flocculation mixers and microsand recirculation pumps without water in the tanks and basins. Always be sure the tanks and basins are full of water before operating this equipment. Operators must perform the following important maintenance tasks:

- Ensure that the drive remains clean
- Check the oil level in the drive on a regular basis
- Change gear oil every 6 months or 2,500 hours, whichever occurs first; refer to mixer O&M manual for exact instructions
- Grease bearings when changing oil
- Keep motors clean and dry
- Tighten terminal connections, assembly screws, bolts, and nuts as required

- Check insulation resistance of motors periodically
- Keep external airway on air-cooled motors clear because obstructions will restrict air passage

3.12.9.3 Scrapers

Operators must perform the following important maintenance tasks:

- Ensure that the drive remains clean
- Inspect oil level in the variable speed reducer every week
- Inspect all reducers every ten years or as indicated by gearbox manufacturer; overhaul as required
- Keep motors clean and dry
- Tighten terminal connections, assembly screws, bolts, and nuts as required
- Check insulation resistance of motors periodically
- Keep external airway on air-cooled motors clear because obstructions will restrict air passage (see the scraper supplier's O&M manual for further information)

3.12.9.4 Microsand Recirculation Pumps

Always operate mixers and microsand recirculation pumps when there is water in the tanks and basins. Be sure the tanks and basins are full of water before operating this equipment. Refer to the pump supplier's O&M manual for further information.

- Ensure that the drive remains clean
- Gland leakage should be checked periodically to ensure that adequate lubrication water is being provided to the seal
- Take oil sample every month and change when necessary
- Grease bearings monthly at a minimum
- Keep motors clean and dry
- Tighten terminal connections, assembly screws, bolts and nuts as required
- Check insulation resistance of motors periodically
- Keep external airway on air-cooled motors clear because obstructions will restrict air passage

3.12.9.5 Hydrocyclones

Refer to the hydrocyclone supplier's O&M manual for further information.

- Inspect hydrocyclone on a monthly basis
- Maintain smooth surfaces on the interior of the cyclone
- Replace parts when interior surfaces become worn or uneven
- Replace apex tip when wear exceeds 1/8 inch (0.32 cm)
- Check cone for wear while replacing apex tip to make sure that internal surfaces are smooth and a lip does not occur

Check Your Understanding

1. What is the main difference between enhanced coagulation and sweep-type coagulation methods?

2. What is the relationship between pH and color in a water sample?

3. How does turbidity affect the measurement of color in a water sample?

4. How are the optimum dosages of acid, alkalinity, and coagulant determined for enhanced coagulation?

5. Why does ballasted flocculation require more operator expertise than conventional coagulant systems?

6. What is the role of hydrocyclones in the ballasted flocculation process?

7. Describe the process to estimate the microsand concentration in a ballasted flocculation system.

8. List the factors that can influence hydrocyclone performance in a ballasted flocculation system.

3.13 Math Assignment

Turn to Appendix A, "Introduction to Basic Math for Operators," and read Section A.6, "Metric System."

Read and work through the example problems in sections:

- "Learning Objectives"
- A.9.2, "Flows" (English System)
- A.10.2, "Flows" (Metric System)
- A.9.3, "Chemical Doses" (English System)
- A.10.3, "Chemical Doses" (Metric System)
- A.9.5, "Coagulation and Flocculation" (English System)
- A.10.5, "Coagulation and Flocculation" (Metric System)

Check the math using a calculator. You should be able to get the same answers.

3.14 Additional Resources

Your local library may be a good source for obtaining these additional resources related to the field of water treatment:

Manual of Water Utilities Operations (Texas Manual). Texas Water Utilities Association, www.twua.org. ASIN: B000KT9EIU. Chapters on water chemistry and coagulation and sedimentation.

Stone, B. G. Notes from "Design of Water Treatment Systems," CE-610, Loyola Marymount University, Los Angeles, CA, 1977.

Water Quality and Treatment, 5th Edition. American Water Works Association (AWWA), www.awwa.org. ISBN: 0-07-001659-3.

Water Treatment Plant Design, 4th Edition. American Water Works Association (AWWA), www.awwa.org. ISBN: 0-07-141872-5.

Operation and Maintenance Manual for Stockton East Water Treatment Plant, James M. Montgomery, Consulting Engineers, Inc., 1979. No longer available.

Chapter Review

3.1 Removing Particulates from Water

1. What is the purpose of coagulation and flocculation?
   1. To control corrosion
   2. To kill disease-causing organisms
   3. To remove leaves, sticks, fish, and other debris
   4. To remove particulate impurities
2. What is coagulation?
   1. Clumping together of fine particles into larger particles with chemicals
   2. Gathering together of particles by gentle stirring
   3. Settling out particles by reducing water velocity
   4. Solidifying large particles by evaporation
3. What is flocculation?
   1. Clumping together of fine particles into larger particles with chemicals
   2. Gathering together of particles by gentle stirring
   3. Settling out of particles by reducing water velocity
   4. Solidifying large particles by evaporation

3.2 Coagulation

4. How long should the coagulation reaction last?
   1. A few days
   2. A few hours
   3. A few minutes
   4. A few seconds
5. Why are coagulant aids used?
   1. To add density and toughness to flocs
   2. To increase the speed of the chemical reaction
   3. To reduce the costs of coagulants
   4. To separate floatables from floc

6. What is the most commonly used coagulation chemical?
   1. Alum
   2. Hypochlorites
   3. Metallic salts
   4. Polymers
7. How can an operator verify the effectiveness of the coagulant chemicals and dosages being applied at the flash mixer?
   1. Determine finished water chlorine demand
   2. Measure raw and finished water temperatures
   3. Measure source water turbidity levels
   4. Perform a series of jar tests
8. What is the disadvantage of mechanical mixers in coagulation?
   1. Higher electricity use than other methods
   2. Less reliable than other methods
   3. More expensive to purchase than other equipment
   4. Slower mixing than other methods

3.3 Flocculation

9. What is the purpose of the flocculation process?
   1. To create floc of good size, density, and toughness
   2. To disinfect floc with chemicals such as metallic salts
   3. To remove floc through sedimentation and filtration
   4. To reverse the clumping that occurred during coagulation

3.5 Process Control

10. How are the proper type and amount of coagulant chemicals determined?
    1. By conducting jar tests at least daily
    2. By reading labels and safety data sheets (SDS)
    3. By spot checking samples at designated sampling sites
    4. By timing the total transit time through the treatment plant

3.6 Normal Operations

11. How can poor process performance be spotted early so that corrective measures can be taken?
    1. Adjust mixer speed often to determine process changes
    2. Monitor water quality throughout the water treatment process
    3. Sample one water quality indicator at one point in the process
    4. Rely on turbidity of filtered water as an indication of overall process performance

12. What is the detention time (minutes) in a rectangular flocculation basin when the flow is 0.9 Mgal/d? The basin is 24 feet long, 12 feet wide, and 8 feet deep.
    1. 27 min
    2. 28 min
    3. 29 min
    4. 30 min

13. What is the basic purpose of the jar test?
    1. Determine whether chemicals from suppliers meet specifications
    2. Duplicate plant detention times and mixing and settling conditions in the laboratory
    3. Prepare water for taste and odor trials with consumers
    4. Test newly developed chemicals for use in treatment plants

14. After a jar test is completed, what does a hazy settled water indicate?
    1. Poor coagulation
    2. Poor flocculation
    3. Poor mixing
    4. Poor turbidity

15. Which of the following factors are not important when evaluating jar test results?
    1. Amount of floc formed
    2. Clarity of water between floc particles
    3. Floc settling rate
    4. Paddle construction

16. Determine the setting on a dry alum feeder in pounds per day when the flow is 1.2 Mgal/d. Jar tests indicate that the best alum dose is 9 mg/L.
    1. 88 lb/day
    2. 90 lb/day
    3. 92 lb/day
    4. 102 lb/day

17. What is a desirable floc appearance?
    1. Bluish tint
    2. Milky appearance
    3. Popcorn flake
    4. Tiny alum floc

18. A water treatment plant used 24 pounds of cationic polymer to treat 1.4 million gallons of water during a 24-hour period. What is the polymer dosage in mg/L?
    1. 1.8 mg/L
    2. 2.0 mg/L
    3. 2.1 mg/L
    4. 2.3 mg/L

19. Why should operators keep polymer powders off of floors?
    1. They become extremely slippery when wet
    2. They become tripping hazards when hardened
    3. They become unsightly when dispersed
    4. They become very sticky when stepped on

20. Liquid polymer is supplied to a water treatment plant as a 15 percent solution. How many gallons (gal) of liquid polymer should be mixed in a barrel with water to produce 100 gallons (gal) of 0.5 percent polymer solution?
    1. 0.03 gal
    2. 0.33 gal
    3. 33 gal
    4. 3.3 gal

3.11 Enhanced Coagulation

21. Which items are adjusted in the enhanced coagulation process?
    1. Detention time and mixing
    2. Dissolved oxygen and turbidity
    3. pH and coagulant dosage
    4. Total suspended solids and temperature

3.12 Ballasted Flocculation

22. What is the purpose of the hydrocyclones in ballasted flocculation?
    1. To separate the dissolved solids from the water
    2. To separate the floc from the coagulant
    3. To separate the microsand from the sludge
    4. To separate the sludge from the slurry

23. A ballasted flocculation process has two hydrocyclones in operation (two trains) with an influent flow of 4,000 gpm and a microsand slurry recirculation rate of 300 gpm. Six V_s samples from the train produced the following volumes of settled microsand: 30 mL, 25 mL, 25 mL, 40 mL, 30 mL, and 30 mL. The six volumes of samples collected in the cone averaged 2,000 mL. Calculate C_m, the microsand concentration in the tanks, in grams of microsand per liter (g/L).
    1. 1.9 g/L
    2. 2.3 g/L
    3. 2.8 g/L
    4. 3.8 g/L

24. Why should operators maintain a diluted feed solution in ballasted flocculation?
    1. To maximize hydrocyclone process efficiency
    2. To maximize inlet pressure to the hydrocyclone
    3. To minimize microsand losses from the hydrocyclone
    4. To minimize roping discharges at the hydrocyclone apex

25. What is the first step in optimizing a ballasted flocculation process?
    1. Optimize the coagulant dose
    2. Optimize the microsand feed solution
    3. Optimize the pH
    4. Optimize the polymer dose

Sedimentation

CHAPTER OUTLINE

LEARNING OBJECTIVES

1. Identify factors affecting the performance of sedimentation basins using results from observations, samples, laboratory tests, and records.

2. Describe various types of sedimentation basins and how they work.

3. Safely start up, operate, maintain, and shut down sedimentation basins and processes.

KEY TERMS

adsorption	launders	septic	surface loading rate
cathodic protection	point source	shock load	turbidity units
density	precipitate	short-circuiting	viscosity

4.1   Presedimentation

The purposes of the sedimentation process are to remove suspended solids (particles) that are denser (heavier) than water and to reduce the load on the filters (Chapter 6, "Disinfection"). The suspended solids may be in their natural state, such as bacteria, clays, or silts; they may be modified (preconditioned) by prior treatment in the coagulation-flocculation process (to form floc); or they may be **precipitated** impurities such as hardness and iron precipitates formed by the addition of chemicals (Figure 4.1).

Sedimentation occurs when the velocity of the water being treated is decreased below the point where it can transport settleable suspended material, thus allowing gravitational forces to remove particles held in suspension. When water is almost still in sedimentation basins, settleable solids will move toward the bottom of the basin.

Larger-sized particles will settle naturally when surface water is stored for a sufficient period of time in a reservoir or a natural lake, as discussed in Chapters 2, "Source Water, Reservoir Management, and Intake Structures," and 3, "Coagulation and Flocculation." Gravitational forces acting in lakes accomplish the same purpose as sedimentation in water treatment plants; larger particles, such as sand and heavy silts, settle to the bottom.

Because source water diverted directly from rivers or streams may carry sand or silt or be contaminated by overland runoff and **point source** waste discharges, presedimentation facilities such as grit basins, debris dams, or sand traps may be installed to remove some of the heavier particles from the source water. These facilities may be located upstream from the reservoir, diversion works, or treatment plant intake or diversion facilities, and serve to protect the municipal intake pipeline from siltation (settling out of solids). Grit basins may be located between the intake structure and the coagulation-flocculation facilities (Figure 4.2). In addition to reducing the solids-removal load at the water treatment plant, these facilities provide equalizing basins to even out fluctuations in the concentration of suspended solids in the source water.

Ideally, surface waters should be stored in a reservoir and transported directly to the water treatment plant in a pipeline. In a reservoir, the heavier solids can settle out before they reach the plant. However, geographical, physical, and economical considerations (such as the lack of a suitable dam site) often make this alternative impractical.

precipitate (pre-SIP-uh-TATE)
(1) An insoluble, finely divided substance that is a product of a chemical reaction within a liquid. (2) The separation from solution of an insoluble substance.

point source
A discharge that comes out the end of a pipe or other clearly identifiable conveyance. Examples of point source conveyances from which pollutants may be discharged include ditches, channels, tunnels, conduits, wells, containers, rolling stock, concentrated animal feeding operations, landfill leachate collection systems, vessels, or other floating craft.

4.2   Process Performance

The size, shape, and weight of the particles to be settled out, as well as physical and environmental conditions in the sedimentation tank, have a significant impact on the type of pretreatment needed and the sedimentation process efficiency.

Factors affecting particle settling include:

1. Particle size and distribution
2. Shape of particles
3. Density of particles

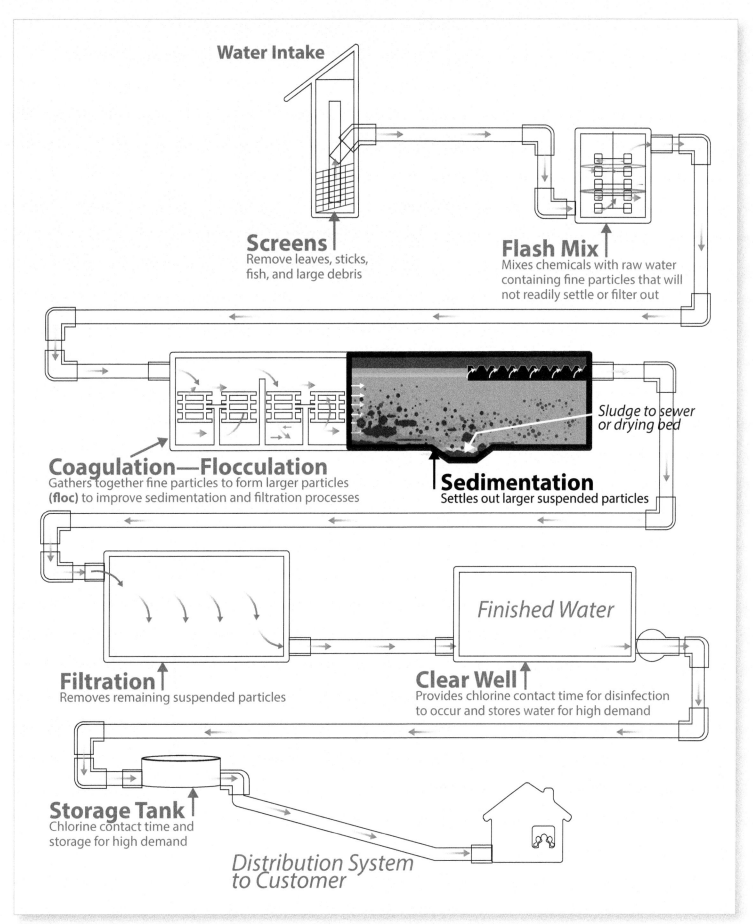

Water Intake

Screens
Remove leaves, sticks,
fish, and large debris

Flash Mix
Mixes chemicals with raw water
containing fine particles that will
not readily settle or filter out

Coagulation—Flocculation
Gathers together fine particles to form larger particles
(**floc**) to improve sedimentation and filtration processes

Sedimentation
Settles out larger suspended particles

*Sludge to sewer
or drying bed*

Filtration
Removes remaining suspended particles

Finished Water

Clear Well
Provides chlorine contact time for disinfection
to occur and stores water for high demand

Storage Tank
Chlorine contact time and
storage for high demand

*Distribution System
to Customer*

Figure 4.1   Flow diagram of a typical plant

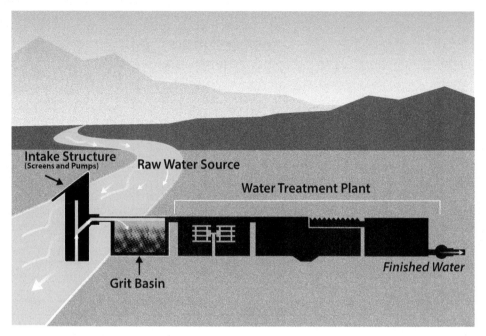

Figure 4.2   Presedimentation facilities

4. Temperature (and related **viscosity** and **density**) of water
5. Electrical charge on particles
6. Dissolved substances in water
7. Flocculation characteristics of the suspended material
8. Environmental conditions (such as wind effects)
9. Sedimentation basin hydraulic and design characteristics (such as inlet conditions and shape of basin)

4.2.1   Nature of Particulate Impurities

Because of their size and density, sand and silt particles greater than 10 microns in diameter (1 micron = 0.001 mm) can be removed from water by sedimentation (simple gravitational settling). In contrast, finer particles do not readily settle and treatment is required to produce larger, denser particles (floc) that are settleable (see Table 4.1).

Table 4.1   Typical size of particles in surface waters

Source	Diameter of Particle (microns)[a]
Coarse Turbidity	1–1,000
Algae	3–1,000
Silt	10
Bacteria	0.3–10
Fine Turbidity	0.1–1
Viruses	0.02–0.26
Colloids	0.001–1

[a]1 micron = 0.0001 cm = 0.00004 of an inch.

viscosity (vis-KOSS-uh-tee)
A property of water, or any other fluid, that resists efforts to change its shape or flow. Syrup is more viscous (has a higher viscosity) than water. The viscosity of water increases significantly as temperatures decrease. Motor oil is rated by how thick (viscous) it is; 20-weight oil is considered relatively thin while 50-weight oil is relatively thick or viscous.

density (DEN-sit-tee)
A measure of how heavy a substance (solid, liquid or gas) is for its size. Density is expressed in terms of weight per unit volume, that is, grams per cubic centimeter or pounds per cubic foot. The density of water (at 39°F or 4°C) is 1.0 gram per cubic centimeter or about 62.4 pounds per cubic foot or 1.0 grams per cubic centimeter.

The shape of particles also influences particle settling. Smooth circular particles will settle faster than irregular particles with ragged edges.

Most particles have a very slight electrical charge. If all of the particles have a negative charge, they will tend to repel each other and not settle. Because alum consists of aluminum with a positive charge, the negatively charged particles are attracted to the positively charged aluminum ions. This causes the clumping together that helps the particles to settle out.

4.2.2 Water Temperature

Another consideration in sedimentation is the effect of water temperature changes. The settling rate (settling velocity) of a particle becomes much slower as the temperature drops because of the increasing viscosity and density of water as it cools. The colder the water temperature becomes, the longer particles take to settle out. Water is similar to syrup in this regard. The colder syrup becomes, the longer it would take a marble to settle to the bottom of the container. This means that longer time periods (lower flows) are required for effective settling at colder water temperatures, or that chemical dosages must be adjusted for the slower settling velocities.

4.2.3 Currents

Several types of currents are found in the typical sedimentation basin:

1. Surface currents caused by winds
2. Density currents caused by differences in suspended solids concentrations and temperature differences
3. Eddy currents produced by the flow of the water coming into and leaving the basin

Currents in the sedimentation basin are beneficial to the extent that they promote flocculation. Collectively, however, these currents distribute the suspended particles unevenly throughout the basin, thereby reducing the expected performance of the sedimentation basin.

Some of these currents can be substantially reduced in the design of a treatment plant by providing baffled inlets and other hydraulic control features (see Section 4.3, "Sedimentation Basins"). Others, such as wind-induced currents, can only be eliminated by providing covers or suitable windbreaks for the sedimentation basins. In most instances, basin covers are not economically feasible nor necessarily desirable from an operations and maintenance standpoint.

4.2.4 Particle Interactions

Suspended particles will continue to clump together (form floc) and other particles will precipitate from solution through flocculation and chemical precipitation in the sedimentation basin. The density and volume of particles will change. As a result, the settling velocities of individual particles will change as larger, denser floc particles are formed when particles of different size and density collide during the sedimentation process. Generally, this results in increased settling velocities.

Check Your Understanding

1. What are the purposes of the sedimentation process?

2. What causes sedimentation to occur?

3. Where are presedimentation facilities installed?

4. List the factors that affect particle settling in a sedimentation basin.

5. What is the purpose of treatment before sedimentation?

6. What types of currents may be found in a typical sedimentation basin?

4.3   Sedimentation Basins

For convenience in discussing sedimentation basins, a typical sedimentation basin can be divided into four zones (see Figure 4.3):

1. Inlet zone
2. Settling zone
3. Sludge zone
4. Outlet zone

The inlet zone to the sedimentation basin should provide a smooth transition from the flocculation basin and should distribute the flocculated water uniformly over the entire cross section of the basin. A properly designed inlet, such as a perforated baffle wall (see Figures 4.4 and 4.5), will significantly reduce **short-circuiting** of water in the basin. An inlet baffle wall will also minimize the tendency of the water to flow at the inlet velocity straight through the basin, minimize density currents due to temperature differences, and minimize wind currents, as previously described.

The settling zone is the largest portion of the sedimentation basin. This zone provides calm, undisturbed storage of the flocculated water for a sufficient time period (3 or more hours) to permit effective settling of the suspended particles in the water being treated.

short-circuiting
A condition that occurs in tanks or basins when some of the flowing water entering a tank or basin flows along a nearly direct pathway from the inlet to the outlet. This is usually undesirable because it may result in shorter contact, reaction, or settling times in comparison with the theoretical (calculated) or presumed detention times.

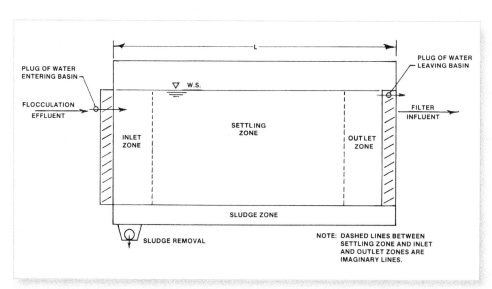

Figure 4.3   Sedimentation basin zones

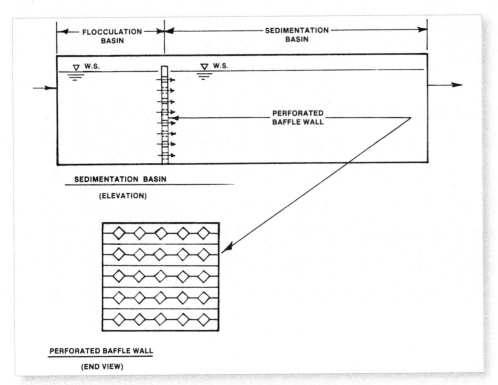

Figure 4.4   Sedimentation basin and inlet baffle wall

Figure 4.5   Sedimentation basin perforated inlet baffle

The sludge zone, located at the bottom of the sedimentation basin, is a temporary storage place for the settled particles (Figure 4.6). Also, the sludge zone is used for compression settling of the sludge; as sludge settles onto the zone, its weight further compacts the sludge below it. If the sludge buildup becomes too great, however, the effective depth of the basin will be significantly reduced, causing localized high flow velocities, sludge scouring, and a decrease in process efficiency. Basin inlet structures should be designed to minimize high flow velocities near the bottom of the sedimentation basin, which could disturb or scour settled particles in the sludge zone, causing them to become resuspended.

Figure 4.6   Sludge buildup in bottom of sedimentation basin

Sludge is removed from the sludge zone by scraper and vacuum devices that move along the bottom of the basin (Figures 4.7, 4.8, and 4.9) as needed or on a regularly scheduled basis. If the removal devices do not operate over the entire length of the basin, it may have to be drained and flushed to remove the sludge.

The outlet zone should provide a smooth transition from the sedimentation basin to the settled water conduit or channel. The outlet can also control the water level in the basin.

In this zone, skimming or effluent troughs, commonly referred to as **launders**, are frequently used to uniformly collect the settled or clarified water (Figures 4.10, 4.11, and 4.12). Adjustable V-notch weirs (Figure 4.12) are generally attached to the launders to enable a uniform draw-off of basin water by controlling the flow. If the water leaving a sedimentation basin flows out unevenly over the weirs or at too high a velocity, floc can be carried over to the filters. The increased loading on the filters may cause shortened filter runs and, therefore, more frequent backwashing.

launders
Sedimentation basin and filter discharge channels consisting of overflow weir plates (in sedimentation basins) and conveying troughs.

Figure 4.7   Traveling bridge (top) and vacuum sweep

Figure 4.8 Sludge scraper

Figure 4.9 Drag-chain and flight

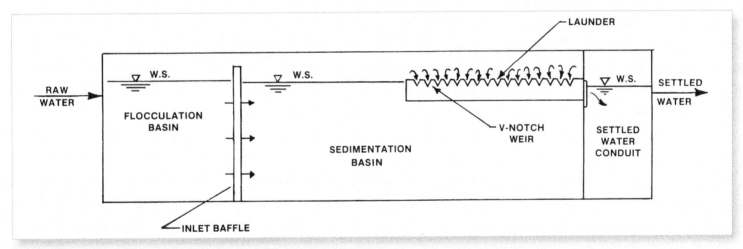

Figure 4.10 Typical sedimentation basin outlet (flow over V-notch weirs)

Figure 4.11 Effluent launders in sedimentation basin

Figure 4.12 Empty sedimentation basin showing sedimentation basin outlet launders and V-notch weirs

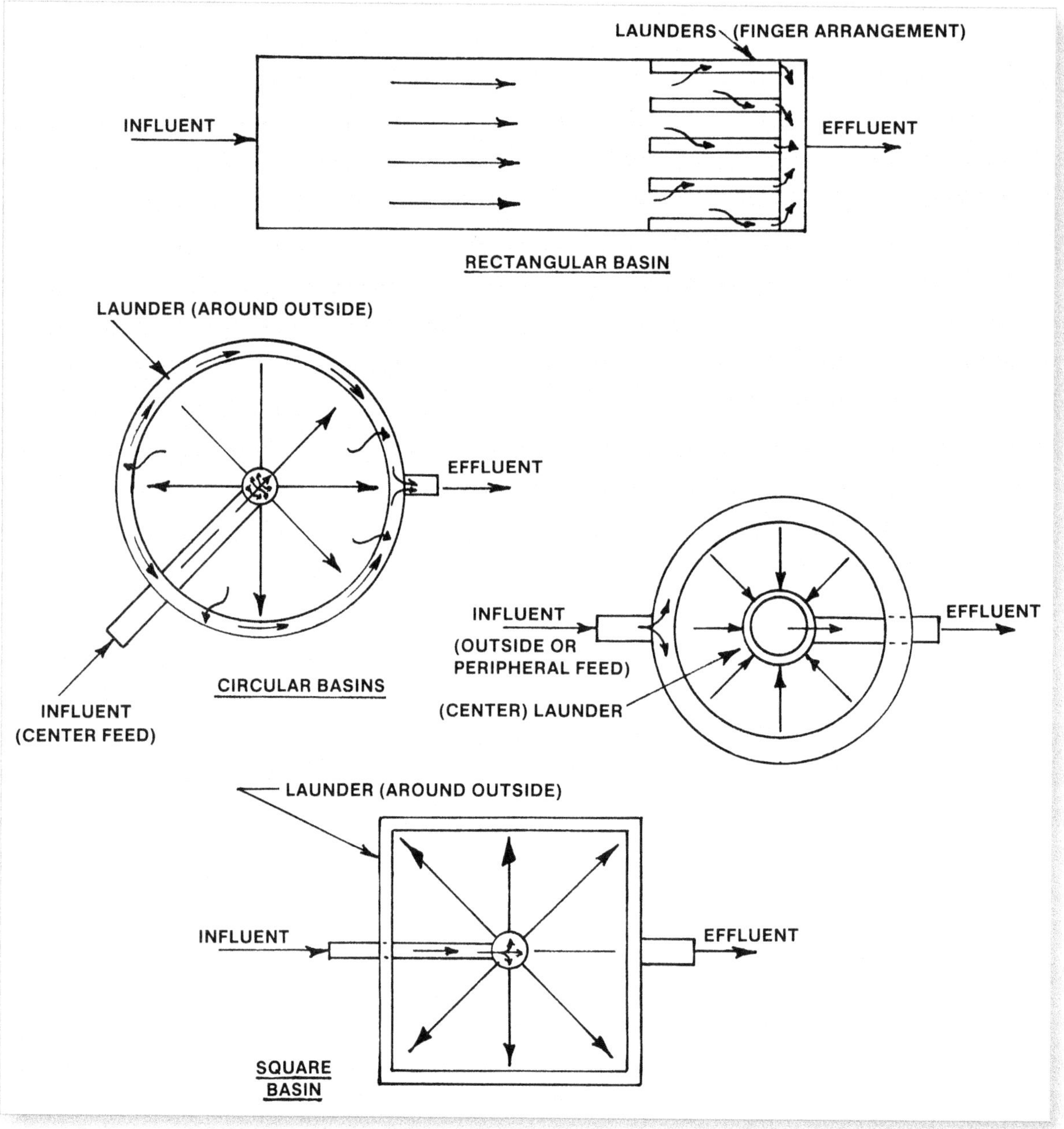

Figure 4.13 Sedimentation basin types (*plan views*)

4.3.1   Basin Types

There are a wide variety of basin types and configurations in use today (Figure 4.13). Several basin types and their major characteristics will be described in the following sections.

4.3.1.1   Rectangular Basins

Rectangular sedimentation basins are commonly found in large-scale water treatment plants. Rectangular basins are popular for the following reasons:

1. High tolerance to **shock loading** (water quality changes)
2. Predictable performance
3. Cost-effectiveness

shock load
The arrival at a water treatment plant of raw water containing unusual amounts of algae, colloidal matter, color, suspended solids, turbidity, or other pollutants.

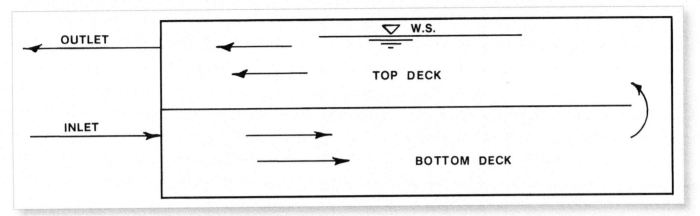

Figure 4.14   Double-deck sedimentation basin (*elevation*)

4. Low maintenance

5. Minimal short-circuiting

4.3.1.2   Double-Deck Basins

Double-deck basins (see Figure 4.14) are an adaptation of the rectangular basin design. By stacking one basin on top of another, double-deck basins provide twice the effective sedimentation surface area of a single basin of equivalent land area. Double-deck basins are designed to conserve land area, but are not in common use owing partially to higher operation and maintenance costs. In this design, sludge removal equipment must operate in both decks, and the entire operation may have to be shut down if an equipment problem develops in either deck.

4.3.1.3   Circular and Square Basins

Circular or square, horizontal-flow basins, as shown in Figures 4.13, 4.15, and 4.16, are often referred to as clarifiers. These basins share some of the advantages of rectangular basin, including predictable performance, cost-effectiveness, and low maintenance. However, they are generally more likely to have short-circuiting and particle removal problems. One of the problem with square and rectangular settling basins is the removal of sludge from the corners. Trained operators will know how to address this challenge.

4.3.1.4   High-Rate Settlers

High-rate or tube settlers were developed to increase the settling efficiency of conventional rectangular sedimentation basins. They also have been successfully installed in circular basins.

Water enters the inclined settler tubes and is directed upward through the tubes as shown in Figures 4.17 and 4.18. Each tube functions as a shallow settling basin. Together, they provide a high ratio of effective settling surface area per unit volume of water. The settled particles can collect on the inside surfaces of the tubes or settle to the bottom of the sedimentation basin.

Parallel plate or tilted plate settlers can also be used to increase the efficiency of rectangular sedimentation basins, and these function in a manner similar to tube settlers.

Clarifier Optimization Package

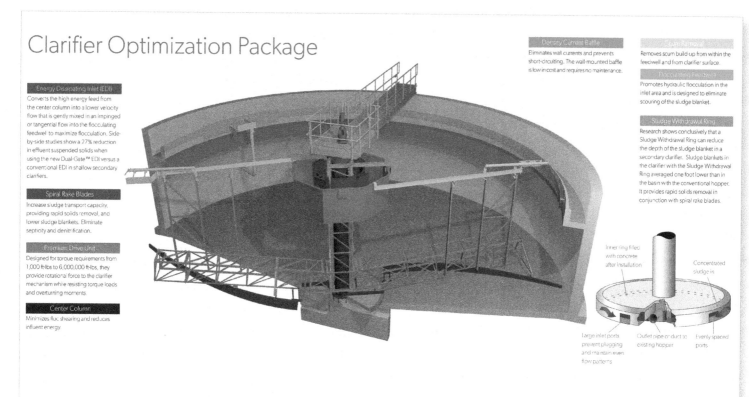

Energy Dissipating Inlet (EDI)
Converts the high energy feed from the center column into a lower velocity flow that is gently mixed in an impinged or tangential flow into the flocculating feedwell to maximize flocculation. Side-by-side studies show a 27% reduction in effluent suspended solids when using the new Dual-Gate™ EDI versus a conventional EDI in shallow secondary clarifiers.

Spiral Rake Blades
Increase sludge transport capacity, providing rapid solids removal, and lower sludge blankets. Eliminate septicity and denitrification.

Premium Drive Unit
Designed for torque requirements from 1,000 ft-lbs to 6,000,000 ft-lbs, they provide rotational force to the clarifier mechanism while resisting torque loads and overturning moments.

Center Column
Minimizes floc shearing and reduces influent energy.

Density Current Baffle
Eliminates wall currents and prevents short-circuiting. The wall-mounted baffle is low in cost and requires no maintenance.

Scum Removal
Removes scum build-up from within the feedwell and from clarifier surface.

Flocculating Feedwell
Promotes hydraulic flocculation in the inlet area and is designed to eliminate scouring of the sludge blanket.

Sludge Withdrawal Ring
Research shows conclusively that a Sludge Withdrawal Ring can reduce the depth of the sludge blanket in a secondary clarifier. Sludge blankets in the clarifier with the Sludge Withdrawal Ring averaged one foot lower than in the basin with the conventional hopper. It provides rapid solids removal in conjunction with spiral rake blades.

Inner ring filled with concrete after installation

Concentrated sludge in

Large inlet ports prevent plugging and maintain even flow patterns

Outlet pipe or duct to existing hopper

Evenly spaced ports

Figure 4.15   Circular clarifier *(Courtesy of WesTech Engineering, Inc.)*

There are several advantages to using high-rate settlers. They are particularly useful for water treatment applications where the site area is limited, in package-type water treatment units, and to increase the capacity of existing sedimentation basins. In existing rectangular and circular sedimentation basins, high-rate settler modules can be conveniently installed

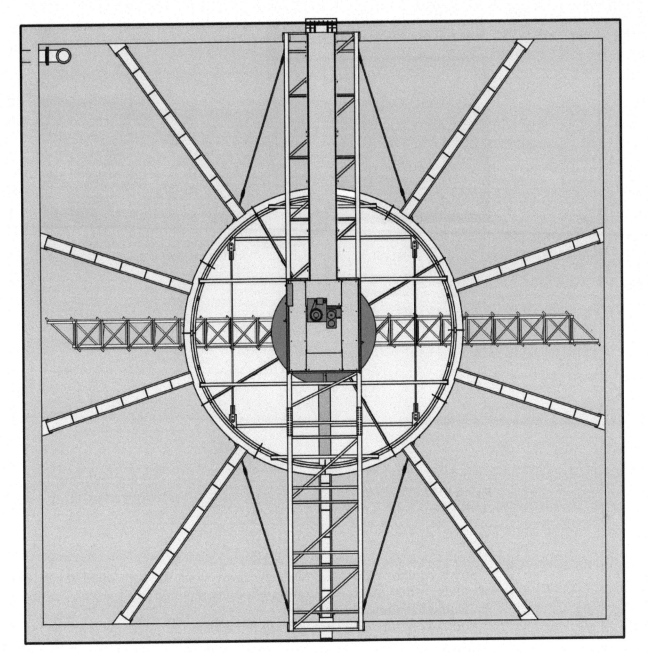

Figure 4.16   Square clarifier basin *(Courtesy of WesTech Engineering, Inc.)*

between the launders. One disadvantage of using tube settlers is that high winds can have an adverse effect on them.

4.3.1.5   Solids-Contact Units

The solids-contact process, also referred to as upflow solids-contact clarification and upflow sludge-blanket clarification, was developed to improve the overall solids removal process under certain design conditions. These units combine the coagulation, flocculation, and sedimentation processes into a single basin, which may be either rectangular or circular in shape. Flow is generally in an upward direction through a sludge blanket or slurry of flocculated, suspended solids as shown in Figure 4.19.

Sedimentation Basin

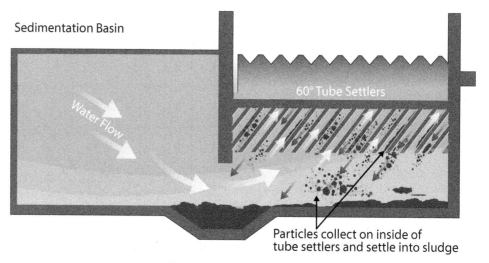

60° Tube Settlers

Water Flow

Particles collect on inside of
tube settlers and settle into sludge

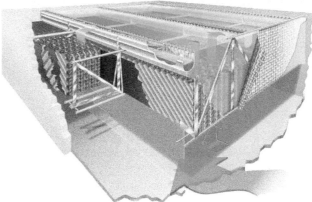

Figure 4.17   Cutaway view of tube settler in rectangular basin *(Bottom figure: Courtesy of Meurer Research, Inc.)*

Solids-contact units generally have provisions for controlled removal of solids so that the concentration of solids retained in the basin can be maintained at some desired level.

Solids-contact units are popular for smaller package-type water treatment plants and in cold climates where the units have to be inside a building. However, care must be exercised in the operation of these units to ensure that a uniform sludge blanket is formed and is subsequently maintained throughout the solids removal process. The sludge blanket is sensitive to changes in water temperature. Temperature density currents tend to upset the sludge blanket. Loss of the sludge blanket will affect the performance of the filters. Other important operational factors include control of chemical dosage and mixing of chemicals.

Under ideal conditions, solids-contact units provide better performance for both turbidity removal and softening processes requiring the precipitation of hardness. With softening processes, chemical requirements are usually lower also. In the case of turbidity removal, coagulant requirements are often higher. In either case, solids-contact units are very sensitive to changes in influent flow or temperature. In these facilities, changes in the rate of flow should be made infrequently, slowly, and with great care.

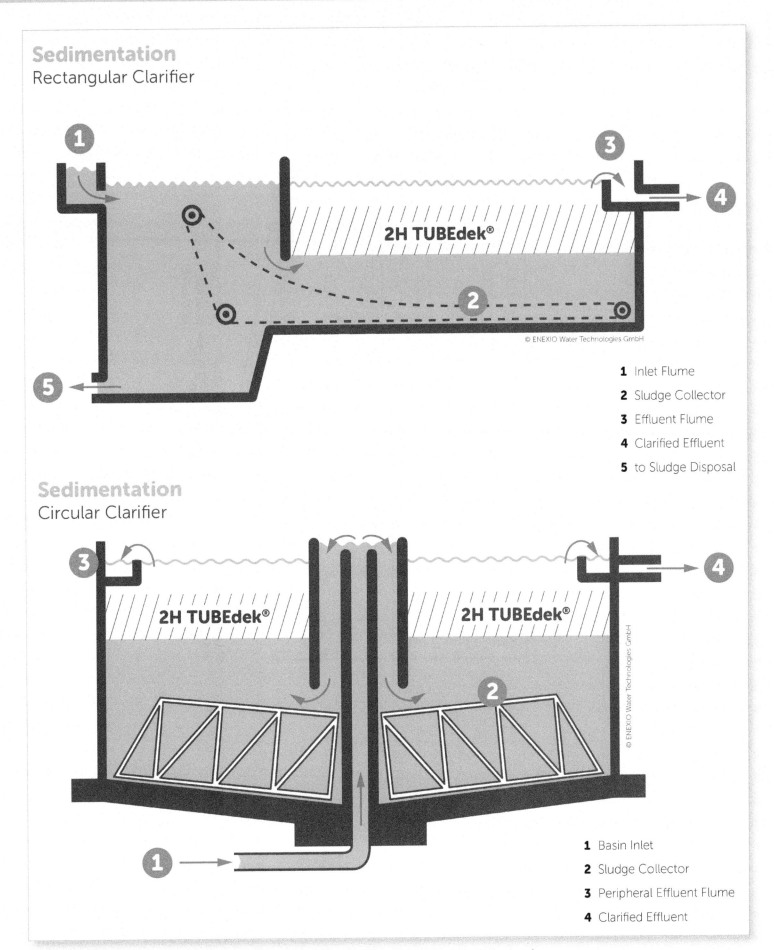

Sedimentation
Rectangular Clarifier

2H TUBEdek®

© ENEXIO Water Technologies GmbH

1 Inlet Flume
2 Sludge Collector
3 Effluent Flume
4 Clarified Effluent
5 to Sludge Disposal

Sedimentation
Circular Clarifier

2H TUBEdek®   2H TUBEdek®

© ENEXIO Water Technologies GmbH

1 Basin Inlet
2 Sludge Collector
3 Peripheral Effluent Flume
4 Clarified Effluent

Figure 4.18   Tube settlers in rectangular and circular clarifiers (*Courtesy of ENEXIO Water Technologies*)

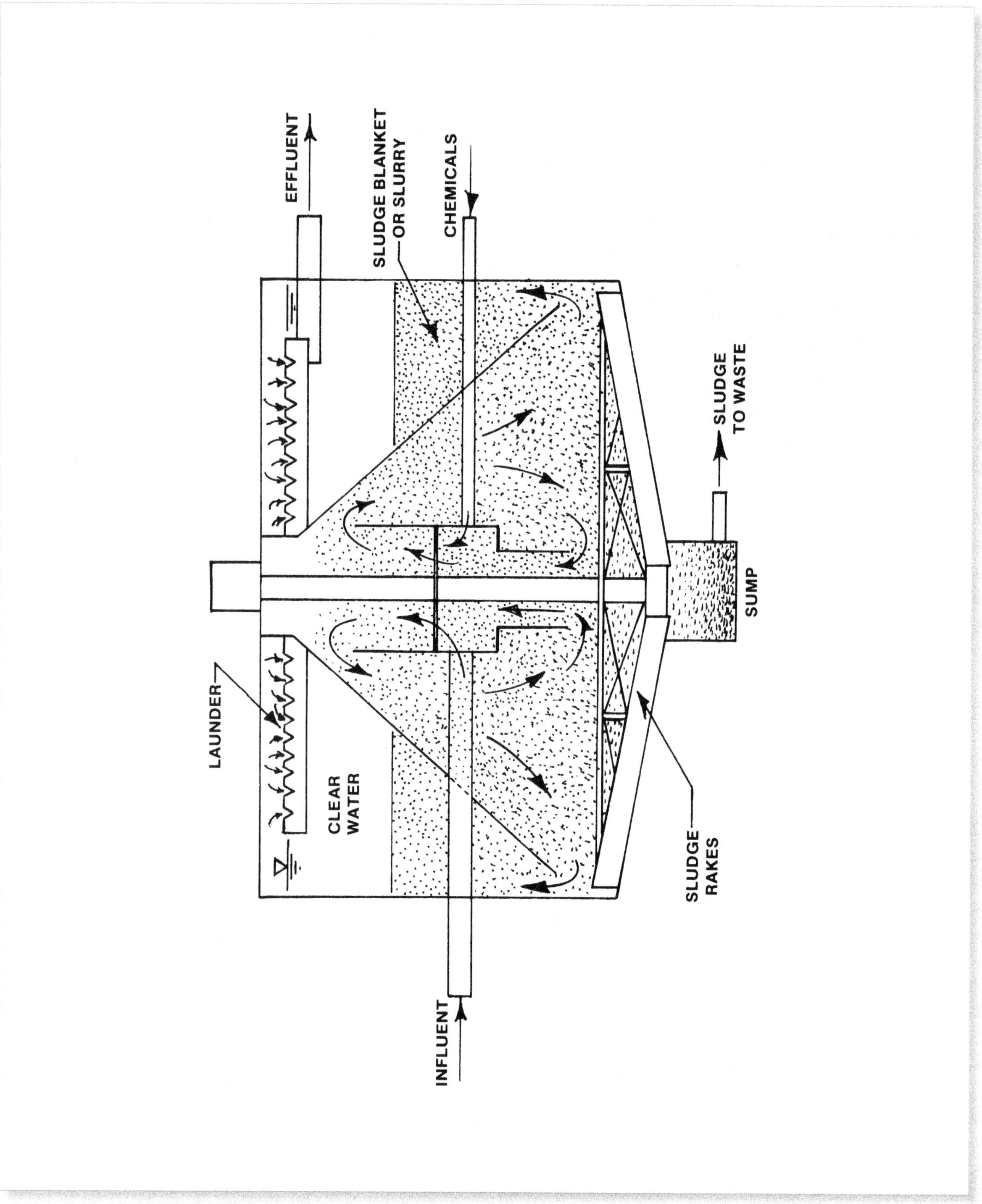

Figure 4.19   Solids-contact unit

For additional information about solids-contact units, see Section 4.5, "Solids-Contact Clarification."

Check Your Understanding

1. List the four zones into which a typical sedimentation basin can be divided.

2. What is the purpose of the settling zone in a sedimentation basin?

3. What are launders?

4. List three possible shapes for sedimentation basins.

5. List the advantages and limitations of double-deck sedimentation basins.

6. Why are rectangular sedimentation basins often preferred over circular basins?

7. During the operation of a solids-contact unit, what items should be of particular concern to the operator?

4.4 Design and Operational Guidelines

Some of the basic guidelines used by engineers to design sedimentation processes are described in this section. Operators should know how these guidelines are obtained and why they are important in order to communicate effectively with design engineers.

By knowing the values used by design engineers and comparing the actual values for the plant, operators may be able to identify the cause of operational problems and poor quality finished water.

4.4.1 Basin Layout

A minimum of two sedimentation basins should be provided in all water treatment plants to allow for maintenance, cleaning, and inspection of a basin without requiring a complete plant shutdown.

A typical rectangular-shaped basin layout is shown in Figure 4.20. A chemical application point is provided in the settled water conduit to permit feeding a filter aid chemical, chlorine, or other chemicals before filtration.

4.4.2 Detention Time

Detention time (or retention time) is the actual time required for a small amount of water to pass through a sedimentation basin at a given rate of flow. Detention time can refer also to the theoretical (calculated) time required for a small amount of water to pass through (or be retained within) a basin at a given rate of flow. The detention time is calculated by dividing the volume of the basin by the flow going into the basin. Actual flow-through times for different small amounts of water in the same basin may vary significantly from the calculated detention time due to short-circuiting, effective exchangeable volume (portion of basin through which the water flows), and other hydraulic considerations such as basin inlet and outlet design. A dye placed at the inlet to a sedimentation basin will produce the curves shown in Figure 4.21 as it leaves the basin. The flatter the curve, the greater the short-circuiting.

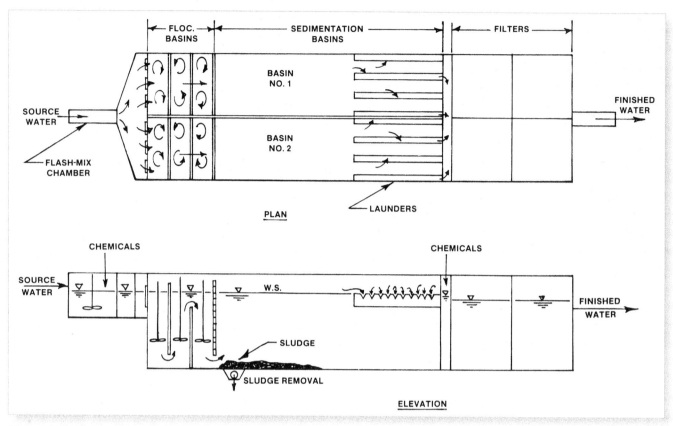

Figure 4.20   Rectangular sedimentation basin

FORMULAS

To calculate the detention time (t) of a sedimentation basin, the volume (V) of the basin and the flow (Q) must be known. The volume of a basin can be calculated from the dimensions of a basin. The dimensions of a basin can be obtained from the plan drawings for the treatment plant. These drawings will have the length, width, and depth for rectangular sedimentation basins and the diameter and depth of circular clarifiers. The flow can be obtained from a flowmeter or flow records.

To calculate the detention time, divide the flow in gallons per day into the tank volume in gallons. To convert this detention time in days to hours, multiply by 24 hours per day.

For rectangular basins:

$$V[\text{gal}] = \text{Length}[\text{ft}] \times \text{Width}[\text{ft}] \times \text{Depth}[\text{ft}] \times \frac{7.48\,\text{gal}}{\text{ft}^3}$$

For circular basins:

$$V[\text{gal}] = 0.785 \times \text{Diameter}^2\,[\text{ft}^2] \times \text{Depth}[\text{ft}] \times \frac{7.48\,\text{gal}}{\text{ft}^3}$$

To calculate the theoretical detention time (t):

$$t[\text{h}] = \frac{V[\text{gal}]}{Q\left[\dfrac{\text{gal}}{\text{d}}\right] \times \dfrac{\text{d}}{24\,\text{h}}}$$

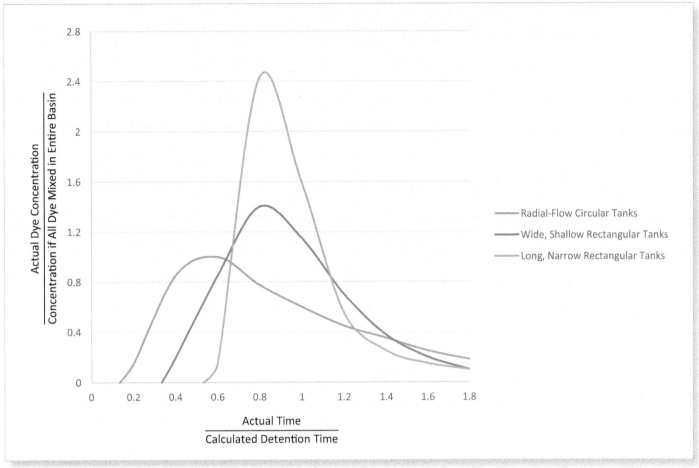

Figure 4.21   Flow-through times for various types of sedimentation basins (*Courtesy of T.R Camp/Water Environment Federation*)

If the size of the basin and design detention time for the sedimentation basin are known, the maximum flow for the basin can be calculated by rearranging the detention time formula:

$$Q\left[\frac{gal}{d}\right] = \frac{V[gal]}{t[h] \times \dfrac{d}{24\,h}}$$

Example 1

A water treatment plant treats a flow of 1.5 MGD. An examination of treatment plant design drawings reveals that the rectangular sedimentation basin is 75 feet (ft) long, 25 feet (ft) wide, and has an effective (water) depth of 12 feet (ft). Calculate the theoretical detention time in hours (h) for the rectangular sedimentation basin.

Known	Unknown
Q = Flow Rate = 1.5 MGD	t = Detention Time, h
L = Length = 75 ft	
W = Width = 25 ft	
H = Depth = 12 ft	

1. Calculate the basin volume (V) in gallons (gal).

$$V[\text{gal}] = L[\text{ft}] \times W[\text{ft}] \times H[\text{ft}] \times \frac{7.48\,\text{gal}}{\text{ft}^3}$$

$$= 75[\text{ft}] \times 25[\text{ft}] \times 12[\text{ft}] \times \frac{7.48\,\text{gal}}{\text{ft}^3}$$

$$= 168{,}300\,\text{gal}$$

2. Determine the theoretical detention time in hours (h).

$$t[\text{hr}] = \frac{V[\text{gal}]}{Q\left[\dfrac{\text{gal}}{\text{d}}\right] \times \dfrac{\text{d}}{24\,\text{h}}}$$

$$= \frac{168{,}300[\text{gal}]}{1{,}500{,}000\left[\dfrac{\text{gal}}{\text{d}}\right] \times \dfrac{\text{d}}{24\,\text{h}}}$$

$$= 2.7\,\text{h}$$

Example 2

What is the maximum flow in MGD for the rectangular sedimentation basin in Example 1 if the theoretical detention time is 2 hours (h)?

Known		Unknown
L = Length	= 75 ft	Q = Maximum Flow
W = Width	= 25 ft	Rate, MGD
H = Depth	= 12 ft	
V = Volume (from Example 1)	= 168,300 gal	
t = Detention Time	= 2.0 h	

Calculate the maximum flow in MGD.

$$Q\left[\frac{\text{gal}}{\text{d}}\right] = \frac{V[\text{gal}]}{t[\text{h}] \times \dfrac{\text{d}}{24\,\text{h}}}$$

$$= \frac{168{,}300[\text{gal}]}{2.0[\text{h}] \times \dfrac{\text{d}}{24\,\text{h}}}$$

$$= 2{,}019{,}600\,\frac{\text{gal}}{\text{d}}$$

$$Q\left[\frac{\text{Mgal}}{\text{d}}\right] = 2.0\,\text{MGD}$$

Example 3

A water treatment plant has a circular clarifier for a sedimentation basin. The treatment plant design drawings indicate that the clarifier has a diameter of 60 feet (ft) and an average water depth of 12 feet (ft). What is the theoretical detention time in hours (h) for the basin when the flow is 2 MGD?

Known	Unknown

Known

D = Diameter = 60 ft
H = Depth   = 12 ft
Q = Flow Rate = 2 MGD

Unknown

t = Detention Time, h

1. Calculate the basin volume (V) in gallons (gal).

$$V[gal] = 0.785 \times D^2[ft^2] \times H[ft] \times \frac{7.48\,gal}{ft^3}$$

$$= 0.785 \times 60^2[ft^2] \times 12[ft] \times \frac{7.48\,gal}{ft^3}$$

$$= 253{,}662\ gal$$

2. Determine the theoretical detention time (t) in hours (h).

$$t[h] = \frac{V[gal]}{Q\left[\dfrac{gal}{d}\right] \times \dfrac{d}{24\,h}}$$

$$= \frac{253{,}662\,[gal]}{2{,}000{,}000\left[\dfrac{gal}{d}\right] \times \dfrac{d}{24\,h}}$$

$$= 3.0\ h$$

Calculating the detention time for a sedimentation basin is not the only way to find problems. Operators can anticipate problems by comparing the actual flow through the water treatment plant with the design flow. Whenever actual flows approach or exceed design flows, problems are likely to develop. Operators do not have to go through any calculations to know that the plant is hydraulically (flow) overloaded.

Also, when water temperature decreases, be prepared to reduce flows if problems should develop. The colder the water, the longer it takes particles to settle out. Reducing flows increases the available detention time.

If the demand for water does not allow for reduced flows, run jar tests with shorter detention times. Adjust the chemical doses as necessary to compensate for the colder water. There is very little an operator can do to control the sedimentation process. Adjusting chemicals and chemical feed rates are the major means by which operators can control water treatment processes.

4.4.3 Surface Loading

Surface loading is one of the most important factors influencing sedimentation. The **surface loading rate** (also called overflow rate) translates into a velocity, and it is equal to the settling velocity of the smallest particle the basin will remove.

Surface loading is determined by dividing the basin flow rate by the basin surface area as follows:

$$\text{Surface Loading} = LR_{surface}\left[\frac{gal}{min \cdot ft^2}\right] = \frac{Q\left[\dfrac{gal}{min}\right]}{A_{surface}[ft^2]}$$

surface loading rate
One of the guidelines for the design of settling tanks and clarifiers in treatment plants. Used by operators to determine if tanks and clarifiers are hydraulically (flow) over- or underloaded. Also called overflow rate or surface settling rate.

$$\text{Surface Loading Rate} = \frac{Flow}{Surface\ Area}$$

Where,

Q = Flow Rate
$A_{surface}$ = Surface Area

Colder water temperatures require lower basin surface loading rates because particle settling velocities also decrease with colder water temperatures. Therefore, plants can effectively treat only lower flow rates during colder weather. The surface loading rate is the same as the theoretical upwelling rate. When this value exceeds the settling rate of a particle, the particle will be unable to settle out and will be carried over to the filters.

Example 4

A water treatment plant treats a flow of 1.5 Mgal/d. An examination of treatment plant design drawings reveals that the rectangular sedimentation basin is 75 feet (ft) long, 25 feet (ft) wide, and has an effective (water) depth of 12 feet (ft). Calculate the surface loading rate in gallons per minute per square foot (gpm/ft²) of surface area.

Known	**Unknown**
Q = Flow Rate = 1.5 Mgal/d	$LR_{surface}$ = Surface Loading, gpm/ft²
L = Length = 75 ft	
W = Width = 25 ft	
D = Depth = 12 ft	

1. Convert flow in Mgal/d to gpm.

$$Q\left[\frac{gal}{min}\right] = Q\left[\frac{Mgal}{d}\right] \times \frac{10^6}{M} \times \frac{d}{24\,h} \times \frac{h}{60\,min}$$

$$= 1.5\left[\frac{Mgal}{d}\right] \times \frac{10^6}{M} \times \frac{d}{24\,h} \times \frac{h}{60\,min}$$

$$= 1,042\ gpm$$

2. Determine the surface loading in gallons per minute per square foot (gpm/ft²) of surface area ($A_{surface}$).

$$LR_{surface}\left[\frac{gal}{min \bullet ft^2}\right] = \frac{Q\left[\frac{gal}{min}\right]}{A_{surface}\left[ft^2\right]} = \frac{1,042\left[\frac{gal}{min}\right]}{75\left[ft\right] \times 25\left[ft\right]} = 0.56\,\frac{gpm}{ft^2}$$

The surface loading rate can be converted to a rise rate in feet per minute by dividing the surface loading by 7.48 gallons per cubic foot (gal/ft³).

$$Rise\ Rate\left[\frac{ft}{min}\right] = LR_{surface}\left[\frac{gal}{min \bullet ft^2}\right] \times \frac{ft^3}{7.48\,gal}$$

Example 5

A solids-contact unit has a surface loading rate of 0.75 gpm/ft². What is the rise rate in feet per minute?

Known	Unknown

$\text{LR}_{\text{surface}}$ = Surface Loading
    = 0.75 gpm/ft²

Rise Rate in feet per minute
(ft/min)

Calculate the rise rate in feet per minute (ft/min) from the surface loading.

$$\text{Rise Rate}\left[\frac{\text{ft}}{\text{min}}\right] = \text{LR}_{\text{surface}}\left[\frac{\text{gal}}{\text{min}-\text{ft}^2}\right] \times \frac{\text{ft}^3}{7.48\,\text{gal}}$$

$$= 0.75\left[\frac{\text{gal}}{\text{min}-\text{ft}^2}\right] \times \frac{\text{ft}^3}{7.48\,\text{gal}} = 0.1\frac{\text{ft}}{\text{min}}$$

4.4.4   Effective Water Depth

In theory, the ideal sedimentation basin would have a very shallow depth and a large surface area. However, practical considerations, such as minimum depth required to accommodate mechanical sludge removal equipment, desired flow velocity, and current and wind effects, must all be considered in the selection of an appropriate basin depth.

Sludge volume storage requirements may also be an important consideration in selecting an appropriate basin depth. This is particularly important in manually cleaned basins where sludge removal requires taking the basin off line for dewatering (draining) to permit removal of accumulated sludge. If the sludge buildup becomes too great, the effective depth of the basin will be significantly reduced. This can cause localized high flow velocities, sludge scouring, and degradation of process efficiency.

4.4.5   Mean Flow Velocity

The flow velocity is a function of the basin cross-sectional area and the flow going into the basin. Flow velocity is calculated by dividing the flow rate by the cross-sectional area of the basin.

The flow velocity is not completely uniform or stable throughout the length of the basin due to variations in water density, currents, and reduction in the cross-sectional area resulting from sludge accumulation in the bottom of the basin.

The main concern about high velocities is scour. When velocities get too high, the flowing water will pick up some of the settled sludge from the bottom of the basin and carry it on to the filters. This scouring reduces sedimentation efficiency.

$$\text{Mean Flow Velocity} = v\left[\frac{\text{ft}}{\text{min}}\right] = \frac{Q\left[\dfrac{\text{gal}}{\text{min}}\right] \times \dfrac{\text{ft}^3}{7.48\,\text{gal}}}{A\left[\text{ft}^2\right]}$$

Where,

    A = Cross-Sectional Area

Example 6

Determine the mean flow velocity in feet per minute (ft/min) for the rectangular sedimentation basin in Example 1.

Known	Unknown
Q = Flow Rate = 1.5 Mgal/d	v = Mean Flow Velocity, ft/min
= 1,042 gpm	
L = Length = 75 ft	
W = Width = 25 ft	
H = Depth = 12 ft	

Determine the mean flow velocity in feet per minute (ft/min).

$$v\left[\frac{ft}{min}\right] = \frac{Q\left[\frac{gal}{min}\right] \times \frac{ft^3}{7.48\,gal}}{A[ft^2]} = \frac{Q\left[\frac{gal}{min}\right] \times \frac{ft^3}{7.48\,gal}}{W[ft] \times H[ft]}$$

$$= \frac{1,042\left[\frac{gal}{min}\right] \times \frac{ft^3}{7.48\,gal}}{25[ft] \times 12[ft]} = 0.46\,\frac{ft}{min}$$

4.4.6   Weir Loading Rate

Launders outfitted with V-notch weirs are generally installed at the basin outlet to provide uniform collection and distribution of the clear water. While insufficient total weir length may reduce settling efficiency due to current effects, operation over a broad range of weir loading rates will generally not interfere with basin performance. If the weir loading rate becomes too high, floc will be carried out of the sedimentation basins and onto the filters. Generally, weir loadings are more important in shallow basins than in deeper basins.

In cold climates, launders use orifices in pipes rather than V-notch weirs on channels. Pipes usually are submerged in 2 feet (0.6 m) of water to avoid ice problems.

$$\text{Weir Loading Rate} = LR_{weir}\left[\frac{gal}{min \cdot ft}\right] = \frac{Q\left[\frac{gal}{min}\right]}{L_{weir}[ft]}$$

Where,
Q   = Flow Rate
L_{weir} = Weir Length

Example 7

Determine the weir loading rate in gallons per minute per foot (gpm/ft) of weir for the rectangular sedimentation basin in Example 1. Four effluent launders 12.5 feet (ft) long with V-notch weirs on both sides of the launder extend into the basin from the outlet end. Therefore, each launder has 25 feet (ft) of weir for a total of 100 feet (ft).

Known	Unknown

Known

Q   = Flow Rate   = 1.5 Mgal/d
                   = 1,042 gpm
L   = Length   = 75 ft
W   = Width   = 25 ft
H   = Depth   = 12 ft
L_{weir} = Weir Length = 100 ft

Unknown

LR_{weir} = Weir Loading
      Rate, gpm/ft

Calculate the weir loading rate in gallons per minute per foot (gpm/ft) of weir.

$$LR_{weir}\left[\dfrac{gal}{min\text{--}ft}\right] = \dfrac{Q\left[\dfrac{gal}{min}\right]}{L_{weir}\,[ft]} = \dfrac{1{,}042\left[\dfrac{gal}{min}\right]}{100\,[ft]} = 10.4\,\dfrac{gpm}{ft}$$

Check Your Understanding

1. How would you calculate the detention time for a sedimentation basin?

2. A rectangular sedimentation basin 50 feet long, 20 feet wide, and 10 feet deep treats a flow of 0.8 MGD. What is the theoretical detention time?

3. During cold weather, why can plants effectively treat only lower flow rates?

4. What problems are caused by reduced effective water depth from excessive sludge buildup?

5. Name two kinds of launders used in basin outlets.

4.5   Solids-Contact Clarification

Solids-contact units were first used in the Midwest as a means of handling the large amounts of sludge generated by water softening processes. It quickly became clear that this compact, single-unit process could also be used to remove turbidity from drinking water.

These types of clarifiers go by several interchangeable names: solids-contact clarifiers, upflow clarifiers, reactivators, and precipitators (Figures 4.22 and 4.23). The basic principles of operation are all the same, even though various manufacturers use different terms to describe how the mechanisms remove solids from water. The settled materials from coagulation or settling are referred to as sludge, and slurry refers to the suspended floc clumps in the clarifier. Sometimes, the terms sludge and slurry are used interchangeably.

The internal mechanism of a clarifier consists of three distinct unit processes that function in the same way as any conventional coagulation-flocculation-sedimentation process chain. Sludge produced by the unit is recycled through the process to act as a coagulant aid, thereby increasing the efficiency of the processes of coagulation, flocculation, and sedimentation. This is the same principle that has been used successfully for many years in the operation of separate coagulation, flocculation, and sedimentation processes.

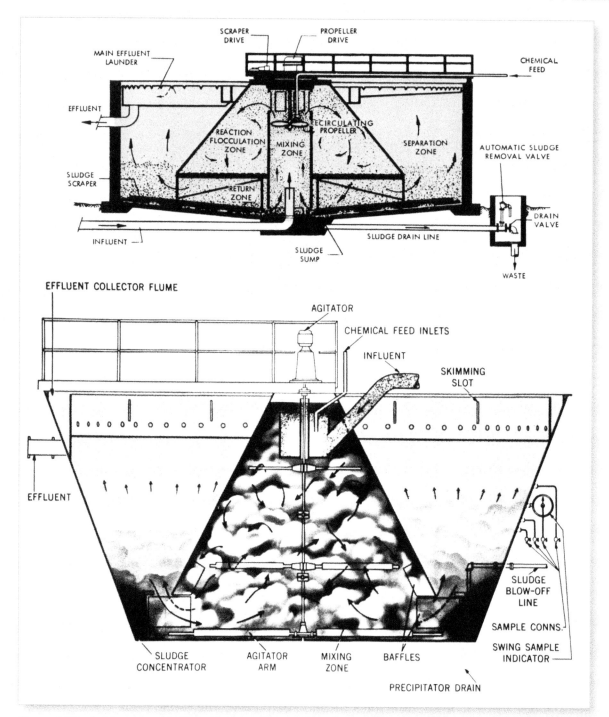

Figure 4.22   Solids-contact clarifiers (*Courtesy of Evoqua Water Technologies LLC.*)

The advantages of using a solids-contact clarifier over operating the same three processes separately are significant, but are sometimes offset by disadvantages. For example, capital and maintenance costs are greatly reduced because the entire chain of processes is accomplished in a single tank. At the same time, though, operation of the single-unit processes requires a higher level of operator knowledge and skill. The operator must have a thorough understanding of how these processes operate and be able to imagine how all of these processes can occur in a small chamber or clarifier at the same time. Operators often have trouble visualizing what is happening in an upflow clarifier and may get frustrated when routine problems associated with solids-contact units occur.

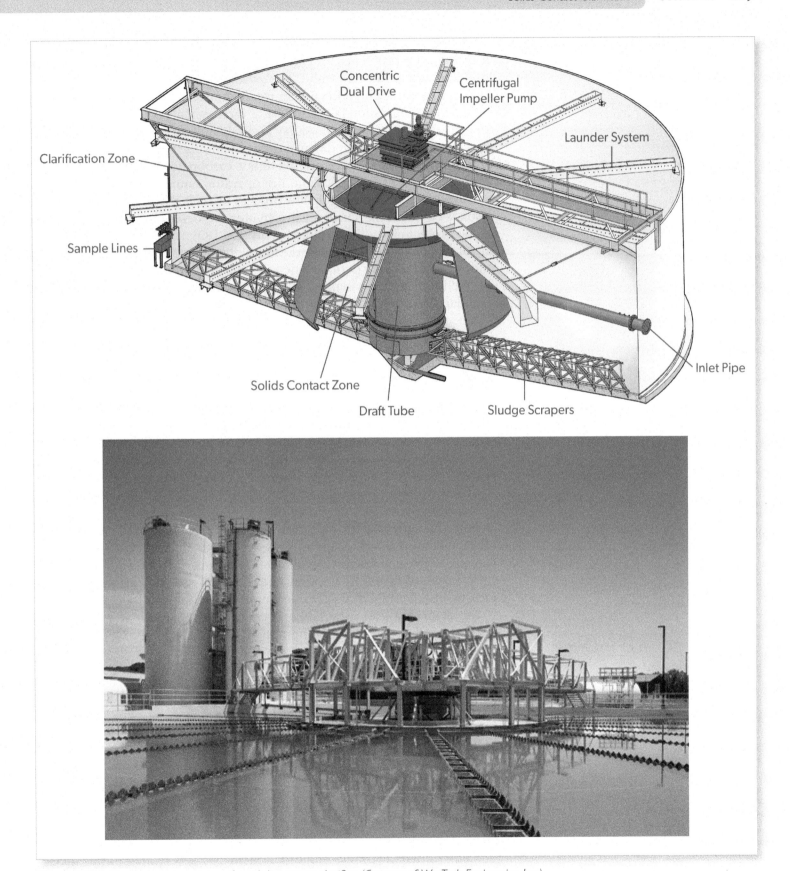

Figure 4.23   Plan and section views of a solids-contact clarifier *(Courtesy of WesTech Engineering, Inc.)*

A tremendous advantage in the use of the solids-contact units is the ability to adjust the volume of slurry (sludge blanket). By proper operational control, the operator can increase or decrease the volume of slurry in the clarifier as needed to cope with certain problems. During periods of severe taste and odor problems, for example, the operator can increase the sludge level and add activated carbon. The **adsorptive** characteristics of activated carbon make it highly effective in treating taste and odor problems. Similarly, when coagulation fails because of increased algal activities, the operator can take advantage of the slurry accumulation to carry the plant through the severe periods of the day when the chemicals will fail to react properly because of changes in the pH, alkalinity, carbonate, and dissolved oxygen (DO). In the conventional plant, the operator cannot respond to this type of breakdown in the coagulation process as well as the operator can with a solids-contact unit. Once algal activities are determined to be the cause of the problem (readily checked by pH and DO), the operator can increase the amount of slurry available during highly functional periods of the day and remove it during periods when the coagulation process is not functioning well. By skillfully making these slurry level adjustments, the operator can maintain a high-quality effluent from the solids-contact unit.

The most serious limitation of solids-contact units is their instability during rapid changes in flow (through-put), turbidity level, and temperature. The solids-contact unit is most unstable during rapid changes in the flow rate. The operator should identify and keep in mind the design flow for which the unit was constructed. To make this easier, convert the design flow to a surface loading rate. This rate may be expressed as the rise rate in inches per minute or feet per minute. For each rise rate, there will be an optimum (best) slurry level to be maintained within the unit. A rising flow rate will increase the depth of the slurry without increasing its volume or density. Conversely, a decrease in flow rate will reduce the level of slurry without changing its volume.

Sampling taps can be installed to enable the operator to monitor changes in slurry depth or concentration. The level of the slurry can be identified by placing sampling taps at various depths along the wall of the solids-contact reactor. The taps should penetrate the wall and extend into the slurry zone. It is often necessary to modify existing sampling taps or install additional sampling pipes to accomplish this.

By observing and measuring slurry depths at frequent intervals, the operator can easily monitor the rise or fall of the slurry levels; this will enable the operator to promptly make appropriate adjustments in the recirculation device or more tightly control the rate of change in the flow rate. This method of operational control is relatively effective in a gravity flow system when the water demands are moderate and the flow rate can be changed slowly. However, operational control in pressure systems is more difficult. Responding to rapid changes in demand or placing pumps into service at full capacity can easily upset an upflow clarifier immediately. In this case, the operator can witness the crisis occurring by observing the sampling taps, but be next to helpless to respond. The slurry will rise very rapidly in the settling zone, approach the overflow weirs, and spill onto the filters with a complete and total breakdown of the plant process.

Solids-contact clarifiers are also sensitive to severe changes in the turbidity of incoming raw water. The operator must be alert to changes in turbidity and must take immediate action. With experience, the operator will learn to accurately forecast when the turbidity may arrive at the reaction zone and will cope with the problem by increasing the chemical dosage before the arrival of excess turbidity. Early application of an

adsorption (add-SORP-shun)
The gathering of a gas, liquid, or dissolved substance on the surface or interface zone of another material.

increased chemical dosage puts the unit in a mode in which the turbidity can be handled successfully. The control of slurry and its influence on operational control during turbidity changes is discussed in greater detail later in this section.

Another factor that exerts a major influence on the operation of a solids-contact unit is temperature. Changes in water temperature will cause changes in the density of the water; changes in density influence the particle settling rate. In extremely cold water, consider using polymers, activated silica, powdered calcium carbonate, or some other weighting agent to aid sedimentation without affecting coagulation. Simple heating by the sun on the wall of the tank or on the flocculant particles within the container will cause a certain amount of carryover of solids to occur. Operators who are not familiar with solids-contact units may become surprised or frustrated by the potential carryover problem. This phenomenon is not a matter of serious concern because, as the position of the sun changes, the convection currents change. The clouds of flocculant particles appear and disappear in response to the currents and there is no real need to control this phenomenon if the overall settled turbidity meets the treatment objectives. So long as the major portion of the sludge blanket lies in the settling zone, the few clouds of flocculant particles (which look like billowing clouds) really do not significantly harm the operation of the unit or the quality of the water produced.

Dramatic changes in temperatures and flow rates may sometimes make it impossible to control or prevent process upsets. If the slurry rises to the weirs and is carried over onto the filters, reduce the flow rate. If possible, use weighting agents before carefully changing flow rates in cold water. The use of weighting agents may cause problems with the slurry requiring changes in recirculation rates. However, too high a recirculation rate may also cause the slurry to overflow onto the filters.

4.5.1   Operational Control

The operator of a solids-contact unit controls the performance of the unit by adjusting three fundamental variables: chemical dosage, recirculation rate, and sludge control.

Operators frequently have trouble distinguishing which of the three variables is the root of the problem because they are interrelated. Another challenge is that one problem may affect the performance of all three fundamentals. However, there are analytical techniques available to identify and separate the three fundamentals into different groups of symptoms and to diagnose the cause of the upset.

First, consider chemical dosage. As in conventional treatment plants, without proper chemical dosage, the entire system of solids-contact clarification will collapse. There must always be sufficient alkalinity in the raw water to react with the coagulant. Assuming for practical purposes the coagulant used is aluminum sulfate, for every mg/L of alum added, 0.45 mg/L of bicarbonate alkalinity is required to complete the chemical reaction. For precipitation to occur, there should be an excess of 20 mg/L of alkalinity present. You may have to add sodium hydroxide (caustic soda), calcium hydroxide (lime), or sodium carbonate (soda ash) to cause sufficient alkalinity to be present. For example, if there is only 30 mg/L of natural alkalinity present, for every mg/L of alum added, add 0.35 mg/L of lime if calcium hydroxide is being used.

All of this information can be verified by jar testing, which is a standard method for determining proper coagulation by chemical dosing. Do not attempt to make changes in solids-contact unit operation without first determining the proper chemical dosage through jar testing. For most solids-contact units, use the chemical dosages that produce floc that gives the lowest turbidity within a 5-minute settling period after stopping the jar tester. Using the above criteria, the operator now can set the chemical feeders to dose the raw water entering the solids-contact unit.

The next control mechanism is recirculation rate. Here, the plant operator may be misled by intuitive judgment. The recirculation rate is established by the speed of the impeller, turbine, pumping unit, or by air injection. Any of these devices causes the slurry to recirculate through the coagulation (reaction) zone.

To help visualize how the slurry should look in the reaction zone, take another look at the lower drawing in Figure 4.22, page 228, and note the cloud-like, billowy appearance of the flocculated slurry in this area. Under normal operating conditions, the entire mass of suspended floc clumps billows and flows within the chamber. Its motion is continually being influenced by the mixing of recirculated sludge and incoming raw water. In principle, operators are attempting to chemically dose the raw water when it enters the reaction zone and is mixed with the recirculated sludge. Coagulation and flocculation occur in the reaction zone and then the water and sludge pass into the settling zone. Some sludge is recirculated and mixed with incoming raw water while the rest of the sludge settles and is removed from the bottom of the settling area. At the point where water and sludge pass from the reaction zone to the settling zone, approximately 1 liter of water should rise and 1 liter of slurry should be returned into the reaction zone.

To sort out the effects of chemical dosages, recirculation rate, and sludge control, keep a log of the speed (RPM) of the recirculation device. If air is used for mixing, then the cubic feet of air applied per minute should be recorded. There is a direct relationship between the percentage of slurry present and the speed at which the mixing device is traveling.

To control the process, the operator must maintain the correct slurry volume in the reaction zone by exercising control of the rate of recirculation. The percentage of slurry can be determined by performing a volume over volume (V/V) test. The test procedure is as follows: using a 100 mL graduated cylinder, collect a sample from the reaction zone. Let the sample sit for 5 minutes and then determine the volume of slurry accumulated (mL) using the formula:

$$V/V[\%] = \frac{\text{Settled Slurry}[mL] \times 100\%}{\text{Total Sample Volume}[mL]}$$

At the same time, observe the clarity of the supernatant (settled water) that remains in the graduated cylinder. The clarity of the water above the slurry (the supernatant) will indicate how well the chemical reaction is proceeding. The percentage of accumulated solids by volume will indicate whether a proper amount of slurry is in the reaction zone. Customarily, such reactors require 5 to 20 percent solids, or a higher percentage in the graduated cylinder at the end of a 5- to 10-minute settling period.

Recordkeeping and experience will help you find an optimum percentage of solids to maintain. Perform the above analyses hourly and more frequently when the raw water quality is undergoing change and keep accurate records.

The final step in controlling the performance of the solids-contact unit is the removal of sludge (sludge control) that has accumulated on the bottom of the clarifier (settling zone). There are several means of sludge collection; some devices are located in areas of clarifiers that hold the sludge and are controlled by opening and closing recirculation gates. Others have scrapers that collect the sludge and move it to a discharge sump. In both cases, the sludge is removed by hydraulic means (water pressure) through a control valve. The sludge removal mechanisms are generally on a timer, which operates periodically for a time duration set by the operator. The time duration is simple to determine. Use a graduated cylinder to collect a sample from the sludge discharge line. The sludge being discharged should be 90 to 98 percent solids in a V/V test, as indicated above. A 5- to 10-minute period should be sufficient to make this determination. When slurries or sludge weaker than 90 percent is pumped, the operator is discharging a considerable amount of water and not leaving enough sludge to be recirculated into the reaction zone. If the percentage is considerably greater than 90 percent, then too much sludge may be accumulating and the recirculation device could become overloaded with too much return slurry.

With increased speed of the recirculation device, a larger amount of slurry can be retained in the unit. At the same time, if this amount becomes too great, it may cause the sludge to rise and ultimately spill over the effluent weirs with the treated water. If the recirculation rate is too slow, the solids may settle too soon and without sufficient recirculation will not return to the reaction zone. The absence of solids in the reaction zone causes improper coagulation. The net result is a failure of the total solids-contact system.

The operational controls described in this section and operator experience will help determine the optimum slurry volume for a solids-contact unit. The goal is to find the correct combination of recirculation rate, chemical dosage, and percentage of solids available for recycling. Keep track of the amount of solids in the reaction zone and, based upon practical experience, approximately the percentage required. This becomes more difficult as the raw water turbidity changes. For instance, in muddy streams carrying silt, sedimentation may occur very rapidly. Even with proper chemical dosage, increased circulation rates and higher sludge removal rates will be required to maintain sufficient slurry in the reaction zone. As the raw water turbidity becomes lighter, the increased circulation rate may cause the slurry blanket to rise to an uncontrolled depth in the settling zone. Removing too much sludge will produce this same effect. All of these problems are readily observed in the V/V test for solids determination in the reaction zone; also, this is cause for increased observations of the V/V during water quality changes.

Another problem may be caused by cold water when the recirculation rate may be too high for the densities of the particles present. A set of recirculation speeds for warmweather operation may be different from those used during cold weather. As a remedy, the operator may select a nonionic polymer as a weighting agent to increase the settling rate in cold waters. Other alternative chemicals are powdered calcium carbonate or the use of activated silica. A note of caution in chemical dosage determination: the reactions in the jar tester should be reasonably rapid to ensure comparable reactions within the solids-contact unit.

Another important point when determining chemical dosage for a solids-contact unit is that a specific set of jar test guidelines will be needed for each plant. For example, determine the volume of the reaction zone and the period of detention of the raw water in that reaction zone. This, along with knowledge of the speed of the recirculation device, is used to determine the detention time and flocculator speed in the jar tester.

In the real world, this means if the flow rate of the solids-contact unit is 10 minutes in the reaction zone and the speed is 2 feet per second (0.6 m/sec), then the jar tester mixer should turn at a speed equal to 2 feet per second with a coagulation period of 10 minutes. Duplicate in the jar tester, as nearly as possible, those conditions of chemical dosage, detention period, and mixing speeds that occur in the solids-contact unit. Using these guidelines, approximate real-world conditions can be achieved in the laboratory resulting in optimized chemical dosages.

4.5.2   Maintenance

The primary maintenance consideration in solids-contact units is the recirculating device, which needs regular inspection of the belt drive and gear boxes, and lubrication. If the unit has a sludge collector, its drive and gear boxes also require the same attention. Inspect the units daily and lubricate following the manufacturers recommendations. The contact unit may also need to be drained periodically to inspect the sludge collectors for wear and corrosion.

Sludge collector devices are usually constructed of steel within a concrete container. If a cathodic protection system is provided with the unit, weekly readings of the amperes and voltage supply should be recorded. Changes in these readings indicate that a problem may be developing. Periodically, inspect the cathodic protection devices. If any defects are detected, correct them.

4.5.3   Calculations

Successful operation of a solids-contact clarification unit requires some math calculations. The volume over volume test provides operators with an indication of the settleability of the slurry or sludge in the sludge blanket. The detention time in the reaction zone is important to ensure that there is sufficient mixing time and time for the chemical reactions (coagulation) to take place.

If the raw water is low in alkalinity and alum is the coagulant, lime may have to be added to provide enough alkalinity. With the necessary information and by following the step-by-step procedures outlined in this section, the setting on the lime feeder can be calculated.

FORMULAS

The volume over volume (V/V) test requires the collection of 100 mL of slurry from a solids-contact unit in a 100 mL graduated cylinder. The slurry is allowed to sit for 10 minutes and the volume of the settled slurry on the bottom of the graduated cylinder is measured and recorded.

cathodic (ca-THOD-ick) protection
An electrical system for prevention of rust, corrosion, and pitting of metal surfaces that are in contact with water, wastewater, or soil. A low-voltage current is made to flow through a liquid (water) or a soil in contact with the metal in such a manner that the external electromotive force renders the metal structure cathodic. This concentrates corrosion on auxiliary anodic parts which are deliberately allowed to corrode instead of letting the structure corrode.

$$V/V[\%] = \frac{\text{Settled Slurry}[mL] \times 100\%}{\text{Total Sample Volume}[mL]}$$

The detention time in the reaction zone of a solids-contact unit is calculated the same way the detention time in any basin is calculated. The flow is divided into the volume of the reaction zone. Any necessary adjustments are made for units such as multiplying by 60 minutes per hour to convert a detention time from hours to minutes.

$$\text{Detention Time}[min] = \frac{\text{Reaction Zone Volume}[gal]}{Q\left[\dfrac{gal}{d}\right] \times \dfrac{d}{24\,h} \times \dfrac{h}{60\,min}}$$

Where,

Q = Flow Rate

The lime dose that must be added to a raw water being treated can be determined from: (1) the alkalinity of the raw water, (2) the alkalinity that must be present to ensure complete precipitation of the alum, (3) the amount of alkalinity that reacts with the alum, and (4) the amount of lime that reacts with the alum; that is:

1. Raw Water Alkalinity, mg/L as HCO_3^-
2. Alkalinity Present for Precipitation, mg/L (at least 30 mg/L)
3. 0.45 mg/L Alkalinity (HCO_3^-) reacts with 1 mg/L Alum
4. 0.35 mg/L Lime ($Ca(OH)_2$) reacts with 1 mg/L Alum

Procedure to Calculate Lime Dose in mg/L
1. Determine the alkalinity available to react with the alum.

$$\text{Alkalinity Available}\left[\frac{mg}{L}\right] = \text{Raw Water Alkalinity}\left[\frac{mg}{L}\right] - \text{Alkalinity Present for Precipitation}\left[\frac{mg}{L}\right]$$

2. Determine the amount of alum that will react with the available alkalinity.

$$\frac{0.45\left[\dfrac{mg}{L}\right]\text{Alkalinity}}{1.0\left[\dfrac{mg}{L}\right]\text{Alum}} = \frac{\text{Alkalinity Available}\left[\dfrac{mg}{L}\right]}{\text{Alum Reacting}\left[\dfrac{mg}{L}\right]}$$

or

$$\text{Alum Reacting}\left[\frac{mg}{L}\right] = \frac{1.0\left[\dfrac{mg}{L}\right]\text{Alum} \times \text{Alkalinity Available}\left[\dfrac{mg}{L}\right]}{0.45\left[\dfrac{mg}{L}\right]\text{Alkalinity}}$$

3. Determine the milligrams per liter (mg/L) of alum that needs additional alkalinity (or is unreacted with). The total alum required is determined by the jar test.

$$\text{Alum Needing Alkalinity}\left[\frac{mg}{L}\right] = \text{Total Alum}\left[\frac{mg}{L}\right]$$

$$- \text{Alum Reacting}\left[\frac{mg}{L}\right]$$

4. Determine the lime dose in milligrams per liter (mg/L).

$$\frac{1\left[\frac{mg}{L}\right]\text{Alum}}{0.35\left[\frac{mg}{L}\right]\text{Lime}} = \frac{\text{Alum Needing Alkalinity}\left[\frac{mg}{L}\right]}{\text{Lime Dosage}\left[\frac{mg}{L}\right]}$$

or

$$\text{Lime Dosage}\left[\frac{mg}{L}\right] = \frac{0.35\left[\frac{mg}{L}\right]\text{Lime} \times \text{Alum Needing Alkalinity}\left[\frac{mg}{L}\right]}{1\left[\frac{mg}{L}\right]\text{Alum}}$$

5. Determine the setting on the lime feeder in pounds per day (lb/d).

$$\begin{aligned}
&\text{Lime Feed}\left[\frac{lb}{d}\right] \\
&= Q\left[\frac{Mgal}{d}\right] \times \text{Lime Dose}\left[\frac{mg}{L}\right] \times \frac{3.785\,L}{gal} \times \frac{10^6}{M} \times \frac{kg}{10^6\,mg} \times \frac{2.205\,lb}{kg}
\end{aligned}$$

6. Determine the setting on the lime feeder in grams per minute.

$$\begin{aligned}
&\text{Lime Feed}\left[\frac{g}{min}\right] \\
&= Q\left[\frac{Mgal}{d}\right] \times \text{Lime Dose}\left[\frac{mg}{L}\right] \times \frac{3.785\,L}{gal} \times \frac{10^6}{M} \times \frac{d}{24\,h} \times \frac{h}{60\,min} \times \frac{g}{10^3\,mg}
\end{aligned}$$

Multiplying by 3.785 liters per gallon cancels out the gallons in MGD and the liters in mg/L. Multiplying by 1,000,000/1 Million cancels out the million units. Dividing by 24 hr/day and 60 min/hr converts the feed rate from days to minutes. Dividing by 1,000 mg/g changes the amount of lime to be fed from milligrams to grams.

Example 8

A graduated cylinder is filled to the 100 mL level with the slurry from a solidscontact unit. After 10 minutes (min), there is 21 mL of slurry on the bottom and 79 mL of clear water remaining in the top part of the cylinder. This is the volume over volume (V/V) test.

Known	Unknown
V_{slurry} = Volume of Settled Slurry = 21 mL	V/V in percentage (%)
V = Total Volume = 100 mL	

Determine V/V as a percent (%).

$$V/V[\%] = \frac{V_{slurry}[mL] \times 100\%}{V[mL]}$$

$$= \frac{21[mL] \times 100\%}{100[mL]}$$

Example 9

The reaction zone in a solids-contact clarifier is 11 feet (ft) in diameter and 4 feet (ft) high. Find the detention time in minutes (min) in the reaction zone if the flow is 2 MGD.

Known	Unknown
D = Diameter = 11 ft	t = Detention Time, min
H = Height = 4 ft	
Q = Flow Rate = 2 MGD	
= 2,000,000 gpd	

1. Calculate the volume of the reaction zone in gallons (gal).

$$Volume[gal] = 0.785 \times D^2[ft^2] \times H[ft] \times \frac{7.48\,gal}{ft^3}$$

$$= 0.785 \times 11^2[ft^2] \times 4[ft] \times \frac{7.48\,gal}{ft^3}$$

$$= 2,842\,gal$$

2. Calculate the detention time in the reaction zone in minutes (min).

$$t[min] = \frac{Reaction\ Zone\ Volume[gal]}{Q\left[\frac{gal}{d}\right] \times \frac{d}{24\,h} \times \frac{h}{60\,min}}$$

$$= \frac{2,842[gal]}{2,000,000\left[\frac{gal}{d}\right] \times \frac{d}{24\,h} \times \frac{h}{60\,min}}$$

$$= 2.05\ min$$

Example 10

A raw water has an alkalinity of 36 mg/L as bicarbonate (HCO_3^-). A chemical dose of 52 mg/L of alum (from a jar test) is needed to reduce the turbidity from 75 TU to 1.0 TU. At least 30 mg/L of alkalinity must be present to ensure complete precipitation of the alum added. Find the dose of lime ($Ca(OH)_2$) in mg/L that is needed to complete this reaction.

Known   **Unknown**

Raw Water Alkalinity = 36 mg/L   D = Lime Dose, mg/L
Total Alum Required = 52 mg/L
Alkalinity Present for
Precipitation   = 30 mg/L

1. Determine the alkalinity available to react with the alum.

$$\text{Alkalinity Available} \left[\frac{mg}{L}\right] = \text{Raw Water Alkalinity} \left[\frac{mg}{L}\right] - \text{Alkalinity Present for Precipitation} \left[\frac{mg}{L}\right]$$

$$= 36 \left[\frac{mg}{L}\right] - 30 \left[\frac{mg}{L}\right]$$

$$= 6 \frac{mg}{L}$$

2. Determine the amount of alum that will react with the available alkalinity.

$$\text{Alum Reacting} \left[\frac{mg}{L}\right] = \frac{1.0 \left[\frac{mg}{L}\right] \text{Alum} \times \text{Alkalinity Available} \left[\frac{mg}{L}\right]}{0.45 \left[\frac{mg}{L}\right] \text{Alkalinity}}$$

$$= \frac{1.0 \left[\frac{mg}{L}\right] \text{Alum} \times 6 \left[\frac{mg}{L}\right] \text{Alkalinity}}{0.45 \left[\frac{mg}{L}\right] \text{Alkalinity}}$$

$$= 13.3 \frac{mg}{L} \text{ Alum}$$

3. Determine the milligrams per liter of alum needing additional alkalinity (or is not reacted with).

$$\text{Alum Needing Alkalinity} \left[\frac{mg}{L}\right]$$

$$= \text{Total Alum} \left[\frac{mg}{L}\right] - \text{Alum Reacting} \left[\frac{mg}{L}\right]$$

$$= 52 \left[\frac{mg}{L}\right] - 13.3 \left[\frac{mg}{L}\right]$$

$$= 38.7 \frac{mg}{L}$$

4. Determine the lime dose in milligrams per liter (mg/L).

$$D \left[\frac{mg}{L}\right] = \frac{0.35 \left[\frac{mg}{L}\right] \text{Lime} \times \text{Alum Needing Alkalinity} \left[\frac{mg}{L}\right]}{1 \left[\frac{mg}{L}\right] \text{Alum}}$$

$$= \frac{0.35 \left[\frac{mg}{L}\right] \text{Lime} \times 38.7 \left[\frac{mg}{L}\right] \text{Alum}}{1 \left[\frac{mg}{L}\right] \text{Alum}}$$

$$= 13.5 \frac{mg}{L}$$

Note: This dose may be verified by using a jar test. Select an alum dose and try different lime doses.

Example 11

If the raw water in Example 10 needs a lime dose of 13.5 mg/L, what should be the setting on the lime feeder in: (1) pounds per day (lb/d), and (2) grams per minute (g/min) when the flow is 2.0 MGD?

Known	Unknown
D = Lime Dose = 13.5 mg/L	Lime Feed, lb/d
Q = Flow Rate = 2.0 MGD	Lime Feed, g/min

1. Determine the setting on the lime feeder in pounds per day (lb/d).

$$\text{Lime Feed}\left[\frac{\text{lb}}{\text{d}}\right]$$

$$= Q\left[\frac{\text{Mgal}}{\text{d}}\right] \times D\left[\frac{\text{mg}}{\text{L}}\right] \times \frac{3.785\,\text{L}}{\text{gal}} \times \frac{10^6}{\text{M}} \times \frac{\text{kg}}{10^6\,\text{mg}} \times \frac{2.205\,\text{lb}}{\text{kg}}$$

$$= 2.0\left[\frac{\text{Mgal}}{\text{d}}\right] \times 13.5\left[\frac{\text{mg}}{\text{L}}\right] \times \frac{3.785\,\text{L}}{\text{gal}} \times \frac{10^6}{\text{M}} \times \frac{\text{kg}}{10^6\,\text{mg}} \times \frac{2.205\,\text{lb}}{\text{kg}}$$

$$= 225\,\frac{\text{lb Lime}}{\text{d}}$$

2. Determine the setting on the lime feeder in grams per minute (g/min).

$$\frac{\text{Lime}}{\text{Feed}}\left[\frac{\text{g}}{\text{min}}\right]$$

$$= Q\left[\frac{\text{Mgal}}{\text{d}}\right] \times D\left[\frac{\text{mg}}{\text{L}}\right] \times \frac{3.785\,\text{L}}{\text{gal}} \times \frac{10^6}{\text{M}} \times \frac{\text{d}}{24\,\text{h}} \times \frac{\text{h}}{60\,\text{min}} \times \frac{\text{g}}{10^3\,\text{mg}}$$

$$= 2.0\left[\frac{\text{Mgal}}{\text{d}}\right] \times 13.5\left[\frac{\text{mg}}{\text{L}}\right] \times \frac{3.785\,\text{L}}{\text{gal}} \times \frac{10^6}{\text{M}} \times \frac{\text{d}}{24\,\text{h}} \times \frac{\text{h}}{60\,\text{min}} \times \frac{\text{g}}{10^3\,\text{mg}}$$

$$= 71\,\frac{\text{g Lime}}{\text{min}}$$

Check Your Understanding

1. List two advantages of solids-contact units.

2. How can the level of the slurry or sludge blanket be determined in solids-contact units?

3. What should be done when a rapid change in turbidity is expected?

4. How is the proper chemical dose selected when operating a solids-contact unit?

5. List the devices that may be used to provide recirculation in a solids-contact unit.

6. How is the percentage of slurry present in the reaction zone determined?

4.6   Sludge Handling

Water treatment plant sludges are typically alum sludges, with solids concentrations varying from 0.25 to 10 percent when removed from the basin. In gravity flow sludge removal systems, the solids concentration should be limited to about 3 percent. If the sludge is to be pumped, solids concentrations as high as 10 percent can be readily transported.

In horizontal-flow sedimentation basins preceded by coagulation and flocculation, over 50 percent of the floc will settle out in the first third of the basin length. Operationally, this must be considered when establishing the frequency for the operation of sludge removal equipment, as well as the volume or amount of sludge to be removed and the sludge storage volume available in the basin.

4.6.1   Sludge Removal Systems

Sludge that accumulates on the bottom of sedimentation basins must be removed periodically to prevent:

1. Interference with the settling process (such as resuspension of solids due to scouring)
2. **Septic** sludge or providing an environment for the growth of microorganisms that can create taste and odor problems
3. Excessive reduction in the cross-sectional area of the basin (reduction of detention time)

In large-scale plants, sludge is normally removed on an intermittent basis with the aid of mechanical sludge removal equipment. However, in smaller plants with low solids loading, manual sludge removal may be more cost effective.

In manually cleaned basins, the sludge is allowed to accumulate until it reduces settled water quality. High levels of sludge reduce the detention time and floc carries over to the filters. The basin is then dewatered (drained), most of the sludge is removed by stationary or portable pumps, and the remaining sludge is removed with squeegees and hoses. Basin floors are usually sloped toward a drain to ease sludge removal. The frequency of shutdown for cleaning will vary from several months to a year or more, depending on source water quality (amount of suspended matter in the water).

In larger plants, a variety of mechanical devices (Figure 4.24) can be used to remove sludge, including:

1. Mechanical rakes
2. Drag-chain and flights
3. Traveling bridges

Circular or square basins are usually equipped with rotating sludge rakes. Basin floors are sloped toward the center, and the sludge rakes progressively push the sludge toward a center outlet.

In rectangular basins, the simplest sludge removal mechanism is the chain and flight system. An endless chain outfitted with wooden flights (scrapers) pushes the sludge into a sump. The disadvantage of this system and of the rotating rakes previously described is high operation and maintenance costs. Most of the moving (wearing) parts are submerged so the basin has to be dewatered to perform major maintenance.

In an attempt to reduce operation and maintenance costs (as well as capital equipment costs), and to improve sludge removal equipment maintainability, the traveling bridge sludge removal system was developed. This bridge looks like an old highway bridge except it has no deck for cars. The traveling bridge spans the width of the sedimentation basin and travels along the length of the basin walls. Movable sludge sweeps,

septic (SEP-tick)
A condition produced by bacteria when all oxygen supplies are depleted. If severe, the bottom deposits produce hydrogen sulfide, the deposits and water turn black, give off foul odors, and the water has a greatly increased chlorine demand.

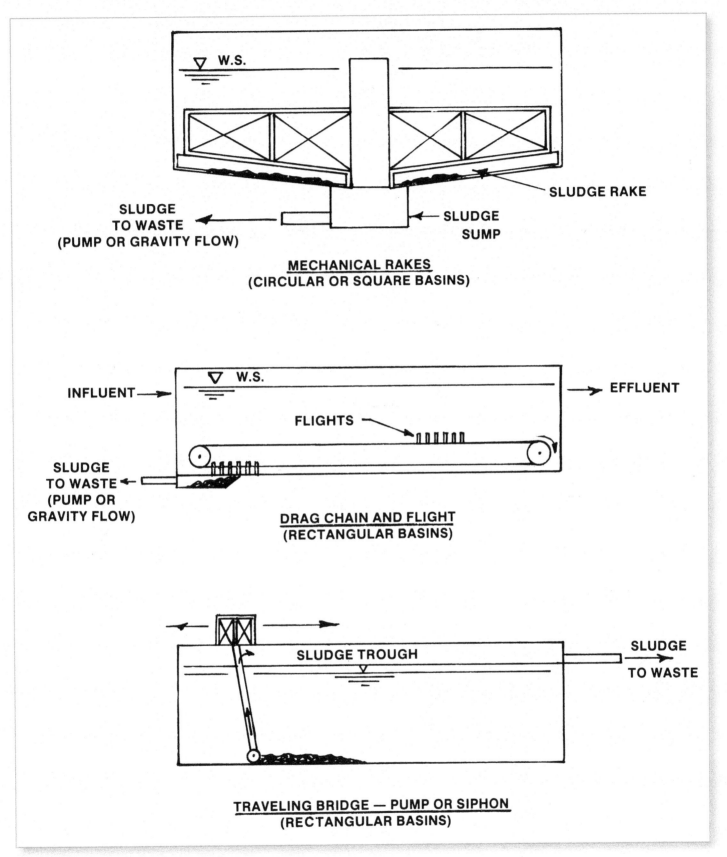

MECHANICAL RAKES
(CIRCULAR OR SQUARE BASINS)

DRAG CHAIN AND FLIGHT
(RECTANGULAR BASINS)

TRAVELING BRIDGE — PUMP OR SIPHON
(RECTANGULAR BASINS)

Figure 4.24 Mechanical sludge removal systems

which are hung from the bridge structure, remove the sludge from the basin floor with suction pumps or by siphon action. There are few submerged parts in this system and these can normally be removed for maintenance without dewatering the basin. Traveling bridge sludge removal systems will operate effectively on the simplest of basin designs.

Sludge may be discharged into sludge basins or ponds for liquid-solids separation. Ultimately, the sludge may be disposed of in a landfill. See Chapter 2, "Source Water, Reservoir Management, and Intake Structures," and the chapter on handling and disposal of process wastes in Volume II of this training manual series for additional details.

4.6.2 Sludge Removal Frequency

Accumulated sludge in the basin bottom or sludge sump is periodically removed for further processing (dewatering) and, ultimately, disposal. The frequency of sludge removal or transfer depends on the rate of sludge buildup and this is directly related to the amount of suspended material and floc removed in the sedimentation process. Other factors influencing the frequency of sludge removal include the size of the sludge sump and the capacity of the sludge pump. In alum coagulation operations, this generally means that sludge removal equipment need only be operated once per shift or perhaps once per day. If polymers are used as the primary coagulant and the source water suspended solids concentration is low (less than 5 mg/L), sludge removal equipment need only be operated once or twice per week.

In some water treatment plants, the operator measures the depth of the accumulated sludge deposit and uses this information to determine the operating frequency of the sludge removal equipment. This measurement can be made with a sludge blanket sounder, a bubbler tube, an aspirator, a clear core sampler, an ultrasonic level indicator, or an infrared level indicator.

Perhaps the simplest sludge blanket measuring device is the sludge blanket sounder. This device consists of a 0.25-inch (6 mm) thick hardware cloth disc about 18 inches (450 mm) in diameter. The disc is suspended from a lightweight chain by three-point suspension. The disc is slowly lowered into the sedimentation basin. When the disc reaches the top of the sludge blanket it stops its descent, and a depth reading is taken from markings on the chain.

Many treatment plants use fixed or portable ultrasonic or infrared level indicators. Fixed level indicators are attached to the basin and can give continuous readings. Portable level indicators can be used at different places in a tank or at several tanks (see Figure 4.25). Ultrasonic level indicators use the echoes from ultrasonic waves to detect the sludge level. Infrared indicators send the light from light emitting diodes (LED) across a gap to detect the increased density of the sludge.

After determining an appropriate time interval for sludge removal equipment operation, the sludge discharge should be periodically checked to determine the concentration of the sludge solids. This is generally done by observation. If

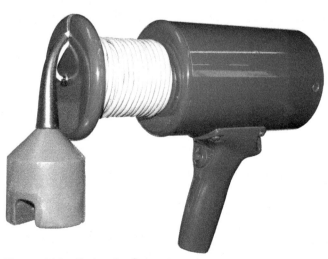

Figure 4.25 Sludge Gun® portable infrared sludge level indicator (*Courtesy of Markland Specialty Engineering Ltd.*)

the sludge is too thick and bulks, the frequency of sludge removal should be increased. Likewise, if the sludge concentration is too low in solids (soupy), decrease the frequency of sludge removal.

Some water treatment plants are equipped with semiautomatic sludge removal equipment that can be adjusted to change the frequency of operation by resetting a time clock.

Check Your Understanding

1. Alum sludge solids concentrations typically vary from ___ to ___ percent when removed from a sedimentation basin.

2. Why must accumulated sludge be removed periodically from the bottom of sedimentation basins?

3. How can the depth of sludge in a sedimentation basin be measured?

4. If the sludge being pumped from a sedimentation basin is too low in solids (soupy), what should the operator do?

4.7   Interaction with Other Treatment Processes

The purposes of the sedimentation process are to remove suspended solids from the water being treated and to reduce the load on the filters. If adequate detention time and basin surface area are provided in the sedimentation basins, solids removal efficiencies greater than 95 percent can be achieved. However, high sedimentation basin removal efficiencies may not always be the most cost-effective way to remove suspended solids.

In low-turbidity source waters (less than about 10 **turbidity units** or TU), effective coagulation, flocculation, and filtration may produce a satisfactory filtered water without the need for sedimentation. In this case, the coagulation-flocculation process is operated to produce a highly filtrable pinpoint floc, which does not readily settle due to its small size. Instead, the pinpoint floc is removed by the filters. However, there is a practical limitation in applying this concept to higher turbidity conditions. If the filters become overloaded with suspended solids, they will quickly clog and need frequent backwashing. This can limit plant production and cause a degradation in filtered water quality.

Thus, the sedimentation process should be operated from the standpoint of overall plant efficiency. If the source water turbidity is only 3 TU, and jar tests indicated that 0.5 mg/L of cationic polymer is the most effective coagulant dosage, then the sedimentation process cannot be expected to remove a significant fraction of the suspended solids. On the other hand, source water turbidities in excess of 50 TU will probably require a high alum dosage (or other primary coagulant) for efficient solids removal. In this case, the majority of the suspended particles and alum floc should be removed in the sedimentation basin.

turbidity units (TU)
Turbidity units are a measure of the cloudiness of water. If measured by a nephelometric (deflected light) instrumental procedure, turbidity units are expressed in nephelometric turbidity units (NTU) or simply TU.

4.8   Normal Operations

From a water quality standpoint, filter effluent turbidity is a good indication of overall process performance. However, the performance of each of the individual water treatment processes, including sedimentation, must still be monitored to anticipate quality or performance changes. Normal operating conditions are conditions within the operating ranges of the plant, while abnormal conditions are unusual or difficult-to-handle conditions. Changes in raw water quality may be considered a normal condition for many plants and an abnormal condition for other water treatment plants.

In the normal operation of the sedimentation process, operators will monitor (Figure 4.26):

1. Turbidity of the water entering and leaving the sedimentation basin
2. Temperature of the entering water

Turbidity of the entering water indicates the floc or solids loading on the sedimentation process. Turbidity of the water leaving the basin reveals the effectiveness or efficiency of the sedimentation process. Low levels of turbidity are desirable to minimize the floc loading on the filters.

The temperature of the water entering the sedimentation basin is very important. Usually, water temperature changes are gradual, depending on time of the year and the weather. As the water becomes colder, the particles will settle more slowly. To compensate for this change, operators should perform jar tests (see Chapter 3, "Coagulation and Flocculation") and adjust the coagulant dosage to produce a denser, faster-settling

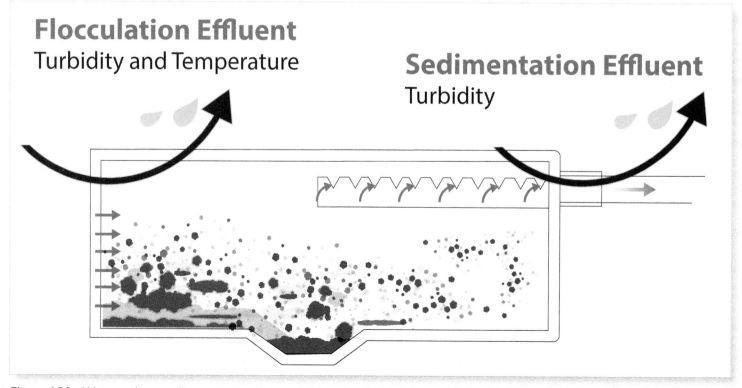

Figure 4.26   Water quality sampling points in sedimentation process monitoring

floc. Another possibility is that if the demand for water decreases during colder weather, the flow to be treated can be reduced, which will produce longer detention times. Longer detention times will allow slower-settling particles or floc to be removed in the sedimentation basins.

Visual checks of the sedimentation process should include observation of floc settling characteristics, distribution of floc at the basin inlet, and clarity of settled water spilling over the launder weirs. An uneven distribution of floc, or poorly settling floc, may indicate that a raw water quality change has occurred or that operational problems may have developed.

4.8.1   Process Control

The actual performance of sedimentation basins depends on the settling characteristics of the suspended particles and the flow rate through the sedimentation basins. To control the settling characteristics of the suspended particles, adjust the chemical coagulant dose and the coagulation-flocculation process (see Chapter 3, "Coagulation and Flocculation"). The flow rate through the sedimentation basin controls the efficiency of the process in removing suspended particles. The higher the flow rate, the lower the efficiency (the fewer suspended particles are removed). Once the actual flow rate becomes greater than the design flow rate, operators can expect an increase in suspended particles flowing over the V-notch weirs.

From a practical standpoint, it is best to operate sedimentation basins near design flows. However, to achieve the intended removal of suspended particles once design flows are exceeded, suspended particles leaving the sedimentation basin may overload the filters with solids and require additional filter backwashing. Study the settling characteristics of the particles by using laboratory jar tests. Then, verify the test results and make adjustments based on actual performance of the water treatment plant.

During periods of low flows, the use of all sedimentation basins may not be necessary. Because the cost to operate a basin is very low, it is common practice to keep all basins in service except during periods of draining for maintenance and repairs.

4.8.2   Process Actions

In rectangular and circular sedimentation basins, operators can generally make a judgment about the performance of the sedimentation process by observing how far the floc is visible beyond the basin inlet. When sedimentation is working well, the floc will only be visible for a short distance. When sedimentation is poor, the floc will be visible for a long distance beyond the inlet.

In upflow or solids-contact clarifiers, the depth of the sludge blanket and the density of the blanket are useful monitoring tools. If the sludge blanket is of normal density (measured as milligrams of solids per liter of water) but is very close to the surface, more sludge should be wasted. If the blanket is of unusually light density, the coagulation-flocculation process (chemical dosage) must be adjusted to improve performance.

With any of the sedimentation processes, it is useful to observe the quality of the effluent as it passes over the launder weir. Flocs coming over at the ends of the basin are indicative of density currents, short-circuiting, sludge blankets that are too deep, or high flows. The clarity of the effluent is also a reliable indicator of coagulation-flocculation efficiency.

Process equipment should be checked regularly to ensure adequate performance. Proper operation of sludge removal equipment should be verified each time the equipment is operated because sludge removal discharge piping systems are subject to clogging. Free-flowing sludge can be readily observed if sight glasses are incorporated in the sludge discharge piping. Otherwise, the outlet of the sludge line should be observed during sludge pumping. Frequent clogging of the sludge discharge line is an indication that the sludge concentration is too high. If this occurs, the frequency of operation of the sludge removal equipment should be increased. This problem can be accurately diagnosed by performing a sludge solids volume analysis in the laboratory. Collect a sludge sample and pour a known volume into a drying dish. Place the sample dish in a drying oven and evaporate the sample to dryness (usually about 1 hour) at 103 to 105°C. Weigh the remaining solids. Use the following formula to calculate the percent of sludge solids.

$$\text{Sludge Solids}[\%] = \frac{\text{Weight of Sample}[\text{mg}] \times 1[\text{mL}] \times 100\%}{\text{Volume of Sample}[\text{mL}] \times 1,000[\text{mg}]}$$

Some treatment plants do not have the capability to perform this analysis, so the samples are sent to outside labs.

The operator should routinely inspect physical facilities and equipment as part of good housekeeping and maintenance practice. Abnormal equipment conditions should be corrected or reported to maintenance personnel, and basin water surfaces and launders should be kept free of leaves, twigs, and other debris that might jam or foul mechanical equipment such as valves and meters.

A summary of routine sedimentation process actions is given in Table 4.2. Actual frequency of monitoring should be based on the source of the water supply and variations in the supply.

4.8.3   Recordkeeping

Recordkeeping is one of the most important administrative functions of the water treatment plant operator. In the routine daily operation of the sedimentation process, operators maintain a daily operations log of process performance and water quality characteristics. Keep the following records:

1. Influent and effluent turbidity and influent temperature
2. Process production inventory (amount of water processed and volume of sludge produced)
3. Process equipment performance (types of equipment in operation, maintenance procedures performed, and equipment calibration)

Entries in logs should be neat and legible, should reflect the date and time of an event, and should be initialed by the operator making the entry. Most plants use computerized logs as part of their O&M recordkeeping system.

Table 4.2   Routine sedimentation process actions

1. Monitor Process Performance and Evaluate Water Quality Conditions	Location	Frequency[a]	Possible Operator Actions
• Turbidity	Influent/Effluent	At least once every 2 hours	1. Increase sampling frequency when process water quality is variable
• Temperature	Influent	Occasionally	2. Perform jar tests if indicated (see procedure in Chapter 3, "Coagulation and Flocculation")
			3. Make necessary process changes
			a. Change coagulant
			b. Adjust coagulant dosage
			c. Adjust flash-mixer/flocculator mixing intensity
			d. Change frequency of sludge removal
			4. Verify response at appropriate time

2. Make Visual Observations	Location	Frequency[a]	Possible Operator Actions
• Floc settling characteristics	First half of basin	At least once per 8-hour shift	1. Perform jar tests if indicated
• Floc distribution	Inlet	At least once per 8-hour shift	2. Make necessary process changes
			a. Change coagulant
• Turbidity (clarity) of settled water	Launders or settled water conduit	At least once per 8-hour shift	b. Change coagulant dosage
			c. Adjust flash-mixer/flocculator mixing intensity
			d. Change frequency of sludge removal
			3. Verify response to process changes at appropriate time

3. Check Process and Sludge Removal Equipment Condition	Location	Frequency[a]	Possible Operator Actions
• Noise	Various	Once per 8-hour shift	1. Correct minor problems
• Vibration	Various	Once per 8-hour shift	2. Notify others of major problems
• Leakage	Various	Once per 8-hour shift	
• Overheating	Various	Once per 8-hour shift	

4. Operate Sludge Removal Equipment	Location	Frequency[a]	Possible Operator Actions
• Perform normal operations sequence	Sedimentation basin	Depends on process conditions (may vary from once per day to several days or more)	1. Change frequency of operation
			a. If sludge is too watery, decrease frequency of operation or pumping rate
• Observe conditions of sludge being removed	Sedimentation basin	Depends on process conditions (may vary from once per day to several days or more)	b. If sludge is too dense, bulks, or clogs discharge lines, increase frequency of operation or pumping rate
			c. If sludge is septic, increase frequency of operation or pumping rate

5. Inspect Facilities	Location	Frequency[a]	Possible Operator Actions
• Check sedimentation basins	Various	Once every 2 hours	1. Report abnormal conditions
• Observe basin water levels and depth of water flowing over launder weirs	Various	Once per 8-hour shift	2. Make flow changes (see Chapter 2, "Source Water, Reservoir Management, and Intake Structures"), or adjust launder weirs
• Observe basin water surface	Various	Once per 8-hour shift	3. Remove debris from basin water surface
• Check for algal buildup on basin walls and launders	Various	Occasionally	

[a] Frequency of monitoring should be based on the source of the water supply and variations in the supply.

Check Your Understanding

1. Under what circumstances are sedimentation basins needed to treat water?

2. What is the difference between normal operations and abnormal operations?

3. The actual performance of sedimentation basins depends on what two major factors?

4. How can an operator determine if sludge lines are free flowing?

5. What should be changed if the sludge line plugs frequently?

6. How often should visual observations of sedimentation process performance be conducted?

7. In the routine operation of the sedimentation process, what types of records should be maintained?

4.9   Abnormal Operations

Sudden changes in source or process water quality indicators, such as turbidity, pH, alkalinity, temperature, chlorine demand (source water), and color, are signals that operators should immediately review the performance of the coagulation-flocculation process and the sedimentation process.

4.9.1   Process Actions

Significant changes in source water turbidity levels, either increases or decreases, require that operators verify the effectiveness of the sedimentation process in removing suspended solids and floc. Measurement of turbidity levels at the sedimentation basin inlet and outlet will give operators a rough idea of process removal efficiency. Grab samples can be used for this determination. Visual observations of floc dispersion and settling characteristics will also help operators evaluate process performance.

Increasing source water turbidity levels may be the result of rainfall and runoff into a river, stream, or impoundment serving the treatment plant. If turbidity levels are increasing rapidly, verify the effectiveness of the coagulant chemicals and dosages being applied at the flash mixer. The efficiency of the coagulation-flocculation process directly affects the performance of the sedimentation process. Performance of jar tests in the laboratory may be used to simulate process performance in the treatment plant (see Chapter 3, "Coagulation and Flocculation"). Use the test results to adjust chemical dosages in the flash mixer.

In the event that higher dosages of alum or other coagulants are required for effective removal of increased suspended solids loads, an increase in the frequency of operation of sludge removal equipment may be required. On the other hand, if source water turbidity levels decrease, less frequent operation of sludge removal equipment may be indicated.

Changes in source water alkalinity and pH caused by storms, waste discharges, or spills have a significant impact on the performance of the sedimentation process as a result of decreased coagulation-flocculation process performance. Again, perform jar tests to assess the impact of source water or process water quality changes.

Water temperature changes may also require a reevaluation of process performance. Decreasing water temperatures lower the rate at which particles settle, while higher water temperatures increase particle settling velocities. Thus, temperature changes may also require that jar tests be performed to establish optimum floc settling rates. Temperature changes usually occur gradually over time, so sudden changes in temperature are unlikely unless a source water change is made.

Sudden increases in settled water turbidity may signal a potential operational issue in the filtration process. Floc carryover from the sedimentation basin will cause premature clogging of filters and may result in the degradation of filtered water quality. Filtration process actions are discussed in Chapter 5, "Filtration."

Table 4.3 gives a summary of sedimentation process problems, how to identify the causes of problems, and how to correct the problems.

Table 4.3   Sedimentation process troubleshooting

1. Source Water Quality Changes	Operator Actions	Possible Process Changes
• Turbidity • Temperature • Alkalinity • pH • Color	1. Perform necessary analyses to determine extent of change 2. Evaluate overall process performance 3. Perform jar tests if indicated 4. Make appropriate process changes (see right-hand column, "Possible Process Changes") 5. Increase frequency of process monitoring 6. Verify response to process changes at appropriate time (allowing sufficient time for change to take effect)	1. Change coagulant 2. Adjust coagulant dosage 3. Adjust flash-mixer/flocculator mixing intensity 4. Change frequency of sludge removal (increase or decrease) 5. Increase alkalinity by adding lime, caustic soda, or soda ash
2. Flocculation Process Effluent Quality Changes	**Operator Actions**	**Possible Process Changes**
• Turbidity • Alkalinity • pH	1. Evaluate overall process performance 2. Perform jar tests if indicated 3. Verify performance of coagulation-flocculation process (see Chapter 3, "Coagulation and Flocculation") 4. Make appropriate process changes 5. Verify response to process changes at appropriate time	1. Change coagulant 2. Adjust coagulant dosage 3. Adjust flash-mixer/flocculator mixing intensity 4. Adjust improperly working chemical feeder
3. Sedimentation Basin Changes	**Operator Actions**	**Possible Process Changes**

• Floc Settling
• Rising or Floating Sludge
	1. Observe floc settling characteristics:
	  a. Dispersion
	  b. Size
	  c. Settling rate
	2. Evaluate overall process performance
	3. Perform jar tests if indicated:
	  a. Assess floc size and settling rate
	  b. Assess quality of settled water (clarity and color)
	4. Make appropriate process changes
	5. Verify response to process changes at appropriate time
		1. Change coagulant
		2. Adjust coagulant dosage
		3. Adjust flash-mixer/flocculator mixing intensity
		4. Change frequency of sludge removal (increase or decrease)
		5. Remove sludge from basin
		6. Repair broken sludge rakes

Table 4.3 Sedimentation process troubleshooting *(continued)*

4. Sedimentation Process Effluent Quality Changes	Operator Actions	Possible Process Changes

• Turbidity
• Color
	1. Evaluate overall process performance
	2. Perform jar test if indicated
	3. Verify process performance:
	   a. Coagulation-flocculation process (see Chapter 3, "Coagulation and Flocculation")
	   b. Floc settling characteristics
	4. Make appropriate process changes
	5. Verify response to process changes at appropriate time
		1. Change coagulant
		2. Adjust coagulant dosage
		3. Adjust flash-mixer/flocculator mixing intensity
		4. Change frequency of sludge removal (increase or decrease)

5. Upflow Clarifier Process Effluent Quality Changes	Operator Actions	Possible Process Changes
• Turbidity • Turbidity Caused by Sludge Blanket Coming to Top Due to Rainfall on Watershed	1. See 4 above 2. Open main drain valve of clarifier	1. See 4 above 2. Drop entire water level of clarifier to bring the sludge blanket down

4.10   Startup and Shutdown Procedures

Startup or shutdown of the sedimentation process is not a routine operating procedure in most water treatment plants that are operated on a continuous basis. Implementation of these procedures is generally associated with a complete plant shutdown for periodic maintenance and cleaning, which is generally performed on an annual basis. In some instances, a shutdown may result from a major process failure. Consider taking photos of special features to provide a visual record of events or conditions that may be difficult to illustrate when a basin is full of water.

4.10.1   Startup Procedure

1. Check operational status and mode of operation (manual or automatic) of equipment and physical facilities.
   a. Check that basin drain valves are closed.
   b. Check that basin isolation gates or stop logs are removed.
   c. Check that launder weir plates are set at equal elevations.
   d. Check to ensure that all trash, debris, and tools have been removed from basin.

2. Test sludge removal equipment.
   a. Check that mechanical equipment is properly lubricated and ready for operation.

3. Operate sludge removal equipment. Check that valves are in proper position (either open or closed).
   a. Observe operation of sludge removal equipment.

4. Fill sedimentation basin with water.
   a. Observe proper depth of water in basin.
   b. Remove floating debris from basin water surface.

5. Start sample pumps (allow sufficient flushing time before securing samples).
6. Perform water quality analyses (make process adjustments as necessary).

4.10.2 Shutdown Procedure

1. Stop flow to sedimentation basin. Install basin isolation gates or stop logs.
2. Turn off sample pumps.
3. Turn off sludge removal equipment.
   a. Shut off mechanical equipment and disconnect where appropriate.
   b. Check that valves are in proper position (either open or closed).
4. Lock out and tag electric switches and equipment.
5. Dewater (drain) basin if necessary.
   a. Verify water table is not high enough to float the empty basin.
   b. Open basin drain valves.
6. Immediately following dewatering, grease and lubricate all gears, sprockets, and mechanical moving parts that were submerged. If this is not done, they can freeze or seize up in a few hours. Frozen parts will require long hours to repair and can result in equipment breakage.

Check Your Understanding

1. What water quality indicator is used as a rough measure of sedimentation basin process removal efficiency?
2. What problems can be created by a sudden increase in settled water turbidity?
3. What actions might an operator take if the sludge is rising or floating in a sedimentation basin?
4. Why should photographs be taken during shutdown and startup procedures?
5. List the steps in the shutdown procedure for a sedimentation basin.

4.11 Laboratory Tests

Critical water quality indicators, such as turbidity, will be routinely monitored, perhaps several times per 8-hour shift. Temperature will be evaluated less frequently, especially under stable water quality conditions.

4.11.1 Sampling Procedures

Process water samples will be either grab samples obtained directly from a specific process monitoring location, or continuous samples that are pumped to the laboratory from various locations in the process (flocculation basin effluent, settled water conduit). Process water samples must be representative of actual conditions in the treatment plant.

Process grab samples should be collected in clean plastic or glass containers and care should be taken to avoid contamination of the sample.

Sample analysis should be performed immediately following sample collection. When this is not possible, care should be taken to preserve samples by proper storage. See Chapter 9, "Laboratory Procedures," for sample preservation and storage procedures.

4.11.2 Sample Analysis

Analyses of certain process control water quality indicators (such as turbidity) are easily performed in the laboratory with the aid of automated analytical instruments such as turbidity meters. The collection of a bad sample or laboratory result is about as useful as obtaining no result. Good results require constant maintenance on and calibration of laboratory equipment and using correct lab procedures.

4.12 Equipment Operation and Maintenance

The operator of a sedimentation process will need to be familiar with the operation and maintenance of a variety of mechanical, electrical, and electronic equipment, including:

1. Sludge removal equipment
2. Sludge pumps
3. Sump pumps
4. Valves
5. Flowmeters and gauges
6. Water quality monitors such as turbidity meters
7. Control systems

Sludge removal equipment will constitute the majority of the electro-mechanical equipment used in the sedimentation process. Because a wide variety of systems can be used to remove sludge from the bottom of the sedimentation basins, the operator will need to be thoroughly familiar with the operation and maintenance instructions for each piece of equipment.

Before starting a piece of mechanical equipment, such as a sludge pump, be sure that the unit is properly lubricated and its operational status is known. After startup and during normal operation, check for excessive noise and vibration, overheating, and leakage (water, lubricants). Check the pump's suction and discharge pressures to be sure the lines are not plugged. When in doubt about the performance of a piece of equipment, always refer to the operation and maintenance instructions.

Many equipment items, such as valves, are simple open or closed devices. Similarly, sump pumps are simple ON/OFF devices. Other equipment, such as sludge pumps, have provisions for flow rate adjustment and may be furnished with sight glasses to visually check sludge flow. Check the flow each time the equipment is operated. Sludge collectors, discharge lines, and troughs should be periodically flushed to maintain a free sludge flow. Calibration of flow rates and measurement of sludge density may require the use of special procedures. Detailed operating, repair, and calibration procedures are usually given in the manufacturer's instructions.

4.12.1 Corrosion Control

The metallic parts of clarifiers and solids-contact units must be protected from corrosion. A good layer of paint or other protective coating over all metal parts exposed to water is one successful approach. Another approach is the use of a cathodic protection system.

Cathodic protection is a process for reducing or eliminating corrosion of metal parts exposed to water, soil, or the atmosphere. In this process, the flow of an electric current from the metal surface into water (the cause of corrosion) is stopped by overpowering it by applying a stronger current from another source (the cathodic protection system).

To maintain a high degree of corrosion control, inspect and test the operation of the cathodic protection system and its parts regularly. Regular and proper maintenance is critical because the amount of protective current required to prevent corrosion can vary with changes in the conditions of the coatings on the metallic surfaces, in the chemical characteristics of the water being treated, and the operation of the clarifier.

For additional information on how to control corrosion, see Chapter 7, "Corrosion Control."

4.12.2 Preventive Maintenance Procedures

Preventive maintenance programs are designed to ensure the continued satisfactory operation of treatment plant facilities by reducing the frequency of breakdown failures. This is accomplished by performing scheduled or routine maintenance of valves, pumps, and other electrical and mechanical equipment.

In the normal operation of the sedimentation process, the operator will be expected to perform routine maintenance functions as part of an overall preventive maintenance program. Typical functions include:

1. Keeping electric motors free of dirt and moisture
2. Ensuring good ventilation (air circulation) in equipment work areas
3. Checking pumps and motors for leaks, unusual noise and vibrations, overheating, or signs of wear
4. Maintaining proper lubrication and oil levels
5. Inspecting for alignment of shafts and couplings
6. Checking bearings for wear, overheating, and proper lubrication
7. Checking for proper valve operation (leakage or jamming)
8. Checking for free flow of sludge in sludge removal collection and discharge systems

Accurate records are a key element in a successful preventive maintenance program. These records provide maintenance and operations personnel with clues for determining causes for equipment breakdowns, and will assist in spotting similar weaknesses in other equipment items. The frequency of performing periodic maintenance on electrical and mechanical equipment items is often based on preventive maintenance records of prior performance and equipment manufacturers' recommendations.

Good housekeeping practices are an important part of every preventive maintenance program.

For additional details on equipment maintenance, see the maintenance chapter in Volume II.

4.12.3 Safety

To avoid accidents or injury when working around sludge removal equipment, such as pumps and motors, follow the safety procedures listed in this section.

ELECTRICAL EQUIPMENT

1. Avoid electric shock by using protective gloves.
2. Avoid grounding yourself in water or on pipes.
3. Ground all electric tools.
4. Use the buddy system.
5. Use a lockout and tag system whenever electrical equipment or electrically driven mechanical equipment is out of service or being worked on.

MECHANICAL EQUIPMENT

1. Keep protective guards on rotating equipment.
2. Do not wear loose clothing around rotating equipment.
3. Keep hands out of valves, pumps, and other pieces of equipment (lock out and tag power switches before cleaning).
4. Clean up all lubricant and sludge spills (slippery surfaces cause falls).
5. Use a lockout and tag system whenever mechanical equipment is out of service or being worked on.

OPEN-SURFACE WATER-FILLED STRUCTURES

1. Use safety devices, such as handrails, ladders, and Coast Guard-approved personal flotation devices (PFDs).
2. Close all openings and replace safety gratings when finished working.
3. Use the buddy system.

VALVE AND PUMP VAULTS, SUMPS

1. Be sure all underground or confined structures are free of hazardous atmospheres by using a gas detector to check for toxic or explosive gases and lack of sufficient oxygen.
2. Only work in well-ventilated structures (use ventilation fans).
3. Use the buddy system.
4. Lock or chain valves when working in an area that could be flooded.

Check Your Understanding

1. How frequently should you monitor the turbidity of the water being treated?
2. How are sedimentation process water samples obtained for analysis?
3. What items should an operator look for after a sludge pump has been started?
4. List typical functions performed as part of a sedimentation process preventive maintenance program.
5. What types of safety hazards are associated with sludge removal equipment?

4.13 Math Assignment

Turn to Appendix A, "Introduction to Basic Math for Operators." Read and work through the example problems in sections:

- "Learning Objectives"
- A.5.5, "Density, Specific Weight, and Specific Gravity"
- A.9.6, "Sedimentation" (English System)
- A.10.6, "Sedimentation" (Metric System)

Check the math using a calculator. You should be able to get the same answers.

4.14 Additional Resources

Your local library may be a good source for obtaining these additional resources related to the field of water treatment:

Manual of Water Utilities Operations (*Texas Manual*). Texas Water Utilities Association, www.twua.org. ASIN: B000KT9EIU. Chapter on coagulation and sedimentation.

Stone, B. G. Notes from "Design of Water Treatment Systems," CE-610, Loyola Marymount University, Los Angeles, CA, 1977.

Water Quality and Treatment, Sixth Edition. American Water Works Association (AWWA), www.awwa.org. ISBN: 9780071630115.

Water Treatment Plant Design, Fifth Edition. American Water Works Association (AWWA), www.awwa.org. ISBN: 9780071745727.

Chapter Review

4.1 Presedimentation

1. What is one purpose of presedimentation facilities?
   1. Avoid need for sedimentation basins
   2. Increase the concentration of suspended solids
   3. Prevent solids settling out before sedimentation
   4. Reduce the solids-removal load at the water treatment plant

4.2 Process Performance

2. What size particles can be removed from water by sedimentation?
   1. 0.3–0.7 microns
   2. 2–8 microns
   3. Greater than 10 microns
   4. Smaller than 10 microns

3. What property of alum attracts negatively charged particles?
   1. Ease of creating alloys
   2. Large particle size
   3. Negative charge
   4. Positive charge

4. What happens to the settling rate (settling velocity) of particles in water when the water temperature drops?
   1. Decreases
   2. Depends on weather
   3. Increases
   4. Stays constant

4.3 Sedimentation Basins

5. What is the largest portion of the sedimentation basin?
   1. Inlet zone
   2. Outlet zone
   3. Settling zone
   4. Sludge zone

4.4 Design and Operational Guidelines

6. Calculate the theoretical detention time for a rectangular sedimentation basin in hours. The basin is 80 feet long, 30 feet wide, 10 feet deep, and treats a flow of 1.8 MGD.
   1. 0.24 hours
   2. 2.4 hours
   3. 24 hours
   4. 240 hours

7. What is the major means by which operators can control water treatment processes?
   1. Adjusting chemicals and chemical feed rates
   2. Controlling sedimentation process
   3. Controlling water temperatures
   4. Modifying demands for water

8. Estimate the surface loading rate in gallons per minute per square foot for a rectangular sedimentation basin 20 feet wide and 40 feet long when the flow is 0.5 MGD.
   1. 0.21 gpm/ft^2
   2. 0.43 gpm/ft^2
   3. 0.64 gpm/ft^2
   4. 0.86 gpm/ft^2

9. Determine the mean flow velocity in feet per minute (ft/min) for a rectangular sedimentation basin with a flow of 2.0 MGD. The basin is 75 feet long and 25 feet wide with a depth of 14 feet.
   1. 0.18 ft/min
   2. 0.36 ft/min
   3. 0.48 ft/min
   4. 0.53 ft/min

10. Determine the weir loading rate in gallons per minute per foot (gpm/ft) of weir for a rectangular sedimentation basin with a flow rate of 2.0 MGD. Four effluent launders 10.5 feet (ft) long with V-notch weirs on both sides of the launder extend into the basin from the outlet end.
    1. 16.5 gpm/ft
    2. 19.4 gpm/ft
    3. 23.2 gpm/ft
    4. 33.1 gpm/ft

4.5 Solids-Contact Clarification

11. _____ combine the coagulation, flocculation, and sedimentation processes in a single basin.

12. Sludge produced by a solids-contact clarification unit is recycled through the process to act as a _____.

13. Changes in which process variables cause instability in solids-contact units?
    1. DO, chlorine demand, and pH
    2. Color, alkalinity, and pesticides
    3. Flow, temperature, and turbidity
    4. Taste, odor, and microbial content

14. What does the volume over volume test result tell an operator about the solids-contact unit?
    1. Chemical feeder setting
    2. Clarity of supernatant
    3. Percentage of slurry
    4. Speed of the recirculation device

4.6 Sludge Handling

15. If the raw water at a treatment plant needs a lime dose of 10.5 mg/L, determine the setting on the lime feeder in pounds per day (lb/d) when the flow is 1.5 MGD.
    1. 126 lb/d
    2. 131 lb/d
    3. 154 lb/d
    4. 175 lb/d

16. What is the primary factor in determining the frequency of sludge removal?
    1. Rate of chemical feed
    2. Rate of flow
    3. Rate of sludge buildup
    4. Rate of weir loading

4.7 Interaction with Other Treatment Processes

17. Low-turbidity source waters may not require_____.

4.8 Normal Operations

18. Which water quality indicators should be monitored in the normal operation of the sedimentation process?
    1. Chlorine residual and coliforms
    2. Suspended solids and dissolved oxygen
    3. Taste, odor, and pH
    4. Temperature and turbidity

19. In solids-contact clarifiers, what should an operator do if the sludge blanket is of normal density but very close to the surface?

  1. Adjust the chemical dose
  2. Install sight glasses
  3. Prevent short-circuiting
  4. Waste more sludge

20. What does frequent clogging of the sludge discharge line indicate?

  1. Sludge blanket is too deep
  2. Sludge blanket is too shallow
  3. Sludge concentration is too high
  4. Sludge concentration is too low

4.9 Abnormal Operations

21. How can operators measure turbidity levels at the sedimentation basin inlet and outlet to get a rough idea of process removal efficiency?

  1. Composite samples
  2. Grab samples
  3. Repeat samples
  4. Representative samples

4.10 Startup and Shutdown Procedures

22. What is the most common reason to shut down a sedimentation basin?

  1. Abnormal amounts of rainfall
  2. Annual maintenance and cleaning
  3. Flow equalization
  4. Intermittent process use

23. Which task should be performed immediately following dewatering of a sedimentation basin?

  1. Grease and lubricate previously submerged mechanical moving parts
  2. Perform necessary laboratory analyses or send samples to outside laboratories
  3. Record all gauge and meter readings and note shutdown in daily diary
  4. Stop flow to sedimentation basin by diverting to holding tank

4.11 Laboratory Tests

24. Process water samples must be _____ of actual conditions in the treatment plant.

4.12 Equipment Operation and Maintenance

25. How can operators determine if sludge pumping lines are clogged?

  1. Align flights on drag-chain and flight systems
  2. Check pump's suction and discharge pressures
  3. Ensure pump is properly lubricated
  4. Perform all scheduled preventive maintenance

Filtration

CHAPTER OUTLINE

LEARNING OBJECTIVES

1. Describe the various types of potable water filters and how they work.

2. Explain how other treatment processes affect the performance of the filtration process.

3. Operate and maintain filters under normal and abnormal process conditions.

4. Start up and shut down filtration processes.

5. Safely perform duties related to the various types of filters.

KEY TERMS

absorption

activated carbon

adsorption

air binding

backwashing

breakthrough

colloids

conventional filtration

diatomaceous earth

diatoms

direct filtration

effective size (ES)

entrain

floc

fluidized

garnet

head loss

inline filtration

micron

mudball

particle counter

particle counting

particulate

permeability

piezometer

pore

sensitivity (particle counters)

slurry

specific gravity

submergence

turbidity meter

uniformity coefficient (UC)

5.1   Filtration Mechanisms

The purpose of filtration is the removal of **particulate** impurities and **floc** from the water being treated. Through filtration, water passes through materials, such as a bed of sand, coal, or other granular substances, to remove floc and particulate impurities. These impurities consist of suspended particles (fine silts and clays), **colloids**, biological forms (bacteria and plankton), and floc in the water being treated.

Filtration preceded by coagulation, flocculation, and sedimentation is commonly referred to as **conventional filtration** (Figure 5.1). In the **direct filtration** process, the sedimentation step is omitted and flocculation facilities are reduced in size or may be omitted.

Filtration is a physical and chemical process. The actual removal mechanisms are interrelated and rather complex, but filter removal of turbidity is based on the following factors:

1. Chemical characteristics of the water being treated (particularly source water quality)
2. Nature of suspension (physical and chemical characteristics of particulates suspended in the water)
3. Types and degree of pretreatment (coagulation, flocculation, and sedimentation)
4. Filter type and operation

A popular misconception is that particles are removed in the filtration process mainly by physical straining. Straining removes particles from a liquid (water) by passing the liquid through a filter whose **pores** are smaller than the particles to be removed. Pores are very small open spaces in a rock or granular material; they are also called interstices, voids, or void spaces. While the straining mechanism does play a role in the overall removal process, especially for larger particles, most of the particles removed during filtration are considerably smaller than the pore spaces in the media. This is particularly true at the beginning of the filtration cycle when the pore spaces are clean (not clogged by particulates removed during filtration).

Thus, a number of interrelated removal mechanisms within the filter media itself are relied upon to achieve high removal efficiencies. These removal mechanisms include the following processes:

1. Sedimentation on media (very important)
2. **Adsorption** (very important)
3. Biological action
4. **Absorption** (not too important after initial wetting)
5. Straining

The relative importance of these removal mechanisms will depend largely on the nature of the water being treated, the degree of pretreatment, and filter characteristics.

Check Your Understanding

1. What is the major difference between conventional filtration and direct filtration?
2. List the particle removal mechanisms involved in the filtration process.

---

particulate
A very small solid suspended in water that can vary widely in size, shape, density, and electrical charge. Colloidal and dispersed particulates are artificially gathered together by the processes of coagulation and flocculation.

floc
Clumps of bacteria and particles or coagulants and impurities that have come together and formed a cluster. Found in flocculation tanks and settling or sedimentation basins.

colloids (KALL-loids)
Very small, finely divided solids (particles that do not dissolve) that remain dispersed in a liquid for a long time due to their small size and electrical charge. When most of the particles in water have a negative electrical charge, they tend to repel each other. This repulsion prevents the particles from clumping together, becoming heavier, and settling out.

conventional filtration
A method of treating water that consists of the addition of coagulant chemicals, flash mixing, coagulation-flocculation, sedimentation, and filtration. Also called complete treatment.

direct filtration
A method of treating water that consists of the addition of coagulant chemicals, flash mixing, coagulation, minimal flocculation, and filtration. The flocculation facilities may be omitted, but the physical-chemical reactions will occur to some extent. The sedimentation process is omitted.

pore
A very small open space in a rock or granular material. Also called an interstice, void, or void space.

adsorption
The gathering of a gas, liquid, or dissolved substance on the surface or interface zone of another material.

absorption
The taking in or soaking up of one substance into the body of another by molecular or chemical action (as tree roots absorb dissolved nutrients in the soil).

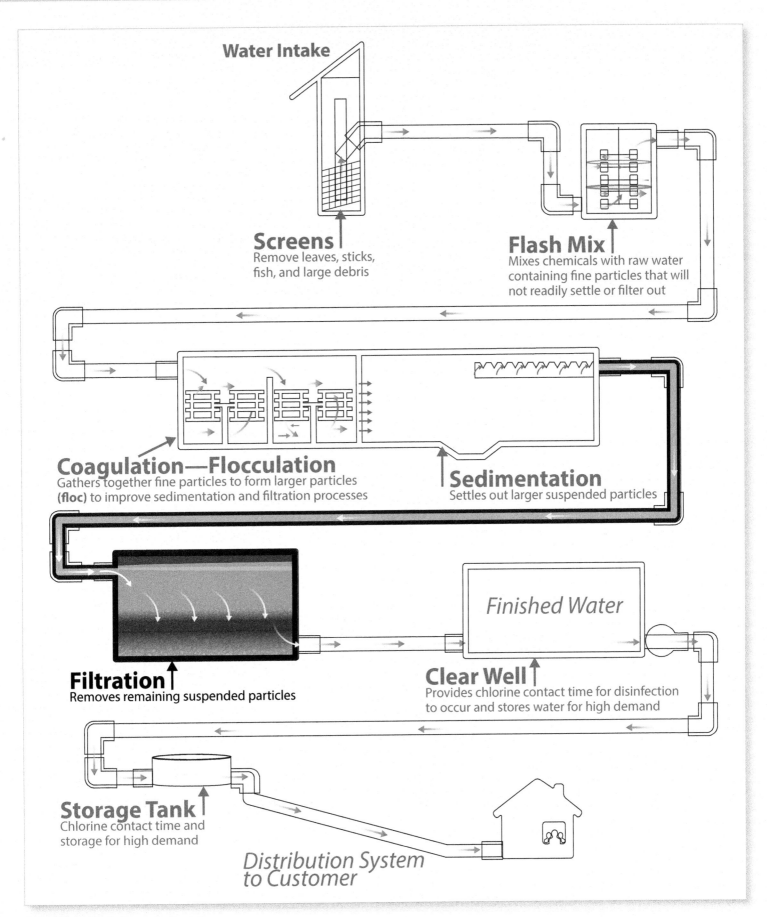

Water Intake

Screens
Remove leaves, sticks, fish, and large debris

Flash Mix
Mixes chemicals with raw water containing fine particles that will not readily settle or filter out

Coagulation—Flocculation
Gathers together fine particles to form larger particles **(floc)** to improve sedimentation and filtration processes

Sedimentation
Settles out larger suspended particles

Finished Water

Filtration
Removes remaining suspended particles

Clear Well
Provides chlorine contact time for disinfection to occur and stores water for high demand

Storage Tank
Chlorine contact time and storage for high demand

Distribution System to Customer

Figure 5.1   Typical process flow diagram

5.2   Types of Filters

Over the years, a distinction has been made between the older slow sand filtration process (used in smaller systems in the United States) and newer rapid sand filtration (used extensively in the United States). This distinction is made because the removal mechanisms that apply to the two sand filtration processes are not directly comparable. However, because of significant improvements in the process, the term "rapid sand filtration" no longer adequately describes the variety of processes used. Thus, a more specific classification system for filter types was developed and includes the following:

1. Gravity filtration (sand, dual media, and mixed media).
2. Pressure filtration (mixed media).
3. Diatomaceous earth (precoat) filtration. Diatomaceous earth is a fine, siliceous (made of silica) earth composed mainly of the skeletal remains of diatoms. Diatoms are single-celled microscopic algae with a rigid, box-like internal structure consisting mainly of silica.
4. Slow sand filtration.

*Note that the numbers for liters per second per square meter are the same as for millimeters per second.

$$1\left[\frac{L}{s \cdot m^2}\right] = 1\left[\frac{L}{s \cdot m^2}\right] \times \frac{m^3}{1,000\ L} \times \frac{1,000\ mm}{m}$$

$$= 1\frac{mm}{s}$$

This occurs because 1 cubic meter equals 1,000 liters and 1 meter equals 1,000 millimeters.

5.2.1   Gravity Filtration

In all gravity filtration systems, the water level or pressure (head) above the media forces the water through the filter media as shown in Figure 5.2. The rate water passes through the granular filter media (commonly referred to as filtration rate) may vary from 2 to about 10 gallons per minute per square foot (gal/min/ft^2) (1.36 to 6.79 liters per second per square meter or L/sec/m^2)*. However, many state health authorities limit the maximum filtration rate to 2 or 3 gal/min/ft^2 (1.36 or 2.04 L/sec/m^2)

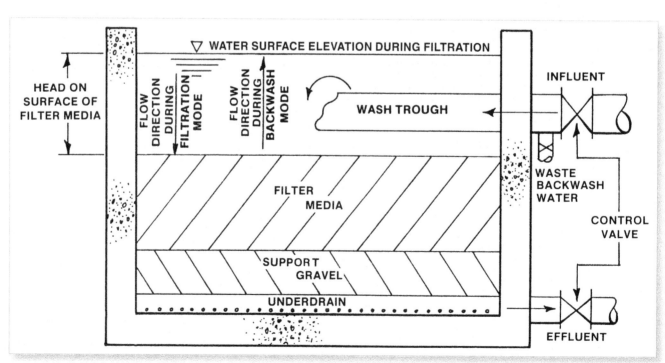

Figure 5.2   Gravity filter module

for gravity filtration. The rate of water flow through the filter is the hydraulic loading or the filtration rate. The filtration rate depends on the type of filter media.

Filter media consists of the following substances:

1. Single media (sand, anthracite coal, or granular activated carbon)
2. Dual media (sand and anthracite coal)
3. Multi- or mixed media (sand, anthracite coal, and **garnet**)

Activated carbon can also be used along with the above filter media to remove tastes, odors, and organic substances.

In gravity filtration, the particulate impurities are removed in/on the media, thus causing the filter to clog after a period of filtration time. In spite of this, gravity filtration is widely used in water treatment plants.

5.2.2   Pressure Filtration

A pressure filter is similar to a gravity sand filter except that the filter is completely enclosed in a pressure vessel such as a steel tank, and is operated under pressure, as shown in Figure 5.3.

Pressure filters frequently offer lower installation and operation costs in small filtration plants; however, they are generally somewhat less reliable than gravity filters (depending upon pumped pressure). Some states do not recommend the use of pressure filters for treating surface waters. Maximum filtration rates for pressure filters are in the 2 to 3 gal/min/ft^2 (1.36 to 2.04 L/sec/m^2) range.

garnet
A group of hard, reddish, glassy, mineral sands made up of silicates of base metals (calcium, magnesium, iron, and manganese). Garnet has a higher density than sand.

activated carbon
Adsorptive particles or granules of carbon usually obtained by heating carbon (such as wood). These particles or granules have a high capacity to selectively remove certain trace and soluble materials from water.

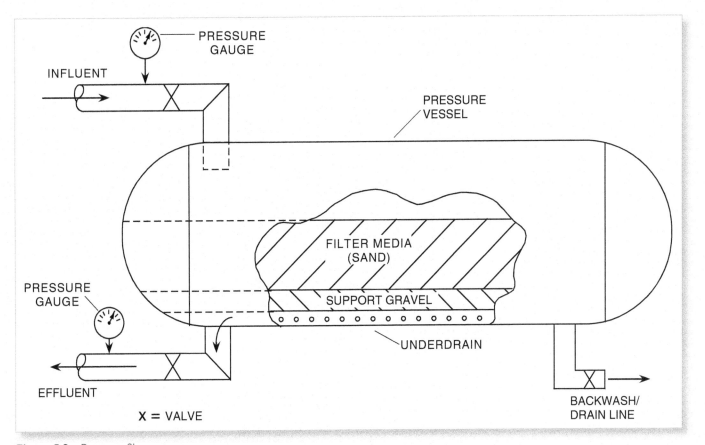

Figure 5.3   Pressure filter

5.2.3   Diatomaceous Earth Filtration

In diatomaceous earth (precoat) filtration, the filter media is added to the water being treated as a **slurry**; it then collects on a septum (a pipe or conduit with porous walls) or other appropriate screening device, as shown in Figure 5.4. After the initial precoat application, water is filtered by passing it through the coated screen. The coating thickness may be increased during the filtration process by gradually adding more media—a body feed. In most water treatment applications, diatomaceous earth is used for both the precoat and body-feed operations.

Diatomaceous earth filtration is primarily a straining process, and finds wide application where very high particle removal efficiencies (high clarity water) are required, such as in the beverage and food industries. Precoat filters can be operated as gravity, pressure, or vacuum filters. They are also commonly used in swimming pool installations due to their small size, efficiency, ease of operation, and relatively low cost. They find limited use in larger water treatment plants due to hydraulic (flow), sludge disposal, and other operational considerations.

5.2.4   Slow Sand Filtration

In slow sand filtration, water is drawn through the filter media (sand) by gravity as it is in the gravity filtration process. However, this is generally where the similarity between these two filtration processes ends.

Small Water System Operation and Maintenance. See section on slow sand filtration. Office of Water Programs. www.owp.csus.edu.

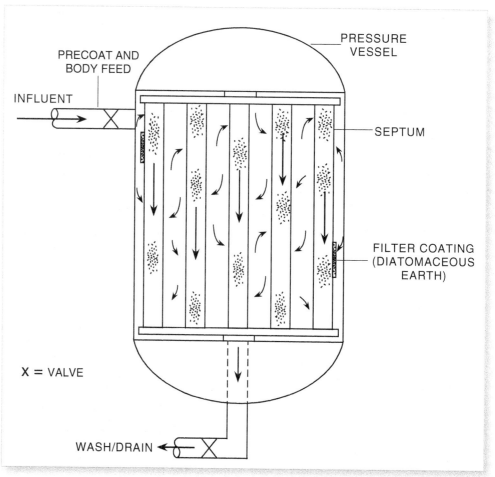

Figure 5.4   Diatomaceous earth (precoat) filter

slurry
A watery mixture or suspension of insoluble (not dissolved) matter; a thin, watery mud or any substance resembling it (such as a grit slurry or a lime slurry).

In the slow sand filtration process, particles are removed by straining, adsorption, and biological action. Filtration rates are extremely low (0.015 to 0.15 gal/min/ft^2 or 0.01 to 0.1 L/sec/m^2). The majority of the particulate material is removed in the top several inches of sand, so this entire layer must be physically removed when the filter becomes clogged.

Slow sand filtration is popular for small systems because of reliability and minimum operation and maintenance (O&M) requirements, but the process has found limited application in larger systems due to the large area required and the need to manually clean the filters.

Check Your Understanding

1. List the four specific classes of filters.

2. What is garnet?

3. What material is used for precoat and body-feed operations?

5.3   Process Performance Considerations

Filter process performance considerations include the filter media, operation mode, cleaning, and control systems. A filter's performance depends on the type and size of the media, the media's characteristics and uniformity coefficient, and if the selected media has associated problems. Operational criteria influence filter performance through the filter layout, filter production and filtration rate, and efficiency. Filter operation depends on the filtration mode, the cleaning or backwashing rates, and the surface washing systems. Filter process performance considerations also include filter control systems that regulate filter flow rates.

5.3.1   Filter Media

Sand, anthracite coal, and garnet are the filter media most often selected by designers in gravity filtration. However, other inert materials can also be used. Gravel is commonly used to support the filter materials. In the diatomaceous earth filtration process, diatomaceous earth is used due to its high strength (very rigid) and high **permeability**. Permeability is the property of a material or soil that permits considerable movement of water through it when it is saturated.

Desirable filter media characteristics include:

1. Good hydraulic characteristics (permeable)
2. Does not react with substances in the water (inert and easy to clean)
3. Hard and durable
4. Free of impurities
5. Insoluble in water

The various filter media are usually classified by the following characteristics:

1. Effective size (ES)
2. Uniformity coefficient

3. **Specific gravity**
4. Hardness

The **effective size (ES)** refers to the diameter of the particles in a granular sample (filter media) for which 10 percent of the total grains are smaller and 90 percent larger on a weight basis. Effective size is obtained by passing granular material through sieves with varying dimensions of mesh and weighing the material retained by each sieve. The effective size is also approximately the average size of the grains. The effective size is referred to as D_{10}.

Effective Size, mm = Diameter of 10% by weight grains, mm

Because some operational problems can be improved by proper selection of media size, two factors are important:

1. The time required for turbidity to break through the filter bed
2. The time required for the filter to reach limiting (terminal) **head loss**

With a properly selected media, these times are about the same.

If the limiting head loss is frequently a problem and turbidity **breakthrough** rarely occurs, then a larger media size may be considered. If turbidity breakthrough is frequently a problem and limiting head loss is rarely encountered, then a smaller media size may be considered.

If both head loss and turbidity breakthrough are constantly a problem, a deeper filter bed with a larger media size should be considered. However, increasing the media depth is not always possible without modification of the filter box or tank. Adequate clearance must be allowed between the top of the media and the bottom of the wash water troughs. Otherwise, filter media will be carried out over the wash troughs during backwash.

The relationship between turbidity breakthrough and limiting head loss is also strongly affected by optimum chemical treatment. Poor chemical treatment can often result in either early turbidity breakthrough or rapid head loss buildup.

Uniformity coefficient (UC) is the ratio of (1) the diameter of a grain (particle) of a size that is barely too large to pass through a sieve that allows 60 percent of the material (by weight) to pass through to (2) the diameter of a grain (particle) of a size that is barely too large to pass through a sieve that allows 10 percent of the material (by weight) to pass through. The resulting ratio is a measure of the degree of uniformity in a granular material, such as filter media. The formula is:

$$UC = \frac{D_{60}}{D_{10}}$$

Where,

UC = Uniformity coefficient

D_{60} = Diameter of particles that 60% of the materials (by weight) are smaller than

D_{10} = Effective size = diameter of particles that 10% of the materials (by weight) are smaller than

For example, if the 60 percent sieve size grain diameter is 1 mm and the 10 percent sieve size grain diameter is 0.5 mm, then the uniformity coefficient would be:

$$UC = \frac{D_{60}}{D_{10}} = \frac{1.0\,mm}{0.5\,mm} = 2.0$$

---

specific gravity
(1) Weight of a particle, substance, or chemical solution in relation to the weight of an equal volume of water. Water has a specific gravity of 1.000 at 39°F (4°C). Particulates with specific gravity less than 1.0 float to the surface and particulates with specific gravity greater than 1.0 sink. (2) Weight of a particular gas in relation to the weight of an equal volume of air at the same temperature and pressure (air has a specific gravity of 1.0). Chlorine gas has a specific gravity of 2.5.

head loss
The head, pressure, or energy (they are the same) lost by water flowing in a pipe or channel as a result of turbulence caused by the velocity of the flowing water and the roughness of the pipe, channel walls, or restrictions caused by fittings. Water flowing in a pipe loses head, pressure, or energy as a result of friction losses. The head loss through a filter is due to friction losses caused by material building up on the surface or in the top part of a filter. Also called friction loss.

breakthrough
A crack or break in a filter bed allowing the passage of floc or particulate matter through a filter. This will cause an increase in filter effluent turbidity. A breakthrough can occur (1) when a filter is first placed in service, (2) when the effluent valve suddenly opens or closes, and (3) during periods of excessive head loss through the filter (including when the filter is exposed to negative heads).

Selection of an appropriate media size and uniformity coefficient depends on the source water quality, filter design, and anticipated filtration rate. Generally, the more uniform the media, the slower the head loss buildup. Media with uniformity coefficients of less than 1.5 are readily available. Media with uniformity coefficients of less than 1.3 are only available at a high cost. Typical filter media characteristics are given in Table 5.1.

Table 5.1 Typical filter media characteristics[a]

Material	Size Range (mm)	Specific Gravity	Hardness (Moh's Scale)
Conventional Sand	0.5–0.6	2.6	7
Coarse Sand	0.7–3.0	2.6	7
Anthracite Coal	1.0–3.0	1.5–1.8	3
Garnet	0.2–0.4	3.1–4.3	6.5–7.5
Gravel	1.0–50	2.6	7

[a]Adapted from:
1. Stone, B. G. Notes from "Design of Water Treatment Systems," CE–610, Loyola Marymount University, Los Angeles, CA, 1977.
2. *Water Quality and Treatment: A Handbook of Community Water Supplies*, by AWWA, 1990.

5.3.2 Operational Criteria

Filter operational criteria include: (1) the filter layout—where the filter media is contained and the number of filter modules; and (2) the filter production and filtration rate. The filtration efficiency—the overall reduction in turbidity—is also an important operational criterion for treatment plants.

5.3.2.1 Filter Layout
In gravity filtration, the filter media is usually contained in concrete (also steel or aluminum) filter modules, which are all the same size; however, the size will vary widely in surface area from plant to plant. In general, the minimum number of filter modules is four. This allows for one filter to be out of service and still have 75 percent capacity available. Typical gravity filter media layouts are shown in Figure 5.5.

5.3.2.2 Filter Production and Filtration Rate
Filter production (capacity) rates are measures of the amount of water that can be processed through an individual filter module in a given time period. Filter production is measured in units of million gallons per day (Mgal/d). Filtration rate (or hydraulic loading) is commonly used to measure the flow of water through a filter in units of gal/min/ft^2 or L/sec/m^2.

Filtration rates used in gravity filtration generally range from about 2 to 10 gal/min/ft^2 (1.36 to 6.79 L/sec/m^2). The higher filtration rates are found in plants with dual-media filters. In larger water treatment plants, filter capacities range from about 5 to 20 Mgal/d (0.2 to 0.8 m^3/sec) for each filter unit.

5.3.2.3 Filtration Efficiency
In the filtration process, filter efficiency is roughly measured by overall plant reduction in turbidity. Reductions of over 99.5 percent can be achieved under optimum conditions, while a poorly operated filter and inadequate pretreatment (coagulation, flocculation, and sedimentation) can result in turbidity removals of less than 50 percent. The best way to ensure

high filtration efficiency is to select an effluent turbidity goal (level) and stay below the target value, such as 0.1 nephelometric turbidity unit (NTU).

Filter removal efficiency depends largely on the quality of the water being treated, the effectiveness of the pretreatment processes in conditioning the suspended particles for removal by sedimentation and filtration, and filter operation itself.

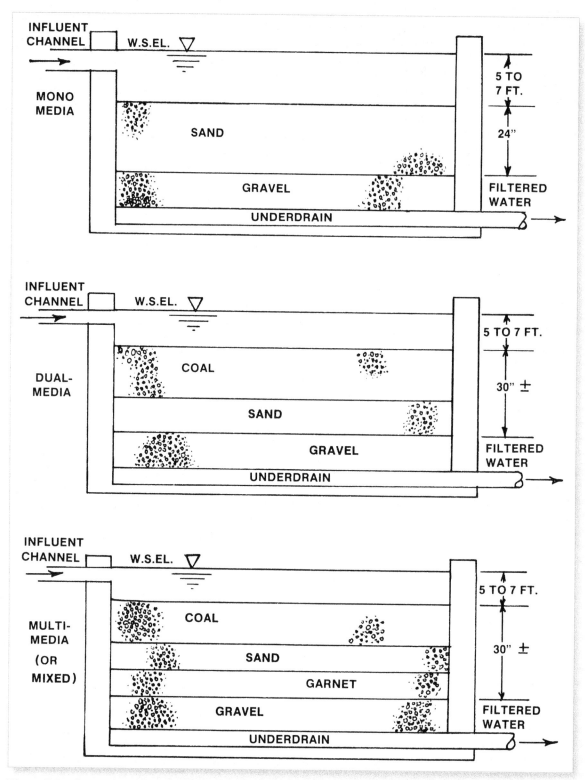

Figure 5.5   Gravity filter media layouts

Filter unit design and filter media type and thickness also play a role in determining filter removal efficiency, but are less important than water quality and pretreatment considerations. Gravity sand filters usually produce a filtered water turbidity comparable to that of a dual-media filter if the applied water quality is similar. However, the operational differences between sand and dual-media filters are significant. Because of their smaller media grain size (see Table 5.1), sand filters tend to clog with suspended matter and floc more quickly than dual-media filters. This means frequent backwashing will be required to keep the sand filter operating efficiently. **Backwashing** is the process of reversing the flow of water back through the filter media to remove the entrapped solids. Sand filters have fine, light grains on the top that stop all floc and particulates at the surface of the filter. Dual-media filters have lighter, larger diameter grains in the top layer of the media that stop the larger particles; the smaller particles are usually stopped farther down in the filter. The larger grain size of the anthracite coal layer (up to 1.5 mm) in the top portion of a dual-media filter permits greater depth penetration of solids into the anthracite coal layer and larger solids storage volume in the filter. The sand layer below the anthracite is used as a protective barrier against breakthrough. Chemical dosages are adjusted to keep the solids in the anthracite. Taps at various depths in the filter are used to observe the depth of solids penetration. These characteristics generally produce filter runs that are several times longer and with higher filtration rates without a high head loss than those achieved by sand filters. Accordingly, dual-media filters are referred to as depth filters while sand filters are known as surface filters.

Multi-media filters (coal, sand, and garnet) are also used to extend filter run times, and these filters generally perform in a manner similar to gravity sand filters, except that the filter media is enclosed in a pressure vessel. You can get consistently satisfactory filtered water quality with pressure filters if you are careful about the routine operation of this type of filtration system. With the filter media fully enclosed, it is difficult to assess the media condition by simple visual observation. In addition, excessive pressure in the vessel will force solids as well as water through the filter media. Obviously, this will result in the deterioration of filtered water quality.

In contrast to gravity filtration, diatomaceous earth filtration is essentially a straining process. In many instances, high particulate removal efficiencies can be achieved in diatomaceous earth filtration without any preconditioning processes (coagulation, flocculation, and sedimentation). However, the application of diatomaceous earth filtration in larger water treatment plants is limited by the following considerations:

1. High head losses across the filter
2. Possible sludge disposal problems
3. Potential decreased reliability (in terms of quality of filtered water)

Check Your Understanding

1. What material is most often used to support granular filter materials?
2. What units are commonly used to express filtration rate?
3. What is the major operational difference between sand and dual-media filters?

5.3.3 Filter Operation

A successful and efficient filter operation depends on the filtration mode and effective backwashing or cleaning, which is often augmented with surface washing. The filtration mode is influenced by the amount of head loss or clogging. A clogged filter or one that has broken through is taken out of service for backwashing or cleaning. This could slow the filtration operation, so the operator must find the right balance of backwashing rates and cycles. Additional surface washing produces the optimum filter media cleaning during backwashing and prevent mudballs.

5.3.3.1 Filtration Mode

In the filtration mode of operation, water containing suspended solids is applied to the surface of the filter media. Depending on the amount of suspended solids in the water being treated and the filtration rate, the filter will exhibit high head loss or clog after a given time period (from several hours to several days).

Clogging is a buildup of head loss (pressure drop) across the filter media until it reaches some predetermined design limit. Total design head loss in gravity filters generally ranges from about 6 to 10 feet (1.8 to 3.0 m), depending mainly on the depth of the water over the media. Clogging of the filter leads to breakthrough, a condition in which solids are no longer removed by the already overloaded filter. The solids pass into the filter effluent where they appear as increased turbidity.

A filter is usually operated until just before clogging or breakthrough occurs, or a specified time period has passed, generally within 72 hours. To save money, energy, and water by maximizing production before backwashing, some operators run their filters until clogging or breakthrough occurs. This is a poor practice; when breakthrough occurs, there will be an increase in filtered water turbidity. See Section 5.9.1 for detailed procedures on how to check out, start up, and shut down filters.

5.3.3.2 Backwashing

After a filter clogs (reaches maximum head loss), or breakthrough occurs, or a specified time period has passed, the filtration process is stopped and the filter is taken out of service for cleaning or backwashing. Backwashing is the process of reversing the flow of water through the filter media to remove the entrapped solids. To remove the trapped solids from the filter media in the gravity and pressure filtration processes, the filter media must be expanded or **fluidized** by reversing the flow of water. Backwash flow rates ranging from 10 to 25 gal/min/ft² (6.8 to 17 L/sec/m²) of filter media surface area are usually required to clean the filter adequately. Insufficient backwash rates may not completely remove trapped solids from the filter media, while too high a backwash rate may cause excessive loss of filter media and media disturbance (mounding). Usually, higher backwash rates are required at high temperatures and with larger media. Higher backwash rates are required at higher temperatures to suspend the media because the water is less viscous (like warmer syrup). Backwashing at too high a rate is much more destructive than at too low a backwash rate. See Section 5.9.1.2, "Backwash Procedures," for details on how to backwash filters.

In diatomaceous earth filtration, accumulated solids are also removed by reversing the flow through the filter. However, the diatomaceous earth

fluidized

A mass of solid particles that is made to flow like a liquid by injection of water or gas is said to have been fluidized. In water and wastewater treatment, a bed of filter media is fluidized by backwashing water through the filter.

Figure 5.6   Backwash pumps

filter coating itself must also be removed; this contributes greatly to the volume of waste sludge produced by this filtration process. An advantage is that less water is required for backwashing.

The pressurized water supply required for backwashing a filter is usually supplied by a backwash pump, which uses filtered water as a supply source. An elevated storage tank can store enough water to backwash large filters and multiple filters. A typical backwash pump and control panel are shown in Figure 5.6.

The backwashing process may use about 2 to 4 percent of the process water for transporting the wash water to dewatering and final solids disposal processes (see the chapter on handling and disposal of process wastes in Volume II of this training manual series). If the wash water can be removed from the water surface in the solids disposal process, a major portion of the water can be recycled through the water treatment plant. In some plants, the wash water is recycled directly to the headworks (ahead of the flash mixer).

Wash water troughs are installed in gravity filters to aid in collecting wash water from the filters (see Figure 5.7).

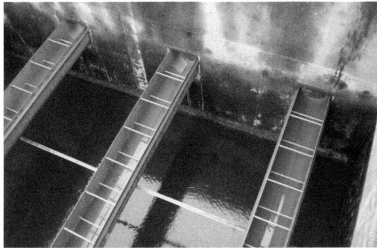

Wash water trough, shown with filter water-surface level drawn down before backwash

Filter shown during backwash process

Figure 5.7   Backwashing in gravity filter process

The Filter Backwash Rule requires that recycled filter backwash water, sedimentation basin sludge thickener supernatant, and liquids from sludge dewatering processes be returned upstream from all conventional or direct filtration treatment systems, including coagulation, flocculation, sedimentation (with conventional filtration only), and filtration. Systems may apply to the state for approval to recycle at an alternate location.

The Filter Backwash Rule aims to improve performance at conventional and direct water filtration plants by reducing the opportunity for recycle practices to adversely affect plant performance that would allow microbes, such as *Cryptosporidium*, to pass through the treatment processes and into the finished drinking water. Surges of recycle flow returned to the treatment processes must also be controlled. Surges may adversely affect treatment processes by creating hydraulically overloaded conditions (when plants exceed design capacity) that lower performance of individual treatment processes within a treatment plant, resulting in lowered *Cryptosporidium* removal efficiency.

5.3.3.3 Surface Wash

To produce optimum cleaning of the filter media during backwashing and to prevent mudballs, surface washing (supplemental scouring) is usually required. **Mudballs** are material, approximately round in shape, that forms in filters and gradually increases in size when not removed by the backwashing process. Mudballs vary from pea-sized up to golf-ball-sized or larger. Surface wash systems provide additional scrubbing action to remove attached floc and other entrapped solids from the filter media. The four types of surface wash systems are Baylis, fixed-grid, rotary, and air scour (Figure 5.8). The first three surface wash systems are mechanical or water-powered, while the air scour system uses air to create the washing action.

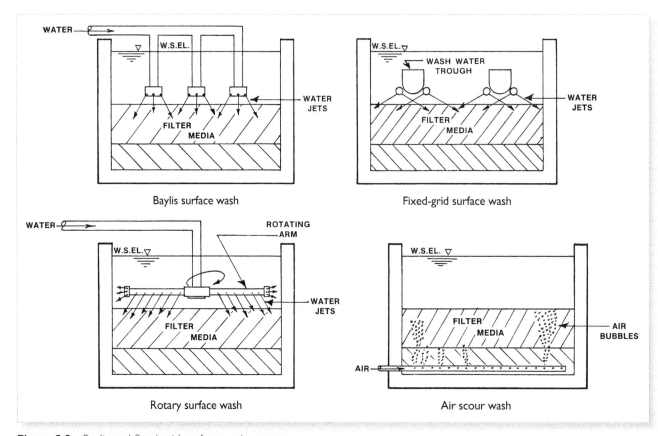

Baylis surface wash

Fixed-grid surface wash

Rotary surface wash

Air scour wash

Figure 5.8 Baylis and fixed grid surface wash systems

5.3.4   Filter Control Systems

Filter control systems regulate flow rates through the filter by maintaining an adequate head above the media surface (see Figure 5.9). This head (**submergence**) forces water through a gravity filter.

Filter operating deck

Control station

Figure 5.9   Filter control systems

submergence
The distance between the water surface and the media surface in a filter.

A filter control system also prevents sudden flow increases or surges that could discharge solids trapped on the filter media. If the solids were suddenly dislodged, they would seriously degrade water quality. One way to prevent incoming water from disturbing (scouring) the media is to maintain an adequate depth of water above the media surface. In this way, the force of the incoming water is absorbed before it reaches the media, thus preventing scouring.

Filter control systems commonly used in gravity filtration installations include the following rate-of-flow, split-flow, declining-rate, and self-back-wash systems. In rate-of-flow control systems, each filter effluent control valve is connected to a flowmeter. As the filter run continues and the media begins to clog, the control valve slowly opens to maintain a constant flow of water through the filter. A master controller monitors the overall plant flow and adjusts the flow rate of each filter accordingly. With this system, all filters operate at the same flow rate, but the rate is variable, dropping each time a new filter is backwashed. Some filters have influent control valves.

In the split-flow control system, the flow to each filter influent is split or divided by a weir. With this system, equal flow is automatically distributed to each filter. The filter effluent valve position is controlled by the water level in the filter. Each filter operates within a narrow water level range.

In declining-rate filters, flow rate varies with head loss. Each filter operates at the same, but variable, water surface level. This system is relatively simple, but requires an effluent control structure (weir) to provide adequate media submergence. Effluent control structures are usually desirable with the other systems described here as well.

In the self-backwash (or Streicher design) system, influent flow to each filter is divided by a weir. The water surface level in each filter varies according to head loss, but the flow rate remains constant for each filter. This system reduces the amount of mechanical equipment required for operation and backwashing, such as wash water pumps, but it requires an effluent control structure and a deeper filter box.

Check Your Understanding

1. What two main factors influence the time period before a filter becomes clogged?
2. Under what conditions is the filtration process stopped and the filter taken out of service for cleaning or backwashing?
3. List four types of surface wash (supplemental scour) systems for filters.
4. What aspects of the filtration process are controlled by the filter control system?

5.4 Activated Carbon Filters

The primary purpose of filtration is to remove suspended particles and floc from the water being treated. Another dimension is added to the filtration process by using activated carbon (granular form) as a filter media. The high adsorptive capacity of activated carbon enables it to remove taste- and odor-causing compounds, as well as other trace organics from the water. However, not all organic compounds are removed with the same degree of efficiency. Other methods of removing volatile organics include air stripping.

While activated carbon filtration is effective in removing taste- and odor-causing compounds, the construction, carbon-handling equipment, and operating costs are generally quite high. Still, activated carbon can be added to existing filter beds or can be incorporated as a separate process. Provisions must be made for regeneration or reactivation of spent carbon (carbon that has lost its adsorptive capacity). Usually, arrangements are made with the carbon manufacturer to regenerate or exchange spent carbon.

An alternative to activated carbon filtration for reducing taste and odor levels is using powdered activated carbon (PAC) before sedimentation and filtration. This option is usually more cost effective than activated carbon filtration, but generally will not yield equivalent results due to operational and physical problems that limit the amount of PAC that can be applied to the water being treated. Powdered activated carbon is difficult to handle due to dust problems.

5.5 Interaction with Other Treatment Processes

How filtration interacts with other treatment processes depends on the pretreatment processes and the type of filtration involved. Pretreatment processes include coagulation, flocculation, and sedimentation. Filtration can be inline, conventional, or direct. Conventional filtration is the most common filtration type used in municipal treatment. This process includes complete pretreatment (coagulation, flocculation, and sedimentation). Direct filtration can be used when source waters are low in turbidity, color, plankton, and coliform organisms. Direct filtration is not preceded by sedimentation. Inline filtration adds chemical coagulants to the filter inlet pipe, but is not as efficient in forming floc as conventional or direct filtration.

5.5.1 Pretreatment Processes

Filtration removes particulate impurities and floc from the water being treated. The filtration process is the final step in the solids removal process, which usually includes the pretreatment processes of coagulation, flocculation, and sedimentation.

The degree of pretreatment applied before filtration depends on the type of treatment plant (pressure filtration, diatomaceous earth filtration, or gravity filtration) and the size of the treatment facility. Most large municipal treatment plants include complete pretreatment facilities and gravity filtration. In any event, the importance of pretreatment before filtration cannot be overemphasized.

Floc particles that are carried over into the filter influent must be small enough to penetrate the upper filter media (depth filtration). Floc that is too large will cause the top portion of the filter bed to clog rapidly, thus leading to short filter runs. In addition, large floc (particularly alum floc) is often weak and easily broken up by water turbulence. This can degrade effluent water quality.

Ideally, floc removal is accomplished by contact with the media grains (sand, coal, and garnet) throughout the upper depths of the filter. After the initial coating or conditioning of the media surfaces with floc at the beginning of the filtration cycle, subsequent applications of floc will build up on the material previously deposited on the media surface. This process is often referred to as the ripening period. Higher filter effluent turbidities may occur during the first few minutes at the beginning of the filter run until the ripening period is completed.

At filtration rates below 4 gal/min/ft² (2.72 L/sec/m²), alum or iron coagulants usually give adequate treatment without assistance.

At filtration rates above 4 or 5 gal/min/ft² (2.72 or 3.40 L/sec/m²), iron and alum floc will shear in the pores of the filter and short filter runs will result because of turbidity breakthrough. Under these conditions, the addition of a polymer coagulant aid or a chemical filter aid is often beneficial.

The polymers most frequently found to be successful in this application are moderate molecular weight cationic polymers (diallyldimethyl-ammonium or DADMA) and relatively high molecular weight nonionic polymers (polyacrylamides). The cationic polymers are best added ahead of the flocculation process to strengthen the floc formed. The nonionic polymers are generally added in the settled water as it moves toward the filters from the sedimentation basins.

Typical doses for cationic polymers range from 0.25 to 2 mg/L. Typical doses for nonionic polymers range from 0.02 to 0.2 mg/L.

5.5.2   Inline Filtration

Inline filtration is the addition of chemical coagulants, called filter aids, directly to the filter inlet pipe. The chemicals are mixed by the flowing water, and flocculation and sedimentation facilities are eliminated (see Figure 5.10). This pretreatment method is commonly used in pressure filter installations.

This process is not as efficient in forming floc as conventional or direct filtration when source water quality has variable turbidity and bacterial levels. Problems may develop with the formation of floc in the water after filtration.

5.5.3   Conventional Filtration (Treatment)

The conventional filtration (treatment) process is used in most municipal treatment plants in the United States. This process includes complete pretreatment (coagulation, flocculation, and sedimentation), as shown in Figure 5.10. This system provides a great amount of flexibility and reliability in plant operation, especially when source water quality is highly variable or is high in suspended solids.

A chemical application point just before filtration permits the application of a filter aid chemical (such as a nonionic polymer) to assist in the solids removal process, especially during periods of pretreatment process upset, or when operating at high filtration rates.

5.5.4   Direct Filtration

Direct filtration is considered a feasible alternative to conventional filtration, particularly when source waters are low in turbidity, color, plankton, and coliform organisms. Direct filtration can be defined as a treatment system in which filtration is not preceded by sedimentation, as shown in Figure 5.10. Many direct filtration plants provide rapid mix, short

Inline Filtration

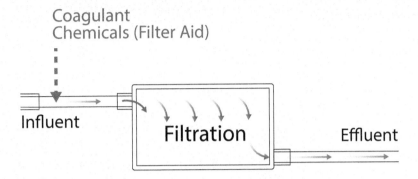

Coagulant
Chemicals (Filter Aid)

Influent

Filtration

Effluent

Conventional Filtration

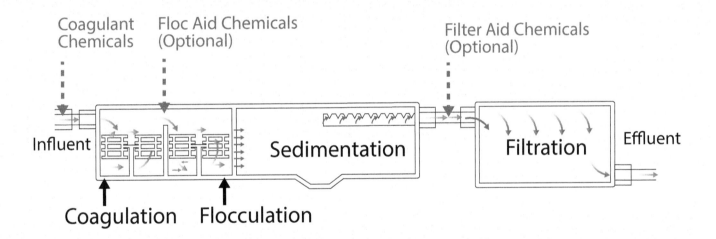

Coagulant
Chemicals

Floc Aid Chemicals
(Optional)

Filter Aid Chemicals
(Optional)

Influent

Sedimentation

Filtration

Effluent

Coagulation   Flocculation

Direct Filtration

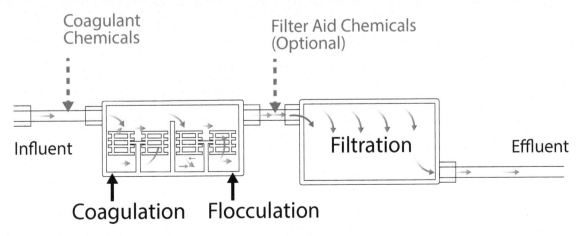

Coagulant
Chemicals

Filter Aid Chemicals
(Optional)

Influent

Filtration

Effluent

Coagulation   Flocculation

Figure 5.10   Filtration systems

detention without agitation (30 to 60 minutes) followed by filtration. As in conventional filtration, a chemical application point just before filtration permits the addition of a filter aid chemical.

5.6   Process Control

In theory, the physical and chemical processes governing solids removal in filtration are rather complex. Yet, from a practical perspective, the treatment plant operator must be provided with the means to measure and control the performance of the filtration process on a day-to-day basis. In this regard, filter influent water quality (turbidity), filter performance (head loss buildup rate and filter run time), and filter effluent water quality (turbidity) are important process control guidelines.

An efficient means of handling wash waters is to recycle the water to the beginning of the plant (before the flash mixer).

Means for measuring and controlling the filtration process are described in the next section.

Check Your Understanding

1. What is the primary purpose of using activated carbon (granular form) as filter media?

2. What is inline filtration?

3. What water quality factors affect the floc formation during inline filtration?

4. When is direct filtration used?

5. What factors must an operator measure to control the performance of the filtration process on a day-to-day basis?

5.7   Operating Procedures Associated with Normal Process Conditions

Filtration is the final and most important step in the solids removal process. From a water quality standpoint, filter effluent turbidity will give you a good indication of overall process performance. However, you must also monitor the performance of each of the preceding treatment processes (coagulation, flocculation, and sedimentation) as well as filter effluent water quality, to anticipate water quality or process performance changes that might affect filter operation.

In the normal operation of the filtration process, the operator should closely monitor filter influent turbidity as well as filter effluent turbidity levels with a **turbidity meter** (Figure 5.11). Filter influent turbidity levels (settled turbidity) can be checked periodically, about every 2 hours, by securing a grab sample either at the filter or from the laboratory sample tap (if such facilities are provided). However, filter effluent turbidity is best monitored and recorded on a continuous basis by an in-process turbidity meter. If the turbidity meter is provided with an alarm feature, process failures can be addressed virtually instantaneously.

Turbidity meter

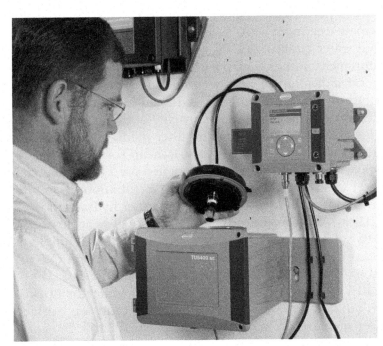

Turbidity meter controls

Figure 5.11   Turbidity meter and controls (*Courtesy of HACH Company*)

A turbidity meter's strong beam of light passes through the sample, providing high sensitivity and accurate turbidity measurements. Sample turbidity causes the light to be scattered, which is then measured by a photocell sensor. The corresponding number of NTUs will be shown on the meter's indicator. If a sample's turbidity is insignificant, light does not reach the meter's photocell sensor.

Other indicators that can be monitored to determine if the filter is performing normally include head loss buildup and filter effluent color.

Develop a written set of process guidelines to assist the operator in evaluating normal process conditions and in recognizing abnormal conditions. These guidelines can be developed based on water quality standards, design considerations, water quality conditions, and most importantly, trial and experience.

5.7.1   Process Actions

In the normal operation of the filtration process, you will perform a variety of functions with an emphasis on maintaining a high-quality filtered water. The quality of the filter effluent is the final product quality that will be distributed to consumers.

In the normal operation of the filtration process, you will be expected to perform the following functions:

1. Monitor process performance
2. Evaluate water quality conditions (turbidity) and make appropriate process changes
3. Check and adjust process equipment (change chemical feed rates)
4. Backwash filters
5. Evaluate filter media condition (media loss, mudballs, cracking)
6. Visually inspect facilities

Monitoring process performance is an ongoing activity. You should look for and anticipate treatment process changes or other problems that might affect filtered water quality such as a chemical feed system failure.

Head loss buildup measurements with a **piezometer** in the filter media will give you a good indication of how well the solids removal process is performing (Figure 5.12). The total designed head loss from the filter influent to the effluent in a gravity filter is usually about 10 feet (3 m). The actual head loss from a point above the filter media to a reference point in the effluent can be monitored as loss-in-head (or loss-of-head). For example, suppose that a gravity filter is designed for a total potential head loss of 10 feet (3 m). At the beginning of the filtration cycle, the actual measured head loss due to clean media and other hydraulic losses might be 3 feet (0.9 m). This would permit an additional head loss of 7 feet (2.1 m) due to solids accumulation in the filter. In this example, a practical cutoff point might be established at an additional 6 feet (1.8 m) of head loss (total of 9 feet or 2.7 m) for backwashing purposes.

The rate of head loss buildup is also an important indicator of process performance. Sudden increases in head loss might indicate surface sealing of the filter media (lack of depth penetration). Early detection of this condition may permit you to make appropriate process changes such as adjustment of the filter aid chemical feed rate or reduction of filtration rate.

piezometer
An instrument used to measure the pressure head in a pipe, tank, or soil. For pipes and tanks, it usually consists of a small pipe or tube. One end is connected or tapped through the wall of the pipe or tank being measured. The other end is most often open to the air, in which case the height of the water in the tube above the connection is the pressure head in inches or feet. For high pressures the tube may be connected to a mercury column in which case the pressure head is expressed in inches of mercury. For soil applications, the tube is connected to a permeable tip which is then buried in or driven into the soil. In this case, the piezometer measures the pore pressure of water in the soil.

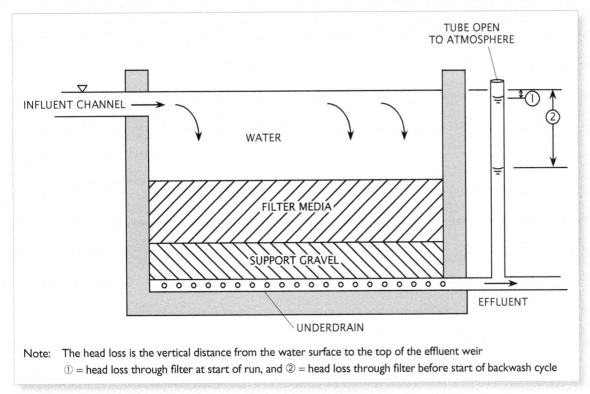

TUBE OPEN
TO ATMOSPHERE

INFLUENT CHANNEL ⟶

WATER

FILTER MEDIA

SUPPORT GRAVEL

EFFLUENT

UNDERDRAIN

Note: The head loss is the vertical distance from the water surface to the top of the effluent weir

① = head loss through filter at start of run, and ② = head loss through filter before start of backwash cycle

Figure 5.12   Measuring head loss buildup in a variable head filter

Monitoring filter effluent turbidity on a continuous basis with an in-process turbidity meter will provide continuous feedback on the filtration process performance (Figure 5.11). In most instances, filter operation should be cut off (terminated) at a predetermined effluent turbidity level. Preset the filter cutoff control at a point where experience and tests show breakthrough will soon occur (Figure 5.13).

In the normal operation of the filter process, calculate when the filtration cycle will be completed on the basis of head loss, effluent turbidity level, and elapsed run time. A predetermined value is established for each guideline as a cutoff point for filter operation. When any one of these levels is reached, the filter is removed from service and backwashed (see Section 5.9, "Startup and Shutdown Procedures," for step-by-step procedures).

Usually, plant operators compare filter performance from season to season and from plant to plant by comparing the filter run length in hours. This is because as the filter run gets shorter, the amount of water wasted in backwash becomes increasingly important when compared to the amount of water produced during the filter run. Percent backwash water statistics are also occasionally collected.

Although of some use, filter run length is not a satisfactory basis for comparing filter runs without considering the filtration rate as well. For example, at a filtration rate of 6 gal/min/ft^2 (4.1 L/sec/m^2), an 18-hour filter run is quite adequate. Whereas, at a filtration rate of 1.5 gal/min/ft^2 (1.0 L/sec/m^2), an 18-hour filter run is not satisfactory.

The best way to compare filter runs is by using the Unit Filter Run Volume (UFRV) technique. The UFRV is the volume of water produced by the filter during the course of the filter run divided by the surface area of the filter. This is usually expressed in gallons per square foot (liters per square meter). UFRVs of 5,000 gal/ft^2 (203,710 L/m^2) or greater are

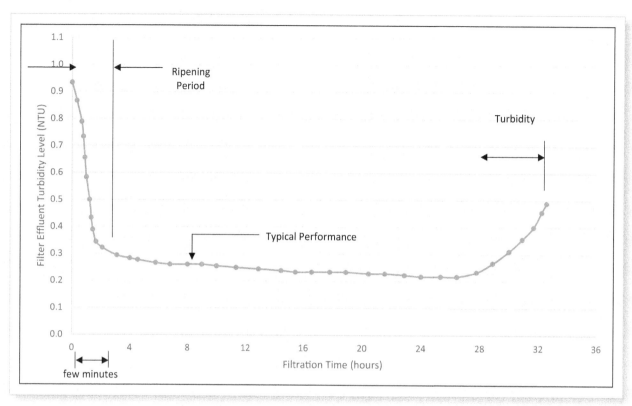

Figure 5.13   Typical filter effluent turbidity data

satisfactory, and UFRVs greater than 10,000 gal/ft² (407,430 L/m²) are desirable. In the examples cited in the previous paragraph, the UFRV for the filter operating at 6 gal/min/ft² (4.1 L/sec/m²) would be 6,480 gal/ft² (264,010 L/m²), and the filter operating at 1.5 gal/min/ft² (1.0 L/sec/m²) would be 1,620 gal/ft² (66,000 L/m²).

Water quality indicators used to assess process performance include turbidity and color (Figure 5.14). Based on an assessment of overall

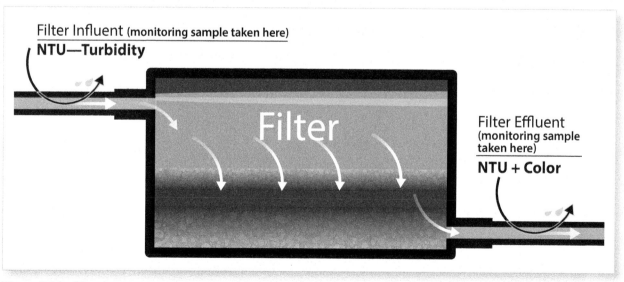

Figure 5.14   Filtration process monitoring guidelines and locations

process performance, you may need to make changes in the coagulation-flocculation process as described in Chapter 3, or in the sedimentation process as described in Chapter 4.

At least once a year, examine the filter media and evaluate its overall condition. Measure the filter media thickness for an indication of media loss during the backwashing process. Measure mudball accumulation in the filter media to evaluate the effectiveness of the overall backwashing operation (see Section 5.7.2, "Mudball Evaluation Procedure").

Routinely, observe the backwash process to qualitatively assess process performance. Watch for media boils (uneven flow distribution) during backwashing, media carryover into the wash water trough, and clarity of the waste wash water near the end of the backwash cycle.

Never bump a filter to avoid backwashing. Bumping is opening the backwash valve during the course of a filter run to dislodge the trapped solids and increase the length of the filter run. This is not a good practice, and in-process turbidity meters on the filter effluent will indicate if bumping is taking place.

Upon completion of the backwash cycle, observe the condition of the media surface and check for filter sidewall or media surface cracks. Corrective actions are described in Section 5.8. Routinely, inspect physical facilities and equipment as part of good housekeeping and maintenance practices. Correct or report abnormal equipment conditions to the appropriate maintenance personnel. Table 5.2 is a summary of routine filtration process actions.

5.7.2 Mudball Evaluation Procedure

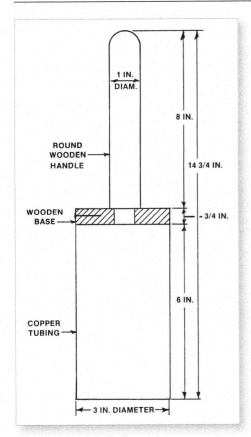

1. Frequency of mudball evaluation. If mudballs in the top of the filter material are a problem, use this procedure on a monthly basis. If mudballs are not a problem, an annual check is sufficient.

2. Sample for mudballs using a mudball sampler as shown in Figure 5.15.

3. Backwash the filter to be sampled and drain the filter to at least 12 inches below the surface of the top media layer (or layer of interest).

4. Push the mudball sampler 6 inches into the media. Tilt the handle until it is nearly level and lift the sampler full of media.

5. Empty the contents of the sampler into a bucket.

Figure 5.15 Mudball sampler

Table 5.2   Routine filtration process actions

Monitor Process Performance and Evaluate Water Quality Conditions	Location	Frequency	Possible Operator Actions
• Turbidity	Influent/Effluent	Influent at least once per 8-hour shift. Effluent, monitor continuously	1. Increase sampling frequency when process water quality is variable 2. Perform jar tests if indicated (see procedure in Chapter 9, "Laboratory Procedures") 3. Make necessary process changes

• Color	Influent/Effluent	At least once per 8-hour shift	(a) Change coagulant
			  (b) Adjust coagulant dosage
			  (c) Adjust flash-mixer/flocculator mixing intensity

• Head loss		At least three times per 8-hour shift	(d) Change chlorine dosage
			  (e) Change filtration rate
			  (f) Backwash filter
			4. Verify response to process changes at appropriate time

Operate Filters and Backwash	Location	Frequency	Possible Operator Actions
• Put filter into service • Change filtration rate • Remove filter from service • Backwash filter • Change backwash rate	Filter module	Depends on process conditions	See operating procedures in Sections 5.7, 5.8, 5.9, and 5.10

Check Filter Media Condition	Location	Frequency	Possible Operator Actions
• Media depth evaluation • Media cleanliness • Cracks or shrinkage	Filter module	At least monthly	1. Replace lost filter media 2. Change backwash procedure 3. Change chemical coagulants

Make Visual Observations of Backwash Operation	Location	Frequency	Possible Operator Actions
• Check for media boils • Observe media expansion • Check for media carryover into wash water trough • Observe clarity of wastewater	Filter module	At least once per day or whenever backwashing occurs when less frequent	1. Change backwash rate 2. Change backwash cycle time 3. Adjust surface wash rate or cycle time 4. Inspect filter media and support gravel for disturbance

Check Filtration Process and Backwash Equipment Condition	Location	Frequency	Possible Operator Actions
• Noise • Vibration • Leakage • Overheating	Various	Once per 8-hour shift	1. Correct minor problems 2. Notify others of major problems

Inspect Facilities	Location	Frequency	Possible Operator Actions
• Check physical facilities • Check for algal buildup on filter sidewalls and on wash water troughs	Various	Once per 8-hour shift	1. Report abnormal conditions 2. Remove debris from filter media surfaces 3. Adjust chlorine dosage to control algae

6. Repeat Steps 4 and 5 four more times from different locations in the filter.

7. Use a 10-mesh sieve for separating the mudballs from the media. Hold the sieve in a bucket or tub of water so the sieve is nearly submerged.

8. Take a handful of media from the bucket containing the samples of media and mudballs and place the material in the sieve.

9. Gently raise and lower the sieve about 0.5 inch at a time until the sand is washed away from the mudballs.

10. Shift the mudballs to one side of the sieve by tipping the submerged sieve and gently shaking the sieve.

11. Repeat Steps 8, 9, and 10 until the entire sample has been washed in the sieve and all of the mudballs have been separated from the sand. If there are so many mudballs that the washing process is hindered, move some of the mudballs to the measuring cylinder described in the next step.

12. Use a 1,000 mL graduated cylinder (a smaller or larger cylinder may be used depending on the volume or amount of mudballs in the sand). Fill the graduated cylinder to the 500 mL mark with water.

13. Allow the water to drain from the mudballs on the sieve. When the draining has stopped (no more dripping), transfer the mudballs to the graduated cylinder.

14. Record the new level of water in the graduated cylinder.

15. Determine the Mudball Volume, mL, by subtracting 500 from the new level of water.

16. Calculate the volume of mudballs as a percent. The total volume of media and mudballs sampled was 3,540 mL if the mudball sampler was full.

$$\text{Mudball Volume}[\%] = \frac{\text{Mudball Volume}[\text{mL}] \times 100\%}{3,540\,[\text{mL}]}$$

*Procedures were adapted from *Water Quality and Treatment.* ISBN: 0-07-001659-3. www.awwa.org.

17. Evaluate the condition of the filtering material using Table 5.3.

18. Mudballs sink more readily in anthracite than sand. Therefore, modify this procedure to collect samples from the bottom 6 inches of anthracite.*

Table 5.3 Condition of filter

Mudball Volume, %	Condition of Filtering Material
0.0–0.1	Excellent
0.1–0.2	Very Good
0.2–0.5	Good
0.5–1.0	Fair
1.0–2.5	Fairly Bad
2.5–5.0	Bad
Over 5.0	Very Bad

Example I

Evaluate the condition of filtering material on the basis of a mudball evaluation. When the drained mudballs were added to the graduated cylinder, the water rose from the 500 mL mark up to the 583 mL mark. The total volume sampled was 3,540 mL.

Known		Unknown
Initial Cylinder Level	= 500 mL	Condition of Filtering Material
Final Cylinder Level	= 583 mL	
V = Total Volume Sample	= 3,540 mL	

1. Determine the mudball volume in milliliters (mL).

   Mudball Volume [mL]

   = Final Cylinder Level [mL] – Initial Cylinder Level [mL]

   = 583 [mL] – 500 [mL]

   = 83 mL

2. Calculate the mudball volume as a percent.

$$\text{Mudball Volume}[\%] = \frac{\text{Mudball Volume}[\text{mL}] \times 100\%}{V}$$

$$= \frac{83[\text{mL}] \times 100\%}{3,540[\text{mL}]} = 2.3\%$$

3. Determine the condition of the filtering material from Table 5.3.

   From Table 5.3, the condition of a sample with a mudball volume of 2.3 percent is fairly bad.

Check Your Understanding

1. What is the most important water quality indicator used to monitor the filtration process?

2. How can filter effluent turbidity be measured on a continuous basis?

3. List some of the typical functions performed by operators in the normal operation of the filtration process.

4. What could cause a sudden increase in head loss through a filter?

5. How would you change the operation of a filter if there was a sudden increase in head loss through the filter?

6. What is meant by bumping a filter?

5.7.3   Process Calculations

In the routine operation of the filtration process, you will be expected to perform a variety of process calculations related to filter operation (flow rate, filtration rate), backwashing (backwash rate, surface wash rate), water production, and percent of water production used to backwash filters.

5.7.3.1   Filter Efficiency

Normally, the efficiency of the filtration process is not calculated as such. The filter effluent turbidity is measured from either a grab sample or an in-process turbidity meter. Whenever the effluent turbidity exceeds 0.5 NTU or some other established value, the filter should be backwashed.

5.7.3.2 Filtration Rate

Filtration rates are measured in gallons per minute per square foot of filter area (gal/min/ft^2), liters per second per square meter of filter area (L/sec/m^2), or rate of rise in millimeters per second. Filtration rates will vary from 2 to 10 gal/m/ft^2 (1.36 to 6.79 L/sec/m^2) with most agencies specifying a maximum of 6 gal/min/ft^2 (4.07 L/sec/m^2) for dual-media units. From a practical standpoint, if your plant has four filters the same size and a plant design flow of 10 Mgal/d, any time more than 2.5 Mgal/d is applied to any filter, you are exceeding the design filtration rate. Problems can develop when design filtration rates are exceeded.

FORMULAS

The filtration rate is described as the flow in gal/min that is filtered by 1 square foot of filter surface area. Obtain the flow from a flowmeter. If the flows are given in Mgal/d, they must be converted to gal/min.

$$Q\left[\frac{\text{gal}}{\text{min}}\right] = Q\left[\frac{\text{Mgal}}{\text{d}}\right] \times \frac{10^6}{\text{M}} \times \frac{\text{d}}{24\,\text{h}} \times \frac{\text{h}}{60\,\text{min}}$$

$$\text{Filtration Rate}\left[\frac{\text{gal}}{\text{min} \cdot \text{ft}^2}\right] = \frac{Q\left[\dfrac{\text{gal}}{\text{min}}\right]}{A_{\text{surface}}\left[\text{ft}^2\right]}$$

Where,

   Q   = Flow Rate

   A_{surface} = Surface Area

Another approach is to turn off the influent valve to a filter and record the time for the water to drop a specified distance. By calculating the velocity of the dropping water and knowing the surface area of the filter, we can calculate the flow rate in gal/min or any other desired units.

1. To calculate the velocity (v) of the water dropping above the filter, measure the time (t) for the water to drop a specific distance.

$$v\left[\frac{\text{ft}}{\text{min}}\right] = \frac{\text{Water Drop}[\text{ft}]}{t[\text{min}]}$$

2. To determine the flow of water through the filter in gal/min, multiply the velocity (ft/min) by the surface area (ft^2) and 7.48 gallons per cubic foot (this converts flow from cubic feet per minute to gal/min).

$$Q\left[\frac{\text{gal}}{\text{min}}\right] = v\left[\frac{\text{ft}}{\text{min}}\right] \times A_{\text{surface}}\left[\text{ft}^2\right] \times \frac{7.48\,\text{gal}}{\text{ft}^3}$$

The flow rate can also be determined in gal/min if the total volume of water filtered between filter runs is known. This is done by obtaining the total flow volume in gallons and dividing this value by the length of filter run in hours. If we divide the results by 60 minutes per hour, we will convert the flow from gallons per hour to gal/min.

$$Q\left[\frac{gal}{min}\right] = \frac{\text{Total Flow}[gal]}{\text{Filter Run}[h] \times \dfrac{60\,min}{h}}$$

The Unit Filter Run Volume (UFRV) is a measure of filter performance. This value is determined by obtaining the total volume of water filtered between filter runs and dividing this number by the surface area of the filter in square feet.

$$\text{UFRV}\left[\frac{gal}{ft^2}\right] = \frac{\text{Volume Filtered}[gal]}{\text{Filter Surface Area}[ft^2]}$$

If the filtration rate in gal/min/ft² is known, this number times the length of filter run in hours times 60 minutes per hour will also give the UFRV.

$$\text{UFRV}\left[\frac{gal}{ft^2}\right] = \text{Filtration Rate}\left[\frac{gal}{min \cdot ft^2}\right] \times \text{Filter Run}[h] \times \frac{60\,min}{h}$$

Example 2

Calculate the filtration rate in gallons per minute per square foot (gal/min/ft²) for a filter with a surface length of 25 feet (ft) and a width of 20 feet (ft) when the applied flow is 2 Mgal/d (2 million gallons during a 24-hour period).

Known	Unknown
L = Length = 25 ft	Filtration Rate, gal/min/ft²
W = Width = 20 ft	
Q = Flow Rate = 2 Mgal/d	

1. Convert the flow rate (Q) from Mgal/d to gal/min.

$$Q\left[\frac{gal}{min}\right] = Q\left[\frac{Mgal}{d}\right] \times \frac{10^6}{M} \times \frac{d}{24\,h} \times \frac{h}{60\,min}$$

$$= 2\left[\frac{Mgal}{d}\right] \times \frac{10^6}{M} \times \frac{d}{24\,h} \times \frac{h}{60\,min}$$

$$= 1{,}389\ gpm$$

2. Calculate the surface area ($A_{surface}$) of the filter in square feet (ft²).

$$A_{surface}\ [ft^2] = L[ft] \times W[ft] = 25[ft] \times 20[ft] = 500\ ft^2$$

3. Calculate the filtration rate in gallons per minute per square foot (gal/min/ft²).

$$\text{Filtration Rate}\left[\frac{gal}{min \cdot ft^2}\right] = \frac{Q\left[\dfrac{gal}{min}\right]}{A_{Surface}\ [ft^2]}$$

$$= \frac{1,389\left[\dfrac{\text{gal}}{\text{min}}\right]}{500\left[\text{ft}^2\right]} = 2.8\,\frac{\text{gpm}}{\text{ft}^2}$$

Example 3

Determine the filtration rate in gallons per minute per square foot (gal/min/ft²) for a filter with a surface length of 30 feet (ft) and a width of 20 feet (ft). With the influent valve closed, the water above the filter dropped 12 inches (in) in 5 minutes (min).

		Known		
L	=	Length	=	30 ft
W	=	Width	=	20 ft
h	=	Water Drop	=	12 in
t	=	Time	=	5 min

Unknown

Filtration Rate, gal/min/ft²

1. Calculate the surface area ($A_{surface}$) of the filter in square feet (ft²).

$$A_{surface}\,[\text{ft}^2] = L[\text{ft}] \times W[\text{ft}] = 30[\text{ft}] \times 20[\text{ft}] = 600\text{ ft}^2$$

2. Calculate the velocity (v) of the dropping water in feet per minute (ft/min).

$$v\left[\frac{\text{ft}}{\text{min}}\right] = \frac{h[\text{ft}]}{t[\text{min}]} = \frac{12[\text{in}] \times \dfrac{\text{ft}}{12\,\text{in}}}{5[\text{min}]} = 0.2\,\frac{\text{ft}}{\text{min}}$$

3. Determine the flow rate (Q) of water through the filter in gallons per minute (gal/min).

$$Q\left[\frac{\text{gal}}{\text{min}}\right] = v\left[\frac{\text{ft}}{\text{min}}\right] \times A_{surface}\left[\text{ft}^2\right] \times \frac{7.48\,\text{gal}}{\text{ft}^3}$$

$$= 0.2\left[\frac{\text{ft}}{\text{min}}\right] \times 600\left[\text{ft}^2\right] \times \frac{7.48\,\text{gal}}{\text{ft}^3}$$

$$= 898\text{ gpm}$$

4. Calculate the filtration rate in gallons per minute per square foot (gal/min/ft²).

$$\text{Filtration Rate}\left[\frac{\text{gal}}{\text{min}\cdot\text{ft}^2}\right] = \frac{Q\left[\dfrac{\text{gal}}{\text{min}}\right]}{A_{surface}\left[\text{ft}^2\right]} = \frac{898\left[\dfrac{\text{gal}}{\text{min}}\right]}{600\left[\text{ft}^2\right]} = 1.5\,\frac{\text{gpm}}{\text{ft}^2}$$

Example 4

A filter with media surface dimensions of 42 feet (ft) long by 22 feet (ft) wide produces a total flow of 18.5 million gallons (Mgal) during a 73.5-hour-long filter run. What is the average filtration rate in gallons per minute per square foot (gal/min/ft²)?

		Known		
L	=	Length	=	42 ft
W	=	Width	=	22 ft
V_{total}	=	Total Flow Volume	=	18.5 Mgal
t	=	Filter Run	=	73.5 hr

Unknown

Average Filtration Rate, gal/min/ft²

1. Find the filter media surface area ($A_{surface}$) in square feet (ft^2).

$$A_{surface} \ [ft^2] = L[ft] \times W[ft] = 42[ft] \times 22[ft] = 924 \ ft^2$$

2. Calculate the average flow rate ($Q_{average}$) in gallons per minute (gal/min).

$$Q_{average} \left[\frac{gal}{min} \right] = \frac{V_{total}[gal]}{t[h] \times \dfrac{60\,min}{h}} = \frac{18,500,000[gal]}{73.5[h] \times \dfrac{60\,min}{h}} = 4,195 \frac{gal}{min}$$

3. Determine the average filtration rate in gallons per minute per square foot (gal/min/ft^2) of surface area.

$$\text{Average Filtration Rate} \left[\frac{gal}{min\text{-}ft^2} \right] = \frac{Q_{average} \left[\dfrac{gal}{min} \right]}{A_{surface} [ft^2]}$$

$$= \frac{4,195 \left[\dfrac{gal}{min} \right]}{924 [ft^2]} = 4.5 \frac{gpm}{ft^2}$$

Example 5

Determine the UFRV for the filter in Example 1. The volume of water filtered between backwash cycles was 3.8 million gallons (Mgal). The filter is 25 feet (ft) long and 20 feet (ft) wide.

	Known		Unknown
L	= Length	= 25 ft	UFRV, gal/ft^2
W	= Width	= 20 ft	
V	= Volume Filtered	= 3,800,000 gal	

Calculate the UFRV in gallons per square foot (gal/ft^2) of filter surface area ($A_{surface}$).

$$UFRV \left[\frac{gal}{ft^2} \right] = \frac{V[gal]}{A_{surface}[ft^2]} = \frac{V[gal]}{L[ft] \times W[ft]} = \frac{3,800,000[gal]}{25[ft] \times 20[ft]} = 7,600 \frac{gal}{ft^2}$$

Example 6

Determine the UFRV for the filter in Example 1. The filtration rate was 2.8 gallons per minute per square foot (gal/min/ft^2) during a 46-hour (h) filter run.

	Known		Unknown
Filtration Rate	=	2.8 gal/min/ft^2	UFRV, gal/ft^2
t	= Filter Run =	46 h	

Calculate the UFRV in gallons per square foot (gal/ft^2) of filter surface area.

$$UFRV \left[\frac{gal}{ft^2} \right] = \text{Filtration Rate} \left[\frac{gal}{min \cdot ft^2} \right] \times t[h] \times \frac{60\,min}{h}$$

$$= 2.8 \left[\frac{gal}{min \cdot ft^2} \right] \times 46[h] \times \frac{60\,min}{h}$$

$$= 7,728 \frac{gal}{ft^2}$$

Note: The method used to calculate the UFRV for your plant will depend on the information available.

5.7.3.3   Backwash Rate

Filter backwash rates are usually given in gal/min/ft² of surface area; or L/sec/m² of surface area; or inches per minute of water "rise rate" (in/min). Backwash rates will vary from 10 to 25 gal/min/ft² (6.8 to 17 L/sec/m²). Usually, the plant O&M instructions will specify the filter backwash rate in gal/min/ft² or the pumping rate for the backwash pump in gal/min. From a practical standpoint, the backwash rate or pumping rate is too low or the length of the backwash cycle too short if the filter is not completely cleaned. This will be obvious if the next filter run is too short or the initial head loss is too high. The backwash pumping or flow rate is too high if excessive amounts of filter media are being lost or disturbed during backwashing. To determine if filter media is being lost, place a burlap bag on the backwash discharge line (if possible) and examine what is caught by the bag. You could also merely observe how much media material remains in the empty wash trough or recovery basin when backwashing is completed.

When filter backwash rates are given in inches per minute (the velocity of the backwash water rising in the filter, such as 24 inches per minute or 10 millimeters per second), you may want to convert this value to other units such as gal/min/ft². By applying the proper mathematical conversion factors, inches per minute can be converted to gal/min/ft². One gal/min/ft² is approximately equal to 1.6 inches per minute of rise. Most operators and designers use gal/min/ft² rather than inches per minute.

FORMULAS

Formulas involving backwash calculations are similar to the formulas for calculating filtration rates. To calculate the backwash pumping or flow rate in gal/min, multiply the filter surface area in square feet by the desired backwash rate in gal/min/ft².

$$Q_{backwash} = A_{surface} \times LR_{backwash}$$

Where,

$Q_{backwash}$ =   Backwash Pumping or Flow Rate, gal/min
$A_{surface}$ =   Filter Surface Area, ft²
$LR_{backwash}$ =   Backwash Loading Rate, gal/min-ft²

To calculate the volume of backwash water needed, multiply the backwash flow in gal/min by the backwash time in minutes.

$$V_{backwash} = Q_{backwash} \times t_{backwash}$$

Where,

$V_{backwash}$ =   Volume of Backwash Water, gal
$t_{backwash}$ =   Backwash Time, min

To convert a backwash loading rate from gal/min/ft² to inches per minute of rise, use the conversion factors 12 inches per foot and 1 cubic foot for every 7.48 gallons.

$$LR_{backwash} \left[\frac{in}{min} \right] = LR_{backwash} \left[\frac{gal}{min \bullet ft^2} \right] \times \frac{12\,in}{ft} \times \frac{ft^3}{7.48\,gal}$$

$$= 1.60 \times LR_{backwash} \left[\frac{gal}{min \bullet ft^2} \right]$$

To determine the percent of water used for backwashing, divide the gallons of backwash water by the gallons of water filtered and multiply by 100 percent.

$$\text{Backwash}[\%] = \frac{V_{backwash}}{V_{filtered}} \times 100\%$$

Where,

$V_{filtered}$ = Volume of Water Filtered, gal

Example 7

Determine the backwash pumping rate in gallons per minute (gal/min) for a filter 30 feet (ft) long and 20 feet (ft) wide if the desired backwash rate is 20 gallons per minute per square foot (gal/min/ft^2).

Known			**Unknown**
L	= Length	= 30 ft	$Q_{backwash}$ = Backwash
W	= Width	= 20 ft	Pumping Rate,
$LR_{backwash}$	= Backwash Loading Rate	= 20 gal/min/ft^2	gal/min

1. Calculate the surface area ($A_{surface}$) of the filter.

$$A_{surface}\ [ft^2] = L[ft] \times W[ft] = 30[ft] \times 20[ft] = 600\ ft^2$$

2. Determine the backwash pumping rate ($Q_{backwash}$) in gal/min.

$$Q_{backwash}\left[\frac{gal}{min}\right] = A_{surface}\left[ft^2\right] \times LR_{backwash}\left[\frac{gal}{min \cdot ft^2}\right]$$

$$= 600\left[ft^2\right] \times 20\left[\frac{gal}{min \cdot ft^2}\right]$$

$$= 12,000\ gpm$$

Example 8

Determine the volume of water in gallons (gal) required to backwash a filter if the filter is backwashed for 8 minutes (min). The backwash pumping rate is 12,000 gallons per minute (gal/min).

Known			**Unknown**
$Q_{backwash}$	= Backwash Pumping Rate	= 12,000 gal/min	$V_{backwash}$ = Volume of Backwash
$t_{backwash}$	= Backwash Time	= 8 min	Water, gal

Calculate the volume of backwash water required ($V_{backwash}$) in gallons (gal).

$$V_{backwash}\left[gal\right] = Q_{backwash}\left[\frac{gal}{min}\right] \times t_{backwash}\left[min\right]$$

$$= 12,000\left[\frac{gal}{min}\right] \times 8\left[min\right] = 96,000\ gal$$

Example 9

How deep must the water be in a backwash tank* if the tank is 50 feet (ft) in diameter and the volume of backwash water required to backwash a filter for 8 minutes (min) is 96,000 gallons (gal)?

*Large filter areas may have backwash tanks instead of pumps because of the large flows required. Variable rate pumps are also used, which are programmed to start with a low backwash rate and then to increase the rate.

Known	Unknown
$V_{backwash}$ = Volume of Backwash Water = 96,000 gal	h = Water Depth, ft
D = Tank Diameter = 50 ft	

1. Convert the volume of backwash water ($V_{backwash}$) from gallons (gal) to cubic feet (ft^3).

$$V_{backwash}\left[ft^3\right] = V\left[gal\right] \times \left[\frac{ft^3}{7.48\ gal}\right]$$

$$= 96,000\left[gal\right] \times \frac{ft^3}{7.48\ gal} = 12,834\ ft^3$$

2. Calculate the depth of water (h) required in the backwash tank.

$$h\left[ft\right] = \frac{V_{backwash}\left[ft^3\right]}{A_{tank}\left[ft^2\right]} = \frac{12,834\left[ft^3\right]}{\frac{\pi}{4} \times 50^2\left[ft^2\right]} = \frac{12,834\left[ft^3\right]}{0.785 \times 50^2\left[ft^2\right]} = 6.54\ ft$$

Where,
A_{tank} = Tank Area

You should have at least 6.6 feet or 6 feet 7 inches of water in the backwash tank.

Example 10

Convert a filter backwash loading rate ($LR_{backwash}$) from 25 gallons per minute per square foot ($gal/min/ft^2$) to inches per minute (in/min) of rise.

Known	Unknown
$LR_{backwash}$ = 25 $gal/min/ft^2$	$LR_{backwash}$, in/min

Convert the backwash rate from gallons per minute per square foot ($gal/min/ft^2$) to inches per minute (in/min).

$$LR_{backwash}\left[\frac{in}{min}\right] = LR_{backwash}\left[\frac{gal}{min \cdot ft^2}\right] \times \frac{12\ in}{ft} \times \frac{ft^3}{7.48\ gal}$$

$$= 25\left[\frac{gal}{min \cdot ft^2}\right] \times \frac{12\ in}{ft} \times \frac{ft^3}{7.48\ gal}$$

$$= 40\frac{in}{min}$$

Example 11

During a filter run, the total volume of water filtered was 18.5 million gallons (Mgal). When the filter was backwashed, 96,000 gallons (gal) of water were used. Calculate the percent of the product water used for backwashing.

	Known		**Unknown**
$V_{filtered}$	= Volume of Water Filtered	= 18,500,000 gal	Backwash, %
$V_{backwash}$	= Volume of Backwash Water	= 96,000 gal	

Calculate the percent of water used for backwashing.

$$\text{Backwash}\,[\%] = \frac{V_{backwash}\,[\text{gal}]}{V_{filtered}\,[\text{gal}]} \times 100\% = \frac{96,000\,[\text{gal}]}{18,500,000\,[\text{gal}]} \times 100\% = 0.5\%$$

5.7.4 Recordkeeping

Recordkeeping is one of the most important administrative jobs required of water treatment plant operators. You will maintain a daily operations log of process performance data and water quality characteristics. Accurate records of the following items should be maintained:

1. Process water quality (turbidity and color)
2. Process operation (filters in service, filtration rates, loss of head, length of filter runs, frequency of backwash, backwash rates, and UFRV)
3. Process water production (water processed, amount of backwash water used, and chemicals used)
4. Percent of water production used to backwash filters
5. Process equipment performance (types of equipment in operation, equipment adjustments, maintenance procedures performed, and equipment calibration)

Entries in logs should be easy to read, should reflect the date and time of an event, and should be initialed by the operator making the entry.

Figure 5.16 is an example of a typical daily operating record for a water treatment plant with 16 filters. If a plant has four filters or even only one filter, similar data should be recorded in the format the plant uses (paper forms or computerized records).

5.7.5 Filter Monitoring Instrumentation

To evaluate filtration process efficiency, you will need to be familiar with the measurement of turbidity. This test can be readily performed in the laboratory with the aid of a turbidity meter. In addition, in-process or continuous water quality monitors, such as turbidity meters, will give you an early warning of process failure and will aid in making a rapid assessment of process performance (Figure 5.11, page 279). Section 5.12, "Particle Counters," describes a technology for monitoring and evaluating filter performance.

You will also need to be familiar with methods used to measure filter media loss and to determine the presence of mudballs in the filter media (see Section 5.7.1).

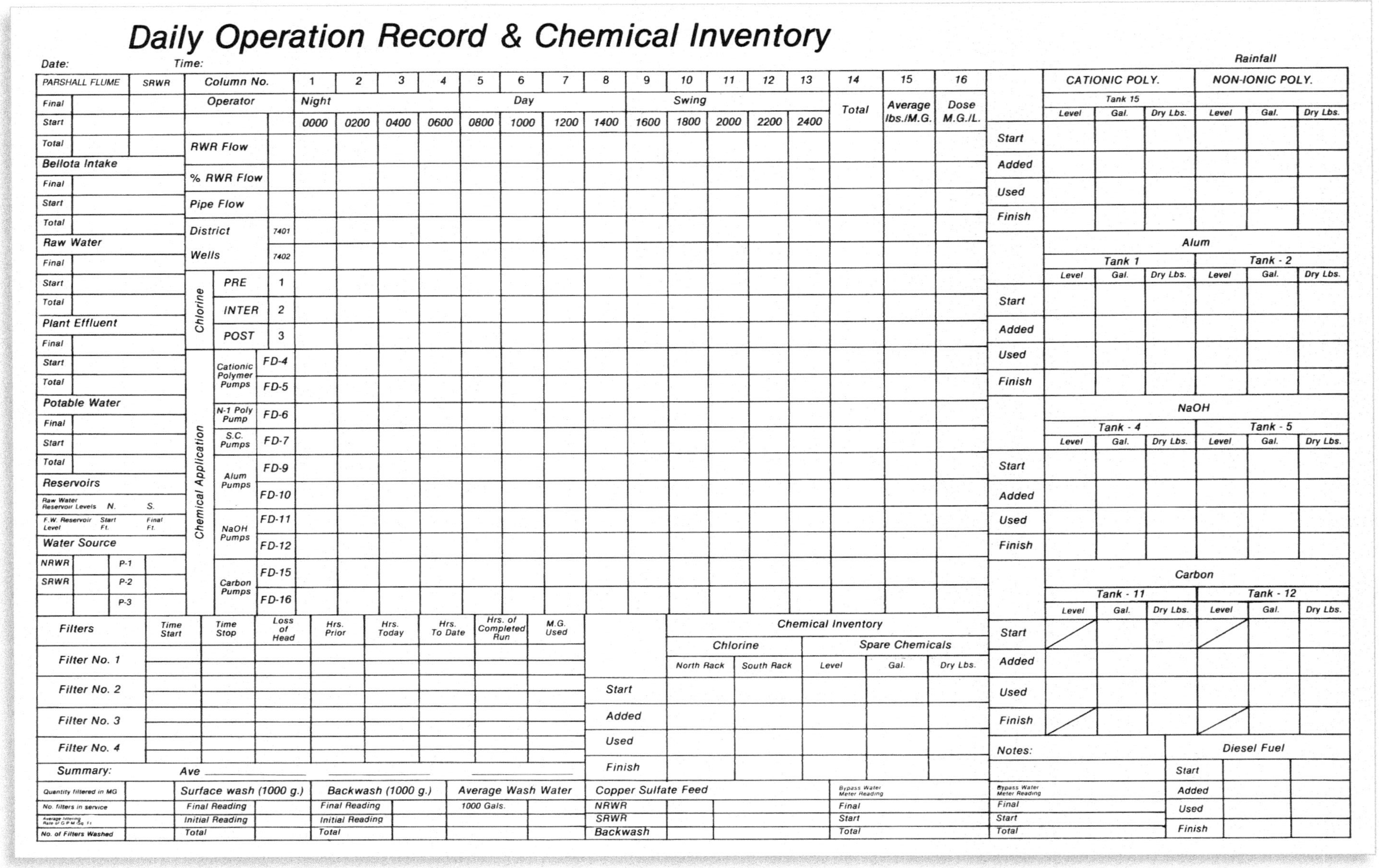

Daily Operation Record & Chemical Inventory

Date: Time: Rainfall

PARSHALL FLUME	SRWR
Final	
Start	
Total	
Bellota Intake	
Final	
Start	
Total	
Raw Water	
Final	
Start	
Total	
Plant Effluent	
Final	
Start	
Total	
Potable Water	
Final	
Start	
Total	
Reservoirs	
Raw Water Reservoir Levels N. S.	
F.W. Reservoir Level Start Ft. Final Ft.	
Water Source	
NRWR P-1	
SRWR P-2	
P-3	

Column No.		1	2	3	4	5	6	7	8	9	10	11	12	13	14	15	16
Operator		Night				Day				Swing					Total	Average lbs./M.G.	Dose M.G./L.
		0000	0200	0400	0600	0800	1000	1200	1400	1600	1800	2000	2200	2400			
RWR Flow																	
% RWR Flow																	
Pipe Flow																	
District	7401																
Wells	7402																
Chlorine PRE	1																
Chlorine INTER	2																
Chlorine POST	3																
Chemical Application: Cationic Polymer Pumps	FD-4																
Cationic Polymer Pumps	FD-5																
N-1 Poly Pump	FD-6																
S.C. Pumps	FD-7																
Alum Pumps	FD-9																
Alum Pumps	FD-10																
NaOH Pumps	FD-11																
NaOH Pumps	FD-12																
Carbon Pumps	FD-15																
Carbon Pumps	FD-16																

	CATIONIC POLY. Tank 15 Level	Gal.	Dry Lbs.	NON-IONIC POLY. Level	Gal.	Dry Lbs.
Start						
Added						
Used						
Finish						

	Alum Tank 1 Level	Gal.	Dry Lbs.	Tank - 2 Level	Gal.	Dry Lbs.
Start						
Added						
Used						
Finish						

	NaOH Tank - 4 Level	Gal.	Dry Lbs.	Tank - 5 Level	Gal.	Dry Lbs.
Start						
Added						
Used						
Finish						

	Carbon Tank - 11 Level	Gal.	Dry Lbs.	Tank - 12 Level	Gal.	Dry Lbs.
Start						
Added						
Used						
Finish						

Filters	Time Start	Time Stop	Loss of Head	Hrs. Prior	Hrs. Today	Hrs. To Date	Hrs. of Completed Run	M.G. Used
Filter No. 1								
Filter No. 2								
Filter No. 3								
Filter No. 4								
Summary:	Ave							

Chemical Inventory	Chlorine North Rack	Chlorine South Rack	Spare Chemicals Level	Gal.	Dry Lbs.
Start					
Added					
Used					
Finish					

Notes:

Diesel Fuel	
Start	
Added	
Used	
Finish	

Quantity filtered in MG	
No. filters in service	
Average filtering Rate of G.P.M./Sq. Ft.	
No. of Filters Washed	

Surface wash (1000 g.)		Backwash (1000 g.)		Average Wash Water		Copper Sulfate Feed		Bypass Water Meter Reading		Bypass Water Meter Reading	
Final Reading		Final Reading		1000 Gals.		NRWR		Final		Final	
Initial Reading		Initial Reading				SRWR		Start		Start	
Total		Total				Backwash		Total		Total	

Figure 5.16 Daily operation record and chemical inventory

Check Your Understanding

1. List the types of process calculations an operator is expected to perform.

2. How are filter backwash rates usually given?

3. Calculate the percent of water filtered used for backwashing if a filtration plant uses 0.12 million gallons for backwashing during a period when a total of 5 million gallons of water was filtered.

4. What types of records should be kept when operating a filtration process?

5.8 Operating Procedures Associated with Abnormal Process Conditions

Abrupt changes in water quality indicators, such as turbidity, pH, alkalinity, threshold odor number (TON), temperature, chlorine demand (source water), chlorine residual (in-process), or color, are signals that the operator should immediately review the performance of the filtration process, as well as pretreatment processes (coagulation, flocculation, and sedimentation).

During a normal filter run, watch for rapid changes in head loss buildup in the filter and turbidity breakthrough. Significant changes in either of these guidelines may indicate an upset or failure in the filtration process or pretreatment processes. Other indicators of abnormal conditions include:

1. Mudballs in filter media
2. Media cracking or shrinkage
3. Media boils during backwash
4. Excessive media loss or visible disturbance
5. Short filter runs
6. Filters that will not come clean during backwash
7. Algae on walls and media

5.8.1 Process Actions

Significant changes in source water turbidity levels, either increases or decreases, require immediate verification of the effectiveness of the filtration process in removing suspended solids and floc. Determine filtration removal efficiency by comparing filter influent and effluent turbidity levels with those of recent records.

If filter turbidity removal efficiency is decreasing, look first at the performance of the coagulation and flocculation processes to determine if the coagulant dosage is correct for current conditions. This may require performing jar tests in the laboratory as described in Chapter 9, "Laboratory Procedures," to properly assess treatment conditions.

Increases in source water turbidity and resultant increases in coagulant feed rates may impose a greater load on the filters if the majority of suspended solids and floc are not removed in the settling basins. This condition may require that you decrease filtration rates (put additional filters into service) or backwash filters more frequently.

If pretreatment processes do not readily respond to source water quality changes, it may be necessary to add filter aid chemicals at the filter influent to improve filtration solids removal efficiency. Filter aid chemicals, such as nonionic polymers, have proven to be effective in improving filtration performance when fed at low dosage rates (parts per billion). However, care must be exercised in selecting the appropriate feed rate because overdosing can seal the filter media resulting in drastically shortened filter runs. Generally, appropriate feed rates are established by trial and experience since this procedure is not easily simulated (duplicated) in the laboratory.

For example, you could begin feeding a polymer as a filter aid at 0.10 mg/L. Note the effectiveness of the results by observing: (1) removal of floc, (2) filter effluent turbidity, and (3) length of filter run. Compare these results to results without the polymer. If there was an improvement in filter effluent quality (reduction of turbidity) but a decrease in length of filter run, decrease the polymer feed. Continue to decrease the polymer feed until there is either a lessening of filter effluent quality or filter run time is maximized.

Changes in source water quality, such as alkalinity and pH, may also affect filtration performance through decreased coagulation-flocculation process performance. This is particularly evident when source water quality changes result from precipitation and runoff or from algal blooms in a source water reservoir. Again, use of filter aids may improve filtration efficiency until other pretreatment processes are stabilized.

Increases in filter effluent turbidity may also result from floc carryover from the sedimentation process. As described in Chapter 3, "Coagulation and Flocculation," the optimum floc size developed in the flocculation process ranges from about 0.1 to 3.0 mm. In conventional filtration, the optimum floc size is closer to 3.0 mm for settling purposes. However, in the direct filtration process (no sedimentation step), the optimum floc size is closer to 0.1 mm to permit depth penetration of the filter media. When larger floc is not removed in sedimentation (too light), it will be carried over into the filters causing rapid media surface clogging. Hydraulic forces in the filter will shear weak flocs, further contributing to turbidity breakthrough. Reevaluation of coagulation-flocculation and sedimentation performance may be required if floc carryover into the filters reduces filtration efficiency. The size of floc can be estimated by observation, as it is seldom necessary to make an accurate measurement of floc size.

Short filter runs may result from increased solids loading or filter aid overdosing, excessively high filtration rates, excessive mudball formation in the filter media, or clogging of the filter underdrain system. Possible corrective actions are summarized in Table 5.4.

Take immediate corrective actions if you encounter backwash problems such as media boils, media loss, or failure of the filter to come clean during the backwash process. Generally, these problems can be solved by adjusting backwash flow rates, surface wash flow rate or duration, or adjusting the time sequence or duration of the backwash cycle. In filters with nozzle-type underdrains, boils are often the result of nozzle failure. In this situation the filter should be taken out of service and the nozzles replaced. Possible operator actions are summarized in Table 5.4.

Problems within the filter itself, such as mudball formation or filter cracks and shrinkage, result from ineffective or improper filter backwashing.

Table 5.4 Filtration process troubleshooting

Source Water Quality Changes	Operator Actions	Possible Process Changes
• Turbidity • Temperature • Alkalinity • pH • Color • Chlorine demand	1. Perform necessary analysis to determine extent of change 2. Assess overall process performance 3. Perform jar tests if indicated 4. Make appropriate process changes 5. Increase frequency of process monitoring 6. Verify response to process changes at appropriate time (allow sufficient time for change to take effect) 7. Add lime or caustic soda if alkalinity is low	1. Change coagulant 2. Adjust coagulant dosage 3. Adjust flash-mixer/flocculator mixing intensity 4. Change frequency of sludge removal (increase or decrease) 5. Change filtration rate (add or delete filters) 6. Start filter aid feed 7. Adjust backwash cycle (rate, duration)

Sedimentation Process Effluent Quality Changes	Operator Actions	Possible Process Changes
• Turbidity or floc carryover	1. Assess overall process performance 2. Perform jar tests if indicated 3. Make appropriate process changes 4. Verify response to process changes at appropriate time	1. Same as for source water quality changes

Filtration Process Changes/Problems	Operator Actions	Possible Process Changes
• Head loss increase • Short filter runs • Media surface sealing • Mudballs • Filter media cracks, shrinkage • Filter will not come clean • Media boils • Media loss • Excessive head loss	1. Assess overall process performance 2. Perform jar tests if indicated 3. Make appropriate process changes 4. Verify response to process changes at appropriate time	1. Change coagulant 2. Adjust coagulant dosage 3. Adjust flash-mixer/flocculator mixing intensity 4. Change frequency of sludge removal (increase or decrease) 5. Decrease filtration rate (add more filters) 6. Decrease or terminate filter aid feed 7. Adjust backwash cycle (rate, duration) 8. Manually remove mudballs (hoses or rakes) 9. Replenish lost media 10. Clear underdrain openings of media, corrosion, or chemical deposits; check head loss indicator

Filter Effluent Quality Changes	Operator Actions	Possible Process Changes

• Turbidity breakthrough
• Color
• pH
• Chlorine demand
	1. Assess overall process performance
	2. Perform jar test if indicated
	3. Verify process performance:
	  a. Coagulation-flocculation process (see Chapter 4, "Coagulation and Flocculation")
	  b. Sedimentation process (see Chapter 5, "Sedimentation")
	  c. Filtration process (see above suggestions)
	4. Make appropriate process changes
	5. Verify response to process changes at appropriate time
		1. Change coagulant
		2. Adjust coagulant dosage
		3. Adjust flash-mixer/flocculator mixing intensity
		4. Change frequency of sludge removal (increase or decrease)
		5. Decrease filtration rate (add more filters)
		6. Start filter aid feed
		7. Change chlorine dosage

Correction of these conditions will require evaluation and modification of backwash procedures. Table 5.4 summarizes filtration process problems, how to identify the causes of problems, and how to correct the problems.

If filter beds are not thoroughly washed, material filtered from the water is retained on the surface of the filter. This material is sufficiently adhesive to form small balls. In time, these balls of material come together in clumps to form larger masses of mudballs. Usually, as time goes on, filter media becomes mixed in and gives the mudballs additional weight. When the mass becomes great enough, it causes the mudballs to sink into the filter bed. These mudballs, if allowed to remain, will clog areas in the filter. Generally, proper surface washing will prevent mudball formation.

5.8.2   Air Binding

Air binding is the clogging of a filter, pipe, or pump due to the presence of air released from water. Air entering the filter media is harmful to both the filtration and backwash processes. Air can prevent the passage of water during the filtration process and can cause the loss of filter media during the backwash process. Shortened filter runs can occur because of air-bound filters. This is caused by the release of dissolved air in saturated cold water due to a decrease in pressure. Air is released from the water when passing through the filter bed by differences in pressure produced by friction through the bed. Subsequently, the released air is entrapped in the filter bed. Air binding will occur more frequently when large head losses develop in the filter. Whenever a filter is operated to a head loss that exceeds the head of water over the media, air will be released. Air-bound filters are a problem because the air prevents water from passing through the filter and causes shortened filter runs. When an air-bound filter is backwashed, the released air can damage the filter media. When air is released during backwashing, the media becomes suspended in the wash water and is carried out of the filter, thus being no longer available for filtration.

5.8.3   Excessive Head Loss

If excessive head losses through a filter remain after backwashing, the filter underdrain system and the head loss measurement equipment should be checked. High head losses can be caused by the reduction in the size and number of underdrain openings. The underdrain openings can be reduced in size or clogged due to media and from corrosion or chemical deposits. Excessive or abnormal head loss readings also may be caused by malfunctioning head loss measurement equipment.

Check Your Understanding

1. How would you identify an upset or failure in the filtration process or pretreatment processes?

2. List the indicators of abnormal filtration process conditions.

3. How could you make a quick determination of filtration removal efficiency?

4. What problems may be encountered during backwash?

5. How does a filter become air bound?

5.9   Startup and Shutdown Procedures

Unlike the previously discussed treatment processes (coagulation, flocculation, and sedimentation), startup or shutdown of the filtration process is a routine operating procedure in most water treatment plants. This is true even if the treatment plant is operated on a continuous basis because it is common practice for a filter module to be brought into service or taken off line for backwashing. Clean filters may be put into service when a dirty filter is removed for backwashing, when it is necessary to decrease filtration rates, or to increase plant production as a result of increased demand for water. However, most plants keep all filters on line except for backwashing and in service except for maintenance. Filters are routinely taken off line for backwashing when the media becomes clogged with particulates, turbidity breakthrough occurs, or demands for water are reduced.

5.9.1   Implementation of Startup/Shutdown Procedures

This section outlines typical actions performed by the operator in the startup and shutdown of the gravity filtration process. These procedures also generally apply to pressure filters. While some of these concepts also apply to diatomaceous earth filtration, the manufacturer's operating procedures will provide more specific instructions.

Figures 5.17 and 5.18 illustrate sectional views of typical gravity filters. The figures show the valve positions and flow patterns in the filtration and backwash modes of filter operation.

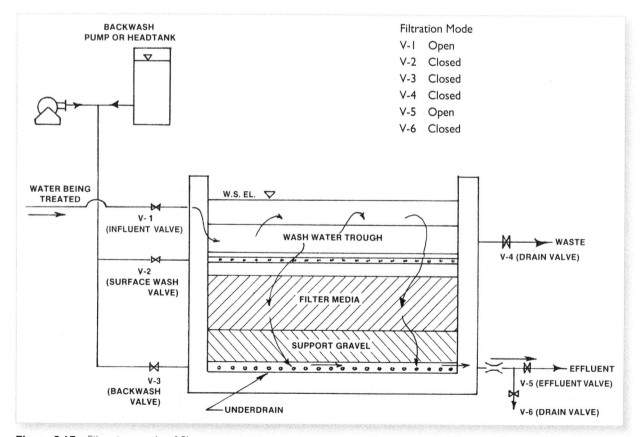

Figure 5.17   Filtration mode of filter operation

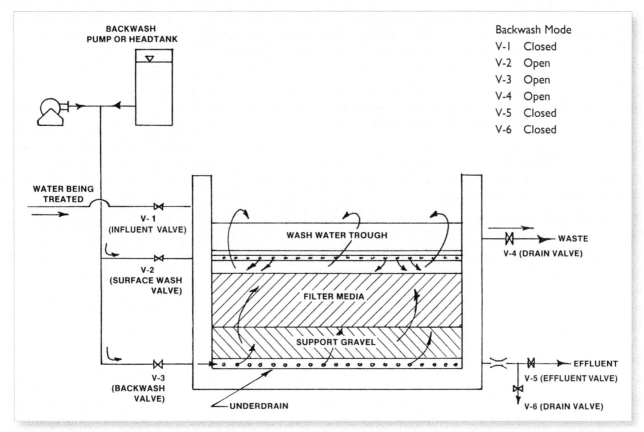

Figure 5.18   Backwash mode of filter operation

5.9.1.1   Filter Checkout Procedures

1. Check operational status of filter.
2. Be sure filter media and wash water troughs are clean of all debris such as leaves, twigs, and tools.
3. Check that all access covers and walkway gratings are in place.
4. Make sure process monitoring equipment, such as head loss and turbidity systems, are operational.
5. Check source of backwash water to ensure it is ready to go. This could be an elevated wash water tank, pumps, or other source.

5.9.1.2   Backwash Procedures

1. Filters should be washed before placing them in service:
   a. If washing filters using automatic equipment, check to be sure length of cycle times set for backwash and surface wash cycles are correct. Correct times vary from plant to plant and time of year. Base these settings on physical observations of actual time required to clean the filter. If filters are usually washed automatically, it is a good idea to occasionally use a manual wash procedure for efficient media cleaning during the wash cycle.
   b. The surface wash system should be activated just before the backwash cycle starts to aid in removing and breaking up solids on the filter media and to prevent the development of mudballs.

The surface wash system should be stopped before completion of the backwash cycle to permit proper reclassification (settling) of the filter media.

c. A filter wash should begin slowly (about 5 gal/min/ft^2 or 3.4 L/sec/m^2) for about 1 minute to permit purging (removing) of any entrapped air from the filter media, and to provide uniform expansion of the filter bed. After this period, the full backwash rate can be applied (15 to 25 gal/min/ft^2 or 10 to 17 L/sec/m^2). Sufficient time should be allowed for cleaning the filter media. Usually, when the backwash water coming up through the filter becomes clear, the media is clean. This generally takes from 3 to 8 minutes. If flooding of wash water troughs or carryover of filter media is a problem, the backwash rate must be reduced. This may be accomplished by adjusting the master backwash valve, thereby throttling the amount of wash water used.

d. In many water treatment plants, waste backwash water is either directly recycled through the plant or is allowed to settle in a tank, pond, or basin and then the supernatant (clear, top portion of water) is pumped back to be recycled through the plant. Usually, it is best to gradually add the waste backwash water to the headworks of the treatment plant (before the flash mixer). This is because a large slug of waste backwash water will require changes in chemical dosages due to the additional flow and increased turbidity.

2. Procedures for the backwash mode of a filter operation includes the following and are shown in Figure 5.18 (which uses "V" to indicate different modes):

a. Log length of filter run since last backwash (see Section 5.7.4, "Recordkeeping").

b. Close filter influent valve (V-1).

c. Open drain valve (V-4).

d. Close filter effluent valve (V-5).

e. Start surface wash system (open V-2).

f. Slowly start backwash system (open V-3).

g. Observe filter during washing process.

h. When wash water from filter becomes clear (filter media is clean), close surface wash system valve (V-2).

i. Slowly turn off backwash water system (close V-3).

j. Close drain valve (V-4).

k. Log length of wash and number of gallons of water used to clean filter.

5.9.1.3 Filter Startup Procedures

The procedures described in this section and shown in Figure 5.17 may include a "V" designation, which indicates different filtration modes. Filter startup procedures include:

1. After washing the filters, they should be eased into service. With automatic equipment, this is generally done by a gradual opening of the filter's effluent valve. Manual operations require a gradual increase of the amount of water treated by the filter. The initial

few hours after a filter is placed in service is a time when turbidity breakthrough can pose a problem. For this reason, filters should be eased into service to avoid hydraulic shock loads.

2. Start filter (Figure 5.17).
   a. Slowly open filter influent valve (V-1).
   b. When proper elevation of water is reached on top of the filter, the filter effluent valve should be gradually opened (V-5). In many systems, the filter effluent valve controls the level of water on the filter. This valve adjusts itself to maintain a constant level of water over the filter media regardless of filtration rate.
   c. Some plants have provisions to waste some of the initial filtered water (open V-6). This provision can be helpful if an initial breakthrough occurs.
   d. Perform turbidity or particle counting analyses of filtered water and make process adjustments as necessary.

5.9.1.4   Filter Shutdown Procedures

1. Remove filter from service by:
   a. Closing influent valve (V-1).
   b. Closing effluent valve (V-5).
2. Backwash filter (see Section 5.9.1.2, "Backwash Procedures").
3. If filter is to be out of service for a prolonged period, drain water from filter to avoid algal growth.
4. Note status of filter in operations log.

Check Your Understanding

1. Under what conditions may (clean) filters be put back into service?
2. When are filters routinely taken off line for backwashing?
3. Why should the surface wash system be activated just before the backwash cycle starts?
4. What should be done if a filter will be out of service for a prolonged period?

5.10   Process and Support Equipment Operation and Maintenance

To run a filtration process, you must be familiar with the O&M instructions for each type a variety of mechanical, electrical, and electronic equipment, and control systems, including:

1. Filter control valves
2. Backwash and surface wash pumps
3. Flowmeters and level/pressure gauges
4. Water quality monitors such as turbidity meters
5. Process monitors (head loss and water level)
6. Mechanical and electrical filter control systems

5.10.1   Equipment Operation

Before starting a piece of mechanical equipment, such as a backwash pump, be sure that the unit has been serviced on schedule and its operational status is positively known.

After startup, always check for excessive noise and vibration, overheating, and leakage (water, lubricants). When in doubt about the performance of a piece of equipment, refer to the manufacturer's instructions.

Much of the equipment used in the filtration process is automated and only requires limited attention by operators during normal operation. However, periodic calibration and maintenance of this equipment is necessary, and this usually involves special procedures. Detailed operating, repair, and calibration procedures are usually described in the manufacturer's literature.

5.10.2   Preventive Maintenance Procedures

Preventive maintenance programs are designed to ensure the continued satisfactory operation of treatment plant facilities by reducing the frequency of breakdown failures. This is accomplished by performing scheduled or routine maintenance on valves, pumps, and other electrical and mechanical equipment items.

In the normal operation of the filtration process, you will be expected to perform routine maintenance functions as part of an overall preventive maintenance program. Typical functions include:

1. Keeping electric motors free of dirt, moisture, and pests (rodents and birds)
2. Ensuring good ventilation (air circulation) in equipment work areas
3. Checking pumps and motors for leaks, unusual noise and vibrations, or overheating
4. Maintaining proper lubrication and oil levels
5. Inspecting for alignment of shafts and couplings
6. Checking bearings for overheating and proper lubrication
7. Checking for proper valve operation (leakage or jamming)
8. Checking automatic control systems for proper operation
9. Checking air/vacuum relief systems for proper functioning, dirt, and moisture
10. Verifying correct operation of filtration and backwash cycles by observation
11. Inspecting filter media condition (look for algae and mudballs and examine gravel and media for proper gradation)
12. Inspecting filter underdrain system (ensure underdrain openings are not becoming clogged due to media, corrosion, or chemical deposits)

Accurate recordkeeping is the most important element of any successful preventive maintenance program. These records provide O&M personnel with clues for determining the causes of equipment failures. They frequently can be used to forecast impending failures thus preventing costly repairs.

5.10.3 Safety Considerations

To avoid accidents or injury when working around filtration equipment such as pumps and motors, follow the safety procedures listed below:

ELECTRICAL EQUIPMENT

1. Avoid electric shock (use protective gloves)
2. Avoid grounding yourself in water or on pipes
3. Ground all electric tools
4. Lock out and tag electric switches and panels when servicing equipment
5. Use the buddy system

MECHANICAL EQUIPMENT

1. Use protective guards on rotating equipment
2. Do not wear loose clothing around rotating equipment
3. Keep hands out of energized valves, pumps, and other pieces of equipment
4. Clean up all lubricant and chemical spills (slippery surfaces cause bad falls)

OPEN-SURFACE FILTERS

1. Use safety devices such as handrails and ladders
2. Close all openings and replace safety gratings when finished working
3. Know the location of all life preservers and other safety devices

VALVE AND PUMP VAULTS, SUMPS, FILTER GALLERIES

1. Be sure all underground or confined structures are free of hazardous atmospheres (toxic or explosive gases, too much or too little oxygen) by checking with gas detectors
2. Only work in well-ventilated structures (use air circulation fans)
3. Use the buddy system

5.11 Drinking Water Regulations

The first US drinking water laws and standards were enacted in 1914 to control waterborne bacteria and viruses that caused diseases such as cholera, dysentery, and typhoid. In 1962, the US Public Health Service revised the national drinking water standards to include limits on certain chemicals. The US Environmental Protection Agency (EPA) replaced the US Public Health Service in 1970 to protect human health and the environment. The EPA sets the standards for drinking water quality in partnership with the states and water suppliers who interpret the laws.

5.11.1 Safe Drinking Water Act (SDWA)

Heightened public awareness and concern regarding cancer became major factors behind the push for additional legislative action on the issue of drinking water contamination, and led to the passage and signing into law of the Safe Drinking Water Act (SDWA) in 1974. The SDWA gave the federal government, through the EPA, the authority to:

1. Set national standards regulating the levels of contaminants in drinking water
2. Require public water systems to monitor and report the levels of their identified contaminants
3. Establish uniform guidelines specifying the acceptable treatment technologies for removing unsafe levels of pollutants from drinking water

The SDWA is the main federal law that ensures the quality of Americans' drinking water. Under this law the EPA sets legal limits on the levels of certain contaminants in drinking water. The legal limits reflect both the levels that safeguard human health and the degree to which water systems must protect drinking water using the best available technology.

In addition to prescribing these legal limits, EPA also develops rules for setting water-testing schedules and methods that water systems must follow. The rules also list acceptable standards or practices to meet the legal limits. EPA's standards include acceptable techniques for treating contaminated water. The SDWA gives individual states the opportunity to set and enforce their own drinking water standards if the standards are at least as strong as EPA's national standards. Most states and territories directly oversee the water systems within their borders.

While the SDWA gave EPA responsibility for promulgating (passing into law) drinking water regulations, it gave state regulatory agencies the opportunity to assume primary responsibility for enforcing those regulations. Implementation of the SDWA has greatly improved basic drinking water purity across the nation. However, EPA surveys of surface water and groundwater indicate the presence of synthetic organic chemicals in 20 percent of the nation's water sources, with a small percentage at levels of concern. In addition, research studies suggest that some naturally occurring contaminants may pose even greater risks to human health than the synthetic contaminants.

In 1986, Congress enacted amendments designed to strengthen the 1974 SDWA. These amendments set deadlines for the establishment of maximum contaminant levels (MCLs), placed greater emphasis on enforcement, authorized penalties for tampering with drinking water supplies, mandated the complete elimination of lead from drinking water, and placed considerable emphasis on the protection of underground drinking water sources. The 1986 SDWA amendments set up a timetable under which EPA was required to develop primary standards for 83 contaminants. Other major provisions of the 1986 SDWA amendments required EPA to do the following:

1. Define an approved treatment technique for each regulated contaminant
2. Specify criteria for filtration of surface water supplies
3. Specify criteria for disinfection of surface and groundwater supplies
4. Prohibit the use of lead products in materials used to convey drinking water

To comply with the provisions of the 1986 SDWA amendments (signed into law in 1996), the EPA, the states, and the water supply industry undertook significant new programs to clean up the country's water supplies. In 1998, two major regulations were signed into federal law: the Disinfectants and Disinfection Byproducts Rule (DBPR) and the Interim Enhanced Surface Water Treatment Rule (IESWTR) (see the chapter on drinking water regulations in Volume II of this training manual series for additional information). The DBPR was developed to protect the public from harmful concentrations of disinfectants and from trihalomethanes (THMs), which could form when DBPs combine with organic matter in drinking water. The goal of the IESWTR is to reduce the occurrence of *Cryptosporidium* and other pathogens in drinking water. This rule was developed to ensure that protection against microbial contaminants is not lowered as water systems comply with the DBPR.

In 2010, EPA announced a new approach to implementing the SDWA. This new approach involves four key elements:

1. Addressing contaminants as a group rather than one at a time to be more cost effective in regulating drinking water protection
2. Promoting the development of new drinking water treatment technologies to address health risks
3. Using the authority of multiple statutes to help protect drinking water
4. Partnering with states to share data collected from monitoring drinking water systems

Operators are urged to work closely with their state regulatory agencies to stay informed about changes in water regulations as they occur in the future.

5.11.2 Surface Water Treatment Rule (SWTR)

In 1989, the EPA developed the Surface Water Treatment Rule (SWTR). The 1996 SDWA amendments imposed deadlines by which EPA must complete work on several major regulations, including an Enhanced Surface Water Treatment Rule (ESWTR). In 1998, the EPA promulgated the Interim Enhanced Surface Water Treatment Rule (IESWTR). In January and December 2002, respectively, the EPA announced the Long Term 1 Enhanced Surface Water Treatment Rule (LT1ESWTR) and the Long Term 2 Enhanced Surface Water Treatment Rule (LT2ESWTR). These regulations established primary drinking water regulations requiring treatment of surface water supplies or groundwater supplies under the direct influence of surface water. These regulations require surface water systems to use a specific water treatment technique (filtration or disinfection) rather than meeting MCLs for *Cryptosporidium*, *Giardia lamblia*, viruses, *Legionella*, and heterotrophic bacteria. Also, the regulations require that all systems must be operated by qualified operators.

The protozoan *Giardia lamblia* is presently the organism most implicated in waterborne disease outbreaks in the United States. These microscopic creatures are found mainly in mountain streams. Once inside the body, they cause a painful and disabling illness. The infection caused by *Giardia* is called Giardiasis. The symptoms of Giardiasis are usually severe diarrhea, gas, cramps, nausea, vomiting, and fatigue.

Giardia and viruses have been added to the traditional coliform and turbidity indicators of microbiological quality. In this case, the MCL goals are zero because the organisms are pathogens, or indicators of pathogens, and should not be present in drinking water.

The regulations define surface water as "all water open to the atmosphere and subject to surface runoff." This would include rivers, lakes, streams, and reservoirs. This surface water definition also includes groundwaters that are directly influenced by surface water. The determination as to whether a supply uses surface water is left up to the state. Generally, if a groundwater source has significant and relatively rapid shifts in water quality such as turbidity, temperature, conductivity, or pH that closely correlate to climatological or nearby surface water conditions, that source can be considered surface water.

At a minimum, the treatment required for surface water would include disinfection systems for clean and protected source waters. These systems would only be required to disinfect to achieve removal of coliforms to meet the requirements.

Turbidity measurements are a key element in determining compliance with both the filtration and the disinfection requirements of the SWTRs. The effectiveness of a filtration system in removing particles and organisms is easily obtained by measuring the water's turbidity before and after filtration. High turbidity also reduces the effectiveness of disinfection processes and, therefore, must be controlled and closely monitored.

Water supply systems that are required to filter can use a variety of treatment technologies, including disinfection, to meet the expected performance levels. These technologies include conventional treatment, direct filtration, slow sand filtration, and diatomaceous earth filtration.

The two-pronged treatment technique of the regulations aims to ensure that harmful microorganisms are either removed by filtration processes or inactivated (and thus made harmless) by disinfection processes. Success, therefore, depends heavily on consistently well-operated filtration and disinfection systems. The general performance standards to be met by all surface water systems and any systems using groundwater under the influence of surface water are as follows:

1. At least 99.9 percent (also called 3 log for the 3 nines) removal or inactivation of *Giardia lamblia* cysts
2. At least 99.99 percent (4 log) removal or inactivation of enteric (intestinal) viruses

In general, compliance by the surface water purveyor could be through one of the following alternatives:

1. Meeting the criteria for which filtration is not required and providing disinfection according to the specific requirements in the regulations
2. Providing filtration and meeting disinfection criteria required for those supplies that are filtered

5.11.3 Turbidity Requirements

Different turbidity monitoring and turbidity MCLs apply for each type of filtration, as follows:

Type of Filtration	Monitoring Frequency	Turbidity Level
Conventional	Every 4 hours	<0.3 NTU
Direct	Every 4 hours	<0.3 NTU
Diatomaceous earth	Every 4 hours	<1.0 NTU
Slow sand	Once per day	<1.0 NTU

Continuous turbidity readings may be substituted for the 4-hour sampling if the meter is periodically calibrated. The regulatory agency may reduce monitoring to once per day for systems serving fewer than 500 people.

CONVENTIONAL OR DIRECT FILTRATION. Systems using either conventional or direct filtration must achieve a filtered water turbidity level of less than or equal to 0.3 NTU in 95 percent of the measurements taken for each month. This limit may be increased by the state to 1 NTU if the system proves it can effectively remove *Giardia* cysts at such turbidity levels. At no time can filtered water turbidity exceed 1 NTU.

SLOW SAND FILTRATION. Systems using slow sand filtration must achieve a filtered water turbidity level of less than or equal to 1 NTU in 95 percent of the measurements for each month. This limit can be increased by the state if there is no interference with disinfection and the turbidity level never exceeds 5 NTU. At no time can filtered water turbidity exceed 5 NTU.

DIATOMACEOUS EARTH FILTRATION. Systems using diatomaceous earth filtration must achieve a filtered water turbidity level of less than or equal to 1 NTU in 95 percent of the measurements for each month. At no time can filtered water turbidity exceed 5 NTU.

See the chapter on drinking water regulations in Water Treatment Plant Operation, Volume II, in this training manual series.

OTHER FILTRATION TECHNOLOGIES. With the approval of the state, systems may use other filtration technologies provided they meet the same criteria for removal of *Giardia* cysts and viruses as are required for the conventional technologies.

Check Your Understanding

1. List the types of equipment used in the filtration process.
2. What should be done before starting a piece of mechanical equipment, such as a backwash pump?
3. What safety hazards may be encountered when working around mechanical equipment?
4. What is the SWTR definition of surface water?
5. Under what conditions would a groundwater source be considered surface water?

5.12 Particle Counters

The SWTR requires that water systems using surface water, or groundwater directly influenced by surface water, achieve a 99.9 percent (3-log) reduction of *Giardia* cysts, and a 99.99 percent (4-log) reduction of viruses through filtration and disinfection. These reductions are typically achieved through a combination of filtration and disinfection. To demonstrate these removals, a water system may use **particle counting** as a substitute for *Giardia* cyst measurement.

particle counting
A procedure for counting and measuring the size of individual particles in water. Particles are divided into size ranges and the number of particles is counted in each of these ranges. The results are reported in terms of the number of particles in different particle diameter size ranges per milliliter of water sampled.

Another threat to public health from drinking water is the exposure to *Cryptosporidium* oocysts. To minimize the potential of public exposure to *Cryptosporidium,* operators try to optimize treatment plant performance. **Particle counters**, devices that count and measure the size of individual particles in water, are one of the best monitoring tools available to optimize plant performance for the removal of particles. Also, while there is no consistently reliable analytical technique for measuring *Cryptosporidium* densities in drinking water, particle counters can be used as an indirect way to indicate the possible removal of *Cryptosporidium*. By monitoring the removal efficiencies of particles in the same size ranges as *Giardia* and *Cryptosporidium,* the operator can estimate the system's *Giardia* and *Cryptosporidium* removal rate.

A particle counter cannot tell the difference between a particle of clay and a microorganism (protozoan cysts or oocysts). While the SWTR permits particle counting to substitute for a portion of microbiological testing for the presence of cysts or oocysts, particle counting does not completely eliminate the need for microbiological testing for the presence, absence, or enumeration (counting) of cysts and oocysts. Particle counting must be viewed as a complementary measurement technique to be used in combination with many other measurements—physical, chemical, and microbiological—not as a replacement for any other technique.

See Standard Methods. 22nd Edition, "2560. Particle Counting and Size Distribution." www.awwa.org. ISBN: 978-0-87553-013-0.

5.12.1   Particle Counting to Monitor Filter Performance

Particle removal is a main concern of municipal plant operation, first at the sedimentation basin, if used, then ultimately at the filter. Monitoring the efficiency of the filtration process is critically important because if filtration is deficient, particulates, and possibly pathogens, may pass through to the consumer.

Particle counters can help operators optimize plant performance for particle removal by enhancing the performance of the filtration process.

5.12.1.1   Filter Ripening

Particle counts and particle size distribution analyses present an informative picture of effluent quality produced during the ripening phase after a filter has been backwashed. In a properly operated system, the total number of particles will increase during the initial few minutes of the ripening phase and then rapidly decrease within a few minutes. Operators should establish operational procedures that make the ripening period as short as possible.

5.12.1.2   Filter Flow Rate

A particle counter can be used to measure the effects of filter flow rate on filter effluent quality. Increases in the flow rate or hydraulic loading on a filter can result in an increase in the particulates detected in the filter effluent. A sudden decrease in filter flow rate can produce a sharp peak of accumulated particulates and then a reduced level of particulates afterward.

5.12.1.3   Filter Run Time

Particle counts show a small spike for a short time immediately after backwash and then drop as ripening is achieved. If a filter gradually accumulates particulates and then releases a large number of particulates at periodic intervals, these changes in performance can be recorded. Cumulative particle counts could be used to initiate backwash before the need for backwash is indicated by effluent turbidity data.

As a filter run progresses, the particle size distribution of particulates in the filter effluent could substantially change. With lengthening filter run time, the percentage of large particles passing through the filter often increases while small particles decrease.

5.12.1.4   Filter Media Selection

Particle counters may be used to evaluate and compare the performance of different types of filter media and different arrangements of filter media.

5.12.1.5   Polymer Application

The effectiveness of polymer addition and polymer dose points to improve filter performance can be measured by analyzing the results recorded by particle counters.

5.12.1.6   Other Uses of Particle Counters

Particle counting, when used in conjunction with other measurement techniques, can be a valuable tool to diagnose filtration problems including improperly graded media, insufficient media, disrupted media, short-circuiting, mudballs, defective underdrains, and excessive filter run loading.

Particle count analyses also can be used to optimize other factors affecting filtration efficiency, such as filter startup rate, the filter-to-waste period, and coagulant type and dosage for sedimentation.

5.12.2   How Particle Counters Work

Two particle counting techniques are used in water treatment plants.

Light-scattering instruments pass particles through a light beam and the light scatters in all directions. The scattered light signal measured by the detector depends on the size, shape, and refractive index of the particle, the refractive index of the liquid medium, the polarization state of the light beam, and the relative orientation between the particle, the light beam, and the detector. Computer software converts the scattering pattern to a count. The light scattering technique is appropriate for measurement of particles less than 1 **micron** (1 μm). While there are sensors available that will measure particles smaller than 0.1 micron, they are not suitable for use in drinking water applications except in a well-controlled research environment. Usually, it is not practical to use a sensor than can measure particles smaller than 0.5 micron in water treatment. As the smallest detectable particle size goes down, the cost, difficulty of maintenance, and unpredictability of the sensor go up. Typically, operators are discouraged from using sensors below 1 micron in water treatment except when they are absolutely the best choice for the given application.

Many water systems today are consistently achieving filter effluent counts (the number of particles detected by the sensor) of less than

micron
μm, Micrometer or Micron. A unit of length. One millionth of a meter or one thousandth of a millimeter. One micron equals 0.00004 of an inch. A typical bacterium cell diameter ranges from 1 to 10 μm.

10 counts (particles) per milliliter for particles greater than 2 microns in size, and it is not unusual to encounter a plant operating day in and day out near one count greater than 2 microns per milliliter.

Light-blocking (sometimes called light obscuration or light interruption) instruments pass particles through a light beam. A photodetector placed directly in the main light beam measures the small drop in the light intensity that occurs when light is blocked. The amount of blockage depends on the size, shape, and refractive index of the particle, the refractive index of the liquid medium, and the relative orientation of the particle within the light beam. Light-blocking instruments with good particle sizing characteristics and useful flow rates are available that can count particles as small as 1.0 to 2.0 μm. This is a good compromise in cost, reliability, and ease of maintenance.

The amount of light blocked from a detector is the sum of the light absorbed by the particle and the light scattered or reflected by the particle. The size and composition of each particle will determine how much light is scattered and how much is absorbed. Carbon particles will absorb most of the light and scatter very little of it. Organic particles have an index of refraction close to the value of water and tend to refract more light. The result of this is that an organic particle 5 μm in size will block less light than an inorganic particle of the same size, and the organic particle will appear smaller to the particle counter. For this reason, *Giardia* and *Cryptosporidium* will be counted in size ranges several microns below their actual size. The orientation of the particle as it passes through the light beam will also affect how much light is blocked. These factors make it necessary to count particles over a range of sizes as opposed to a single size.

Particle analyzers will produce differential or cumulative counts, or both. A cumulative particle count is the total of all particles greater than the sensor's lower limit. A differential particle count expresses the number of particles between two size limits. For example, a sensor with 2-micron **sensitivity** will detect and count all the particles larger than 2 μm and report a cumulative number.

For example, suppose you have a water sample containing the particles shown in Figure 5.19. As shown, there is one 30 μm particle, five with sizes between 5 and 20 μm, and 10 with sizes between 2 and 5 μm. In water samples, there are many more small particles than large ones. As you can see in the table below the picture, differential counts are the numbers in each individual size range (often called channels or bins). For a given lower limit, cumulative counts are the sum of the particles in all the channels larger than the given lower limit. As shown in the figure, the cumulative count of particles greater than 5 μm is 1 + 5 = 6, and the cumulative count of particles greater than 2 μm is 1 + 5 + 10 = 16.

The choice of differential vs. cumulative counts depends on your interest. For example, if you want to know how the total number of particles changed by passing the water through a filter, you would use cumulative data. If you want to see which sizes of particles are taken out best by the filter, you would use the differential data. In coagulation/flocculation, small particles are combined into larger flocs. A differential count would show how the size distribution changed in the treatment process.

sensitivity (particle counters)
The smallest particle a particle counter will measure and count reliably.

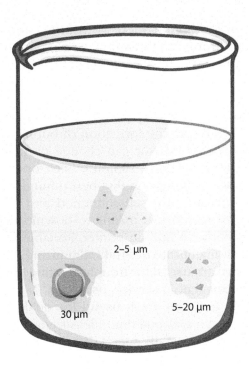

Size Range	Differential Counts	Size Range	Cumulative Counts
> 20	1	> 20	1
5–20	5	> 5	6
2–5	10	> 2	16

Figure 5.19   Illustration of cumulative particle counts and differential particle counts

5.12.3   Grab vs. In-Process Particle Counters

Accurate particle count measurements, whether they are based on grab samples or in service readings, depend on achieving a representative sample. While this is true of all analytical measurements, achieving this is critical for particle counting. The sample lines must be as short as possible and the instrument should be located as close as possible to the measurement point. Some operators argue that a sample cannot even be transported from one room to another without changing the sample.

5.12.3.1   Grab Samples

Two types of particle counters are available: dedicated laboratory models that are installed at a fixed location in a laboratory, or portable units. Use of a fixed-location laboratory particle counter complicates achieving a representative sample without much delay in analysis or the use of long sample lines (both are potential sources of significant error). One benefit of the dedicated lab units is that they make it easier for the lab technician to remove **entrained** air in solution either by pressure or by vacuum degassing. (Entrained air can form bubbles, which can be counted.)

A portable unit, on the other hand, can be transported to the sampling location, thus minimizing errors due to sample transport and long sample lines. However, it is more difficult to deal with entrained air when using

entrain
To trap bubbles in water either mechanically through turbulence or chemically through a reaction; or to trap one substance or material by another substance or material.

a portable unit. In addition, if there is insufficient sample pressure to deliver the sample through the instrument, a pump must be used (preferably downstream of the sensor). Pumping can introduce errors and thus should be avoided whenever possible for lab, portable, or in-process measurements. Placing a pump downstream from the sensor complicates interference due to air bubbles because the pressure reduction that occurs in the cell tends to promote the formation of bubbles. However, placing a pump just ahead of the sensor can change particle size distribution (resulting in a nonrepresentative sample). Each operator must examine the trade-offs and select the option that best minimizes errors.

Portable units normally are self-contained and include a means of measurement, digital memory, a means to download data directly to a computer, and sometimes a built-in printer. Portable units are convenient where offsite measurements are desired (minimizing the need to transport the sample), and they are useful for troubleshooting parts of the treatment process where a permanent in-service instrument is not installed.

Lab and portable units are most useful for quality assurance/quality control (QA/QC) checks and one-time measurements to gather baseline data. The operator must normally spend time and effort to download data to a software package (spreadsheet or database) and then manipulate the data. Software for downloading and manipulating data from a particular model of particle counter may be available from the particle counter manufacturer.

Lab or portable particle counting usually involves use of a sample vessel. Any laboratory glassware or equipment used for particle counting must be extremely well cleaned and, if possible, used only for particle counting.

In summary, both fixed-location particle counters and portable models have advantages and limitations. With both types of devices, however, the reliability of the measurements obtained is dependent on the use of representative samples and on the laboratory analyst's consistent use of good technique. Because particle counting requires rigorous attention to technique, the labor costs can be significant.

5.12.3.2 In-Process Measurements

In-process units are permanently mounted on pipes so that they can make continuous measurements of water flowing past. The actual instruments should be located as close as possible to the sample point to provide representative sampling and fast response to process changes. Once a proper sampling location has been chosen and installation has been completed, the in-process particle count measurement is much less technique-dependent than measurements made with laboratory or portable instruments. The laboratory technician only needs to perform periodic maintenance consisting mainly of cleaning the device (maintenance for in-process devices is usually less than maintenance needed for devices installed in the lab or grab sample devices). Most in-process instruments are designed to transmit data directly to a central computer system that provides automated data accumulation and manipulation (thus further reducing labor costs).

Cost per unit is significantly less for in-process instruments than for portable or dedicated laboratory units. Therefore, unless the few benefits of the lab or portable units are required, it is often more economical to install in-process instruments.

5.12.4   Particle Counters Compared to Turbidity Meters

Almost any application in which turbidity measurement is useful will be an appropriate application for a particle counter. Turbidity is a measure of the relative clarity or cloudiness of a water sample. Turbidity is determined by shining a light into a sample of water and measuring the amount of light scattered by the particles suspended in the sample. The detection device converts light energy into electrical voltage, which is scaled to output values in NTUs.

Because light scattering depends on several factors, turbidity readings are not reliable indicators of the number or size of particles. A relatively small number of large particles can produce exactly the same NTU value as many small particles. For this reason, the particle counter is far superior in applications where particle size is used as an indicator of *Giardia-* and *Cryptosporidium*-sized particle removals. On the other hand, turbidity meters can sense the presence of particles that are too small for particle counters to detect.

The ability of particle counters to detect specific sizes can be very valuable in some cases. For example, near the end of a filter run, the first particles to break through will be the smaller ones. Because turbidity measurements only indicate the average of all the particles, these smaller particles may not sufficiently affect the overall average for quite some time. A particle counter set to count particles in that range will detect the change immediately. It is common to see indications of filter breakthrough with a particle counter several minutes or even hours before turbidity measurements change.

Particle counters can detect carbon particles that do not scatter light and cannot be detected by turbidity meters using a single 90-degree light-scatter detector. This is a useful feature when carbon addition is part of the treatment process or when diagnosing problems with carbon filters.

In settled or applied water, particle counters can be used to measure the distribution of particle sizes, and can thereby aid in optimizing chemical feed and/or flocculation. Polymers have been known to cause strange readings from turbidity meters, resulting in over- or underfeeding. Particle distribution can be correlated with flows to treatment processes to achieve more precise process control.

Particle counters will not replace turbidity meters in water treatment. A turbidity measurement is preferable in some applications, mainly when dealing with high-turbidity waters. Since the particle counter is counting individual particles, there is a physical limit on the particle concentration that can be accurately measured. As a general guideline, a turbidity between 5 and 10 NTUs may be over the range of a particle counter. However, actual measurements in the field have identified situations where 1 NTU was over the range of a particle counter and other situations where 20 NTUs was not over the range of a particle counter. Experience indicates that a typical limit for particle counters is in the range of 5 to 10 NTUs; however, operators will have to test on a site-by-site basis to determine at what NTU point the particle counter will be beyond its range for a particular plant.

Operators must realize that particle counters and turbidity meters are tools to help them do a better job, similar to Phillips and slotted screwdrivers—both are important but they do different jobs.

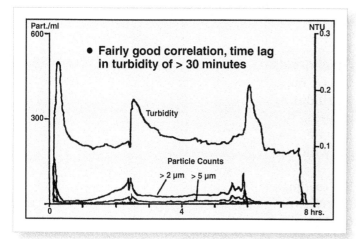

Figure 5.20   Data from pilot filter at Grants Pass, Oregon (*Courtesy of Met One Inc, a division of HACH Company*)

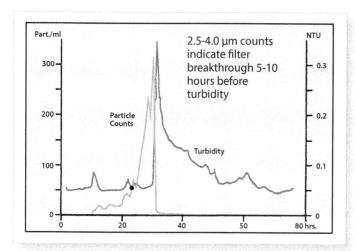

Figure 5.21   Data from full-scale filter in Utah (*Courtesy of Met One Inc, a division of HACH Company*)

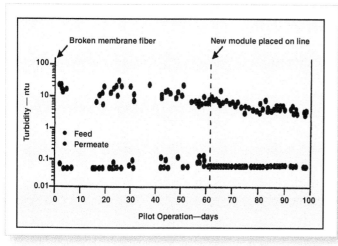

Figure 5.22   Loss of membrane integrity monitored by turbidity (*Courtesy of Met One Inc, a division of HACH Company*)

Scenario 1, Sensing Filter Breakthrough

Figure 5.20 shows two pilot filter breakthrough events. Both turbidity readings and particle counts sensed the breakthroughs; however, increasing particle counts warned of the impending turbidity breakthrough by more than 30 minutes. A second pilot filter was switched on line after the first breakthrough (at about 2.5 hours).

Notice that after the switchover, the counts for particles larger than 2 µm dropped much faster than the turbidity. Particles smaller than 2 µm were sensed by the turbidity meter even though they were too small to be counted.

Figure 5.21 shows the same advanced warning of turbidity breakthrough afforded by particle counting from a full-scale filter at a different plant site. In this case, the particle counts reveal reduced filter performance 5 to 10 hours before turbidity.

Notice the comparative slopes of the particle counts and turbidity responses. Characteristically, particle counts show a gradual upward trend leading to breakthrough, with a steep drop-off after backwash, whereas turbidity shows a sudden steep increase at breakthrough with a long, gradual drop-off after backwash. Turbidity meters actually see a lot of very small particles that are not measured by particle counters because the sensitivity of the particle counter selected was purposely limited to 2 microns. A particle counter with a higher sensitivity (a sensitivity to less than 2 microns) would be capable of seeing these smaller particles.

Scenario 2, Damaged Membrane Filter

Figures 5.22, 5.23, and 5.24 compare the feed and permeate values of turbidity, particle counts, and *Giardia* cysts for both a compromised (damaged) and an intact membrane filter.

During the first 61 days of the test, the ultrafiltration (UF) membrane filter had two broken fibers. The effluent turbidity always remained below 0.1 NTU, but the permeate particle count read as high as 5,000 particles per mL and *Giardia* cyst recovery was as high as 400 per sample. After day 61, a new, intact UF membrane was installed. Thereafter, all measurements showed minimal effluent detection levels. These results show that even two broken membrane fibers were revealed by a 3 order of magnitude change in particle counts, while the turbidity meter was unable to conclusively detect the failure.

5.12.5   Operation and Maintenance

For a particle counter to be useful, it needs a reliable sample handling and delivery system and a smooth flow delivery system. This requires particle counters to be maintained and operated correctly. Carefully checking procedures during the initial startup and preventing overconcentration will ensure particle counters operate properly. Knowledgeable troubleshooting and good QA/QC procedures also reduce operation and maintenance problems and increase particle counter efficiency and accuracy. Other ways to increase the dependable use of particle counters include using the recommended sample tubing and preventing or minimizing bubbles in the sample line.

5.12.5.1 Sampling

The high sensitivity of the particle counter makes sample handling and delivery critical for proper operation. Every sample collected must be representative of the process stream. Particles can be added to the sample stream if the sample tap is located at the bottom of a pipe where sediment accumulates, or particles can be lost from the sample stream in long sample lines where particles may settle out due to low flow rates and flow velocities. Particle distribution can be altered by sample pumps chopping up larger particles and creating more small particles.

5.12.5.2 Flow Control

Because all particle count data must be based on sample volume, flow control is critical for accurate and repeatable performance. The simplest and most effective way to achieve constant flow is with an overflow weir (Figures 5.25 and 5.26). As long as enough flow is delivered to maintain some overflow in the weir, a constant flow will be present in the sensor. If this requirement is maintained, flow will only be altered by clogging of the sensor flow cell. The flow rate through a particle counter should be monitored daily to ensure that the flow is within 5 percent of the calibrated flow rate specified by the manufacturer. Where flow measurement and control are better than 5 percent of the calibrated flow rate, operators can use electronic flow monitoring to meter the flow and to activate a low flow alarm.

5.12.5.3 Sample Tubing

Black nylon tubing with a Hytril liner is recommended for sample tubing used with in-process instruments. Hytril has the slickness of Teflon, thus particle accumulation in the tubing is minimized. The black outer shell of the tubing discourages biological growths in the tube. Black tubing is definitely preferred when monitoring very clean samples with fewer than 10 particles per milliliter. Hytril is expensive, is somewhat stiff, and loses some flexibility with use.

When using grab samplers and for some lab applications, operators often use short lengths of Tygon tubing because it is easy to work with, flexible, and inexpensive. Also, operators can see the sample flow to the particle counter.

5.12.5.4 Bubbles

Bubbles can be introduced into the sample if air is pulled into the sample line or if the temperature of the sample water is allowed to increase while

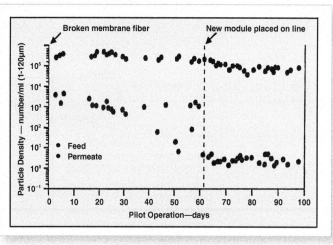

Figure 5.23 Loss of membrane integrity monitored by particle counts (*Courtesy of Met One Inc, a division of HACH Company*)

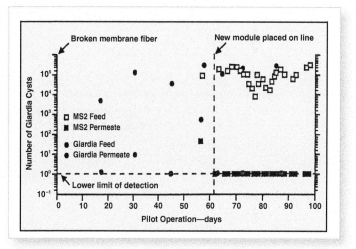

Figure 5.24 Loss of membrane integrity monitored by *Giardia* cysts (*Courtesy of Met One Inc, a division of HACH Company*)

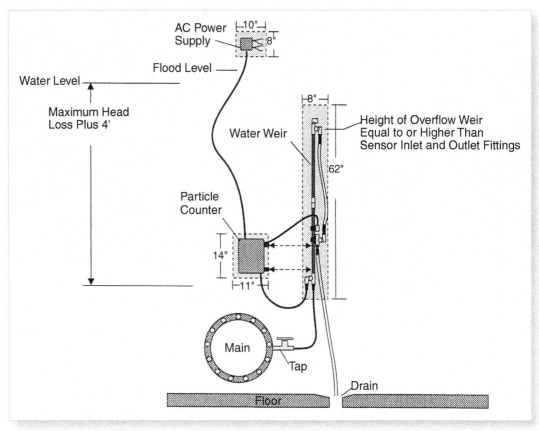

Figure 5.25   Particle counter mounting diagram   (*Courtesy of Met One Inc, a division of HACH Company*)

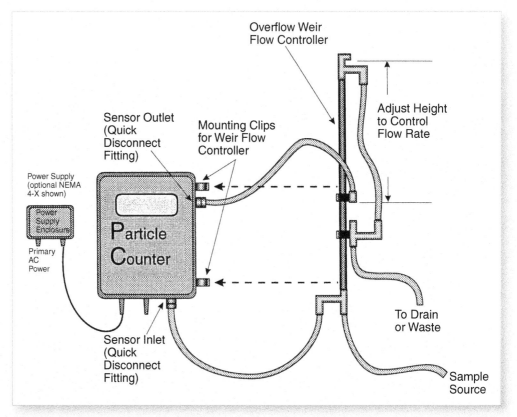

Figure 5.26   Particle counter connection to flow control weir (*Courtesy of Met One Inc, a division of HACH Company*)

in the sample line. The particle counter will count bubbles if they are large enough, just as a turbidity meter will.

Overflow weirs must be specifically designed to exhaust gas bubbles in all but the most severe cases. Usually a "debubbling" weir design is sufficient. Operators may encounter some samples with such severe air entrainment that other measures are necessary. For example, some operators provide high pressure to the sensor and all flow downstream from the sensor, thus minimizing pressure drop across the cell and minimizing the corresponding tendency of gas bubbles to come out of solution in the sensing area. However, this might involve eliminating the weir altogether. Problems such as these need to be resolved by working with the particle counter supplier.

5.12.5.5 Initial Startup

Be sure that all fittings used in making connections in the sample lines are noncorrosive. Never allow dissimilar metals to be connected. For example, ductile iron, galvanized steel, and brass bushings should never be combined in a single fitting or coupling, When looking for just a few particles per mL, the slightest amount of corrosion from the contact of any dissimilar metals can cause a sample error.

Install or place the particle counter as close as practical to the sample point in the flow stream to minimize the length of sample tubing and to ensure rapid response time of the counting of the particles in the sample.

Be sure to flush out all taps and valves before connecting the particle sensor. Sometimes taps or valves have not been used recently and can release a slug of debris that will clog the sensor. Once the installation is complete, all sample lines should be flushed for a short time. If the particle counter output is being recorded, it is easy to determine when the sample lines have been sufficiently flushed because the particle counts per mL will level out or become steady.

5.12.5.6 Overconcentration

Always monitor the total particle count over the entire monitoring range of the particle counter to ensure that the total particle count is not exceeding the limit of the sensor. If the limit of the sensor is being exceeded, dilution will be necessary. Dilution of samples for in-process particle counters may be difficult. Contact the instrument manufacturer for assistance. An automated in-process dilution system may be available from some manufacturers.

5.12.5.7 Troubleshooting

The first step is to check that the power is ON. Nearly all difficulties with a particle counter can be solved by cleaning the sample lines. If the particle concentration appears low, check for clogging of the sample line. If the particle concentration appears high, check for contamination, air bubbles, or an increased flow rate through the sensor.

5.12.5.8 Quality Assurance/Quality Control

Frequently, verify the baseline of the particle counter. When analyzing particle-free water, the particle counter should indicate less than one particle per mL in the 2 to 150 μm size range. Instrument calibration should be verified by the manufacturer annually. With multiple-sensor systems, verify that the sensors give similar results (within plus or minus 10 percent) with the same water source. This can be a time-consuming process until operators gain experience and confidence in particle counters.

Check Your Understanding

1. Particle counters can be used as a substitute for indicating the potential removal of what two microorganisms that are a threat to public health when found in drinking water?

2. Particle counters can be used to enhance the performance of which water treatment process?

3. What is the difference between turbidity meter and particle counter measurements?

5.13   Math Assignment

Turn to Appendix A, "Introduction to Basic Math for Operators." Read and work through the example problems in sections:

- "Learning Objectives"
- A.5.8, "Force and Pressure"
- A.5.9, "Work, Head, and Power"
- A.9.7, "Filtration" (English System)
- A.10.7, "Filtration" (Metric System)

Check the math using a calculator. You should be able to get the same answers.

5.14   Additional Resources

Your local library may be a good source for obtaining these additional resources related to the field of water treatment:

Manual of Water Utilities Operations (Texas Manual). Texas Water Utilities Association, www.twua.org. ASIN: B000KT9EIU. Chapter on filtration.

Key Term Matching

_____ 1. Breakthrough

_____ 2. Mudball

_____ 3. Diatomaceous Earth

_____ 4. Direct Filtration

_____ 5. Turbidity Meter

_____ 6. Head Loss

_____ 7. Adsorption

_____ 8. Backwashing

_____ 9. Piezometer

_____ 10. Submergence

A. The gathering of a gas, liquid, or dissolved substance on the surface or interface zone of another material.

B. The process of reversing the flow of water back through the filter media to remove the entrapped solids.

C. A crack or break in a filter bed allowing floc or particulate matter to pass through a filter, causing increased filter effluent turbidity.

D. Fine, siliceous (made of silica) earth composed mainly of the skeletal remains of diatoms.

E. A method of treating water that consists of the addition of coagulant chemicals, flash mixing, coagulation, minimal flocculation, and filtration.

F. The head, pressure, or energy lost by water flowing in a pipe or channel due to turbulence from the velocity of the flowing water and the roughness of the pipe, channel walls, or fittings restrictions.

G. Material round in shape that forms in filters and gradually increases in size when not removed by the backwashing process.

H. An instrument used to measure the pressure head in a pipe, tank, or soil. Can measure the pore pressure of water in the soil.

I. The distance between the water surface and the media surface in a filter.

J. An instrument that measures and compares the turbidity of liquids by passing light through them and determining how much light is reflected by the particles in the liquid.

5.2 Types of Filters

11. Why is diatomaceous earth filtration used in some water treatment applications and not in others?
    1. Because it is expensive
    2. Because it is explosive
    3. Limited use in larger treatment plants due to hydraulic (flow), sludge disposal, and other operational conditions
    4. Limited use in smaller treatment plants due to hydraulic (flow), sludge disposal, and other operational conditions

5.3 Process Performance Considerations

12. What is a desirable characteristic for filter media?
    1. Hard and durable
    2. Reacts with substances in water
    3. Soft and flexible
    4. Soluble in water
13. What is one solution if both head loss and turbidity breakthrough are constantly a problem due to improperly sized media?
    1. A deeper filter bed with a larger media size
    2. A deeper filter bed with a smaller media size
    3. A shallower filter bed with a larger media size
    4. A shallower filter bed with a smaller media size
14. What problems can develop from improper backwash rates?
    1. An insufficient backwash may not completely remove trapped solids from the filter media
    2. An insufficient backwash rate may cause excessive loss of filter media and media disturbance
    3. Too high a backwash rate may add prohibitive expense to the filter process due to water costs
    4. Too high a backwash rate may not clean the filter adequately

5.4 Activated Carbon Filters

15. What is the primary purpose of the filtration process?
    1. To add contact time with disinfectants
    2. To add particulates and floc to the water being treated
    3. To keep expenses down
    4. To remove particulate impurities and floc from water being treated

5.7 Operating Procedures Associated with Normal Process Conditions

16. Which instrument would you use to monitor a filtration process?
    1. Backwash pump
    2. Gas detector
    3. Surface washer
    4. Turbidity meter
17. How frequently should the performance of the filtration process be evaluated?
    1. Daily
    2. Effluent, every 2 hours; influent, continuously
    3. Influent, every 2 hours; effluent, continuously
    4. Weekly
18. What would you look for in evaluating the filter media condition after a backwash cycle?
    1. Backwash pump overheating
    2. Cracking, media loss, and mudballs
    3. Higher backwash pump rates
    4. Lower piezometer readings

19. Calculate the filtration rate in gallons per minute per square foot (gal/min/ft²) for a filter with a surface length of 50 ft and a width of 25 ft when the applied flow is 4 Mgal/d.
    1. 2.1 gpm/ft²
    2. 2.2 gpm/ft²
    3. 2.3 gpm/ft²
    4. 2.4 gpm/ft²

5.8 Operating Procedures Associated with Abnormal Process Conditions

20. Which item is an indicator of an upset or failure in the filtration process or pretreatment process?
    1. Excessive media loss
    2. Filters that come clean during backwash
    3. Long filter runs
    4. Steady head loss
21. How could you adjust the filter operation if increases in source water turbidity and resultant increases in coagulant feed rates impose a greater load on the plant's active filters?
    1. Backwash filters less frequently
    2. Decrease filtration rates
    3. Increase filtration rates
    4. Take some filters out of service
22. What backwash process adjustments would you make to solve such problems as media boils, media loss, or incomplete cleaning of media?
    1. Adjust the backwash flow rates, surface wash flow rate, or duration
    2. Adjust the chemical additions involved
    3. In the case of a nozzle failure, adjust the nozzle for full force so the blockage is cleared
    4. Replace the backwash pump fittings

5.10 Process and Support Equipment Operation and Maintenance

23. What should you not check for after starting any piece of equipment?
    1. Be sure that the unit has been serviced on schedule
    2. Be sure that the unit's operational status is positively known
    3. Check for excessive noise and vibration, overheating, and leakage
    4. Check that the manufacturer's warranty is still in effect

5.11 Drinking Water Regulations

24. Name a type of filtration that can be used to meet the requirements of the Surface Water Treatment Rule.
    1. Diatomaceous earth filtration
    2. Fast sand filtration
    3. Indirect filtration
    4. Unconventional filtration

5.12 Particle Counters

25. How can particle counters be used to monitor filter performance?
    1. Measure amount of finished water
    2. Measure effects of filter flow rate
    3. Present an informative picture of influent quality
    4. Review the disinfection process

CHAPTER 6

Disinfection

LEARNING OBJECTIVES

1. Explain the disinfection process using chlorine, hypochlorite, chlorine dioxide, and chloramines, as well as ultraviolet (UV) and ozone systems.
2. Describe the breakpoint chlorination process.
3. Identify the points of disinfectant application and select the proper dosage.
4. Start up, shut down, troubleshoot, and maintain disinfection equipment and systems.
5. Handle disinfectants safely and develop and conduct a chlorine safety program.

KEY TERMS

air gap
ambient temperature
amperometric titration
bacteria
ballast
breakpoint chlorination
buffer capacity
carcinogen
chloramination

chloramines
chlorination
chlorine demand
chlorine residual
chlorophenolic
chlororganic
coliform
combined available
     chlorine

combined chlorine
   residual
Cryptosporidium
DPD method
dew point
diatoms
disinfection
eductor
ejector

electron
enzymes
free available residual
   chlorine
HTH
heterotrophic
hypochlorite
IDLH
injector water

nitrification
oxidation
oxidizing agent
Pascal (Pa)
pathogenic organisms
phenolic compounds

postchlorination
potable water
prechlorination
precursor, THM
reducing agent
reduction

reliquefaction
residual chlorine
rotameter
sterilization
titrate
total chlorine

total chlorine residual
trihalomethanes (THMS)
turbidity
ultraviolet (UV)

6.1   Drinking Water Safety

Our single most important natural resource is water. Without water we could not exist. Unfortunately, safe water is becoming harder to find. In the past, safe water could be found in remote areas, but with population growth and related pollution of waters, there are few natural waters left that are safe to drink without treatment of some kind.

Water is the universal solvent and, therefore, carries all types of dissolved materials. Water also carries biological life forms that can cause diseases. These waterborne pathogenic organisms can cause the types of diseases listed in Table 6.1. Most of these organisms and the diseases they transmit are no longer a problem in the United States due to proper water protection, treatment, and monitoring. However, many developing regions of the world still experience serious outbreaks of various waterborne diseases.

One of the cleansing processes in the treatment of safe water is called disinfection. Disinfection is the process designed to kill or inactivate most microorganisms in water or wastewater, including essentially all pathogenic (disease-causing) bacteria. There are several ways to disinfect, with chlorination being the most frequently used in water and wastewater treatment plants. Do not confuse disinfection with sterilization. Sterilization is the complete destruction of all organisms. Sterilization is not necessary in water treatment and is quite expensive.

Table 6.1   Diseases transmitted by pathogenic organisms in water

Bacteria-Caused	Internal Parasite-Caused
Anthrax	Amoebic dysentery
Bacillary dysentery	Ascariasis
Cholera	Cryptosporidiosis
Gastroenteritis	Giardiasis
Leptospirosis	**Virus-Caused:**
Paratyphoid	Gastroenteritis
Salmonellosis	Heart anomalies
Shigellosis	Infectious hepatitis
Typhoid fever	Typhoid fever
	Poliomyelitis

pathogenic (PATH-o-JEN-ick) organisms
Organisms, including bacteria, viruses, protozoa, or internal parasites, capable of causing disease (such as giardiasis, cryptosporidiosis, typhoid fever, cholera, or infectious hepatitis) in a host (such as a person). Nonpathogenic organisms do not cause disease.

6.1.1   Safe Drinking Water Laws

In the United States, the US Environmental Protection Agency (EPA) is responsible for setting drinking water standards and for ensuring their enforcement. This agency sets federal regulations that all state and local agencies must enforce. The Safe Drinking Water Act (SDWA) and its amendments contain specific maximum allowable levels of substances known to be hazardous to human health. In addition to describing maximum contaminant levels (MCLs), these federal drinking water regulations also give detailed instructions on what to do when the MCL for a particular substance is exceeded.

The Surface Water Treatment Rule (SWTR) requires disinfection of all surface water supply systems as protection against exposure to viruses, **bacteria**, and *Giardia*. Table 6.2 shows the total coliform sampling requirements according to the population served. Table 6.3 shows an example of the regulations for **coliform** bacteria and viruses to be killed by disinfection. See the chapter on drinking water regulations in Volume II for details on the SDWA's rules and regulations.

Table 6.2   Total coliform sampling requirements according to population served (Federal Register, February 13, 2013, Part II, Environmental Protection Agency, 40 CFR, Parts 141 and 142, National Primary Drinking Water Regulations: Revisions to the Total Coliform Rule; Final Rule. www.gpo.gov. [p. 10361])

Population Served			Minimum Number of Routine Samples Per Month	Population Served			Minimum Number of Routine Samples Per Month
1,001	to	2,500	2	70,001	to	83,000	80
2,501	to	3,300	3	83,001	to	96,000	90
3,301	to	4,100	4	96,001	to	130,000	100
4,101	to	4,900	5	130,001	to	220,000	120
4,901	to	5,800	6	220,001	to	320,000	150
5,801	to	6,700	7	320,001	to	450,000	180
6,701	to	7,600	8	450,001	to	600,000	210
7,601	to	8,500	9	600,001	to	780,000	240
8,501	to	12,900	10	780,001	to	970,000	270
12,901	to	17,200	15	970,001	to	1,230,000	300
17,201	to	21,500	20	1,230,001	to	1,520,000	330
21,501	to	25,000	25	1,520,001	to	1,850,000	360
25,001	to	33,000	30	1,850,001	to	2,270,000	390
33,001	to	41,000	40	2,270,001	to	3,020,000	420
41,001	to	50,000	50	3,020,001	to	3,960,000	450
50,001	to	59,000	60	3,960,001	or	more	480
59,001	to	70,000	70				

Drinking water regulations are constantly changing. The Interim Enhanced Surface Water Treatment Rule (IESWTR) and the Disinfectant/ Disinfection By-Products (D/DBP) Rule were passed in 1998 and further modifications of these rules have been developed. The goal of the IESWTR was to increase public protection from illness caused by the *Cryptosporidium* organism. The D/DBP Rule, which applies to water systems using a disinfectant during treatment, limits the amount of certain potentially harmful disinfection byproducts that may remain in drinking water after treatment.

Water systems serving 10,000 or more people had 3 years (from December 16, 1998) to comply with both the D/DBP and the IESWTR regulations. Small systems (serving fewer than 10,000 people) had 5 years to comply.

bacteria (back-TEAR-e-ah)
Bacteria are living organisms, microscopic in size, that usually consist of a single cell. Most bacteria use organic matter for their food and produce waste products as a result of their life processes.

coliform (COAL-i-form)
A group of bacteria found in the intestines of warmblooded animals (including humans) and in plants, soil, air, and water. The presence of coliform bacteria is an indication that the water is polluted and may contain pathogenic (disease-causing) organisms. Fecal coliforms are those coliforms found in the feces of various warm-blooded animals, whereas the term coliform also includes other environmental sources.

Cryptosporidium
A waterborne intestinal parasite that causes a disease called cryptosporidiosis (KRIP-toe-spo-rid-ee-O-sis) in infected humans. Symptoms of the disease include diarrhea, cramps, and weight loss. *Cryptosporidium* contamination is found in most surface waters and some groundwaters. Commonly referred to as crypto.

Table 6.3 Microbial standards and sampling requirements (Federal Register, Volume 78, No. 30, February 13, 2013, Environmental Protection Agency, 40 CFR, Parts 141 and 142, National Primary Drinking Water Regulations: Revisions to the Total Coliform Rule; Final Rule. www.gpo.gov. [p. 10269])

Constituent	Maximum Contaminant Level (MCL)
Microbial	
• Total coliform	1 per 100 mL
	<40 samples/mo—no more than 1 positive
	>40 samples/mo—no more than 5% positive
• *Giardia lamblia*	3-log (99.9%) removal[a]
• Legionella	treatment technique[a]
• Enteric viruses	4-log (99.99%) removal[a]
• Heterotrophic bacteria	treatment technique[a]

Routine Sampling Requirements

- Total coliform samples must be collected by Public Water Systems (PWSs) at sites which are representative of water quality throughout the distribution system according to a written sample siting plan subject to state review and revision.
- For PWSs collecting more than one sample per month, collect total coliform samples at regular intervals throughout the month, except that groundwater systems serving 4,900 or fewer people may collect all required samples on a single day if the samples are taken from different sites.
- Each total coliform-positive (TC+) routine sample must be tested for the presence of *E. coli*.
- If any TC+ sample is also *E. coli*-positive (EC+), then the EC+ sample result must be reported to the state by the end of the day that the PWS is notified.
- If any routine sample is TC+, repeat samples are required.

  PWSs on quarterly or annual monitoring must take a minimum of three additional routine samples (known as additional routine monitoring) the month following a TC+ routine or repeat sample.
- Reduced monitoring may be available for PWSs using only groundwater and serving 1,000 or fewer persons that meet certain additional PWS criteria.

Repeat Sampling Requirements

• *Within 24 hours of learning of a TC+ routine sample result, at least 3 repeat samples must be collected and analyzed for total coliform:*	• One repeat sample must be collected from the same tap as the original sample.
	• One repeat sample must be collected from within five service connections upstream.
	• One repeat sample must be collected from within five service connections downstream.
	• The PWS may propose alternative repeat monitoring locations that are expected to better represent pathways of contamination into the distribution system.
• *If one or more repeat samples is TC+:*	• The TC+ sample must be analyzed for the presence of *E. coli*.
	• If any repeat TC+ sample is also EC+, then the EC+ sample result must be reported to the state by the end of the day that the PWS is notified.
	• The PWS must collect another set of repeat samples, unless an assessment has been triggered and the PWS has notified the state.

[a]Applies to systems using surface water or groundwater under the influence of surface water.

The Long Term 2 Enhanced Surface Water Treatment Rule (LT2ESWTR) builds upon earlier rules to address higher risk public water systems for protection measures beyond those required for previous regulations. The purpose of the LT2ESWTR is to reduce illness linked with *Cryptosporidium* and other pathogenic microorganisms in drinking water. The LT2ES-WTR was implemented simultaneously with the Stage 2 Disinfection By-Products (Stage 2 D/DBP) Rule to address concerns about risk trade-offs between pathogens and DBPs.

> Drinking water rules and regulations are continually changing. Keep in contact with your state drinking water agency to obtain the rules and regulations that currently apply to your water utility. For additional information or answers to specific questions about the regulations, phone EPA's Safe Drinking Water Hotline at (800) 426-4791.

Check Your Understanding

1. What are pathogenic organisms?

2. What is disinfection?

3. Drinking water standards are established by what agency of the United States government?

4. MCL stands for what words?

6.2   Factors Influencing Disinfection

Many factors influence a successful disinfection program at a water treatment plant. These factors include pH, temperature, turbidity, reducing agents, and microorganisms.

6.2.1   pH

The pH of water being treated can alter the efficiency of disinfectants. Chlorine, for example, disinfects water much faster at a pH around 7.0 than at a pH over 8.0.

6.2.2   Temperature

Temperature conditions also influence the effectiveness of the disinfectant. The higher the temperature of the water, the more efficiently it can be treated. Water near 70 to 85°F (21 to 29°C) is easier to disinfect than water at 40 to 60°F (4 to 16°C). Longer contact times are required to disinfect water at lower temperatures. To speed up the process, operators often simply use larger amounts of chemicals. Be aware, though, that the higher the chlorine concentration, the greater the dissipation rate of chlorine into the atmosphere. This can produce odors and wastes chlorine.

6.2.3   Turbidity

Under normal operating conditions, the **turbidity** level of water being treated is very low by the time the water reaches the disinfection process. Excessive turbidity will greatly reduce the efficiency of the disinfecting chemical or process. Studies in water treatment plants have shown that when water is filtered to a turbidity of one unit or less, most of the bacteria have been removed.

turbidity (ter-BID-it-tee)
The cloudy appearance of water caused by the presence of suspended and colloidal matter. In the waterworks field, a turbidity measurement is used to indicate the clarity of water. Technically, turbidity is an optical property of the water based on the amount of light reflected by suspended particles. Turbidity cannot be directly equated to suspended solids because white particles reflect more light than dark-colored particles and many small particles will reflect more light than an equivalent large particle.

The suspended matter itself may also change the chemical nature of the water when the disinfectant is added. Some types of suspended solids can create a continuing demand for the chemical, thus changing the effective germicidal (germ-killing) properties of the disinfectant.

6.2.3.1   Organic Matter

Organics found in the water can consume great amounts of disinfectants while forming unwanted compounds. **Trihalomethanes (THMs)** are an example of undesirable compounds formed by reactions between chlorine and certain organics. Disinfecting chemicals often react with organics and **reducing agents** (Section 6.2.4). Then, if any of the chemical remains available after this initial reaction, it can act as an effective disinfectant. The reactions with organics and reducing agents, however, will have significantly reduced the amount of chemical available for disinfection.

6.2.3.2   Inorganic Matter

Inorganic compounds, such as ammonia (NH_3) in the water being treated, can create special problems. In the presence of ammonia, some oxidizing chemicals form side compounds causing a partial loss of disinfecting power. Silt can also create a chemical demand. It is clear, then, that the chemical properties of the water being treated can seriously interfere with the effectiveness of disinfecting chemicals.

6.2.4   Reducing Agents

Chlorine combines with a wide variety of materials, especially reducing agents. Most of the reactions are rapid, though others are much slower. These side reactions complicate the use of chlorine for disinfection. The demand for chlorine by reducing agents must be satisfied before chlorine becomes available to disinfect. Examples of inorganic reducing agents present in water that will react with chlorine include hydrogen sulfide (H_2S), ferrous ion (Fe^{2+}), manganous ion (Mn^{2+}), ammonia (NH_3), and the nitrite ion (NO_2^-). Organic reducing agents in water also will react with chlorine and form chlorinated organic materials of potential health significance.

6.2.5   Microorganisms

The concentration of microorganisms is important because the higher the number of microorganisms, the greater the demand for a disinfecting chemical. The resistance of microorganisms to specific disinfectants varies greatly. Non-spore-forming bacteria are generally less resistant than spore-forming bacteria. Cysts and viruses can be resistant to certain types of disinfectants.

6.2.5.1   Removal Processes

Pathogenic organisms can be removed from water, killed, or inactivated by various physical and chemical water treatment processes. These processes are:

1. Coagulation—Chemical coagulation followed by sedimentation and filtration will remove 90 to 95 percent of the pathogenic organisms, depending on which chemicals are used. Alum usage can increase virus removals up to 99 percent.

---

trihalomethanes (THMs)
(tri-HAL-o-METH-hanes)
Derivatives of methane, CH_4, in which three halogen atoms (chlorine or bromine) are substituted for three of the hydrogen atoms. Often formed during chlorination by reactions with natural organic materials in the water. The resulting compounds (THMs) are suspected of causing cancer.

reducing agent
Any substance, such as base metal (iron) or the sulfide ion (S^{2-}), that will readily donate (give up) electrons. The opposite is an *oxidizing agent*.

2. Sedimentation—Properly designed sedimentation processes can effectively remove 20 to 70 percent of the pathogenic microorganisms. This removal is accomplished by allowing the pathogenic organisms (as well as non-pathogenic organisms) to settle out by gravity, assisted by chemical floc.

3. Filtration—Filtering water through granular filters is an effective means of removing pathogenic and other organisms from water. The removal rates vary from 20 to 99+ percent, depending on the coarseness of the filter media and the type and effectiveness of pretreatment.

4. Disinfection—Disinfection chemicals, such as chlorine, are added to water to kill or inactivate pathogenic microorganisms.

In previous chapters, you have already studied the first three processes. The fourth, disinfection, is the subject of this chapter.

Check Your Understanding

1. How does pH influence the effectiveness of disinfection?

2. How does the temperature of the water influence disinfection?

3. What does chlorine produce when it reacts with organic matter?

4. What two factors influence the effectiveness of disinfection on microorganisms?

6.3   Disinfection Process

Disinfection destroys harmful organisms. This can be accomplished either physically or chemically. Physical methods may: (1) physically remove the organisms from the water, or (2) introduce motion that will disrupt the cells' biological activity and kill or inactivate them.

Chemical methods alter the cell chemistry causing the microorganism to die. The most widely used disinfectant chemical is chlorine. Chlorine is easily obtained, relatively inexpensive, and most importantly, leaves a **residual chlorine** that can be measured. Other disinfectants are also used. There has been increased interest in disinfectants other than chlorine because of the **carcinogenic** compounds that chlorine may form (trihalomethanes or THMs).

This chapter will focus primarily on the use of chlorine as a disinfectant. However, let us take a brief look first at other disinfection methods and chemicals. Some of these are being more widely applied today because of the potential adverse side effects of **chlorination**.

6.3.1   Physical Means of Disinfection

1. *Ultraviolet rays* can be used to destroy pathogenic microorganisms. To be effective, the rays must come in contact with each microorganism. The ultraviolet energy disrupts various organic components of the cell causing a biological change that is fatal to the microorganisms.

residual chlorine
The concentration of chlorine present in water after the chlorine demand has been satisfied. The concentration is expressed in terms of the total chlorine residual, which includes both the free and combined or chemically bound chlorine residuals. Also called chlorine residual.

carcinogen (CAR-sin-o-JEN)
Any substance that tends to produce cancer in an organism.

chlorination (KLOR-uh-NAY-shun)
The application of chlorine to water, generally for the purpose of disinfection, but frequently for accomplishing other biological or chemical results (aiding coagulation and controlling tastes and odors).

ultraviolet (UV)
Pertaining to a band of electromagnetic radiation just beyond the visible light spectrum. Ultraviolet radiation is used in water treatment to disinfect the water. When ultraviolet radiation is absorbed by the cells of microorganisms, it damages the genetic material in such a way that the organisms are no longer able to grow or reproduce, thus ultimately killing them.

This system has not had widespread acceptance because of the lack of measurable residual and the cost of operation. Currently, use of ultraviolet rays is limited to small or local systems and industrial applications. Oceangoing ships have used these systems for their water supply.

Advances in UV technology and concern about disinfection byproducts (DBPs) produced by other disinfectants have prompted a renewed interest in UV disinfection. See Section 6.10, "Ultraviolet (UV) Systems," for a more detailed description of this process.

2. *Heat* has been used for centuries to disinfect water. Boiling water for about 5 minutes will destroy essentially all microorganisms. This method is energy intensive and thus expensive. The only practical application is in the event of a disaster when individual local users are required to boil their water.

3. *Ultrasonic* Waves have been used to disinfect water on a limited scale. Sonic waves destroy microorganisms by vibration. This procedure is not yet practical and is expensive.

6.3.2   Chemical Disinfectants (Other Than Chlorine)

1. *Iodine* has been used as a disinfectant in water since 1920, but its use has been limited to emergency treatment of water supplies. Although it has long been recognized as a good disinfectant, iodine's high cost and potential physiological effects (pregnant women can suffer serious side effects) have prevented widespread acceptance. The recommended dosage is two drops of iodine (tincture of iodine which is 7 percent available iodine) in a liter of water.

2. *Bromine* has been used only on a very limited scale for water treatment because of its handling difficulties. Bromine causes skin burns on contact. Because bromine is a very reactive chemical, residuals are hard to obtain. This also limits its use. Bromine can be purchased at swimming pool supply stores.

3. *Bases*, such as sodium hydroxide and lime, can be effective disinfectants but the high pH leaves a bitter taste in the finished water. Bases can also cause skin burns when left too long in contact with the skin. Bases effectively kill all microorganisms (they sterilize rather than just disinfect water). Although this method has not been used on a large scale, bases have been used to sterilize water pipes.

4. *Ozone* has been used in the water industry since the early 1900s, particularly in France. In the United States, it has been used primarily for taste and odor control. The limited use in the United States has been due to its high costs, lack of residual, difficulty in storing, and maintenance requirements.

Although ozone is effective in disinfecting water, its use is limited by its solubility. The temperature and pressure of water being treated regulate the amount of ozone that can be dissolved in the water. These factors tend to limit the disinfectant strength that can be made available to treat the water.

Many scientists claim that ozone destroys all microorganisms. Unfortunately, significant residual ozone does not guarantee that a water is safe to drink. Organic solids may protect organisms from the disinfecting action of ozone and increase the amount of ozone needed for disinfection. In addition, ozone residuals cannot be maintained in metallic conduits for any period of time because of ozone's reactive nature. The inability of ozone to provide a residual in the distribution system is a major drawback to its use. However, recent information about the formation of THMs by chlorine compounds has resulted in renewed interest in ozone as an alternative means of disinfection. See Section 6.11, "Ozone Systems," for a description of disinfection using ozone.

Check Your Understanding

1. List the physical agents that have been used for disinfection.

2. List the chemical agents other than chlorine that have been used for disinfection.

3. What is a major limitation to the use of ozone?

6.3.3   Chlorine

Chlorine (Cl_2) is a greenish-yellow gas with a penetrating and distinctive odor. The gas is two-and-a-half times heavier than air. Chlorine has a very high coefficient of expansion. If there is a temperature increase of 50°F (28°C) (from 35°F to 85°F or 2°C to 30°C), the volume will increase by 84 to 89 percent. This expansion could easily rupture a cylinder or a line full of liquid chlorine. For this reason, no chlorine containers should be filled to more than 85 percent of their capacity. One liter of liquid chlorine can evaporate and produce 450 liters of chlorine gas.

Chlorine by itself is nonflammable and nonexplosive, but it will support combustion. When the temperature rises, so does the vapor pressure of chlorine. This means that when the temperature increases, the pressure of the chlorine gas inside a chlorine container will increase. This property of chlorine must be considered when:

1. Feeding chlorine gas from a container
2. Dealing with a leaking chlorine cylinder

6.3.3.1   Chlorine Disinfection Action

The exact mechanism of chlorine disinfection action is not fully known. One theory holds that chlorine exerts a direct action against the bacterial cell, thus destroying it. Another theory is that the toxic character of chlorine inactivates the enzymes, which enable living microorganisms to use their food supply. As a result, the organisms die of starvation. From the point of view of water treatment, the exact mechanism of chlorine disinfection is less important than its demonstrated effects as a disinfectant.

When chlorine is added to water, several chemical reactions take place. Some involve the molecules of the water itself, and some involve organic and inorganic substances suspended in the water. We will discuss these chemical reactions in more detail in the next few sections of this chapter. First, however, there are some terms associated with chlorine disinfection that you should understand.

enzymes (EN-zimes)
Organic substances (produced by living organisms) that cause or speed up chemical reactions. Organic catalysts or biochemical catalysts.

When chlorine is added to water containing organic and inorganic materials, it will combine with these materials and form chlorine compounds. If you continue to add chlorine, you will eventually reach a point where the reaction with organic and inorganic materials stops. At this point, you have satisfied what is known as the **chlorine demand**.

The chemical reactions between chlorine and these organic and inorganic substances produce chlorine compounds. Some of these compounds have disinfecting properties; others do not. In a similar fashion, chlorine reacts with the water itself and produces some substances with disinfecting properties. The total of all the compounds with disinfecting properties plus any remaining free (uncombined) chlorine is known as the **chlorine residual**. The presence of this measurable chlorine residual indicates to the operator that all possible chemical reactions have taken place and that there is still sufficient available residual chlorine to kill the microorganisms present in the water supply.

If you add together the amount of chlorine needed to satisfy the chlorine demand (Cl_{demand}) and the amount of chlorine residual ($Cl_{residual}$) needed for disinfection, you will have the chlorine dose (Cl_{dose}). This is the amount of chlorine you will have to add to the water to disinfect it.

$$Cl_{dose}\left[\frac{mg}{L}\right] = Cl_{demand}\left[\frac{mg}{L}\right] + Cl_{residual}\left[\frac{mg}{L}\right]$$

or

$$Cl_{demand}\left[\frac{mg}{L}\right] = Cl_{dose}\left[\frac{mg}{L}\right] - Cl_{residual}\left[\frac{mg}{L}\right]$$

and

$$Cl_{residual}\left[\frac{mg}{L}\right] = Cl_{combined\ forms}\left[\frac{mg}{L}\right] + Cl_{free}\left[\frac{mg}{L}\right]$$

Where,
$Cl_{combined}$ = Combined Chlorine Forms
Cl_{free} = Free Chlorine

6.3.3.2   Reaction with Water

Free chlorine combines with water to form hypochlorous and hydrochloric acids:

Chlorine + Water ⇌ Hypochlorous Acid + Hydrochloric Acid
$$Cl_2 + H_2O \rightleftharpoons HOCl + HCl$$

In solutions that are dilute (low concentration of chlorine) and have a pH above 4, the formation of HOCl (hypochlorous acid) is almost complete and leaves little free chlorine (Cl_2).

Depending on the pH, some hypochlorous acid will disassociate (break up) and produce a hydrogen ion and a **hypochlorite** ion. Hypochlorous acid is a weak acid and hence is poorly dissociated at pH levels below 6. As shown in Figure 6.1, below pH 6, the free chlorine (HOCl + OCl⁻)is

chlorine demand
Chlorine demand is the difference between the amount of chlorine added to water or wastewater and the amount of chlorine residual remaining after a given contact time. Chlorine demand may change with dosage, time, temperature, pH, and nature and amount of the impurities in the water.

Chlorine Demand = Chlorine Applied − Chlorine Residual

chlorine residual
The concentration of chlorine present in water after the chlorine demand has been satisfied. The concentration is expressed in terms of the total chlorine residual, which includes both the free and combined or chemically bound chlorine residuals. Also called residual chlorine.

hypochlorite (HI-poe-KLOR-ite)
Chemical compounds containing available chlorine; used for disinfection. They are available as liquids (bleach) or solids (powder, granules, and pellets) and are packaged and shipped by various methods.

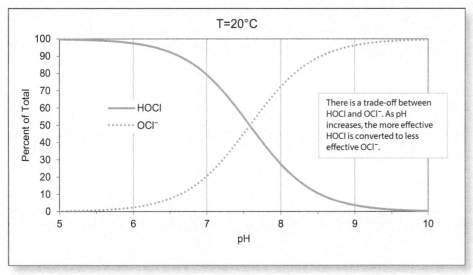

Figure 6.1 Relationship between hypochlorous acid (HOCl), hypochlorite ion (OCl⁻), and pH

almost all in the HOCl form. Above pH 9, almost all the free chlorine is in the OCl⁻ (hypochlorite) form and almost none is HOCl.

Hypochlorous Acid $\rightleftharpoons$ Hydrogen Ion + Hypochlorite Ion

$$HOCl \rightleftharpoons H^+ + OCl^-$$

Normally, in water with a pH of 7.5, approximately 50 percent of the chlorine present will be in the form of HOCl and 50 percent in the form of OCl⁻. This is extremely important since HOCl and OCl⁻ differ in disinfection ability. HOCl has a much greater disinfection potential than OCl⁻.

6.3.3.3 Reaction with Impurities in Water

Most waters that have been processed still contain some impurities. In this section, we will discuss some of the more common impurities that react with chlorine and we will examine the effects of these reactions on the disinfection ability of chlorine.

1. Hydrogen sulfide (H_2S) and ammonia (NH_3) are two inorganic substances that may be found in water when it reaches the disinfection stage of treatment. Their presence can complicate the use of chlorine for disinfection purposes. This is because hydrogen sulfide and ammonia are what is known as reducing agents. That is, they give up **electrons** easily. Chlorine reacts rapidly with these particular reducing agents producing some undesirable results.

    Hydrogen sulfide produces an odor that smells like rotten eggs. It reacts with chlorine to form sulfuric acid and elemental sulfur (depending on temperature, pH, and hydrogen sulfide concentration). Elemental sulfur is objectionable because it can cause odor problems and will precipitate as finely divided white particles that are sometimes colloidal in nature.

    The chemical reactions between hydrogen sulfide and chlorine are as follows:

Hydrogen Sulfide	+ Chlorine +	Oxygen Ion	→	Elemental Sulfur	+ Water +	Chloride Ions

$$H_2S + Cl_2 + O^{2-} \rightarrow S \downarrow + H_2O + 2\,Cl^-$$

electron
(1) A very small, negatively charged particle that is practically weightless. According to the electron theory, all electrical and electronic effects are caused either by the movement of electrons from place to place or because there is an excess or lack of electrons at a particular place. (2) The part of an atom that determines its chemical properties.

The chlorine required to oxidize hydrogen sulfide to sulfur and water is 2.08 mg/L chlorine to 1 mg/L hydrogen sulfide. The complete **oxidation** of hydrogen sulfide to the sulfate form is as follows:

Hydrogen Sulfide + Chlorine + Water → Sulfuric Acid + Hydrochloric Acid

$$H_2S + 4\,Cl_2 + 4\,H_2O \rightarrow H_2SO_4 + 8\,HCl$$

Thus, 8.32 mg/L of chlorine are required to oxidize 1 mg/L of hydrogen sulfide to the sulfate form. Note that in both reactions, the chlorine is converted to the chloride ion (Cl^- or HCl), which has no disinfecting power and produces no chlorine residual. In waterworks practice, we always chlorinate to produce a chlorine residual; therefore, the second reaction (complete oxidation of hydrogen sulfide) occurs before we have any chlorine residual in the water we are treating.

When chlorine is added to water containing ammonia (NH_3), it reacts rapidly with the ammonia and forms **chloramines**. This means that less chlorine is available to act as a disinfectant. As the concentration of ammonia increases, the disinfectant power of the chlorine drops off at a rapid rate.

2. When organic materials are present in water being disinfected with chlorine, the chemical reactions that take place may produce suspected carcinogenic compounds (THMs). The formation of these compounds can be prevented by limiting the amount of **prechlorination** and by removing the organic materials before chlorination of the water.

Check Your Understanding

1. How is the chlorine dosage determined?
2. How is the chlorine demand determined?
3. List two inorganic reducing chemicals with which chlorine reacts rapidly.

6.3.4 Hypochlorite

The use of hypochlorite (OCl^-) to treat **potable water** achieves the same result as chlorine gas. Hypochlorite may be applied in the form of calcium hypochlorite ($Ca(OCl)_2$) or sodium hypochlorite (NaOCl). The form of calcium hypochlorite most frequently used to disinfect water is known as High Test Hypochlorite (**HTH**).

6.3.4.1 Reactions with Water

The chemical reactions of hypochlorite in water are similar to those of chlorine gas.

CALCIUM HYPOCHLORITE

Calcium Hypochlorite + Water → Hydrochlorous Acid + Calcium Hydroxide

$$Ca(OCl)_2 + 2\,H_2O \rightarrow 2\,HOCl + Ca(OH)_2$$

oxidation (ox-uh-DAY-shun)
Oxidation is a chemical reaction that results in the addition of oxygen, removal of hydrogen, or the removal of electrons from an element or compound; in the environment and in wastewater treatment processes, organic matter is oxidized to more stable substances. The opposite of *reduction*.

reduction (re-DUCK-shun)
Reduction is a chemical reaction that results in the addition of hydrogen, removal of oxygen, or the addition of electrons to an element or compound. Under anaerobic conditions (no dissolved oxygen present), sulfur compounds are reduced to odor-producing hydrogen sulfide (H_2S) and other compounds. In the treatment of metal finishing wastewaters, hexavalent chromium (Cr^{6+}) is reduced to the trivalent form (Cr^{3+}). The opposite of *oxidation*.

chloramines (KLOR-uh-means)
Compounds formed by the reaction of hypochlorous acid (or aqueous chlorine) with ammonia.

prechlorination
The addition of chlorine at the headworks of the plant before other treatment processes mainly for disinfection and control of tastes, odors, and aquatic growths. Also applied to aid in coagulation and settling.

potable (POE-tuh-bull) water
Water that does not contain objectionable pollution, contamination, minerals, or infective agents and is considered satisfactory for drinking.

HTH
High Test Hypochlorite. Calcium hypochlorite or $Ca(OCl)_2$.

SODIUM HYPOCHLORITE

Sodium Hypochlorite + Water → Hydrochlorous Acid + Sodium Hydroxide

$$NaOCl + H_2O \rightarrow HOCl + NaOH$$

HTH is used by a number of small water supply systems. A problem occurs in these systems when sodium fluoride is injected at the same point as the hypochlorite. A severe crust forms when the calcium and fluoride ions combine.

6.3.4.2 Differences Between Chlorine Gas and Hypochlorite Compound Reactions

The only difference between the reactions of the hypochlorite compounds and chlorine gas is the side reactions of the end products. The reaction of chlorine gas tends to lower the pH (increases the hydrogen ion, H^+, concentration) by the formation of hydrochloric acid, which favors the formation of hypochlorous acid (HOCl). The hypochlorite tends to raise the pH with the formation of hydroxyl ions (OH^-) from the calcium or sodium hydroxide. At a high pH of around 8.5 or higher, the hypochlorous acid (HOCl) is almost completely dissociated to the ineffective hypochlorite ion (OCl^-) (Figure 6.1). This reaction also depends on the **buffer capacity** (amount of bicarbonate, HCO_3, present) of the water.

Hypochlorous Acid ⇌ Hydrogen Ion + Hypochlorite Ion

$$HOCl \rightleftharpoons H^+ + OCl^-$$

6.3.4.3 Onsite Chlorine Generation

Small water systems are generating chlorine on site for their water treatment processes. Onsite generation (OSG) of process chlorine is attractive due to the lower safety hazards and costs involved. Onsite generated chlorine systems produce 0.8 percent sodium hypochlorite. This limited solution strength (about $\frac{1}{15}$ the strength of commercial bleach and $\frac{1}{7}$ that of household bleach) is below the lower limit deemed a hazardous liquid, with obvious economic and safety advantages.

Operators' only duties with onsite generation systems are to observe the control panel daily for proper operating guidelines and to dump bags of salt every few weeks. Since the assemblies (which are quite small) include an ion exchange water softener, mineral deposits forming with the electrolytic cell are minimal, with an acid cleaning being necessary only every few months. Cell voltage is controlled at a low value to maximize electrode life, which is about 3 years. Process brine strength and cell current determine chlorine production at the anode, with hydrogen gas continually vented from the cathode. The units include provisions for storing the chlorine solution to deliver chlorine for several days in the event of a power failure or other problems causing equipment failure.

6.3.5 Chlorine Dioxide

Chlorine dioxide (ClO_2) may be used as a disinfectant. Chlorine dioxide does not form carcinogenic compounds that may be formed by other chlorine compounds. Also, it is not affected by ammonia, and is a very effective disinfectant at higher pH levels. In addition, chlorine dioxide reacts with sulfide compounds; thus, helping to remove them and eliminate

buffer capacity
A measure of how much a solution or liquid can neutralize acids or bases. For example, this measures the capacity of water or wastewater to resist changes in pH due to the addition of an acid or base.

their characteristic odors. Phenolic tastes and odors can be controlled by using chlorine dioxide.

6.3.5.1 Reaction in Water

Chlorine dioxide reacts with water to form chlorate and chlorite ions in the following manner:

Chlorine Dioxide + Water → Chlorate Ion + Chlorite Ion + Hydrogen Ions

$$2\,ClO_2 + H_2O \rightarrow ClO_3^- + ClO_2^- + 2\,H^+$$

6.3.5.2 Reactions with Impurities in Water

1. Inorganic compounds—Chlorine dioxide is an effective **oxidizing agent** with iron and manganese and does not leave objectionable tastes or odors in the finished water. Because of its oxidizing ability, chlorine dioxide usage must be monitored and the dosage will have to be increased when treating waters with iron and manganese.

2. Organic compounds—Chlorine dioxide does not react with organics in water. Therefore, there is little danger of the formation of potentially dangerous THMs.

Check Your Understanding

1. How do chlorine gas and hypochlorite influence pH?
2. How does pH influence the relationship between HOCl and OCl⁻?

6.3.6 Breakpoint Chlorination

In determining how much chlorine you will need for disinfection, remember you will be attempting to produce a certain chlorine residual in the form of **free available chlorine residual**. Chlorine in this form has the highest disinfecting ability. **Breakpoint chlorination** is the name of this process of adding chlorine to water until the chlorine demand has been satisfied. Further additions of chlorine will result in a chlorine residual that is directly proportional to the amount of chlorine added beyond the breakpoint. Public water supplies are normally chlorinated past the breakpoint.

Look at the breakpoint chlorination curve in Figure 6.2. Assume the water being chlorinated contains some manganese, iron, nitrite, organic matter, and ammonia. Now add a small amount of chlorine. The chlorine reacts with (oxidizes) the manganese, iron, and nitrite. That is all that happens—no disinfection and no chlorine residual (Figure 6.2, region A). Add a little more chlorine, enough to react with the organics and ammonia; **chlororganics** and chloramines will form. The chloramines produce a **combined available chlorine residual**—chlorine combined with other substances, mainly ammonia (Figure 6.2, region B). Combined residuals have rather poor disinfecting power and may cause tastes and odors.

With just a little more chlorine, the chloramines and some of the chlororganics are destroyed, which results in a drop in combined chlorine residual (Figure 6.2, region C). When all of the chloramines are gone, adding more chlorine produces free available residual chlorine (Figure 6.2, region D)—free in the sense that it has not reacted with anything and available in that it can and will react if need be. Free available residual chlorine is the best

oxidizing agent
Any substance, such as oxygen (O_2) or chlorine (Cl_2), that will readily add (take on) electrons. When oxygen or chlorine is added to water or wastewater, organic substances are oxidized. These oxidized organic substances are more stable and less likely to give off odors or to contain disease-causing bacteria. The opposite is a *reducing agent*.

breakpoint chlorination
Addition of chlorine to water or wastewater until the chlorine demand has been satisfied. At this point, further additions of chlorine will result in a free chlorine residual that is directly proportional to the amount of chlorine added beyond the breakpoint.

free available chlorine residual
That portion of the total available chlorine residual composed of dissolved chlorine gas (Cl^2), hypochlorous acid (HOCl), or hypochlorite ion (OCl⁻) remaining in water after chlorination at the end of a specified contact period. This does not include chlorine that has combined with ammonia, nitrogen, or other compounds.

chlororganic (klor-or-GAN-ick)
Organic compounds combined with chlorine. These often toxic compounds generally originate from chlorinating water containing dissolved organic substances or microbes such as algae.

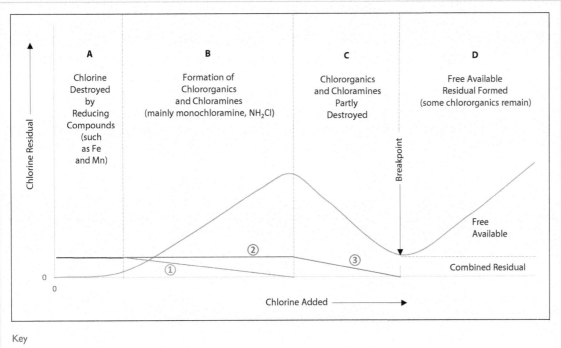

Key
1. Free (or excess) ammonia-N. In this region, free ammonia is converted to chloramines.
2. Total ammonia-N (free ammonia plus chloramines). Ammonia is converted to chloramines, but the total does not change.
3. Total ammonia-N. In this region, the chloramines are converted to nitrogen gas, which leaves the water.

Figure 6.2   Breakpoint chlorination curve

residual for disinfection. It disinfects faster and without the swimming-pool odor of combined residual chlorine. The point at which the chlorine residual curve bottoms out is called the breakpoint, and chlorinating beyond this point is called breakpoint chlorination. In water treatment plants today it is common practice to go past the breakpoint. This means that the treated water will have a very effective disinfectant because it is in the form of free available residual chlorine.

In plants where THMs are not a problem, sufficient chlorine to get past the breakpoint is added to the raw water (prechlorination). The chlorine residual will aid coagulation, control algal problems in basins, reduce odor problems in treated water, and provide sufficient chlorine contact time to effectively kill or inactivate pathogenic organisms. Therefore, the treated water will have a very low chlorine residual, but the residual will be a very effective disinfectant.

Let us look more closely at some of the chemical reactions that take place during chlorination. When chlorine is added to waters containing ammonia (NH_3), the ammonia reacts with hypochlorous acid (HOCl) to form monochloramine, dichloramine, and trichloramine. The formation of these chloramines depends on the pH of the solution and the initial chlorine-ammonia ratio.

Ammonia must be present to produce the breakpoint chlorination curve. Sources of ammonia in raw water include fertilizer in agricultural runoff and discharges from wastewater treatment plants. High-quality raw water without any ammonia will not produce a breakpoint curve. Therefore, if there is no ammonia present in the water, and chlorinated water smells like chlorine, and a chlorine residual is present, do not add more chlorine.

Ammonia + Hypochlorous Acid → Chloramine + Water
$NH_3 + HOCl → NH_2Cl + H_2O$ Monochloramine
$NH_2Cl + HOCl → NHCl_2 + H_2O$ Dichloramine
$NHCl_2 + HOCl → NCl_3 + H_2O$ Trichloramine (more commonly
                                     called nitrogen trichloride)

As the chlorine-to-ammonia-nitrogen ratio increases, the ammonia molecule becomes progressively more chlorinated. At Cl_2:NH_3-N weight ratios higher than 7.6:1, all available ammonia is theoretically oxidized to nitrogen gas and chlorine residuals are greatly reduced. Therefore, 7.6:1 is the theoretical ratio of the breakpoint. The actual Cl_2:NH_3-N ratio for breakpoint for a given source water will usually be greater than 7.6:1 (typically 10:1 for most water), depending on the levels of other substances present in the water (such as nitrite and organic nitrogen). Once this point is reached, additional chlorine dosages yield an equal and proportional increase in free available chlorine.

At the pH levels that are usually found in water treatment plants (pH 6.5 to 7.5), monochloramine and dichloramine exist together. At pH levels below 5.5, dichloramine exists by itself. Below pH 4.0, trichloramine is the only compound found. The mono- and dichloramine forms have definite disinfection powers and are of interest in the measurement of chlorine residuals. Dichloramine has a more effective disinfecting power than monochloramine. However, dichloramine is not recommended as a disinfectant because of taste and odor problems. Chlorine reacts with **phenolic compounds** and salicylic acid (both are leached into water from leaves and blossoms) to form **chlorophenol**, which has an intense medicinal odor. This reaction goes much slower in the presence of monochloramine.

6.3.7 Chloramination

Chloramination has been used as an alternative disinfection process by water utilities for over 80 years. An operator's decision to use chloramines depends on several factors, including the quality of the raw water, the ability of the treatment plant to meet various regulations, operational practices, and distribution system characteristics. Chloramines have proven effective in accomplishing the following objectives:

1. Reducing the formation of THMs and other DBPs
2. Maintaining a detectable residual throughout the distribution system
3. Penetrating the biofilm (the layer of microorganisms on pipeline walls) and reducing the potential for coliform regrowth
4. Killing or inactivating **heterotrophic** plate count bacteria
5. Reducing taste and odor problems

6.3.7.1 Methods for Producing Chloramines
There are three primary methods by which chloramines are produced: (1) preammoniation followed by later chlorination, (2) addition of chlorine and ammonia at the same time, and (3) prechlorination/postammoniation.

1. *Preammoniation Followed by Later Chlorination*. In this method, ammonia is applied at the rapid-mix unit process and chlorine is added downstream at the entrance to the flocculation basins. This approach usually produces lower THM levels than the postammoniation method. Preammoniation to form chloramines (monochloramine) does not produce phenolic tastes and odors, but this method may not be as effective as postammoniation for controlling tastes and odors associated with **diatoms** and anaerobic bacteria in source waters.

phenolic (fee-NO-lick) compounds
Organic compounds that are derivatives of benzene. Also called phenols (FEE-nolls).

chlorophenolic (klor-o-FEE-NO-lick)
Chlorophenolic compounds are phenolic compounds (carbolic acid) combined with chlorine.

chloramination (KLOR-ah-min-NAY-shun)
The application of chlorine and ammonia to water to form chloramines for the purpose of disinfection.

heterotrophic (HET-er-o-TROF-ick)
Describes organisms that use organic matter for energy and growth. Animals, fungi, and most bacteria are heterotrophs.

diatoms (DYE-uh-toms)
Unicellular (single cell), microscopic algae with a rigid, box-like internal structure consisting mainly of silica.

2. *Concurrent Addition of Chlorine and Ammonia.* In this method, chlorine is applied to the plant influent and, at the same time or immediately thereafter, ammonia is introduced at the rapid-mix unit process. Concurrent chloramination produces the lowest THM levels of the three methods.

3. *Prechlorination/Postammoniation.* In prechlorination/postammoniation, chlorine is applied at the head of the plant and a free chlorine residual is maintained throughout the plant processes. Ammonia is added at the plant effluent to produce chloramines. Because of the longer free chlorine contact time, this application method will result in the formation of more THMs, but it may be necessary to use this method to meet the disinfection requirements of the Surface Water Treatment Rule (SWTR). A major limitation of using chloramine residuals is that chloramines are less effective as a disinfectant than free chlorine residuals.

6.3.7.2 Chlorine-to-Ammonia-Nitrogen Ratios

After a method of chloramine application has been selected, the best ratio of chlorine to ammonia-nitrogen (by weight) and the desired chloramine residual for each system must be determined. A dosage of three parts of chlorine to one part ammonia (3:1) will form monochloramines. This 3:1 ratio provides an excess of ammonia-nitrogen, which will be available to react with any chlorine added in the distribution system to boost the chloramine residual.

Higher chlorine-to-ammonia-nitrogen weight ratios such as 4:1 and 5:1 also have been used successfully by many water agencies. However, the higher the chlorine-to-ammonianitrogen ratio, the less excess ammonia will be available for rechlorination. Some agencies have found it necessary to limit the amount of excess available ammonia to prevent incomplete nitrification.

Monochloramines form combined residual chlorine (Figure 6.2, region B) as the chlorine dose is increased in the presence of ammonia. As the chlorine dose increases, the combined residual increases and excess ammonia decreases (line 2 in Figure 6.2). The maximum chlorine-to-ammonia ratio that can be achieved is 5:1. At a chlorine dose above the 5:1 ratio, the combined residual actually decreases and the total ammonia-nitrogen also begins to decrease as it is oxidized by the additional chlorine (Figure 6.2, region C and line 3). Dichloramines form during this oxidation and may cause unwanted tastes and odors. As the chlorine dose is further increased, breakpoint chlorination will eventually occur. Trichloramines are formed past the breakpoint and may form unpleasant tastes and odors. Further additions of chlorine will result in a chlorine residual that is proportional to the amount of chlorine added beyond the breakpoint (Figure 6.2, region D).

Calculating the chlorine-to-ammonia-nitrogen ratio on the basis of actual quantity of chemicals applied can lead to incorrect conclusions regarding the finished water quality. In applications in which chlorine is injected before the ammonia, chlorine demand in the water will reduce the amount of chlorine available to form the combined residual (Figure 6.2, region A). In such applications the applied ratio will be greater than the actual ratio of chlorine to ammonia-nitrogen leaving the plant.

As an example, assume that an initial dosage of 5.0 mg/L results in a free chlorine residual of 3.5 mg/L at the ammonia application point; it

nitrification (NYE-truh-fuh-KAY-shun) An aerobic process in which bacteria oxidize the ammonia and organic nitrogen in water into nitrite and then nitrate.

can be concluded that a chlorine demand of 1.5 mg/L exists. If ammonia-nitrogen is applied at a dose of 1.0 mg/L, the applied chlorine-to-ammonia-nitrogen ratio is 5:1, whereas the actual ratio in the water leaving the plant is only 3.5:1.

6.3.7.3 Special Water Users

Although chloramines are nontoxic to healthy humans, they can have a weakening effect on individuals with kidney disease who must undergo kidney dialysis. Chloramines must be removed from the water used in the dialysis treatments. Granular activated carbon and ascorbic acid are common substances used to reduce chloramine residuals. All special water users should be notified before chloramines are used as a disinfectant in municipal waters.

Also, like free chlorine, chloramines can be deadly to fish. They can damage gill tissue and enter the red blood cells causing a sudden and severe blood disorder. For this reason, all chloramine compounds must be removed from the water before any contact with fish.

6.3.7.4 Blending Chloraminated Waters

Care must be taken when blending chloraminated water with water that has been disinfected with free chlorine. Depending on the ratio of the blend, these two different disinfectants can cancel each other out, resulting in very low disinfectant residuals. When chlorinated water is blended with chloraminated water, the chloramine residual will decrease after the excess ammonia has been combined (this is equivalent to moving from region B to region C in the breakpoint chlorination curve). Knowing the amount of uncombined ammonia available is important in determining how much chlorinated water can be blended with a particular chloraminated water without significantly affecting the monochloramine residual. Knowing how much uncombined ammonia-nitrogen is available is also important before you make any attempt to boost the chloramine residual by adding chlorine.

6.3.7.5 Chloramine Residuals

When measuring combined chlorine residuals (chloramines) in the field, analyze for total chlorine. Total chlorine is the total concentration of chlorine in water, including the combined chlorine (such as inorganic and organic chloramines) and the free available chlorine. No free chlorine should be present at chlorine-to-ammonia-nitrogen ratios of 3:1 to 5:1 (these are the ratios in region B of Figure 6.2). Care must be taken when attempting to measure free chlorine with chloraminated water because the chloramine residual will interfere with the DPD method for measuring free chlorine. (See Section 6.8, "Measurement of Chlorine Residual," for more specific information about the methods commonly used to measure chlorine residuals, including the DPD method.)

6.3.8 Nitrification

Nitrification is an important and effective microbial process in the oxidation of ammonia in both land and water environments. Two groups of organisms are involved in the nitrification process: ammonia-oxidizing bacteria (AOB) and nitrite-oxidizing bacteria (see Figure 6.3). Nitrification has been well recognized as a beneficial treatment for the removal of ammonia in municipal wastewater.

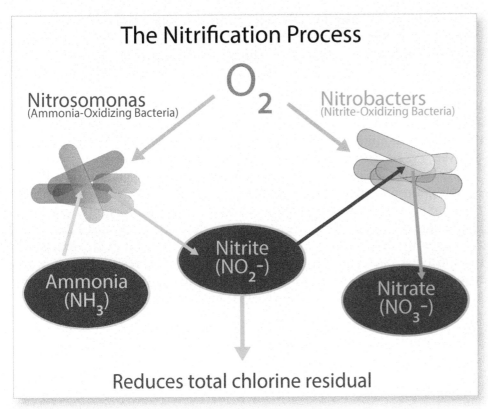

Figure 6.3   The nitrification process

When nitrification occurs in chloraminated drinking water, however, the process may lower the water quality unless the nitrification process reaches completion. Incomplete or partial nitrification causes the production of nitrite from the growth of AOB. This nitrite, in turn, rapidly reduces free chlorine and can interfere with the measurement of free chlorine. The end result may be a loss of total chlorine and ammonia and an increase in the concentration of heterotrophic plate count bacteria.

Factors influencing nitrification include the water temperature, the detention time in the reservoir or distribution system, excess ammonia in the water system, and the chloramine concentration used. The conditions most likely to lead to nitrification when using chloramines are a pH of 7.5 to 8.5, a water temperature of 77 to 86°F (25 to 30°C), a free ammonia concentration in the water, and a dark environment. The danger in allowing nitrification episodes to occur is that you may be left with very low or no **total chlorine residual**. Total chlorine residual is the total concentration of chlorine in water, including the combined chlorine (such as inorganic and organic chloramines) and the free available chlorine.

6.3.8.1   Nitrification Prevention and Control

When using chloramines for disinfection, an early warning system should be developed to detect the signs that nitrification is beginning to occur so that you can prevent or at least control the nitrification process. The best way to do this is to set up a regularly scheduled monitoring program. The warning signs to watch for include decreases in ammonia level, total chlorine level, and pH; increases in nitrite level; and an increase in heterotrophic plate count bacteria. In addition, action response levels should be established for chloraminated distribution systems and reservoirs. Normal background levels of nitrite should be measured and then alert levels should be established so that increasing nitrite levels will not be overlooked.

An inexpensive way to help keep nitrite levels low is to reduce the detention times through the reservoirs and the distribution system, especially during warmer weather. Adding more chlorine to reservoir inlets and increasing the chlorine-to-ammonia-nitrogen ratio from 3:1 to 5:1 at the treatment plant effluent will further control nitrification by decreasing the amount of uncombined ammonia in the distribution system. However, at a chlorine-to-ammonia-nitrogen ratio of 5:1, it is critical that the chlorine and ammonia feed systems operate accurately and reliably because an overdose of chlorine can reduce the chloramine residual.

Other strategies for controlling nitrification include establishing a flushing program and increasing the chloramine residual. A uniform flushing program should be a key component of any nitrification control program. Flushing reduces the detention time in low-flow areas, increases the water velocity within pipelines to remove sediments and biofilm that would harbor nitrifying bacteria, and draws higher disinfectant residuals into problem areas. Increasing the chloramine residual in the distribution system to greater than 2.0 mg/L is also effective in preventing the onset of nitrification.

Check Your Understanding

1. What is breakpoint chlorination?

2. An operator's decision to use chloramines depends on what factors?

3. What are the three primary methods by which chloramines are produced?

4. Why is the applied chlorine-to-ammonia-nitrogen ratio usually greater than the actual chlorine-to-nitrogen ratio leaving the plant?

5. Incomplete nitrification causes the production of nitrite, which produces what problems in the disinfection of water?

6.3.9   Chlorine Residual Testing

Many small system operators attempt to maintain a chlorine residual throughout the distribution system. Chlorine is effective in biological control and especially in the elimination of coliform bacteria that might reach the water in the distribution system through cross connections or leakage into the system. A chlorine residual also helps to control any microorganisms that could produce slimes, tastes, or odors in the water in the distribution system.

Adequate control of coliform aftergrowth is usually obtained only when chlorine residuals are carried to the farthest points of the distribution system. To ensure that this is taking place, perform daily chlorine residual tests. A chlorine residual of about 0.2 mg/L measured at the extreme ends of the distribution system is usually a good indication that a free chlorine residual is present in all other parts of the system. This small residual can destroy a small amount of contamination, so a lack of chlorine residual could indicate the presence of heavy contamination. If routine checks at a given point show measurable residuals, any sudden absence of a residual at that point should alert the operator to the possibility that a potential problem has arisen that needs prompt investigation. Immediate action that can be taken includes retesting for chlorine residual, then checking chlorination equipment, and finally searching for a contamination source that could cause an increase in the chlorine demand.

6.3.9.1   Chlorine Residual Curve

The chlorine residual curve procedure is a quick and easy way for an operator to estimate the proper chlorine dose, especially when surface water conditions are changing rapidly such as during a storm.

Fill a clean 5-gallon bucket from a sample tap located at least two 90 degree elbows downstream from the chlorine injection point (or where chlorine is completely mixed with the water in the pipe). Immediately measure the chlorine residual and record this value on the time zero line of your record sheet (Figure 6.4). This is the initial chlorine residual. At 15-minute intervals, vigorously stir the bucket using an up and down motion. (A large plastic spoon works well for this purpose.) Collect a sample from the bucket from 1 or 2 inches below the water surface and measure the chlorine residual. Record this chlorine residual value on the record sheet. Repeat this procedure for at least 1 hour at 15-minute intervals. Plot these recorded values on a chart or graph paper as shown in Figure 6.4. Connect the plotted points to create a chlorine residual curve. If the chlorine residual after 1 hour is not correct, increase or decrease the initial chlorine dose so that the final chlorine residual will be at the desired value for the water distribution system. Repeat this procedure as needed.

When performing this test, be sure the 5-gallon plastic test bucket is clean and used only for this purpose. A new bucket does not need to be used for every test, but the bucket should be new when the first test is performed. The stirrer should also be clean. Do not use the stirrer for the chlorine solution mixing and holding tank. During the test, keep the bucket cool to minimize the loss of chlorine to the atmosphere. The purpose of the chlorine residual curve is to show the effects of chemical reactions on chlorine residual. These effects will be overestimated if the chlorine is lost to the air.

Groundwater quality changes slowly, or not at all; therefore, the initial or target chlorine residual does not have to be checked more than once a month. In contrast, surface water quality can change rapidly, especially

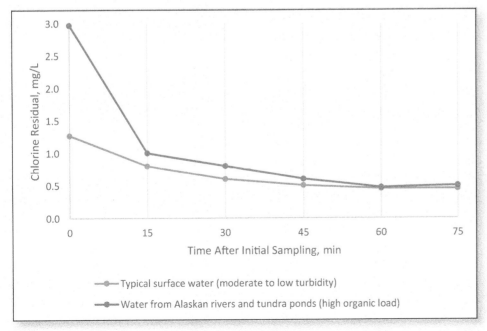

Figure 6.4   Chlorine residual curves

during storms and the snow melt season. Experience has proven that the required initial or target chlorine residual at time zero is directly tied to the turbidity of the finished (treated) water. The higher the finished water turbidity, the higher the initial chlorine residual value will have to be to ensure the desired chlorine residual in the distribution system. If you carefully document this information in your records, you will eventually be able to estimate the required initial chlorine residual based on the turbidity. This will reduce the time and effort needed to fine-tune the chlorine feed rate to achieve the desired residual in the distribution system.

Bear in mind that the chlorine residual curve is not a substitute for checking the actual chlorine residual in the distribution system. This should be done daily to assure that the water you are delivering to your customers is safe.

6.3.9.2 Critical Factors

Both chlorine residual and contact time are essential for effective killing or inactivation of pathogenic microorganisms. Complete initial mixing is important. Changes in pH affect the disinfection ability of chlorine and you must reexamine the best combination of contact time and chlorine residual when the pH fluctuates. Critical factors influencing disinfection are summarized as follows:

1. Effectiveness of upstream treatment processes. The lower the turbidity (suspended solids, organic content, reducing agents) of the water, the better the disinfection.
2. Injection point and method of mixing to get disinfectant in contact with water being disinfected. This depends on whether using prechlorination or postchlorination.
3. Temperature. The higher the temperature, the more rapid the rate of disinfection.
4. Dosage and type of chemical. Usually, the higher the dosage, the faster the disinfection rate. The form (chloramines or free chlorine residual) and type of chemical also influence the disinfection rate.
5. pH. The lower the pH, the better the disinfection.
6. Contact time. With good initial mixing, the longer the contact time, the better the disinfection.
7. Chlorine residual.

6.3.10 CT Values

The Surface Water Treatment Rule (SWTR) ensures that pathogenic organisms are removed or inactivated by the treatment process. To meet this goal, all systems are required to disinfect their water supplies. For some water systems using very clean source water and meeting the other criteria to avoid filtration, disinfection alone can achieve the 3-log (99.9%) Giardia and 4-log (99.99%) virus inactivation levels required by the SWTR. For extremely clean source waters, there may be virtually no Giardia or viruses and achieving 3-log or 4-log inactivation levels will be impossible and unnecessary.

Several methods of disinfection are in common use, including free chlorination, chloramination, use of chlorine dioxide, and application of ozone. The concentration of chemical needed and the length of contact time needed to ensure disinfection are different for each disinfectant. Therefore, the efficiency of the disinfectant is measured by the time (T) in minutes of the disinfectant's contact in the water and the concentration (C) of the disinfectant residual in mg/L measured at the end of the contact time. The product of these two factors (C × T) provides a measure of the degree of pathogenic inactivation (also known as the "kill"). The required CT value to achieve inactivation is dependent upon the organism in question, the type of disinfectant, and the pH and temperature of the water supply.

Time (T) is measured from the point of application to the point where C is determined. T must be based on peak hour flow rate conditions. In pipelines, T is calculated by dividing the volume of the pipeline in gallons by the flow rate in gallons per minute (GPM). In reservoirs and basins, dye tracer tests must be used to determine T. In this case, T is the time it takes for 10 percent of the tracer to pass the measuring point.

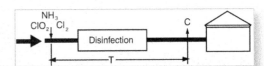

A properly operated filtration system can achieve limited removal or inactivation of microorganisms. Because of this, systems that are required to filter their water are permitted to apply a factor that represents the microorganism removal value of filtration when calculating CT values to meet the disinfection requirements. The factor (removal credit) varies with the type of filtration system. Its purpose is to take into account the combined effects of both disinfection and filtration in meeting the SWTR microbial standards.

6.3.11   Process Calculations

There are two basic chlorination process calculations:

1. Chlorine dose (Cl_{dose}), mg/L
2. Chlorine demand (Cl_{demand}), mg/L

To calculate the chlorine dosage (dose) of the water being treated, we need to know the chlorine fed to the water being treated in pounds of chlorine per day and the amount of water treated in million gallons per day. With this information, we can calculate the chlorine dosage in milligrams of chlorine per liter of water. Start with the basic chemical feed formula:

$$Cl_{feed}\left[\frac{lb}{d}\right] = Q\left[\frac{Mgal}{d}\right] \times Cl_{dose}\left[\frac{mg}{L}\right] \times \frac{3.785\ L}{gal} \times \frac{10^6}{M} \times \frac{kg}{10^6\ mg} \times \frac{2.205\ lb}{kg}$$

$$= Q\left[\frac{Mgal}{d}\right] \times Cl_{dose}\left[\frac{mg}{L}\right] \times 8.34$$

Where,

Cl_{feed} = Chlorine Feed

Q   = Flow Rate

Rearranging the above formula to determine chlorine dose (Cl_{dose}), we obtain:

$$Cl_{dose}\left[\frac{mg}{L}\right] = \frac{Cl_{feed}\left[\frac{lb}{d}\right]}{8.34 \times Q\left[\frac{Mgal}{d}\right]} = 0.120 \times \frac{Cl_{feed}\left[\frac{lb}{d}\right]}{Q\left[\frac{Mgal}{d}\right]}$$

When we determine the chlorine dose in milligrams of chlorine per liter for the water we are treating, we must add enough chlorine to the water to meet the chlorine demand (Cl_{demand}) and the desired chlorine residual ($Cl_{residual}$).

$$Cl_{dose}\left[\frac{mg}{L}\right] = Cl_{demand}\left[\frac{mg}{L}\right] + Cl_{residual}\left[\frac{mg}{L}\right]$$

If we wish to know the chlorine demand (Cl_{demand}), then we rearrange the terms by subtracting the chlorine residual ($Cl_{residual}$) from the chlorine dose (Cl_{dose}).

$$Cl_{demand}\left[\frac{mg}{L}\right] = Cl_{dose}\left[\frac{mg}{L}\right] - Cl_{residual}\left[\frac{mg}{L}\right]$$

Example 1

A chlorinator is set to feed 20 pounds (lb) of chlorine in 24 hours (h) to a flow of 0.85 million gallons per day (MGD). Find the chlorine dose in mg/L.

Known		Unknown	
Cl_{feed}	= Chlorine Feed	Cl_{dose}	= Chlorine Dose, mg/L
	= 20 lb Cl/24 h (20 lb Cl/d)		
Q	= Flow Rate = 0.85 MGD		

Calculate the chlorine dose in mg/L.

$$Cl_{dose}\left[\frac{mg}{L}\right] = 0.120 \times \frac{Cl_{feed}\left[\frac{lb}{d}\right]}{Q\left[\frac{Mgal}{d}\right]} = 0.120 \times \frac{20\left[\frac{lb}{d}\right]}{0.85\left[\frac{Mgal}{d}\right]} = 2.8 \frac{mg}{L}$$

Example 2

Find the chlorine demand in mg/L for the water being treated in Example 1 with a chlorine dose of 2.8 mg/L. The chlorine residual after 30 minutes (min) of contact time is 0.5 mg/L.

Known		Unknown	
Cl_{dose}	= Chlorine Dose	Cl_{demand}	= Chlorine Demand, mg/L
	= 2.8 mg/L		
$Cl_{residual}$	= Chlorine Residual		
	= 0.5 mg/L		

Find the chlorine demand in mg/L.

$$Cl_{demand}\left[\frac{mg}{L}\right] = Cl_{dose}\left[\frac{mg}{L}\right] - Cl_{residual}\left[\frac{mg}{L}\right]$$

$$= 2.8\left[\frac{mg}{L}\right] - 0.5\left[\frac{mg}{L}\right]$$

$$= 2.3 \frac{mg}{L}$$

Check Your Understanding

1. What actions should an operator take when there is a sudden absence of chlorine residual in the distribution system?

2. How does the length of chlorine contact time affect the disinfection process?

3. How is the efficiency of a disinfectant measured?

6.4   Points of Chlorine Application

Chlorine may be applied at several points in the treatment plant or the distribution system. The two most common points (locations) of chlorination in a water treatment plant are prechlorination and postchlorination (Figure 6.5).

6.4.1   Prechlorination

Prechlorination is the application of chlorine ahead of any other treatment processes. While prechlorination may increase the formation of trihalomethanes in raw water containing organic **precursor (THM)** compounds and tastes and odors when phenolic compounds are present, it provides the following benefits:

1. Control of algal and slime growths
2. Control of mudball formation
3. Improved coagulation
4. Reduction of tastes and odors
5. Increased chlorine contact time
6. Increased safety factor in disinfection of heavily contaminated waters

6.4.2   Postchlorination

Postchlorination is the addition of chlorine to the effluent after water treatment but before it enters the distribution system. This is the primary point of disinfection and it is normally the last application of any disinfectant.

6.4.3   Rechlorination

Rechlorination is the practice of adding chlorine in the distribution system. This practice is common when the distribution system is long or complex. The application point could be any place where adequate mixing is available.

6.4.4   Wells

Chlorination of wells is required in some areas and is a good practice whenever wells are used for public water supplies. This is usually accomplished with a small system and can be automated for ease of operation.

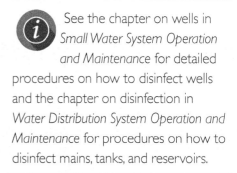

See the chapter on wells in *Small Water System Operation and Maintenance* for detailed procedures on how to disinfect wells and the chapter on disinfection in *Water Distribution System Operation and Maintenance* for procedures on how to disinfect mains, tanks, and reservoirs.

precursor, THM (pre-CURSE-or)
Natural, organic compounds found in all surface and groundwaters, which may react with halogens (such as chlorine) to form trihalomethanes (THMs); they must be present in order for THMs to form.

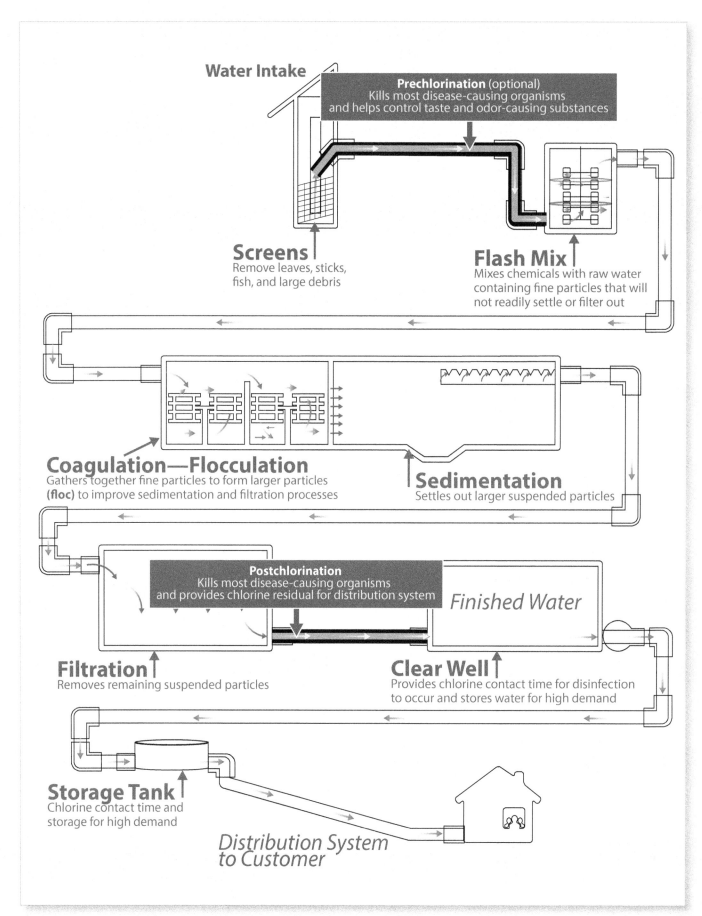

Water Intake

Prechlorination (optional)
Kills most disease-causing organisms
and helps control taste and odor-causing substances

Screens
Remove leaves, sticks,
fish, and large debris

Flash Mix
Mixes chemicals with raw water
containing fine particles that will
not readily settle or filter out

Coagulation—Flocculation
Gathers together fine particles to form larger particles
(floc) to improve sedimentation and filtration processes

Sedimentation
Settles out larger suspended particles

Postchlorination
Kills most disease-causing organisms
and provides chlorine residual for distribution system

Finished Water

Filtration
Removes remaining suspended particles

Clear Well
Provides chlorine contact time for disinfection
to occur and stores water for high demand

Storage Tank
Chlorine contact time and
storage for high demand

*Distribution System
to Customer*

Figure 6.5   Typical flow diagram for a water treatment plant

6.4.5   Mains

Mains are usually not a problem and are not chlorinated except in long pipelines and in complex systems. Mains must be chlorinated after initial installation and any repairs.

6.4.6   Tanks and Reservoirs

Usually, tanks and reservoirs are not chlorinated unless specific problems develop that cannot be solved by other means. Tanks must be chlorinated after initial installation, repairs, maintenance, repainting, and cleaning. In other words, any time a tank has been drained or entered, it should be chlorinated.

Check Your Understanding

1. List the two most common points (locations) of chlorination in a water treatment plant.
2. Under what conditions should waters not be prechlorinated?
3. What are the benefits of prechlorination?

6.5   Operation of Chlorination Equipment

Chlorination equipment feeds liquid and gas chlorine at specific rates for disinfection activities at a water treatment plant. When operating chlorination equipment, an operator must be protected from chlorine using personal protective equipment, including specialized breathing equipment and chemical suits. An operator needs to know how to correctly and safely remove chlorine from plastic and steel containers by becoming familiar with the plant's operation and maintenance (O&M) instructions for the chlorination equipment. A water treatment plant operator also must be able to recognize normal and abnormal operations at the plant and to troubleshoot problems that occur.

Figure 6.6 Typical small hypochlorinator installation *(Courtesy of Clean Water Systems and Stores, Inc.)*

6.5.1   Hypochlorinators

A hypochlorinator is a piece of equipment used to feed liquid chlorine (bleach) solutions. Hypochlorinators used on small water systems are simple and easy to install. Typical installations are shown in Figures 6.6 and 6.8. Hypochlorinator systems usually consist of a chemical solution tank for the hypochlorite, pump (Figure 6.7), power supply, water pump, pressure switch, and water storage tank.

There are two methods of feeding the hypochlorite solution into the water being disinfected. The hypochlorite solution may be pumped directly into the water (Figure 6.9). In the other method, the hypochlorite solution is pumped through an **ejector** (also called an eductor or injector), which draws in additional water for dilution of the hypochlorite solution (Figure 6.10).

ejector
A device used to disperse a chemical solution into water being treated.

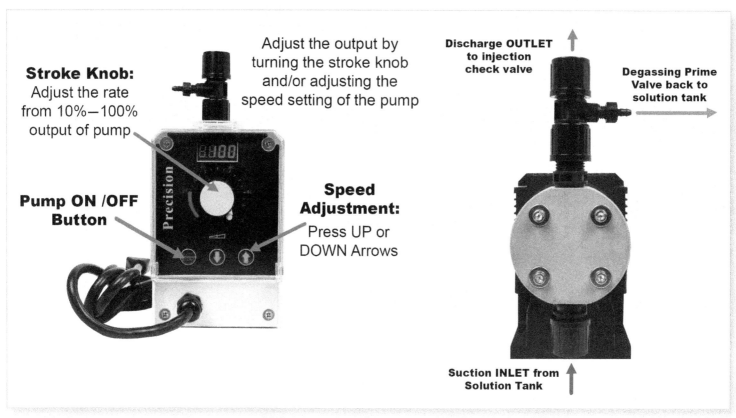

Stroke Knob:
Adjust the rate
from 10%–100%
output of pump

Adjust the output by
turning the stroke knob
and/or adjusting the
speed setting of the pump

Discharge OUTLET
to injection
check valve

Degassing Prime
Valve back to
solution tank

**Pump ON /OFF
Button**

**Speed
Adjustment:**
Press UP or
DOWN Arrows

Suction INLET from
Solution Tank

Figure 6.7   Hypochlorinator pump *(Courtesy of Clean Water Systems and Stores, Inc.)*

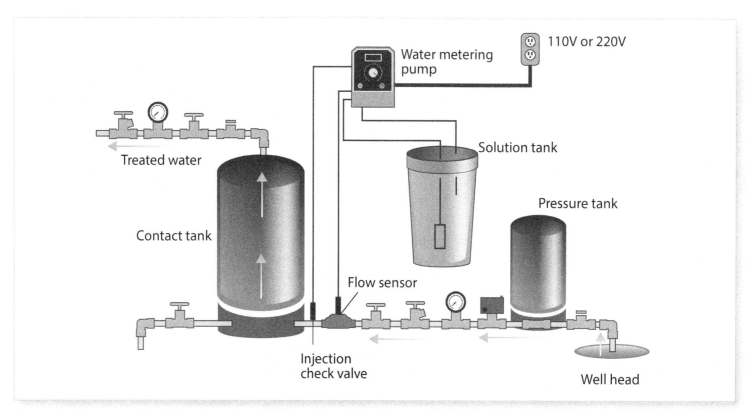

110V or 220V

Water metering
pump

Solution tank

Treated water

Pressure tank

Contact tank

Flow sensor

Injection
check valve

Well head

Figure 6.8   Proportional-feed flow meter hypochlorinator system *(Adapted from Clean Water Systems and Stores, Inc.)*

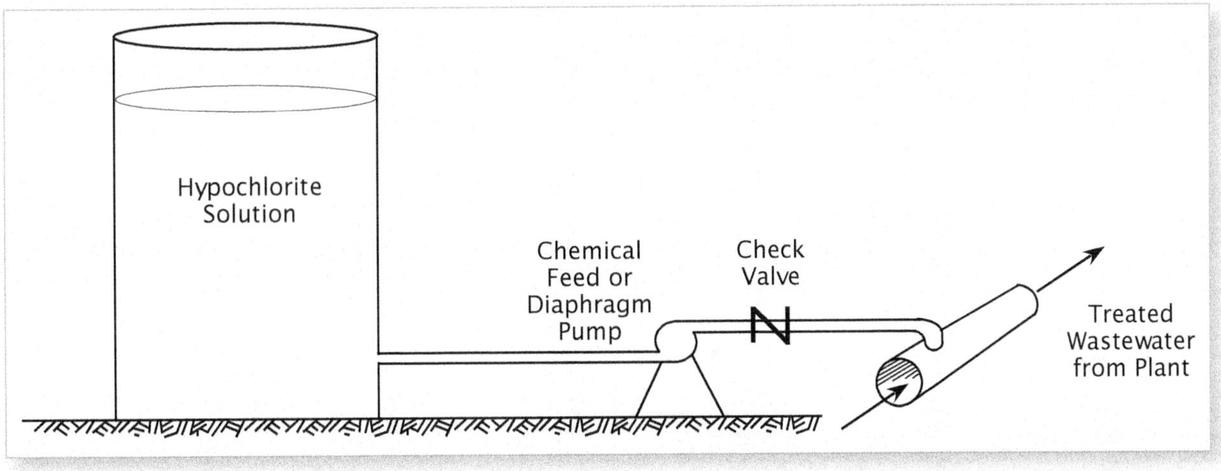

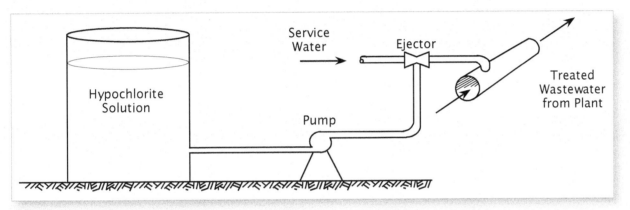

Figure 6.9 Hypochlorinator direct pumping system

Figure 6.10 Hypochlorinator ejector feed system

Check Your Understanding

1. List the major parts of a typical hypochlorinator system.

2. What are the two common methods of feeding hypochlorite to the water being disinfected?

6.5.2 Chlorinators

A typical installation for gaseous chlorine is shown in Figure 6.11 (see Section 6.6.2, "Installation," Item 8, for details about chlorine room ventilation systems and requirements). Chlorine gas may be removed from chlorine containers by a valve and piping arrangement to the chlorinators (Figure 6.11) and delivered by vacuum-fed gas chlorinators (Figures 6.11 and 6.12). The chlorine gas is regulated, metered, introduced into a stream of **injector water**, and then is conducted as a solution to the point of application. Some of the components of a typical vacuum-fed chlorinator are shown in Figures 6.12 and 6.13. The purpose of chlorinator parts is summarized in Table 6.4.

injector water
Service water in which chlorine is added (injected) to form a chlorine solution.

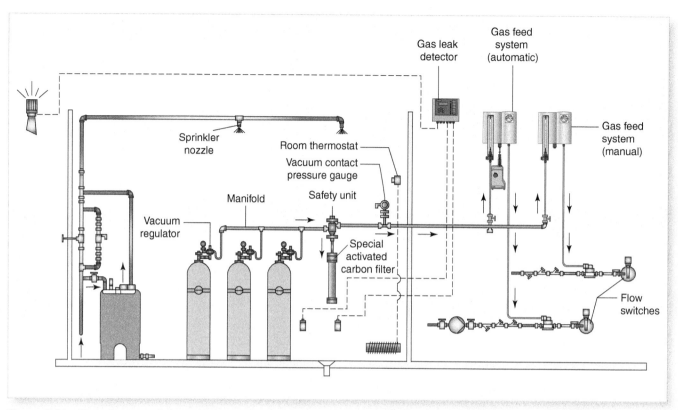

Figure 6.11　Typical vaccum-fed gas chlorinator installation.

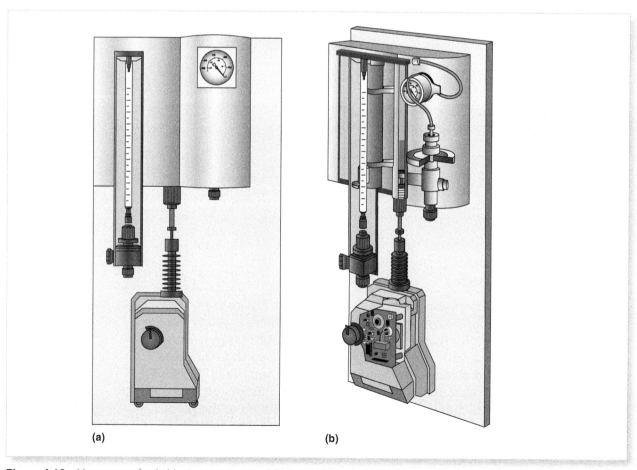

Figure 6.12　Vacuum gas feed chlorinator system, wall-mounted front and cutaway views

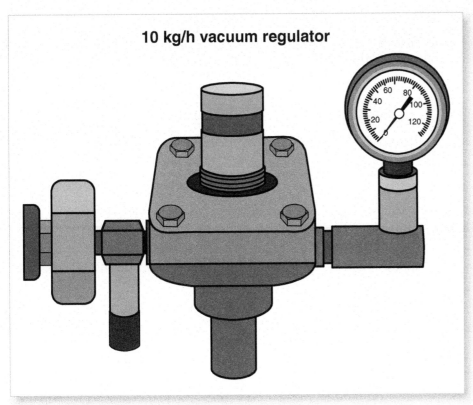

Figure 6.13   Examples of chlorinator parts: 10 kg/h vacuum regulator

Table 6.4   Typical chlorinator parts and purpose

Part	Purpose
Pressure Gauge	Indicates chlorine gas pressure at chlorinator system from chlorine manifold and supply (20 psi minimum and 40 psi maximum or 2.4 kg/cm^2 minimum and 4.8 kg/cm^2 maximum).
Gas Supply	Provides source of chlorine gas from containers to chlorinator system.
Vacuum Regulator—Check Unit	Maintains a constant vacuum on chlorinator.
Standby Pressure Relief	Relieves excess gas pressure on chlorinator.
Vent	Discharges any excess chlorine gas (pressure) to atmosphere outside of chlorination building.
Gas Inlet	Allows entrance of chlorine gas to chlorinator. Gas flows from chlorine container through supply line and gas manifold to inlet.
Heater	Prevents **reliquefaction** of chlorine gas.
Vacuum Gauge	Indicates vacuum on chlorinator system.
Rotameter Tube and Float	Indicate chlorinator feed rate by reading top of float or center of ball for rate marked on tube.
Differential Regulating (Reducing) Valve	Regulates (reduces) chlorinator chlorine gas pressure. Serves to maintain a constant differential across the orifice to obtain repeatable chlorine gas flow rates at a given V-notch orifice opening regardless of fluctuations in the injector vacuum.

reliquefaction (re-LICK-we-FACK-shun)
The return of a gas to the liquid state; for example, a condensation of chlorine gas to return it to its liquid form by cooling.

Table 6.4   Typical chlorinator parts and purpose *(continued)*

Part	Purpose
V-notch Plug and Variable Orifice	Control chlorine feed rate by regulating flow of chlorine gas. A wide V-notch in the plug allows high feed rates through the orifice and a small V-notch in the plug provides low feed rates.
Vacuum Relief Valve	Relieves excess vacuum by allowing air to enter system and reduce vacuum.
Vacuum Relief	Provides source of air to reduce excess vacuum.
Injector Vacuum Gauge	Indicates vacuum at the injector.
Diaphragm Check Valve	Regulates chlorinator vacuum, which in turn adjusts chlorinator feed rate. Receives signal from chlorine feed rate controls and then adjusts feed rate by regulating vacuum.
Manual Feed Rate Adjuster	Regulates chlorine feed rate manually. Most chlorination systems have automatic feed rate controls with a manual override.
Injector Water Supply	Provides source of water for chlorine solution. Must provide sufficient pressure and volume to operate injector.
Injector	Mixes or injects chlorine gas into water supply. Creates sufficient vacuum to operate chlorinator and to pull metered amount of chlorine gas.
Solution Discharge	Discharges solution mixture of chlorine and water.

In many smaller systems, chlorine gas is withdrawn with equipment installed directly on the cylinder (Figure 6.14). A direct-mounted chlorinator meters prescribed (preset or selected) doses of chlorine gas from a chlorine cylinder, conveys it under a vacuum, and injects it into the water supply. Direct cylinder mounting is the safest and simplest way to connect the chlorinator to the chlorine cylinder. The valves on the cylinder and chlorinator inlet are connected by a positive metallic yoke, which is sealed by a single lead or fiber gasket.

rotameter (RODE-uh-ME-ter)
A device used to measure the flow rate of gases and liquids. The gas or liquid being measured flows vertically up a tapered, calibrated tube. Inside the tube is a small ball or bullet-shaped float (it may rotate) that rises or falls depending on the flow rate. The flow rate may be read on a scale behind or on the tube by looking at the middle of the ball or at the widest part or top of the float.

6.5.2.1   Chlorinator Flow Path
Chlorine gas flows from a chlorine container to the gas inlet. After entering the chlorinator, the gas passes through a spring-loaded pressure regulating valve, which maintains the proper operating pressure. A **rotameter** is used to indicate the rate of gas flow. The rate is controlled by an orifice. The gas then moves to the injector where it is dissolved in water. This mixture leaves the chlorinator as a chlorine solution (HOCl) ready for application.

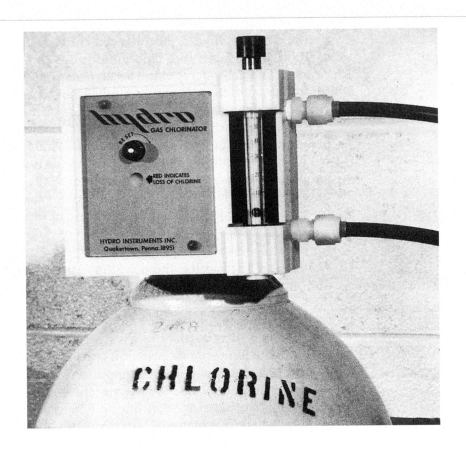

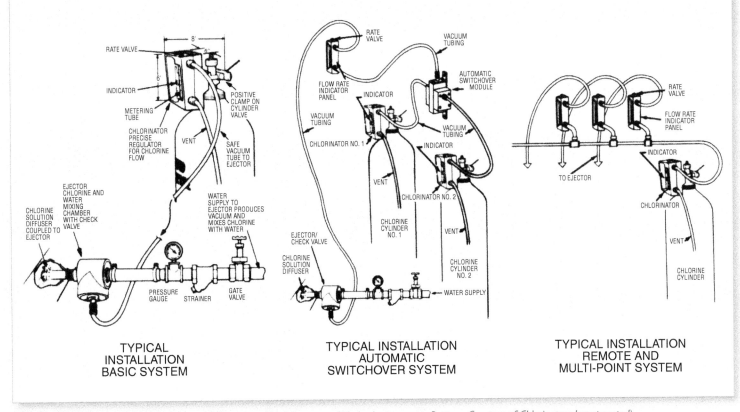

TYPICAL
INSTALLATION
BASIC SYSTEM

TYPICAL INSTALLATION
AUTOMATIC
SWITCHOVER SYSTEM

TYPICAL INSTALLATION
REMOTE AND
MULTI-POINT SYSTEM

Figure 6.14   Cylinder-mounted gas chlorinators *(Top: Courtesy of Hydro Instruments.; Bottom: Courtesy of Chlorinators Incorporated)*

The operating vacuum is provided by a hydraulic injector. The water supplied by this injector absorbs the chlorine gas. The resulting chlorine solution is conveyed to a chlorine diffuser through a corrosion-resistant conduit. A vacuum regulating valve dampens fluctuations and gives smoother operation. A vacuum relief valve prevents excessive vacuum within the equipment.

The primary advantage of a vacuum operation is safety. If a failure or breakage occurs in the vacuum system, the chlorinator either stops the flow of chlorine into the equipment or allows air to enter the vacuum system rather than allowing chlorine to escape into the surrounding atmosphere. In case the chlorine inlet shutoff fails, a vent valve discharges the incoming gas to a chlorine treatment facility outside of the chlorinator building.

6.5.2.2   Chlorinator Parts and Their Purpose

The Ejector: The ejector, fitted with a Venturi nozzle, creates the vacuum that moves the chlorine gas. Water supplied by a pump moves across the Venturi nozzle creating a differential pressure, which establishes the vacuum. The gas chlorinator is able to transport the chlorine gas to the water supply by reducing the gas pressure from the chlorine cylinder to less than the atmospheric pressure (vacuum).

In the past, it was not uncommon to find the ejector and the vacuum regulator mounted inside some type of cabinet. However, it makes better sense to locate the ejector at the site where the chlorine is to be applied, eliminating the necessity of pumping the chlorine over long distances and the associated problems inherent with gas pressure lines. Also, by placing the ejector at the application point, any tubing break will cause the chlorinator to shut down. This halting of operation stops the flow of gas and any damage that could result from a chlorine solution leak.

Check Valve Assembly: The vacuum created by the ejector moves through the check valve assembly. This assembly prevents water from back-feeding, that is, entering the vacuum-regulator portion of the chlorinator.

Rate Valve: The rate valve controls the flow rate at which chlorine gas enters the chlorinator. The rate valve controls the vacuum level and thus directly affects the action of the diaphragm assembly in the vacuum regulator. A reduction in vacuum lets the diaphragm close, causing the needle valve to reduce the inlet opening, which restricts chlorine gas flow to the chlorinator. An increase in the rate valve setting applies more vacuum to the diaphragm assembly, pulling the needle valve back away from the inlet opening and permitting an increased chlorine gas flow rate.

Diaphragm Assembly: This assembly connects directly to the inlet valve of the vacuum regulator, as described above. A vacuum (of at least 20 inches or 508 mm of water column) exists on one side of the diaphragm; the other side is open to atmospheric pressure through the vent. This differential in pressure causes the diaphragm to open the chlorine inlet valve allowing the gas to move (under vacuum) through the rotameter, past the rate valve and through the tubing to the check valve assembly, into the ejector nozzle area, and then to the point of application. If for some reason the vacuum is lost, the diaphragm will seat the needle valve on the inlet, stopping chlorine gas flow to the chlorinator.

Interconnection Manifold: If several gas cylinders provide the chlorine gas, direct cylinder mounting is not possible. An interconnection manifold made of seamless steel pipe and flexible connectors of cadmium-plated copper fitted with isolation valves must be used as the bridge between the chlorinator and the various cylinders.

The steel gas manifold with chlorine valve is mounted to the chlorinator. The flexible connector links the rigid manifold and the chlorine cylinder. The isolation valve between the flexible connector and the cylinder valve provides a way to close off the flexible connector when a new gas cylinder must be attached. This limits the amount of moisture that enters the system. Moisture in the system will combine with the chlorine gas to produce hydrochloric acid and cause corrosion. Corrosion can cause the manifold to fail.

The chlorine is usually injected directly into the water supply pipe and there may not be contact chambers or mixing units. The location of the injection point is important. The injection should never be on the intake side of the pump as it will cause corrosion problems. There should be a check valve and a meter to monitor the chlorine dose.

On most well applications, a chlorine booster pump is needed to overcome the higher water pump discharge pressures. If the well produces sand in the water, this pump will wear rapidly and become unreliable. In this situation, the chlorine solution should be introduced down the well through a polyethylene tube.

The polyethylene tube (½ inch or 12 mm) must be installed in the well so as to discharge a few inches below the suction screen. The chlorinator should operate only when the pump is running. The chlorine solution flowing through the polyethylene tube is extremely corrosive. If the tube does not discharge into flowing water, the effect of the solution touching the metal surface can be disastrous. Wells have been destroyed by corrosion from chlorine.

Check Your Understanding

1. How is the rate of gas flow in a chlorinator measured?
2. What is the primary advantage of vacuum system chlorinators?

6.5.3   Chlorine Containers

Chlorine containers are made of plastic or steel and range from small portable cylinders to 1-ton steel tanks.

6.5.3.1   Plastic

Plastic containers are commonly used for storage of hypochlorite solution (Figure 6.6). The container size depends on usage. Plastic containers should be large enough to hold a 2- or 3-days' supply of hypochlorite solution. The solution should be prepared every 2 or 3 days. If a larger amount of solution is mixed, the solution may lose its strength and thus affect the chlorine feed rate. Normally, a week's supply of hypochlorite should be in storage and available for preparing hypochlorite solutions. Store the hypochlorite in a cool, dark place. Sodium hypochlorite can lose from 2 to 4 percent of its available chlorine content per month at room temperature. Therefore, manufacturers recommend a maximum shelf life of 60 to 90 days.

Figure 6.15 Chlorine cylinder *(Courtesy of Hhakim/iStock)*

Steel Cylinders

Cylinders containing 100 to 150 pounds (45 to 68 kg) of chlorine are convenient for very small treatment plants with capacities of less than 0.5 MGD (1,890 m³/day) (Figure 6.15). A fusible plug is placed in the valve (Figure 6.16) below the valve seat. This plug is a safety device. The fusible metal softens or melts at 158 to 165°F (70 to 74°C) to prevent buildup of excessive pressures and the possibility of rupture due to a fire or high surrounding temperatures. Cylinders will not explode under normal conditions and can be handled safely.

The following are safe procedures for handling and storing chlorine cylinders:

1. Move cylinders with a properly balanced hand truck with clamp supports that fasten about two-thirds of the way up the cylinder.
2. Cylinders (100- and 150-pound or 45 to 68 kg) can be rolled in a vertical position. Avoid lifting these cylinders except with approved equipment. Use a lifting clamp, cradle, or carrier. Never lift with homemade chain devices, rope slings, or magnetic hoists. Never roll, push, or drop cylinders off the back of trucks or loading docks.
3. Always replace the protective cap when moving a cylinder.
4. Keep cylinders away from direct heat (steam pipes or radiators) and direct sun, especially in warm climates.
5. Transport and store cylinders in an upright position.
6. Firmly secure cylinders to an immovable object.

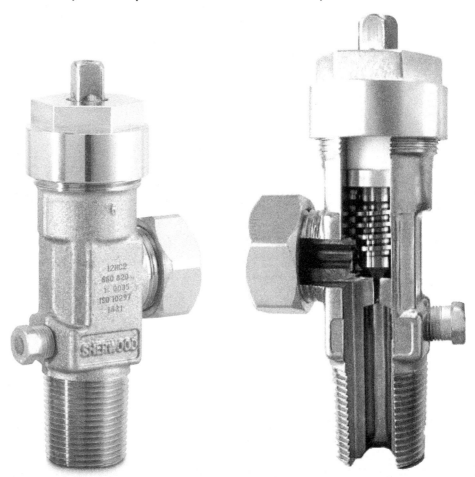

Figure 6.16 Standard chlorine cylinder valve *(Courtesy of Sherwood Valve LLC.)*

7. Store empty cylinders separately from full cylinders. All empty chlorine cylinders must be tagged as empty.

    Note: Never store chlorine cylinders near turpentine, ether, anhydrous ammonia, finely divided metals, hydrocarbons, or other materials that are flammable in air or will react violently with chlorine.

8. Remove the outlet cap from the cylinder and inspect the threads on the outlet. Cylinders having outlet threads that are corroded, worn, cross-threaded, broken, or missing should be rejected and returned to the supplier.

9. The specifications and regulations of the US Interstate Commerce Commission require that chlorine cylinders be tested at 800 pounds per square inch (psi) (5,516 kPa or 56.24 kg/cm²) every 5 years. The date of testing is stamped on the dome of the cylinder. Cylinders that have not been tested within that period of time should be rejected and returned to the supplier.

6.5.3.3 Ton Tanks

1. Ton tanks are of welded steel construction and have a loaded weight of as much as 3,700 pounds (1,680 kg). They are about 80 inches (200 cm) in length and 30 inches (75 cm) in outside diameter. The ends of the tanks are crimped inward to provide a substantial grip for lifting clamps.

2. Ton tanks have 6 openings for fusible plugs and valves. Generally, two operating valves are located on one end near the center. There are six fusible metal safety plugs, three on each end. These are designed to melt within the same temperature range as the safety plug in the cylinder valve.

For safe handling of ton tanks, follow these procedures:

1. Ship ton tanks by rail in multiunit cars or by truck or semitrailer.

2. Handle ton tanks with a suitable lift clamp in conjunction with a hoist or crane of at least 2-ton capacity.

3. For storage and for use, lay ton tanks on their sides, above the floor or ground, on steel or concrete supports. They should not be stacked more than one high and should be separated by 30 inches (0.75 m) for access in case of leaks.

4. Place ton tanks on trunnions (pivoting mounts) that are equipped with rollers so that the withdrawal valves may be positioned one above the other. The upper valve will discharge chlorine gas, and the lower valve will discharge liquid chlorine. The ability to rotate tanks is also a safety feature. In case of a liquid leak, the container can be rolled so that the leaking chlorine escapes as a gas rather than a liquid.

5. Use trunnion rollers that do not exceed 3½ inches (9 cm) in diameter so that the containers will not rotate too easily and be turned out of position.

6. Equip roller shafts with a zerk-type lubrication fitting. Roller bearings are not advised because of the ease with which they rotate.

7. Use locking devices to prevent ton tanks from rolling while connected.

 It is very important that fusible plugs should not be tampered with under any circumstances and that the tank should not be heated. Once this plug opens, all of the chlorine in the tank will be released.

Pascal (Pa)
The pressure or stress of one newton per square meter.
1 psi = 6,895 Pa = 6.895 kN/m²
= 0.0703 kg/cm²

⚠ Do not use canister-type gas masks where chlorine leaks occur. Self-contained air or supplied-aid types of breathing apparatus (Figure 6.17) are the best option. Chemical cartridge respirators may be used for escape only.

⚠ Small amounts of chlorine cause large problems. Leaks never get better if ignored.

Figure 6.17   Self-contained breathing apparatus *(Courtesy of rmfox/iStock)*

6.5.4   Protect Yourself from Chlorine

Every person working with chlorine should know the proper ways to handle it, should be trained in the use of appropriate respiratory protective devices and methods of detecting hazards, and should know what to do in case of emergencies.

Self-contained air supply and demand-breathing equipment must fit properly and be used properly. Pressure-demand and rebreather kits may be safer. Pressure-demand units use more air from the air bottle, which reduces the time a person may work on a leak. There are certain physical constraints when using respiratory protection. Confirm requirements with your local safety regulatory agency.

Before entering an area with a chlorine leak, wear protective clothing. A chemical suit will prevent chlorine from contacting the sweat on your body and forming hydrochloric acid. Chemical suits are very cumbersome, but should be worn when the chlorine concentration is high. A great deal of practice is required to perform effectively while wearing a chemical suit.

The best protection that one can have when dealing with chlorine is to respect it. Each individual should practice rules of safe chlorine handling and preventive safety strategies.

6.5.5   Removing Chlorine from Containers

Whenever you do any work or maintenance involving the removal of chlorine from containers, a self-contained breathing apparatus (Figure 6.17) should be worn, or at least be readily available. This is especially true when searching for chlorine leaks.

The maximum rate of chlorine removal from a 150-pound (68 kg) cylinder is 40 pounds (18 kg) of chlorine per day. If the rate of removal is greater, freezing can occur and less chlorine will be delivered.

6.5.5.1   Connections

The outlet threads on container valves are specialized threads; they are not ordinary tapered pipe threads. Use only the fittings and gaskets furnished by your chlorine supplier or chlorinator equipment manufacturer when making connections to chlorine containers. Do not try to use regular pipe thread fittings. Whenever you make a new connection, always use a new gasket. The outlet threads on container valves should always be inspected before being connected to the chlorine system. Containers with outlet threads that are badly worn, cross threaded, or corroded should be rejected and returned to the supplier. The connecting nut on the chlorine system should also be inspected and replaced if it develops any of these problems. Since the threads on the cylinder connection may become worn, yoke-type connectors (Figure 6.18) are recommended.

Flexible ⅜-inch 2,000-pound (psi) (0.95 cm, 13,790 kPa or 140 kg/cm²) annealed (toughened) copper tubing (pigtail, or coiled horizontally at least three times) is recommended for connection between chlorine containers and stationary piping. Care should be taken to prevent sharp bends in the tubing because this will weaken it and eventually the tubing will start leaking. Many operators recommend use of a sling to hold the tubing when disconnecting it from an empty cylinder to prevent the tubing from flopping around and getting kinked or getting dirt inside it. Cap or plug the connectors and valves when they are not connected to prevent entry of dirt/debris and moisture.

To simplify changing containers, you will also need a shutoff valve located just beyond the container valve or at the beginning of the stationary piping.

Figure 6.18   Yoke and adapter-type connection (*Courtesy of the Chlorine Institute*)

6.5.5.2 Valves

Do not use wrenches longer than 6 inches (15 cm), pipe wrenches, or wrenches with an extension on container valves. If you do, you could exert too much force and break the valve. Use only a square end-open or box wrench, which can be obtained from your chlorine supplier (see Figure 6.18). To unseat the valve, strike the end of the wrench with the heel of your hand to rotate the valve stem in a counterclockwise direction. Then open slowly. One complete turn permits maximum discharge. Do not force the valve beyond the full open position or you may strip the internal valve stem threads. If the valve is too tight to open in this manner, loosen the packing gland nut slightly to free the stem. Open the valve, then retighten the packing nut. If you are uncertain how to loosen the nut, you should return the container to the supplier. Do not use organic lubricants on valves and lines because further chemical reactions can block the lines.

6.5.5.3 Ton Tanks

One-ton tanks must be placed on their sides with the valves in a vertical position so either chlorine gas or liquid chlorine may be removed. Connect the flexible tubing to the top valve to remove chlorine gas from a tank. The bottom valve is used to remove liquid chlorine and is used only with a chlorine evaporator. The valves are similar to those on the smaller chlorine cylinders (fusible plugs are not located at valves on ton containers) and must be handled with the same care.

Check Your Understanding

1. When are mouthbit chemical cartridge respirators used?
2. How can copper tubing be prevented from getting kinks or dirt inside the tubing?
3. Why are 1-ton tanks placed on their sides with the valves in a vertical position?

6.5.6   Performance of Chlorination Units

Before attempting to start or stop any chlorination system, read the manufacturer's literature and your plant's operation and maintenance (O&M) instructions to become familiar with the equipment. Review the plans and drawings of the facility. Determine what equipment, pipelines, pumps, tanks, and valves are to be placed into service or are in service.

The current status of the entire system must be known before starting or stopping any portion of the system. This section provides the procedures for a typical system and will give you ideas for your system.

6.5.6.1   Hypochlorinators

1. *Startup of Hypochlorinators*
   a. Prepare the chemical solution. Most agencies buy commercial or industrial hypochlorite at a strength of 12 to 15 percent chlorine. This solution is usually diluted down to a 2 percent solution. If using commercially prepared solutions, you will need to calculate feed rates.
   b. Lock out the electric circuit and then inspect it. Normally, no adjustments are needed. Look for frayed wires. Turn power back on. Leave solution switch off.
   c. Turn on the chemical pump. Make adjustments while the pump is running. Never adjust the pump while it is off because damage to pump will occur.
   d. Calibrate the pump to ensure accurate delivery of chlorine solution. See Chapter 3, "Coagulation and Flocculation," Section 3.10.2, "Calibration Method."
   e. Make sure the solution is being fed into the system. Measure the chlorine residual just downstream from where the solution is being fed into the system. You may have a target residual you wish to maintain at the beginning of the system such as 2.0 mg/L.
   f. Check the chlorine residual in the system. The residual should be measured at the most remote test location within the distribution system and should be at least 0.2 mg/L free residual chlorine. This chlorine residual is necessary to protect the treated water from any recontamination. Adjust the chemical feed as needed.

2. *Shutdown of Hypochlorinators*
   a. Short Duration
      (1) Turn the water supply pump off. You do not want to pump any unchlorinated water and possibly contaminate the rest of the system. Never pump any unchlorinated water into your distribution system.
      (2) Turn the hypochlorinator off.
      (3) When making repairs, lock out the circuit or pull the plug from an electric socket.
   b. Long Duration
      (1) Obtain and place another hypochlorinator in service.

3. *Normal Operation of Hypochlorinators*
   Normal operation of a hypochlorinator requires routine observation and preventive maintenance.

   DAILY
   a. Inspect the building to make sure only authorized personnel have been there.
   b. Read and record the level of the solution tank at the same time every day.
   c. Read the meters and record the amount of water pumped.
   d. Check the chlorine residual (at least 0.2 mg/L) in the system and adjust the chlorine feed rate as necessary. Try to maintain a

chlorine residual of 0.2 mg/L at the most remote point in the distribution system. The suggested free chlorine residual for treated water or well water is 0.5 mg/L at the point of chlorine application provided the 0.2 mg/L residual is maintained throughout the distribution system an d coliform test results are negative.

e. Check chemical feed pump operation. Most hypochlorinators have a dial with a range from 0 to 10 that adjusts the chlorine feed rate. Start with a setting around 6 or 7 on the dial and use a 2 percent hypochlorite solution. The pump should operate in the upper ranges of the dial so that the strokes or pulses from the pump will be close together. In this way, the chlorine will be fed continuously to the water being treated. Adjust the feed rate after testing residual levels.

WEEKLY

a. Clean the building.

b. Replace the chemicals and wash the chemical storage tank. Try to have a 15- to 30-day supply of chlorine in storage for future needs. When preparing hypochlorite solutions, prepare only enough for a 2- or 3-day supply.

MONTHLY

a. Check the operation of the check valve.

b. Perform any required preventive maintenance suggested by the manufacturer.

c. Cleaning
Commercial sodium hypochlorite solutions (such as Clorox) contain an excess of caustic (sodium hydroxide or NaOH). When this solution is diluted with water containing calcium and carbonate alkalinity, the resulting solution becomes supersaturated with calcium carbonate. This calcium carbonate tends to form a coating on the poppet valves in the solution feeder. The coated valves will not seal properly and the feeder will fail to feed properly.

Use the following procedure to remove the carbonate scale:

(1) Fill a 1-quart (1-liter) Mason jar half full of tap water.

(2) Place 1 fluid ounce (20 mL) of 30 to 37 percent hydrochloric acid (swimming pool acid) in the jar. Always add acid to water, never the reverse.

(3) Fill the jar with tap water.

(4) Place the suction hose of the hypochlorinator in the jar and pump the entire contents of the jar through the system.

(5) Return the suction hose to the hypochlorite solution tank and resume normal operation.

You can prevent the formation of the calcium carbonate coatings by obtaining the dilution water from an ordinary home water softener.

NORMAL OPERATION CHECKLIST

a. Check chemical usage. Record the solution level and the water pump meter reading or number of hours of pump operation. Calculate the amount of chemical solution used and compare with the desired feed rate. See Example 3.

b. Determine if every piece of equipment is operating.

    c. Inspect the lubrication of the equipment.
    d. Check the building for any possible problems.
    e. Clean up the area.

FORMULAS

When operating a hypochlorinator, you should compare the actual chlorine dose applied to the water being treated with the desired chlorine dose in milligrams per liter. The actual dose is calculated by determining the amount of chlorine actually used and the amount of water treated. The amount of chlorine used is found by measuring the amount of hypochlorite solution used and knowing the strength of the hypochlorite solution. The amount of water used is determined from a flowmeter.

To calculate the amount of water treated, determine the amount in gallons from a flowmeter and convert this amount from gallons to pounds.

$$\text{Water}\,[\text{lb}] = \text{Water Treated}\,[\text{gal}] \times \frac{8.34\ \text{lb}}{\text{gal}}$$

To calculate the amount of hypochlorite used in gallons, determine the volume of hypochlorite ($V_{\text{hypochlorite}}$) used in gallons.

$$V_{\text{hypochlorite}}\,[\text{gal}] = 0.785 \times \text{Diameter}^2\,[\text{ft}^2] \times \text{Depth}\,[\text{ft}] \times \frac{7.48\ \text{gal}}{\text{ft}^3}$$

To determine the pounds of chlorine used to disinfect the water being treated, we have to convert the hypochlorite used from gallons to pounds of chlorine (Cl) by considering the strength of the hypochlorite solution.

$$\text{Cl}\,[\text{lb}] = V_{\text{hypochlorite}}\,[\text{gal}] \times \frac{8.34\ \text{lb}}{\text{gal}} \times \frac{\text{Hypochlorite}\,[\%]}{100\,[\%]}$$

Finally, to estimate the actual chlorine dose in milligrams of chlorine per liter of water treated, we divide the pounds of chlorine used (Cl_{used}) by the millions of gallons (Mgal) of water treated (V_{water}).

$$\text{Cl}_{\text{dose}}\left[\frac{\text{mg}}{\text{L}}\right] = \frac{\text{Cl}_{\text{used}}\,[\text{lb}] \times \dfrac{\text{kg}}{2.205\ \text{lb}} \times \dfrac{10^6\ \text{mg}}{\text{kg}}}{V_{\text{water}}\,[\text{Mgal}] \times \dfrac{10^6}{\text{M}} \times \dfrac{3.785\ \text{L}}{\text{gal}}} = \frac{\text{Cl}_{\text{used}}\,[\text{lb}]}{8.34 \times V_{\text{water}}\,[\text{Mgal}]}$$

$$= 0.120 \times \frac{\text{Cl}_{\text{used}}\,[\text{lb}]}{V_{\text{water}}\,[\text{Mgal}]}$$

Example 3

Water pumped from a well is disinfected by a hypochlorinator. A chlorine dose of 1.2 mg/L is necessary to maintain an adequate chlorine residual throughout the system. During a 1-week time period, the water meter indicated that 2,289,000 gallons (gal) of water were pumped. A 2 percent (%) sodium hypochlorite solution is stored in a 3-foot-diameter plastic tank. During this 1-week period, the level of hypochlorite in the tank dropped 2 feet 8 inches (2.67 feet). Does the chlorine feed rate appear to be too high, too low, or about OK?

Known		Unknown	
Desired Chlorine Dose	= 1.2 mg/L	$V_{hypochlorite}$	= Volume of Hypochlorite Used, gal
V_{water} = Volume of Water Pumped	= 2,289,000 gal		
Hypochlorite Content	= 2%	Cl_{dose}	= Actual Chlorine Dose, mg/L
D = Chemical Tank Diameter	= 3 ft	Is Actual Dose OK?	
h = Chemical Drop in Tank	= 2.67 ft		

1. Calculate the pounds (lb) of water disinfected.

$$\text{Water}[\text{lb}] = V[\text{gal}] \times \frac{8.34 \text{ lb}}{\text{gal}}$$

$$= 2,289,000[\text{gal}] \times \frac{8.34 \text{ lb}}{\text{gal}}$$

$$= 19,090,000 \text{ lb}$$

$$= 19.09 \text{ Mlb}$$

2. Calculate the volume of 2 percent (%) sodium hypochlorite solution used in gallons (gal).

$$V_{hypochlorite}[\text{gal}] = 0.785 \times D^2[\text{ft}^2] \times h[\text{ft}] \times \frac{7.48 \text{ gal}}{\text{ft}^3}$$

$$= 0.785 \times 3^2[\text{ft}^2] \times 2.67[\text{ft}] \times \frac{7.48 \text{ gal}}{\text{ft}^3}$$

$$= 141.1 \text{ gal}$$

3. Determine the pounds (lb) of chlorine (Cl) used to disinfect the water.

$$Cl[\text{lb}] = V_{hypochlorite}[\text{gal}] \times \frac{8.34 \text{ lb}}{\text{gal}} \times \frac{\text{Hypochlorite}[\%]}{100\%}$$

$$= 141.1[\text{gal}] \times \frac{8.34 \text{ lb}}{\text{gal}} \times \frac{2\%}{100\%} = 23.5 \text{ lb}$$

4. Convert the volume of water from gallons (gal) to million gallons (Mgal).

$$V_{water}[\text{Mgal}] = V_{water}[\text{gal}] \times \frac{M}{10^6} = 2,289,000[\text{gal}] \times \frac{M}{10^6} = 2.289[\text{Mgal}]$$

5. Estimate the chlorine dose in mg/L.

$$Cl_{dose}\left[\frac{\text{mg}}{\text{L}}\right] = 0.120 \times \frac{Cl_{used}[\text{lb}]}{V_{water}[\text{Mgal}]} = 0.120 \times \frac{23.5[\text{lb}]}{2.289[\text{Mgal}]} = 1.23\left[\frac{\text{mg}}{\text{L}}\right]$$

Because the actual estimated chlorine dose (1.23 mg/L) was similar to the desired dose of 1.2 mg/L, the chlorine feed rate appears OK.ç

4. *Abnormal Operation of Hypochlorinators*
   a. Inform your supervisor of the problem.
   b. If the hypochlorinator malfunctions, it should be repaired or replaced immediately. See the shutdown operation (Step 2 in this section).
   c. Solution tank level.
      (1) If Too Low: Check the adjustment of the pump.
                        Check the hour meter on the water pump.
      (2) If Too High: Check the chemical pump.
                        Check the hour meter on the water pump.
   d. Determine if the chemical pump is not operating.

      Troubleshooting Guidelines

      (1) Check the electrical connection.
      (2) Check the circuit breaker.
      (3) Check for stoppages in the flow lines.

      Corrective Measures

      (1) Shut off the water pump so that no unchlorinated water is pumped into the system.
      (2) Check for a blockage in the solution tank.
      (3) Check the operation of the check valve.
      (4) Check the electric circuits.
      (5) Replace the chemical feed pump with another pump while repairing the defective unit.
   e. The solution is not being pumped into the water line.

      Troubleshooting

      (1) Check the solution level.
      (2) Check for blockages in the solution line.

5. *Maintenance of Hypochlorinators*
   Hypochlorinators on small systems may be sealed systems and cannot be repaired so replacement is the only solution. Some units are repairable and can be serviced by following the manufacturer's instructions. Maintenance requirements are normally minor such as oil changes and lubrication of moving parts. Review the manufacturer's specifications and instructions for maintenance requirements.

Check Your Understanding

1. What would you do before attempting to start any chlorination system?

2. What should be the chlorine residual in the most remote part of the distribution system?

3. Why should a hypochlorite feed pump be operated in the upper end of its range (at 6 or 7 in a range of 0 to 10)?

6.5.6.2 Chlorinators

1. *Safety Equipment Required and Available Outside the Chlorinator Room*
   a. Protective clothing
      (1) Gloves
      (2) Rubber suit

b. Self-contained pressure-demand air supply system (Figure 6.17)

c. Chlorine leak detector/warning device (Figure 6.20, page 385) should be located outside the chlorine room and should have a battery backup in case of a power failure. The chlorine sensor unit should be in the chlorine room and connected to the leak detector/warning device, which is located outside the chlorine storage room.

2. *Chlorinator Startup*

Procedures for startup, operation, shutdown, and troubleshooting are outlined in this section and are intended to be typical procedures for all types of chlorinators. For specific directions, see the manufacturer's literature and O&M instructions for your plant. During emergencies, you must act quickly and may not have time to check out each of the steps outlined below, but you still must follow established procedures. Work in pairs. Never work alone when hooking up a chlorine system.

a. Gas Chlorinators

Startup procedures for chlorinators using chlorine gas from containers are outlined in this section.

(1) Be sure the chlorine gas valve at the chlorinator is closed. This valve should already be closed since the chlorinator is out of service.

(2) All chlorine valves on the supply line should have been closed during shutdown. Be sure they are still closed. If any valves are required to be open for any reason, this exception should be indicated by a tag on the valve.

(3) Inspect all tubing, manifolds, and valve connections for potential leaks and be sure all joints are properly gasketed.

(4) Check chlorine solution distribution lines to be sure that system is properly valved to deliver chlorine solution to desired point of application.

(5) Open the chlorine metering orifice slightly by adjusting chlorine feed-rate control.

(6) Start the injector water supply system. This is usually a potable water supply protected by an **air gap** or air-break system. Injector water is pumped at an appropriate flow rate and the flow through the injector creates sufficient vacuum in the injector to draw chlorine. Chlorine is absorbed and mixed in the water at the injector. This chlorine solution is conveyed to the point of application.

(7) Examine injector water supply system.

(a) Note the reading of injector supply pressure gauge. If the reading is abnormal (different from usual reading), try to identify the cause and correct it. Injectors should operate at not less than 50 psi inlet pressure.

(b) Note the reading of injector vacuum gauge. If the system does not have a vacuum gauge, have one installed. If the vacuum reading is less than normal, the machine may function at a lower feed rate, but will be unable to deliver at rated capacity.

(8) Inspect chlorinator vacuum lines for leaks.

(9) Crack open the chlorine container valve and allow gas to enter the line. Inspect all valves and joints for leaks by placing an

air gap
An open, vertical drop, or vertical empty space, between a drinking (potable) water supply and the nonpotable point of use. This gap prevents the contamination of drinking water by backsiphonage because it stops wastewater from reaching the drinking water supply. Air gap devices are also used to provide adequate space above the top of a manhole and the end of the hose from the fire hydrant.

*Use a concentrated ammonia solution containing 28 to 30 percent ammonia as NH_3 (this is the same as 58 percent ammonium hydroxide, NH_4OH, or commercial 26° Baumé).

ammonia-soaked rag* near each valve and joint. A polyethylene squeeze bottle filled with ammonia water to dispense ammonia vapor may also be used. Care should be taken to avoid spraying ammonia water on any leak or touching the soaked cloth to any metal. The formation of a white cloud of vapor will indicate a chlorine leak. Start with the valve at the chlorine container, move down the line and check all joints between this valve and the next one downstream. Never apply ammonia solutions directly to any valve because an acid will form that will eat away the valve fittings. If the downstream valve passes the ammonia test, open the valve and continue to the next valve. If there are no leaks to the chlorinator, open the cylinder valve approximately one complete turn to obtain maximum discharge and continue with the startup procedure.

(10) Inspect the chlorinator.

(a) Chlorine gas pressure at the chlorinator should be between 20 and 30 psi (137.9 kPa and 206.85 kPa or 1.4 and 2.1 kg/cm²). However, in the summer on very hot days, the pressure in a chlorine cylinder may exceed 150 psi (1,035 kPa or 10.5 kg/cm²).

(b) Operate the chlorinator over the complete range of feed rates.

(c) Check the operation of manual and automatic settings.

(11) The chlorinator is ready for use. Set the desired feed rate. Log in the time the system is placed into operation and the application point.

b. Chlorinators with Evaporators (Figure 6.19)

Startup procedures for chlorinators using liquid chlorine from containers are outlined in this section. Only the very largest water treatment plants require chlorinators equipped with evaporators. In most plants, liquid chlorine is delivered in 1-ton containers. However, liquid chlorine may be delivered in railroad tank cars to very large plants.

(1) Inspect all joints, valves, manifolds, and tubing connections in the chlorination system, including application lines, for proper fit and for leaks. Make sure that all joints have gaskets.

(2) If the chlorination system has been broken open or exposed to the atmosphere, verify that the system is dry. Usually, once a system has been dried out, it is never opened again to the atmosphere. However, if moisture enters the system, from the air or by other means, it readily mixes with chlorine and forms hydrochloric acid, which will corrode the pipes, valves, joints, and fittings. Corrosion can cause leaks and require that the entire system be replaced.

To verify that the system is dry, determine the **dew point** (must be lower than –40°F or –40°C). If not dry, turn the evaporator on, pass dry air through the evaporator and force this air through the system. If this step is omitted and moisture remains in the system, serious corrosion damage can result and the entire system may have to be repaired.

(3) Start up the evaporators. Fill the water bath and adjust the device according to the manufacturer's directions.

dew point
The temperature to which air with a given quantity of water vapor must be cooled to cause condensation of the vapor in the air.

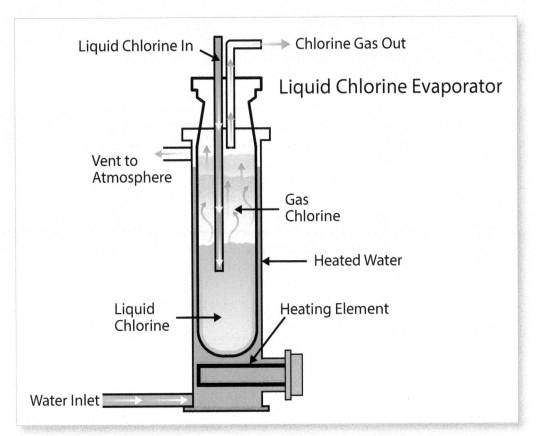

Figure 6.19   Liquid chlorine evaporator

Water baths or evaporators should be equipped with low-water alarms and automatic shutoffs in case of excessive heat. Meltdown can occur with lack of water and chlorine is explosive at high temperatures.

(4) Turn on the heaters on the evaporators.

(5) Wait until the temperature of the evaporators reaches 180°F (82°C). This may take over an hour on large units.

(6) Inspect and close all valves on the chlorine supply line. One chlorine valve on the evaporator (the inlet valve) should be open to allow for expansion of the chlorine if heated.

(7) Open the chlorine metering orifice slightly by adjusting the chlorine feed-rate knob. This is to prevent damage to the rotameter.

(8) Start the injector water supply system.

(9) Examine the injector water supply system.

 (a) Note the reading of the injector water supply pressure gauge. If the gauge reading is abnormal (different than usual reading), try to identify the cause and correct it. Injectors should operate at not less than 50 psi (345 kPa or 3.5 kg/cm²) inlet pressure.

 (b) Note the reading of the injector vacuum gauge. If the system does not have a vacuum gauge, have one installed. If the vacuum reading is less than normal, the machine may function at a lower feed rate, but will be unable to deliver at the maximum rated capacity.

(10) Inspect chlorinator vacuum lines for leaks.

(11) On the system connected to the gas side of the supply container, crack open the chlorine gas line at the chlorine container. All liquid chlorine systems should be checked by using gas because of the danger of leaks (1 liter of liquid will produce 450 liters of gas). Inspect the joints between this valve and the next one downstream. If this valve passes the ammonia leak test, continue to the next valve down the line. Follow this procedure until the evaporator is reached. Before allowing chlorine to enter the evaporator and the chlorinator, make sure that all valves between the evaporator and the chlorinator are open. Heat in the evaporator will expand the gas and, if the system is closed, excessive pressure can develop. Chlorine should never be trapped in a line, an evaporator, or a chlorinator. After the gas test, reconnect to the liquid side and retest at connection and pigtail valve.

(12) If no problems develop, the gas line can be put in service by opening the valve one complete turn.

(13) Check the operation of the chlorinator.

(a) Operate over the complete range of chlorine feed rates.
(b) Check the operation of manual and automatic settings.

(14) If the chlorinator is operating properly, close the gas valve on the container, wait a few minutes to evacuate the pigtail connector, close the valve from the pigtail to the manifold, disconnect the pigtail, replace the gas valve cap, remove the liquid valve cap, clean the valve face and threads, install a new gasket to connect the pigtail, crack the valve open and then close it. Check for leaking at the new connection. If there is no leak, open the liquid valve two turns (usually one turn open is sufficient).

(15) After admitting liquid chlorine to the system, wait until the temperature of the evaporator again reaches 180°F (82°C) and full working pressure (100 psi, 690 kPa or 7 kg/cm²). Inspect the evaporator by looking for leaks around pipe joints, unions, and valves.

(16) The system is ready for normal operation.

3. *Chlorinator Shutdown (when operating from container gas connection)*

Work in pairs. A plan should be used where both people are not exposed to the chlorine at the same time.

a. Short-Term Shutdown

The following is a typical procedure for shutting down a chlorinator for a time period of less than 1 week.

(1) Have safety equipment available in the event of a chlorine leak.

(2) Close chlorine container gas outlet valve.

(3) Allow chlorine gas to completely evacuate the system through the injector. Chlorine gas pressure gauges will fall to 0 psi on the manifold and the chlorinator.

(4) Close chlorinator gas discharge valve. The chlorinator may remain in this condition indefinitely and is ready to be placed back in service by reopening the chlorinator discharge valve and the chlorine container gas outlet valve. After these valves have been reopened, inspect for chlorine leaks throughout the chlorination system.

  b. Long-Term Shutdown
     (1) Perform Steps 1 through 4 above for short-term shutdown.
     (2) Turn off the chlorinator power switch, lock out, and tag.
     (3) Secure the chlorinator gas manifold and chlorinator valve in closed position.

Check Your Understanding

1. When starting a gas chlorinator, how is the system checked for chlorine gas leaks?

2. List the steps to follow when shutting down a chlorinator for a long time period.

6.5.7   Normal and Abnormal Operation

Normal operation of the chlorination process requires regular observation of facilities and a regular preventive maintenance program. When something abnormal is observed or discovered, corrective action must be taken. This section outlines normal operation procedures and possible operator responses to abnormal conditions. Exact procedures will depend on the type of equipment in your plant.

6.5.7.1   Container Storage Area

DAILY

  1. Inspect building or area for ease of access by authorized personnel to perform routine and emergency duties.
  2. Be sure fan and ventilation equipment are operating properly.
  3. Read scales, charts, or meters at the same time every day to determine use of chlorine and any other chemicals. Notify plant superintendent when chlorine supply is low.
  4. Look at least once per shift for chlorine and chemical leaks.
  5. Maintain the temperature of the storage area below the temperature of the chlorinator room.
  6. Determine manifold pressure before and after the chlorine pressure regulating valve.
  7. Be sure all chlorine containers are properly secured.

WEEKLY

  1. Clean the building or storage area.
  2. Check the operation of the chlorine leak-detector alarm.

MONTHLY

  1. Exercise all valves, including flex connector's auxiliary, manifold, filter bypass, and switchover valves.
  2. Inspect all flex connectors and replace any that have been kinked or flattened.
  3. Inspect hoisting equipment.
     a. Cables: frayed or cut.
     b. Container hoisting beam and hooks: cracked or bent.

c. Controls: operate properly, do not stick or respond sluggishly, cords not frayed, safety chains or cables in place.

d. Hoist travels on monorail easily and smoothly.

4. Examine building ventilation.

a. Ducts and louvers: clean and operate freely.

b. Fans and blowers: operate properly, guards in place, equipment properly lubricated.

5. Perform preventive maintenance as scheduled. These duties may include:

a. Lubricating equipment (monthly to quarterly).

b. Repacking valves and regulators (6 to 12 months).

c. Cleaning and replacing valve seats and stems (annually).

d. Cleaning filters and replacing glass wool (annually).

e. Painting equipment if needed (annually).

f. Inspecting condition and parts of all repair kits and safety equipment (monthly).

Glass wool soaked with liquid chlorine or chlorine impurities may burn your skin or give off sufficient chlorine gas to be dangerous.

6.5.7.2 Evaporators

Evaporators are used to convert liquid chlorine to gaseous chlorine for use by gas chlorinators (Figure 6.19).

DAILY

1. Check the evaporator water bath to be sure the water level is at the midpoint of the sight glass.

2. In most evaporators the water bath temperature is between 160 and 195°F (71 and 91°C). The low alarm should sound at 160°F (71°C) and the high alarm should sound at 200°F (93°C).

3. Determine the chlorine inlet pressure to evaporator. The pressure should be the same as on the supply manifold from containers (20 to 100 psi, 138 to 690 kPa, or 1.4 to 7 kg/cm²).

4. Determine the chlorine outlet temperature from the evaporator, typically from 90 to 105°F (32 to 41°C). The high alarm should sound at 110°F (43°C). At low temperatures the chlorine pressure-reducing valve (PRV) will close due to the low temperature in the water bath.

5. Check the chlorine PRV operation.

6. If the evaporator is equipped with a water bath recirculation pump at the back of the evaporator, determine if the pump is operating properly.

7. Look for leaks and repair any discovered.

ABNORMAL EVAPORATOR CONDITIONS

1. The evaporator water level is low. The water level is not visible in the sight glass.

Troubleshooting Measures

a. Determine the actual level of the water.

b. Monitor the temperature of the water.

c. Check the temperature and pressure of chlorine in the evaporator system and feed lines back to containers and chlorinators for possible overpressure of the system (pressure should not exceed 100 psi, 690 kPa, or 7 kg/cm²).

Corrective Action

a. If the chlorine pressure on the system is near or over 100 psi (690 kPa or 7 kg/cm^2), close the supply container valves to stop chlorine addition to the system, increase the chlorinator feed rate to use the chlorine in the system, and drop the pressure back down to a safe range. *Note*: The alarm system is usually set to sound at 110 psi (758 kPa or 7.7 kg/cm^2). If the system pressure was at or over 110 psi (758 kPa or 7.7 kg/cm^2), inspect the alarm circuit to determine why the alarm failed.

b. If the temperature of the water bath is set at an abnormal level, find the cause. Water bath levels are usually set in the following sequence:

    (1) 160°F or 71°C: low temperature alarm.
    (2) 185 to 195°F or 85 to 91°C: normal operating range.
    (3) 185°F or 85°C: actuates PRV to open position.
    (4) 200°F or 93°C: high temperature alarm.

      If the temperature is above 200°F (93°C), the alarm should have sounded. Open the control switch on the evaporator heaters to stop the current flow to the heating elements. If the temperature is in the normal range, return to correcting the original problem of a low water level.

c. Evaporator water levels are controlled by a solenoid valve. First check to see if the drain valve is fully closed and then use the following steps:

    (1) Override the solenoid valve and fill the water bath to proper level. Note: The installation of a manual bypass on the water line would allow continued operation of the evaporator in the event of a solenoid failure and would facilitate solenoid valve replacement with the unit in service.

    (2) If water cannot be added, take the evaporator out of service:
        (a) Try to remove all liquid chlorine left in the line.
        (b) Switch to another evaporator.
        (c) Switch the supply to the gas side of containers. Connect more containers to supply sufficient chlorine gas, if necessary.

2. Low water temperature in evaporator.

a. Check the chemical (chlorine) flow-through rate. The rate may have exceeded the unit's capacity and may require two evaporators to be in service to handle the chemical feed rate.

b. Inspect the immersion heaters for proper operation. First, examine the control panel for thermal overload on the breaker. Most evaporators are equipped with two to three heating elements. An inspection of the electrical system will indicate if the breakers are shorted or open and will locate the problem. Replace any heating elements that have failed.

c. If no spare evaporators are available, operate from the valves on the chlorine gas supply. If necessary, reduce the chlorine feed rate to keep the chlorination system working properly.

3. No chlorine gas flow to the chlorinator.

a. Inspect the pressure-reducing valve downstream from the evaporator and determine if the valve is in the open or closed position.

    (1) The valve may be closed due to low water temperature in the evaporator (less than 185°F or 85°C).

⚠️ When frost appears on valves and flex connectors, the chlorine gas may condense to liquid chlorine (reliquefy). The liquid chlorine may plug the chlorine supply lines (sometimes this is referred to as chlorine ice or frozen chlorine). If you disconnect the chlorine supply line to unplug it, be very careful. The liquid chlorine in the line could reevaporate, expand as a gas, build up pressure in the line, and cause liquid chlorine to come shooting out the open end of a disconnected chlorine supply line.

(2) The valve may be closed due to loss of vacuum on the system or loss of continuity of electrical control circuits, which may have been caused by a momentary power drop. Correct the problem and reset the valve.

(3) The valve may be out of adjustment and restricting gas flow through the valve due to a low pressure setting.

b. Inspect the supply containers and manifold. Possible sources of a lack of chlorine gas flow to chlorinators include:

(1) Containers are empty.

(2) Container chlorine supply lines are incorrectly connected to gas instead of liquid side of containers. High flow rates of gas will remove gas from the container faster than it can change from liquid to gas. This will cause a reduced flow of chlorine gas. When this happens, a frost may appear on valves and flex connectors. Reduce the flow of water being treated or connect more chlorine containers.

(3) The chlorine manifold filters are plugged. Check the pressure upstream and downstream from the filter. The pressure drop should not exceed 10 psi (6.9 kPa or 0.7 kg/cm²). Frost on the manifold may indicate an excessive flow of chlorine gas through the filters.

(4) Inspect manifold and system for closed valves. Most systems operate properly with all chlorine valves at only one turn open position.

MONTHLY

1. Exercise all valves, including inlet, outlet, PRV, water, drain, and fill valves.

2. Inspect the evaporator cathodic protection meter (if so equipped). Cathodic protection protects the metal water tank and piping from corrosion due to electrolysis. Electrolysis is the flow of electric current and is the reverse of metal plating. In electrolysis, the flow of certain compounds away from the metal causes corrosion and holes in a short time. This type of corrosion is controlled by either a sacrificial anode made of magnesium and zinc or by applying small electric currents to suppress or reverse the normal corroding current flow.

3. Check the PRV setting to maintain the desired pressure of chlorine gas to chlorinators.

4. Inspect heating and ventilating equipment in the chlorinator area.

5. Perform scheduled routine preventive maintenance.
   a. Drain and flush the water bath.
   b. Clean the evaporator tank.
   c. Check heater elements.
   d. Repack the gasket and reseat PRVs (annually).
   e. Replace anodes (annually).
   f. Paint the system (annually).

6.5.7.3   Chlorinators, Including Injectors

DAILY

1. Check the injector water supply pressure. Pressures will range from 40 to 90 psi (276 to 520 kPa or 2.8 to 6.3 kg/cm²) depending on system.

2. Determine the injector vacuum. Values will range from 15 to 25 inches (38 to 64 cm) of mercury.

3. Check the chlorinator vacuum. Values will range from 5 to 10 inches (13 to 25 cm) of mercury.

4. Determine the chlorinator chlorine supply pressure. Values will range from 20 to 40 psi (138 to 276 kPa or 1.4 to 2.8 kg/cm^2) after the pressure regulating valve.

5. Read the chlorinator feed rate on the rotameter tube. Is the feed rate at the required level? Record the rotameter reading and time.

6. Calculate the chlorine usage.

7. Examine and record the mode of control.
   a. Manual
   b. Single-input automatic
   c. Dual-input automatic

8. Measure the chlorine residual at the application point.

9. Inspect the system for chlorine leaks.

10. Inspect auxiliary components.
   a. Flow signal input. Does the chlorinator feed rate change when flow changes? Chlorinator response is normally checked by biasing (adjusting) the flow signal, which may drive the dosage control unit on the chlorinator to full open or closed position. When the switch is released, the chlorinator will return to the previous feed rate. During this operation, the unit should have responded smoothly through the change. If the response was not smooth, look for mechanical problems of binding, lubrication, or vacuum leaks.
   b. If the chlorinator also is controlled by a residual analyzer, be sure the analyzer is working properly. Check the following items on the residual analyzer. Be sure to follow the manufacturer's instructions.
      (1) Actual chlorine residual is properly indicated.
      (2) Recorder alarm set point.
      (3) Recorder control set point.
      (4) Sample water flow.
      (5) Sample water flow to cell block after dilution with fresh water.
      (6) Adequate flow of dilution water.
      (7) Filter system and drain.
      (8) Run comparison tests of chlorine residual. Do the tests match with analyzer output readings?
      (9) If the residual analyzer samples two streams, start the other stream flow and compare tested residual of that stream with the analyzer output readings. Standardize the analyzer output readings against tested residuals. Enter the changes and corrections in the log.
      (10) Change the recorder chart daily or as necessary. If a strip chart is used, date the chart for recordkeeping purposes.
      (11) Check the recorder output signal controlling the chlorinator for control responses on the feed rate. Correct the feed rates through the ratio controller.

WEEKLY

1. Put the chlorinator on manual control. Operate feed-rate adjustment through a full range from 0 to full scale (250, 500, 1,000, 2,000, 4,000, 6,000, 8,000, or 10,000 pounds/day). At each end of the scale, check:
   a. Chlorinator vacuum
   b. Injector vacuum
   c. Solution line pressure
   d. Chlorine pressure at the chlorinator

   If any of the readings do not produce normal set points, make proper adjustments.
   a. Injector should produce necessary vacuum at the chlorinator (5 to 10 inches or 13 to 25 centimeters of mercury).
   b. Adjust the PRV to obtain sufficient pressure and chemical feed for full feed-rate operation of the chlorinator. This chlorine PRV is on the main chlorine supply line and is usually located near or in the chlorine container area.

2. If the unit performs properly through a complete range of feed rates, return the unit to automatic control. If any problems develop, locate the source and make corrections.

3. Clean the chlorine residual analyzer (see Section 6.8, "Measurement of Chlorine Residual"), including the following tasks:
   a. Clean the filters.
   b. Clean the sample line.
   c. Clean the hydraulic dilution wells and baffles.
   d. Flush the discharge hoses and pipes.
   e. Clean and flush the cell block.
   f. Fill the buffer reservoirs.
   g. Check the buffer pump and feed rate.
   h. Wipe the machine clean and keep it clean.

4. Calculate chlorine usage so that replacement supply containers can be ordered and constant chlorination can be maintained. Try to have a 15- to 30-day supply of chlorine in storage.

MONTHLY

1. Exercise all chlorine valves.
2. Inspect heaters and room ventilation equipment.
3. Check the chlorinator vent line to the outside of the structure for any obstructions that could prevent free access to the atmosphere. Bugs and wasps like vent lines for nests. If this becomes a problem, install a fine wire mesh over the open end of the vent line.
4. Inspect the unit for vacuum leaks.
5. Clean the rotameter sight glass.
6. Inspect all drain lines and hoses.
7. Perform scheduled routine maintenance.
   a. Repack the seats and stems of valves.
   b. Inspect tubing and fittings for leaks. Wash and dry thoroughly before reassembling.
   c. Inspect control system.
      (1) Electrical and electronics
      (2) Pneumatics

(3) Lubrication

(4) Calibration of total system

d. Chlorine analyzer.

(1) Lubricate chart drives, filter drives, and pumps.

(2) Clean and flush all piping and hoses, filters, tubing, cell blocks, and hydraulic chambers.

(3) Clean acid and iodide reservoirs.

(4) Calibrate the unit with known standards.

(5) Repaint the unit if needed.

e. Inspect safety equipment, including self-contained breathing equipment and repair kits.

ANNUALLY

1. Disassemble, clean, and regasket the chlorinator.

POSSIBLE ABNORMAL CONDITIONS

1. Chlorine leak in the chlorinator.
   Shut off the gas flow to the chlorinator. Leave the injector in service. Place another chlorinator in service, if available, until repairs are accomplished. Allow the chlorinator to operate under a vacuum with 0 psi showing on the chlorine pressure gauge for 3 to 5 minutes to remove chlorine gas. If the system must be opened to repair the leak, the vacuum on the chlorinator may minimize the potential for residual chlorine to be released to the atmosphere. If effective evacuation of the unit cannot be ensured, appropriate respiratory protection must be used.

2. Gas pressure is too low, less than 20 psi (138 kPa or 1.4 kg/cm²).
   Alarm indicated. Check chlorine supply:

   a. Empty containers: switch to standby units.

   b. Evaporator shut down. See Section 6.5.7.2, "Evaporators."

   c. Inspect the manifold for closed valves or restricted filters. If necessary, switch to another manifold, clean/replace filters, or set valves and controls to the proper positions.

3. Injector vacuum is too low.

   a. Adjust the injector to achieve required vacuum.

   b. Inspect the injector water supply system.

      (1) Start the pump if it is off.

      (2) Clean dirty strainers.

      (3) If the pump is worn out and will not deliver appropriate flow and pressure to injector, use another unit and repair or replace the pump.

      (4) Inspect the valves in the system. Place the valves in proper positions.

   c. Inspect the solution line discharge downstream from injector. Check for the following items:

      (1) The valve is closed or partially closed.

      (2) The line is broken or a restriction is reducing the flow or increasing back pressure.

      (3) The diffuser is plugged, thus restricting flow and creating a higher back pressure on the discharge line and injector. Clean the diffuser and flush the pipe.

Note: The installation of a pressure gauge on the discharge side of the injector or on the solution line would alert the operator to abnormal system back pressure.

4. Low chlorine residual. The alarm indicator is on from the chlorine residual analyzer.

   Determine the actual chlorine residual and compare it with the residual reading from the chlorine analyzer. If the residual analyzer is indicating a different chlorine residual, recalibrate the analyzer and readjust. If the chlorine residual analyzer is correct and the chlorine residual is low, check the following items:

   a. Sample pump.
   (1) Operation, flow, and pressure.
   (2) Sample lines are clean and free of solids or algae that could create a chlorine demand.
   (3) The strainer is dirty and restricting flow, thus preventing adequate pressure (15 to 20 psi, 103 to 138 kPa, or 1.0 to 1.4 kg/cm^2) at the analyzer.

   b. Control system if the chlorinator is on automatic control. If the chlorine feed rate remains too low, switch the chlorinator from automatic control to manual control. Set the chlorinator to the proper feed rate as determined by previous adequate feed rates.

   c. Chlorine demand is higher than the amount one chlorinator can supply. Place additional chlorinator in service.

TROUBLESHOOTING DIRECT-MOUNT CHLORINATORS

Table 6.5   Direct-mount chlorinator troubleshooting guide

Operating Symptoms	Probable Cause	Remedy
1. Water in the chlorine metering tube.	Check valve failure, deposits on seat of check valve, or check valve seat distorted by high pressure.	Clean deposits from check ball and seat with dilute muriatic acid. Badly distorted check valve may have to be replaced.
2. Water venting to atmosphere.	Excess water pressure in the vacuum regulator.	Remove vacuum regulator from chlorine cylinder and allow chlorinator to pull air until dry.
3. No indication on flowmeter when vacuum is present.	Vacuum leak due to bad or brittle vacuum tubing, connections, rate valve O-rings, or gasket on top of flowmeter.	Check the vacuum tubing, rate valve O-rings, and flowmeter gasket for vacuum leaks. Replace bad tubing connectors, O-rings, or gasket.
4. Indication on flowmeter but air present, not chlorine gas.	Connection below meter tube gasket leaks.	Check connections and replace damaged elements.

Check Your Understanding

1. Normal operation of a chlorinator includes daily inspection of what facilities or areas?

2. What is the purpose of evaporators?

3. What abnormal conditions could be encountered when operating an evaporator?

4. What are possible chlorinator abnormal conditions?

5. How can you determine if the chlorine residual analyzer is working properly?

6.5.7.4 Summary, Daily Operation

Actual procedures for operating chlorination equipment will vary from plant to plant, region to region, season to season, and water to water. The following procedures are provided to serve as guidelines to help you develop your own procedures.

1. *Prechlorination*

   If THMs are not a problem, prechlorination is a very cost-effective means of disinfecting water. Prechlorination should be at a chlorine dosage that will produce a free chlorine residual (past the breakpoint) of 0.5 to 1.5 mg/L before the flocculation basins. If THMs are a problem, see the section on trihalomethanes in the chapter on specialized treatment processes in Volume II. If phenolic compounds are present, discontinue prechlorination, adsorb phenols with activated carbon and remove them by filtration, and disinfect by postchlorination.

2. *Postchlorination*

   Regardless of whether prechlorination is practiced, a free chlorine residual of at least 0.5 mg/L should be maintained in the clear well or distribution reservoir immediately downstream from the point of postchlorination.

3. *Distribution System*

   Postchlorination dosages should be adequate to produce a free chlorine residual of 0.2 mg/L at the farthest point in the distribution system at all times.

Very often when consumers complain about chlorine tastes in their drinking water, the chlorine dose has been inadequate. When the chlorine dose is inadequate, the measured chlorine residual is frequently **combined available chlorine**. By increasing the chlorine dose, you chlorinate past the breakpoint and the chlorine residual may be lower, but the residual contains free available chlorine. One way to determine if you have reached the chlorine breakpoint is to increase the chlorine dose rate. If the residual chlorine increases in proportion to the increased dose, then you are chlorinating past the breakpoint. For a discussion of breakpoint chlorination, see Section 6.3.6, "Breakpoint Chlorination."

The objective of disinfection is the destruction or inactivation of pathogenic organisms, and the ultimate measure of the effectiveness is the bacteriological result (negative coliform test results). The measurement of

combined available chlorine
The total chlorine, present as chloramine or other derivatives, that is present in a water and is still available for disinfection and for oxidation of organic matter. The combined chlorine compounds are more stable than free chlorine forms, but they are somewhat slower in disinfection action.

chlorine residual does supply a tool for practical control. If the chlorine residual value commonly effective in most water treatment plants does not yield satisfactory bacteriological kills in a particular plant, the residual chlorine that does produce satisfactory results must be determined and used as a control in that plant. In other words, the 0.5 mg/L chlorine residual, while generally effective, is not a rigid standard but a guide that may be changed to meet local requirements.

FORMULAS

To determine if the chlorinator setting is high enough to produce a free available residual chlorine past the breakpoint, increase the chlorinator feed rate. If all of the chlorine fed at the new or higher setting is converted to chlorine residual in milligrams per liter, the original setting was at or past the breakpoint. The expected increase in chlorine residual can be calculated by using the formula:

$$\text{Expected Increase } Cl_{residual}\left[\frac{mg}{L}\right] = \frac{\left(Cl_{feed}\right)_{new}\left[\frac{lb}{d}\right] - \left(Cl_{feed}\right)_{old}\left[\frac{lb}{d}\right]}{Q\left[\frac{Mgal}{d}\right] \times \frac{8.34\,lb}{gal}}$$

$$= 0.120 \times \frac{\left(Cl_{feed}\right)_{new}\left[\frac{lb}{d}\right] - \left(Cl_{feed}\right)_{old}\left[\frac{lb}{d}\right]}{Q\left[\frac{Mgal}{d}\right]}$$

Where,

$\left(Cl_{feed}\right)_{old}$ = Chlorine Feed at the Old Setting

$\left(Cl_{feed}\right)_{new}$ = Chlorine Feed at the New Setting

To determine the actual increase in free chlorine residual, find the difference in milligrams per liter between the new and old free available residual chlorine.

$$\text{Actual Increase } Cl_{residual}\left[\frac{mg}{L}\right] = \left(Cl_{residual}\right)_{new}\left[\frac{mg}{L}\right] - \left(Cl_{residual}\right)_{old}\left[\frac{mg}{L}\right]$$

Where,

$\left(Cl_{residual}\right)_{old}$ = Chlorine Residual at the Old Setting

$\left(Cl_{residual}\right)_{new}$ = Chlorine Residual at the New Setting

Example 4

A chlorinator is set to feed chlorine to a treated water at a dose of 27 pounds of chlorine per 24 hours. This dose rate produces a free chlorine residual of 0.4 mg/L. When the chlorinator setting is increased to 30 pounds per 24 hours, the free chlorine residual increases to 0.6 mg/L. If the average 24-hour flow is 1.6 MGD, is the water being chlorinated past the breakpoint?

	Known		Unknown
$(Cl_{feed})_{old}$	= Old Setting	= 27 lb/d	Increase in Residual
$(Cl_{feed})_{new}$	= New Setting	= 30 lb/d	in mg/L
$(Cl_{residual})_{old}$	= Old Residual	= 0.4 mg/L	Is the Water Being
$(Cl_{residual})_{new}$	= New Residual	= 0.6 mg/L	Chlorinated past
Q	= Flow Rate	= 1.6 MGD	the Breakpoint?

1. Calculate the expected increase in chlorine residual if chlorination is at the breakpoint.

$$\text{Expected Increase Cl}_{residual}\left[\frac{mg}{L}\right] = 0.120 \times \frac{(Cl_{feed})_{new}\left[\frac{lb}{d}\right] - (Cl_{feed})_{old}\left[\frac{lb}{d}\right]}{Q\left[\frac{Mgal}{d}\right]}$$

$$= 0.120 \times \frac{30\left[\frac{lb}{d}\right] - 27\left[\frac{lb}{d}\right]}{1.6\left[\frac{Mgal}{d}\right]} = 0.2\frac{mg}{L}$$

2. Determine the actual increase in free chlorine residual in mg/L.

$$\text{Actual Increase Cl}_{residual}\left[\frac{mg}{L}\right] = (Cl_{residual})_{new}\left[\frac{mg}{L}\right] - (Cl_{residual})_{old}\left[\frac{mg}{L}\right]$$

$$= 0.6\left[\frac{mg}{L}\right] - 0.4\left[\frac{mg}{L}\right] = 0.2\frac{mg}{L}$$

3. Is the water being chlorinated past the breakpoint? Yes, because the calculated expected increase in chlorine residual (0.2 mg/L) is the same as the actual increase in the chlorine residual (0.2 mg/L), we can conclude that we are chlorinating past the breakpoint.

6.5.7.5 Laboratory Tests

1. Chlorine Residual in System
    a. Daily chlorine residual tests using the **DPD method** should be taken at various locations in the system. A remote tap is ideal for one sampling location. Take the test sample from a tap as close to the main as possible. Allow the water to run at least 5 minutes before sampling to ensure a representative sample from the main.
       Operators using the DPD colorimetric method to test water for a free chlorine residual need to be aware of a potential error that may occur. If the DPD test is run on water containing a combined chlorine residual, a precipitate may form during the test. The particles of precipitated material will give the sample a turbid appearance or the appearance of having color. This turbidity can produce a positive test result for free chlorine residual when there is actually no chlorine present. Operators call this error a false positive chlorine residual reading.
    b. Chlorine residual test kits are available for small systems.
2. Bacteriological Analysis (Coliform Tests)
    Samples should be taken routinely in accordance with EPA and health department requirements. Take samples according to

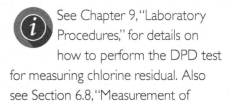

See Chapter 9, "Laboratory Procedures," for details on how to perform the DPD test for measuring chlorine residual. Also see Section 6.8, "Measurement of Chlorine Residual," later in this chapter.

DPD (pronounce as separate letters) method
A method of measuring the chlorine residual in water. The residual may be determined by either titrating or comparing a developed color with color standards. DPD stands for N,N-diethyl-p-phenylenediamine.

(i) See Chapter 9, "Laboratory Procedures," for proper procedures for collecting and analyzing samples for chlorine residuals and coliform tests.

approved procedures. Be sure to use a sterile plastic or glass bottle. If the sample contains any chlorine residual, sufficient sodium thiosulfate (thio) should be added to neutralize all of the chlorine residual. Usually, 0.1 milliliter of 10 percent thio in a 120 mL (4 oz) bottle is sufficient for distribution systems. The thio should be added to the sample bottle before sterilization.

6.5.8 Troubleshooting Gas Chlorinator Systems

Table 6.6 Troubleshooting gas chlorinator systems

Operating Symptoms	Probable Cause	Remedy
1. Injector vacuum reading low	1. Hydraulic system	1. Check injector water supply system
	2. Flow restricted	2. Adjust injector orifice
	3. Low pressure	3. Close throat
	4. High pressure	4. Open throat
	5. Back pressure	5. Change injector or increase water supply to injector
	6. Low flow of water	6. Increase pump output
2. Leaking joints	1. Missing gasket	1. Repair joint
3. Chlorinator will not reach maximum point	1. Faulty injector (no vacuum)	1. Repair injector system
	2. Restriction in supply	2. Find restriction in supply system
	3. Faulty chlorinator	3. Check for vacuum leaks
	4. Leaks	4. Repair leaks
	5. Wrong orifice	5. Install proper orifice
4. Chlorinator will feed OK at maximum output, but will not control at low rates	1. Vacuum regulating valve	1. a. Repair diaphragm b. Check valve capsule
	2. Dirty chlorine pressure-reducing valve (PRV)	2. Clean chlorine PRV cartridge, diaphragm, and gaskets
5. Chlorinator does not feed	1. Supply	1. Renew Cl$_2$ supply
	2. Piping	2. a. Open valve b. Clean filter
6. Variable vacuum control, formerly working well, now will not go below 30% feed. Signal OK	1. Chlorine PRV	1. Clean chlorine PRV
7. Variable vacuum control reaches full feed, but will not go below 50% feed. Chlorine PRV OK	1. Signal vacuum too high	1. a. Hole in diaphragm b. Clean dirty filter disks c. Clean converter nozzle
8. Variable vacuum control will not go to full feed. Gas pressure OK. Chlorine PRV OK	1. Plugged restrictor	1. Clean restrictor
	2. Air leak in signal	2. Repair air leak
9. Freezing of manometer	1. Rate too high	1. Lower rate
	2. Restriction in manometer orifice	2. Clean piping

6.5.9   Disinfection Troubleshooting

Table 6.7   Disinfection troubleshooting guide

Operating Symptoms	Probable Cause	Remedy
1. Increase in coliform level	Low chlorine residual	Raise chlorine dose
2. Drop in chlorine level	a. Increase in chlorine demand b. Drop in chlorine feed rate	Raise chlorine dose and find out why chlorine demand increased or chlorine feed rate dropped

6.5.10   Chlorination System Failure

If your chlorination system fails, do not allow unchlorinated water to enter the distribution system. Never allow unchlorinated water to be delivered to your consumers. If your chlorination system fails and cannot be repaired within a reasonable time period, notify your supervisor and officials of the health department. To prevent this problem from occurring, your plant should have backup or standby chlorination facilities.

For hypochlorinator maintenance, see Section 6.5.6.1, "Hypochlorinators".

For chlorinator maintenance, see Section 6.5.6.2, "Chlorinators", and Section 6.5.7, "Normal and Abnormal Operation".

Check Your Understanding

1. What is the suggested free chlorine residual for treated water (measured at a point just beyond postchlorination)?
2. What is the suggested free chlorine residual for the farthest point in the distribution system?
3. How would you determine if you were chlorinating at the breakpoint?

6.6   Maintenance

Necessary maintenance of chlorination equipment requires a thorough understanding of the manufacturer's literature. Generally, the daily, weekly, and monthly operating procedures also contain the appropriate maintenance procedures. Do not attempt maintenance tasks that you are not qualified to perform. There are too many possibilities for serious accidents when you are not qualified.

6.6.1   Chlorine Leaks

Your sense of smell can detect chlorine concentrations as low as 3 ppm. Portable and permanent automatic chlorine detection devices (Figure 6.20) can detect chlorine concentrations of 1 ppm or less. Whenever you must deal with a chlorine leak, always follow the safe procedures outlined in Section 6.9, "Chlorine Safety Program."

Chlorine leaks must be taken care of immediately or they will become worse. Corrective measures should be undertaken only by trained operators wearing proper safety equipment. All operators should be trained to

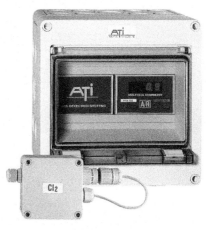

Figure 6.20   Chlorine leak detector
(Courtesy of Analytical Technology, Inc.)

repair chlorine leaks. Always work in pairs when looking for and repairing leaks. All other persons should leave the danger area during repairs until conditions are safe again.

If the leak is large, all persons in the adjacent areas should be warned and evacuated. Obtain help from your fire department. They have self-contained breathing equipment and can help evacuate people. The police department can help control curious sightseers. Repair crews and drivers of emergency vehicles must realize that vehicle engines will quit operating in the vicinity of a large chlorine leak because of the lack of oxygen. You must always consider your neighbors—people, animals, and plants.

1. Before any new system is put into service, it should be cleaned, dried, and tested for leaks. Clean and dry pipelines by flushing and steaming from the high end to allow condensate and foreign materials to drain out, or by the use of commercially available cleaning solvents compatible with chlorine. After the empty line is heated thoroughly, blow dry air through the line until it is dry. After drying, test the system for tightness with 150 psi (1,034 kPa or 10.5 kg/cm^2) dry air. Apply soapy water to the outside of joints to detect leaks. Small quantities of chlorine gas may now be introduced into the line, the test pressure built up with air, and the system tested for leaks with ammonia. Whenever a new system is tested for leaks, at least one chlorinator should be in service to withdraw chlorine from the system in case of a leak. The same is true in case of an emergency leak at any installation. If a chlorinator is not running, at least one should be started. Preferably, all available chlorinators should be put in service.

2. To find a chlorine leak, dip a 1-inch (2.5 cm) paintbrush or a rag tied on a stick into a strong ammonia solution,* and hold the paintbrush or rag near the suspected leaks. A polyethylene squeeze bottle filled with ammonia water to dispense ammonia vapor may also be used. Take care to avoid spraying ammonia water on any leak or touching the brush or cloth to any metal. White fumes will indicate the exact location of the leak. Locating leaks by this method may not be possible for large leaks, which diffuse the gas over large areas. Do not use an ammonia spray bottle because the entire room could turn white if it is full of chlorine gas. Also, any chlorine deposits will draw water from the ammonia and form an acid, which will eat away any material it contacts.

3. If the leak is in the equipment in which the chlorine is being used, close the valves on the chlorine container at once. Repairs should not be attempted while the equipment is in service. All chlorine piping and equipment that is to be repaired by welding should be flushed with water or steam. Before returning equipment to use, it must be cleaned, dried, and tested as previously described.

4. If the leak is in a chlorine cylinder or container, use the emergency repair kit supplied by most chlorine suppliers (Figure 6.21). These kits can be used to stop most leaks in a chlorine cylinder or container and can usually be delivered to a plant within a few

*Use a concentrated ammonia solution containing 28 to 30 percent ammonia as NH$_3$ (this is the same as 58 percent ammonium hydroxide, NH$_4$OH, or commercial 26° Baumé).

Emergency kit "A" for chlorine cylinders

Emergency kit "B" for chlorine ton containers

Figure 6.21   Chlorine emergency repair kits *(Courtesy of the Chlorine Institute)*

hours if one is not already at the site of the leak. It is advisable to have emergency repair kits available at your plant at all times and to train personnel frequently in their use. Respiratory protective equipment should be located outside chlorine storage areas. The repair kits may be located within chlorine storage areas because during an emergency requiring their use, you will already be wearing approved respiratory protection and hence they will be accessible. Refer to Figure 6.16 in Section 6.5.3.2, "Steel Cylinders," for typical locations of cylinder valve leaks.

5. If chlorine is escaping as a liquid from a cylinder or a ton tank, turn the container so that the leaking side is on top. In this position, the chlorine will escape only as a gas, and the amount that escapes will be only $1/15$ as much as if the liquid chlorine were leaking. Keeping the chlorinators running also will reduce the amount of chlorine gas leaking out of a container. Increase the feed rate to cool the supply tanks as much as possible.

6. For situations in which a prolonged or unstoppable leak is encountered, emergency disposal of chlorine should be provided. Chlorine may be absorbed in solutions of caustic soda (sodium hydroxide), soda ash (sodium carbonate), or agitated hydrated lime (calcium hydroxide) slurries (Table 6.8). Chlorine should be passed into the solution through an iron pipe or a properly weighted rubber hose to keep it immersed in the absorption solution. The container should not be immersed because the leaks will be aggravated due to the corrosive effect, and the container may float when partially empty. In some cases, it may be advisable to move the container to an isolated area. Discuss the details of such precautions with your chlorine supplier.

7. Never put water on a chlorine leak. A mixture of water and chlorine will increase the rate of corrosion of the container and make the leak larger. Besides, water may warm the chlorine, thus increasing the pressure and forcing the chlorine to escape faster.

8. Leaks around valve stems can often be stopped by closing the valve or tightening the packing gland nut. Tighten the nut or stem by turning it clockwise.

9. Leaks at the valve discharge outlet can often be stopped by replacing the gasket or adapter connection.

10. Leaks at fusible plugs and cylinder valves usually require special handling and emergency equipment. Call your chlorine supplier immediately and obtain an emergency repair kit for this purpose if you do not have a kit readily available.

11. Pinhole leaks in the walls of cylinders and ton tanks can be stopped by using a clamping pressure saddle with a turnbuckle available in repair kits. This is only a temporary measure, and the container must be emptied as soon as possible.

   If a repair kit is not available, use your ingenuity. One operator stopped a pinhole leak temporarily until a repair kit arrived by placing several folded layers of neoprene packing over a leak, a piece of scrap steel plate over the packing, wrapping a chain around the cylinder and steel plate, and applying leverage pressure with a crowbar.

Table 6.8   Chlorine absorption solutions (Source: Courtesy of the Chlorine Institute)

Absorption Solution	Container Size (lb net)	Chemical (lb)	Water (gal)
Caustic Soda (Sodium Hydroxide), (100%)	100	125	40
	150	188	60
	2,000	2,500	800
Sodium Ash (Sodium Carbonate)	100	300	100
	150	450	150
	2,000	6,000	2,000
Hydrated Lime[a] (Calcium Hydroxide)	100	125	125
	150	188	188
	2,000	2,500	2,500

[a]Hydrated lime solution must be continuously and vigorously agitated while chlorine is to be absorbed.

Dry ice has been applied to chlorine containers to cool the liquid and thus reduce the amount of gas escaping through a pin-hole leak.

12. A leaking container must not be shipped. If the container leaks or if the valves do not work properly, keep the container until you receive instructions from your chlorine supplier for returning it. If a chlorine leak develops in transit, keep the vehicle moving until it reaches an open area.

13. Do not accept delivery of containers showing evidence of leaking, stripped threads, or abuse of any kind.

14. If a chlorine container develops a leak, be sure your supplier does not charge you for the unused chlorine.

15. Chlorine leaks may be detected by chlorine gas detection devices (Figure 6.20). Alarm systems may be connected to these devices. Be sure to follow the manufacturer's recommendations regarding frequency of checking and testing detection devices and alarm systems.

Check Your Understanding

1. If chlorine is escaping from a cylinder, what would you do?

2. How can chlorine leaks around valve stems be stopped?

3. How can chlorine leaks at the valve discharge outlet be stopped?

6.6.2   Installation

The following are some features of importance when working with chlorine facilities. Also, examine these items when reviewing plans and specifications.

1. Chlorinators should be located as near the point of application as possible.

2. There must be a separate room for chlorinators and chlorine container storage (above ground) to prevent chlorine gas leaks from damaging equipment and harming personnel. There should be no access to this room from a room containing equipment or where personnel work.

3. Ample working space around the equipment and storage space for spare parts should be provided.

4. There should be an ample supply of water to operate the chlorinator at required capacity under maximum pressure conditions at the chlorinator injector discharge.

5. The building should be adequately heated. The temperature of the chlorine cylinder and chlorinator should be above 50°F (10°C). Line heaters may be used to keep chlorine piping and the chlorinator at higher temperatures to prevent condensing of gas into liquid in the pipelines and chlorinator. In general, a temperature difference of 5 to 10°F (3 to 6°C) above surroundings is recommended. The maximum temperature at which a chlorine cylinder is stored should not exceed 100°F (38°C).

*For procedures on how to calculate maximum withdrawal rates, see "Maximum Withdrawal Rates from Chlorine, Sulfur Dioxide, and Ammonia Cylinders," by Robert J. Baker in the October, 1984, issue of *Opflow*, published by AWWA.

6. It is not advisable to draw more than 40 pounds (18 kg) of chlorine from any one 100 to 150-pound (45 and 68 kg) cylinder in a 24-hour period. At higher chlorine withdrawal rates, the chlorine gas is removed from the cylinder faster than the liquid chlorine is being converted to chlorine gas. When this happens, the actual flow of chlorine gas will be reduced and become less than the desired rate. The maximum allowable chlorine withdrawal rate varies with temperature with the maximum withdrawal rate increasing as the temperature increases. With ton containers, the limit of chlorine gas withdrawal is about 8 pounds of chlorine per day per °F (6.4 kg/°C) **ambient temperature**. When evaporators are provided, these limitations do not apply.*

7. There should be adequate light.

8. Adequate ventilation is important in a chlorinator room to remove any leaking chlorine gas that would be hazardous to personnel and damaging to equipment. Many state and local fire codes require proper ventilation of chlorine storage rooms and rooms where chlorine is used. Mechanical exhaust systems must draw air from the room at a point no higher than 12 inches (30.5 cm) above the floor at a rate of not less than 1 cubic foot of air per minute per square foot (0.00508 m³/s/m²) of floor area in the storage area. (The system should not draw air through the fan itself because chlorine gas can damage the fan motor.) Normally, ventilation from chlorine storage rooms is discharged to the atmosphere. When a chlorine leak occurs, the ventilated air containing the chlorine should be routed to a treatment system for processing. A caustic scrubbing system can be used to treat air containing chlorine from a leak. The treatment system should be designed to reduce the maximum allowable discharge concentration of chlorine to one-half the **IDLH** (Immediately Dangerous to Life or Health) at the point of discharge to the atmosphere. The IDLH for chlorine is 10 ppm. A secondary standby source of power is required for the chlorine detection, alarm, ventilation, and treatment systems.

9. Adequate measuring and controlling of chlorine dosage are required. Scales and recorders indicating loss in weight are desirable as a continuous check and as a record of the continuity of chlorination. Weigh chlorine containers and record the weights at the same time every day. Compare the actual weight of chlorine used with the calculated use based on the chlorinator setting. Also, compare the results of chlorine residual tests with the calculated dosage.

10. There should be continuity of chlorination. When chlorination is practiced for disinfection, it is needed continuously when the plant is operating for the protection of the water consumers. To ensure continuous chlorination, the chlorine gas lines from cylinders should feed to the manifold so that the cylinders can be removed without interrupting the feed of gas. Duplicate units with automatic cylinder switchover should be provided. Hypochlorinators are sometimes used during emergencies.

ambient (AM-bee-ent) temperature
Temperature of the surrounding air (or other medium). For example, temperature of the room where a gas chlorinator is installed.

IDLH
Immediately Dangerous to Life or Health. The atmospheric concentration of any toxic, corrosive, or asphyxiant substance that poses an immediate threat to life or would cause irreversible or delayed adverse health effects or would interfere with an individual's ability to escape from a dangerous atmosphere.

Check Your Understanding

1. Why should chlorinators be located in a separate room?

2. Why is adequate ventilation important in a chlorinator room?

3. How can chlorination rates be checked against the chlorinator setting?

4. When and how often should the weights of chlorine containers be recorded?

6.7   Chlorine Dioxide Facilities

Use of chlorine dioxide as a disinfecting chemical instead of chlorine is of considerable interest to operators. A major reason for this interest is that trihalomethanes are not formed when disinfecting with chlorine dioxide. Other reasons for considering chlorine dioxide include the fact that chlorine dioxide is effective in killing bacteria and viruses. Chlorine dioxide is much more effective than chlorine in killing bacteria in the pH range from 8 to 10. Chlorine dioxide does not combine with ammonia and is more selective in its reaction with many organics than chlorine, therefore less chlorine dioxide is required to achieve equivalent chlorine residuals and bacterial kills in waters containing such contaminants.

Most existing chlorination units may be used to produce chlorine dioxide. In addition to the existing chlorination system, a diaphragm pump, solution tank, mixer, chlorine dioxide generating tower, and electrical controls are needed (Figure 6.22). The diaphragm pump and

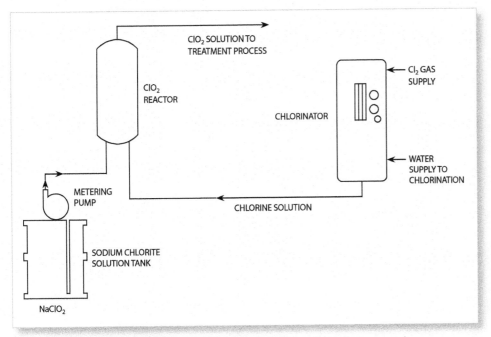

Figure 6.22   Chlorine dioxide facility *(Courtesy of the United States Environmental Protection Agency)*

piping must be made of corrosion-resistant materials because of the corrosive nature of chlorine dioxide. Usually, PVC or polyethylene pipe is used.

Chlorine dioxide cannot be compressed or stored commercially as a gas because it is explosive under pressure. Therefore, it is never shipped but is generated on site (Figures 6.23 and 6.24). The chemical used to generate chlorine dioxide for drinking water application is sodium

Figure 6.23 Chlorine dioxide generator, wall-mounted *(Courtesy of Dioxide Pacific Inc.)*

chlorite. Special precautions must be taken when handling sodium chlorite. Sodium chlorite is usually supplied as a salt and is very combustible around organic compounds. Whenever spills occur, sodium chlorite must be neutralized with anhydrous sodium sulfite. Combustible materials (including combustible gloves) should not be worn when handling sodium chlorite. If sodium chlorite comes in contact with clothing, the clothes should be removed immediately and soaked in water to remove all traces of sodium chlorite or they should be burned immediately. If you follow safe procedures, you can safely handle sodium chlorite. Chlorine dioxide has not been widely used to treat drinking water because costs are higher than other more commonly used disinfection methods.

6.7.1 Safe Handling of Chemicals

Due to the corrosive nature of the chemicals involved in chlorine dioxide generation (sodium chlorite and chlorine), certain precautions should be taken in the handling of these chemicals.

1. Sodium chlorite solutions have very strong bleaching capabilities. Always wear protective clothing, face shield or goggles, and chemical-resistant, noncombustible gloves and boots.

2. An emergency shower and eye wash station should be located in the immediate vicinity. If sodium chlorite comes in contact with eyes or skin, flush immediately with large amounts of fresh water. Soak contaminated clothing in water or burn it.

3. Dried sodium chlorite presents a fire hazard, particularly if allowed to come in contact with organic materials, for example, cotton, leather, wood, or oils. Do not allow the solution to dry. Flush spills with large amounts of fresh water.

4. Thoroughly rinse rags and mops used for cleanup before disposal or storage.

5. Chlorine gas is a respiratory irritant. If a leak occurs, evacuate the vicinity immediately.

Refer to the Safety Data Sheets (SDSs) for more information on the safe handling of sodium chlorite, chlorine dioxide solutions, and chlorine gas.

6.7.2 Operation

Operating a chlorine dioxide generator requires strict adherence to prestart, startup, and shutdown procedures to ensure the equipment operates safely and efficiently.

6.7.2.1 Prestart Procedures

The following section provides a comprehensive list of prestart procedures that must be taken to ensure safe and efficient operation of the chlorine dioxide generator.

Debris such as PVC or metal filings can seriously damage or plug the eductor assembly and rotameters, resulting in poor performance and delays in startup. If the water and chemical lines have not been properly

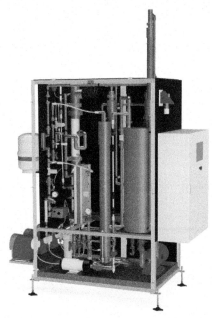

Figure 6.24 Chlorine dioxide generator, platform-mounted (*Courtesy of Dioxide Pacific Inc.*)

eductor (e-DUCK-ter)
A hydraulic device used to create a negative pressure (suction) by forcing a liquid through a restriction, such as a Venturi. An eductor or aspirator (the hydraulic device) may be used in the laboratory in place of a vacuum pump. As an injector, it is used to produce vacuum for chlorinators. Sometimes used instead of a suction pump.

flushed during installation, follow the procedures outlined in Steps 1 through 4.

1. Eductor Water Supply Line
   a. Close the shutoff valve.
   b. Remove the filter from the Y-strainer.
   c. Disconnect the water line from the generator inlet.
   d. Fully open the shutoff valve and allow the water to flow for 2 or 3 minutes.
   e. Close the shutoff valve.
   f. Replace the Y-strainer filter.
   g. Reconnect the water line to the generator.
   h. Fully open the shutoff valve.

2. Chlorite Supply Line
   a. Disconnect the tubing from the supply line to the chlorite rotameter.
   b. Remove the filter element from the chemical filter.
   c. Disconnect the upper union of the ball check valve.
   d. Connect a water source to one end of the chlorite supply line. Be sure the water supply you choose is clean and has sufficient pressure to properly flush debris from the line.
   e. Open both ball valves and flush the line for at least 2 to 3 minutes.
   f. Turn off the water supply and close both ball valves.
   g. Reconnect the chemical line to the ball check valve.
   h. Replace the filter element of the chemical filter.
   i. Reconnect the tubing from the chlorite supply line to the chlorite rotameter.
   j. Fully open both ball valves.

3. Chlorine Supply Line
   a. Ensure that the shutoff valve on the chlorine cylinder is fully closed.
   b. Disconnect the chlorine supply line from the vacuum demand regulator.
   c. Disconnect the tubing from the supply line to the chlorine rotameter.
   d. Connect a source of dry compressed air to one end of the line. To prevent damaging the line and fittings, be sure the air pressure does not exceed 50 psi.
   e. Apply several short bursts of air through the supply line.
   f. Disconnect the compressed air.
   g. Reconnect the tubing from the chlorine supply line to the chlorine rotameter.

   Note: Do not reconnect the chlorine supply line to the vacuum demand regulator at this time.

4. Chlorine Dioxide Distribution Lines
   a. Disconnect the chlorine dioxide distribution line from the generator output.
   b. Disconnect the distribution line from the injection point.

c. Connect a clean water source to the generator end of the distribution line and flush for several minutes.

d. Disconnect the water source from the distribution line.

e. Reconnect the chlorine dioxide distribution line to the injection point.

f. Reconnect the distribution line to the generator output.

5. Generator Cycle

Water testing is the final prestart test and will determine the readiness of the system for operation.

a. Ensure the shutoff valves for both the chlorite and chlorine supplies are in the OFF position.

b. Fully open the eductor water supply valve.

c. Ensure any shutoff valves on the distribution line are open.

d. With the generator running and the chemical shutoff valves closed, the generator should pull full vacuum. The meter floats will fall to and remain at zero. If the floats bounce off the bottom of the meters, there is a leak in the system that must be found before you proceed. Open the chlorine shutoff valve; the rotameter float will begin to rise indicating air flow.

e. Close the eductor water supply valve.

This completes the system prestart tests.

6.7.2.2   Startup

Once the prestart checks have been completed and any problems that were detected have been resolved, the generator is ready for startup.

1. Reconnect the chlorine supply line to the vacuum demand regulator.

2. Open all the shutoff valves for both the chlorine and chlorite supply lines.

3. Open both the chlorine and the sodium chlorite rate valves, which are located at the top of the rotameters.

4. Determine the chlorine dioxide demand. Open the eductor supply valve. The generator should begin to run.

5. Use the chlorine rate valve to adjust the feed rate to correspond to the feed-rate chart located on the generator.

6. Adjust the chlorite rate valve until the rotameter reads 100. Use the manual sodium chlorite valve to adjust the feed rate to correspond to the feed-rate chart. The production of chlorine dioxide will be evident by a yellow-green solution in the sight glass.

 Do not open the rate valves all the way against the stops. This can stress the valve and eventually cause vacuum leaks.

It will take 15 to 20 seconds for the system to reach full vacuum and stabilize. No further adjustments are required to the generator unless the production demand changes.

6.7.2.3   Shutdown

1. Short Term

A short-term shutdown would be less than 24 hours, as in the case of maintenance or troubleshooting.

a. Open the eductor supply valve, allowing the generator to run.

b. Close the sodium chlorite supply line shutoff valve. This will allow the full system vacuum to be applied to the chlorine line.

c. Close the chlorine shutoff valve at the vacuum demand regulator. Allow the generator to run until the float in the chlorine rotameter falls to zero. Disconnect the feed line from the vacuum regulator. This will allow air to be drawn in and will purge the chemical feed line.

d. Disconnect the Teflon tubing from the chemical filter on the chlorite supply line and immerse the tubing in a container of clean, warm water. This will flush the chlorite line. Allow the generator to flush in this fashion for at least 3 or 4 minutes.

e. Close the eductor supply valve.

2. Long Term

If the system is to be shut down for more than 24 hours, as in a seasonal application, additional precautions should be taken.

a. Follow all of the steps outlined in Item 1, Short Term.

b. Remove the filter from the Y-strainer and allow the water to drain.

c. Disconnect the Teflon tubing from the water and two chemical inlets. Loosen the two unions at the bottom of the reaction column. Open the plastic retaining clamps and remove both check valve assemblies. Use low pressure air to dry the assemblies by placing the air nozzle on the inlet side of the assemblies.

d. Loosen the union on the water inlet and remove the connecting pipe. Empty the water from this section and dry with compressed air.

e. Loosen the union on the solution outlet and dry the rest of the generator assembly using compressed air.

f. Open the rate valves on the rotameters and allow them to drain.

Note: Do not use pressurized air to dry these components.

g. If distribution lines are located in areas subject to freezing, they must also be drained and properly dried. Disconnect the line at the injection point and open all shutoff valves in the distribution line.

Once all of the assemblies are thoroughly dry, the system should be reassembled to prevent the loss of any components. The unit is now ready for storage.

6.7.3   Maintenance

Use of a periodic maintenance schedule is the best safeguard against production losses of chlorine dioxide and unscheduled downtime. The recommended maintenance schedule shown in Table 6.9 is for systems operating under optimum conditions. Factors such as dirty water, hard water, or wet chlorine can increase the frequency of maintenance that is required. Keeping accurate maintenance logs will help you to establish a schedule specific to your system.

Table 6.9   Periodic maintenance schedule

Service Interval	Bimonthly	Monthly	Quarterly	Yearly
Water Filter				
Inspect	X			
Clean	X			
Rebuild			X	
Ball Checks				
Inspect		X		
Rebuild			X	
Generator				
Inspect	X			
Clean			X	
Tubing and Connectors				
Inspect			X	
Replace				X
Rotameters				
Inspect	X			
Rebuild				X
Sight Glass				
Inspect	X			
Clean			X	
Rebuild				X
Nozzle and Throat				
Inspect			X	
Clean			X	
Replace				X
Reaction Column				
Inspect	X			
Clean			X	
Check Valve				
Inspect	X			
Clean		X		
Teflon Solenoid				
Inspect	X			
Clean		X		
Rebuild			X	
Chemical Filter				
Inspect		X		
Clean		X		
Replace			X	

6.7.4   Troubleshooting

Table 6.10 lists the symptoms of several problems that sometimes develop with chlorine dioxide generators, the probable causes of these symptoms, and some possible solutions.

Table 6.10 Chlorine dioxide generator troubleshooting

Symptoms	Probable Cause	Remedy
1. Low water flow	1. Input supply restricted	1. Eliminate restriction
	2. Y-strainer blocked	2. Clean filter element
	3. Nozzle and throat plugged or damaged	3. Clean or replace
	4. Output blocked	4. Fully open outlet shutoff valve or clear blockage
2. Low sodium chlorite flow	1. Water flow low	1. Refer to Item 1 in this table
	2. Rotameter setting incorrect	2. Adjust setting
	3. Manual rate valve setting incorrect	3. Adjust setting
	4. Chemical filter blocked	4. Clean or replace filter
	5. Chemical feed check valve failure	5. Repair or replace
	6. Rotameter inlet plugged	6. Disassemble and clean
	7. Teflon solenoid plugged	7. Clean or replace
	8. Reaction column blocked	8. a. Open water bleed and flush column b. Clean reaction column
	9. Nozzle and throat plugged or damaged	9. Clean or replace
3. Low chlorine flow	1. Rotameter setting incorrect	1. Adjust setting
	2. Vacuum regulator failure	2. Repair or replace
	3. Rotameter inlet plugged	3. Disassemble and clean
	4. Chemical check valve failure	4. Clean or replace
	5. Teflon solenoid plugged	5. Clean or replace
	6. Reaction column blocked	6. a. Open water bleed and flush column b. Clean reaction column
	7. Nozzle and throat plugged or damaged	7. Clean or replace
	8. Vacuum leak	8. Find and repair leak
4. No sodium chlorite flow	1. Out of chemical	1. Replenish supply
	2. Chemical shutoff valve closed	2. Open valve
	3. Rotameter rate valve closed	3. Open valve
	4. Manual rate valve closed	4. Open valve
	5. Teflon solenoid failure	5. Clean or replace
5. No chlorine flow	1. Out of chemical	1. Replenish supply
	2. Chemical shutoff valve closed	2. Open valve
	3. Rotameter rate valve closed	3. Open valve
	4. Teflon solenoid failure	4. Clean or replace
	5. Check valve salted out	5. Clean
	6. Vacuum leak	6. Find and repair leak
6. Bubbles in sodium chlorite meter	1. Vacuum leak	1. Find and repair leak
	2. Chemical level low	2. Replenish supply
	3. Bottom O-ring in rotameter leaking	3. a. Rotate meter tube b. Replace O-ring
7. Low vacuum	1. Low water flow	1. Refer to Item 1 in this table
	2. Low water pressure	2. Check water supply line pressure
	3. Water bleed setting too high	3. Adjust setting
	4. Reaction column blocked	4. Clean reaction column

Table 6.10   Chlorine dioxide generator troubleshooting (*continued*)

Symptoms	Probable Cause	Remedy
8. Falling meter ball	1. Vacuum leak	1. Find and repair leak
	2. Unstable water supply	2. Correct condition
	3. Chemical feed line incorrectly installed	3. Reinstall feed line
	4. Chemical feed line incorrectly sized	4. Re-size feed line
	5. Feed rate below 10 percent of system capacity	5. Adjust sodium chlorite feed rate
	6. Rate valve incorrectly adjusted	6. Adjust rate valve
9. Chlorite in chemical lines	1. Chemical check valve failure	1. Clean or replace
	2. Teflon solenoid failure	2. Clean or replace

Check Your Understanding

1. What additional equipment is necessary to use an existing chlorination unit to produce chlorine dioxide?

2. What hazards are associated with the handling of sodium chlorite?

3. What factors could increase the maintenance needed on chlorine dioxide generators?

6.8   Measurement of Chlorine Residual

Systems are required to monitor the disinfectant residual leaving the plant and at various points in the distribution system. Residuals are measured to ensure that the treated water is being adequately disinfected. The water leaving the plant must have at least 0.2 mg/L of the disinfectant (0.5mg/L recommended), and samples taken in the distribution system must have a detectable residual (0.2 mg/L recommended).

Residual chlorine measurements of treated water leaving the plant should be taken at least three times per day on small systems and once every 2 hours on large systems. If at any time the disinfectant residual leaving the plant is too low, the system is allowed up to four hours to correct the problem. If the problem is corrected within this time, it is not considered a violation but the regulatory agency must be notified.

6.8.1   Methods of Measuring Chlorine Residual

There are multiple methods for measuring residual chlorine. The most commonly used methods involve amperometric titration and DPD tests. These are described in more detail in Chapter 9, "Laboratory Procedures."

Amperometric titration is the standard for comparison because it is sensitive, relatively free of interferences and provides the most repeatable chlorine residual results. It does, however, require specialized amperometric titration equipment, and some degree of operator skill.

amperometric (am-PURR-o-MET-rick) titration
A means of measuring concentrations of certain substances in water (such as strong oxidizers) based on the electric current that flows during a chemical reaction.

DPD tests are based on a chemical reaction that produces a color whose intensity is proportional to the concentration of chlorine. Comparing the color to a standard color wheel by eye will give an approximate concentration reading. A more accurate method is to **titrate** the sample with a chemical that destroys the color-producing compound. In recent years, affordable electronic readers have become commercially available that can read the color intensity and determine the chlorine concentration from an internal calibration curve.

Inline instruments based on these chemical principles can be used to provide continuous measurements of chlorine residual as the water leaves the plant. Guidance for adopting these kinds of instruments has been published by EPA (Method 334.0, EPA 815-B-09-013, September 2009).

6.8.2   ORP Probes

ORP (oxidation-reduction potential) probes are being used in some plants to optimize chlorination processes. ORP (also called the redox potential) is a direct measure of the disinfecting power of a chlorine residual. Chlorine forms that are toxic to microorganisms (including coliforms) are missing one or more electrons in their molecular structure. They satisfy their need for electrons by taking electrons from organic substances or microorganisms present in the water being treated. When microorganisms lose electrons they become inactivated and can no longer transmit a disease or reproduce.

The ability of chlorine to take electrons (the electrical attraction or electrical potential) is the ORP and is measurable in millivolts. The strength of the millivoltage (or the redox measurement) is directly proportional to the oxidative disinfection strength of the chlorine in the treatment system. The higher the concentration of chlorine disinfectant, the higher the measured ORP voltage. Conversely, the higher the concentration of organics (chlorinedemanding substances), the lower the measured ORP voltage. The ORP probe (redox sensing unit) measures the voltage present in the water being treated and thus provides a direct measure of the disinfecting power of the disinfectant present in the water.

In a typical installation, a high resolution redox (HRR) chlorine controller monitors the chlorine residual using an ORP probe suspended in the chlorine contact chamber downstream from the chlorine injection point. The controller converts the ORP signal to a 4- to 20-milliamp (mA) signal that automatically adjusts the chlorine feed rate from the chlorinator.

HRR units can control the chlorination chemical feed rate to account for the chlorine demand in the water. These HRR units can be set to automatically maintain chemical residuals in the ideal ranges, regardless of changes in the chlorine demand or water flow.

Maintenance for the ORP probe consists of cleaning the unit's sensor once a month. Chlorine demand is discussed in Chapter 9, "Laboratory Procedures."

titrate (TIE-trate)
To titrate a sample, a chemical solution of known strength is added drop by drop until a certain color change, precipitate, or pH change in the sample is observed (end point). Titration is the process of adding the chemical reagent in small increments (0.1–1.0 milliliter) until completion of the reaction, as signaled by the end point.

Check Your Understanding

1. How often should treated water residual chlorine measurements be made?

2. What methods are used to measure chlorine residual in treated water?

3. What does an ORP probe measure in a disinfection system?

4. What happens to a microorganism when it loses an electron?

5. What maintenance is required on ORP probes?

6.9   Chlorine Safety Program

Safety is everyone's priority. Every good safety program begins with cooperation between the employee and the employer, with the employee taking an active part in the overall program. The employee must be responsible and should take all necessary steps to prevent accidents. The employer also must take an active part by supporting safety programs. There must be funding to purchase equipment and to enforce safety regulations required by OSHA and state industrial safety programs. The following items should be included in all safety programs.

1. Establishment of a formal safety program.
2. Written rules and specific safety procedures.
3. Periodic hands-on training using safety equipment.
    a. Leak-detection equipment
    b. Self-contained breathing apparatus (Figure 6.17)
    c. Atmospheric monitoring devices
4. Establishment of emergency procedures for chlorine leaks and first aid.
5. Establishment of a maintenance and calibration program for safety devices and equipment.
6. Provide police and fire departments with tours of facilities to locate hazardous areas and provide chlorine safety information.

All persons handling chlorine should be thoroughly aware of its hazardous properties. Personnel should know the location and use of the various pieces of protective equipment and be instructed in safety procedures. In addition, an emergency procedure should be established and each individual should be instructed how to follow the procedures. An emergency checklist also should be developed and available.

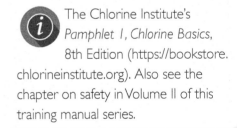

The Chlorine Institute's *Pamphlet 1, Chlorine Basics,* 8th Edition (https://bookstore.chlorineinstitute.org). Also see the chapter on safety in Volume II of this training manual series.

6.9.1   Chlorine Hazards

Chlorine is a gas that is 2.5 times heavier than air, extremely toxic, and corrosive in moist atmospheres. Dry chlorine gas can be safely handled in steel containers and piping, but with moisture must be handled in corrosion-resistant materials such as silver, glass, Teflon, and certain other plastics. Chlorine gas at container pressure should never be piped in silver, glass, Teflon, or any other material that cannot handle the pressure. Even in dry atmospheres, the gas is very irritating to the mucous membranes of the nose, to the throat, and to the lungs; a very small percentage in the air causes severe coughing. Heavy exposure can be fatal (see Table 6.11).

Table 6.11   Physiological response to concentrations of chlorine gas (Adapted from data in the 1955 US Bureau of Mines *Technical Paper 248*)

Effect	Parts of Chlorine Gas Per Million Parts of Air by Volume (ppm)[a]
Slight symptoms after several hours' exposure	1
Detectable odor	0.3 to 3.5
Noxiousness	5
Throat irritation	15
Coughing	30
Dangerous from 30 minutes to 1 hour	40
Death after a few deep breaths	1,000

[a] The current Occupational Safety and Health Administration (OSHA) permissible exposure limit (PEL) for chlorine is 1 ppm (part per million) as a ceiling limit. A worker's exposure to chlorine shall at no time exceed this ceiling level. The National Institute for Occupational Safety and Health (NIOSH) has established a recommended exposure limit (REL) for chlorine of 0.5 ppm as a time-weighted average (TWA) for up to a 10-hour workday and a 40-hour workweek and a short-term exposure limit (STEL) of 1 ppm. The revised Immediately Dangerous to Life or Health (IDLH) limit is 10 ppm. IDLH is the atmospheric concentration of any toxic, corrosive, or asphyxiate substance that poses an immediate threat to life or would cause irreversible or delayed adverse health effects or would interfere with an individual's ability to escape from a dangerous atmosphere.

When entering a room that may contain chlorine gas, open the door slightly and check for the smell of chlorine. Never go into a room containing harmful concentrations of chlorine gas in the air without a self-contained air supply, protective clothing, and help standing by. Help may be obtained from your chlorine supplier and your local fire department.

Most people can usually detect concentrations of chlorine gas above 0.3 ppm and you should not be exposed to concentrations greater than 1 ppm. However, chlorine gas can deaden your sense of smell and cause a false sense of security. Never rely on your sense of smell to protect you from chlorine because your sense of smell might not be able to detect harmful levels of chlorine.

6.9.2   Why Chlorine Must Be Handled with Care

You must always remember that chlorine is a hazardous chemical and must be handled with respect. Concentrations of chlorine gas in excess of 1,000 ppm (0.1 percent by volume in air) may be fatal after a few breaths.

Because the characteristic sharp odor of chlorine is noticeable even when the amount in the air is small, it is usually possible to get out of the gas area before serious harm is suffered. So, chlorine exposure is more avoidable than exposure to other gases, such as carbon monoxide, which is odorless, and hydrogen sulfide, which impairs your sense of smell in a short time.

Inhaling chlorine causes general restlessness, panic, severe irritation of the throat, sneezing, and production of much saliva. These symptoms are followed by coughing, retching and vomiting, and difficulty in breathing. Chlorine is particularly irritating to persons suffering from asthma and certain types of chronic bronchitis. Liquid chlorine causes severe irritation and blistering on contact with the skin.

6.9.3   Protect Yourself from Chlorine

Every person working with chlorine should know the proper ways to handle it, should be trained in the use of self-contained breathing apparatus (SCBA), methods of detecting hazards, and should know what to do in case of emergencies. The clothing of persons exposed to chlorine can be saturated with chlorine, which will irritate the skin if exposed to moisture or sweat. These people should not enter confined spaces before their clothing is purged of chlorine (stand out in the open air for awhile). This is particularly applicable to police and fire department personnel who leave the scene of a chlorine leak and ride back to their stations in closed vehicles. Suitable protective clothing for working in an atmosphere containing chlorine includes a disposable rainsuit with hood to protect your body, head, and limbs, rubber gloves to protect your hands, and rubber boots to protect your feet.

Self-contained air supply and positive pressure/demand breathing equipment must fit properly and be used properly. Pressure demand and rebreather kits may be safer. Pressure demand units use more air from the air bottle, which reduces the time a person may work on a leak. There are certain physical constraints when using respiratory protection. Contact your local safety regulatory agency to determine these requirements.

Before entering an area containing a chlorine leak, wear protective clothing. A chemical suit will prevent chlorine from contacting the sweat on your body and forming hydrochloric acid. Chemical suits are very cumbersome, but should be worn when the chlorine concentration is high. A great deal of practice is required to perform effectively while wearing a chemical suit.

The best protection that one can have when dealing with chlorine is to respect it. Each individual should practice rules of safe handling and good preventive maintenance.

Prevention is the best emergency tool you have. Because small amounts of chlorine cause large problems, plan ahead.

> Canister-type gas masks are usually inadequate and ineffective in situations where chlorine leaks occur and are therefore not recommended for use under any circumstances. Self-contained air- or oxygen supply-type breathing apparatus are recommended. Operators serving on emergency chlorine teams must be carefully selected and receive regular approved training. They must be provided the proper equipment that receives regular maintenance and is ready for use at all times.

1. Have your fire department and other available emergency response agencies tour the area so that they know where the facilities are located. Give them a clearly marked map indicating the location of the chlorine storage area, chlorinators, and emergency equipment.

2. Have regularly scheduled practice sessions in the use of respiratory protective devices, chemical suits, and chlorine repair kits. Involve all personnel who may respond to a chlorine leak.

3. Have a supply of ammonia available to detect chlorine leaks.

4. Write emergency procedures. Prepare a chlorine emergency list of names of companies and phone numbers of persons to call during an emergency and ensure that all involved personnel are trained in notification procedures. This list should be posted at plant telephones and should include:
   a. Fire department
   b. Chlorine emergency personnel
   c. Chlorine supplier
   d. Police department

5. Follow established procedures during all emergencies.
   a. Never work alone during chlorine emergencies.
   b. Obtain help immediately and quickly repair the problem. Problems do not get better.
   c. Only authorized and properly trained persons with adequate equipment should be allowed in the danger area to correct the problem.
   d. If you are caught in a chlorine atmosphere without appropriate respiratory protection, shallow breathing is safer than breathing

deeply. Recovery depends upon the duration and amount of chlorine inhaled, so it is important to keep that amount as small as possible.

e. If you discover a chlorine leak, leave the area immediately unless it is a very minor leak. Small leaks can be found by using a rag soaked with ammonia. A white gas will form near the leak so it can be located and corrected.

f. Use approved respiratory protection and wear disposable clothing when repairing a chlorine leak.

g. Notify your police department that you need help if it becomes necessary to stop traffic on roads and to evacuate persons in the vicinity of the chlorine leak.

6. Develop emergency evacuation procedures for use during a serious chlorine leak. Coordinate these procedures with your police department and other officials. Ensure that all facility personnel are thoroughly trained in any evacuation procedure developed.

7. Post emergency procedures in all operating areas.

8. Inspect equipment and routinely make any necessary repairs.

9. At least twice weekly, inspect areas where chlorine is stored and where chlorinators are located. Remove all obstructions from these areas.

10. Schedule routine maintenance on all chlorine equipment at least once every 6 months.

11. Have health appraisals for employees on chlorine emergency duty. All those who have heart or respiratory problems should not be allowed on emergency teams. There may be other physical constraints. Contact your local safety regulatory agency for details.

FIRST-AID MEASURES

Operators may be exposed to chlorine in spite of taking appropriate precautions. Be prepared to implement first-aid measures for yourself or others.

Mild Cases. Whenever you have a mild case of chlorine exposure (which does happen from time to time around chlorination equipment), you should first leave the contaminated area. Move slowly, breathe lightly without exertion, remain calm, keep warm, and resist coughing. Notify other operators and have them repair the leak immediately.

If clothing has been contaminated, remove it as soon as possible; otherwise, the clothing will continue to give off chlorine gas, which will irritate the body even after leaving the contaminated area. Immediately wash any area affected by chlorine. Shower and put on clean clothes.

If the victim has slight throat irritation, immediate relief can be accomplished by drinking milk. A mild stimulant such as hot coffee or hot tea is often used for coughing. See a physician.

Extreme Cases. Call 911 immediately. Then begin treating the victim.

1. Check for breathing. Tilt the victim's head back (this opens the airway and may restore breathing), put your ear over the victim's mouth and nose, and listen and feel for air. Look at the victim's chest to see if it is rising and falling. Watch for breathing for 3 to 5 seconds.

a. If the victim is breathing, place the person face up with head and back in a slightly elevated position. Keep victim warm and comfortable.

b. If there is no breathing, perform hands-only cardiopulmonary resuscitation (CPR).* With the victim face up, push hard and fast in the center of the chest. Continue hands-only CPR until breathing starts or help arrives.

2. Eye irritation caused by chlorine gas should be treated by flushing the eyes with generous amounts of water for at least 15 minutes. Hold eyelids apart to ensure maximum flushing of exposed areas. Do not attempt to neutralize with chemicals. Do not apply any medication (except as directed by a physician).

3. Minor throat irritation can be relieved by drinking milk. Do not give the victim any drugs (except as directed by a physician).

6.9.4 Hypochlorite Safety

Hypochlorite does not present the hazards that gaseous chlorine does and, therefore, is safer to handle. When spills occur, wash with large volumes of water. The solution is messy to handle. Hypochlorite causes damage to your eyes and skin upon contact. Immediately wash affected areas thoroughly with water. Consult a physician if the area appears burned. Hypochlorite solutions are very corrosive. Hypochlorite compounds are nonflammable; however, they can cause fires when they come in contact with organics or other easily oxidizable substances.

6.9.5 Chlorine Dioxide Safety

Chlorine dioxide is generated in much the same manner as chlorine and should be handled with the same care. Of special concern is the use of sodium chlorite to generate chlorine dioxide. Sodium chlorite is very combustible around organic compounds. Whenever spills occur, sodium chlorite must be neutralized with anhydrous sodium sulfite. Combustible materials (including combustible gloves) should not be worn when handling sodium chlorite. If sodium chlorite comes in contact with clothing, the clothes should be removed immediately and soaked in water to remove all traces of sodium chlorite or the clothes should be burned immediately.

6.9.6 Operator Safety Training

Training is a concern to everyone, especially when your safety and perhaps your life is involved. Every utility agency should have an operator chlorine safety training program that introduces new operators to the program and updates previously trained operators. As soon as a training session ends, obsolescence begins. People will forget what they have learned if they do not use and practice their knowledge and skills. Operator turnover can dilute a well-trained staff. New equipment and new techniques and procedures can dilute the readiness of trained operators. An ongoing training program could include a monthly lunch seminar, a monthly safety bulletin that is to be read by every operator, and outside speakers who reinforce and refresh specific elements of safety training.

*If you are trained in conventional CPR using chest compressions and mouth-to-mouth breathing, use that method instead of hands-only CPR. Training in both hands-only and conventional CPR is available from local Red Cross and American Heart Association chapters as well as many public health agencies. Video training for hands-only CPR and other information about CPR is available from the American Heart Association at cpr.heart.org.

The Chemical Transportation Emergency Center (CHEMTREC) is a public service of the American Chemistry Council dedicated to assisting emergency responders deal with incidents involving hazardous materials. Their 24-hour emergency phone number is (800) 424-9300.

6.9.7   CHEMTREC

Safely handling chemicals used in daily water treatment is an operator's responsibility. However, if the situation ever gets out of hand, there are emergency teams that will respond with help anywhere there is an emergency. If an emergency does develop in your plant and you need assistance, call the Chemical Transportation Emergency Center (CHEMTREC) at (800) 424-9300 for assistance. CHEMTREC will provide immediate advice for those at the scene of an emergency and then quickly alert experts whose products are involved for more detailed assistance and appropriate follow-up.

Check Your Understanding

1. What properties make chlorine gas so hazardous?
2. What type of breathing apparatus is recommended when repairing chlorine leaks?
3. What first-aid measures should be taken if a person comes in contact with chlorine gas?

6.10   Ultraviolet (UV) Systems

Just beyond the visible light spectrum there is a band of electromagnetic radiation, which we commonly refer to as ultraviolet (UV) light. When ultraviolet radiation is absorbed by the cells of microorganisms, it damages the genetic material in such a way that the organisms are no longer able to grow or reproduce, thus ultimately killing them. This ability of UV radiation to disinfect water has been understood for almost a century, but technological difficulties and high energy costs prevented widespread use of UV systems for disinfection. Today, however, with growing concern about the safety aspects of handling chlorine and the possible health effects of chlorination byproducts, UV disinfection is gaining in popularity. UV technology can also provide inactivation of *Cryptosporidium and Giardia*, which are resistant to common disinfectants like chlorine or ozonation. However, a chlorination stage is still required to provide disinfection residual in the distribution system. The combination of both UV technology and chlorination allows an efficient disinfection system by killing or inactivating a larger range of microorganisms than using only one disinfectant. The UV disinfection is particularly adapted to water with a good quality. The efficiency of the UV disinfection depends on the quality of water and on the treatment stages upstream. A raw water with a low turbidity and with a low level of color favors the penetration of UV light and improves the disinfection efficiency. Corrosive water can also damage the UV system. Technological advances are being made and several manufacturers now produce UV disinfection systems for water and wastewater applications. As operating experience with installed systems increases, UV disinfection may become a practical alternative to the use of chlorination at water treatment plants.

6.10.1   UV Lamp Types

Each UV lamp assembly consists of a UV lamp enclosed in an individual quartz sleeve with the ends appropriately sealed using an O-ring and a quartz end plug. All lamps within a UV system are identical (type, length, diameter, power, and output).

There are three types of electrode-type lamps used to produce UV radiation or UV light. These types are based on their internal operating design:

1. Low-pressure, low-intensity
2. Low-pressure, high-intensity
3. Medium-pressure, high-intensity

UV lamp technology is currently changing as manufacturers strive to improve their products and search for potential new technologies.

A **ballast** is a type of transformer that is used to limit the current to a UV lamp. Because UV lamps are arc-discharging devices, the more current in the arc, the lower the resistance becomes. Without a ballast to limit current, the lamp would destroy itself. Therefore, matching the lamp and ballast is very important in the design of UV disinfection systems.

Check Your Understanding

1. What happens when ultraviolet radiation is absorbed by the cells of microorganisms?

2. What are the three types of electrode-type lamps used to produce UV radiation or UV light?

3. What is a ballast?

6.10.2   UV System Types

The usual source of the UV radiation for disinfection systems is from low-pressure mercury vapor UV lamps that have been made into multi-lamp assemblies. Each lamp is protected by a quartz sleeve and each has watertight electrical connections. The lamp assemblies are mounted in a rack (or racks) and these racks are immersed in the flowing water. The racks may be mounted either within an enclosed vessel or in an open channel. Most of the UV installations in North America are of the open channel configuration.

When UV lamps are installed in open channels, they are typically placed either horizontal and parallel to the flow or vertical and perpendicular to the flow. In the horizontal and parallel-to-flow open channel lamp configuration, the lamps are arranged into horizontal modules of evenly spaced lamps. The number of lamps per module establishes the water depth in the channel. For example, 16 lamps could be stacked 3 inches apart to provide disinfection for water flowing through a 48-inch-deep open channel.

Each horizontal lamp module has a stainless-steel frame. Each module is fitted with a waterproof wiring connector to the power distribution

ballast
A type of transformer that is used to limit the current to an ultraviolet (UV) lamp.

center. The connectors allow each module to be disconnected and removed from the channel separately for maintenance. The horizontal lamp modules are arranged in a support rack to form a lamp bank that covers the width of the UV channel and there may be several such lamp banks along the channel. The number of UV banks per channel is determined by the required UV dosage to achieve the target effluent quality.

In the vertical and perpendicular-to-flow lamp configuration, rows of lamps are grouped together into vertical modules. Each vertical lamp module has a stainless-steel support frame and can be removed individually from the channel for cleaning or inspection. The electrical wirings for the lamps are located within the frame above the water level. Each individual lamp can be removed from the top of the frame without removing the entire module from the channel. The length of the UV lamps establishes the depth of water in the channel. One or more vertical modules are installed to cover the width of the channel. As with the horizontal arrangement, the number of vertical lamp modules per channel will depend on the UV dosage needed to achieve the desired effluent quality.

When it is necessary to maintain pressure within the water transmission system, UV lamps can be installed in a closed pressure vessel.

Another type of UV system, the thin film type, uses a chamber with many lamps spaced one-quarter inch (6 mm) apart. This system has been used in the water industry for a 9 MGD (0.35 m³/s) treatment plant.

Operators may also encounter a Teflon tube UV disinfection system, although this design is not in common use. Water flows in a thin-walled Teflon tube past a series of UV lamps. UV light penetrates the Teflon tube and is absorbed by the fluid. The advantage to this system is that water never comes in contact with the lamps. However, scale does eventually build up on the pipe walls and must be removed, or the Teflon tube must be replaced. This type of system has generally been replaced by the quartz sleeve systems described earlier.

6.10.3 Safety

The light from a UV lamp can cause serious burns to your eyes and skin. Always take precautions to protect them. Never look into the uncovered parts of the UV chamber without proper protective glasses. Do not plug a UV unit into an electrical outlet or switch a unit on without having the UV lamps properly secured in the UV water chamber and the box closed.

UV lamps contain mercury vapor, a hazardous substance that will be released if a lamp is broken. Handle UV lamps with care and be prepared with the proper equipment to clean up any spills.

Check Your Understanding

1. How is the number of UV banks per channel determined?
2. How are UV lamps installed when it is necessary to maintain pressure within the water transmission system?
3. What kinds of damage can the light from a UV lamp do to operators?

6.10.4   Operation

The operation of ultraviolet disinfection systems requires very little operator attention. To prevent short-circuiting and ensure that all microorganisms receive sufficient exposure to the UV radiation, the water level over the lamps must be maintained at the appropriate level. Water levels in channels can be controlled by weirs or automatic control gates.

Proper water depth must be maintained in the UV channel to ensure acceptable disinfection levels over the entire range of design flows. The UV channel water level control device must be regulated by the operator to:

1. Minimize variation of the channel's water level
2. Maintain the channel's water level at a defined level
3. Keep the UV lamps submerged at all times
4. Prevent excessive water layer thickness above the top lamp row

6.10.4.1   UV Light Intensity Effectiveness

To disinfect the water, UV light must be intense enough to penetrate the cell walls of the pathogens. The UV light intensity that reaches the pathogens is affected by the condition of the UV lamps and the quality of the water. The UV unit automatically adjusts UV dose according to the light transmission and effluent flow.

The UV lamp condition is affected by the lamp's age and the amount of slime on its surface. An old or dirty lamp has a reduced UV light intensity. The UV unit periodically cleans the lamps by mechanical means. The operator can adjust the frequency and the number of wiping cycles of the cleaning process. The UV lights can be cleaned manually in the cleaning tanks that contain the chemical solution formulated for this purpose.

Upstream processes affect the quality of the water, measured as turbidity and total suspended solids (TSS). High turbidity inhibits light transmission through the water, thereby reducing the disinfecting power of the light in proportion to its distance from the light source. High TSS, besides inhibiting light transmission, shields bacteria and thus protects them from the UV radiation. Insufficient UV light intensity can initiate a chain of events leading to ineffective disinfection and noncompliance.

Low UV light intensity will produce a low level of disinfection. This will result in high total coliform bacteria or high virus counts in the plant effluent, insufficient disinfection, and noncompliance.

The lamp must be replaced when its output at maximum power is insufficient to disinfect.

6.10.4.2   Minimum UV Dose Management

The primary control function of a programmable logic controller (PLC) is to manage the minimum UV dose applied to a UV channel. The actual UV dose control (dose pacing) is controlled for each UV channel and is based on maintaining a minimum dosing level.

The maintenance of a minimum dosing level is usually done by flow pacing. The applied dose is calculated from the flow rate and the end-of-lamp-life intensity at the specified transmittance multiplied by ballast power (that is, 50 percent power results in half the intensity). The formulas on page 410 show how the values needed to determine the necessary UV dose are computed based on UV intensity.

The dose calculation is based on received dose, derived from flow and input from the intensity sensors.

The PLC also controls the UV intensity in the UV channel. Each channel has its own separate ballast control loop using the intensity set point as its target. The loop controls the intensity by calculating the lowest intensity of the banks in use. Then, it adjusts the ballast/lamp power (changes intensity) to achieve the intensity set point. The action of the UV system PLC can be summarized as follows:

- Receives a minimum UV intensity from a preset value, which is field adjustable
- Receives a target UV intensity from a preset value, which is field adjustable
- Compares the target UV intensity with the actual UV intensity (lowest intensity of running banks)
- Makes an adjustment to the output power of all the ballast cards in operation, accordingly

If the actual UV intensity goes below minimum UV intensity for more than 2 minutes, the PLC should activate a low intensity channel bank alarm and flag the bank as failed. This will start the next assist bank, and after a preset time (10 minutes) will shut down the failed bank.

The level of ballast output power is identical for all of the ballast cards within a UV bank.

The UV system operates to maintain a minimum UV dose at all times, with a safety margin to accommodate operational changes and change-over procedures:

- As flow increases or transmission reduces, the UV dose will thus be reduced and the PLC will increase ballast output to compensate. It will start up the assist bank when the duty bank is at 85 percent of full power or the dose is less than the design level plus the safety factor. Once the assist bank is at 100 percent power, output will be reduced on both banks to get the correct dose.
- As flow decreases or transmission increases, the UV dose will thus be increased and the PLC will reduce ballast output to compensate. It will shut down the assist bank when ballast output is at 50 percent and the dose is greater than the design level plus the safety factor. Output on the duty bank will be increased to 100 percent before shutting down the assist bank.

FORMULAS

UV Dose Calculation

The intensity of the UV radiation and the contact time determine the UV dose received by the bacteria and, hence, the effectiveness of the process. UV dose is the standard indicator of the UV effectiveness and is expressed as follows:

$$UV_{dose}\left[\frac{mJ}{cm^2}\right] = UV_{intensity}\left[\frac{mW}{cm^2}\right] \times t[s]$$

Where,

t = Retention Time, s

Note: Dosage is typically expressed as milli-joules per square centimeter (mJ/cm²) and intensity as milliwatt seconds per square centimeter (mW-s/cm²).

In water disinfection systems, molecules or suspended solids in the water absorb UV energy. Thus, the worst-case intensity (at the farthest point from the UV source) is used in the calculation.

Channel Volume Calculation

The UV channel volume refers to the irradiated volume of the UV reactor, in other words, the volume in which the bacteria are exposed to UV radiation. This is a fixed value calculated in the following manner:

UV Channel Volume per Bank[ft³] = (UV Channel Width [ft] × Top Water Level [ft] × Lamp Arc Length [ft]) − Volume of Quartz Sleeves [ft³]

Retention Time Calculation

The retention time is the amount of time that the bacteria are in contact with the UV radiation. Head loss and velocity calculations ensure that an optimal hydraulic condition exists in the channel. The retention time is calculated by dividing the UV channel volume by the flow rate within the UV channel as follows:

$$t[s] = \frac{V_{reactor}[ft^3]}{Q\left[\dfrac{ft^3}{s}\right]}$$

Where,

$V_{reactor}$ = Volume of the UV Reactor

Flow Rate Calculation

The total inflow rate is supplied by the flowmeter. This value is scaled by using the maximum flow, which transforms the inflow rate to tenths of MGD, then is displayed on a screen as a flow rate. A programmable logic controller (PLC) calculates the rate per channel. This flow rate is for use in the control of the UV system exclusively and should not be used for any other purpose.

Check Your Understanding

1. The UV light intensity that reaches the pathogens in the water is affected by which factors?

2. The UV unit automatically adjusts UV dose according to which factors?

3. How is the UV dose calculated?

6.10.4.3   Routine Operations Tasks

The main routine operations tasks for a UV disinfection system include:

- Check UV monitors for UV transmission.
- Routinely clean the UV lamps.

6.10.4.4 Wiping System

Operators need to periodically observe the wiping system process to ensure proper operation of the wiping action of a bank and the proper wiping cycle. The wiping action of a bank is operated on a regular timed basis in the following manner:

- Wiping solenoid is energized.
- Wiper rings reach the far end (as opposed to the rest end).
- Wiping solenoid is de-energized.
- The system waits until all the wipers in the bank reach the rest end before repeating the cycle on the next bank.
- If the wipers do not complete the cycle in 1 minute, an alarm is generated and the system goes on to the next bank.
- Once all of the banks have been wiped, the wipe counter is incremented and the cycle resets to the first bank again.
- The cycle is repeated until the wipe counter (number of wipes) exceeds the preset value (wiping sequence is complete).

The wiping cycle is carried out whether or not the bank is operational; however, it must be in position and submerged in water.

6.10.4.5 Equipment Shutdown/Startup Preliminary Steps

The following preliminary steps provide general guidance for shutdown and startup of the process units and equipment.

1. Plan ahead
   - Choose a system configuration that involves the unit to be started up or shut down.
   - Determine how unit startup or shutdown will affect other units. If necessary, calculate how the change in flow distribution will affect flow rates, unit capacity, solids concentration, and effluent quality. Prepare upstream, downstream, and parallel units accordingly.
   - Determine how unit startup or shutdown will affect process control. Modify processcontrol strategies as needed.
2. Check switches
   - When starting a piece of equipment, check to ensure that all local and panel switches connected with the unit and its associated equipment are in the OFF position.
   - When shutting down a piece of equipment, open and label all breakers and local power disconnects for all equipment and instrumentation associated with the unit. Label using proper lockout/tagout procedure.
3. Clean units
   - Before startup and after shutdown, it may be necessary to clean the unit. When entering a unit, always follow the prescribed safety practices, including, where applicable, the confined space entry procedures.
4. Prepare equipment
   - Close all breakers and local power disconnects for all equipment and instrumentation associated with a piece of equipment.
   - Prepare the equipment for startup according to instructions in the manufacturer's O&M manual.

5. Set controls
   - For normal operation, set equipment controls as specified in the O&M manual.
   - For alternative operation, set equipment controls, as needed, to implement the chosen process-control strategy.

6.10.4.6 Shutdown Sequence

The following sequence is used to bring one UV channel off line:

1. Follow the preliminary steps listed under Section 6.10.4.5.
2. Close the respective influent channel slide gate.
3. Turn off all UV banks in the UV channel.
4. Drain the UV channel by opening the mud valve.
5. Wash the UV channel and the UV equipment.
6. Repair the UV equipment or channel and place it back into service as soon as possible.

6.10.4.7 Cleaning the Tank

UV light bulbs may be cleaned in an acid bath in a cleaning tank. Refer to the manufacturer's O&M manual for details. When a module is removed from a bank, the rest of the modules in the bank will be turned off.

6.10.4.8 Startup Sequence

The following sequence is used to bring one UV channel on line:

1. Follow the preliminary steps listed under Section 6.10.4.5.
2. Install the UV banks, if removed previously.
3. Close the UV channel mud valve.
4. Open the corresponding influent channel slide gate.
5. Return the UV banks to service (see the manufacturer's O&M manual).

Check Your Understanding

1. Why do operators need to periodically observe the UV wiping system process?
2. What preliminary steps should be followed when shutting down or starting up a UV disinfection process?
3. How can UV light bulbs be cleaned?

6.10.4.9 Monitoring Lamp Output Intensity

Lamp output declines with use so the operator must monitor the output intensity and replace lamps that no longer meet design standards, as well as any lamps that simply burn out. Lamp intensity monitors can be installed to assist the operator in monitoring the level of light output. Lamp failure indicators connected to the main UV control panel will alert the operator when a lamp burns out and requires replacement. In addition, computerized systems are available to monitor and record the age (burn time) of each lamp.

6.10.4.10 Monitoring Influent and Effluent Characteristics

Care must be taken not to exceed the maximum design turbidity levels and flow velocities when using this type of equipment. Suspended particles will shield microorganisms from the UV light and thus protect them from its destructive effects. Flows should be somewhat turbulent to ensure complete exposure of all organisms to the UV light, but flow velocity must be controlled so that the water is exposed to UV radiation long enough for the desired level of disinfection to occur.

Because ultraviolet rays leave no chemical residual like chlorine does, bacteriological tests must be made frequently to ensure that adequate disinfection is being achieved by the ultraviolet system. In addition, the lack of residual disinfectant means that no protection is provided against recontamination after the treated water has left the disinfection facility. When the treated water is exposed to visible light, the microorganisms can be reactivated. Microorganisms that have not been killed have the ability to heal themselves when exposed to sunlight. The solution to this problem is to design UV systems with a high efficiency for killing microorganisms.

6.10.5   Emergency Alarms

UV systems require extensive alarm systems to ensure continuous complete disinfection of the water being treated. Typical emergency alarms on UV systems include:

- Inlet level high
- Inlet turbidity high
- Sample flow low
- Inlet gate motorized gate control failure
- Inlet channel high transmittance
- Inlet channel low transmittance
- Diversion gate failure
- Isolation gate failure
- Inlet gate failure
- System flow rates
- Unit power failure
- Channels inlet gates failure
- Channel level sensor low/high
- Channel level sensor low/low
- Low dose

Operators need to inspect these instruments and be sure they are performing as intended.

Check Your Understanding

1. Why must operators monitor the lamp output intensity?
2. What influent and effluent characteristics must be monitored?
3. Why do UV systems require extensive alarm systems?

6.10.6 Maintenance

A UV system is capable of continuous use if a simple maintenance routine is performed at regular intervals. By checking the following items regularly, the operator of a UV system can determine when maintenance is needed.

1. Check the UV monitor for significant reduction in lamp output.
2. Monitor the process for major changes in normal flow conditions such as incoming water quality.
3. Check for fouling of the quartz sleeves and the UV intensity monitor probes.
4. Check the indicator light display to ensure that all of the UV lamps are energized.
5. Monitor the elapsed time meter, microbiological results, and lamp log sheet to determine when UV lamps require replacement.
6. Check the quartz sleeves for discoloration. This effect of UV radiation on the quartz is called solarization. Excessive solarization is an indication that a sleeve is close to the end of its useful service life. Solarization reduces the ability of the sleeves to transmit the necessary amount of UV radiation to the process.

Maintenance on UV systems requires two tasks: cleaning the quartz sleeves and changing the lamps.

Algae and other attached biological growths may form on the walls and floor of the UV channel. This slime can slough off, potentially hindering the disinfection process. If this condition occurs, the UV channel should be dewatered and hosed out to remove accumulated algae and slimes.

6.10.6.1 Quartz Sleeve Cleaning

Fouling of the quartz sleeves occurs when cations such as calcium, iron, or aluminum ions attach to protein and colloidal matter that crystallizes on the quartz sleeves. As this coating builds up on the sleeves, the intensity of the UV light decreases to the point where the buildup has to be removed for the system to remain effective. The rate at which fouling of the quartz sleeves occurs depends on several factors, including:

1. Types of treatment processes before UV disinfection.
2. Quality of water being treated.
3. Chemicals used in the treatment processes.
4. Length of time that the lamps are submerged.
5. Velocity of the water through the UV system. Very low or stagnant flows are especially likely to permit the settling of solids and the resulting fouling problems.

How often quartz sleeves need to be cleaned will depend on the quality of the water being treated and the treatment chemicals used before disinfection. Dipping the UV modules for 5 minutes in a suitable cleaning solution will completely remove scale that has deposited on the quartz sleeves. Cleaning is best done using an inorganic acid solution with a pH between 2 and 3. The two most suitable cleaning solutions are nitric acid in strengths to approximately 50 percent concentration and a 5 percent or 10 percent solution of phosphoric acid. To clean the system while still continuing to disinfect normal flows, single modules can be removed from the channel, cleaned, and then reinstalled. The other modules remaining on line while one is being cleaned should still be able to provide for continuous disinfection.

In-channel cleaning of UV lamps is another option, but it has some disadvantages. A backup channel is required and a much greater volume of acid solution is needed. Also, additional equipment and storage tanks for chemicals are required. Precautions must be taken to prevent damage to concrete channels from the acid cleaning solution. Epoxy coatings normally used to protect concrete from acid attack are not used in UV disinfection systems because the epoxy tends to break down under high UV-light intensities.

The type and complexity of the cleaning system will depend on the size of the system and the required cleaning frequency. Table 6.12 offers some guidelines.

Table 6.12   Recommended UV lamp cleaning methods

Peak Flow, MGD	Location of Lamps	Type of Cleaning	Work Hours per MGD
<5	out of channel	manual wipe	1
5–20	out of channel	immersion in cleaning tank	0.5
>20	out of channel	remove bank of lamps	0.25
>20	in channel	isolate a channel	0.25

6.10.6.2 Lamp Maintenance

The lamps are the only components that have to be changed on a regular basis. Their service life can be from 7,500 hours to 20,000 hours. This considerable variation can be attributed to three factors:

1. The level of suspended solids in the water to be disinfected and the fecal coliform level to be achieved. Better-quality effluents or less-stringent fecal coliform standards require smaller UV doses. Since lamps lose intensity with age, the smaller the UV dose required, the greater the drop in lamp output that can be tolerated.

2. The frequency of ON/OFF cycles. High cycling rates contribute to lamp electrode failure, the most common cause of lamp failure. Limiting the number of ON/OFF cycles to a maximum of 4 per 24 hours can considerably prolong lamp life.

3. The operating temperature of the lamp electrodes. System temperature usually depends on system conditions. Systems with both lamp electrodes operating at the same temperature (both electrodes submerged in the water) operate up to three times longer than systems where the two electrodes operate at different temperatures. This can occur in systems with lamps protruding through a bulkhead where only one electrode is immersed in the water and the other electrode is surrounded by air if the air temperature is routinely higher than the water temperature.

The largest drop in lamp output occurs during the first 7,500 hours. This decrease is between 30 and 40 percent. Thereafter, the annual decrease in lamp output (usually 5 to 10 percent) is caused by the decreased volume of gases within the lamps and by a compositional change of the quartz (solarization), which makes it more opaque to UV light.

Contact your appropriate regulatory agency to determine the proper way to dispose of used UV lamps. Do not throw used lamps in a garbage can to get rid of them because of hazardous mercury in the lamps.

6.10.7   Troubleshooting

Potential causes of poor performance by UV disinfection systems are described in this section. Operators should use this information when troubleshooting poor disinfection performance.

6.10.7.1   System Hydraulics

Erratic or reduced inactivation performance is often caused by poor system hydraulics. System short-circuiting, poor entry and exit flow conditions, and dead spaces or dead zones in the reactor all can be sources of poor performance.

6.10.7.2   Biofilms on UV Channel Walls and Equipment

Biofilms are typically fungal and filamentous bacteria that develop on exposed surfaces and are especially troublesome on areas exposed to light. Biofilms can contain and shield bacteria. When biofilms break away from surfaces, they protect the bacteria in the clumps as they pass through the UV disinfection system. Operators should periodically remove biofilms using a hypochlorite disinfecting solution.

6.10.7.3   Particles Shielding Bacteria

Particles can shield bacteria and reduce the effectiveness of the UV disinfection process. These particles should be removed by upstream treatment processes such as improved clarifier performance or some type of filtration.

Check Your Understanding

1. What tasks are included in the routine maintenance of UV disinfection systems?
2. How often should quartz sleeves be cleaned?
3. What factors influence the service life of UV lamps?
4. How can operators determine the proper way to dispose of used UV lamps?

6.11   Ozone Systems

Ozone is an alternative treatment process for disinfecting water. Ozone is produced when oxygen (O_2) molecules are exposed to an energy source and converted to the unstable gas, ozone (O_3), which is used for disinfection. Ozone is a very strong oxidant and virucide (kills viruses).

The effectiveness of ozone disinfection depends on the susceptibility of the target organisms, the contact time, and the concentration of the ozone. After generation, ozone is fed into a down-flow contact chamber containing the water to be disinfected. The purpose of the contact chamber is to transfer ozone from the gas bubble to the water while providing sufficient contact time for disinfection. Because ozone is consumed quickly, it must be exposed to the water uniformly in a plug-flow-type contactor.

The key process control guidelines are dose, mixing, and contact time. An ozone disinfection system strives for the maximum solubility of ozone in water because disinfection depends on the transfer of ozone into the water. The amount of ozone that will dissolve in water at a constant temperature is a function of the partial pressure of the gaseous ozone above the water or in the gas feed stream. All ozone disinfection systems should be pilot tested and calibrated before installation to ensure they meet the disinfection requirements for their particular sites.

6.11.1 Equipment

Ozone is normally generated on site because it is very unstable and decomposes to elemental oxygen in a short time after generation. Ozonation equipment (Figures 6.25, 6.26, and 6.27) consists of four major parts:

1. Gas preparation unit
2. Electrical power unit
3. Ozone generator
4. Contactor

Gauges, controls, safety equipment, and housing are also required.

6.11.2 Gas Preparation

The gas preparation unit to produce dry air usually consists of a commercial air dryer with a dew point monitoring system. This is the most critical part of the system.

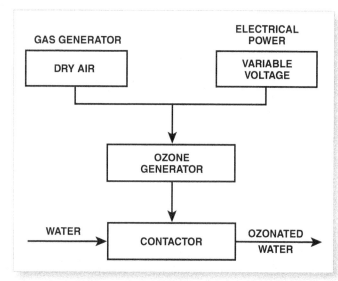

Figure 6.25 Ozonation equipment

Figure 6.26 Ozone disinfection system

OZONE GENERATION

The basic system for ozone production is presented in the diagram below. The ozone-producing unit, or ozonator, is fed an oxygen-containing gas that has first been cleaned and dried (A). As oxygen molecules (O_2) travel through the discharge gap (B) and are subjected to corona discharge, some are separated into free oxygen atoms (O) which combine with intact molecules to form ozone (O_3). Power (C) is required to generate the corona discharge, and the heat given off during ozonation is dissipated by some form of cooling process (D).

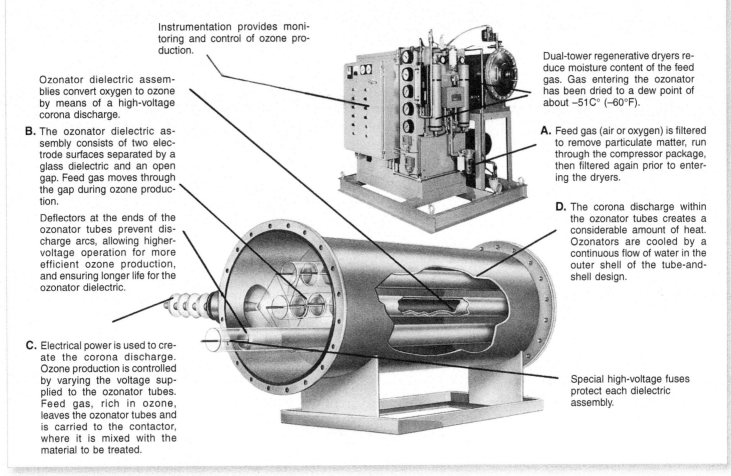

Instrumentation provides monitoring and control of ozone production.

Ozonator dielectric assemblies convert oxygen to ozone by means of a high-voltage corona discharge.

B. The ozonator dielectric assembly consists of two electrode surfaces separated by a glass dielectric and an open gap. Feed gas moves through the gap during ozone production.

Deflectors at the ends of the ozonator tubes prevent discharge arcs, allowing higher-voltage operation for more efficient ozone production, and ensuring longer life for the ozonator dielectric.

C. Electrical power is used to create the corona discharge. Ozone production is controlled by varying the voltage supplied to the ozonator tubes. Feed gas, rich in ozone, leaves the ozonator tubes and is carried to the contactor, where it is mixed with the material to be treated.

Dual-tower regenerative dryers reduce moisture content of the feed gas. Gas entering the ozonator has been dried to a dew point of about –51C° (–60°F).

A. Feed gas (air or oxygen) is filtered to remove particulate matter, run through the compressor package, then filtered again prior to entering the dryers.

D. The corona discharge within the ozonator tubes creates a considerable amount of heat. Ozonators are cooled by a continuous flow of water in the outer shell of the tube-and-shell design.

Special high-voltage fuses protect each dielectric assembly.

Figure 6.27   Ozone generation *(Courtesy of Ozone Engineering)*

6.11.3   Electrical Supply Unit

This unit is normally a very special electrical control system. The most common electrical supply unit provides low frequency, variable voltage. For large installations, medium frequency, variable voltage is used to reduce power costs and because it allows for higher output of ozone.

6.11.4   Ozone Generator

This unit consists of a pair of electrodes separated by a gas space and a layer of glass insulation (Figure 6.27). An oxygen-containing gas (air) is passed through the empty space as a high-voltage alternating current is applied. An electrical discharge occurs across the gas space and ozone is formed when a portion of the oxygen is ionized and then becomes associated with nonionized oxygen molecules.

$$\text{Oxygen from Air + High Electrical Voltage} \rightarrow \text{Ionized Oxygen + Heat}$$
$$O_2 \qquad\qquad\qquad\qquad \rightarrow \qquad 2(O)$$

and

$$\text{Ionized Oxygen + Non-Ionized Oxygen} \rightarrow \text{Ozone}$$
$$2(O) \qquad + \qquad 2(O_2) \qquad \rightarrow 2(O_3)$$

6.11.5   Ozone Contactor

This unit is a mixing chamber for the ozone-rich material and the process water. The objective is to dissolve enough ozone in the water to achieve disinfection at the lowest possible cost. These units are available in many configurations.

1. Multi-Stage Porous Diffuser
   a. Single application of an ozone-rich stream
   b. Application of ozone to second state
2. Eductor System
   a. Total flow through eductor
   b. Partial plant flow through the eductor
3. Turbine
   a. Positive pressure
   b. Negative pressure
4. Packed bed
   a. Concurrent ozone-rich flow
   b. Countercurrent ozone-rich flow
5. Two-level diffuser
   a. Lower chamber off gases applied to upper chamber
   b. Application of ozone-rich gas to lower chamber

For disinfection purposes, the diffuser-type ozone contactor (#5 above) is the most commonly used design. The off gases must be treated before release to the atmosphere. The most common method of treatment is the use of activated carbon and dilution.

6.11.6   Ozone Residuals

Residual ozone is measured by the iodometric method. The procedure is as follows:

1. Collect an 800 mL sample in a 1-liter wash bottle.
2. Pass pure air or nitrogen through the sample and then through an absorber containing 400 mL KI solution. Continue for 5 to 10 minutes at a rate of 1.0 liter/minute to purge all ozone from the sample.
3. Transfer KI solution to another vessel.
4. Add 20 mL 1 N H_2SO_4 to reduce the pH to 2.
5. Titrate with a 0.005-N sodium thiosulfate solution.
6. Add several drops of starch.

7. The end point is reached when the purple color is discharged (solution becomes colorless).

8. Repeat this test using a blank or distilled water.

9. Calculation

$$O_3\left[\frac{mg}{L}\right] = \frac{(A \pm B) \times N \times 24{,}000}{V_{sample}\,[mL]}$$

Where,

A   = mL of Titrant for Sample

B   = mL of Titrant for Blank (positive if turned blue and negative if had to back titrate blank)

N = Normality of Sodium Thiosulfate

V_{sample} = Volume of the Sample

Continuous inline ozone residual analyzers are available similar to the continuous inline chlorine residual analyzers (see Chapter 9, "Laboratory Procedures").

6.11.7   Safety

Ozone is a toxic gas that is a hazard to plants and animals. Ozone irritates nasal passages in low concentrations. When ozone breaks down in the atmosphere as a result of photochemical reactions (reactions taking place in the presence of sunlight), the resulting atmospheric pollutants can be very harmful. However, ozone is less of a hazard than gaseous chlorine because chlorine is normally manufactured and delivered to the plant site. Ozone is produced on the site, it is used in low concentrations, and it is not stored under pressure. Problem leaks can be stopped by turning the unit off.

Ozone production equipment has various fail-safe protection devices that will automatically shut off the equipment when a potential hazard develops.

6.11.8   Maintenance

Electrical equipment and pressure vessels should be inspected monthly by trained operators. A yearly preventive maintenance program should be conducted by factory representatives or by an operator trained by the manufacturer. Lubrication of the moving parts should be done according to the manufacturer's recommended schedule.

6.11.9   Applications of Ozone

In addition to using ozone after filtration for bacterial disinfection and viral inactivation, ozone may be used for several other purposes in treating drinking waters. Ozone may be used before coagulation for treating iron and manganese, helping flocculation, and removing algae (Figure 6.28). When ozone is applied before filtration, it may be used for oxidizing organics, removing color, or treating tastes and odors.

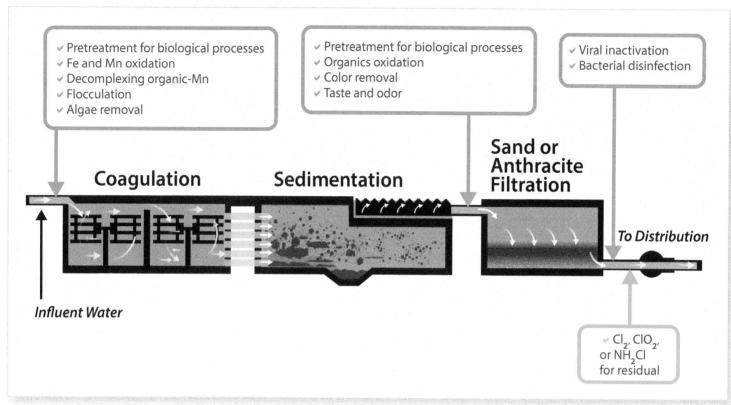

Figure 6.28 Typical ozone application points in drinking water processes

6.11.10   Advantages and Limitations of Ozone

The advantages of ozone include:

- Ozone is more effective than chlorine in destroying viruses and bacteria.
- The ozonation process uses a short contact time (10 to 30 minutes).
- There are no harmful residuals that need to be removed after ozonation.
- After ozonation, there is no regrowth of microorganisms, except for those protected by the particulates in the water stream.
- Ozone is generated on site, thus there are fewer safety problems associated with shipping and handling.
- Removes color, odor, and tastes (phenols).
- Oxidizes iron, manganese, sulfide, and organics.

The limitations of ozone include:

- Low dosage may not effectively inactivate some viruses, spores, and cysts.
- Ozonation is a more complex technology than chlorination or UV disinfection, requiring complicated equipment and efficient contacting systems.
- Ozone is very reactive and corrosive, thus requiring corrosion-resistant materials such as stainless steel.
- Ozonation is not economical for water with high levels of suspended solids (SS), biochemical oxygen demand (BOD), chemical oxygen demand (COD), or total organic carbon (TOC).

- Ozone is extremely irritating and possibly toxic, so off gases from the contactor must be destroyed to prevent exposure.
- The cost of treatment can be relatively high in capital and power costs.

In summary, ozone can be an effective disinfectant, however, the capital costs and the O&M costs of ozone may not be competitive with available disinfection alternatives.

Check Your Understanding

1. What factors influence the effectiveness of ozone disinfection?

2. What are the key process control guidelines for ozone disinfection?

3. Why is ozone generated on site?

6.12   Mixed-Oxidants (MIOX) Systems

Mixed-oxidants disinfection systems provide good protection from *Cryptosporidium and Giardia*. Field studies using mixed oxidants have achieved 99.99 percent reduction of *Giardia* after 30 minutes, and eliminated *E. coli* and cholera after 30 minutes. Mixed-oxidants treatment has reduced total THMs/DBPs produced when compared with conventional chlorination.

The mixed-oxidants disinfection system works by placing salt and water in a membraneless electrolytic cell (Figure 6.29) to produce a concentrated solution of the oxidants that work together as a disinfectant. The system produces on demand a liquid stream of disinfectant that is safe to handle but can be immediately injected into raw water. No mixing of chemicals or gases is required. The disinfectant is safe to store, produces no toxic fumes, and poses no explosive threat.

The mixed-oxidant cell generates a mixed-oxidant solution electrolytically from a sodium chloride (NaCl) brine. The mixed-oxidant cell separates solutions generated at the anode (oxidants) and at the cathode (reductants) into two fluid streams.

The composition of the mixed-oxidant solution can be varied by changing the brine salt (NaCl) concentration, voltage, and brine flow rate. For example, the cell operating guidelines may be adjusted to produce less chlorine/hypochlorite, more chlorine dioxide and ozone, and a higher oxidation reduction potential by lowering the salt concentration and raising the voltage.

The mixed-oxidant solution is more effective than chlorine/hypochlorite, chlorine dioxide, and ozone used individually. The fast-reacting oxidants (chlorine dioxide and ozone) satisfy the oxidant demand while providing exceptional disinfection and leaving chlorine/hypochlorite as a disinfectant residual. Disinfection occurs within minutes of treatment. These disinfection systems produce no chemical aftereffects and the treated water has no tastes or odors.

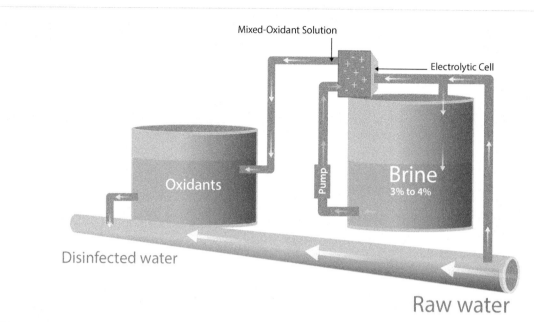

Features | Benefits
Strong Water Disinfection | Kills virus, *Giardia*, cholera, and *Cryptosporidium*. Two to six times more effective than chlorine.
Excellent Water Quality | No chemical aftereffects—MIOX-treated water appears, smells, and tastes good.
Meets EPA Standards | Properly dosed water maintains required chlorine residual.
No Harmful Products Used | No toxic gas or liquid chemicals required; uses only salt water.
Low Operating Cost | Operating cost is about 2–4¢/1,000 gallons of treated water.
Wide Range of Capacities | Standard cells treat up to 6,000 gallons per hour.
 | Large cells treat 20,000 gallons per hour.
Simple Operation/Low Maintenance | Operated and maintained by water system operators with minimal training.
Remote Operation | Operates on site anywhere, using any salt, and is adaptable to virtually any power source.

Figure 6.29   Mixed-oxidants system

6.13   Typical Chlorination Math Problems

This section provides practical examples of calculations typically needed to operate chlorination equipment at a water treatment plant.

6.13.1   Chlorinators

Example 5

At 8:00 am on Monday, a chlorine cylinder weighs 83 pounds (lb). At 8:00 am Tuesday the same cylinder weighs 69 pounds (lb). What is the chlorinator feed rate in pounds per 24 hours (lb/24 h)?

Known	Unknown
M_{Monday} = Monday Weight = 83 lb	F = Chlorinator Feed Rate, lb/24 h
$M_{Tuesday}$ = Tuesday Weight = 69 lb	
t = Time Period = 24 h	

Calculate the chlorinator feed rate in pounds per 24 hours.

$$F\left[\frac{lb}{24\ h}\right] = \frac{\text{Initial Weight}[lb] - \text{Final Weight}[lb]}{t[h]}$$

$$= \frac{M_{\text{Monday}}[lb] - M_{\text{Tuesday}}[lb]}{t[h]}$$

$$= \frac{83[lb] - 69[lb]}{24[h]}$$

$$= 14\frac{lb}{24\ h}$$

or

$$= 14\frac{lb}{d}$$

Note: The chlorinator setting should read 14 pounds (lb) per 24 hours (h).

Example 6

Estimate the chlorine dose in mg/L for the chlorinator in Example 5. The flow totalizer read 12,982,083 gallons (gal) at 8:00 am on Monday and 13,528,924 at 8:00 am on Tuesday. Fourteen pounds (lb) of chlorine were fed during the 24-hour (hr) period.

Known	Unknown
V_{Monday} = Monday Flow = 12,982,083 gal	Cl_{dose} = Chlorine Dose
V_{Tuesday} = Tuesday Flow = 13,528,924 gal	in mg/L
Cl_{feed} = Chlorine Feed = 14 lb/d	
t = Time Period = 1 d	

Calculate the chlorine dose in mg/L.

$$Cl_{\text{dose}}\left[\frac{mg}{L}\right] = \frac{Cl_{\text{feed}}\left[\frac{lb}{d}\right] \times t[d]}{\left(V_{\text{Tuesday}}[Mgal] - V_{\text{Monday}}[Mgal]\right) \times 8.34\frac{lb}{gal}}$$

$$= \frac{14\left[\frac{lb}{d}\right] \times 1[d]}{(13.53[Mgal] - 12.98[Mgal]) \times 8.34\frac{lb}{gal}} = \frac{14[lb]}{0.55[Mgal] \times 8.34\frac{lb}{gal}}$$

$$= 3.1\frac{lb}{Mlb} = 3.1\ ppm = 3.1\frac{mg}{L}$$

Example 7

A centrifugal pump delivers approximately 200 gpm (gallons per minute) against typical operating heads. If the desired chlorine dosage is 3 mg/L, what should be the setting on the rotameter for the chlorinator (lb chlorine per 24 h)?

Known	**Unknown**
Q = Pump Flow Rate = 200 gpm	F = Rotameter Setting,
Cl_{dose} = Chlorine Dose = 3 mg/L	lb Chlorine/24 h

1. Convert pump flow to million gallons per day (MGD).

$$Q\left[\frac{Mgal}{d}\right] = 200\left[\frac{gal}{min}\right] \times \frac{M}{10^6} \times \frac{60\ min}{h} \times \frac{24\ h}{d} = 0.288\ MGD$$

Note: When we multiply or divide an equation by $10^6/M$ we do not change anything except the units. This is just like multiplying an equation by 12 in/ft or 60 min/h; all we are doing is changing units.

2. Calculate the rotameter setting in pounds (lb) of chlorine per 24 hours (h). Refer to Section 6.3.11, "Process Calculations," for details.

$$F\left[\frac{lb}{d}\right] = Q\left[\frac{Mgal}{d}\right] \times Cl_{dose}\left[\frac{mg}{L}\right] \times \frac{3.785\ L}{gal} \times \frac{10^6}{M} \times \frac{kg}{10^6\ mg} \times \frac{2.205\ lb}{kg}$$

$$= 0.288\left[\frac{Mgal}{d}\right] \times 3\left[\frac{mg}{L}\right] \times \frac{3.785\ L}{gal} \times \frac{10^6}{M} \times \frac{kg}{10^6\ mg} \times \frac{2.205\ lb}{kg}$$

$$= 7.2\frac{lb}{d} = 7.2\frac{lb}{24\ h}$$

Note: We use the pounds per day formula regularly in our work to calculate the setting on a chemical feeder (lb chlorine per day) and the loading on a treatment process (pounds BOD per day). An explanation of how the units in the formula cancel helps us understand how to use and apply the formula to problems such as this one.

Example 8

Using the results from Example 7 (a chlorinator setting of 7.2 pounds per day [lb/d]), how many pounds (lb) of chlorine would be used during 1 week (wk) if the pump hour meter showed 100 hours (h) of pump operation? If the chlorine cylinder contained 78 pounds (lb) of chlorine at the start of the week, how many pounds of chlorine (lb) should remain at the end of the week?

Known	**Unknown**
F = Chlorinator Setting = 7.2 lb/d	Cl_{used} = Chlorine Used, lb/wk
$Cl_{initial}$ = Mass of Chlorine = 78 lb	
t = Time = 100 h	$Cl_{remaining}$ = Chlorine Remaining, lb

1. Calculate the chlorine used in pounds per week.

$$Cl_{used}\left[\frac{lb}{wk}\right] = Cl_{feed}\left[\frac{lb}{d}\right] \times t\left[\frac{h}{wk}\right]$$

$$= 7.2\left[\frac{lb}{d}\right] \times 100\left[\frac{h}{wk}\right] \times \frac{d}{24\ h} = 30\frac{lb}{wk}$$

2. Determine the amount of chlorine that should be in the cylinder at the end of the week.

$$Cl_{remaining}\left[lb\right] = Cl_{initial}\left[lb\right] - Cl_{used}\left[lb\right] = 78\left[lb\right] - 30\left[lb\right] = 48\ lb$$

Example 9

Given the pumping and chlorination system in Examples 7 and 8, if 30 pounds (lb) of chlorine are used during an average week, how many 150-pound (lb) chlorine cylinders (cyl) will be used per month (assume 30 days per month)?

<table>
<tr><td align="center">Known</td><td align="center">Unknown</td></tr>
<tr><td>Cl_{used} = Chlorine Used = 30 lb/wk
Mass of Chlorine   = 150 lb/cyl</td><td>M = Amount of Chlorine
      Used per Month, lb
N = Number of 150 lb Cylinders
      Used per Month, cyl/month</td></tr>
</table>

1. Calculate the amount of chlorine used in pounds (lb) of chlorine per month.

$$M\left[\frac{lb}{month}\right] = Cl_{used}\left[\frac{lb}{wk}\right] \times \frac{wk}{7\ d} \times \frac{30\ d}{month}$$

$$= 30\left[\frac{lb}{wk}\right] \times \frac{wk}{7\ d} \times \frac{30\ d}{month} = 129\frac{lb}{month}$$

2. Determine the number of 150-pound (lb) chlorine cylinders used per month.

$$N\left[\frac{number}{month}\right] = \frac{M\left[\frac{lb}{month}\right]}{Mass\ of\ Chlorine\left[\frac{lb}{cyl}\right]}$$

$$= \frac{129\left[\frac{lb}{month}\right]}{150\left[\frac{lb}{cyl}\right]}$$

$$= 0.86\frac{cyl}{month}$$

This installation requires less than one 150-pound chlorine cylinder per month.

6.13.2 Hypochlorinators

Example 10

Water from a well is being treated by a hypochlorinator. If the hypochlorinator is set at a pumping rate of 50 gallons per day (gpd) and uses a 3 percent (%) available chlorine solution, what is the chlorine dose in mg/L if the pump delivers 350 gpm?

Known	**Unknown**
$Q_{hypochlorinator}$ = Flow Rate of Hypochlorinator = 50 gpd	Cl_{feed} = Chlorine Feed Rate, lb/d
Hypochlorite Content = 3%	Cl_{dose} = Chlorine Dose, mg/L
Q_{pump} = Flow Rate of Pump = 350 gpm	

1. Convert the pumping rate to MGD.

$$Q_{pump}\left[\frac{Mgal}{d}\right] = 350\left[\frac{gal}{min}\right] \times \frac{M}{10^6} \times \frac{60\ min}{h} \times \frac{24\ h}{d} = 0.50\ MGD$$

2. Calculate the chlorine feed rate, Cl_{feed}, in pounds per day (lb/d).

$$Cl_{feed}\left[\frac{lb}{d}\right] = \frac{Q_{hypochlorinator}\left[\frac{gal}{d}\right] \times Hypochlorite\ Content[\%] \times \frac{8.34\ lb}{gal}}{100\%}$$

$$= \frac{50\left[\frac{gal}{d}\right] \times 3\% \times \frac{8.34\ lb}{gal}}{100\%}$$

$$= 12.5\frac{lb}{d}$$

3. Estimate the chlorine dose in mg/L.

$$Cl_{dose}\left[\frac{mg}{L}\right] = 0.120 \times \frac{Cl_{feed}\left[\frac{lb}{d}\right]}{Q_{pump}\left[\frac{Mgal}{d}\right]} = 0.120 \times \frac{12.5\left[\frac{lb}{d}\right]}{0.50\left[\frac{Mgal}{d}\right]} = 3\frac{mg}{L}$$

You can get the same answer by converting the water into million pounds and then using the property of water that parts per million is the same as mg/L.

$$Cl_{dose}\left[\frac{mg}{L}\right] = \frac{Cl_{feed}\left[\frac{lb}{d}\right]}{Q_{pump}\left[\frac{Mgal}{d}\right]} = \frac{12.5\left[\frac{lb}{d}\right]}{0.50\left[\frac{Mgal}{d}\right] \times \frac{8.34\ lb}{gal}}$$

$$= 3\frac{lb}{Mlb} = 3\ ppm = 3\frac{mg}{L}$$

Example 11

Water pumped from a well is disinfected by a hypochlorinator. During a 1 week time period, the water meter indicated that 1,098,000 gallons (1.098 Mgal) of water were pumped. A 2 percent (%) sodium hypochlorite solution is stored in a 2.5-foot-diameter plastic tank. During this 1-week time period, the level of hypochlorite in the tank dropped 18 inches (1.50 ft). What was the chlorine dose in mg/L?

Known	Unknown
V = Volume of Water Treated = 1.098 Mgal	Cl_{used} = Chlorine Used, lb
Hypochlorite Content = 2.0%	Cl_{dose} = Chlorine Dose,
D = Hypochlorite Tank	mg/L
Diameter = 2.5 ft	
h = Hypochlorite Used = 1.5 ft	

1. Calculate the gallons of hypochlorite solution used.

$$\text{Hypochlorite}\left[\text{gal}\right] = \frac{\pi}{4} \times D^2\left[\text{ft}^2\right] \times h\left[\text{ft}\right] \times \frac{7.48\ \text{gal}}{\text{ft}^3}$$

$$= 0.785 \times 2.5^2\left[\text{ft}^2\right] \times 1.5\left[\text{ft}\right] \times \frac{7.48\ \text{gal}}{\text{ft}^3}$$

$$= 55.0\ \text{gal}$$

2. Determine the pounds of chlorine (Cl) used to treat the water.

$$Cl_{used}\left[\text{lb}\right] = \text{Hypochlorite}\left[\text{gal}\right] \times \frac{\text{Hypochlorite}\left[\%\right]}{100\%} \times \frac{8.34\ \text{lb}}{\text{gal}}$$

$$= 55.0\left[\text{gal}\right] \times \frac{2\%}{100\%} \times \frac{8.34\ \text{lb}}{\text{gal}}$$

$$= 9.17\ \text{lb}$$

3. Estimate the chlorine dose in mg/L.

$$Cl_{dose}\left[\frac{\text{mg}}{\text{L}}\right] = \frac{Cl_{used}\left[\text{lb}\right]}{V\left[\text{Mgal}\right]} = \frac{9.17\left[\text{lb}\right]}{1.098 \times \left[\text{Mgal}\right] \times \frac{8.34\ \text{lb}}{\text{gal}}}$$

$$= 1.0\ \frac{\text{lb}}{\text{Mlb}} = 1.0\ \text{ppm}$$

$$= 1.0\ \frac{\text{mg}}{\text{L}}$$

You will get the same answer by converting the chlorine use into mg and converting the million gallons into liters as in the previous example.

Example 12

Estimate the required concentration of a hypochlorite solution (%) if a pump delivers 600 gpm from a well. The hypochlorinator can deliver a maximum of 120 gpd and the desired chlorine dose is 1.8 mg/L.

Known   **Unknown**

Q_{pump} = Pump Flow Rate = 600 gpm   Hypochlorite
$Q_{hypochlorinator}$ = Hypochlorinator   Strength, %
                Flow Rate   = 120 gpd
Cl_{dose} = Chlorine Dose   = 1.8 mg/L

1. Calculate the flow of water treated in million gallons per day (MGD).

$$Q_{pump}\left[\frac{Mgal}{d}\right] = 600\left[\frac{gal}{min}\right] \times \frac{60\ min}{h} \times \frac{24\ h}{d} \times \frac{M}{10^6} = 0.86\ MGD$$

2. Determine the pounds of chlorine required ($Cl_{required}$) per day (lb/d).

$$Cl_{required}\left[\frac{lb}{d}\right] = Q_{pump}\left[\frac{Mgal}{d}\right] \times Cl_{dose}\left[\frac{mg}{L}\right] \times \frac{3.785\ L}{gal} \times \frac{10^6}{M} \times \frac{kg}{10^6\ mg} \times \frac{2.205\ lb}{kg}$$

$$= 0.86\left[\frac{Mgal}{d}\right] \times 1.8\left[\frac{mg}{L}\right] \times \frac{3.785\ L}{gal} \times \frac{10^6}{M} \times \frac{kg}{10^6\ mg} \times \frac{2.205\ lb}{kg}$$

$$= 12.9\ \frac{lb}{d}$$

3. Calculate the hypochlorite solution strength as a percent.

$$Hypochlorite\ Strength[\%] = \frac{Cl_{required}\left[\frac{lb}{d}\right] \times 100\%}{Q_{hypochlorinator}\left[\frac{gal}{d}\right] \times \frac{8.34\ lb}{gal}}$$

$$= \frac{12.9\left[\frac{lb}{d}\right] \times 100\%}{120\left[\frac{gal}{d}\right] \times \frac{8.34\ lb}{gal}} = 1.3\%$$

Example 13

A hypochlorite solution for a hypochlorinator is being prepared in a 55-gallon drum. If 10 gallons of 5 percent hypochlorite is added to the drum, how much water should be added to the drum to produce a 1.3 percent hypochlorite solution?

Sketch the problem.

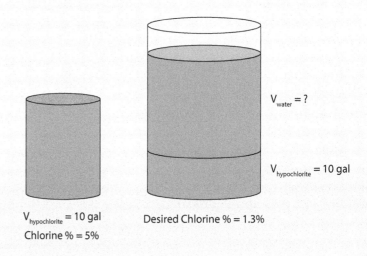

V_{water} = ?

$V_{hypochlorite}$ = 10 gal

$V_{hypochlorite}$ = 10 gal   Desired Chlorine % = 1.3%

Chlorine % = 5%

	Known			**Unknown**
V_{drum}	= Drum Capacity	= 55 gal	V_{water}	= Volume of
$V_{hypochlorite}$	= Volume of			Water Added, gal
	Hypochlorite	= 10 gal		
Actual Hypochlorite Content	= 5%			
Desired Hypochlorite Content	= 1.3%			

Start with the formula to determine the desired hypochlorite content.

$$\text{Desired Hypochlorite Content}[\%]$$

$$= \frac{V_{hypochlorite}[\text{gal}] \times \text{Actual Hypochlorite Content}[\%]}{V_{hypochlorite}[\text{gal}] + V_{water}[\text{gal}]}$$

$$1.3[\%] = \frac{10[\text{gal}] \times 5[\%]}{10[\text{gal}] + V_{water}[\text{gal}]}$$

Rearrange by cross multiplying.

$$10[\text{gal}] + V_{water} = \frac{10[\text{gal}] \times 5[\%]}{1.3[\%]}$$

$$10 + V_{water} = 38.5$$

$$V_{water} = 38.5 - 10 = 28.5 \text{ gal}$$

Check Your Understanding

1. Estimate the chlorine dose in mg/L for a chlorinator that fed 15 lb chlorine in a 24-hr period. The flow totalizer read 14.4 Mgal at 8:00 am on Monday and 16.1 Mgal at 8:00 am on Tuesday.

2. What should be the setting on a chlorinator (lbs chlorine per 24 hours) if a pump usually delivers 600 GPM and the desired chlorine dosage is 4.0 mg/L?

3. Water from a well is being disinfected by a hypochlorinator. If the hypochlorinator is set at a pumping rate of 60 gallons per day (GPD) and uses a 2 percent available chlorine solution, what is the chlorine dose rate in mg/L? The pump delivers 400 GPM.

6.14 Math Assignment

Turn to Appendix A, "Introduction to Basic Math for Operators." Read and work through the example problems in sections:

- "Learning Objectives"
- A.5.7, "Velocity and Flow Rate"
- A.9.8, "Disinfection" (English System)
- A.10.8, "Disinfection" (Metric System)

Check the math using a calculator. You should be able to get the same answers.

6.15   Additional Resources

Your local library may be a good source for obtaining these additional resources related to the field of water treatment:

Chlorine Basics, 8th Edition, Pamphlet 1, (The Chlorine Institute, https://bookstore.chlorineinstitute.org).

Chlorine Safety for Water and Wastewater Operators (DVD AVM_DVD_W) (The Chlorine Institute, https://bookstore.chlorineinstitute.org).

Handbook of Chlorination and Alternative Disinfectants, by George C. White, Fifth Edition, 2010. ISBN: 978-0-470-18098-3.

Manual of Water Utilities Operations (Texas Manual). Texas Water Utilities Association, www.twua.org. ASIN: B000KT9EIU. Chapter on disinfection of water.

Water Chlorination/Chloramination Practices and Principles (M20), Second Edition, 2006. ISBN: 978-1-583-21408-4. American Water Works Association (AWWA) (www.awwa.org).

Key Term Matching

_____ 1. Buffer Capacity

_____ 2. Oxidation

_____ 3. Reducing Agent

_____ 4. Total Chlorine

_____ 5. Combined Chlorine Residual

_____ 6. Enzymes

_____ 7. Nitrification

_____ 8. Residual Chlorine

_____ 9. Turbidity

_____ 10. Chloramines

A. A measure of the capacity of a solution or liquid to neutralize acids or bases. This is a measure of the capacity of water or wastewater for offering a resistance to changes in pH.

B. Compounds formed by the reaction of hypochlorous acid (or aqueous chlorine) with ammonia.

C. Chlorine combined with other substances, mainly ammonia.

D. Organic substances (produced by living organisms) that cause or speed up chemical reactions. Organic catalysts or biochemical catalysts.

E. An aerobic process in which bacteria oxidize the ammonia and organic nitrogen in water into nitrite and then nitrate.

F. The addition of oxygen, removal of hydrogen, or the removal of electrons from an element or compound.

G. Any substance, such as base metal (iron) or the sulfide ion (S^{2-}), that will readily donate (give up) electrons.

H. The concentration of chlorine present in water after the chlorine demand has been satisfied.

I. The total concentration of chlorine in water, including the combined chlorine (such as inorganic and organic chloramines) and the free available chlorine.

J. The cloudy appearance of water caused by the presence of suspended and colloidal matter.

6.1 Drinking Water Safety

11. _____ is the process designed to kill or inactivate most microorganisms in water or wastewater, including essentially all pathogenic bacteria. _____ is the complete destruction of all organisms.

6.2 Factors Influencing Disinfection

12. Organic matter in water can consume great amounts of _____ while forming unwanted _____.

6.3 Disinfection Process

13. What is not a type of physical disinfection?
    1. Chlorine
    2. Heat
    3. Ultrasonic waves
    4. Ultraviolet rays
14. What is not a chemical disinfectant?
    1. Bromine
    2. Carmine
    3. Ozone
    4. Sodium hydroxide
15. Chlorine dioxide is an effective disinfectant because it does not form _____ that may be formed by other chlorine compounds; it is not affected by _____; it is effective at higher pH levels; it reacts with _____, helping to remove them and eliminate their characteristic odors; and it can control _____.

6.4 Points of Chlorine Application

16. What is a limitation of prechlorination?
    1. May control algal and slime growths
    2. May control mudball formation
    3. May increase chlorine contact time
    4. May increase the formation of trihalomethanes in raw water

6.5 Operation of Chlorination Equipment

17. Care should be taken to prevent sharp bends in copper tubing because this will _____ and eventually the tubing will_____.

18. Why do you not use wrenches longer than 6 inches (15 cm), pipe wrenches, or wrenches with an extension on to open chlorine container valves?
    1. These wrenches are only available in metric sizes
    2. These wrenches are too costly
    3. These wrenches do not fit on chlorine container valves
    4. You could exert too much force and break the valve
19. What should be done to the water supply pump when the hypochlorinator is shut down?
    1. Operate the water supply pump as usual
    2. Start the pump slowly to ease into full operation
    3. Turn water supply pump off
    4. Turn water supply pump on at full operation
20. Liquid chlorine piping systems must be _____ because if moisture enters the system, it readily mixes with chlorine and forms _____, which will corrode the pipes, valves, joints, and fittings. Corrosion can cause leaks and require that the entire system be replaced.
21. If consumers complain about chlorine tastes in their drinking water, what is the most likely cause of this problem?
    1. The chlorine dose is inadequate
    2. The chlorine dose is the wrong mix of chemicals
    3. The chlorine dose is too high
    4. The chlorine dose was added too late in the treatment process

6.6 Maintenance

22. The maximum temperature at which a chlorine cylinder is stored should not exceed _____°F (_____°C).
23. When a chlorine leak occurs, the _____ containing the _____ should be routed to a treatment system for processing.

434

24. When chlorination is practiced for disinfection, it should be _____ to protect the _____.

6.7 Chlorine Dioxide Facilities

25. What is a disadvantage of using chlorine dioxide for disinfection?
    1. Chlorine dioxide costs are higher than other more commonly used disinfection methods
    2. Chlorine dioxide is effective in killing bacteria and viruses
    3. Chlorine dioxide is much more effective than chlorine in killing bacteria in the pH range from 8 to 10
    4. Trihalomethanes are not formed when disinfecting with chlorine dioxide

6.9 Chlorine Safety Program

26. Which item is not included in a chlorine safety program?
    1. Atmospheric monitoring devices
    2. Leak-detection equipment
    3. Oral rules and safety procedures
    4. Self-contained breathing apparatus

6.10 Ultraviolet (UV) Systems

27. _____ and _____ previously prevented widespread use of UV systems for disinfection, but it is gaining in popularity.
28. UV disinfection systems are gaining in popularity today because of growing concern about the _____ of handling chlorine and the possible _____ of chlorination byproducts.

29. Which item is not a goal of regulating the UV channel water level control device?
    1. Maximize variation of the channel's water level
    2. Maintain the channel's water level at a defined level
    3. Keep the UV lamps submerged at all times
    4. Prevent excessive water layer thickness above the top lamp row
30. _____ for the UV transmission and _____ are routine operations tasks for a UV disinfection system.
31. What is a potential cause of poor UV disinfection system performance?
    1. Clean UV channel walls and equipment
    2. Limited ON/OFF cycles
    3. Improved clarifier performance
    4. Short-circuiting

6.11 Ozone Systems

32. Why is ozone not used more extensively in the disinfection of water?
    1. After ozonation, there is no regrowth of microorganisms, except for those protected by the particulates in the water stream
    2. Ozonation is a more complex technology than is chlorination or UV disinfection, requiring complicated equipment and efficient contacting systems
    3. The ozonation process oxidizes iron, manganese, sulfide, and organics
    4. The ozonation process uses a short contact time (10 to 30 minutes)

6.13 Typical Chlorination Math Problems

33. At 8:00 am on Monday, a chlorine cylinder weighs 94 pounds (lb). At 8:00 am Tuesday the same cylinder weighs 76 pounds (lb). What is the chlorinator feed rate in pounds per 24 hours (lb/24 h)?

    1. 16 lb/24 hr
    2. 18 lb/24 hr
    3. 20 lb/24 hr
    4. 22 lb/24 hr

34. For a chlorinator set at 6.7 lb/d, the pump hour meter shows 100 hours (h) of pump operation. If the chlorine cylinder contained 83 pounds (lb) of chlorine at the start of the week, how many pounds of chlorine (lb) should remain at the end of the week?

    1. 55.1 lb
    2. 62.3 lb
    3. 65.1 lb
    4. 68.4 lb

35. Estimate the required concentration of a hypochlorite solution (%) if a pump delivers 800 gpm from a well. The hypochlorinator can deliver a maximum of 145 gpd and the desired chlorine dose is 1.8 mg/L.

    1. 1.6 %
    2. 1.3 %
    3. 1.5 %
    4. 1.4 %

Corrosion Control

LEARNING OBJECTIVES

1. Recognize adverse effects of corrosion and describe how a pipe corrodes.

2. Determine if corrosion problems exist in your system.

3. Determine if a water is saturated with calcium carbonate.

4. Control corrosion using proper chemicals and dosages and cathodic protection.

5. Prevent soil corrosion (external corrosion).

6. Troubleshoot and solve corrosion problems.

7. Explain and implement the Lead and Copper Rule.

KEY TERMS

alkalinity	compound	electrolytic cell	oxidation
amperage	corrosion	electromotive force (EMF)	pH
ampere	corrosion inhibitors	electromotive series	polarization
anaerobic	corrosive gases	electron	representative sample
anion	corrosivity	element	sacrificial anode
annular space	coupon	galvanic cell	saturation
anode	current	galvanic series	slake
base metal	dead end	hardness, water	slurry
buffer capacity	depolarization	ion	stray current corrosion
calcium carbonate	dielectric	Langelier Index (LI)	supersaturated
($CaCO_3$) equivalent	electrochemical reaction	molecule	titrate
cathode	electrochemical series	NPDES permit	tubercle
cathodic protection	electrolysis	noble metal	tuberculation
cation	electrolyte	ohm	

7.1   Adverse Effects of Corrosion

Corrosive water can cause serious problems in both water supply facilities and water treatment plants. Many hundreds of millions of dollars in damage occurs each year due to corrosive conditions in water systems. Water main replacement is often required when **tuberculation** reduces the carrying capacity of a main. Tuberculation increases pipe roughness, which causes an increase in pump energy costs and may reduce distribution system pressures. Leaks in water mains are usually caused by corrosion and may eventually require the replacement of a water main. Figure 7.1 shows corrosion damage to pipes due to corrosive soils.

Many other serious problems are caused by corrosive water. Corrosive water causes materials to deteriorate and go into solution (be carried by the water). Corrosion of toxic metal pipe materials, such as lead, can create a serious health hazard. Corrosion of iron may produce a flood of unpleasant telephone calls from consumers complaining about rusty water, stained laundry, and bad tastes. Corrosive drinking water causes internal corrosion (the inside of the pipe corrodes) and corrosive soils and moisture cause external corrosion (the outside of the pipe corrodes).

tuberculation
(too-BURR-kyoo-LAY-shun)
The development or formation of small mounds of corrosion products (rust) on the inside of iron pipe. These mounds (tubercles) increase the roughness of the inside of the pipe thus increasing resistance to water flow (decreases the C Factor).

7.2   Corrosion Process

Corrosion is a complex topic. This chapter aims to make this subject as understandable as possible, without oversimplifying all of the factors that influence corrosion. There are many possible solutions to corrosion problems and a combination of solutions may be necessary to solve any corrosion problems in your water treatment and water distribution facilities.

Corrosion is a complex problem, involving numerous chemical, electrical, physical, and biological factors. Corrosion can occur both on the interior and exterior of metal piping and equipment. A few basic concepts describe the major principles of corrosion. Selecting and implementing an effective corrosion-control program requires an understanding of these basic concepts.

Corrosion is the gradual decomposition or destruction of a material by chemical action, often due to an electrochemical reaction. Corrosion may be caused by (1) stray current electrolysis, (2) galvanic corrosion caused by dissimilar metals, or (3) differential-concentration cells. Corrosion starts at the surface of a material and moves inward. The severity and type of corrosion depend on the chemical and physical characteristics of the water and the material.

electrochemical reaction
Chemical changes produced by electricity (electrolysis) or the production of electricity by chemical changes (galvanic action). In corrosion, a chemical reaction is accompanied by the flow of electrons through a metallic path. The electron flow may come from an external source and cause the reaction, such as electrolysis caused by a DC (direct current) electric railway, or the electron flow may be caused by a chemical reaction, as in the galvanic action of a flashlight dry cell.

current
A movement or flow of electricity. Electric current is measured by the number of coulombs per second flowing past a certain point in a conductor. A coulomb is equal to about 6.25×10^{18} electrons (6,250,000,000,000,000,000 electrons). A flow of one coulomb per second is called one ampere, the unit of the rate of flow of current.

7.2.1   Electrochemical Corrosion: The Galvanic Cell

This section explains the corrosion reaction for you to understand what causes corrosion and determine whether a corrosion problem exists in your system.

Metallic (metal) corrosion in potable water is always the result of an **electrochemical reaction**. An electrochemical reaction is a chemical reaction where the flow of electric **current** itself is an essential part of

Corrosion of Plain Carbon Steel N Exposed
14 Years at 5 Test Sites

SOIL 51   53   55   56   58

Corrosion of Plain Cast Iron G
Exposed 14 Years at 5 Test Sites

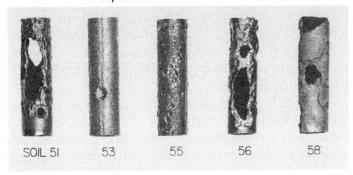

SOIL 51   53   55   56   58

Figure 7.1   Exterior corrosion due to corrosive soils: Corrosion of Plain Carbon Steel N Exposed 14 Years at 5 Test Sites (*Courtesy of AK Steel Corporation*)

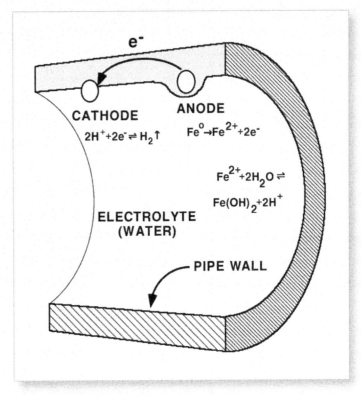

Figure 7.2   Simplified anode and cathode reactions of iron in contact with water. Source of H^+ ions is the normal dissociation of water, $H_2O \rightleftharpoons H^+ + OH^-$

the reaction. If the electric current is stopped by breaking the circuit, the chemical reaction will stop. Also, if the chemical reaction is stopped by removing one of the reacting chemicals, the flow of electric current will stop. For corrosion to occur, both of the factors—electric current and chemical reaction—must be present.

The deterioration of metal during corrosion is called an electrochemical reaction because both a chemical and an electrical process are occurring. Figure 7.2 shows a simplified electrochemical corrosion reaction with the following components:

- **Anode**
  Point from which metal is lost and electric current begins. The anode (positive pole or electrode of an electrolytic system, such as a battery) attracts negatively charged particles or ions (anions) in an electrolyte solution. **Anions** are attracted to the anode under the influence of a difference in electrical potential. The chloride **ion** (Cl^-) is an anion.

- **Cathode**
  Point where electric current leaves the metal and flows to the anode through the electrolyte. A cathode is the negative pole or electrode of an **electrolytic cell** or system. It will attract positively charged particles or ions (**cations**) in an electrolyte solution. Cations are attracted to the cathode under the influence of a difference in electrical potential. The sodium ion (Na^+) is a cation.

ion
An electrically charged atom, radical (such as SO_4^{2-}), or molecule formed by the loss or gain of one or more electrons.

electrolytic (ee-LECK-tro-LIT-ick) cell
A device in which the chemical decomposition of material causes an electric current to flow. Also, a device in which a chemical reaction occurs as a result of the flow of electric current. Chlorine and caustic (NaOH) are made from salt (NaCl) in electrolytic cells.

- **Electrolyte**
  Conducting solution (usually water with dissolved salts). A substance that separates into two or more ions when it is dissolved in water.

In addition, the anode and the cathode must be joined. In Figure 7.2, the only metal present is iron. At the anode, a **molecule** of iron dissolves into the water as a ferrous (Fe^{2+}) ion (chemical reaction), and two **electrons** ($2e^-$) flow to the cathode (electrical reaction). At the cathode, the electrons leave the metal at the point of contact with the electrolyte (water) and react with hydrogen ions (H^+) in the water to form hydrogen gas (H_2). The hydrogen ions are always present in the water from the normal separation of water ($H_2O \rightleftharpoons H^+ + OH^-$). The electrolyte (water) is in contact with both the anode and the cathode, completing the circuit.

At the anode, the dissolved iron reacts with oxygen and the water, forming a rust film composed initially of ferrous hydroxide, $Fe(OH)_2$, as shown in Figure 7.3. Additional water and oxygen then react with the ferrous hydroxide to form ferric hydroxide, $Fe(OH)_3$, which becomes a second layer over the ferrous hydroxide.

This multilayered rust deposit is known as a **tubercle** (a crust of rust that builds up over a pit caused by the loss of metal due to corrosion). Tubercles can grow to the point that the carrying capacity of the pipe is significantly reduced. Also, during periods of high flow rates, the tubercles may dislodge, resulting in rusty or red-colored water. The formation of a rust coating on the pipe has another important effect on the rate of corrosion. As the rust film forms, it begins to cover and protect the anode, slowing the rate of corrosion. If the rust film is flushed away, the corrosion reaction accelerates again.

A more complex form of electrochemical corrosion is caused by the joining of dissimilar metals. This type of corrosion is called galvanic corrosion. Figure 7.4 shows a galvanic corrosion model resulting from the joining of sections of copper and iron pipe. Figure 7.5 contains photos of corrosion damage caused by galvanic corrosion. Like the corrosion cell in Figure 7.2, the galvanic corrosion cell in Figure 7.4 also has an anode, a cathode, and electrolyte, as well as a connection between the anode and cathode. This **galvanic cell**, however, has two dissimilar metals, copper for the cathode and iron for the anode. The electrochemical corrosion reaction is otherwise similar.

The degree to which a particular metal will become anodic (corrode) in a galvanic reaction is related to its tendency to enter into solution. Another way to view this is the tendency of a metal to revert to its natural ore state (ferric hydroxide, for example) from a refined, or finished, metal state (steel, for example). The relative tendency of various metals to revert to an ore state can be shown by their positions on an **electromotive series** or **galvanic series** in which the most active metals are listed at the top as is shown in Table 7.1. An electromotive or galvanic series is a list of metals and alloys presented in the order of their tendency to corrode (or go into solution). This is a practical application of the theoretical **electrochemical series**, which is a list of metals with the standard electrode potentials given in volts. The size and sign of the electrode potential indicates how easily these **elements** will take on or give up electrons, or corrode. Hydrogen is conventionally assigned a value of zero.

The higher the level of activity, the greater the tendency for that metal to corrode. Also, the farther two metals are apart on the galvanic series,

molecule
The smallest division of a compound that still retains or exhibits all the properties of the substance.

electron
(1) A very small, negatively charged particle that is practically weightless. According to the electron theory, all electrical and electronic effects are caused either by the movement of electrons from place to place or because there is an excess or lack of electrons at a particular place. (2) The part of an atom that determines its chemical properties.

galvanic cell
An electrolytic cell capable of producing electric energy by electrochemical action. The decomposition of materials in the cell causes an electric (electron) current to flow from cathode to anode.

element
A substance that cannot be separated into its constituent parts and still retain its chemical identity. For example, sodium (Na) is an element.

the greater the galvanic corrosion potential. The more active metal of any two in the galvanic series will always become the anode. When iron and copper water pipes are joined, the iron will corrode if water contains dissolved oxygen (DO) and the copper will be protected. Because of their active positions on the galvanic series, zinc and magnesium make excellent anodes and are commonly used as **sacrificial anodes** in water tanks or for buried pipelines. These reactive metals, called **base metals**, will corrode preferentially to aluminum and iron, for example. A metal, such as iron, will react with dilute hydrochloric acid to form hydrogen.

On the cathodic side of the galvanic series, the least reactive metals are called **noble metals** (chemically inactive metals that do not corrode easily and are much scarcer and more valuable than the base metals). One of the most noble metals is gold, which has been known for millennia to be resistant to corrosion, even when worn as jewelry in continuous contact with oils and acids on the skin. Stainless steel is also cathodic to most other metals, which, although quite expensive, is why it is often used in critical chemical industry applications where corrosion potential is great. Stainless steel is commonly used in high-pressure reverse osmosis desalination systems due to its excellent corrosion resistance in the presence of highly conductive seawater.

Table 7.1   Galvanic series

		Galvanic Series
ANODE	(Most Active)	Magnesium
		Zinc
		Aluminum (2S)
		Cadmium
		Aluminum (175T)
		Steel or Iron
		Cast Iron
		Lead—Tin Solder
		Lead
		Nickel
		Brasses
		Copper
		Bronzes
		Stainless Steel (304)
		Monel Metal
		Stainless Steel (316)
		Silver
		Graphite
CATHODE	(Least Active)	Gold

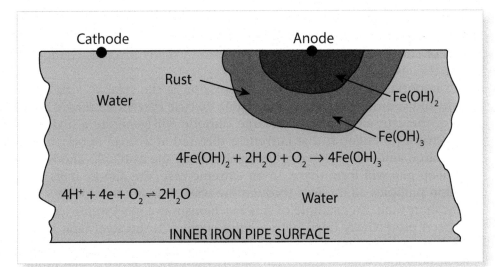

Figure 7.3 Role of oxygen in iron corrosion

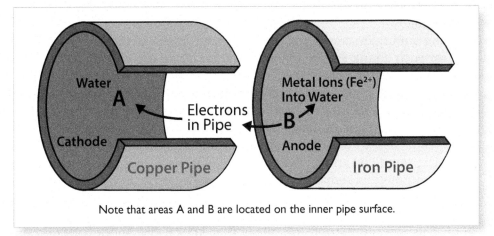

Note that areas A and B are located on the inner pipe surface.

Figure 7.4 Simplified galvanic cell

sacrificial anode
An easily corroded material deliberately installed in a pipe or tank. The intent of such an installation is to give up (sacrifice) this anode to corrosion while the water supply facilities remain relatively corrosion free.

Local galvanic cell

Corrosion at corporation stop hole caused by joining two dissimilar metals

Figure 7.5 Galvanic cells in water mains (*Permission of ARMCO*) (*Courtesy of AK Steel Corporation*)

If a steel nail is immersed in a solution of a copper salt, such as a solution of bluestone or copper sulfate ($CuSO_4 \bullet 5 H_2O$), metallic copper will be formed or plate out on the surface of the nail. In the same manner, if a reservoir has been treated with bluestone to control algal growths, there is a possibility that copper will plate out on steel pipes in the distribution system. The metallic copper might then act as a cathode in a galvanic cell and cause corrosion of the steel.

A similar phenomenon occurs in the so-called dezincification of brass. Brass is an alloy of zinc and copper. When brass corrodes, the result is a mass of spongy copper having nearly the same shape as the original brass. For a long time people believed that the zinc dissolved and left the copper. Today we know that the alloy itself dissolves and, since the copper is more noble (less reactive) than the brass, the copper plates out in more or less its original position.

As the galvanic series shows, when copper and lead solder are in contact, the lead becomes the anode and will corrode in preference to the copper. The relatively high toxicity of lead and its anodic tendencies are why lead has been banned from use in potable water distribution systems by the US Government under the 1986 Safe Drinking Water Act Amendments and the Lead and Copper Rule (LCR). Other factors, such as water chemistry, biological films, and physical characteristics (temperature and flow rate), all play a role in the severity of the corrosion reaction. These other factors can be used to help lessen the effects of galvanic and other forms of corrosion. Control of lead corrosion usually involves changes in water chemistry.

An important feature of galvanic corrosion is the relative size of the anode and cathode. The level of galvanic electric current increases as the area of the cathode increases. A large cathode will generate a high level of electrical potential. If that current is directed at a small anode, a relatively large amount of metal will dissolve from the available anode area and deep pits will form there. In an extreme case, the anodic metal may develop pinholes all the way through the wall of a tank or pipe in a relatively short time. An example of this mechanism is a steel water tank that has been protectively lined with a paint coating. If the steel tank is connected to a large copper distribution system line, for example, the corrosion potential will be high; the tank is the anode and the distribution system is the cathode. If the tank coating is imperfect, the entire energy potential of the cathode will be directed to the holiday, or imperfection, in the coating and severe pitting leading to penetration of the tank wall may result.

Using our understanding of galvanic electrochemistry, corrosion protection for such a tank can be gained by uncoupling the copper distribution system line from the tank. Specially designed **dielectric** couplings, which do not conduct an electric current, and that are partially constructed of ceramic, plastic, or other nonconductive materials, are used to separate the copper and steel. Additionally, sacrificial zinc or magnesium anodes can be attached to the tank wall and suspended in the water inside the tank. The sacrificial anodes will dissolve first, protecting the steel if some corrosion potential remains.

Corrosion inhibitors that slow the cathodic reaction also slow corrosion (in this example, the reaction on the copper electrode) and are known as safe inhibitors. Inhibitors that act on the anode reaction are called dangerous inhibitors because if there is not quite enough inhibitor added, severe pitting will result and corrosion will be worse than if none had been added.

In the ordinary large gate valve used in water distribution systems, brass or bronze parts often make contact with the valve's cast-iron body. The brass or bronze and cast iron create a galvanic cell, but the area of the brass or bronze is so small, compared to the area of the cast iron, that galvanic corrosion is minimal.

We encounter many similar situations in the waterworks field. For example, copper in a valve can be the cathode and the steel or cast iron in the valve the anode. As long as the cathode area is small relative to the anode, corrosion will not be a problem. However, if a copper service line is connected to a steel main, the cathode area is large relative to the anode and corrosion will occur at the anode. The opposite will occur with a brass valve in a steel water line; the cathode area is small and the anode area is large so there will be no problem.

Check Your Understanding

1. List the problems that can be created by corrosive waters.

2. What is an electrochemical reaction?

3. What happens to an iron water pipe when the pipe is connected to a copper pipe?

4. What is the dezincification of brass?

5. What happens when copper and lead solder are in contact?

6. What is a dangerous corrosion inhibitor?

7. What will happen if a copper service line is connected to a steel water main?

7.3   Factors Influencing Corrosion

As we discussed earlier, corrosion is a complex phenomenon with many possible variables. The essential elements of a corrosion cell, including the specific case of a galvanic cell, have been presented in the previous section. In this section, we will discuss how other factors influence the corrosion cell, either increasing or decreasing the rate and severity of the electrochemical reaction. The factors to be considered are physical, chemical, and biological.

7.3.1   Physical Factors

We will start by considering physical factors. Physical factors that influence corrosion include the type and arrangement of materials used in the system, system pressure, soil moisture, the presence of stray electric currents, temperature, and the water flow velocity.

7.3.1.1   System Construction

Some of the effects of the type of materials that make up the anode and cathode were discussed earlier. Also discussed was the influence of the relative size of the anode and cathode on each other, particularly with regard to the corrosion rate and development of pitting. The types of coatings or other protective measures used on both the inside and outside of pipes and tanks play a major role in corrosion activity.

7.3.1.2   System Pressure

If a system is under high pressure, corrosion can be affected greatly. An entire field of corrosion control is concerned with servicing boilers, steam lines, and similar industrial systems under high pressure. This special set of circumstances is not the topic of discussion of this manual, but some of the effects of high pressure on the rate of corrosion are applicable. Higher pressure, for example, increases the maximum concentration of **corrosive gases** that can exist in water, such as carbon dioxide and oxygen. When dissolved in water, these gases can oxidize construction materials (steel and concrete).

7.3.1.3   Soil Moisture

For buried distribution system lines, contact with moist soil can cause external pipe corrosion. The moisture functions as the electrolyte, the same as water inside the pipe does in the case of internal corrosion.

7.3.1.4   Stray Electric Current

Grounding of electric circuits to water pipes is a common practice, but one that is not recommended because it can lead to corrosion of the pipes. **Stray current corrosion** is much more pronounced from direct current (DC) grounding than from alternating current (AC) grounding. As a rule of thumb, AC current effects are only about 1 percent as great as DC current effects. Both types of stray current effects are most pronounced on the outside of pipes and fixtures because water inside pipes has a much greater resistance to the rate of current flow than does the metal.

Homeowners may notice the effect of stray current corrosion when pitting penetrates fixtures or pipes, resulting in leaks. Unfortunately, due to the nature of stray current corrosion, the anode can occur some distance from the point where the ground contacts the pipe. This may add to the difficulty of identifying this type of corrosion problem.

Electrolysis is the decomposition of a material by an outside electric current. When a DC current flows (plus to minus) from a metal into soil, the metal is corroded, except for very noble metals such as platinum. Before 1940, electrolysis from stray electrical **amperage** (the strength of an electric current measured in amperes) from streetcar power systems caused a great deal of damage to metallic water mains, but presently there are few such sources of direct current. Modern electric transit systems, however, may be a problem again in the future. Electric current flow is similar to the flow of water in gallons per minute. An **ampere** is the unit used to measure current strength; it is the current produced by an **electromotive force** (also called voltage) of 1 volt acting through a resistance of 1 **ohm** (the unit of electrical resistance).

Another form of electrolysis of water mains that can result from stray currents may be generated by **cathodic protection** systems installed by utilities other than the water utility. Cathodic protection systems are discussed in Section 7.5, "Methods of Controlling Corrosion."

stray current corrosion
A corrosion activity resulting from stray electric current originating from some source outside the plumbing system such as DC grounding on phone systems.

electromotive force (EMF)
The electrical pressure available to cause a flow of current (amperage) when an electric circuit is closed.

cathodic (kath-ODD-ick) protection
An electrical system for prevention of rust, corrosion, and pitting of metal surfaces that are in contact with water, wastewater, or soil. A low-voltage current is made to flow through a liquid (water) or a soil in contact with the metal in such a manner that the external electromotive force renders the metal structure cathodic. This concentrates corrosion on auxiliary anodic parts, which are deliberately allowed to corrode instead of letting the structure corrode.

7.3.1.5 Temperature

The rate of chemical reactions usually increases as temperatures rise. Because chemical reactions are involved in corrosion, temperature generally has the effect of increasing the corrosion rate. However, there are several exceptions to this general rule.

As the temperature of water increases, the amount of calcium carbonate that can remain dissolved in water is reduced. This means that protective calcium carbonate film formation may be improved under this circumstance. The deposition may be so enhanced, however, that severe scaling may result, clogging hot water lines and fittings, and coating hot water heating elements. Heavy scaling of heating elements will increase the amount of energy used, and may result in their failure due to overheating.

Higher temperatures can also alter the form of corrosion. In cold water, pitting may be the dominant form of attack on anodic metals, resulting in short service life. Although the rate of corrosion may increase as the temperature rises, the form of corrosion in the same water may change to a uniform or more generalized metal loss, resulting in a longer service life for the pipe.

Another special case of temperature effects is the influence on zinc/iron galvanic corrosion. In hot water heaters with temperatures exceeding 140°F (60°C), the normal condition of zinc being anodic to iron may reverse, resulting in the zinc becoming cathodic and the iron corroding. Penetration of the steel tank wall can result.

7.3.1.6 Flow Velocity

Flow velocity has several significant influences on corrosion. Moderate flow rates are often beneficial while very high or low flow rates usually increase the rate of corrosion.

- Negative effects: Under stagnant water flow conditions, corrosion is usually more severe, particularly in the form of pitting and tuberculation in iron pipes. However, highly oxygenated water can become even more corrosive under higher flow conditions as the movement of water increases the contact of oxygen with the pipe surface.

  In the extreme case of very high velocities, erosion corrosion can occur, particularly in copper pipes. At rates exceeding 5 ft/sec (1.5 m/sec), copper tubing will erode rapidly. This is usually noticeable at joints or elbows, and results in structural damage to the pipe. This problem often becomes evident by the occurrence of persistent leaks in even relatively new copper tubing. Circulating hot water systems in large buildings are particularly susceptible to erosion corrosion due to both flow velocities and high temperature effects.

- Beneficial effects: Water that has protective properties, such as the tendency to deposit calcium carbonate films, or to which a corrosion inhibitor has been added, will be less corrosive under moderate flow conditions. Film formation requires deposition of calcium carbonate or the inhibitor (usually a phosphate or silicate compound) on the surface of the metal. In stagnant water, deposition is limited. Under very high flows, it may be scoured off the pipe as it forms, or erosion corrosion may occur faster than the film deposits. If the flow rate is moderate, less than 5 ft/sec (1.5 m/sec),

compound
A pure substance composed of two or more elements whose composition is constant. For example, table salt (sodium chloride, NaCl) is a compound.

deposition of protective films is enhanced. We will discuss the effect of film formation as a corrosion-control measure in later sections of this chapter.

7.3.2 Chemical Factors

A very important part of understanding the theory of corrosion is an understanding of the chemistry involved. Various chemical factors influence corrosion, such as **pH**; alkalinity; chlorine residual; levels of dissolved solids; dissolved gases, such as oxygen and carbon dioxide; and the types and concentrations of various minerals present in the water. The major chemical factors are discussed in this section.

7.3.2.1 Alkalinity

Alkalinity is the capacity of water or wastewater to neutralize acids. This capacity is caused by the water's content of carbonate, bicarbonate, hydroxide, and occasionally borate, silicate, and phosphate. Alkalinity is expressed in milligrams per liter of equivalent calcium carbonate. Alkalinity is not the same as pH because water does not have to be strongly basic (high pH) to have a high alkalinity. Alkalinity is also a measure of how much acid must be added to a liquid to lower the pH to 4.5.

Alkalinity is a measure of the buffering capacity, or the ability of a particular quality of water to resist a change in pH. Alkalinity is primarily composed of carbonate (CO_3^{2-}) and bicarbonate (HCO_3^-) ions. Acids, compounds that contain free H^+ ions, react with carbonate and bicarbonate:

Carbonate Ion + Hydrogen Ion (Acid) $\rightarrow$ Bicarbonate Ion

$$CO_3^{2-} \quad + \quad H^+ \quad \rightarrow \quad HCO_3^-$$

Bicarbonate Ion + Hydrogen Ion (Acid) $\rightarrow$ Carbonic Acid

$$HCO_3^- \quad + \quad H^+ \quad \rightarrow \quad H_2CO_3$$

Conversely, bases (OH^- ions) can react with bicarbonate to form carbonate:

Bicarbonate Ion + Hydroxide Ion (Base) $\rightarrow$ Carbonate Ion + Water

$$HCO_3^- \quad + \quad OH^- \quad \rightarrow \quad CO_3^{2-} \quad + \quad H_2O$$

In either case, the acid or the base is neutralized by the carbonate or bicarbonate. The effect of a high concentration of carbonate and bicarbonate in water (high alkalinity) is that the water has a strong tendency to resist a change in pH. Water with low alkalinity will have little **buffer capacity** and may become acidic very easily. The buffer capacity is a measure of the capacity of a solution or liquid to neutralize acids or bases. This measures the capacity of water or wastewater for offering a resistance to changes in pH. Carbonate is also necessary to react with calcium in the water to form protective calcium carbonate ($CaCO_3$) films on the inside of pipes:

Carbonate Ion + Calcium Ion $\rightarrow$ Calcium Carbonate

$$CO_3^{2-} \quad + \quad Ca^{2+} \quad \rightarrow \quad CaCO_3$$

The simplest form of corrosion control in many water systems is to simply add more alkalinity in the form of lime, soda ash, or caustic soda, or

pH (pronounce as separate letters)
pH is an expression of the intensity of the basic or acidic condition of a liquid. Mathematically, this is equivalent to the negative of the base 10 logarithm. If $\{H^+\} = 10^{-6.5}$, then pH = 6.5. The pH may range from 0 to 14, where 0 is most acidic, 14 most basic, and 7 neutral.

$$pH = Log\frac{1}{\{H^+\}} \text{ or } -Log(\{H^+\})$$

directly as calcium carbonate in the form of crushed limestone. This will be discussed further in Section 7.4.3.3, "Calcium Carbonate Saturation."

7.3.2.2 pH

The hydrogen ion is extremely active (corrosive) at pH values below 4. Neither chlorine nor hydrogen ions are usually present in sufficient concentrations in potable water to have a significant effect on corrosion. Low pH (<7.0) water tends to be corrosive and high pH (>7.5) water is protective of pipe materials. Very high pH water may have a tendency to deposit excessive amounts of scale. The pH of the water is influenced by the level of alkalinity present. The ability of an acid or a base to change the pH when added to a given amount of water is a direct function of the alkalinity.

Certain pH ranges offer less protection than would otherwise be assumed. For example, at pH values very near 8.3, the transition point between carbonate and bicarbonate, the buffering system is weak, and slightly lower pH values (such as 7.8 to 8.0) may be more protective.

7.3.2.3 Dissolved Oxygen

Oxygen is often considered the most corrosive component of water chemistry. Oxygen plays an important role in corrosion both at the anode and at the cathode. The hydrogen gas released at the cathode coats the cathode area, slowing the rate of corrosion (Figure 7.2). This phenomenon is called **polarization**. However, when DO reacts with the hydrogen gas (H_2) to form water (H_2O), the gas is removed from circulation and the rate of corrosion at the cathode accelerates. The removal of hydrogen from the cathode is called **depolarization**, which is the removal or depletion of ions in the thin boundary layer adjacent to a membrane or pipe wall.

At the anode, dissolved oxygen reacts with iron as it dissolves into the water to form ferric hydroxide (rust). This process is known as **oxidation**. It produces red water and, when rust particles attach to the anode area, tubercles are formed. The formation of a rust coating can actually slow the rate of corrosion, until rapid flushing of the water lines clears the coating and again exposes the anode to corrosive oxygen.

7.3.2.4 Dissolved Solids

Solids dissolved in water are present as ions that increase the electrical conductivity of the water. Generally, the higher the dissolved solids or salt content of the water, the greater the potential for corrosion to occur due to the increased conductivity. Some dissolved solids are involved in scale formation, possibly slowing the rate of corrosion if a protective film is formed. All scale-forming components, such as iron oxide (rust) and calcium carbonate (limestone), are present first as dissolved solids in the water before they deposit on the surface of pipes and fixtures.

7.3.2.5 Hardness

The dissolved form of some of the principal scale-forming components in water is referred to as hardness. **Hardness** is composed primarily of calcium and magnesium ions, but may also include such ions as iron, manganese, and strontium. All hardness ions have the common property of forming a scale on the inside of pipes or fixtures under conditions of high enough concentration, and at elevated pH and temperature levels.

polarization
The concentration of ions in the thin, boundary layer adjacent to a membrane or pipe wall.

oxidation
Oxidation is a chemical reaction that results in the addition of oxygen, removal of hydrogen, or the removal of electrons from an element or compound; in the environment and in wastewater treatment processes, organic matter is oxidized to more stable substances. The opposite of reduction.

hardness, water
A characteristic of water caused mainly by the salts of calcium and magnesium, such as bicarbonate, carbonate, sulfate, chloride, and nitrate. Excessive hardness in water is undesirable because it causes the formation of soap curds, increased use of soap, deposition of scale in boilers, damage in some industrial processes, and sometimes causes objectionable tastes in drinking water.

Planned deposition of calcium carbonate film is one of the most common corrosion-control measures used in water systems. There are several methods of measuring the relative level of calcium carbonate saturation in water. One of the simpler methods is called the Marble Test, which directly measures whether a water sample will increase in hardness and pH when dosed with an excess of calcium carbonate. The Marble Test and a more extensive determination of the saturation level of calcium carbonate called the **Langelier Index** are discussed in Section 7.4.3.3, "Calcium Carbonate Saturation."

If hardness levels are too high (a condition referred to as oversaturation), calcium carbonate deposits can clog pipes and reduce flows. They can also coat hot water heating elements, increasing power consumption. Conversely, water containing very little hardness tends to be more aggressive because it does not deposit a protective calcium carbonate film.

7.3.2.6   Chloride and Sulfate

Chloride (Cl^-) and sulfate (SO_4^{2-}) ions in water may inhibit the formation of protective scales by keeping hardness ions in solution. The relative amount of alkalinity compared to chloride and sulfate greatly affects this tendency. It is recommended that the alkalinity, expressed as **calcium carbonate ($CaCO_3$) equivalent**, be five times higher than the sum of chloride and sulfate ions, also expressed as $CaCO_3$ equivalents.

7.3.2.7   Phosphate and Silicate

These compounds have a tendency to form protective films in water systems when present in high enough concentrations and when in the correct chemical form for the particular conditions of the water. Phosphate and silicate compounds are frequently added at the water treatment works as a corrosion-control method. The selection and use of phosphate and silicate inhibitors are discussed in Section 7.5.4, "Zinc, Silica, and Polyphosphate Compounds."

7.3.2.8   Trace Metals

Trace metals of significance in corrosion control include copper, iron, lead, and zinc. These metals are referred to as trace because they are normally present in relatively low concentrations. When present at high levels in distribution system water, they are usually indicators of corrosion of the pipes and fittings. Copper and lead usually indicate corrosion of copper pipe and lead solder or service lines. Iron usually results from the corrosion of iron or steel pipe and fittings, and zinc may result from corrosion of galvanized pipe.

Due to their relatively high toxicity to the consumer, copper and lead have been specifically identified by the US EPA in a set of regulations known as the Lead and Copper Rule (LCR), finalized in 1991. These regulations have a significant impact on all water utilities in the United States. Control of lead and copper in water systems involves corrosion monitoring and, if necessary, corrosion control. The Lead and Copper Rule is discussed in Section 7.7 of this chapter.

Iron and zinc may be involved extensively in the formation of protective films, limiting the rate of corrosion. Zinc is also a common corrosion-control additive, usually in a compound formed of zinc and phosphate.

Langelier Index (LI)
An index reflecting the equilibrium pH of a water with respect to calcium and alkalinity. This index is used in stabilizing water to control both corrosion and the deposition of scale.
Langelier Index = pH − pH$_s$
    where pH = actual pH of the water
        pH$_s$ = pH at which water having the same alkalinity and calcium content is just saturated with calcium carbonate

calcium carbonate ($CaCO_3$) equivalent
An expression of the concentration of specified water constituents in terms of their equivalent value to calcium carbonate. For example, the hardness in water that is caused by calcium, magnesium, and other ions is usually described as calcium carbonate equivalent. Alkalinity test results are usually reported as mg/L $CaCO_3$ equivalents.

When used as an additive for corrosion control, zinc levels may rise to several mg/L in the water. The use of zinc compounds in corrosion control is discussed later in this chapter.

7.3.3   Biological Factors

Two types of microorganisms that can play an important role in corrosion of water distribution systems are iron bacteria and sulfate-reducing bacteria. Both can increase the rate of corrosion and the formation of undesirable corrosion byproducts. Iron bacteria use dissolved iron as an energy source, and sulfate-reducing bacteria use sulfate for their energy. Because both types of bacteria can grow in dense masses, they may be relatively tolerant to disinfection by chlorine. Both types of bacteria can be particularly troublesome in low-flow areas of distribution systems.

7.3.3.1   Iron Bacteria

Iron bacteria are often present in a filamentous (string-like) form, resulting in slimy, reddish or brown-colored masses. These may be present on well screens, inside pipes or water storage tanks, or on the inside of fixtures. Some of the types of iron bacteria common to water systems include *Crenothrix, Sphaerotilus,* and *Gallionella,* all of which are filamentous forms.

Iron bacteria convert ferrous (dissolved) iron into ferric hydroxide precipitate, and deposit the rust particles on or in the slime sheaths surrounding the cells. The deposited iron can be released during periods of high water flow velocity, contributing to red water conditions. An unpleasant musty odor is often associated with the presence of iron bacteria. Corrosion can be accelerated underneath deposits of iron bacteria on pipes and tanks, resulting in pitting and tuberculation.

7.3.3.2   Sulfate-Reducing Bacteria

The presence of sulfate-reducing bacteria can often be readily distinguished by the characteristic rotten egg odor of hydrogen sulfide. Using sulfate as an energy source, sulfate-reducing bacteria produce sulfide as a byproduct. Some of the byproduct sulfide may be released as gaseous hydrogen sulfide, generating the rotten egg odor, or it may react with metals producing black metal sulfide deposits on the inside of pipes and fixtures. Free hydrogen sulfide (H_2S) can react with water forming sulfuric acid (H_2SO_4), which is extremely corrosive to metals.

Check Your Understanding

1.   What effect of stray current corrosion may be evident to homeowners?

2.   What causes erosion corrosion in copper tubing?

3.   List the chemical factors that influence corrosion.

4.   Why does water with a higher dissolved solids content have a greater potential for corrosion?

5.   What is the impact of calcium carbonate on corrosion?

7.3.4   Oxygen Concentration Cell

Although galvanic cells are responsible for some corrosion problems, by far the most common corrosion cell is the oxygen concentration cell. To understand an oxygen concentration cell, think in terms of the dead end created by a 6-inch (150 mm) dry barrel fire hydrant installed on an 8-inch (200 mm) water main (Figure 7.6). A **dead end** is the end of a water main that is not connected to other parts of the distribution system by means of a connecting loop of pipe. An oxygen concentration cell can be started in the dead end of a water main. The fire hydrant assembly will normally consist of an 8″ × 8″ × 6″ (200 mm × 200 mm × 150 mm) tee, 6-inch (150 mm) nipple, a 6-inch (150 mm) gate valve, 10 to 20 feet (3 to 6 m) of 6-inch (150 mm) pipe, and the fire hydrant. When corrosion starts in the 6-inch (150 mm) pipe, the following chemical reaction occurs:

$$\text{Iron} \rightarrow \text{Ferrous Ion} + \text{Electrons}$$

$$\text{Fe} \rightarrow \text{Fe}^{2+} + 2\,\text{e}^-$$

This is the anode reaction of the galvanic cell. The ferrous ions (Fe^{2+}) formed by this reaction will in turn react with dissolved oxygen and water as follows:

$$\underset{\text{Ions}}{\text{Ferrous}} + \underset{\text{Oxygen}}{\text{Dissolved}} + \text{Water} \rightarrow \underset{\text{(Solid)}}{\text{Ferric Hydroxide}} + \underset{\text{Ions}}{\text{Hydrogen}}$$

$$4\,\text{Fe}^{2+} + \text{O}_2 + 10\,\text{H}_2\text{O} \rightarrow 4\,\text{Fe(OH)}_3 + 8\,\text{H}^+$$

This is chemical shorthand for saying that four ferrous ions and one molecule of oxygen react with 10 water molecules to form four molecules of solid (precipitate) ferric hydroxide and eight hydrogen ions.

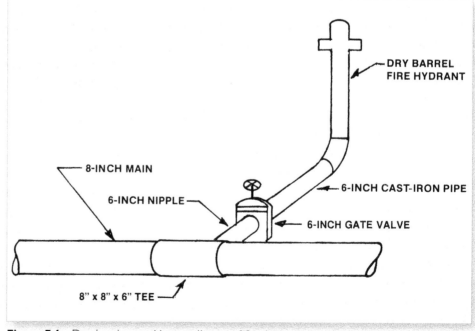

Figure 7.6   Dead end caused by installation of fire hydrant

Just as they did in the galvanic cell, the electrons from the anode reaction flow through the metallic path (pipe) to the 8-inch (200 mm) main where they react with dissolved oxygen, which is continually being replenished by the flowing water.

Oxygen + Electrons + Water → Hydroxide Ions

$$O_2 \ + \ 4\,e^- \ + 2\,H_2O \rightarrow \ 4\,OH$$

This is the cathode reaction.

The continued production of ferrous ions within the 6-inch (150 mm) pipe completely removes dissolved oxygen from the water in the dead end and scavenges any DO that may diffuse in from the flowing water in the 8-inch (200 mm) pipe. The hydrogen ions produced by this reaction can lower the pH in the water to around 5.2 to 5.8.

Also, the production of positively charged ferrous ions in the 6-inch (150 mm) pipe requires that there be an inflow of negatively charged ions to maintain electrical neutrality in the water. The common negative ions present in water are bicarbonate (HCO_3^-), chloride (Cl^-), and sulfate (SO_4^{2-}).

The absence of oxygen and the low pH value in the 6-inch (150 mm) pipe make conditions ideal for the growth of **anaerobic** bacteria. An anaerobic condition is one in which atmospheric or dissolved oxygen is not present in water. The action of these bacteria on traces of organic matter and on reducing the sulfate ions to sulfide are responsible for the foul odors usually found in the dead ends of water mains.

The typical distribution system contains many dead ends. In addition to those at the end of the system, other dead-end conditions include the **annular spaces** (circular cavities) created by mechanical couplings, tapping sleeves, and even the bonnets (covers) of gate valves, but the number of dead ends from pits (holes) in the pipe surfaces far exceeds the number of dead ends from all other causes combined (Figure 7.7).

Pits may be started by anything that will shield the metal surface from DO in the water, such as bits of clay, dirt, sand, or a colony of bacteria. Also, impurities in the metal may cause a local anode to form.

As the iron ions from the anode react with the dissolved oxygen in the water, the resulting ferric hydroxide forms a membrane over the anode. As the reactions continue and the membrane ages, it turns into a crust. The membrane, and later the crust, protect the anode area from dissolved oxygen in the water, thus causing the oxygen concentration cell to intensify itself. As the iron ions leave the metal, a pit is formed that grows deeper and deeper. As this occurs, the crust becomes thicker and thicker. In this way, a tubercle, a mound of iron rust, is built up over the pit.

As the ferric hydroxide ($Fe(OH)_3$) ages, it forms other minerals such as ferric oxide (Fe_2O_3) or iron rust. Eventually, the crust becomes so thick that negative ions cannot enter the pit, nor can iron ions escape, and the corrosion stops. At this point, the pit is said to be inactive.

The reaction of dissolved oxygen with ferrous ions is very slow at low pH values. When the pH is less than seven, the reaction is so slow that tubercles do not form and the pits are not self-perpetuating. New pits keep starting in different places, so that corrosion appears to be uniform over the surface of the pipe.

annular (AN-yoo-ler) space
A ring-shaped space located between two circular objects. For example, the space between the outside of a pipe liner and the inside of a pipe.

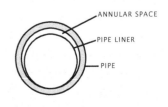

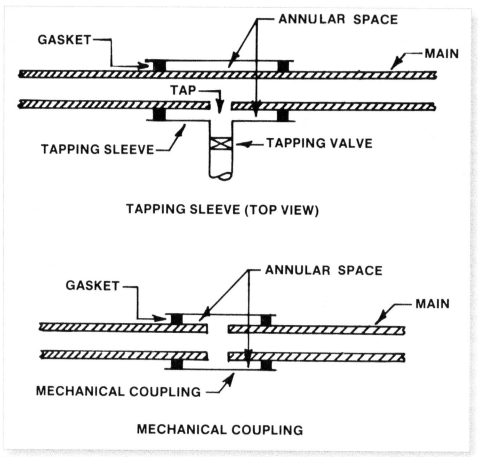

Figure 7.7   Creation of annular spaces in distribution systems

Check Your Understanding

1. What is the most common type of corrosion cell?

2. Where can an oxygen concentration cell be started?

3. How can pits be started on a metallic surface under water?

4. What is a tubercle?

7.4   How to Determine if Corrosion Problems Exist

The LCR requires water systems to control corrosion to protect the public from the harmful effects of lead, copper, or other toxic metals in drinking water. To determine if the water you are treating is causing corrosion problems, you can examine materials removed from your distribution system for signs of corrosion damage. Chemical tests on the water can be used to indicate the corrosiveness of the water. The most common indicators of corrosion problems are an increasing number of leaks in the distribution system and consumers are complaining about dirty or red water.

7.4.1 Examine Materials Removed from Distribution System

Corrosion rates may be measured by inserting special steel specimens called **coupons** in the water mains (Figure 7.8). After a period of time, usually a month or two, the coupons are removed and the loss of weight and nature of corrosion damage are measured. The rate of corrosion is measured by the loss of weight of the coupon between weighing time intervals. Be sure to scrape off all encrustations before weighing the coupons. These tests should be made under the supervision of an experienced corrosion engineer because standard procedures must be used to obtain results that can be compared with other water supply systems.

Much can be learned about the corrosiveness of a water by prowling through the scrap heap and examining sections of water mains that have been taken out of service for various reasons. When examining internal corrosion damage on a pipe, pay particular attention to the maximum depth of the pits. When pit depth equals the wall thickness of a pipe, a leak develops. As a rule of thumb, you can assume (for internal corrosion only) that pit depth increases with the cube root of time $\left(\sqrt[3]{\text{Time}}\right)$. Thus, if a pit depth reaches a certain value in 1 year, it will about double this depth in 8 years ($2 \times 2 \times 2 = 8$).

Leaks are often detected by the observation of wet spots above a pipeline. All reports of leaks should be recorded. If the number of leaks is large, plot the location of the leaks on a map to identify trouble spots.

If effervescence (bubbles) occurs when a drop of dilute hydrochloric (HCl) acid is placed on an obvious cathodic area (such as a brass ring on a gate valve), this indicates the presence of a calcium carbonate ($CaCO_3$) film that may be too thin to see. This indicates that the water is, at worst, only moderately corrosive.

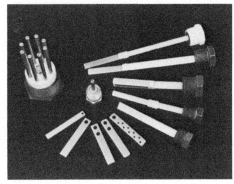

Pipe plug assembly coupons (*Courtesy of Metal Sales Co, Inc.*)

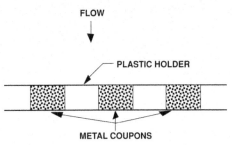

Figure 7.8   Coupons

Check Your Understanding

1. Why is corrosion of water system facilities a public health concern?
2. How can corrosion rates be measured?
3. How can leaks in pipes be detected?
4. How can you detect a film of calcium carbonate that is too thin to see?

7.4.2 Flow Tests

Corrosivity is an indication of the corrosiveness of a water. The corrosiveness of a water is described by the water's pH, alkalinity, hardness, temperature, total dissolved solids (TDSs), DO concentration, and the Langelier Index. A very simple and useful test is to measure the change in water flow through a 20-foot (6-meter) length of half-inch (12.5 mm) standard black iron pipe under a constant head of 1 foot (0.3 m). The initial flow rate will be small enough that the flow rate can be determined by measuring the time required to fill a 1-quart (or 1-liter) container. If the water is highly corrosive, the flow rate will be reduced by as much as 50 percent in 2 weeks. In other words, the fill time will be twice as long.

coupon
A steel specimen inserted into water or wastewater to measure corrosiveness. The rate of corrosion is measured as the loss of weight of the coupon or change in its physical characteristics. Measure the weight loss (in milligrams) per surface area (in square decimeters) exposed to the water or wastewater per day. 1 meter = 10 decimeters = 100 centimeters.

If the water is relatively noncorrosive, the flow reduction will be only 10 percent or less after 2 weeks. This test may be run for longer time periods, but the changes in flow caused by tuberculation are greatest during the first few weeks. Usually, there is little change after 26 weeks.

7.4.3   Chemical Tests on Water

Chemical tests on water include tests for dissolved oxygen, toxic heavy metals, and calcium carbonate saturation. Tests for DO show the operator if the water contains oxidizable organic matter or if gross corrosion is occurring. Testing a customer's plumbing for toxic heavy metals, such as lead, copper, cadmium, and chromium, can help determine whether the water is corrosive. Tests for calcium carbonate saturation include the calcium carbonate precipitation potential (CCPP) and the Langelier Index.

7.4.3.1   Dissolved Oxygen

Certain chemical tests on water may be helpful to indicate its corrosiveness. As you have learned, corrosion occurs most readily in the presence of oxygen. Other oxidants (electron acceptors), such as chlorine, nitrate, and nitrite, can contribute to corrosion. If no oxygen is present, the water can still be corrosive even though rust will not form.

Water supplies taken from lakes or streams will contain dissolved oxygen. If your source water is drawn from wells, measure the DO* concentration in each well. By measuring the dissolved oxygen at various points in the distribution system, you can calculate how much oxygen is used up as water passes through the system. Loss of dissolved oxygen indicates either that the water contains oxidizable organic matter or that gross corrosion is occurring.

When taking DO samples, you must avoid exposing the sample to air because air contains oxygen. You will need the following equipment and reagents:

1. A 4-foot (1.2 m) length of polyethylene tubing with fittings to connect to a hose bib (faucet).
2. BOD (biochemical oxygen demand) bottle. A glass bottle with 300 mL capacity with a tapered, pointed, ground-glass stopper and a flared mouth.
3. Three small (100 mL) plastic bottles with screw-on rubber bulb dispensing pipets. These pipets should deliver approximately 0.5 mL when the bulb is squeezed (an ordinary eye dropper is satisfactory). The bottles are filled with standard reagents available from chemical supply houses.

    a. Manganous sulfate.
    b. Alkaline potassium iodide.
    c. Concentrated sulfuric acid. This is a very dangerous chemical and must be handled with great care.

To obtain a **representative sample**, select a hose bib as close as possible to the meter of a customer on a short water service. Flush the service line at a rate of approximately half a gallon per minute (2 L/min) for 5 minutes for each 100 feet (30 m) of pipe between the hose bib and the main so that you can be sure you are getting your sample from the main. Do not flush the service line at a high water velocity because this

*See Chapter 9, "Laboratory Procedures," and the advanced laboratory procedures chapter in Volume II.

representative sample
A sample portion of material, water, or wastestream that is as nearly identical in content and consistency as possible to that in the larger body being sampled.

may disturb sediment in the pipe. Therefore, take a long time to flush the line. After the line has been flushed, connect the plastic tube to the hose bib, place the end of the plastic tube at the bottom of the BOD bottle, and turn the hose bib on again at a rate of flow no higher than was used for flushing the service line. Allow the flow to continue until the BOD bottle has overflowed at least three times its volume. Withdraw the plastic tube taking care not to introduce any air bubbles while removing the tube.

Next, fix (stop any chemical reactions involving dissolved oxygen) the sample so it can be transported to the laboratory where the DO can be measured.

1. Gently, but rapidly, add 1 mL of manganous sulfate reagent (two squirts with dropper) followed by 1 mL of alkaline potassium iodide. Each time, hold the tip of the pipet below the surface of the water so you will not add any dissolved oxygen.

    A heavy brown floc of manganese hydroxide will form at this point if DO is present. A white floc will indicate that there is no dissolved oxygen in the water.
2. Insert the glass stopper without trapping any air bubbles. Mix sample and reagents by rapidly inverting the bottle back and forth.
3. Allow the floc to settle halfway down in the bottle. Invert the bottle and allow the floc to settle halfway again.
4. Carefully remove the stopper and add 1 mL of concentrated sulfuric acid. Allow the acid to run down the neck of the bottle and into the sample.
5. Mix the sample by inverting again.

The sample now contains an iodine solution that is chemically equivalent to the initial DO concentration in the sample. The final solution is stable and can be transported to the laboratory where the iodine solution can be measured by **titrating** with a standard solution of sodium thiosulfate. See the laboratory chapter in Volume II for details.

7.4.3.2 Toxic Heavy Metals

Testing for toxic heavy metals in water samples from customers' plumbing has been recommended as a way of determining whether the delivered water is corrosive. However, the results of such samples can be so variable that a great many samples must be taken to obtain a meaningful average.

A common source of lead from plumbing is copper tubing with soldered fittings. A water sample taken shortly after a period of heavy use will contain much less lead than a sample taken after standing overnight. Experiments have shown that on new copper tube systems, significant amounts of lead are found in water samples regardless of the water quality. With noncorrosive water, no lead can be detected from copper plumbing 1 month old, and even with moderately corrosive water, lead is low after approximately 2 years. With highly corrosive water, excessive lead concentrations may be found after more than 10 years.

Copper is a toxic heavy metal that can leach into drinking water from copper and brass materials such as piping, tubing, and fittings. High levels of copper in corrosive waters can produce a bitter or metallic taste in water and cause green stains on plumbing fixtures.

titrate (TIE-trate)
To titrate a sample, a chemical solution of known strength is added drop by drop until a certain color change, precipitate, or pH change in the sample is observed (end point). Titration is the process of adding the chemical reagent in small increments (0.1–1.0 milliliter) until completion of the reaction, as signaled by the end point.

Cadmium is another toxic metal found in samples from plumbing systems; it is found only in very small amounts. Cadmium is a contaminant found in zinc used for galvanizing steel pipes. Cadmium-plated water-works fittings are not used in the United States. However, they have been used in Europe and have been responsible for serious cases of cadmium poisoning.

Chromium is used for external decorative plating and has virtually no chance to get into the water. Arsenic, antimony, and silver are found in copper and zinc ores, but the quantities found in the refined metals used for plumbing are too small to be significant sources of contamination of drinking water.

Check Your Understanding

1. What does a loss of dissolved oxygen in the water flowing in a distribution system indicate?

2. What toxic metals may enter drinking waters from the customer's plumbing due to corrosive water?

Standard Methods, Part 2330 C, "Indices Predicting the Quantity of $CaCO_3$ That Can Be Precipitated or Dissolved." www.awwa.org. ISBN: 978-0-87553-013-0. Order No. 10085.

7.4.3.3   Calcium Carbonate Saturation

A water is considered stable when it is just **saturated** with calcium carbonate. In this condition, the water will neither dissolve nor deposit calcium carbonate. Water treatment plant operators commonly use two approaches to determine the calcium carbonate saturation level of their water: the CCPP and the Langelier Index.

The CCPP indicates the tendency to either precipitate or dissolve $CaCO_3$. It can be determined by several experimental methods, two of which are discussed below.

MARBLE TEST AND ENSLOW COLUMN

To conduct a Marble Test for calcium carbonate saturation, first measure the pH, alkalinity, and hardness of the water sample. Add a pinch of powdered calcium carbonate and then stir the water for at least five minutes. The water should be stirred in a nearly-full stoppered flask to avoid the introduction of carbon dioxide from the air. Also, the water being stirred should be at the same temperature as the water in the distribution system. (See the laboratory chapter in Volume II for detailed procedures on how to perform the Marble Test.) If the pH, alkalinity, or calcium increase during the test, the water was undersaturated with respect to calcium carbonate; if they decrease, the water was **supersaturated**. See Table 7.2 for guidance on interpreting experimental results. The CCPP equals the quantitative change in alkalinity (or calcium) due to the water being exposed to the powdered calcium carbonate.

If you need to perform the Marble Test frequently because of changing water conditions, it is convenient to use an Enslow column. This is a column packed with calcium carbonate granules. The pH, alkalinity, and calcium are measured on a sample stream of water before and after passing through the column. The results are interpreted in the same way as in the Marble Test (Table 7.2). See Section 7.5.2 for a more complete description of the Marble Test procedures, and Section 7.5.3 for a detailed description of an Enslow column.

---

saturation
The condition of a liquid (water) when it has taken into solution the maximum possible quantity of a given substance at a given temperature and pressure.

supersaturated
An unstable condition of a solution (water) in which the solution contains a substance at a concentration greater than the saturation concentration for the substance.

Table 7.2   Interpreting marble test or Enslow column results

Change in pH, Alkalinity, or Hardness[a]	What's Happening in the Test	Saturation State of the Water Sample
Increase	Water sample is dissolving the $CaCO_3$ (marble)	Undersaturated
No change		Saturated
Decrease	Scale is being formed on the suspended $CaCO_3$ or in the column bed	Supersaturated

[a]Change = (Concentration or pH values after the test) − (initial values).

LANGELIER INDEX (LI)

As previously mentioned, a water is considered stable when it is just saturated with calcium carbonate; in this condition, it will neither dissolve nor deposit calcium carbonate. Thus, in a stable water, the calcium carbonate is in equilibrium with the hydrogen ion concentration. If the pH is higher than the equilibrium point (LI positive), the water is scale-forming and will deposit calcium carbonate. If the pH is lower than the equilibrium point (LI negative), the water is considered corrosive.

The Langelier Index (LI) (also known as the Saturation Index or SI) is the most common index used to indicate how close a water is to the equilibrium point, or the corrosiveness of the water. This index is based on the equilibrium pH of a water with respect to calcium and alkalinity. The Langelier Index can be determined by using the equation:

$$\text{Langelier Index} = pH - pH_S$$

> where pH = actual pH of the water
>
> pH_S = pH at which water having the same alkalinity and calcium content is just saturated with calcium carbonate

In this equation, pH_S is defined as the pH value at which water of a given calcium content and alkalinity is just saturated with calcium carbonate (i.e., at the equilibrium point). For some waters of low calcium content and alkalinity, there is no pH value that satisfies this definition; however, for most waters, there will be two values for pH_S. These difficulties can be avoided by defining pH_S as that pH where a water of given calcium and bicarbonate concentrations is just saturated with calcium carbonate.

T. E. Larson's method* for calculating pH_S is a satisfactory approximation when the value of pH_S is calculated using the equation:

$$pH_S = A + B - \log(Ca^{2+}) - \log(\text{Alkalinity})$$

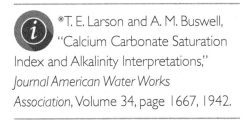

*T. E. Larson and A. M. Buswell, "Calcium Carbonate Saturation Index and Alkalinity Interpretations," *Journal American Water Works Association*, Volume 34, page 1667, 1942.

The values for A and B are found in Tables 7.3 and 7.4; (Ca^{2+}) is the calcium hardness in mg/L as $CaCO_3$; and (Alkalinity) is the alkalinity in mg/L as $CaCO_3$.

This calculation is accurate enough for practical purposes up to a pH_S value of 9.3. Above this value, errors are large.

Example 1

Find the pH_S and Langelier Index of a water at 15°C having a TDS of 200 mg/L, alkalinity of 100 mg/L, and a calcium hardness of 50 mg/L. The pH is 8.6.

Table 7.3 Values of "A" for Larson's method

Temperature, °C	A
0	2.34
5	2.27
10	2.20
15	2.12
20	2.05
25	1.98
30	1.91
40	1.76
50	1.62
60	1.47
80	1.18
100	0.88

Table 7.4 Values of "B" for Larson's method

TDS, mg/L	B
0	9.63
50	9.72
100	9.75
200	9.80
400	9.86
800	9.94
1,600	10.04

Langelier Index = pH − pHS

where $pH_S = A + B - \log(Ca^{2+}) - \log(Alky)$

Known		**Unknown**
T = Water Temp = 15°C		pH_S
TDS = 200 mg/L		Langelier Index
Alkalinity = 100 mg/L as $CaCO_3$		
Ca Hardness = 50 mg/L as $CaCO_3$		
pH = 8.6		

1. Find the formula values from the tables.
   From Table 7.3 for a water temperature of 15°C,

   A = 2.12

   From Table 7.4 for a TDS of 200 mg/L,

   B = 9.80

   $\log(Ca^{2+})$ = $\log(50)$ = 1.70
   $\log(Alkalinity)$ = $\log(100)$ = 2.00

2. Calculate pH_S.

   pH_S = A + B − $\log(Ca^{2+})$ − $\log(Alkalinity)$
        = 2.12 + 9.80 − 1.70 − 2.00
        = 8.22

3. Calculate the Langelier Index.

   Langelier Index = pH − pH_S
                    = 8.6 − 8.22
                    = 0.38

   A positive Langelier Index (pH greater than pH_S) indicates that the water is supersaturated with calcium carbonate ($CaCO_3$) and will tend to form scale. That is, the water is not corrosive. A negative Langelier Index means that the water is corrosive.

   Soft, low-alkalinity waters having excessively high pH values are corrosive even though the calculated LI may indicate a noncorrosive tendency. No protective calcium carbonate film can form, in this instance, due to the insufficient amount of calcium ions and alkalinity.

   The corrosive tendencies of water to particular metals, such as the ones used in distribution systems, are also significantly influenced by the amount of total dissolved solids (TDS). Waters containing TDS exceeding 50 mg/L may exhibit corrosive tendencies in spite of a positive Langelier Index. The presence of various ions, such as sulfate and chloride ions in water, may interfere with the formation and maintenance of a uniform protective calcium carbonate layer on metal surfaces. In addition, the presence of these ions will accelerate the corrosion process.

   Because of the various water quality indicators involved, the LI should only be used to determine the corrosive tendencies of water within a pH range of 6.5 to 9.5 provided that a sufficient amount of calcium ions and alkalinity over 40 mg/L as $CaCO_3$ are present in the water.

OTHER CORROSIVITY INDICES

Three other indices for calcium carbonate saturation have been used: (1) Driving Force Index (DFI), (2) Ryznar Index (RI), and (3) Aggressive Index (AI). These indices will not be described because they are not as widely used as the Langelier Index.

7.4.4   Complaints

Accurate records should be maintained of all complaints. If more than a few rusty or red water complaints are received, they should be plotted on a map of the system. The distribution of the plots can tell you where corrosion problems are occurring and indicate how the problem can be corrected. See Chapter 1, "Introduction to Water Treatment," Section 1.5.4.2, "Water Quality Complaints," for more details.

Check Your Understanding

1. When is a water considered stable?

2. How can water be tested to determine if it is undersaturated or supersaturated with calcium carbonate?

3. The Langelier Index is determined by what equation?

4. What is the meaning of pH_s?

5. Why do some waters not have a meaningful pH_s value?

6. Find the pH_s of a water at 10°C having a TDS of 100 mg/L, alkalinity of 80 mg/L as $CaCO_3$, and a calcium hardness of 40 mg/L as $CaCO_3$.

7. What do rusty or red water complaints indicate?

7.5   Methods of Controlling Corrosion

To control corrosion, you must select the correct chemicals to treat the water, calculate the correct chemical dosage, and determine the proper chemical feeder settings. The water treatment plant operator must understand cathodic protection to control corrosion and the compounds used for cathodic protection. In addition, the operator must know that about external corrosion, including soil corrosion, corrosion of steel in concrete, and stray electric current.

7.5.1   Selection of Corrosion Control Chemicals

If the water is corrosive, steps should be taken to reduce the corrosivity. Reducing corrosivity is almost always accomplished by treating the water with chemicals so that the water is saturated or slightly supersaturated with calcium carbonate. Chemicals should be fed after filtration (Figure 7.9). A slight excess of chemical could result in a supersaturated solution that could cement together the filter sand grains. (Samples for turbidity measurements should be taken after filtration but before chemical feed. Small amounts of turbidity could be introduced with the

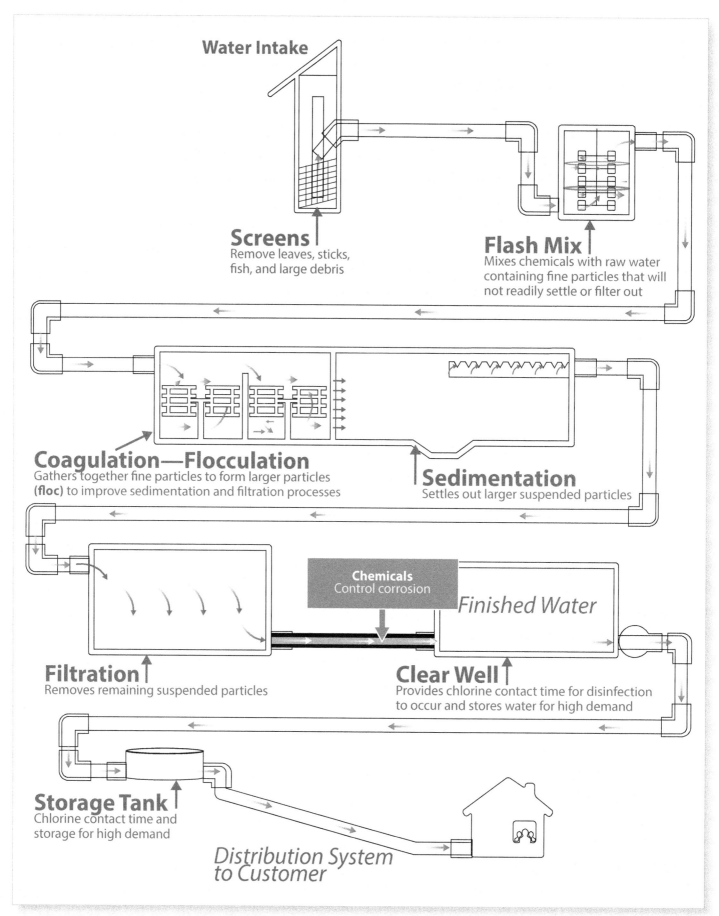

Water Intake

Screens
Remove leaves, sticks,
fish, and large debris

Flash Mix
Mixes chemicals with raw water
containing fine particles that will
not readily settle or filter out

Coagulation—Flocculation
Gathers together fine particles to form larger particles
(floc) to improve sedimentation and filtration processes

Sedimentation
Settles out larger suspended particles

Chemicals
Control corrosion

Finished Water

Filtration
Removes remaining suspended particles

Clear Well
Provides chlorine contact time for disinfection
to occur and stores water for high demand

Storage Tank
Chlorine contact time and
storage for high demand

*Distribution System
to Customer*

Figure 7.9   Typical process flow diagram

chemical and might produce misleading results suggesting poor filter performance.) The chemical feed can take place before, after, or along with postchlorination; however, samples should be taken only after post-chlorination because the chlorine may react with the chemicals used to reduce corrosivity. For example, chlorine gas will lower the pH of the water, while hypochlorite compounds will raise the pH.

The selection of a chemical to achieve calcium carbonate saturation will depend on the water quality characteristics of the water and the cost of the chemicals. The central idea is that all three chemical characteristics—calcium (measured as hardness), carbonate (measured as alkalinity), and pH—must be present in the right proportions to achieve calcium carbonate saturation. For waters that have low hardness and low alkalinity, quicklime and hydrated lime should be added to increase the calcium content and the pH. For waters with sufficient calcium but low alkalinity, soda ash may be used. For waters in which the calcium and alkalinity are sufficient but the pH is too low, caustic soda would be indicated, although soda ash can also be used because it raises pH. See Table 7.5 for a summary of water characteristics and remedies.

Quicklime is calcium oxide (CaO); it is the least expensive of the four chemicals. However, lime requires expensive equipment to **slake** or hydrate it. This procedure is cost effective only for very large water treatment plants.

Hydrated lime is calcium hydroxide ($Ca(OH)_2$); it is slightly more costly than unslaked lime. Hydrated lime is only slightly soluble in water so it cannot be fed as a true solution. Lime **slurry** reacts with carbon dioxide to form limestone (calcium carbonate or $CaCO_3$) so pipes, pumps, and solution feeders tend to become plugged with scale or deposits very rapidly. For this reason, hydrated lime is best fed using a dry feeder. Use of either form of lime will add hardness to the water. This extra hardness may be a slight disadvantage where the water already contains too much calcium or hardness. However, the lime is advantageous for waters that contain so little calcium that they cannot otherwise be saturated with calcium carbonate.

Table 7.5   Chemical summary table

Water Characteristics	Hardness × Alkalinity (both as mg/L as $CaCO_3$)	Remedy	Chemical
Very low hardness and alkalinity	< 1,000	Increase either or both calcium and carbonate	(Seek expert advice)
Low hardness and moderate to high alkalinity	1,000–5,000	Increase calcium	Quicklime (CaO) or hydrated lime ($Ca(OH)_2$)
Sufficient calcium (≥ 30 mg/L as $CaCO_3$) but low alkalinity	1,000–5,000	Increase carbonate	Soda ash (Na_2CO_3)
High hardness and alkalinity	> 5,000	Increase pH	Caustic soda (NaOH) or lime (CaO or $Ca(OH)_2$)[a]

[a]Soda ash (Na_2CO_3) will raise the pH but is normally too expensive.

slake
To mix with water so that a true chemical combination (hydration) takes place, such as in the slaking of lime.

slurry
A watery mixture or suspension of insoluble (not dissolved) matter; a thin, watery mud or any substance resembling it (such as a grit slurry or a lime slurry).

Lime should be fed to the water after the water has passed through the filters, but before it enters the clear well. Provisions should be made to collect turbidity samples or measure turbidity between the filters and the location of the lime feed because lime will cause an increase in the turbidity of the filtered water.

Lime can be very difficult to handle. If you are having problems with calcium carbonate forming in the pipes that are delivering lime to the point of application, consider the construction of open channels so the deposits of lime can be cleaned from the channel with a hoe. Other possibilities include the use of flexible pipelines (rather than rigid pipe) such as plastic hose for ease in breaking loose the calcium carbonate deposits. The outlet of any pipe or hose should not be submerged. To reduce the problem of cleaning either pipes or open channels, minimize the length of the solution lines by placing the chemical feeder as close as possible to the point of application.

Caustic soda (NaOH), also known as sodium hydroxide, is more expensive than lime, but is available in a 50 percent solution that can be fed directly with less expensive solution feeders. The purpose of adding caustic soda is to raise the pH. There are some waters, however, with initial calcium and alkalinity levels so low that calcium carbonate ($CaCO_3$) saturation cannot be reached by feeding caustic soda. When considering the use of caustic soda, problems associated with chemical feed systems must be considered. Caustic soda crystallization (freezing) occurring at temperatures below 50°F (10°C) can be a problem. At locations where caustic soda is fed to the water being treated, the caustic should not be fed in a closed conduit or pipe because this will encourage clogging.

Caustic soda must be handled with great care because it: (1) dissolves human skin, (2) produces heat when mixed with water, and (3) reacts with amphoteric metals (such as aluminum) generating hydrogen gas, which is flammable and may explode if ignited. When handling caustic soda, control the mists with good ventilation. Protect your nose and throat with an approved respiratory system. Wear chemical worker's goggles or a full-face shield to protect your eyes. Protect your body by being fully clothed, and by using impervious gloves, boots, apron, and face shield.

Soda ash is sodium carbonate (Na_2CO_3). Sodium carbonate will dissolve in water up to approximately 1⅓ pounds per gallon (0.16 kg/L) and can thus be fed with solution feeders. Soda ash can be used in waters of low alkalinity provided the calcium hardness is greater than about 30 mg/L as $CaCO_3$. Because such waters are rare, it is not a common method of treatment. Both soda ash and caustic soda can increase alkalinity. To decide which chemical to use, calculate the cost of increasing the alkalinity by 1 mg/L. This will require consideration of the cost per pound of chemical, the percent purity of the chemical, and the change in alkalinity or pH resulting from the application of the chemical.

Utilities should exercise caution when applying compounds containing sodium if the added sodium will increase the level in the water to more than 20 mg/L. Public health officials are concerned because evidence indicates that sodium is an important factor in the development of high blood pressure in susceptible individuals.

Once you have analyzed your water and determined that you have a potential corrosion problem, the next step is to select the appropriate chemical or chemicals. Selection of chemicals depends on the characteristics of the water, where in the treatment process the chemicals can be applied, how they can be applied and mixed with the water, and the costs of the chemicals. You want to solve your corrosion problem by the most cost-effective means (see Table 7.5).

When you multiply the calcium hardness by the alkalinity (both in mg/L as $CaCO_3$) and the product is less than 1,000, then the treatment required may be complicated. For example, both lime and carbon dioxide may be required. A qualified expert's advice should be sought to determine the proper chemical doses.

If the calcium hardness multiplied by the alkalinity is between 1,000 and 5,000, either lime or soda ash (Na_2CO_3) will be satisfactory. The decision regarding which chemical to use will depend on the cost of equipment and the cost of chemicals.

If the calcium hardness times the alkalinity is greater than 5,000, either lime or caustic soda (NaOH) may be used. Soda ash will be ruled out because of expense.

7.5.2   Determination of Chemical Dose

The chemical dose required to saturate the water with calcium carbonate may be determined graphically or by a trial-and-error calculation that is practical only with the help of a computer.

In any event, the calculations are not exact and the chemical dose must be checked by the Marble Test. Only slightly more time is required to find the proper chemical dose by experiment than by using the graphical approach or a computer, and the results obtained by these methods should be verified by the Marble Test anyway.

To determine the chemical dose experimentally, first calculate the value of pH_S using Larson's method (page 457). Prepare a solution containing 1 gram per liter of the chemical to be used (1 mL will then contain 1 mg of chemical). Be sure the dilution water does not contain any carbon dioxide. Carbon dioxide may be removed from distilled water by boiling. Deionized, instead of distilled water, is usually satisfactory. Treat a 1-liter sample of the water to be tested with 1 mL portions of the chemical solution (mixing well) until the calculated value of pH_S is reached. As 1 mL portions of the chemical solution are added, the pH of the sample will gradually increase until the pH_S value is reached. The degree of saturation is then measured by the Marble Test. Compare the total hardness (mg/L as $CaCO_3$) of the water before and after the Marble Test. If the total hardness (as $CaCO_3$) is reduced by more than 10 mg/ L (i.e., the water is supersaturated with $CaCO_3$), try a smaller chemical dose and repeat the procedure until the total hardness decreases by only 0 and 10 mg/L as $CaCO_3$.

To conduct the Marble Test, first measure the pH, alkalinity, and hardness. Stir a 0.5-liter sample of the water being tested with approximately 0.5 gram of pulverized marble ($CaCO_3$) for 5 minutes. Filter the water through filter paper and again measure the pH, alkalinity, and hardness. A decrease in all three values means the water was supersaturated; an increase in all three values means the water was undersaturated; and no change indicates the water was just saturated with $CaCO_3$. When stirring the sample, use a magnetic stirrer and a nearly full stoppered bottle to prevent a loss or gain of carbon dioxide from the air.

For information on determining the chemical dose graphically, see D. T. Merril and R. L. Sanks, "Corrosion Control by Deposition of $CaCO_3$," *Journal American Water Works Association*, Part 1, page 592, November 1977; Part 2, page 634, December 1977; and Part 3, page 12, January 1978.

See the laboratory procedures chapter in Volume II for detailed procedures on how to perform the Marble Test.

To be sure that the temperature of the water stays nearly constant, execute the test quickly.

Water that is just saturated will form a $CaCO_3$ scale only on the cathodic corrosion areas, but water that is supersaturated will form a scale on all surfaces exposed to flowing water. The thickest scale will form on the surfaces where the water velocity is the highest (up to 5 ft/sec or 1.5 m/sec) because these have the greatest contact of calcium carbonate with the surface. The thickness of scale in water mains seldom exceeds ⅛ inch (3 mm). However, ridges of scale may form transverse (perpendicular) to the flow and very seriously reduce the carrying capacity of the water main. When the velocities are higher (greater than 5 ft/sec or 1.5 m/sec), the scale can be washed or eroded away.

It is important to note that calcium carbonate, unlike most salts, is less soluble in hot water than in cold water. Therefore, even if a water is just saturated as it enters the distribution system, it may become highly supersaturated in hot water systems. Scale has a strong tendency to form on heat transfer surfaces in hot water heaters. Because calcium carbonate does not conduct heat as well as steel, this results in a lowered heating efficiency and in overheating of the heat-transferring metal. Hot water pipes are sometimes found almost completely plugged with a calcium carbonate deposit. Scale formation can be inhibited by feeding a 1 to 2 mg/L solution of sodium trimetaphosphate (commercially known as sodium hexametaphosphate). This chemical eventually reverts to the less effective chemical called orthophosphate in the distribution system. The chemical change occurs slowly in cold waters, but quite rapidly in hot water systems.

7.5.3   Determination of Chemical Feeder Setting

The desirable feeder setting must be established by analysis of the results of your corrosioncontrol program. If the chemical composition of the water is fairly constant, periodic Marble Tests may be used. If the chemical composition of the water is quite variable and frequent Marble Tests are required, an Enslow column might be more practical.

There are no standards for constructing an Enslow column, nor are any ready-made columns commercially available. You can construct your own column based on Enslow's design (see Figure 7.10). The important parameter is the detention time, which must be long enough for the water to become saturated with calcium carbonate in the first column. Enslow recommends 2 hours, although that time may vary, depending on water characteristics. Longer contact times are not a problem. To achieve a 2-hour contact time, and assuming a porosity of 60 percent, 3-foot-tall columns that are 2, 3, and 4 inches in diameter would require flows of about 10, 20, and 37 mL/min, respectively. *Standard Methods* (Part 2330C) suggests constructing this device using a laboratory separatory funnel. The shape is not important so long as there is good contact between the water and the chalk.

The interpretation of Enslow column data was previously discussed in Section 7.4.3.3, "Calcium Carbonate Saturation." If the pH increases across the column, the water is undersaturated (marble dissolving) and the chemical feed should be increased. If the pH decreases across the column, the water is supersaturated (scale forming) and the chemical

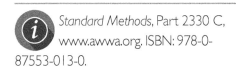

Standard Methods, Part 2330 C, www.awwa.org. ISBN: 978-0-87553-013-0.

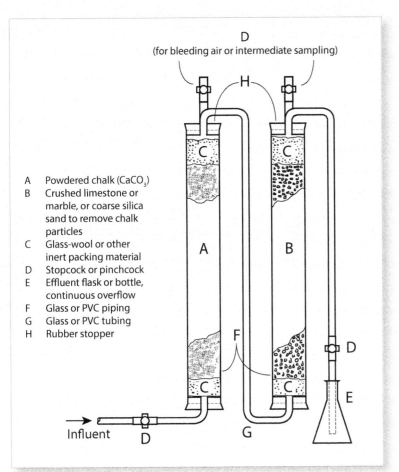

A   Powdered chalk (CaCO₃)
B   Crushed limestone or
      marble, or coarse silica
      sand to remove chalk
      particles
C   Glass-wool or other
      inert packing material
D   Stopcock or pinchcock
E   Effluent flask or bottle,
      continuous overflow
F   Glass or PVC piping
G   Glass or PVC tubing
H   Rubber stopper

Figure 7.10   Enslow column (*Courtesy of the American Water Works
Association*)

feed can be decreased. A slight drop in pH is acceptable, but to control corrosion the chemical feed should be maintained to prevent
any increase in pH. If the alkalinity is measured before and after the
water passes through the columns, the amount of increase in alkalinity
would indicate the amount of increase required in the chemical feed
system.

Check Your Understanding

1.   How can the corrosivity of a water be reduced?

2.   What chemicals may be added to waters to reduce the corrosivity?

3.   What chemicals may be required for corrosion control if the product of calcium hard-
     ness multiplied by the alkalinity (both in mg/L as CaCO₃) is less than 1,000?

4.   What chemicals will be required for corrosion control if the product of calcium
     hardness multiplied by the alkalinity (both in mg/L as CaCO₃) is greater than 5,000?

5.   How can the proper chemical dose be determined to produce water that is just
     saturated with calcium carbonate (CaCO₃)?

7.5.4   Zinc, Silica, and Polyphosphate Compounds

Certain zinc compounds, such as zinc phosphate, are capable of forming effective cathodic films that will control corrosion. These zinc compounds are largely proprietary (can only be bought from the owner) and the companies that market these compounds usually supply technical advice and assistance without charge. The zinc compound treatments are generally more expensive than treatment with lime or caustic, but they have the advantage that scale is less apt to be a problem. Do not use zinc phosphate compounds to control corrosion caused by water that will be stored in an open reservoir. The phosphate may cause algal blooms. Residual chlorine lasts longer in distribution systems using zinc orthophosphate (ZOP). This has been attributed to decreased chlorine demand as a result of reducing iron sediments in distribution piping.

Sodium silicate has been used to treat corrosive waters. A solution of sodium silicate ($Na_2O : 3 \ SiO_2$) fed at a rate of approximately 12 mg/L as silica is used for the first month, after which the rate is reduced to 8 mg/L. This method of treatment is used by individual customers, such as apartment houses and large office buildings, but is not commonly used by water utilities.

Sodium polyphosphates, usually either tetrasodium pyrophosphate ($Na_4P_2O_7$) or sodium hexametaphosphate, have been used for corrosion control. Solutions of these compounds may form protective films but, because they react with calcium, they reduce the effective calcium concentration and thereby actually increase corrosion rates. The major use of these chemicals in water treatment is to control scale formation in waters that are supersaturated with calcium carbonate.

The deterioration of asbestos-cement (AC) pipe may be prevented by maintaining calcium carbonate saturation. There is evidence that the zinc treatments are also effective for this purpose as are treatments using traces of iron, manganese, or silica in the water. Any deterioration of AC pipe will cause an increase in pH and calcium content of water as it passes through the pipe. Tests for pH and calcium should be performed after the pipe has been in service for 2 months or longer because all AC pipe contains at least traces of free lime, which will result in an initial increase in water pH when the pipe is placed in service.

7.5.5   Cathodic Protection

Mixers, tanks, flocculators, clarifiers, and filter troughs are frequently constructed of steel and require some sort of corrosion protection. Both manual and automatic cathodic protection systems are available. Automatic cathodic protection systems are preferred because the conductivity (total dissolved solids or TDS) in water can change. Cathodic protection systems are very costly to install. However, it is also very costly to shut down a water treatment process, drain the facility, sandblast metal surfaces, apply paint or a protective coating, and put the facility back on line.

7.5.5.1   How the Protection System Works

Cathodic protection is a process used to reduce or inhibit corrosion of metal exposed to water or soil. This process consists of the deliberate act of reversing the electrochemical force to check the destruction that naturally occurs to metals whenever they are buried.

The technique introduces into the natural corrosion cycle an external DC (direct current) electric current sufficiently strong enough to offset and cancel out the corrosion-producing action. Key to the system is the use of an auxiliary anode of expendable metal that is immersed or buried in the soil or water (called the electrolyte) a predetermined distance from the metal to be protected. Electric current flowing from the anode to the structure (pipeline, flocculator, or clarifier) in precisely the proper flow can exactly counteract corrosion losses.

The application of cathodic protection is an involved process. Factors affecting a corrosion problem include soil conductivity (which varies considerably even within a limited area), soil moisture content, soil and water characteristics, DO content, temperature, seasonal variations of environment (weather), protective coatings, dissimilar metals to be protected, position of other metallic structures, and stray currents already present in the ground or water.

All cathodic protection systems pass current through the soil or water from anodes connected to the structure that is to be protected. Two basic methods are used. Sacrificial anode material, such as magnesium or zinc, is used to create a galvanic cell. Such anodes are selfenergized and are connected directly to the structure to be protected. These anodes are commonly used to apply small amounts of current at many locations.

The other basic method uses anodes energized by a DC power supply such as a rectifier (Figure 7.11). This method, commonly referred to as impressed current, uses relatively inert anodes (usually graphite or high-silicon cast iron) connected directly to the positive terminal of a DC power supply or rectifier, with the pipe or structure being protected connected to the negative terminal. Such systems are generally used where large amounts of current are required at relatively few locations.

7.5.5.2 Equipment

The equipment to be used in any given cathodic protection system is very important. A wide selection of rectifiers is available, including air-cooled, oil-immersed, and automatic units. The range of voltage and amperage output is almost infinite. Selection should be based on the particular requirements of the pipe or facilities being protected from corrosion.

Anodes, which serve to distribute the direct current into the earth or water, are manufactured from various metals. Graphite, carbon, high-silicon cast iron, platinum, magnesium, aluminum, and zinc alloys are commonly used. Each has its own particular application. Proper usage is a determining factor in the success or failure of a cathodic protection system.

7.5.5.3 Protection of Flocculators, Clarifiers, and Filters

Cathodic protection of flocculators, clarifiers, and filters is a very effective means of controlling corrosion because maintenance and repair of protective coatings on these facilities is very difficult (Figure 7.12). Automatic cathodic protection control devices can provide precise corrosion control under nearly all conditions. Anodes with a 10-year design life are usually installed below the low-water level.

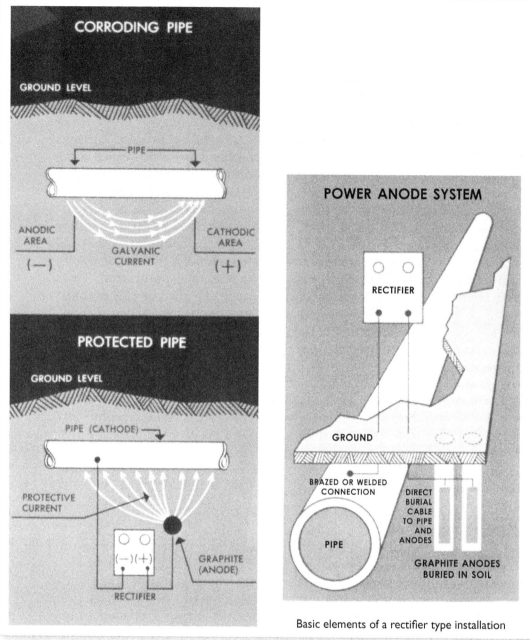

Figure 7.11   Cathodic protection for buried pipe: Basic elements of a rectifier type installation
(*Courtesy of Harco Corporation*)

7.5.5.4   Maintenance

To achieve and maintain a high degree of corrosion control, you must regularly inspect and test the operation of the cathodic protection system and its parts. Regular and proper maintenance is critical because the amount of protective current required to prevent corrosion can vary with changes in the condition of the coatings on the metallic surfaces, in the chemical characteristics of the water being treated, and the operation of the facilities.

An annual maintenance inspection should include a visual examination of all anodes, wiring, electrical splices and connections, power units, meters, and reference cells. In addition, a complete potential profile should be taken inside the structure to determine the proper automatic controller setting to ensure that corrosion control will be maintained automatically on all submerged surfaces throughout the year.

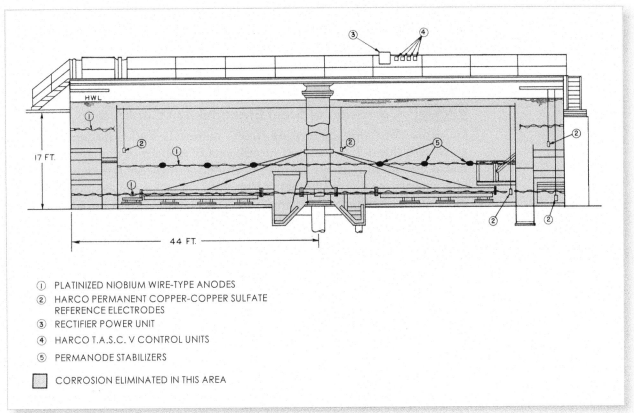

① PLATINIZED NIOBIUM WIRE-TYPE ANODES
② HARCO PERMANENT COPPER-COPPER SULFATE
   REFERENCE ELECTRODES
③ RECTIFIER POWER UNIT
④ HARCO T.A.S.C. V CONTROL UNITS
⑤ PERMANODE STABILIZERS

▨ CORROSION ELIMINATED IN THIS AREA

Figure 7.12   Cathodic protection for a clarifier (*Courtesy of Harco Corporation*)

7.5.6   Removal of Oxygen

Other methods of water treatment for corrosion control are not practical for domestic water systems. Removal of oxygen is used in boilers and in other water heating systems.

7.5.7   External Corrosion

The water treatment plant operator should understand the complications of external corrosion, including soil corrosion, corrosion of steel embedded in concrete, and stray electric currents.

7.5.7.1   Soil Corrosion

Although corrosion on the outside of the water mains is not the responsibility of the water treatment plant operator, there are many smaller systems where the operator is the person who comes closest to being the corrosion engineer. For this reason, some of the factors influencing soil corrosion will be discussed very briefly. The best measure of the corrosivity of soil is the soil resistivity, which is easily measured using a soil resistance meter. The fourpoint type is the most useful because it can measure average resistivity down to the depth of the pipeline. Some water systems use soil resistivity as an indication of the kind of pipe to install. If the soil resistivity is greater than 5,000 ohms/cm, serious corrosion is unlikely. Steel pipe, with its superior strength and flexibility, may be used under these conditions. If the soil resistivity is below 500 ohms/cm, nonmetallic pipe, such as AC or PVC, is used. Cast-iron pipe or pipe that is lined and coated with cement mortar is used in the intermediate ranges of soil resistivity.

The chemical reactions involved in external corrosion are electrochemical in nature just like those in internal corrosion. The current paths (flows) are not confined to the inside surfaces of the pipe; therefore, galvanic corrosion and electrolysis are relatively more important and cathodic protection is usually practical.

7.5.7.2   Corrosion of Steel Embedded in Concrete

Galvanic corrosion of ferrous (iron) materials under homes with concrete slab floors has resulted in millions of dollars in damage before the cause was identified. When steel is embedded in concrete, it assumes the characteristics of a noble metal. Concrete slabs are always poured over a steel mesh or other reinforcing iron. The electrical resistance of the steel is very small compared to that of concrete, so the entire slab behaves as if it were a solid sheet of a noble metal. Since the area of the slabs in a subdivision of slab floor homes is much larger than the area of the pipe, the flow of electrons from the pipes in the area may be fairly large; thus, the pipes may be seriously corroded. This problem may be avoided by using insulating fittings at the service cock or by using plastic service pipe. After the concrete has cured for a few years, the electrical resistance becomes so high that corrosion ceases.

A corrosion problem can develop where a steel water transmission main many miles long runs alongside a second main that is coated with cement. If the two mains are connected at pump stations, corrosion problems can develop. At these pump stations, a galvanic current of more than 1.0 ampere might be measured. This instance of corrosion may sometimes be controlled by installing insulating fittings to interrupt the electric circuit.

Aluminum should never be embedded directly in concrete without protection such as a zinc chromate primer. Aluminum and, to a lesser extent, zinc may be oxidized by hydroxyl ions:

$$2\ Al + 2\ OH^- + 4\ H_2O \rightarrow 2\ H_2AlO_3^- + 3\ H_2$$

For this reason, aluminum is unsatisfactory in highly alkaline aquatic (water) environments.

7.5.7.3   Stray Electric Currents

As explained previously in this chapter, electrolysis is the decomposition of a substance by the passage of an exterior source of DC electric current (Figure 7.13). Internal corrosion caused by electrolysis is practically impossible in a system with properly made joints because the resistance of the water in the pipe is about a billion times as great as the resistance of the pipe itself so that corrosive electric currents must be extremely small by comparison.

Alternating current (AC) electrolysis will also corrode metals, but AC-caused electrolysis is a rare occurrence and seldom a serious problem.

Electrolysis of water mains can result from stray currents generated by cathodic protection installed by the gas company or other utilities, but engineers who install cathodic protection systems are aware of this possibility and can avoid these problems (Figure 7.14). Electrolysis caused by defective grounding of a customer's piping may be avoided by using insulating fittings or using plastic service pipe.

As in the past, electric transit systems may pose a serious threat in the future. Although contemporary transit systems are designed with

Figure 7.13 Stray-current electrolysis
(*Courtesy of AK Steel Corporation*)

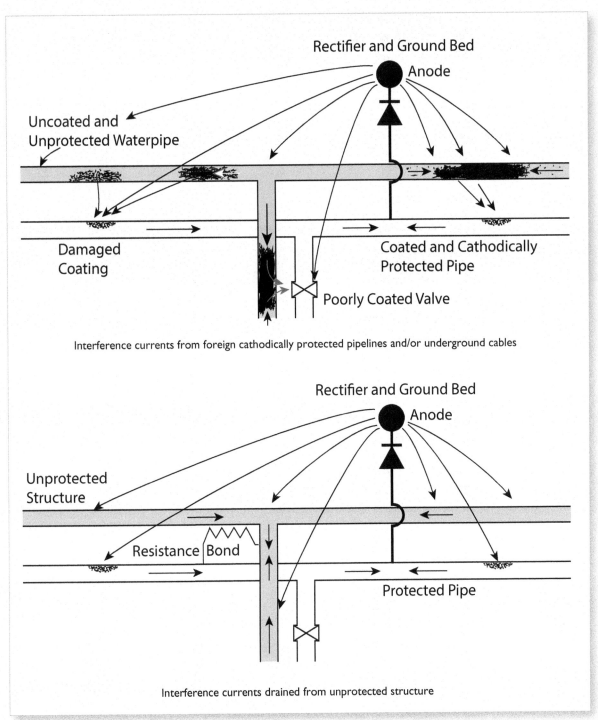

Figure 7.14   Electrolysis caused by interference currents (*Courtesy of AK Steel Corporation*)

many effective features to control stray current, it still can cause costly problems for neighboring infrastructure, including buried pipelines and cables. Controlling stray current corrosion requires regular inspection and testing during the life of the transit system. In addition to periodic inspections, tests for stray current include taking resistance measurements and ground current surveys, and measuring for possible current leakage. Stray current corrosion should be addressed during transit system design, operation and maintenance.

7.6   Troubleshooting

Troubleshooting at the water treatment plant involves understanding both internal and external pipe corrosion and applying the necessary controls or solutions such as pipe replacement.

7.6.1   Internal Pipe Corrosion

Internal corrosion can be detected through rusty water complaints and by examining the insides of pipes for pitting, tubercles, and other evidence of corrosion. To control internal corrosion, treat the water to achieve calcium carbonate saturation.

Select a target pH_S. Dose to this pH. Run a Marble Test on a sample of the treated water. If the pH does not change by more than ±0.2 pH before and after the Marble Test, the target pH_S is satisfactory. If the pH increases during the Marble Test, the target pH_S should be increased by the amount of the pH increase during the Marble Test. Reset the chemical feeder to dose to the new pH. Repeat this procedure until satisfactory results are obtained.

7.6.2   External Pipe Corrosion

External corrosion is indicated by pinhole leaks or rust on the outside of pipes. If the rusting pipes were installed with the proper bonds and insulating fittings, cathodic protection can be effective. If cathodic protection is not feasible, the pipe will have to be replaced with PVC, plastic pipe, or a cement-coated line.

7.7   The Lead and Copper Rule

The Lead and Copper Rule (LCR) describes the health concerns and regulations for waters affected by lead and copper. The water treatment plant operator must understand the monitoring and treatment requirements for these waters, and must also know the public education and reporting requirements.

7.7.1   Health Concerns

The health concerns about exposure to lead are best described by the EPA:

Lead is a common, natural and often useful metal found throughout the environment in lead-based paint; air; soil; household dust; food; certain types of pottery, porcelain, and pewter; and water. Lead can pose a significant risk to your health if too much of it enters your body. Lead builds up in the body over many years and can cause damage to the brain, red blood cells, and kidneys. The greatest risk is to young children and pregnant women. Amounts of lead that will not hurt adults can slow down normal mental and physical development of growing bodies. In addition, a child at play often comes into contact with sources of lead contamination—like dirt and dust—that rarely affect an adult.

Lead in drinking water, although rarely the sole cause of lead poisoning, can significantly increase a person's total lead exposure, particularly the exposure of infants who drink baby formulas and concentrated juices that are mixed with water. The EPA estimates that drinking water can make up 20 percent or more of a person's total exposure to lead.

Lead is unusual among drinking water contaminants in that it seldom occurs naturally in water supplies like rivers and lakes. Lead enters drinking water primarily as a result of the corrosion, or wearing away, of materials containing lead in the water distribution system and household plumbing (see Figure 7.15). These materials include lead-based solder used to join copper pipe, brass and chrome-plated brass faucets, and, in some cases, pipes made of lead that connect your house to the water main (service lines).

When water stands in lead pipes or plumbing systems containing lead for several hours or more, the lead may dissolve into your drinking water. This means the first water drawn from the tap in the morning, or later in the afternoon after returning from work or school, can contain fairly high levels of lead.

The health effects of copper include stomach and intestinal distress. Prolonged doses result in liver damage. Excess intake of copper or the inability to metabolize copper is called Wilson disease.

7.7.2   Regulations

As part of the Safe Drinking Water Act Amendments of 1986, the US Congress directed the Environmental Protection Agency to develop regulations for monitoring and control of lead and copper in drinking water. On June 7, 1991, EPA published the final LCR. The rule and its implications are both complex and significant. Major features of the rule are explained in this section. More detailed information can be obtained from the EPA

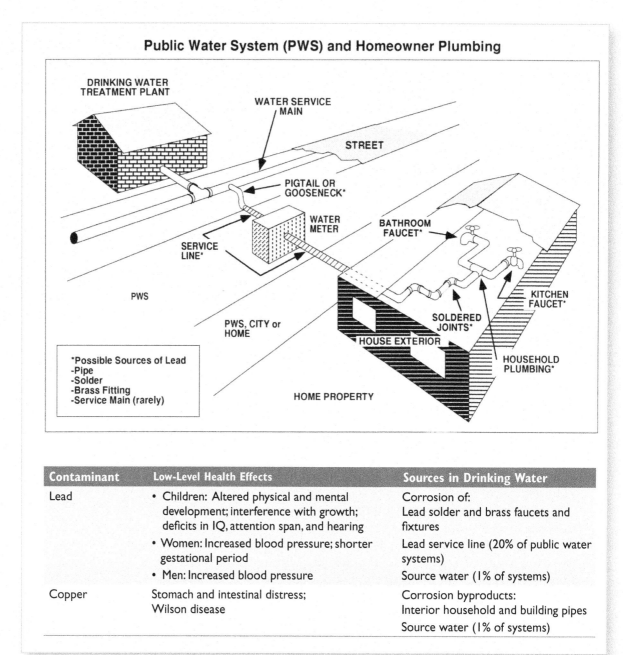

Public Water System (PWS) and Homeowner Plumbing

Contaminant	Low-Level Health Effects	Sources in Drinking Water
Lead	• Children: Altered physical and mental development; interference with growth; deficits in IQ, attention span, and hearing • Women: Increased blood pressure; shorter gestational period • Men: Increased blood pressure	Corrosion of: Lead solder and brass faucets and fixtures Lead service line (20% of public water systems) Source water (1% of systems)
Copper	Stomach and intestinal distress; Wilson disease	Corrosion byproducts: Interior household and building pipes Source water (1% of systems)

Figure 7.15 Health effects and sources of lead and copper

(www.epa.gov), your state regulatory agency, and the American Water Works Association (AWWA) (www.awwa.org).

The issue of lead contamination in water supplies and in other environmental media (air, paint, food) continues to be a subject of considerable activity on the part of individual state governments, Congress, and the EPA. Future amendments to the rules or regulations governing lead contamination in water, monitoring requirements, and mitigation measures may affect how the June 7, 1991, LCR is implemented. Investigate the most current status of lead contamination regulations from the regulatory agencies in your state before starting a monitoring or mitigation program to ensure that your program meets all applicable requirements. The information provided in this section applies to the final Lead and Copper Rule of June 7, 1991, published in the *Federal Register*.

The 1991 LCR includes the following elements:

- Maximum contaminant level goals (MCLGs)—nonenforceable health-based targets—and action levels established for lead and copper
- Monitoring requirements for lead, copper, and other corrosion analysis constituents; analytical methods; and laboratory certification requirements
- Treatment techniques for lead and copper (required if action levels are exceeded during monitoring) including optimal corrosion-control treatment, source water treatment, and lead service line replacement
- Public notification and public education program requirements
- Utility system recordkeeping and reporting requirements
- Variances and exemptions from the regulations and compliance schedules based on the size of the population served by the utility system

On October 10, 2007, EPA published revisions and clarifications to the LCR. These revisions are as follows:

- Enhance the implementation of the LCR in the areas of monitoring, treatment, customer awareness, and lead service line replacement
- Improve compliance with the public education requirements of the LCR and ensure drinking water consumers receive meaningful, timely, and useful information needed to help them limit their exposure to lead in drinking water

7.7.3   Monitoring Requirements

The first requirement of the LCR is for monitoring to determine if either of these metals exceeds the levels at which further action must be taken. All systems regardless of size had to begin their initial monitoring programs by July 1993. An unusual provision of the LCR is that the monitoring samples must be collected at the consumers' taps, rather than only at the water treatment plant or in the distribution system. The samples must be taken from locations identified by the utility as high risk, including:

- Homes with lead solder installed after 1982
- Homes with lead pipes
- Homes with lead service lines

7.7.3.1   Monitoring Frequency

The number of samples to be collected for lead and copper analysis is based on the size (population served) of the distribution system; the sampling frequency is every 6 months for the initial monitoring program. There are two monitoring periods each year, January to June, and July to December. If the system is in compliance, either as demonstrated by monitoring or after installation of corrosion controls, a reduced monitoring frequency can be initiated. The number of sampling sites required based on system size for initial and reduced monitoring are listed in Table 7.6.

7.7.3.2   Sampling Procedure

The samples are to be collected as first-draw samples from the cold water tap in either the kitchen or bathroom, or from a tap routinely used for consumption of water if in a building other than a home. A first-draw sample is defined as the first liter of water collected from a tap that has not been used

Table 7.6   Sampling sites required for lead and copper analysis

System Size (Population)	Sampling Sites Required (Base Monitoring)	Sampling Sites Required (Reduced Monitoring)
>100,000	100	50
10,001–100,000	60	30
3,301–10,000	40	20
501–3,300	20	10
101–500	10	5
≤100	5	5

Reduced Monitoring:
- All public water systems that meet the lead and copper action levels or maintain optimal corrosion-control treatment for two consecutive 6-month monitoring periods may reduce the number of tap water sampling sites as shown above and their collection frequency to once per year.
- All public water systems that meet the lead and copper action levels or maintain optimal corrosion-control treatment for 3 consecutive years may reduce the number of tap water sampling sites as shown above and their collection frequency to once every 3 years.

for at least 6 hours, but preferably unused no more than 12 hours. Faucet aerators should be removed before sample collection. The LCR allows homeowners to collect samples for the utility as long as the proper sample collection instructions have been provided. The EPA specifically prohibits the utility from disputing the accuracy of a sample collected by a resident.

7.7.3.3   Maximum Contaminant Level Goals (MCLGs)

Under the 1977 National Interim Primary Drinking Water Regulations, a maximum contaminant level (MCL) was established for lead, and in 1979, under the National Secondary Drinking Water Regulations, a secondary maximum contaminant level (SMCL) was set for copper. The LCR replaces those with MCLGs, and with treatment technique requirements when action levels are exceeded during system monitoring. The MCLGs and action levels for lead and copper, compared to the older interim limits, are:

Contaminant	Interim Limits	1991 MCLG	1991 Action Level
Lead	0.05 mg/L	Zero	0.015 mg/L
Copper	1 mg/L	1.3 mg/L	1.3 mg/L

Action levels are defined in this rule as the value measured in the 90th percentile at the consumer's tap. This means that when all of the samples measured at consumer taps are arranged from lowest to highest value in a list, the value at the point $^9/_{10}$ (90 percent) of the way up from the lowest number is the value used to determine compliance with the action level. For example, in the following series of lead and copper test results, the 90th percentile is shown:

	Lead	Copper
	0.002	0.387
	0.002	0.396
	0.008	0.411
	0.010	0.430
	0.012	0.465
	0.014	0.534
	0.015	0.653
	0.021	0.758
90th Percentile:	**0.025**	**0.806**
	0.031	0.950

In the previous example, the lead value at the 90th percentile exceeds the 0.015 mg/L action level, requiring the utility to begin implementation of a corrosion-control program. All utilities with populations larger than 50,000 are required to initiate corrosion-control studies unless they can prove through the routine monitoring program that any increase between the source water lead level and the 90th percentile tap samples is less than 0.005 mg/L for two consecutive 6-month monitoring periods. For small systems with only five sampling sites, the 90th percentile is defined as the average of the 4th and 5th highest values in the series.

7.7.3.4 Other Water Quality Monitoring

In addition to the required lead and copper monitoring program, all large systems and any medium or small systems that exceed the action levels must also monitor for other corrosion-related water quality indicators. These indicators are:

- Alkalinity
- Conductivity
- pH
- Temperature
- Corrosion Inhibitor Concentration (Calcium, Orthophosphate, or Silica)
- Corrosion Inhibitor Dosage Rate (if added)

Specified limits for these water quality indicators are to be set by each state. The monitoring locations include both the point(s) of entry to the distribution system and consumer taps. The sampling program requirements are:

- All large systems, and those medium and small systems that exceed the lead or copper action levels, must collect two samples for each applicable water quality indicator at each lead and copper tap water sample site every 6 months, and one sample for each applicable water quality indicator at each entry point to the distribution system every 2 weeks.
- All large systems, and those medium and small systems that exceed the lead or copper action levels, even after installing optimal corrosion control treatment, must continue to collect two samples for each applicable water quality indicator at each lead and copper tap water sample site every 6 months, and one sample for each applicable water quality indicator at each entry into the distribution system every 2 weeks.
- The same number of reduced monitoring stations as are allowed with the lead and copper monitoring are allowed for systems in compliance or after successful implementation of corrosion-control treatment.

7.7.3.5 Analytical Methods and Certification Requirements

US EPA-approved methods must be used for laboratory analysis of all samples tested in Lead and Copper Rule monitoring programs. Methods are listed as either EPA-approved methods or in *Standard Methods for the Examination of Water and Wastewater,* 22nd Edition (www.awwa.org. ISBN: 978-0-87553-013-0). Laboratories are not required to be certified to test for alkalinity, calcium, conductivity, orthophosphate, pH, silica, or temperature, but they must be certified to test for copper and lead.

7.7.4   Treatment Requirements

Treatment requirements for lead and copper include action levels for lead and copper. If a large system does not meet those action levels, corrosion-control studies must be conducted. If a system is not in compliance, possible solutions include source water treatments, lead service line replacement, and treatment for control of lead and copper.

7.7.4.1   Corrosion Treatment Studies

All large water systems (>50,000 population) that do not meet the action levels for copper or lead, or that cannot show that the difference between source water and 90th percentile tap water samples is less than 0.005 mg/L, must conduct corrosion-control studies. These studies should include comparison of the following accepted potable water corrosion-control treatments:

- pH and alkalinity adjustment (to reduce acidity)
- Calcium adjustment (to form protective calcium carbonate films inside plumbing)
- Phosphate or silica-based inhibitor addition (to form protective films inside plumbing)

The corrosion studies can be conducted using pipe rig/loop tests, metal coupon tests, partial-system tests, or documented analogous treatment. Pipe rig/loop and metal coupon tests directly evaluate the performance of corrosion inhibitors on the corrosion rate of materials similar to those used in the distribution system. Partial-system tests determine the effectiveness of inhibitors on an isolated portion of a distribution system as compared to other sections with other treatments or with no treatment. Documented analogous treatment refers to using data from other similar water systems that have already conducted corrosion studies.

Small or medium-sized systems that do not meet the action levels are also required to install corrosion-control treatment. States have some flexibility in providing smaller systems with standard recommended treatment practices for corrosion control, and in allowing such systems to recommend treatments without extensive field studies. Some states may require full studies of any size system, however. In either case, all water systems are required to implement corrosion treatment within 24 months once approved by the state, and to begin collecting follow-up samples within 1 year of completion of installation.

7.7.4.2   Source Water Treatment

A small number of systems may not be in compliance with the lead and copper action levels if these metals occur in the source water. If it is discovered during the routine monitoring that a lead or copper problem exists in the source water feeding the distribution system, the utility must specify a treatment method to remove that contaminant from the source water. One of the following treatment methods or an alternative that is equally effective must be selected:

- Ion exchange
- Reverse osmosis
- Lime softening
- Coagulation/filtration

As with corrosion treatment, once the state has approved the recommended treatment, installation must be accomplished within 24 months, and collection of follow-up samples must begin within 1 year of completion of installation. See *Water Treatment Plant Operation*, Volume II, for procedures on how to operate and maintain the first three source water treatment processes listed above.

7.7.4.3 Lead Service Line Replacement

Special requirements also apply to distribution systems with lead service lines into the homes. If corrosion-control and source water treatment programs fail to achieve compliance with lead action levels for such systems, the following requirements take effect:

- All public water systems that continue to exceed the lead action level after installing optimal corrosion-control treatment and source water treatment must replace lead service lines that contribute in excess of 0.015 mg/L of lead to the tap water.
- Utilities must replace 7 percent of their lead service lines per year, or demonstrate that the lines not replaced contribute less than 0.015 mg/L of lead to the tap water. Samples to determine lead service line contribution levels can either be taken directly from a tap in the service line, or from the consumer's tap after the water has been run sufficiently long to ensure that the water being sampled is drawing directly from the sample line. This may be indicated by a change in temperature at the tap.
- A system must replace the entire lead service line unless it can demonstrate that it does not control the entire line. Water systems must offer to replace the owner's portion of the service line (at the owner's expense).
- A system that exceeds the lead action level after installing optimal corrosion-control treatment and source water treatment has a maximum of 15 years to complete the lead service line replacement.

7.7.4.4 Treatment for Control of Lead and Copper

A water's pH/alkalinity combination determines that water's tendency to dissolve lead and copper. Corrosion-control treatment often involves adjusting pH and alkalinity to make water less corrosive.

The inhibitors used to control lead and copper act by forming a protective coating over the site of corrosion activity, thus inhibiting corrosion. The success of inhibitor addition depends on the ability of the inhibitor to provide a continuous coating throughout the distribution system.

Phosphates are by far the most common inhibitors used in water treatment for corrosion control. Silicates have a more limited application and may be most suitable for small systems with iron and manganese problems. The control of lead using a phosphate inhibitor occurs when a lead-phosphate compound is formed.

Orthophosphates are primarily used to control lead, not copper. Under optimal conditions, orthophosphate treatment is usually more effective in reducing lead than pH/alkalinity adjustment. The optimal pH for orthophosphate treatment is between 7.2 and 7.8. The orthophosphate residual must be maintained continuously throughout the distribution system. Systems that have a raw water naturally in the optimal pH range from

7.2 to 7.8 can treat with orthophosphate alone, provided the water has sufficient alkalinity for a stable pH.

Optimal corrosion-control treatment minimizes lead and copper concentrations at users' taps while ensuring that the treatment does not cause the water system to violate any drinking water regulations. When applying corrosion-control treatment, a substantial amount of time may elapse between the time treatment changes are made and the reduction of lead and copper is detected through tap water sample analyses. Be very cautious because by changing the chemistry of water, conditions may get worse before they get better.

When using corrosion inhibitors, operators must be aware of both the positive and negative side effects that may result from the use of inhibitors. Phosphate-based inhibitors may stimulate biofilms in the distribution system. These biofilms may deplete (reduce) disinfection residuals within the distribution system. Consumer complaints regarding red water, dirty water, color, and sediment may result from the action of the inhibitor on existing corrosion byproducts within the distribution system. The use of zinc orthophosphate may present problems for wastewater facilities with zinc or phosphorus limits in their **NPDES permits**. Customers with specific water quality needs, such as health care facilities, should be advised of any treatment changes at the plant. The use of sodium-based chemicals will increase total sodium levels in the finished water. Since silicates precipitate rapidly at higher water temperatures, the use of silicates may reduce the useful life of domestic hot water heaters. The precipitate (a glass-like substance) eventually fills up the water heater and reduces volume or capacity. The term for this effect is glassification.

For systems using alum or other products containing aluminum, the aluminum will bind with the orthophosphate at a ratio of 1:4 and will interfere with the formation of an effective inhibitor film. For example, if the target orthophosphate residual is 0.5 mg/L and the water has 0.1 mg/L of aluminum, only 0.1 mg/L of the orthophosphate will be available to coat the pipes while 0.4 mg/L will be bound to the aluminum. To solve this problem, measure the aluminum level at the entry to the distribution system, multiply the amount by four, and add the result to the desired operating orthophosphate residual. Calculating the orthophosphate dose in this manner will give you a sufficient residual of orthophosphate for effective corrosion control in the distribution system.

7.7.5　Public Education and Reporting Requirements

The LCR has public education and reporting and recordkeeping requirements. An EPA public education program must be conducted if a utility does not meet the lead action level. Under the LCR, utilities must produce reports for various programs, treatment, and activities, and keep complete records of sampling and testing data and other applicable information.

7.7.5.1　Public Education

If a utility fails to meet the lead action level, an EPA-developed public education program must be initiated. The program for community water supplies includes mail-out notices in the utility bills;

NPDES permit

A National Pollutant Discharge Elimination System permit is the regulatory agency document issued by either a federal or state agency that is designed to control all discharges of potential pollutants from all point sources of pollution and stormwater runoff into US waterways. NPDES permits regulate discharges from industries, municipal wastewater treatment plants, sanitary landfills, large animal feedlots, and return irrigation flows, among others.

announcements in major local daily and weekly newspapers; distribution of pamphlets to schools, health care, and day care facilities; and submittal of public service announcements to at least five of the largest radio and television stations in the area. Each of these actions must be repeated every 12 months as long as the lead action level is exceeded, with the exception of the radio and television announcements, which must be distributed every 6 months. For noncommunity systems that exceed the lead action level, information posters must be placed in public locations and brochures must be distributed to each person served by the system.

In any routine 6-month monitoring period that the lead action levels are complied with, no public education activity is required. The public education programs are designed to provide the consumer with information on how to minimize their exposure to lead in the water through such activities as:

- Flushing the taps before use
- Cooking with cold rather than hot tap water
- Checking new plumbing for lead solder
- Testing their water for lead

Utilities that do not meet the lead action level must offer to sample the tap water of any consumer who requests it, although the consumer can be charged for the cost of the sample collection and analysis. Mandatory health effects language (a portion of which was quoted in Section 7.7.1) is provided by the EPA and required to be used in the public education program. Copies of the public education materials can be obtained from the EPA, state regulatory agencies, or AWWA.

The revised regulation requires that all utilities must provide a notification of tap water monitoring results for lead to owners and/or occupants of homes and buildings who consume water from taps that are part of the utility's sampling program.

7.7.5.2 Reporting and Recordkeeping Requirements

Under the provisions of the Lead and Copper Rule, utilities must provide separate reports for tap water monitoring programs, source water monitoring, corrosion-control treatment, source water treatment, lead service line replacement activities, and public education programs. All utilities are required to keep complete, original records of sampling data, test results, reports, surveys, letters, evaluations, schedules, state determinations, and any other information required to comply with the LCR for at least 12 years.

Check Your Understanding

1. List the important elements of the 1991 Lead and Copper Rule.
2. What is the definition of an action level in the Lead and Copper Rule?
3. What are the accepted potable water corrosion-control treatments?
4. How do chemical inhibitors control lead and copper in water distribution systems?
5. How must a utility attempt to educate the public if it fails to meet the lead action level?

7.8   Math Assignment

Turn to Appendix A, "Introduction to Basic Math for Operators." Read and work through the example problems in sections:

- "Learning Objectives"
- A.7, "Pumps"
- A.9.9, "Corrosion Control" (English System)
- A.10.9, "Corrosion Control" (Metric System)

Check the math using a calculator. You should be able to get the same answers.

7.9   Additional Resources

Your local library may be a good source for obtaining these additional resources related to the field of water treatment:

Manual of Water Utilities Operations (Texas Manual). Texas Water Utilities Association, www.twua.org. ASIN: B000KT9EIU. Chapter on special water treatment.

External Corrosion: Introduction to Chemistry and Control (M27). American Water Works Association (AWWA), www.awwa.org. ISBN: 1-58321-300-7.

Standard Methods for the Examination of Water and Wastewater, 22nd Edition. American Public Health Association (APHA), American Water Works Association (AWWA), and the Water Environment Federation (WEF). www.awwa.org. ISBN: 978-0-87553-013-0.

"Corrosion Control by Deposition of $CaCO_3$ Films, A Practical Approach for Plant Operators," by Douglas T. Merrill and Robert L. Sanks. *Journal American Water Works Association;* Part 1, November 1977, pages 592–599; Part 2, December 1977, pages 634–640; and Part 3, January 1978, pages 12–18.

Another excellent source of information about all aspects of corrosion control is the National Association of Corrosion Engineers (NACE). NACE's website (www.nace.org) offers a variety of information, including technical papers on corrosion-related topics, testing and materials standards, abstracts of publications, and training and certification opportunities.

Key Term Matching

_____ 1. Galvanic Cell

_____ 2. Amperage

_____ 3. Coupon

_____ 4. NPDES Permit

_____ 5. Electrolytic Cell

_____ 6. Sacrificial Anode

_____ 7. Electromotive Force

_____ 8. Electrolyte

_____ 9. Langelier Index

_____ 10. Representative Sample

A. The strength of an electric current measured in amperes.

B. A steel specimen inserted into water or wastewater to measure corrosiveness.

C. A special solution capable of conducting electricity.

D. A device in which the chemical decomposition of material causes an electric current to flow.

E. An electrolytic cell capable of producing electric energy by electrochemical action.

F. An index reflecting the equilibrium pH of a water with respect to calcium and alkalinity.

G. The regulatory agency document issued by either a federal or state agency that is designed to control all discharges of potential pollutants from all point sources of pollution and stormwater runoff into US waterways.

H. A sample portion of material, water, or waste stream that is as nearly identical in content and consistency as possible to that in the larger body being sampled.

I. An easily corroded material deliberately installed in a pipe or tank.

J. The electrical pressure available to cause a flow of current (amperage) when an electric circuit is closed.

7.1 Adverse Effects of Corrosion

11. What is not an adverse effect caused by corrosive waters?

  1. Health hazards from corrosion of toxic metal pipe materials such as lead

  2. Increase in distribution system pressures

  3. Increase in pump energy costs from tuberculation

  4. Leaks in water mains and water main replacement

12. Corrosion in the water treatment industry is the gradual decomposition or destruction of a material by _____, often due to an electrochemical reaction.

7.2 Corrosion Process

13. If a reservoir has been treated with bluestone (copper sulfate) to control _____, the copper might plate out on steel pipes in the distribution system. The metallic copper could act as a cathode in a galvanic cell and _____ the pipes.

14. What is the corrosion potential of a a steel water tank that has been protectively lines with a paint coating connected to a large copper distribution system line?

    1. High
    2. Low
    3. Moderate
    4. Variable

7.3 Factors Influencing Corrosion

15. Which of the following is not a physical factor that influences corrosion?

    1. Alkalinity
    2. System pressure
    3. The presence of stray electric currents
    4. Type and arrangement of materials used in the system

16. Which item is not a problem caused by scaling in hot water systems?

    1. Clogged hot water lines and fittings
    2. Heating element failure due to overheating
    3. Increased energy use
    4. Stronger hot water heating elements

17. Some dissolved solids are involved in scale formation, possibly _____ the rate of corrosion if a protective film is formed.

18. Where can corrosion-causing bacteria be particularly troublesome in distribution systems?

    1. Building connections
    2. High-flow areas
    3. Low-flow areas
    4. Water meters

7.4 How to Determine if Corrosion Problems Exist

19. Which activity is not a way for operators to determine if the water from their treatment plant is causing corrosion problems?

    1. Determine if the distribution system has a decreasing number of leaks
    2. Examine materials removed from your distribution system for signs of corrosion damage
    3. Notice that consumers are complaining about dirty or red water
    4. Run chemical tests on the water to indicate the corrosiveness of a water

20. A flow test measures the _____ in water flow through a 20-foot (6-meter) length of half-inch (12.5 mm) standard black iron pipe under a constant head of 1 foot (0.3 m). The flow rate of highly corrosive water will be _____ by up to 50 percent in 2 weeks.

21. What chemical test is not helpful to determine the corrosiveness of water?

    1. Measure the calcium carbonate saturation using the Langelier Index
    2. Measure the calcium carbonate saturation using the Marble Test
    3. Measure the copper sulfate concentration
    4. Test for heavy metals

22. What is the meaning of a negative Langelier Index?

    1. A negative Langelier Index means that the water is corrosive
    2. A negative Langelier Index means that the water is not corrosive
    3. A negative Langelier Index means that the water is supersaturated with calcium carbonate
    4. A negative Langelier Index means that the water is potable

23. Find the Langelier Index of a water at 25°C having a TDS of 100 mg/L, alkalinity of 100 mg/L, and a calcium hardness of 50 mg/L. The pH is 8.2.
    1. 0.17
    2. 0.27
    3. 0.37
    4. 0.47

7.5 Methods of Controlling Corrosion

24. What is not a reason to handle caustic soda with great care?
    1. Dissolves human skin
    2. Expensive
    3. Produces heat when mixed with water
    4. Reacts with amphoteric metals (such as aluminum) generating hydrogen gas, which is flammable and may explode

25. When selecting corrosion-control chemicals, which factor is not important?
    1. Ambient temperature
    2. Characteristics of the water
    3. Costs of the chemicals
    4. Where the chemicals can be applied

26. Cathodic protection is a process used to _____ or _____ corrosion of metal exposed to water or soil. This process consists of _____ the electro-chemical force to check the natural destruction to buried metals.

27. Galvanic corrosion of _____ under homes with _____ occurs when steel is embedded in concrete and it assumes the characteristics of a noble metal.

7.7 The Lead and Copper Rule

28. Which monitoring location is not a requirement for the Lead and Copper Rule?
    1. Homes with lead pipes
    2. Homes with lead service lines
    3. Homes with lead solder installed after 1982
    4. Homes with lead solder installed before 1982

29. If the _____ value in drinking water exceeds the action level, the utility must begin implementation of a _____.

30. What activities could a utility suggest in its public education program to minimize the consumers' exposure to lead in their drinking water?
    1. Checking new plumbing for lead solder
    2. Cooking with cold rather than hot tap water
    3. Flushing the taps before use
    4. Testing their water for copper sulfate

Taste and Odor Control

LEARNING OBJECTIVES

1. Identify causes and locate sources of tastes and odors.
2. Prevent the development of tastes and odors.
3. Treat or eliminate tastes and odors.
4. Develop a taste and odor control strategy with a taste and odor monitoring program.

KEY TERMS

adsorption	decomposition	**NPDES permit**	septic
aeration	degasification	olfactory fatigue	slurry
algae	diatoms	organic	superchlorination
anaerobic	eutrophication	oxidation	supernatant
bacteria	flagellates	ozonation	supersaturated
biological growth	hydrophobic	percolation	threshold odor number
carcinogen	imhoff cone	phenolic compounds	(TON)
chlorophenolic	inorganic	photosynthesis	toxic
chlorganic	Kjeldahl nitrogen	plankton	volatile
cross-connection	microbial growth	reduction	
decant water	microorganisms	respiration	
dechlorination	nephelometric	saturation	

8.1 Effects of Taste and Odor Problems

Tastes and odors in drinking water are among the most common and difficult problems that confront waterworks operators. In a nationwide survey, 20 percent of the people served by municipal water systems rate their water as having an objectionable taste or odor. A 1989 American Water Works Association (AWWA) survey found that the primary objectionable odors were described as chlorinous, earthy, fishy, and medicinal (AWWA, 1996). Taste and odor (T&O) problems may occur locally on a persistent, seasonal, occasional, or infrequent basis. Regardless of the frequency or type of T&O problem that a utility may face, these survey results indicate that T&O problems are widespread. Most water treatment plant operators will have to deal with T&O problems during their careers.

Taste and odor, along with colored water complaints, are the most common types of water quality complaints received by a water utility. This is because the average consumer uses three senses to evaluate water: sight, smell, and taste. If the water looks dirty or colored, smells bad, or has an objectionable taste, the consumer will rate it as poor-quality water. This is true whether or not any health-related problem exists.

Taste and odor problems are likely to have significant effects on a water utility when they occur. First, numerous complaints must be handled. This requires a great deal of staff time, creates a sense of frustration for both the consumer and for office personnel, and may well require the time of operators who could better be used to solve the problem rather than deal with the effects of the problem.

Water with an objectionable taste or odor is not desirable and may cause a number of consumers to purchase bottled drinking water. Other consumers may switch to alternative water supplies, such as old, poorly maintained private wells, which may not be as safe as the public supply. People could use water that has a more pleasing taste or smell than the public water supply, not realizing that this water may be hazardous to their health.

Perhaps the most damaging effect of a T&O episode is the loss of public confidence in the water utility's ability to provide a safe, high-quality water. Serious loss of confidence may later result in funding restrictions and increased public relations problems for the utility.

Because of the widespread occurrence of T&O problems in public drinking water supplies and the serious effects of these types of problems to the utility, T&O control is a vital area in water treatment operator training.

The key to successful T&O control is to prevent tastes and odors from ever developing by treating them before a problem occurs. This means control of **algae** and other **microorganisms** in source waters and treatment plants through preventive treatment.

algae (AL-jee)
Microscopic plants containing chlorophyll that live floating or suspended in water. They also may be attached to structures, rocks, or other submerged surfaces. Excess algal growths can impart tastes and odors to potable water. Algae produce oxygen during sunlight hours and use oxygen during the night hours. Their biological activities appreciably affect the pH, alkalinity, and dissolved oxygen of the water.

microorganisms (MY-crow-OR-gan-is-ums)
Living organisms that can be seen individually only with the aid of a microscope.

Check Your Understanding

1. How frequently may taste and odor problems occur at a water treatment plant?
2. What are the most common types of water quality complaints received by a water utility?
3. What is the most damaging effect of T&O problems for a water utility?

8.2   Causes of Tastes and Odors

Taste and odor problems may arise from diverse causes and may also be the result of a combination of factors. Nonetheless, a general understanding of the conditions that can contribute to T&O problems is useful in trying to prevent and treat water of objectionable quality. According to a 1989 AWWA survey, the major causes of T&O problems are algal blooms in source waters, disinfectants used, and the water distribution system (AWWA, 1996). Determining the cause of a T&O episode may be extremely difficult and, in many cases, no definite answers are ever found.

It is helpful to have a basic understanding of common taste- and odor-producing compounds even though most prevention and treatment efforts are not aimed at those compounds but at the conditions and activities that produce the compounds. Natural conditions or human activities anywhere within the total water supply system can produce tastes and odors. Raw water sources, conveyance facilities, treatment plants, chlorination stations, finished water storage reservoirs, distribution systems, and consumer plumbing have all been identified as potential sources of tastes and odors. Each water system must be evaluated individually because each will have unique characteristics that may significantly affect water quality.

8.2.1   Common Taste and Odor Compounds

Considerable headway has been made in the identification of the T&O compounds that routinely affect the aesthetic quality of treated water. The following are the most common T&O compounds:

- Geosmin—Geosmin is a natural chemical byproduct of various species of *Actinomycetes* and blue-green algae, including *Oscillatoria, Aphanizomenon,* and *Phormidium,* that have been identified as potent producers of geosmin. This compound is associated with earthy odors in the water.
- 2-Methylisoborneol (MIB)—MIB is a natural chemical compound produced by various species of *Actinomycetes* and blue-green algae, including species of *Oscillatoria, Synechococcus,* and *Phormidium,* that have been identified as producers of MIB. This compound imparts a musty odor to the water.
- Chlorine—Chlorine is the most common disinfectant used in the water industry. It is also a common source of T&O complaints that range from bleach to chlorinous and medicinal tastes and odors.
- Chloramines—When ammonia is added to the treatment process, the formation of odoriferous inorganic chloramines is a potential problem. Inorganic materials are chemical substances of mineral origin, such as sand, salt, iron, and calcium salts, whereas organic substances are usually of animal or plant origin. Organic substances always contain carbon. The three main chloramine compounds that can be formed before breakpoint chlorination are monochloramine, dichloramine, and trichloramine. Monochloramines rarely cause T&O problems unless levels exceed 5 milligrams per liter (mg/L). However, dichloramine, which has a swimming pool or bleach odor, has a lower detection level and becomes a problem at levels of 0.9 to 1.3 mg/L. Trichloramine produces a geranium-like odor at levels of 0.02 mg/L or above.

- Aldehydes—The most common odor problem associated with aldehydes is the fruity odor that develops in ozonated waters. Aldehydes are formed as a result of the oxidation of amino acids and nitriles during the treatment process.
- Phenols and Chlorophenols—Phenols in the water supply react with chlorine to form chlorophenols. Pharmaceutical and medicinal tastes and odors are most often associated with chlorophenols in drinking water.

Check Your Understanding

1. List the major causes of tastes and odors in a water system.
2. Where could T&O conditions develop in a water supply system?
3. What two compounds can produce earthy and musty odors in the water?
4. Which inorganic chloramine causes the least odor problems in treated water?

8.2.2   Natural Causes

Natural causes for T&O problems stem from two sources (1) biological growth in source waters, and (2) environmental conditions. Taste and odor problems from biological growth can result from various types of **bacteria** and algae. Taste and odor problems related to environmental conditions can occur from oxygen depletion in the bottom layers of water bodies, from the inflow of pollutants, from algal blooms and diurnal oxygen fluctuations in water, and from natural runoff.

8.2.2.1   Biological Growth in Source Waters

Biological growth is the activity and growth of any and all living organisms. Microscopic organisms that grow in water or in the sediments of lakes, reservoirs, and rivers are major contributors to the tastes and odors experienced by water utilities. Various types of *Actinomycetes*, algae, and bacteria have all been reported as the cause of these problems at waterworks across the nation.

Actinomycetes are a group of filamentous bacteria that grow in sediments, water, and aquatic plant life. They have been identified as one of the sources of earthy-musty tastes and odors in both source waters and distribution systems. Geosmin, an earthy-smelling odor compound, and 2-methylisoborneol (MIB), a musty-smelling odor compound, were first isolated from *Actinomycetes* cultures. Correlation between *Actinomycetes* and T&O events may be difficult to prove because *Actinomycetes* are not detected with routine microbiological water quality tests.

Numerous types of algae are known to produce T&O compounds. The major T&O producing algal groups include blue-green algae (*Cyanophyta*), yellow-green algae (*Chrysophyceae*), diatoms (*Bacillariophyceae*), and dinoflagellates (*Pyrrophyta*). **Flagellates** are microorganisms that move by the action of tail-like projections. The most common odor producers belong to the blue-green algae class, including *Anabaena, Aphanizomenon, Oscillatoria,* and *Microcystis*. The odors produced by these organisms range from earthy-musty to **septic** and are

bacteria (back-TEER-e-uh)
Bacteria are living organisms, microscopic in size, that usually consist of a single cell. Most bacteria use organic matter for their food and produce waste products as a result of their life processes.

septic (SEP-tick)
A condition produced by bacteria when all oxygen supplies are depleted. If severe, the bottom deposits produce hydrogen sulfide, the deposits and water turn black, give off foul odors, and the water has a greatly increased chlorine demand.

dependent on species, density, and physiological state of the population. Blue-green algae usually develop into blooms that float on the top of the water and are visually noticeable. The yellow-green algae, most notably *Dinobryon, Mallomonas,* and *Synura,* produce compounds with odors ranging from cucumbery to fishy. These algae do not need to be present in large numbers and are often overlooked as the source of T&O events. In large quantities, diatoms can also produce noticeable odors. *Asterionella, Cyclotella,* and *Tabellaria* are known to produce fishy odors while *Melosira* and *Fragellaria* produce musty odors. The most common dinoflagellates are *Ceratium* and *Peridinium,* which can produce rotten, septic, or fishy odors when found in large quantities.

Although planktonic algae are usually associated with T&O events, attached algae (periphyton) can also be a significant source of T&O compounds. Attached beds of *Oscillatoria* have been identified in several California lakes as the source of both geosmin and MIB.

Other microorganisms can also produce tastes and odors. Most notably, sulfate-reducing bacteria, a group of anaerobic bacteria, reduces sulfate to hydrogen sulfide, which produces the rotten egg odor associated with highly polluted waters. These bacteria can also create T&O problems in thermally stratified lakes and reservoirs when the bottom layer becomes anaerobic and sulfate in the sediments and decaying organic matter is reduced by these bacteria to hydrogen sulfide.

Microbial populations can contribute to unpleasant tastes and odors in water in two general ways. As microorganisms grow and multiply, they produce metabolic byproducts. These byproducts are released into the water and some may lead to the deterioration of T&O quality. The concentration of compounds produced by microorganisms is generally very low and may be measured in nanograms per liter (parts per trillion parts). However, even such extremely low levels of these materials may result in T&O complaints.

Cellular material of common aquatic microorganisms present in a drinking water supply can also be responsible for T&O complaints. As the microorganisms grow, organic matter accumulates within the cells. As long as the organisms are healthy, these cellular components are retained and usually do not affect the taste or odor of a water. When the population begins to die off, either as a result of natural processes or treatment, the cells rupture and the cellular materials are released into the water. This is one reason why, in some cases, water has better T&O qualities before being treated than after treatment.

This phenomenon (die-off causing tastes and odors) must be considered in developing plans for copper sulfate treatment of reservoirs. Even relatively high plankton counts do not necessarily indicate that treatment is either required or that it will be beneficial. A large total plankton count may be the result of low numbers of many different types of organisms. This can indicate a well-balanced, diversified plankton community that does not pose a threat to T&O quality. Treating with copper sulfate may disrupt this natural balance, allowing an objectionable organism to gain dominance, and may create a T&O problem by causing the release of cellular components from killed organisms that otherwise would not have had a detrimental effect on water quality.

Microbial decomposition of organic matter in a water supply may also create offensive tastes and odors through a combination of both metabolic byproduct formation and release of cellular materials. Following an algal bloom, significant natural die-off of the predominant organism

diatoms (DYE-uh-toms)
Unicellular (single cell), microscopic algae with a rigid, box-like internal structure consisting mainly of silica.

anaerobic (AN-air-O-bick)
A condition in which atmospheric or dissolved oxygen (DO) is not present in the aquatic (water) environment.

plankton
(1) Small, usually microscopic, plants (phytoplankton) and animals (zooplankton) in aquatic systems. (2) All of the smaller floating, suspended, or self-propelled organisms in a body of water.

decomposition
The conversion of chemically unstable materials to more stable forms by chemical or biological action. Also called decay.

may result in the release of cellular materials with objectionable qualities. The resulting tastes and odors can be further worsened by the growth of other microorganisms feeding on the dying algal mass. This secondary growth may lead to the production of obnoxious metabolic byproducts.

While decomposition is a critical step in the continuous cycling of nutrients through nature, it can have serious consequences on water quality. For this reason, raw water treatment programs are only useful if operated to prevent massive algal blooms. If treatments are begun only after large populations have developed, the effect may be to accelerate the decomposition process and worsen a T&O outbreak. Frequent monitoring of both the type and density of plankton populations in source waters will provide the early warning needed for preventive treatment such as the application of algicide. The application of an algicide to source water should be considered when: (1) the dominant algae is identified as a T&O producer, (2) algae is not a dominant organism in the population but is known for its ability to produce potent T&O compounds (for example, *Synura*), and (3) sensory or chemical analysis indicates an increase in the odor of the water.

Check Your Understanding

1. Name two groups of bacteria that are known to produce T&O compounds in water.

2. What common blue-green algae are associated with odors ranging from earthy-musty to septic?

3. What are the two general ways in which microbial populations can contribute to unpleasant tastes and odors in water?

4. Why does water sometimes have better T&O qualities before treatment than after treatment?

5. When should the application of an algicide be considered?

8.2.2.2 Environmental Conditions

The effects of lake and reservoir stratification on water quality were discussed in Chapter 2, "Source Water, Reservoir Management, and Intake Structures." The depletion of oxygen in the bottom layers of reservoirs provides suitable conditions for the growth of microorganisms capable of producing compounds such as hydrogen sulfide, which are very objectionable to consumers. The information in Chapter 2 should be reviewed as it relates to the causes of T&O problems in drinking water supplies drawn from large reservoirs.

The inflow of many types of pollutants, particularly of organic matter and compounds containing nitrogen and phosphorus, may indirectly cause oxygen-poor conditions in reservoirs, ponds, rivers, and canals. Increased nutrient levels (called **eutrophication**) may result from either natural conditions or from human activity within the watershed.

Pollutants themselves normally do not cause the oxygen concentration in the water to decrease. Microorganisms capable of growing on the organic materials (pollutants) are responsible for depletion of the oxygen following nutrient enrichment from runoff. Microbial populations are

eutrophication (YOO-tro-fi-KAY-shun) The increase in the nutrient levels of a lake or other body of water; this usually causes an increase in the growth of aquatic animal and plant life.

relatively low in unpolluted waters, and the rate of oxygen transfer from the atmosphere to the water is sufficient to prevent oxygen-poor conditions from developing. When available nutrient concentrations increase, rapid microbial growth consumes dissolved oxygen at a rate faster than it can be replaced from the air. Oxygen-poor conditions in water following increased nutrient loading are the result of this rapid **microbial growth**.

Blooms of photosynthetic algae resulting from increased nutrient concentrations, suitable water temperatures, and favorable sunlight can cause both oxygen depletion and oxygen supersaturation in water during a 24-hour period. **Photosynthesis** is a process in which organisms, with the aid of chlorophyll, convert carbon dioxide and inorganic substances into oxygen and additional plant material, using sunlight for energy. The process of photosynthesis occurs during daylight hours and results in oxygen being released into the water. Large populations of algae can produce oxygen faster than it can escape to the atmosphere, leading to afternoon dissolved oxygen levels higher than would normally occur in the absence of major algal activity. **Saturation** is the condition of a liquid (water) when it has taken into solution the maximum possible quantity of a given substance at a given temperature and pressure. When dissolved oxygen levels in water exceed normal saturation levels, the condition is known as **supersaturation**.

At night, algal photosynthesis stops and **respiration** begins. In the dark hours, the algae use the available oxygen at a rate faster than it can be replenished from the air. By the early morning hours, almost all of the dissolved oxygen may have been consumed. This pattern of supersaturation and depletion is known as the diurnal dissolved oxygen cycle and is shown in Figure 8.1. (The graph in Figure 8.1 shows the percent of oxygen saturation during a 24-hour period because oxygen saturation changes with both temperature and elevation. At sea level, for example,

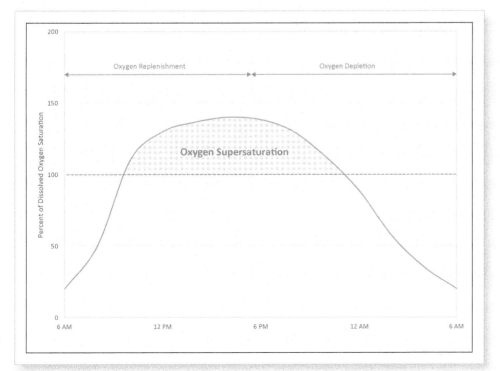

Figure 8.1   Diurnal variation in dissolved oxygen concentrations

microbial (my-KRO-bee-ul) growth
The activity and growth of microorganisms, such as bacteria, algae, diatoms, plankton, and fungi.

respiration
The process in which an organism takes in oxygen for its life processes and gives off carbon dioxide.

oxygen saturation at 50°F or 10°C is approximately 11 mg/L, while at 68°F or 20°C it is only about 9 mg/L.)

The importance of diurnal oxygen fluctuations in water is that significantly different conditions can and do exist from day to night. These differences may have a major effect on T&O quality. Oxygen depletion during the night may result in fish kills and die-off of aquatic organisms and vegetation, which will produce foul tastes and odors in water. Oxygen-poor conditions during darkness may also allow anaerobic organisms to become established and contribute to a general degradation of the aesthetic qualities of the water.

From an operations viewpoint, significant dissolved oxygen fluctuations caused by algae in raw water will also be accompanied by changes in the pH. When algae produce oxygen, carbon dioxide (CO_2) is removed from the water and the pH will increase during the daylight hours. At night, during the respiration process, algae will consume oxygen and release carbon dioxide, which will lower the pH. These changes in pH caused by carbon dioxide fluctuations will influence the chemical doses required to effectively treat the water during coagulation-flocculation, disinfection, and corrosion control processes.

A special case of nutrient enrichment can occur at water treatment plants designed to recycle water used in filter backwashing and settling basin sludge removal. Typically, some type of settling basin or lagoon is provided to allow the sludge to settle before the water is decanted off (separated) and returned to the plant influent. These sludge lagoons produce a water supply in which nutrients previously removed by treatment are concentrated. Microbial populations can flourish in these facilities and, unless careful management is practiced, severe T&O problems can originate from sludge settling lagoons. This is especially true when the **decant water** passes through the plant as a high percentage of total plant flow over extended periods.

Frequent treatment with copper sulfate and rotation of sludge lagoons where multiple units are available, or removal of sludge to separate drying facilities are appropriate methods for controlling the T&O problems associated with these water recovery systems.

Natural runoff may also lead to tastes and odors by substantially increasing flow velocities in rivers and canals. Sediments deposited on the bottom of channels during low-flow periods may be resuspended by scouring action and transported into the treatment plant. When present in the plant influent, bottom sediments may impart septic, musty, or earthy tastes and odors to the drinking water supply. Problems associated with high-flow conditions may subside as runoff decreases, or may persist for several weeks until natural settling redeposits the suspended material on the channel bottom.

decant (de-KANT) water
Water that has separated from sludge and is removed from the layer of water above the sludge.

Check Your Understanding

1. What types of materials in runoff waters can lead to oxygen depletion?

2. How can the nutrient levels in a water supply be increased?

3. What causes the diurnal dissolved oxygen cycle?

8.2.3   Human Causes

Many of our activities in the environment lead to objectionable tastes and odors in public water systems. Increased awareness of environmental degradation has led to many new pollution control regulations at both the state and federal levels. These regulations have significantly improved conditions in many of the nation's waterways, but discharges upstream from municipal intake facilities or into raw water storage reservoirs continue to be contributing factors in many T&O experiences.

Types of pollution that may enter a municipal water supply upstream from the water treatment plant and result in water quality degradation include inadequately treated municipal wastewaters, domestic wastes from individual homes, industrial discharges, urban runoff, chemical spills, agricultural wastes (manure), and irrigation runoff. Tastes and odors may be created by these discharges directly, or may develop because of microbial activities associated with the pollution.

8.2.3.1   Municipal Wastewaters

Inadequately treated municipal wastewaters may enter a water supply due to a process failure at the wastewater treatment plant, mechanical breakdowns, or overloading of the facility. Some older municipal wastewater treatment plants discharge inadequately treated effluent during periods of heavy precipitation because of high flows (inflow and infiltration or I&I) from combined sanitary sewer and storm drain systems. Other wastewater works simply do not have the extensive facilities required to maintain the effluent quality necessary to prevent degradation of the receiving water.

Individual wastewater disposal systems in rural areas of the watershed also contribute pollutants to the water supply. Septic tanks and leach fields may be located too close to rivers for adequate protection of the water supply. Improper siting of individual systems in soils with inadequate percolation rates can create a situation in which nearly untreated wastewater reaches a municipal source.

Poor maintenance and substandard installation are also problems of septic systems that may lead to tastes and odors in community water supplies. Contamination of both surface and groundwater is often associated with individual wastewater disposal system deficiencies in relatively high-density resort and vacation developments.

Municipal and individual wastewater discharges can contribute to tastes and odors in two ways: (1) directly adding odoriferous compounds, such as phenols and aromatic hydrocarbons, to the water, and (2) adding nutrients that result in T&O causing algal blooms. Wastewater discharges have been identified as one of the major sources of phosphorus in surface waters. It is also a significant source of nitrogen. Both nutrients contribute to the eutrophication of a water body and subsequent T&O episodes produced by algal blooms.

8.2.3.2   Industrial Wastes

Industrial discharges sometimes present significant T&O problems for downstream municipal water suppliers. While each individual discharger is required to limit concentrations of chemicals in the effluent according to levels specified by its **National Pollutant Discharge Elimination System (NPDES) permit**, water treatment plants downstream from

percolation (purr-ko-LAY-shun)
The slow passage of water through a filter medium; or, the gradual penetration of soil and rocks by water.

NPDES permit
A National Pollutant Discharge Elimination System permit is the regulatory agency document issued by either a federal or state agency that is designed to control all discharges of potential pollutants from all point sources of pollution and stormwater runoff into US waterways. NPDES permits regulate discharges from industries, municipal wastewater treatment plants, sanitary landfills, large animal feedlots, and return irrigation flows, among others.

heavily industrialized areas may encounter problems arising from the total effects of all facilities that discharge into the supply. Furthermore, industrial wastewater treatment works are just as likely to experience process and mechanical failures as municipal treatment works are.

8.2.3.3 Chemical Spills

Chemical spills into municipal raw water sources can have detrimental effects on T&O quality, but the primary concern of water utility operators in such cases must be the health-related effects due to the toxicity of spilled chemicals. Most spills are the result of accidents at industrial plants, chemical storage facilities, or during transportation. Because chemical spills are unpredictable events, which can lead to a large amount of contamination reaching the water treatment plant in a short time, every water utility should have an Emergency Response Plan to deal with this problem.

(i) For detailed information about planning for emergencies and handling the threat of contaminated water supplies, see the emergency response section of the administration chapter in Volume II of this training manual series.

Check Your Understanding

1. What factors can cause contamination of a water supply by septic tank and leach field systems in rural areas and in resort and vacation developments?

2. How can industrial waste discharges cause taste and odor problems?

3. What are the sources of most chemical spills?

8.2.3.4 Urban Runoff

Urban runoff contributes to tastes and odors, especially in areas where precipitation occurs only during a limited portion of the year. During dry periods, oil, grease, gasoline, and other residues accumulate on paved surfaces. When storms begin, this material is washed into the local receiving water from roadway storm drainage systems. Urban runoff also contains animal droppings from pets and fertilizers used for landscaping. Taste and odor complaints may be received by systems served by water taken downstream from this urban runoff. Usually the complaints slow down after the storm passes, and subsequent storms during the same wet season will typically not create the same degree of problem as the first storm. Nitrate concentrations in runoff can indirectly lead to tastes and odors by increasing nutrient levels in the receiving water bodies, which in turn lead to T&O producing algal blooms.

8.2.3.5 Agricultural Wastes

The contribution to municipal drinking water T&O problems by agricultural runoff depends on the nature and extent of farming in the watershed, precipitation patterns, and local irrigation practices. Many municipal water intake facilities are located upstream from major agricultural areas to avoid possible contamination of the water supply by fertilizers, microbial contaminants (*Cryptosporidium* and *Giardia*), pesticides, and herbicides. This siting of intake facilities above heavily cultivated agricultural lands also helps protect against tastes and odors.

Grazing lands are not usually a major source of tastes and odors in municipal supplies. Significant amounts of waste material that could reach raw water sources cannot accumulate in areas where precipitation occurs regularly throughout the year. Those regions that experience limited seasons of precipitation cannot support high herd densities on grazing lands, so the annual quantity of waste material generated is less per acre. Furthermore, long dry periods allow for the drying out of animal wastes, rendering them far less offensive. When storms do occur, the dried material is generally diluted by the heavy precipitation and wet season river flows.

High-density animal feeding and dairy operations can cause problems if located near surface supplies. The high concentration of animal wastes in a confined area can contribute to significant nutrient loading if runoff is allowed to drain into reservoirs or rivers during a storm. Feedlots and dairies are required to control their discharges under the NPDES program, and utilities should work with the health department and water pollution control agency to prevent serious water quality degradation due to runoff from such operations. Water systems faced with concentrated animal waste runoff into the water source need to practice careful reservoir management and river monitoring programs to prevent massive T&O problems.

Runoff from cultivated fields can contribute both nutrients and objectionable materials to water supplies. Modern, high-intensity farming requires the use of a wide variety of chemicals to achieve maximum crop production. Precipitation or irrigation in excess of the water-holding capacity of the field will lead to runoff that may contain residues of previously applied fertilizers, pesticides, herbicides, and the spreading agent used to apply them.

Often, it will take days for irrigation water to return to the stream or canal. During this time, microbial activity may create very high concentrations of objectionable byproducts. Even if these return flows represent only a small portion of the total supply, the presence of these microbial byproducts in the finished water can lead to consumer complaints of tastes and odors.

8.2.3.6 Treatment Plant and Distribution System Conditions

Inadequate or incomplete maintenance of water treatment plants and distribution systems will result in water quality deterioration no matter how clean the raw water supply may be. Debris and sediments transported to the plant accumulate during the year in areas such as influent conduits and flocculator basins that are not equipped with sludge removal systems. Sludge removal from settling basins with pumping equipment is never 100 percent complete, and deposits will build up over a period of time. Good housekeeping in and around water treatment plants is required to keep the plant in a clean and sanitary condition.

Microorganisms will grow in plant debris and sludges even in the presence of a strong chlorine residual. The conditions that lead to foul, septic, musty, or other types of tastes and odors in raw water supplies may be duplicated on a smaller scale in treatment plants that are not kept clean. In treatment plants with little or no oxidant at the headworks, such as in plants using biological activated carbon filters (BAC), algal growth on the sedimentation basin and filter walls and weirs can cause severe odor problems within the plant. Filamentous algae, such as *Oscillatoria* and *Phormidium,* can colonize these areas and release both geosmin and MIB into the waters. When odors are detected in the treated waters and not in the source waters, a sensory profile of the plant will usually determine the location of the colonized area. Periodic inspection of plant facilities will also help to identify colonized areas within the plant and is a vital part of good water treatment practice and a necessary part of an effective taste and odor prevention program.

Distribution system maintenance is also an important part of T&O prevention. Debris, which accumulates in distribution system mains and laterals, provides an environment for bacterial growth. Especially susceptible are low-flow zones and dead ends in which no chlorine residual is maintained. These areas allow for abundant bacterial regrowth in distribution lines, which results in stagnant, septic, or foul tastes and odors. Comprehensive flushing programs should be used by utilities as part of a system water quality maintenance effort.

8.2.3.7 Household Plumbing

Sometimes a T&O problem is traceable directly to the consumer's plumbing system. The age and types of plumbing materials in older homes may contribute to unpleasant-tasting water. The plastic household plumbing in new housing subdivisions may require several days, or longer, before the plastic taste of the water disappears. Low flows in some homes, or inadequate flushing of lines and cleaning of strainers and aerators may also contribute to water quality degradation in the consumer's plumbing. However, it is poor practice to attribute widespread complaints about the taste and odor of the water to conditions in consumers' plumbing. If a large number of complaints are received from throughout the system, the chances are very good the problem is with the water supply and not the result of a large number of individual problems.

Check Your Understanding

1. Why are many municipal water intake facilities located upstream from major agricultural areas?

2. List some sources of agricultural wastes that may cause T&O problems in a water supply.

3. How do debris and sludge cause tastes and odors?

4. What parts of the plant are likely locations for algal colonization?

8.3   Locating Taste and Odor Sources

A wide variety of conditions that can occur in any portion of a water supply system may cause objectionable tastes and odors at the consumer's tap. When evaluating a T&O complaint, no segment of the system, from the raw water supply to the consumer's household plumbing, should be ignored as a potential source of the problem.

The first step in determining the cause of tastes and odors should be to locate where in the overall system the problem is originating. Once the point of origin is known, it is usually easier to determine an underlying cause and to develop plans for correcting the situation. Locating a T&O source is often a time-consuming process of elimination that may not yield any conclusive information. Nonetheless, the benefits of successfully identifying sources and causes of T&O problems are well worth the effort required.

8.3.1 Raw Water Sources

The most commonly reported problem faced by water facilities is the development of tastes and odors in the raw water supply (a lake, reservoir, river, or canal) or in the raw water transmission facilities that deliver water from the source to the treatment plant. Any parts of the system that are used to store, transport, or regulate untreated water may provide a suitable habitat for organisms that produce objectionable tastes and odors in the drinking water due to an absence of chlorine residual.

When investigating a T&O problem, divide the system into its major component parts based on each component's primary function such as storage, open conveyance channel, and transmission pipelines. Also, consider each component's accessibility for sampling, time required for sampling, and the number of samples that can be reasonably analyzed in a timely manner. You must choose sampling locations where the water is representative of the water consumers will receive. Collecting and evaluating surface samples in a strongly stratified reservoir is useless if the water being treated is released from the lower layers.

Some examples of sampling points that would allow you to test major components of raw water supplies include: outlet works of major reservoirs and regulating basins, inlets and outlets of transmission channels and pipelines, and the plant influent upstream from any chemical additions. Analyze samples from these locations for plankton levels and predominant type, turbidity, pH, taste threshold test, **threshold odor number (TON)**, geosmin, and MIB.* Major changes in any of these water quality indicators between sample locations may be the result of conditions contributing to the T&O problem. Resampling and inspection (if possible) of that portion of the system between sample points showing remarkably different characteristics should be conducted as soon as possible.

Once the origin of the problem is identified, a sanitary survey of the source water and its watershed should be conducted. The purpose of this survey is to locate the biological or industrial sources of the offending taste or odor or the nutrients responsible for a taste- and odor-producing algal bloom. When resampling and inspecting the segment of the system suspected of being the origin of the T&O problem, look for new or expanded residential, commercial, or industrial activity, as well as new or altered tributary streams, which could contribute poor quality inflow to rivers and canals. Also, look for runoff from agricultural areas if the problem occurred after a heavy rain. Examine pipelines for unauthorized or unintentional **cross-connections** that could provide a route for contamination to enter the supply. Analyses for various nutrients such as total phosphorus, total **Kjeldahl nitrogen** (nitrogen in the form of organic proteins or their decomposition product ammonia, as measured by the Kjeldahl Method), ammonia, nitrate, and nitrite may be useful in determining the extent of nutrient inflow into the water body and the potential

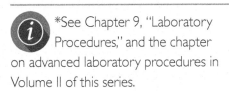

*See Chapter 9, "Laboratory Procedures," and the chapter on advanced laboratory procedures in Volume II of this series.

threshold odor number (TON)
The greatest dilution of a sample with odorfree water that still yields a just-detectable odor.

cross-connection
(1) A connection between a drinking (potable) water system and an unapproved water supply. (2) A connection between a storm drain system and a sanitary collection system. (3) Less frequently used to mean a connection between two sections of a collection system to handle anticipated overloads of one system.

for future algal blooms. This information is valuable in developing a source water management plan.

8.3.2   Treatment Plant

Accumulated debris and sludge in treatment plant facilities will lead to T&O deterioration as the water is processed. Algal growth due to poor housekeeping practices is both unsightly and a potential source of tastes and odors in the finished water. Routine inspection and cleaning of all facilities are necessary elements of treatment plant operation.

Collecting laboratory samples from various points throughout the treatment plant is usually not as productive in locating T&O sources as it is in the raw water system. Treatment chemicals, especially chlorine and powdered activated carbon (PAC), tend to mask any changes in T&O quality that may be occurring within the plant. Comparing T&O quality of raw and finished water may be useful in indicating a problem of plant origin, but there is a natural tendency to always rate treated water as of better quality than untreated. Additionally, the presence of strong, easily treated odors in the raw water may invalidate any direct comparison with finished water. However, if a taste threshold test is conducted, a sensory profile of the treatment plant may reveal the source of a plant-related problem. Plant profiles of geosmin and MIB levels can also be helpful with musty, earthy odors.

Visually inspect basin and filter walls, channels, and weirs for algal or slime (bacterial) growth. Plants without an oxidant at the headworks will often develop problems with attached algae on the sedimentation and filter walls and weirs. As noted earlier, *Oscillatoria* and *Phormidium* are two blue-green algae that readily colonize these areas and can produce both geosmin and MIB. Remove any slime material and send samples to the laboratory for identification. Future growths can sometimes be prevented by regularly washing affected areas with a high-pressure hose. Shock treatment with chlorine through the plant or direct application of a strong chlorine solution to points that prove to be especially difficult to keep clean can also help.

Conduct an evaluation of plant facilities, looking for potential zones of debris accumulation in areas such as influent conduits and flocculator and settling basins. Review records of previous plant inspections and cleanings to see if some areas have a history of particularly heavy sludge buildup. Seldom is it practical to dewater a treatment plant during a T&O episode, but plans should be made for regular (yearly) dewatering operations to allow inspection and cleaning of suspected or potential problem facilities.

Windows in structures over filter galleries may allow sunlight to encourage algal and slime growths. Some plants have corrected this problem by covering the windows.

8.3.3   Distribution System

Conditions within a municipal water supply distribution system can significantly affect water quality received by the public. Just as adequate maintenance and housekeeping are required at the treatment plant, they are a necessary part of distribution system operation. The main causes of tastes and odors in the distribution system are: microbiological activity, disinfection residuals and their byproducts, organic or mineral compounds from system materials, and external contaminants from cross-connections.

Taste and odor complaints originating within the distribution system are usually confined to limited areas or zones. Dead ends, low-flow zones, and areas subject to wide flow variations or changes in supply source may experience higher than normal numbers of T&O complaints. Records of complaints should be reviewed so such areas can be identified and preventive measures, such as more frequent flushing, can be implemented. You should recognize, however, that any change in your source water may result in taste or odor complaints from an area, even in the absence of any actual quality problems.

Cross-connections in the distribution system are potentially very hazardous and can be a source of T&O complaints. A variety of contaminants have been introduced to drinking water through cross-connections, and often the first warning the supplier has of such a condition are the complaints about a musty, septic, or other strange odor in the water. Because of the potential public health hazard, complaints of this nature should receive prompt and careful attention.

Backflow-prevention devices that have been improperly installed or bypassed at industrial plants have resulted in the contamination of numerous water systems. Cross-connections are also made by contractors, landscape workers, and, with disturbing frequency, municipal employees. Careless or unthinking homeowners and amateur repair persons have contaminated systems by creating cross-connections. In many cases, the problem was located because of the number, location, and type of T&O complaints received by the water utility.

Cross-connections commonly occur in the following situations:

1. A sprinkler system using nonpotable water is connected to a potable water supply.
2. A potable water source used as a seal supply is connected to a pump delivering unapproved or nonpotable water.
3. A hose connected to the house is left in a swimming pool. When water is drawn from indoor taps, the hose sucks the pool water into the potable water supply.
4. A hose connected to the house is used to apply chemical fertilizers or pesticides. Without a vacuum breaker or other backflow-prevention device at the house connection, the chemicals will enter the potable water supply.
5. A hose connected to the house is used to flush a car radiator. If the hose is left in the radiator after it has been filled with antifreeze and if there is no backflow protection at the house connection, the antifreeze will enter the potable water supply.

Distribution-related T&O problems can be reduced by making sure that the water exiting the plant is stable and that the disinfectant used does not deteriorate quickly or produce odoriferous disinfection byproducts.

Check Your Understanding

1. What are the most likely sources for the development of tastes and odors?

2. What kind of survey is used to identify the sources of tastes and odors in raw water?

3. How do the chemicals used in a water treatment plant interfere with a search for the source of tastes and odors?

4. Where are the potential sources of tastes and odors within a water treatment plant?

5. What are the main causes of tastes and odors in the distribution system?

6. What types of complaints would alert you to the potential of a cross-connection?

8.4   Taste and Odor Prevention and Monitoring

An important aspect of any taste and odor control program is prevention. When dealing with a T&O problem, or for that matter any water quality problem, an ounce of prevention is worth far more than a pound of treatment. Preventing problems is usually both more economical and more effective than trying to treat for tastes and odors at the plant. Development of an effective T&O monitoring program is a valuable asset, especially to those systems that experience frequent T&O episodes. Such a program is important in predicting the onset of T&O episodes, linking odors to specific algal populations, and assessing the effectiveness of control measures.

A taste and odor monitoring program should include, at a minimum, the following components: (1) routine counting and identification of source water algal populations and attached algae in the plant, and (2) sensory analysis, such as a taste threshold test, of source and distribution system waters, and plant profiles when needed. If appropriate, chemical analysis (closed-loop stripping analysis) of the odor-producing compound can also be a valuable tool in the investigation of T&O problems. The information developed through such a program can be incorporated into a more extensive source water management program.

The taste threshold test is a method that uses a trained panel of individuals to identify, describe, and judge the intensities of T&O qualities in the water. This is the only method that provides a description of the type of taste or odor. This method is being used in addition to the more widely used threshold odor number (TON) and has been found to be a better tool for monitoring and evaluating water treatment processes and distribution waters.

8.4.1   Raw Water Management

As explained in Chapter 2, "Source Water, Reservoir Management, and Intake Structures," the role of raw water management in providing high-quality water to the consumer cannot be ignored. Once a supply has deteriorated, it usually will take a change of seasons before conditions return to a desirable level. If no alternative water supplies are available to a community, deterioration of the raw water source may mean an extended period of poor-quality water, or an extended period of significantly increased treatment costs, or both.

The major techniques and considerations in developing and implementing a raw water management system were described in detail in Chapter 2. If the water supply is a river rather than a reservoir, many of the treatment approaches described for reservoirs will obviously not work. However, monitoring rivers for changing raw water quality and for sources of pollution that may adversely affect a community water supply are both activities that should be conducted as part of a river management strategy.

8.4.2 Plant Maintenance

Settled sludges and other debris, which may be transported to the plant by the raw water, need to be removed on a regular basis. These materials provide local environments in which organisms grow and multiply. If left for an extended time, these areas will become septic and impart a foul taste and odor to the water. Depending on the type of material, the quality of the water, and the source and nature of debris that accumulates, rotten egg (hydrogen sulfide), bitter, musty, earthy, swampy, fishy, or grassy tastes and odors will result.

Many treatment plants schedule annual shutdowns to allow inspection and thorough cleaning of all facilities, especially those that are normally submerged. Other, generally larger, plants are designed so that one part of the plant may be taken out of service for inspection and cleaning while another is still operating. This routine cleaning is commonly done during the winter months when flows are lower and full plant capacity is not needed to meet the community's water demands.

Another aspect of plant maintenance that relates to the prevention of T&O episodes at the consumer's tap is the use of ongoing programs that will ensure that those facilities that may be required to treat a T&O problem are functioning properly at all times. No matter how well managed a surface supply is, or how well the treatment plant is maintained in sanitary condition, surface water conditions will change from time to time and will result in water quality deterioration. If facilities such as PAC feed systems are not in functioning condition when they are needed, they are of no value. Periodic inspection and testing of such equipment is necessary to ensure that it will work properly when needed and that emergency repairs will not have to be made just at the time when the equipment is needed most.

8.4.3 Distribution System Maintenance

The quality of water delivered to the consumer is the result of both adequate treatment and maintenance of that quality through the distribution system. Without procedures to ensure that the distribution system is capable of maintaining water quality, the consumer will not receive high-quality water no matter how well the treatment plant is operated.

Many systems have specified stations within the distribution system that are routinely flushed throughout the year to prevent the development of problems. The locations of these stations and the frequency at which they are flushed is determined from records of complaints and water quality tests. This indicates the importance of good recordkeeping as a tool for water quality management. Records can be used to evaluate the effectiveness of spot flushing, the frequency of flushing, and the need to add or rotate stations during the year.

Annual flushing programs also play a part in maintenance of water quality in the distribution system. Again, records should be used to guide the program. In many cases, it is both impossible and unnecessary to flush all parts of the system every year. Review of records may indicate that most zones of a system only need complete flushing every 3 to 5 years. Such a program, in combination with routine flushing of problem zones, is more economical and as effective as a complete flushing every year.

Flushing alone does not provide an adequate level of protection against the development of tastes and odors in a municipal distribution system. Routine collection of samples for T&O analysis using a sensory method such as the taste threshold test, especially in systems subject to seasonal outbreaks, can provide an early warning of quality deterioration. These samples should be collected when coliform bacteria samples in the distribution system are collected. This is especially important since tests at the treatment plant may not accurately indicate water quality in the distribution system. The higher chlorine residual and shorter chlorine contact time at the plant tend to give the water at the plant better T&O properties than may be encountered in the distribution system.

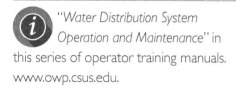

"Water Distribution System Operation and Maintenance" in this series of operator training manuals. www.owp.csus.edu.

Check Your Understanding

1. What two components should be incorporated into a taste and odor monitoring program?

2. What happens when settled sludges and other debris are allowed to accumulate in the bottoms of channels and tanks in a water treatment plant?

3. When are portions of water treatment plants usually taken out of service for inspection and cleaning while the remainder of the plant continues to operate?

4. Why should T&O treatment equipment be capable of operating properly at all times?

5. What sensory method uses a trained panel to identify tastes and odors in both the treatment process and distribution system?

8.5   Taste and Odor Treatment

No single treatment will be applicable to all T&O problems because objectionable tastes and odors are the result of so many different causes and because each water system has its own unique characteristics. Each utility must develop procedures to deal with its problems on an individual basis. Successful problem solving depends on understanding some of the important general properties of the most commonly used T&O treatment methods.

Taste and odor treatment methods can be divided into two broad categories: removal and destruction. Often, both techniques are used at the same time. Multiple treatments, perhaps three or four, may be necessary to produce water of acceptable quality. Removal techniques include optimum coagulation/flocculation/sedimentation, degasification, and adsorption. Destruction of tastes and odors is accomplished by various methods of oxidation.

8.5.1   Improved Coagulation/Flocculation/ Sedimentation

Depending on the type of taste and odor, and on the raw water quality, improving sedimentation and associated processes may produce a better tasting water. This is especially true if the T&O quality has deteriorated during a period when changes in raw water turbidity, color, or pH have suddenly occurred. Such changes might occur during spring or fall turnover of a lake or reservoir, or during high flows in rivers and canals due to storm runoff. Increases in color and pH may also be the results of an algal bloom, in which case algal levels in the raw water can be expected to increase.

As discussed in Chapter 4, "Sedimentation," turbidity, color, and pH can all have a significant effect on coagulation. In addition to removing suspended particulates, inorganic salts, such as alum, have been shown to reduce the organic content of water, though turbidity and organic removal may occur at different dosages. To determine if increased coagulant dosages will reduce objectionable tastes and odors, the standard jar test procedure can be used followed by tests for both settled water turbidity and threshold odor number.

In cases where tastes and odors increase because of increased algal populations, successful attempts to improve coagulation and sedimentation may produce longer filter runs. In plants where chlorine is added at the headworks, an increase in odor levels may occur due to the destruction of algal cells by the chlorine. The reaction of chlorine on algal cells results in the release of objectionable cellular materials into the water. Further reaction between these cellular products and chlorine may produce chlororganic compounds, which impart an even more objectionable taste to the water. When possible, suspension of chlorine at this location and removal of algae in the sedimentation basin will reduce the overall odor levels going on to the filters. Removal of the algae in the sedimentation basin rather than by filtration will also increase filter runs. In plants where chlorine is applied to the water just upstream from the filters, removal of the organisms by sedimentation may improve T&O quality by reducing the action of chlorine on the algal cells that would otherwise be trapped on the filters.

The degree of success that may be obtained by improving coagulation may be difficult to assess, especially if the problem is of short duration. However, if it solves the problem, improving coagulation may be the simplest and most economical approach available to the water treatment plant operator for controlling tastes and odors.

degasification (DEE-gas-if-uh-KAY-shun)
A water treatment process that removes dissolved gases from the water. The gases may be removed by either mechanical or chemical treatment methods or a combination of both.

adsorption (add-SORP-shun)
The gathering of a gas, liquid, or dissolved substance on the surface or interface zone of another material.

oxidation
Oxidation is the addition of oxygen, removal of hydrogen, or the removal of electrons from an element or compound; in the environment and in wastewater treatment processes, organic matter is oxidized to more stable substances. The opposite of *reduction*.

chlororganic (klor-or-GAN-ick)
Organic compounds combined with chlorine. These often toxic compounds generally originate from chlorinating water containing dissolved organic substances or microbes such as algae.

Check Your Understanding

1. List the two broad categories of taste and odor treatment methods.

2. How does the use of chlorine influence taste and odors when water contains algae?

3. How can tastes and odors caused by algae be removed most economically in a water treatment plant?

8.5.2 Aeration Processes and Systems

The use of **aeration** in T&O control is effective only in removing gases and organic compounds that are relatively **volatile**. In general, volatile compounds will be noticed as objectionable odors, while less volatile compounds are more often associated with objectionable tastes. Aeration is somewhat more successful in treating an odor problem than in treating water with objectionable tastes.

Removal of odor-producing substances that are volatile, as well as other volatile compounds, is known as degasification. Because the compounds being removed exist at a lower concentration in the air than in the water, they will tend to leave the water and move into the air. The more air that is circulated through the water, the greater the amount of objectionable volatile compounds that will be removed from the water. By increasing the aeration rates, the concentration of the objectionable compound present in the water may be reduced to a level at which it no longer causes a problem. As stated above, this process only works if the compound is sufficiently volatile.

Aeration can also destroy some compounds by oxidation. While this can be quite effective for treating **reduced** inorganic compounds, such as ferrous iron or manganous manganese, it usually is not very effective in the treatment of tastes and odors resulting from the presence of nonvolatile organic compounds. Aeration normally does not provide enough oxidation to attack the taste- and odor-producing organic compounds.

Aeration systems are designed to operate in one of two ways: some systems pass air through the water; other systems pass the water through the air. That is to say, in one type of system air is pumped into the flow of water by some type of air pump, while in the other the water is distributed through the air by nozzles or cascades. A process called air stripping combines elements of both techniques by flowing water over columns of support medium while air is introduced into the water through openings at many points within the support system.

When chemicals are used to treat T&O problems, additional treatment by aeration may reduce the chemical dosage needed. The additional benefit to improve T&O quality from aeration is usually a minor consideration.

8.5.2.1 Air Blowers

Air blowers are basically compressors that supply air under pressure to the water. Large volumes of air are pumped into the water, generally through diffusers along the bottom of a trough or channel, and the air

aeration (air-A-shun)
The process of adding air to water. Air can be added to water by either passing air through water or passing water through air.

volatile (VOL-uh-tull)
(1) A volatile substance is one that is capable of being evaporated or changed to a vapor at relatively low temperatures. Volatile substances can be partially removed from water or wastewater by the air stripping process. (2) In terms of solids analysis, volatile refers to materials lost (including most organic matter) upon ignition in a muffle furnace for 60 minutes at 1,022°F (550°C). Natural volatile materials are chemical substances usually of animal or plant origin. Manufactured or synthetic volatile materials, such as plastics, ether, acetone, and carbon tetrachloride, are highly volatile and not of plant or animal origin.

reduction (re-DUCK-shun)
Reduction is a chemical reaction that results in the addition of hydrogen, removal of oxygen, or the addition of electrons to an element or compound. Under anaerobic conditions (no dissolved oxygen present), sulfur compounds are reduced to odor-producing hydrogen sulfide (H_2S) and other compounds. In the treatment of metal finishing wastewaters, hexavalent chromium (Cr^{6+}) is reduced to the trivalent form (Cr^{3+}). The opposite of *oxidation*.

is allowed to rise to the surface. Along the way, gases are exchanged between the air bubbles and the surrounding water.

Efficient degasification requires bubbles of very small size to achieve maximum gas transfer. The small bubble size requires that air diffuser orifices be very small. For a specific volume of air, the small orifice size means that the compressor must supply high delivery pressures. To make air blowers effective as direct aeration units, a large unit may be needed.

Air blowers can serve as very effective mixing devices because small bubble size is not an important consideration for mixing. Air diffusers located on the bottom of a channel can produce significant turbulence. This turbulence can be used to achieve some degree of degasification by exchanging water between the bottom of the channel and the water surface where some volatile compounds escape to the atmosphere. For odor problems caused by highly volatile substances such as hydrogen sulfide, this type of aerator may provide adequate control. However, general application of air blowers for T&O control is usually not a very effective technique and it is not widely used today.

8.5.2.2   Cascades and Spray Aerators

Cascades and spray aerators are termed waterfall devices because they aerate water in a manner similar to waterfalls in rivers. These systems pass the water through the air, as opposed to blower devices that introduce air into the water. Both systems are limited to the removal of readily oxidizable or highly volatile compounds just as are air blowers.

Cascade systems are essentially a series of small waterfalls. The water flows down over a series of tiers, which may have some type of medium to increase turbulence and improve aeration efficiency. A simple cascade system design consists of a series of concrete steps over which the water flows. When the water reaches the bottom, it flows into a collection basin and is routed on through the treatment plant. A simplified diagram of a cascade system is shown in Figure 8.2.

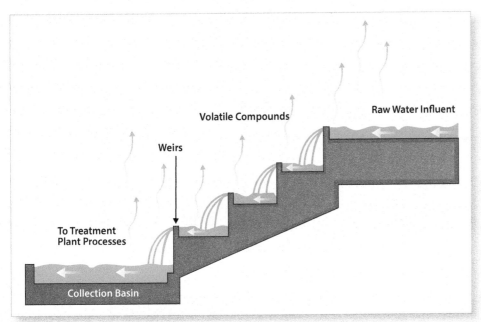

Figure 8.2   Simple cascade aerator system

Spray aerators, as the name implies, spray water through the air to achieve aeration. To be effective, the water must be dispersed into fine droplets. Again, efficiency depends on extensive exposure of the water to air for gas transfer. The number of spray nozzles necessary will depend on plant flow rates, available head, and space limitations. Evaporation losses are a disadvantage of these systems.

Waterfall devices can provide a visual appeal similar to public fountains, which they resemble. However, they may encourage biological growths that may contribute to T&O problems. Continuous copper sulfate treatment has been applied to control such growth at some facilities using these open-water systems.

8.5.2.3 Air Stripping

Air stripping is a process that is used to remove dissolved gases and volatile substances from water. It combines elements of both air blowers and waterfall devices to achieve aeration. While not commonly used at conventional water treatment plants, this method may find future application at facilities that must treat both tastes and odors and other trace organic contamination in drinking water supplies.

Air stripping is achieved by flowing water over a support medium, or packing, contained in a tower while pumping air through the packing in the opposite direction (Figure 8.3). This arrangement of countercurrent (opposite direction) flow provides increased aeration and, therefore, improved removal of volatile substances. Water flow is downward through the support medium. As with other aeration devices, air stripping will only be effective in removing compounds that are highly volatile.

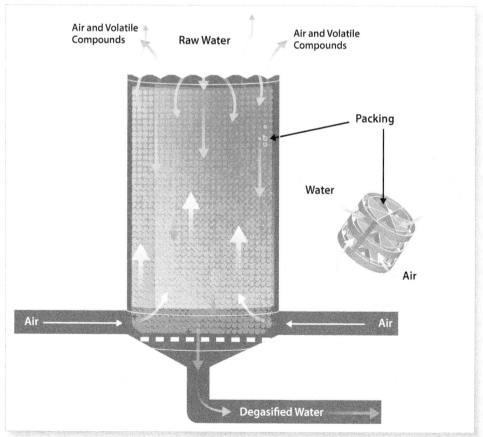

Figure 8.3 Updraft air stripping tower

Check Your Understanding

1. Aeration is best suited for treating what type of taste and odor problems?

2. What are the two basic ways that aeration systems can be designed?

8.5.3   Oxidative Processes

Chemical oxidation is a destructive technique used to control tastes and odors. By application of a strong oxidant, objectionable compounds are chemically modified or broken down into less objectionable byproducts. This method is perhaps the most common T&O control process because of the widespread use of chlorine, a strong oxidant, for disinfection in water treatment.

In addition to chlorine, there are three other chemicals that deserve some attention as oxidants for the destruction of T&O compounds. These are potassium permanganate ($KMnO_4$), ozone (O_3), and chlorine dioxide (ClO_2). All are receiving increased attention in the water industry as alternatives to chlorine at plants that must deal with levels of trihalomethanes in excess of the maximum contaminant level.

8.5.3.1   Chlorine

Chlorine has been used in water treatment in this country for over 100 years, primarily because it is such an effective and relatively inexpensive disinfectant. Because facilities for chlorination are already in place at most water treatment plants, adjustments of chlorine dosages are often used to improve T&O quality. In normal cases, where chlorine is applied in a disinfection dosage range, it will often be necessary to increase the normal chlorine dosage to obtain optimal T&O treatment.

Many odors in raw water are readily treated by the use of chlorine. Easily detectable odors such as fishy, grassy, or flowery odors can often be decreased significantly with normal chlorination dosages or slight overdoses. Iron and sulfide can also be treated effectively by chlorination. In plants where prechlorination is practiced, it may not even be necessary to change routine treatment to produce a high-quality water. In cases where prechlorination is not routinely applied, some tastes and odors may be eliminated by adding chlorine early in the treatment processes. The success of chlorination as a T&O treatment will depend on the type of odor, the seriousness of the problem, the dose applied, and the contact time between chlorine and the water before delivery to the consumer.

Increased chlorine doses, including **superchlorination**, have been used successfully to treat some difficult T&O problems. Higher doses are most beneficial when applied to the plant influent. This allows the greatest contact time and the use of higher doses without adversely affecting consumers located near the treatment plant. Some plants routinely use superchlorination as the initial step in T&O control. This process, which uses chlorine doses far in excess of the doses required to meet the chlorine demand, increases the likelihood of adequate oxidation of the target compound. Superchlorination is followed by **dechlorination** as part of the treatment for tastes and odors. Dechlorination is the deliberate removal of chlorine from water. The use of high chlorine doses in

superchlorination (SOO-per-KLOR-uh-NAY-shun)
Chlorination with doses that are deliberately selected to produce free or combined residuals so large as to require dechlorination.

pretreatment permits later use of powdered or granular activated carbon to remove excess chlorine.

Despite widespread use and success, chlorine may sometimes be the wrong choice of treatment for a T&O problem. Some compounds may become more objectionable after chlorination than before treatment. This is usually the result of incomplete oxidation or substitution of chlorine atoms onto a taste- and odor-causing molecule. For example, runoff from streets and highways contains gasoline residues, which are **phenolic compounds** (phenols). Chlorination of water containing phenols often results in the production of highly objectionable **chlorophenolic compounds**. They cause noticeable tastes and odors even in concentrations 40 to 200 times less than the original phenol. In such cases, further laboratory testing is needed to determine if the chlorine dosage should be increased or decreased. The standard jar test procedure followed by tests for odor quality can be used to indicate the direction to proceed (usually the dosage can be reduced).

8.5.3.2   Potassium Permanganate

Potassium permanganate ($KMnO_4$) has been used for a number of years in both water and wastewater treatment. Permanganate is a strong chemical oxidizer that can be used to destroy many organic compounds, both natural and manufactured, present in water supplies. Permanganate is also used to oxidize iron, manganese, and sulfide compounds and is often used in conjunction with aeration for the control of these and other taste- and odor-producing substances.

A critical aspect of potassium permanganate treatment is color control. This material produces an intense purple color when mixed with water. As the permanganate ion is reduced during its reaction with compounds that it oxidizes, the color changes from purple to yellow or brown. The final product formed is manganese dioxide (MnO_2), an insoluble precipitate that can be removed by sedimentation and filtration. All of the $KMnO_4$ applied must be converted to the MnO_2 form before filtration. If the purple-to-pink color reaches the filters, it will pass into the clear well or distribution system. This may result in the consumer finding pink tap water, or the reaction may continue in the system and the same conditions as exist with naturally occurring manganese will result (staining of plumbing fixtures).

To prevent this situation, it is necessary to determine the maximum $KMnO_4$ dose that can be safely applied in the plant before starting a permanganate treatment program. To do this, measure the time required for the last trace of pink to disappear from a series of jar tests (see Chapter 9, "Laboratory Procedures," for instructions on conducting jar testing). Mix the test samples for a time similar to the detention time in the sedimentation basin. You should allow for a margin of error in this determination; for example, use a time equal to one-half to three-quarters of the actual detention time when running these laboratory trials. In actual plant operation, many operators control permanganate dose by adjusting the feed rate so that the pink or purple color is visible only to a specific point in the basin. This involves a simple process of observation, and is important when treating water that exhibits major fluctuations in permanganate demand. At night, however, visual observations are more difficult and you may want to use a reduced daytime dosage. Some plants use in-line permanganate analyzers to assist with permanganate dose control (Figure 8.4).

phenolic (fee-NO-lick) compounds
Organic compounds that are derivatives of benzene. Also called phenols (FEE-nolls).

chlorophenolic (KLOR-o-fee-NO-lick)
Chlorophenolic compounds are phenolic compounds (carbolic acid) combined with chlorine.

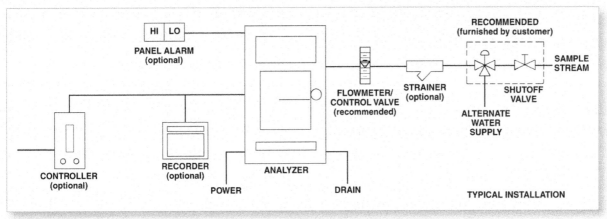

Figure 8.4   Permanganate in-line analyzer *(Courtesy of HACH Company)*

If raw water is transported to the plant through a pipeline or canal, it is best to apply the potassium permanganate at the inlet. This allows an extended contact time for the reaction to produce manganese dioxide and allows somewhat higher permanganate doses (if required for adequate T&O control). This also provides the opportunity to monitor the plant influent for color or soluble manganese, giving the operator an early warning system for a permanganate overdose.

If a permanganate overdose does occur, powdered activated carbon can be used to control the problem until the permanganate dosage rate has been adjusted to the correct level. In some extreme cases, it may be necessary to increase the pH of the water for a short time to increase the rate at which the manganese is precipitated. While it is better to maintain a permanganate dose that will not cause a problem, or to respond by the addition of PAC, pH adjustment to precipitate manganese in the clear well is still preferable to manganese deposition in the distribution system or in the consumers' sinks.

Experience at various water treatment plants has shown that typical permanganate application rates for T&O control are about 0.3 to 0.5 mg/L, though the range reported is from 0.1 to 5 mg/L. The cost of this comparatively expensive chemical must be considered when you are trying to set cost-effective dosage ranges.

At one water treatment plant, potassium permanganate ($KMnO_4$) in the range of 0.3 to 0.5 mg/L has been applied at a river intake structure approximately 13 miles upstream from the plant for T&O control. Permanganate treatment has resulted in a 65 percent reduction in complaints at this facility, which has a seasonal problem with a musty-earthy taste in the water. Detention time between $KMnO_4$ application and the treatment plant is normally more than 6 hours. This allows plenty of time for complete conversion of the permanganate to the manganese dioxide form before the water enters the sedimentation basin.

Potassium permanganate is a dry, crystalline product that is best delivered from a dry feeder into a special mixer immediately before application. Permanganate storage facilities must be dry and well ventilated as moisture in the atmosphere can cause caking of the material. This will clog the feeder and prevent accurate delivery of the permanganate.

Ventilation in a permanganate feeding and storage area is important for both operator safety and equipment protection. Potassium permanganate will produce a very fine dust during loading and handling. This dust is irritating to the eyes, mucous membranes, and even to the skin of some sensitive individuals. When handling permanganate, wear goggles, a dust mask, and gloves. Protective outer clothing (such as rain gear) should also be worn to prevent discoloration* of skin and clothes.

Adequate dust control and ventilation are also important for equipment protection. Permanganate and moisture create a very corrosive mixture that will attack metal, including electrical connections. By ventilating the area during loading and maintaining as dry an environment as possible, the service life and reliability of permanganate feed equipment will be improved.

Never store permanganate in the same area where activated carbon is stored because they are both highly flammable.

*To remove permanganate stains from the skin, a strong solution of sulfite compound (sodium sulfite/meta-bisulfite) works very well.

Check Your Understanding

1. What types of odors can often be decreased significantly by chlorination?

2. Under what circumstances might the use of chlorine be the wrong treatment for a taste and odor problem?

3. How would you respond to a permanganate overdose?

4. Why must permanganate storage facilities be dry and well ventilated?

5. Why are adequate permanganate dust control and ventilation important for equipment protection?

8.5.3.3   Ozone

Ozone (O_3) is an unstable form of oxygen that is an extremely powerful oxidant. The compound in its pure state is a bluish gas that has a pungent, penetrating odor; it can sometimes be detected around large electric motors. Ozone is produced by passing dry air, or oxygen, through a high energy ionizing unit known as an ozonator. Because it is unstable, ozone must be generated on site for use in water treatment.

Ozone is a stronger oxidant than chlorine and, therefore, destroys a wider range of organic compounds. Ozone has a distinct advantage over chlorine treatment for taste and odor in that objectionable byproducts of the reaction do not normally form. This is of particular advantage when the taste and odor is of industrial origin because the combination of chlorine and some industrial pollutants can lead to more intense tastes and odors than those caused by the original compounds. In addition, ozone has received much attention in recent years as an agent for oxidizing organic contaminants present in some water supplies and as a means of disinfecting water without producing trihalomethanes (THMs).

Production of ozone on site for water treatment requires specialized equipment. The basic elements of an ozonation system include a source of dry air or oxygen, a condenser and dryer to remove traces of moisture in the feed gas, the ozonator, an enclosed contactor unit where the ozone is mixed with the water, and a means for venting or recycling waste gas from the contactor. These units are shown schematically in Figure 8.5.

Because of the specialized equipment requirements and costs, laboratory and pilot-scale testing of the effectiveness of ozonation as a T&O

ozonation (O-zoe-NAY-shun)
The application of ozone to water for disinfection or for taste and odor control.

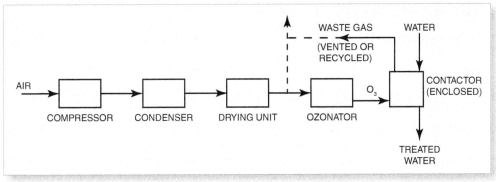

Figure 8.5   Processes in ozonation

control process or for elimination of THMs should be conducted before full-scale installation. Utilities considering purchase of an ozonation system should contact manufacturers or consulting engineering firms for assistance in the evaluation and design of ozone systems.

8.5.3.4   Chlorine Dioxide

Chlorine dioxide (ClO_2) is another chemical oxidant that has received increased attention in recent years as a result of concern about the formation of trihalomethanes in water following chlorination. For more information, see Chapter 6, "Disinfection," Section 6.3.5, "Chlorine Dioxide (ClO_2)," and Section 6.7, "Chlorine Dioxide Facilities." Chlorine dioxide, like ozone, is a strongly oxidizing, unstable compound. Chlorine dioxide is formed by reacting sodium chlorite and chlorine in a special ClO_2 generator. Because it is unstable, it too must be generated on site at the time it is to be applied to the treatment process. Chlorine dioxide has been used to treat tastes and odors caused by industrial pollution, especially in cases where chlorine has intensified the problem.

While chlorine dioxide has been reported to reduce T&O complaints in some cases, other agencies have experienced more severe problems following ClO_2 treatment. The Louisville (Kentucky) Water Company experienced sharp increases in T&O complaints during initial trials with ClO_2 as a method for lowering distribution system trihalomethane (THM) levels. The problem was resolved by using ClO_2 at the plant influent followed by combined chlorine treatment of water flowing into the clear well.

A further consideration in the use of chlorine dioxide is the formation of unwanted chlorite and chlorate ions. There is concern over the possible health-related effects of these reaction products. Limitations for allowable concentrations have been imposed that may prevent the use of chlorine dioxide at doses adequate to control tastes and odors.

Check Your Understanding

1. Why should laboratory and pilot-scale testing of the effectiveness of ozonation as a taste and odor control process be conducted before full-scale installation?

2. Why must chlorine dioxide be generated on site?

3. What types of taste- and odor-producing wastes have been treated successfully by using chlorine dioxide?

8.5.4   Adsorption Processes

Adsorption is the gathering of a gas, liquid, or dissolved substance on the surface or interface zone of another material. This process removes T&O-producing compounds because the compounds attach themselves to a material added to the water for that purpose. In water treatment, this is accomplished in one of two ways: the addition of powdered activated carbon (PAC) to the treatment process, usually at the influent; or the use of granular activated carbon (GAC) as a filter medium. The material being removed by adsorption is known as the adsorbate, and the material responsible for the removal is known as the adsorbent.

The primary adsorbents for water treatment are the two types of activated carbon mentioned above: powdered and granular. These materials are activated by a process involving high temperature and high-pressure steam treatment. The original source of the carbon may be wood, coal, coconut shells, or even bones. The purpose of the activation process is to significantly increase the surface area of the particles so that more adsorption can take place per pound of carbon. The surface area is increased during activation by the formation of holes and crevices in the carbon resulting in particles that have a very porous structure. The surface of activated carbon may range from 2 million to 9 million square feet per pound (400 to 2,000 square meters per gram). This large surface area is responsible for the high degree of effectiveness that can be achieved with this remarkable substance.

When evaluating carbon treatment, consider that the nature of the porous structure will exert a significant effect on the success of the treatment. If the pores of the carbon are too small, the compounds that are being treated will not be able to enter the structure and only a small portion of the available surface area will be used. Activation processes produce carbons with different surface areas. Because adsorption is a surface phenomenon, carbons with greater surface area generally provide greater adsorptive capacity.

Activated carbons are typically rated on the basis of a phenol number or an iodine number. The higher the value, the greater the adsorption capacity of the carbon for phenol or iodine. This is an excellent approach for evaluating the effectiveness of a carbon for removal of phenol or iodine, but there may be no direct relationship between the compounds causing T&O problems and the phenol or iodine of the test. Only by testing various carbons for effective removal of the objectionable taste or odor can a good comparison be made for a particular application.

8.5.4.1   Powdered Activated Carbon

Powdered activated carbon adsorption is the most common technique used specifically for T&O control at water treatment plants in this country. This widespread use is due largely to its nonspecific action over a broad range of taste- and odor-causing compounds. While useful in treating many T&O problems, PAC treatment does have limitations, and its effectiveness and required dose rate vary widely from plant to plant.

Powdered activated carbon may be applied to the water at any point in the process before filtration. Because carbon must contact the material to be removed for adsorption to occur, it is advantageous to apply PAC at plant mixing facilities. Powdered activated carbon is often less effective at removing compounds after chlorination, so application upstream from chlorine treatment is desirable. Chlorine will react with carbon and neutralize the effects of both. From an economic standpoint, therefore, it is not good practice to apply chlorine and PAC near the same location.

Powdered activated carbon is often applied at the plant flash mixer. This location provides high-rate initial mixing and the greatest contact time through the plant. Again, both thorough mixing and long contact time improve the effectiveness of PAC. Another common location for PAC application is the filter influent. While contact time and mixing are drastically reduced compared to flash-mix application, this method ensures that all of the water passes through a PAC layer before release into the distribution system. This procedure is sometimes used with $KMnO_4$ treatment to prevent colored water problems in the distribution system.

When PAC is used, dosages may range from 1 to 15 mg/L. Some reports in the literature have indicated that as much as 100 mg/L have been required to adequately treat some serious T&O problems. At very high doses, treatment costs become prohibitive, and consideration of granular activated carbon treatment in its own contactor (filter) is warranted.

POWDERED ACTIVATED CARBON FEED SYSTEMS

Powdered activated carbon feed systems may be either dry-type feeders or **slurry** feeders. For small-scale applications, the dry-type feed system may be satisfactory, especially if it is only used for short-term, occasional incidents. More frequent PAC treatment requirements or larger applications are usually best accomplished with slurry systems.

Carbon slurry tanks require continuous mixing to prevent settling and caking of the carbon. Typical installations provide a vertical bi-level (two-blade) paddle mixer and vertical wall-mounted baffles for mixing. One mixer blade is located near the bottom of the tank and the other is situated about mid-depth. Carbon tanks should have drain sumps, but the suction side of the carbon intake should not extend into the sump to avoid foreign material that, during loading operations, can find its way into the tank through the dump chute or in the material itself. At one water treatment plant, carbon tanks have been found to contain sand, wood chips, plastic, and paper after having been thoroughly cleaned before reloading.

Because PAC has a tendency to cake, even when continuously mixed, and because of the variety of miscellaneous materials found in carbon tanks, standard practice at some plants is to clean each carbon slurry tank before receipt of a new shipment of carbon. The cleaning procedure allows a visual inspection of the tank, including the mixer blades and baffles.

Powdered activated carbon is a somewhat **hydrophobic** (has a dislike for water) material and does not mix readily with water. When carbon is being loaded into slurry tanks, an overhead spray system in the tank should be used. This spray accomplishes some degree of dust control as well as an initial wetting of the carbon. To load a PAC tank, fill it about onequarter to one-third full of water before starting, and operate the mixers during loading.

Proceed slowly enough to allow complete mixing and wetting of the carbon as it is introduced into the tank. If the mixing capability of the tank is exceeded, the PAC will form a cake on the surface of the water. The caked material will prevent additional carbon from being properly mixed, and a thick, tough layer of floating carbon will form in the tank.

slurry
A watery mixture or suspension of insoluble (not dissolved) matter; a thin, watery mud or any substance resembling it (such as a grit slurry or a lime slurry).

Once formed, caked PAC is difficult to disperse into a useful slurry. If PAC does not mix rapidly during loading, stop the operation, check mixers and sprayers, and break up any cake that has formed. The cake can be broken up using breaker bars (crowbars) or wooden paddles (2×4s). If all equipment is operating properly, the loading rate must be reduced to the level at which the system can adequately mix the carbon into a slurry.

Figure 8.6 shows a typical PAC feeder.

Check Your Understanding

1. What two forms of activated carbon are used in water treatment?
2. What terms are used to describe the adsorptive capacities or ratings of activated carbons?
3. Why should PAC be applied at the plant flash-mixing facilities?
4. What would you do if a caked layer of carbon starts to form on the surface of the water in the slurry tank?

Figure 8.6 Powdered activated carbon (PAC) feeder *(Courtesy of Eagle Microsystems, Inc.)*

olfactory (all-FAK-tore-ee) fatigue
A condition in which a person's nose, after exposure to certain odors, is no longer able to detect the odor.

POWDERED ACTIVATED CARBON DOSE DETERMINATION

The appropriate PAC dose for any particular problem will vary depending upon the nature of the problem, the concentration of the material to be removed, the mixing available, the contact time, and the location of application points. Jar tests should be used to determine the necessary PAC doses required to treat the specific T&O problem that exists. By simulating mixing speeds and detention times of the various locations at which the carbon might be applied, and by varying the doses of carbon, the jar test can be used to indicate the most effective range of carbon treatment. (See Chapter 9, "Laboratory Procedures," for instructions on conducting jar testing.)

Accurate determination of the threshold odor number (TON) of a sample is difficult to make when several jar tests are run all at once. A person's sense of smell becomes rapidly fatigued and after just two or three TON tests individuals tend to become desensitized to the odors present. Because only a limited number of operators are usually available for odor testing, a single operator will often have to do all of the tests. Misleading results may be obtained if more than a few TON determinations are made by the same person in a short time. Also, if the air around the plant contains the odor of concern, none of the people at the plant site will be able to perform the TON test. When your nose smells the same odor for a long time, eventually you will be unable to detect that odor due to **olfactory fatigue**.

A more practical approach, especially during initial testing, is to compare the relative (subjective) odor of jars dosed at different PAC levels. (This same method can be applied to the preliminary evaluation of all T&O treatments.) After mixing for an appropriate time, collect undiluted samples from each jar. Starting with the sample from the jar test treated at the highest level, test (smell) each sample in order of descending treatment dose and note your initial reaction to any odor present using descriptive terms such as none, slightly fishy, or foul. After having tested all samples, rate the various treatments from best to worst in terms of odor removal. An example of the results from this type of comparative test are shown in Figure 8.7.

The test shown in Figure 8.7 included alum and polymer treatment. This test was conducted to evaluate PAC application at the flash mixer

DATE _____ TIME _____ Operator: _____

SOURCE _____
TEMP _____ °C pH _____
TURBIDITY _____ NTU COLOR _____
ALKALINITY _____ mg/L HARDNESS _____ mg/L
_____ _____

	RPM	TIME
1.	Max	2 min
2.	85	8 min
3.	40	16 min
4.	___	___

	Chemical (mg/L)	JAR 1	JAR 2	JAR 3	JAR 4	JAR 5	JAR 6
1.	Alum Liquid / Dry	6.0	6.0	6.0	6.0	6.0	6.0
2.	Cat Polymer	1.0	1.0	1.0	1.0	1.0	1.0
3.	PAC	0	2	4	6	8	10
4.							
5.							
6.							
Floc Characteristics	After flash mix						
	After rapid mix						
	5 min. slow mix						
	10 min. slow mix						
	15 min. slow mix						
Floc Settling	5 min.						
	10 min.						
	20 min.						
	30 min.						
Settled Water Quality	Turbidity						
	pH						
	Color						
Comments: Odor:		Grassy/ Stale · Dirty	Grass · Dirty/ Dirty	Slightly Dirty	None	None	None
Subjective Rating: (Scale of 1 to 6)		6 (worst)	5	4	1	1 (all similar)	1

Figure 8.7 Jar test for relative odor

where the coagulants are added. If chlorine is added near the PAC delivery point, the jar samples should be treated to reflect this. As in all jar testing, the laboratory trials should simulate plant conditions as closely as possible.

This technique may indicate that PAC treatment alone will not be enough to control the T&O problem, which is certainly valuable information. This method is most useful, however, for deciding whether PAC treatment above a certain level provides any additional benefit to water quality. In the example, all jars treated at 6 mg/L and higher gave similar results, indicating that a higher PAC dose would not be useful. Results such as these provide guidelines for establishing the most effective treatment at the most economical costs.

One difficulty often overlooked in the use of PAC is the problem of measuring the actual concentration (lbs/gal) of carbon in the slurry tank. For convenience in calculating feed rates, powdered activated carbon slurry tanks are normally loaded at a rate of 1 pound of PAC per gallon of water. Unfortunately, off-loading from a bulk delivery into multiple tanks usually results in an unequal division of the carbon among the tanks. The following simple laboratory procedures are used to determine the actual carbon content of the PAC slurry.

Collect a 1-liter sample of the carbon slurry from a recently loaded tank after several hours of mixing. Allow this sample to settle for approximately 2 hours in an **Imhoff cone** and record the amount of settled carbon as mL carbon/liter of **supernatant.**

After the sample has settled, dry portions of the settled carbon in a laboratory drying oven and weigh them. In this manner, you can calculate the dry weight of carbon per milliliter of settled carbon. From this information, it is easy to calculate the amount of carbon per gallon in the slurry (see Example 2 on page 525). Results from several years of testing at a water treatment plant indicated that different carbon manufacturers' carbons have their own characteristic values on a gram-per-settled-mL basis. Generally, multiplying the mL/L of settled carbon by a factor, which ranges from about 0.0022 to 0.0028 at the plant, gives the weight of carbon in the tank, expressed as pounds per gallon.

Powdered activated carbon is an abrasive material. Equipment used to feed PAC often requires more frequent inspection, cleaning, and maintenance than other chemical feed equipment at a water treatment plant to ensure proper dosage delivery. Pumping equipment should be routinely inspected for signs of wear. Check valves and slurry suction lines can become clogged with carbon that has caked in the tank. Frequent inspection during actual carbon delivery is necessary when using this equipment.

FORMULAS

There are two approaches to determining the dosage of PAC in either pounds per gallon or pounds per million gallons. With the first approach, convert the milligrams of PAC per liter of water dose from the jar test to pounds per gallon using conversion factors.

$$\text{PAC}_{desired}\left[\frac{lb}{Mgal}\right] = \text{PAC}_{conc}\left[\frac{mg}{L}\right] \times \frac{3.785\,L}{gal} \times \frac{kg}{10^6\,mg} \times \frac{2.205\,lb}{kg} \times \frac{10^6}{M}$$

$$= 8.34 \times \text{PAC}_{conc}\left[\frac{mg}{L}\right]$$

Imhoff cone
A clear, cone-shaped container marked with graduations. The cone is used to measure the volume of settleable solids in a specific volume (usually one liter) of water or wastewater.

supernatant (soo-per-NAY-tent)
The relatively clear water layer between the sludge on the bottom and the scum on the surface of a basin or container. Also called clear zone.

Where,

$$PAC_{desired} = \text{Desired PAC}$$
$$PAC_{conc} = \text{PAC Concentration}$$

Multiply by 3.785 liters per gallon to convert liters to gallons. Divide by a million (10^6) to convert milligrams into kilograms. Then multiply by 2.205 to convert kilograms into pounds. Finally, multiply by a million (10^6) to convert the gallons into million gallons. Operators typically use pounds of PAC per million gallons because pounds per gallon is a very small number.

The second approach is to refer to the basic loading or chemical feed equation.

$$PAC_{feed}\left[\frac{lb}{d}\right] = Q\left[\frac{Mgal}{d}\right] \times PAC_{conc}\left[\frac{mg}{L}\right] \times \frac{3.785L}{gal} \times \frac{10^6}{M} \times \frac{kg}{10^6 mg} \times \frac{2.205\,lb}{kg}$$

$$= 8.34 \times Q\left[\frac{Mgal}{d}\right] \times PAC_{conc}\left[\frac{mg}{L}\right]$$

Where,

$$PAC_{feed} = \text{PAC Feed}$$
$$Q = \text{Flow Rate}$$

Example 1

Results of jar tests indicate that 5 mg/L of PAC is the most effective dosage for treating a T&O problem. What is the desired concentration in pounds per million gallons (lb/Mgal)?

Known

PAC_{conc} = PAC Concentration = 5 mg/L

Unknown

$PAC_{desired}$ = Desired PAC, lb/Mgal

Convert the PAC concentration from mg/L to pounds/million gallons (lb/Mgal).

$$PAC_{desired}\left[\frac{lb}{Mgal}\right] = 8.34 \times PAC_{conc}\left[\frac{mg}{L}\right] = 8.34 \times 5\left[\frac{mg}{L}\right] = 42\frac{lb}{Mgal}$$

Example 2

A sample of PAC slurry was collected from a slurry tank. The 1-liter (L) sample was allowed to settle for 24 hours (h). The amount of settled carbon was recorded as 50 mL carbon per liter (L) of supernatant. Ten mL of the settled carbon was dried and found to weigh 20 mg. Calculate the amount of carbon in the slurry in pounds of dry carbon per gallon (lb/gal) of water.

Known

V_{sample} = Sample Volume = 1 L
$V_{settled}$ = Carbon Settled = 50 mL
V_{dried} = Portion Dried = 10 mL
M = Dried Sample Weight = 20 mg

Unknown

PAC Slurry, lb/gal

1. Calculate the amount of settled carbon as mg/mL.

$$\text{Settled Carbon}\left[\frac{mg}{mL}\right] = \frac{M[mg]}{V_{dried}[mL]}$$

$$= \frac{20[mg]}{10[mL]}$$

$$= 2\frac{mg}{mL}$$

2. Calculate the PAC concentration in the slurry as mg/L.

$$\text{PAC Slurry}\left[\frac{mg}{L}\right] = \frac{V_{settled}[mL] \times \text{Settled Carbon}\left[\frac{mg}{mL}\right]}{V_{sample}[L]}$$

$$= \frac{50[mL] \times 2\left[\frac{mg}{mL}\right]}{1[L]}$$

$$= 100\frac{mg}{L}$$

3. Convert PAC slurry from mg/L to pounds PAC per gallon (lb/gal).

$$\text{PAC}_{desired}\left[\frac{lb}{Mgal}\right] = 8.34 \times \text{PAC}_{conc}\left[\frac{mg}{L}\right]$$

$$= 8.34 \times 100\left[\frac{mg}{L}\right] = 834\frac{lb}{Mgal}$$

FILTRATION CONSIDERATIONS WITH POWDERED ACTIVATED CARBON

Use of powdered activated carbon for T&O removal may interfere with filter performance at a water treatment plant. Caking of PAC on the surface of filters may cause substantially shorter filter runs than otherwise expected. If this occurs, adjustments to improve PAC removal in the settling process may increase the effective length of filtration. An added advantage of optimizing the settling process is the physical removal of T&O components with the settled sludge.

A second and more difficult problem to detect can occur when applying PAC. Because the particle size of PAC is nearly microscopic, PAC can penetrate through filters before either head loss or turbidity breakthrough indicate the need for backwashing. Because the PAC particles are black, they absorb light. Therefore, standard **nephelometric** turbidity measurements will not warn of the carbon passing through filters. Penetration of PAC through filters can cause dirty water complaints in the distribution system and is particularly apparent to those consumers who have installed home filtration units on their faucets.

Studies conducted at a water treatment plant indicate that carbon penetration through filters is related to both the PAC dose and the hydraulic loading on the filters. At higher flow rates and increased carbon loading on the filters, PAC penetrates through the filter beds much earlier in the filter run than at lower flows or carbon doses.

nephelometric (neff-el-o-MET-rick)
A means of measuring turbidity in a sample by using an instrument called a nephelometer. A nephelometer passes light through a sample and the amount of light deflected (usually at a 90-degree angle) is then measured.

Powdered activated carbon penetration through a filter bed occurs in approximately the same manner as turbidity breakthrough. A freshly backwashed filter may show the presence of PAC during the recovery period. Following recovery, PAC filtration is excellent, though the length of time this performance is maintained will vary. Depending on the severity of the PAC penetration problem, special treatment may be necessary. At one plant, filter-aid dosage of a nonionic polymer in the range of 0.005 to 0.010 mg/L had been used to control PAC penetration during recovery. However, continued application of this level of filter aid throughout the run increased the rate of head loss, so filters required backwashing at about the same rate as before the filter aid was applied.

One simple method of determining carbon penetration through a filter is to collect a 1 liter sample of filter effluent and filter it through a 0.45 µm membrane filter. The presence of PAC in the effluent can be determined from a darkening of the white membrane filter surface.

POWDERED ACTIVATED CARBON HANDLING

Powdered activated carbon may be purchased either in individual bags (usually 50 pounds or 23 kg) or by bulk truck load (up to 40,000 lb or 18,180 kg per load).* Regardless of loading arrangement, by bulk or individual container, PAC is usually an unpleasant material to work with at a water treatment plant. Hand loading of a carbon storage tank from bags is a time-consuming, dirty job. Off-loading of bulk loads is less time-consuming but not necessarily any less dirty.

*Large, 1-ton (910 kg) bladders or carboys of PAC are available in some areas. Special facilities are required for the use of these containers.

Various types of equipment have been used to control the dust generated during PAC handling. The success of any dust collection system depends on how well it meets the particular needs of the facility for carbon handling and on adequate maintenance. Several different approaches to dust control have been used, with more elaborate (and occasionally more successful) designs usually found at larger treatment plants. No matter what type of equipment is used, dust collectors must be clean and functional at the start of loading operations and further cleaning may be required during the operation to maintain dust collection effectiveness.

Operators assigned to carbon loading, especially when bags are used, must wear protective clothing. At the very least, goggles, dust masks, and gloves are required. Protective outer clothing is also highly recommended, and showering facilities should be provided for cleaning up after unloading the carbon.

Because powdered carbon will scavenge (actively remove) oxygen from the air, safety procedures for working in confined spaces must be carefully followed when cleaning tanks. Open the tank fully, set up some means of circulating air in the tank, and work in teams—*never work alone in a powdered carbon tank.* Arrange for a spotter (operator) overhead (outside) to watch the cleaning operation closely and be prepared to assist if the need arises. The operator in the tank should wear a safety harness and the oxygen level in the tank must be monitored continuously.

If bags of PAC are to be stored, elevate them off the floor to prevent caking due to spilled or sprayed water. Wooden pallets work well for this purpose. Carbon is a combustible material, but tends to smolder rather than burn fiercely. A smoldering pallet may be removed by dragging it out of the area, thereby saving the rest of the material from water and fire damage. For this reason, bags should not be stacked more than three or four high on an individual pallet.

Because carbon is combustible, it should be stored away from other materials, especially flammable materials like potassium permanganate ($KMnO_4$) and high test hypochlorite (HTH) compounds. If at all possible, an isolated storage facility is highly desirable. Electrical wiring and switches in a carbon storage area should be of special design for flammable storage area use. Whenever there is dust in the air, a spark could cause an explosion. The area should be prominently marked with No Smoking signs and this regulation should be strictly enforced.

8.5.4.2   Granular Activated Carbon

Granular activated carbon is activated in much the same way as PAC, although the raw material is usually a lignite coal. Granular carbon, as the name implies, is made up of larger particles than PAC. Typical surface area for GAC is in the range of 2 million to 4 million square feet per pound (400 to 800 square meters per gram). The basic adsorption process with granular carbon is the same as with powdered carbon, except that the granular carbon is placed in a stationary bed through which the water flows.

Granular carbon filters are capable of filtering water in the same manner as rapid-sand or dual-media filters, and can produce low-turbidity water when operated under similar conditions. GAC filters, which are also referred to as biologically activated carbon filters because of the high levels of bacteria that will develop within the filter bed, have received attention as a means of treating trace organic contamination of drinking water. Generally, the effective life of a GAC filter being used for T&O control will be much greater than one used for removal of potentially toxic or carcinogenic trace organics.

Operational procedures for GAC filters closely resemble procedures used with other typical water treatment filters. Two unique considerations for GAC filtration are empty bed contact time (EBCT) and the regeneration interval of the carbon. Empty bed contact time is the time that the water is actually in the filter bed (use calculated volume not including carbon when determining contact time). Successful application of GAC filtration for T&O control usually requires a minimum of five minutes, although some cases have required contact times greater than 15 minutes. The EBCT is an important design consideration in GAC units to be used for T&O control because the filters must be large enough to provide adequate contact time as the water is filtered.

Periodic regeneration or replacement of GAC is necessary as the capacity of the filter to adsorb and retain organic compounds decreases with time. The time between regenerations or replacement will vary with the type of material being removed and the volume of water treated. Regeneration is accomplished essentially in the same manner as initial activation, and most utilities replace spent carbon with fresh rather than attempting to regenerate. Regeneration or replacement intervals of 3 to 7 years have been reported by utilities using GAC filters for the control of tastes and odors.

Retrofitting existing filters with GAC is an expensive modification. Initial installation costs are high and long-term expense for regeneration or replacement must be considered. However, in situations where high doses of PAC must be used, or where T&O control is required during a substantial portion of the year, evaluation of GAC filtration may indicate that it is a more economical long-term alternative. Because of the expense, thorough investigation in the laboratory and in pilot-scale tests should be made before plant installation.

toxic
A substance that is poisonous to a living organism.

carcinogen (kar-SIN-o-jen)
Any substance that tends to produce cancer in an organism.

Check Your Understanding

1. What precautions must operators take because powdered activated carbon is so abrasive?

2. What adjustments would you make if PAC tends to cake on the surface of the filters?

3. How can carbon penetration through a filter be detected?

4. What is empty bed contact time in a granular activated carbon filter?

8.6   Identifying Taste and Odor Problems

Before an effective control strategy for a T&O problem can be implemented, the fact that a problem exists must be recognized. This may seem obvious, but delayed response (or no response at all) to the onset of a T&O problem in water systems frequently occurs when the problem is not identified at the time it develops. This is especially true in organizations that have had no previous experience with T&O problems, or in agencies where complaints from consumers are received at one location while operation and water quality activities are conducted at another.

Every water utility must have an effective communications network to deal with all types of consumer complaints. No utility can expect to respond in a timely manner to legitimate consumer dissatisfaction with the water unless it operates under adequate complaint recording and notification procedures.

A good practice is to supply any personnel who deal with the public, either by telephone or in person, with a standard water quality complaint form (Figure 8.8). The form should include, at a minimum, a place for recording the date and time of the complaint, the location (address), and as complete a description of the problem given by the consumer as is possible to obtain.

Treat dissatisfied consumers courteously and respectfully, and consider them a valuable resource for the utility in its efforts to provide a high-quality water for public consumption. If the problem has already been identified by the utility, personnel involved in public contact should be provided with a clear, easily understood explanation of the problem and the actions being taken to correct the situation. If the problem is likely to be corrected in the near future, a date at which the consumer can expect an improvement in the water should be provided.

Notification of the proper personnel within an organization when a complaint has been received is vital to the utility's ability to respond to the problem. When T&O complaints are received, it is important that the operators and those responsible for water quality control be provided with the information. Unless the people who are in a position to take action are aware of a problem, no solution is likely to be found. A standard notification procedure for complaints should be established and followed for providing the necessary information to the appropriate units within the organization.

In many cases, it is not necessary to rely on consumer complaints to identify a T&O problem or to anticipate the onset of a T&O episode. Seasonal tastes and odors in the source water plague many utilities. By keeping complete and accurate records of T&O outbreaks, it is often possible to predict the start of problems and take corrective action before the

WATER QUALITY COMPLAINT FORM			
DATE CALL RECEIVED	CALL TAKEN BY		TIME
TYPE OF COMPLAINT (BACTERIA, COLOR, ODOR, PLANKTON, TASTE, TURBIDITY)			
NAME & ADDRESS OF CUSTOMER MAKING COMPLAINT			
LOCATION OF COMPLAINT			
DESCRIPTION OF COMPLAINT			
COMMENTS OF FIELD INVESTIGATOR			
RESPONSE ACTIONS (INCLUDE DATE, YOUR NAME, & NAME OF PERSON YOU TALKED TO)			
CONCLUSIONS			

Send completed copy to _____.

Figure 8.8   Typical water quality complaint form

consumers are affected. Records must be maintained from 1 year to the next and organized to allow easy retrieval of information.

Routine testing of the raw and finished water for T&O quality is an important, ongoing activity of every water supplier. A taste and odor monitoring program as outlined in Section 8.4 should be implemented. Even if no previous record of T&O problems exists, testing should be carried out as part of a preventive program. If plant operators identify a problem before it reaches the distribution system, action to control objectionable tastes and odors can be taken without exposing the consuming public to unpalatable water. Taste tests of the finished water and threshold odor number determinations of the raw and finished water should be conducted on a regular basis. The frequency of these tests should be increased during periods of the year when tastes and odors have occurred in the past and during any time that T&O treatments are being used.

Check Your Understanding

1. What information should be recorded on a water quality complaint form?

2. How should operators be kept informed of taste and odor problems?

3. What tests can operators conduct to identify T&O problems before they reach the consumer?

8.7   Developing a Taste and Odor Control Strategy

The first steps in developing a T&O control strategy are to recognize that a problem exists and determine the extent of the problem. If only one complaint has been received, the problem may exist within the consumer's home plumbing and no action at the treatment plant can be expected to improve the situation. If several complaints have been received from one area of the distribution system but the rest of the system has not experienced problems, the problem is probably the result of local conditions in the distribution system. Factors that may affect one area of a system include chlorine residuals above or below normal, low flows due to low demands, or a cross-connection. However, because other, more serious water quality problems may be associated with a localized T&O problem, prompt response to such complaints is required.

In systems that use multiple water sources, T&O complaints may arise when a new source of water is provided to an area. The complaints do not necessarily mean one water is better than the other—only that the two are different. This frequently occurs in systems that use both a surface supply and groundwater, although use of two or more different surface supplies may also lead to complaints. When the utility switches back to the original source, more complaints can be expected. In cases where water from two different sources is occasionally introduced into the distribution system at opposite ends of the system, the point in the system where these waters meet can also be a source of odor complaints. Utility personnel who handle complaints must be made aware of the situation so they can explain to consumers that a change in supply does not necessarily mean a change in water quality.

If the entire system is being affected by a T&O problem, it is important to try to determine just where the problem originates. If complaints can be associated with the use of a particular raw water supply, changing to an alternative source may resolve the problem. If an alternative water supply is not available, then an evaluation of conditions that might be responsible for the deterioration of water quality is necessary.

Reviewing water quality test results from different locations within the supply system may indicate that a problem has developed at a particular point. Conduct a physical inspection of the suspected area and correct any conditions that appear to be adversely affecting water quality. The raw water source, storage facilities, raw water transmission systems, and facilities at the treatment plant all should be evaluated as potential sources of the problem. Finished water storage facilities within systems have also been sources of tastes and odors for many water facilities.

Collect samples at a number of locations from both the source and distribution system for analysis in the laboratory. Examine raw water samples for algal composition and density and, using a sensory analysis method such as a taste threshold test, determine whether there has been an increase in source water odors. If the offensive odor is identified in the source water but no taste- and odor-producing planktonic algae is detected, an examination of attached algae may be necessary, especially if the odor is earthy or musty. Sensory analysis should also be conducted on distribution system samples to determine the extent of the problem.

Investigations to determine the cause of a T&O episode may take several weeks before providing any useful information. While the laboratory is conducting tests to try to determine the cause, the plant operators are faced with the question of how to treat the problem. Evaluation of different treatment techniques should be done in cooperation with the laboratory, but testing needs to be done in such a manner that the results are useful to plant operation, especially during the initial stages of a T&O problem.

Do not waste time evaluating treatment programs that are not available for use at the treatment plant. Concentrate the initial efforts in treatment evaluation in those areas that the plant can most readily use; write up an inventory of available treatment techniques. (Ideally, this information should be collected and updated before a T&O problem begins.) Review the operational status of installed equipment during this inventory process. For example, PAC carbon feed systems that are not operable because of repair or maintenance work will not be much help in controlling current tastes and odors.

In addition to looking at readily available treatment processes, identify alternative methods for T&O control at other locations. An example would be the capacity to superchlorinate the plant influent, or possible locations upstream from the plant at which potassium permanganate could be applied. Collecting this information in the early stages of a T&O outbreak can be useful in developing control programs that are both effective and realistic for a particular facility.

Once you have identified the options available for treating the taste and odor, determine what treatment or combination of treatments is effective in reducing or eliminating the objectionable qualities of the water. Test each available treatment in the laboratory for its ability to improve water quality; combinations of treatments should also be investigated. Most treatment plants faced with T&O problems use a combination of two or three treatments to produce finished water of the best possible quality. In one particular situation, for example, it may be possible to use $KMnO_4$ at the raw water intake; powdered activated carbon at the flash mixer; improved coagulation, flocculation, and sedimentation; and above-normal chlorination before filtration to reduce the T&O properties of the water to a level that the consumers find acceptable. The most important aspect of a taste and odor control evaluation program is to test as many possible combinations as you reasonably can to determine the most effective approach to solving the problem.

Each potential treatment scheme may require changes in the various steps involved. Dose control at each step of the process is critical so that the individual processes work in harmony and do not create additional problems while trying to control the original source of the consumer complaints. In most cases, for example, it is both ineffective and uneconomical to superchlorinate and add PAC at the same point in the process stream. But at the same time, it may be unwise to completely shut down prechlorination just because carbon is being fed at a point close to the chlorine diffuser. Without prechlorination, algae may rapidly develop in downstream facilities at the treatment plant. The savings associated with reduced chlorine use may well be offset by the increase in problems associated with the algal growth in the treatment works.

In developing a taste and odor control treatment program, it is useful to remember that many of the available methods require some minimum dose level to be effective. Application of amounts greater than this minimum dose may provide no further benefit to water quality and, in some cases, may actually impair water quality. For example, tests show that a water supply required 10 mg/L PAC for T&O control. To save money, only 3 mg/L PAC is used. However, at 3 mg/L there is almost no improvement in the water quality so this dosage level may actually be a waste of money. The water requires at least 10 mg/L PAC to show any significant improvement. At the same time, treating the water with 15 mg/L PAC does not improve the quality over what would be achieved by the 10 mg/L treatment. Similarly, treatment with 0.5 mg/L of $KMnO_4$ may reduce the objectionable qualities of the water while 0.3 mg/L may do no good and 1.0 mg/L may well create colored water problems in addition to the existing T&O problem.

Development of a taste and odor monitoring program that includes both routine counting and identification of algal populations in source waters, and sensory analysis of raw, treated, and distribution system waters is essential. Routine monitoring of the algal populations in source waters, in conjunction with the use of sensory analysis in the form of a taste threshold test panel, can provide a robust early warning system. These analyses will provide valuable information for evaluating the source and cause of a T&O problem and the effectiveness of various treatment techniques in controlling the problem.

Sophisticated laboratory analytical methods using a gas chromatography-mass spectrometry instrument have been developed to measure and identify taste- and odor-causing compounds at the part-per-trillion (ppt) level. The Metropolitan Water District of Southern California applied specialized analytical methods to the detection of geosmin and methylisoborneol, compounds known to cause earthy-musty odors and tastes in water. Other laboratories may develop similar capabilities for these and other compounds. However, for laboratories that cannot afford the costs of maintaining this type of instrumentation, an odor panel, experienced in flavor profile analysis, should be the first step in dealing with T&O events.

Increased public awareness of potential health hazards associated with drinking water supplies and the legitimate concern of consumers for the quality of the water that they receive have emphasized the need for prompt and effective action during T&O incidents. Constant testing of both raw and treated water and adequate training of operators responsible for water treatment are required to provide the public with high-quality, safe, and pleasing supplies of drinking water.

Prevention of taste and odor in public water supplies is usually more effective and more economical than treatment. To prevent the development of T&O problems, operators should: (1) keep a clean and well-maintained water treatment plant and distribution system, and (2) study plant records to anticipate changes that might lead to T&O problems. However, every water system is subject to unusual and uncontrollable changes in source quality. Familiarity with available treatment alternatives and training in the operation and application of these alternatives is an important aspect of operator development and improvement.

Check Your Understanding

1. What factors could cause taste and odor complaints in a local area of a distribution system?

2. Which locations in a water supply system might contribute to a T&O problem?

3. Why is it important to write out and update a list of options available for treating T&O problems?

4. What two analyses can be used as an early warning system for taste and odor events?

8.8   Math Assignment

Turn to Appendix A, "Introduction to Basic Math for Operators," and read the "Learning Objectives" section. Next, read and work through the example problems in Section A.3, "Basic Algebra (Solving Equations)." Check the math using a calculator. You should be able to get the same answers.

8.9   Additional Resources

Your local library may be a good source for obtaining these additional resources related to the field of water treatment:

Manual of Water Utilities Operations (Texas Manual). Texas Water Utilities Association, www.twua.org. ASIN: B000KT9EIU. Chapter on tastes and odors in surface water supplies.

Chapter Review

Key Term Matching

——— 1. Ozonation

——— 2. Anaerobic

——— 3. Saturation

——— 4. Decant Water

——— 5. Olfactory Fatigue

——— 6. Superchlorination

——— 7. Degasification

——— 8. Adsorption

——— 9. Supernatant

——— 10. Eutrophication

A. The gathering of a gas, liquid, or dissolved substance on the surface or interface zone of another material.

B. A condition in which atmospheric or dissolved oxygen (DO) is not present in the aquatic (water) environment.

C. Water that has separated from sludge and is removed from the layer of water above the sludge.

D. A water treatment process that removes dissolved gases from the water. The gases may be removed by either mechanical or chemical treatment methods or a combination of both.

E. The increase in the nutrient levels of a lake or other body of water; this usually causes an increase in the growth of aquatic animal and plant life.

F. A condition in which a person's nose, after exposure to certain odors, is no longer able to detect the odor.

G. The application of ozone to water for disinfection or for taste and odor control.

H. The condition of a liquid (water) when it has taken into solution the maximum possible quantity of a given substance at a given temperature and pressure.

I. Chlorination with doses that are deliberately selected to produce free or combined residuals so large as to require dechlorination.

J. The relatively clear water layer between the sludge on the bottom and the scum on the surface of a basin or container.

8.1 Effects of Taste and Odor Problems

11. Taste and odor problems are likely to have _____ effects on a water utility when they occur.

12. Water with an objectionable taste or odor may cause consumers to purchase _____ or switch to _____.

8.2 Causes of Tastes and Odors

13. If algicide treatments are begun only after large algal populations have developed, the effect may be to _____ the decomposition process and _____ a taste and odor outbreak.

14. Oxygen depletion during the _____ may result in fish kills and die-off of aquatic organisms and vegetation, which will produce foul tastes and odors in water.

15. _____ may also lead to tastes and odors by substantially increasing flow velocities in rivers and canals.

16. What types of pollution may enter a municipal water supply upstream from a water treatment plant?
    1. Fully treated municipal wastewaters
    2. Industrial discharges
    3. Ocean current inflow
    4. Precipitation as rain or snow

17. What is one factor that influences the contribution of agricultural runoff to taste and odor problems?
    1. Irregular coastline formations
    2. Irrigation practices in neighboring watersheds
    3. Precipitation patterns
    4. The nature and extent of farming in neighboring states

18. No matter the condition of the raw water supply, tastes and odors will develop within a water treatment plant as a result of _____.

8.3 Locating Taste and Odor Sources

19. The first step in determining the cause of tastes and odors should be to _____.

20. _____, _____, and _____ may experience higher than normal numbers of T&O complaints.

8.4 Taste and Odor Prevention and Monitoring

21. Any change in the _____ may result in taste and odor complaints, even in the absence of any actual quality problems.

8.5 Taste and Odor Treatment

22. Use the _____, followed by tests for both settled water turbidity and threshold odor number, to determine if increased coagulant dosages will reduce objectionable tastes and odors.

23. What problems may be created by the use of cascades and spray aerators?
    1. They may encourage biological growths that may contribute to T&O problems
    2. They may encourage children to play in the water, causing safety issues
    3. They may encourage corrosion, leading to an iron taste in the water
    4. They may interrupt the sedimentation process

24. How would you prevent color from potassium permanganate reaching a consumer's tap?
    1. Adjust the feed rate so that the orange or brown color is visible only to a specific point in the basin
    2. Adjust the feed rate so that the pink or purple color is visible throughout the basin
    3. Determine the maximum $KMnO_4$ dose that can be safely applied in the plant
    4. Determine the minimum $KMnO_4$ dose that can be safely applied in the plant

25. What is not a necessary safety precaution when handling potassium permanganate?
    1. Keep permanganate storage facilities dry and well ventilated
    2. Provide adequate dust control and ventilation for workers and equipment
    3. Wear goggles, a dust mask, and gloves when handling permanganate
    4. Wear steel-toed boots when working with permanganate

26. Ozone has a distinct advantage over chlorine treatment for taste and odor in that _____.

27. You can determine which type of activated carbon would be effective in removing specific tastes and odors only by _____ for effective removal of the objectionable taste or odor.

28. Why should an overhead spray system be used when powdered activated carbon is being loaded into slurry tanks?
    1. To clean the carbon
    2. To clean the equipment
    3. To control dust
    4. To control insects and rodents

29. Results of jar tests indicate that 11 mg/L of PAC is the most effective dosage for treating a T&O problem. What is the desired concentration in pounds per million gallons (lb/Mgal)?
    1. 91.7 lb/Mgal
    2. 93.8 lb/Mgal
    3. 94.6 lb/Mgal
    4. 96.3 lb/Mgal

30. How does powdered activated carbon influence the performance of filters?
    1. PAC may bypass the filters and clog the system
    2. PAC may cake on the surface of filters and allow tastes and odors through the process
    3. PAC may cake on the surface of filters and cause longer filter runs
    4. PAC may cake on the surface of filters and cause shorter filter runs

31. Because powdered carbon will remove oxygen from the air, what safety precaution should be exercised when cleaning a PAC tank?
    1. Follow procedures for working in confined spaces
    2. Keep the tank closed so air will not circulate
    3. Stack PAC bags at least five high
    4. Store PAC near $KMnO_4$ and HTH

8.6 Identifying Taste and Odor Problems

32. Every water utility must have an _____ to deal with all types of consumer complaints.

8.7 Developing a Taste and Odor Control Strategy

33. The first step in developing a T&O control strategy is to _____ and _____.

34. Why should a taste and odor control evaluation program test as many combinations of treatment schemes as possible?
    1. To determine the most effective approach to solving the problem
    2. To determine who would be responsible for repair costs
    3. To fulfill state contracts
    4. To satisfy federal regulations

35. What is the most important step an operator can take to prevent the development of taste and odor problems?
    1. Follow federal or state regulations, whichever is easier to comply with
    2. Make sure the plant is clean with the utility stockholders are visiting
    3. Run the water through the plant as quickly as possible to save money
    4. Study plant records to anticipate changes that might lead to T&O problems

Laboratory Procedures

CHAPTER OUTLINE

LEARNING OBJECTIVES

1. Safely operate laboratory equipment.

2. Collect representative samples and preserve and transport the samples.

3. Prepare samples for analysis.

4. Describe lab test limitations, recognize precautions for those tests, and record lab test results.

5. Perform the following field or laboratory tests—alkalinity, chlorine residual, chlorine demand, coliform, hardness, jar test, pH, temperature, and turbidity.

KEY TERMS

acidic

aliquot

alkaline

amperometric

amperometric titration

aseptic

bacteria

blank

buffer

calcium carbonate
   equivalent

carcinogen

chlororganic

colorimetric measurement

composite (proportional)
   sample

compound

DPD method

disinfection

element

endpoint

facultative bacteria

flame polished

grab sample

indicator

inorganic

ion

M or molar or mol/L

MPN

meniscus

milligrams per liter, mg/l

mole

molecular weight

N (normal)

nephelometric

OSHA

organic

organism

pathogenic organisms

pH

potable water

precipitate

reagent

reduction

representative sample

safety data sheet (SDS)

solution

specific gravity

standard solution

standardize

sterilization

supernatant

titrate

volatile liquids

volumetric

9.1   Basic Laboratory Concepts, Equipment, and Techniques

(i) *Standard Methods,* 22nd Edition. 2012. American Public Health Association, American Water Works Association, and Water Environment Federation. www.awwa. org. ISBN: 978-0-87553-013-0.

Laboratory tests provide the necessary information to monitor and properly operate water treatment processes to ensure a safe and pleasant-tasting product water. You should think of laboratory testing like you do the instruments on the dashboard of your car. You need to be able to read the speedometer and gas gauge to operate your car. You need to be able to do the tests described in this chapter to properly operate your treatment plant. Moreover, relating laboratory results to treatment operations allows operators to select the most effective operational procedures and to identify potential problems before they affect finished water quality or threaten public health. For these reasons, a clear understanding of laboratory procedures is a must for every waterworks operator.

In this chapter we present laboratory procedures in some detail so that you get an accurate feel for the effort and intricacies involved in doing these tests. While adequate for educational purposes, these procedures are, in fact, somewhat simplified versions of the procedures you should be following on the job. The official versions of common analytical procedures can be found in *Standard Methods for the Examination of Water and Wastewater. Standard Methods* also contains details on chemical preparation and troubleshooting that are not included in this chapter. Most treatment plants and analytical laboratories follow the procedures in *Standard Methods,* but your local regulatory agency may specify alternatives that you need to follow.

9.1.1   Laboratory Units: The Metric System

In the laboratory, the metric system is used to express units of length, volume, weight (mass), concentration, and temperature. The metric system is based on decimals. All units of length, volume, and weight use factors of 10 to express larger or smaller quantities of these units. Below is a summary of metric and English unit names and their abbreviations.

Type of Measurement	US Customary Units	Abbreviation	Metric System Units	Metric Abbreviation
Length	inch	in	meter	m
	foot	ft		
	yard	yd		
Temperature	Fahrenheit	°F	Celsius	°C
Volume	quart	qt	liter	L
	gallon	gal		
Weight	ounce	oz	gram	g
	pound	lb		
Concentration	lb/gal		milligrams per liter	mg/L
	strength, %			

Many times in the water laboratory, we use smaller amounts than a meter, a liter, or a gram. To express these smaller amounts, prefixes are added to the names of the base metric unit. There are many prefixes in use; however, we commonly use two or three prefixes more than any others in the laboratory.

Prefix	Abbreviation	Meaning
centi-	c	1/100 of; or 0.01 times
milli-	m	1/1,000 of; or 0.001 times
micro-	μ	1/1,000,000 of; or 0.000001 times

One centimeter (cm) is 1/100 (one hundredth) of a meter, 1 milliliter (mL) is 1/1,000 (one thousandth) of a liter and 1 microgram (μg) is 1/1,000,000 (one millionth) of a gram.

Larger amounts than a meter, liter, or gram can be expressed using such prefixes as kilomeaning 1,000. A kilogram is 1,000 grams.

Additional information on the metric system can be found in Appendix A.

Example 1

Convert 3 grams (g) into milligrams (mg).

$$1 \text{ milligram} = 1 \text{ mg} = \frac{1}{1,000}\text{g}$$

therefore, 1 g = 1,000 mg

$$3[\text{g}] \times \frac{1,000 \text{ mg}}{\text{g}} = 3,000 \text{ mg}$$

Example 2

Convert 750 milliliters (mL) to liters (L).

$$1 \text{ mL} = \frac{1}{1,000}\text{L}$$

therefore, 1 liter = 1,000 mL

$$750[\text{mL}] \times \frac{1 \text{ L}}{1,000 \text{ mL}} = 0.750 \text{ L}$$

Example 3

Convert 50 micrograms (μg) to grams (g).

$$1 \mu\text{g} = \frac{1}{1,000,000}\text{g}$$

therefore, 1 g = 1,000,000 μg

$$50[\mu\text{g}] \times \frac{1 \text{ g}}{1,000,000 \,\mu\text{g}} = 0.00005 \text{ g}$$

In the lab, the Celsius (or centigrade) temperature scale is used rather than the Fahrenheit scale.

	Fahrenheit (°F)	Celsius (°C)
Freezing point of water	32	0
Boiling point of water	212	100

To convert Fahrenheit to Celsius, you can use the following formula:

$$\text{Temperature[°C]} = (°F - 32°) \times \frac{5}{9}$$

Example 4

Convert 68°F to °C.

$$\text{Temperature[°C]} = (°F - 32°) \times \frac{5}{9}$$

$$= (68° - 32°) \times \frac{5}{9}$$

$$= 36° \times \frac{5}{9}$$

$$= 20°C$$

To convert Celsius to Fahrenheit, the following formula can be used:

$$\text{Temperature[°F]} = \left(°C \times \frac{9}{5}\right) + 32°$$

Example 5

Convert 35°C to °F.

$$\text{Temperature[°F]} = \left(°C \times \frac{9}{5}\right) + 32°$$

$$= \left(35° \times \frac{9}{5}\right) + 32°$$

$$= 63° + 32°$$

$$= 95°F$$

9.1.2   Chemical Names and Formulas

element
A substance that cannot be separated into its constituent parts and still retain its chemical identity. For example, sodium (Na) is an element.

In the laboratory, chemical symbols are used as shorthand for the names of the **elements**. The full listing is contained in the Periodic Table of Elements. The names and symbols for elements commonly encountered in water treatment are listed below.

Chemical Name	Symbol	Chemical Name	Symbol
Aluminum	Al	Lead	Pb
Calcium	Ca	Magnesium	Mg
Carbon	C	Manganese	Mn
Chlorine	Cl	Nitrogen	N
Copper	Cu	Oxygen	O
Fluorine	F	Potassium	K
Hydrogen	H	Sodium	Na
Iron	Fe	Sulfur	S

A **compound** is a pure substance composed of two or more elements whose composition is constant. For example, table salt (sodium chloride, NaCl) is a compound. Generally, all chemical compounds can be divided into two main groups, **organic** and **inorganic**. Organic compounds are those that contain the element carbon (C). A few substances containing carbon are however considered to be inorganic compounds. These are carbon dioxide (CO_2), carbon monoxide (CO), bicarbonate (HCO_3^-), and carbonate (CO_3^{-2}) as in calcium carbonate ($CaCO_3$).

Many different compounds can be made from the same two or three elements. Therefore, you must carefully read the formula and name to prevent errors and accidents. A chemical formula is a shorthand or abbreviated way to write the name of a chemical compound. For example, the name sodium chloride (common table salt) can be written NaCl. Table 9.1 lists chemical compounds commonly found in the water laboratory.

Poor results and safety hazards are often caused by using a chemical from the shelf that is not exactly the same as the chemical called for in a particular procedure. The mistake usually occurs when the chemicals are not properly labeled or have similar names or formulas. This problem can be eliminated if you use both the chemical name and formula as a double check. The spellings of many chemical names are quite similar. These slight differences are critical because the chemicals do not behave alike. For example, the chemicals potassium nitrate (KNO_3) and potassium nitrite (KNO_2) are quite different chemically.

Finally, many compounds dissociate (separate) into **ions** when dissolved in water. Ions are molecules with an electric charge, positive or negative, due to extra electrons sticking to one of the ions.

For example, table salt (NaCl) dissociates into a sodium ion (Na^+) and a chloride ion (Cl^-). In this case, an extra negatively charged electron has stayed with the chlorine atom, giving it a negative charge (shown by the negative superscript in the symbol). The sodium atom, being short an electron, has a positive charge. We can write this dissolution like a chemical reaction:

$$NaCl \rightarrow Na^+ + Cl^-$$

Table 9.1 Names and formulas of chemicals commonly used in water analyses

Chemical Name	Chemical Formula
Acetic Acid	CH_3COOH
Aluminum Sulfate (alum)[a]	$Al_2(SO_4)_3 \cdot 14.3\ H_2O$
Ammonium Hydroxide	NH_4OH
Calcium Carbonate	$CaCO_3$
Chloroform	$CHCl_3$
Copper Sulfate	$CuSO_4$
Ferric Chloride	$FeCl_3$
Hydrochloric Acid	HCl
Nitric Acid	HNO_3
Phenylarsine Oxide	C_6H_5AsO
Potassium Iodide	KI
Sodium Bicarbonate	$NaHCO_3$
Sodium Hydroxide	NaOH
Sulfuric Acid	H_2SO_4

[a]Alum in the dry form based on 17% Al_2O_3. The 14.3 H_2O nomenclature means that there is some water contained in the crystals of alum at a ratio of 14.3 water molecules to each aluminum sulfate molecule.

organic
Describes chemical substances that come from animal or plant sources. Organic substances always contain carbon though not every compound that contains carbon is organic. Carbon dioxide (CO_2), bicarbonate (HCO_3^-), and carbonate (CO_3^{-2}) are considered to be inorganic.

inorganic
Describes materials such as sand, salt, iron, calcium salts, and other minerals. Inorganic materials are chemical substances of mineral origin, whereas organic substances are usually of animal or plant origin.

Many of the compounds dissociate like this. Some examples include:

Ammonium hydroxide: $NH_4OH \rightarrow NH_4^+ + OH^-$
Calcium carbonate: $CaCO_3 \rightarrow Ca^{2+} + CO_3^{2-}$
Ferric chloride: $FeCl_3 \rightarrow Fe^{3+} + 3Cl^-$
Sodium bicarbonate: $NaHCO_3 \rightarrow Na^+ + HCO_3^-$
Sulfuric acid: $H_2SO_4 \rightarrow 2H^+ + SO_4^{2-}$

Notice that because the compounds on the left side of the reactions start out as electrically neutral, the net charge (sum of the positives and negatives) of the ions must be zero. For example, the ions that result from sulfuric acid dissociation are two hydrogen ions with one charge each (total of two positive charges) and one sulfate ion with a −2 charge (total of two negative charges). The net charge on both sides of the arrow is zero.

Check Your Understanding

1. Why are laboratory tests important?
2. What does the prefix milli- mean?
3. What is the proper name of this chemical compound: $CaCO_3$?

9.2   Laboratory Equipment and Techniques

This section features specialized equipment for water laboratories, including containers, measuring equipment, and special-purpose laboratory glassware. When working in a laboratory, cuts and burns might occur unless safety precautions are followed. An operator must know how to mix and safely use chemical solutions, and know the importance of proper data recording and recordkeeping and following laboratory quality control.

9.2.1   Water Laboratory Equipment

In any laboratory, there are certain basic pieces of equipment that are used routinely to perform water analysis tests. The items of equipment in a water laboratory are some of the operator's tools of the trade. Common items of glassware and equipment that you should be familiar with are shown in Table 9.2. Other specialized pieces of equipment will be discussed in association with their functions later in the chapter.

Table 9.2   Common laboratory glassware and equipment

Vessels for Containing Liquids (Approximate Volumes)

Beakers are used for mixing, heating, and weighing chemicals. They come in many sizes and often have volume-indicating lines (graduations) imprinted on their sides. These are only approximate volumes. Do not use a beaker to try to make an accurate volume measurement.

(Source: Courtesy of Africa Studio/Shutterstock)

Flasks are used for the same purposes as beakers. There are many different sizes and shapes for particular purposes, but the most common style for everyday use is the Erlenmeyer flask (see the figure on the right). Like beakers, they can measure approximate volumes only.

(Source: Courtesy of pedrosala/Shutterstock)

Test Tubes are used for holding and mixing small quantities of chemicals. They are also used as containers for bacterial testing (culture tubes).

Test Tube

(Source: Courtesy of Africa Studio/Shutterstock)

Culture Tube Without Lip

(Source: Courtesy of HUANSHENG XU/Shutterstock)

Bottles are used to store chemicals, to collect samples, and to dispense liquids. They are used for containing, not measuring, liquids.

Bottle, Reagent

(Source: Courtesy of Tsz-shan Kwok/Pearson Education)

Glassware for Measuring Volumes

Graduated Cylinders are used to measure volumes more accurately than beakers, but not as accurately as pipets or volumetric glassware (see below). Graduated cylinders should never be heated in an open flame because they will break.

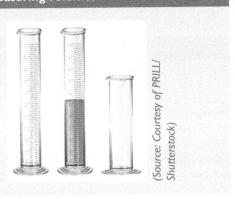

(Source: Courtesy of PRILL/Shutterstock)

Table 9.2   Common laboratory glassware and equipment *(continued)*

Glassware for Measuring Volumes *(continued)*

Burets are similar to graduated cylinders except that they can be emptied through the bottom through a regulating valve called a stopcock. The stopcock controls the amount of liquid that will flow from the buret. A glass stopcock must be lubricated with stopcock grease and should not be used with alkaline solutions which will cause the stopcock to freeze. Teflon stopcocks do not need lubrication. Burets are especially useful in a procedure called titration which is discussed later.

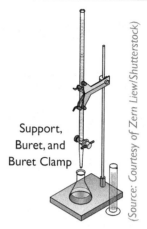

Support, Buret, and Buret Clamp

(Source: Courtesy of Zern Liew/Shutterstock)

Buret

(Source: Courtesy of mazarekic/Shutterstock)

Pipets are small tubes used to measure and deliver accurate (but small) volumes. They are the most accurate way to measure a volume of liquid. Liquids are drawn up into the pipet using a suction device (not your mouth). Graduated pipets (also called Mohr pipets) and serological pipets look similar except for their tips. Verify which type you need to use. See the discussion of pipetting technique below.

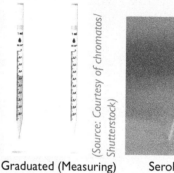

Graduated (Measuring) or Mohr Pipet

(Source: Courtesy of chromatos/Shutterstock)

Serological Pipet

(Source: Courtesy of Terry Leung/Pearson Education)

Volumetric Flasks are used to prepare solutions because of their high accuracy. Each flask is individually calibrated and there is only one line on the neck corresponding to the desired volume. Consequently, they are more accurate than graduated cylinders. Volumetric flasks should never be heated. It is not a good practice to store liquid chemicals in volumetric flasks; the chemicals should be transferred to storage bottles.

(Source: Courtesy of chromatos/Shutterstock)

Volumetric Pipet. Like volumetric flasks, volumetric pipets are used to deliver a single volume very accurately.

(Source: Courtesy of Pearson Education)

Special-Purpose Glassware

Filter Flasks are shaped like Erlenmeyer flasks except for a stem in the neck. They are usually used as the base of a filtering device, particularly for filter paper. Suction tubing is attached to the stem and the vacuum created in the flask pulls the liquid sample through the filter paper. The liquid is captured in the flask, and is not sucked into the vacuum system.

(Source: Courtesy of chromatos/Shutterstock)

Table 9.2   Common laboratory glassware and equipment *(continued)*

Special-Purpose Glassware *(continued)*

Membrane Filter Holder (below). This is a device that fits on top of a filter flask and supports membrane or paper filters. They are commonly made of stainless steel screen or porous glass. Create a *Membrane Filtration Apparatus* (right) by clamping a funnel over the membrane filter holder and attaching a vacuum line to the filter flask.

(Source: Courtesy of Pearson Education)

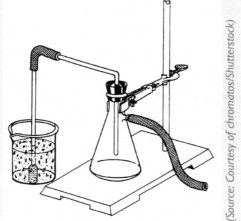

(Source: Courtesy of chromatos/Shutterstock)

Petri (Culture) Dish. These are shallow glass or plastic dishes with lids that are used to culture bacteria and other microorganisms on solid media.

(Source: Courtesy of Photographee.eu/Shutterstock)

General Laboratory Equipment

Tongs and Clamps. There are a variety of different types of tongs and clamps used for moving glassware, particularly when it is hot.

(Source: Courtesy of Trevor Clifford/Pearson Education)

Magnetic Stirrers are used to mix liquids. They contain magnets that can be spun at variable speeds. The mixing is provided by a *Magnetic Stir Bar* placed in the liquid that rotates in synch with the magnet in the base. Often magnetic stirrers are combined with *Hot Plates* so that samples can be heated and stirred simultaneously.

(Source: Courtesy of Dikiiy/Shutterstock)

Table 9.2   Common laboratory glassware and equipment *(continued)*

General Laboratory Equipment *(continued)*

Bunsen Burners are used to provide a small open flame. Typically they operate using natural gas.

(Source: Courtesy of Martyn F Chillmaid/Pearson Education)

Analytical Balances are sophisticated instruments for determining masses with high resolution (down to 1/10,000 of a gram). Newer balances are electronic and are easy to operate. Older versions are mechanical and require a bit of skill. Although analytical balances are accurate, they do not have a large capacity. *Platform Balances* (electronic or mechanical) will have larger capacities, but lesser resolutions (usually only down to 1/100 of a gram).

Autoclaves are specialized devices for sterilizing glassware and other objects through a combination of heat and pressure. They are used in bacterial testing.

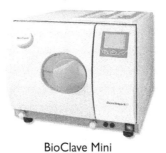

BioClave Mini

BioClave 16

(Source: Courtesy of Benchmark Scientific, Inc.)

meniscus (meh-NIS-cuss)
The curved surface of a column of liquid (such as water, oil, or mercury) in a small tube. When the liquid wets the sides of the container (as with water), the curve forms a valley. When the confining sides are not wetted (as with mercury), the curve forms a hill or upward bulge.

9.2.2   Using Laboratory Glassware

Accurate laboratory test results depend not only on using the proper equipment, but using proper techniques as well.

9.2.2.1   Reading Volumes

When a buret or cylinder is filled with liquid, the surface of the liquid is curved. This surface curve is called the **meniscus** (meh-NIS-cuss). Depending on the liquid, the curve forms a valley, as with water, or forms a hill, as with mercury. Since most **solutions** used in the laboratory are water based, always read the bottom of the meniscus with your eye at the same level (Figure 9.1). If you have the meniscus at eye level, graduation marks that completely circle the glassware should appear as straight lines. If they look like ovals, then your eyes are not at the right level and your reading will contain a parallax error.

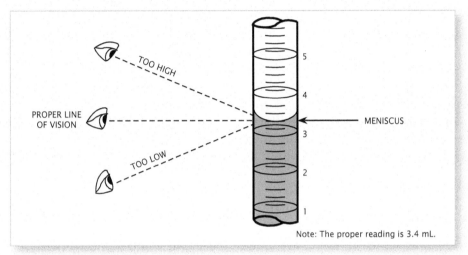

Note: The proper reading is 3.4 mL.

Figure 9.1   How to read a meniscus

9.2.2.2 Using Pipets

Accuracy in pipetting is important for accurate test results. This depends on both the instrument and the operator's technique. A regular calibration program can minimize the risk of error from faulty equipment. Contact your pipet supplier for calibration procedures. Incorrect pipetting technique can also lead to errors.

There are three types of pipets commonly used in the laboratory—**volumetric** pipets, graduated (measuring or Mohr) pipets, and serological pipets (see Figure 9.2). In addition, pipets are manufactured to deliver (TD) or to contain (TC) a specified volume.

Let's consider TD and TC first. In using a pipet to transfer liquids, pipets should be held in a vertical or near-vertical position and the outflow should be unrestricted. The tip should be touched to the wet surface of the receiving vessel and kept in contact with it until the emptying is complete. A small drop of liquid will stick in the tip. A TD pipet is manufactured to deliver the correct volume not including that droplet. In other words, it will deliver the correct volume with that droplet remaining in the tip. So, for TD pipets, do NOT blow that droplet out. TC pipets are manufactured to contain the specified volume, including the droplet in

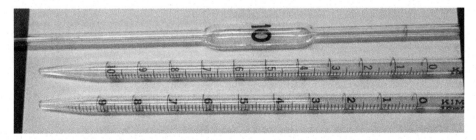

Figure 9.2   The three types of pipets (from top to bottom): volumetric, graduated (measuring or Mohr), and serological

solution
A liquid mixture of dissolved substances. In a solution it is impossible to see all the separate parts.

volumetric
A measurement based on the volume of some factor. Volumetric titration is a means of measuring unknown concentrations of water quality indicators in a sample by determining the volume of titrant or liquid reagent needed to complete particular reactions.

the tip. So, for TC pipets, do blow that droplet out. The TD and TC markers will be printed at the top of the pipet.

Volumetric pipets are used to deliver a single volume very accurately.* These pipets can be recognized by a bulb in the middle of the tube. They will have only one marking for a specific volume (see Figure 9.2). Graduated (Mohr) and serological pipets are designed to deliver fractions of the total volume indicated on the pipet. Their tips, though, are different (see Figure 9.3). The graduated marks on a Mohr pipet stop above the tip; those on a serological pipet extend into the tip. So to deliver 10 mL from a 10 mL Mohr pipet, you should stop the flow when the meniscus reaches the bottom line; but for a serological pipet, you sould allow the pipet to drain completely. (Whether or not you blow out the final drop depends on whether you are using a TD or TC pipet.)

Use a pipet filler or pipet bulb (Figure 9.4) to draw liquid into a pipet. Never pipet chemical **reagent** solutions or unknown water samples by mouth. For best results, use the following techniques:

1. Draw liquid up into the pipet past the zero mark.
2. Quickly remove the bulb and place a dry fingertip over the top end of the pipet.

*Volumetric glassware will deliver accurately only when the inner surface is so scrupulously clean that water wets the surface immediately and forms a uniform film on the surface upon emptying. For best results, keep your pipets clean.

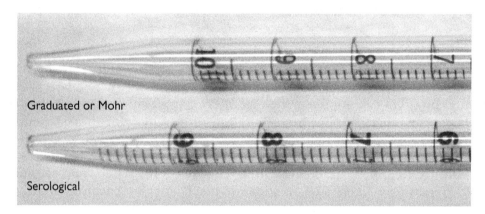

Graduated or Mohr

Serological

Figure 9.3   The tips of graduated (or Mohr) and serological pipets

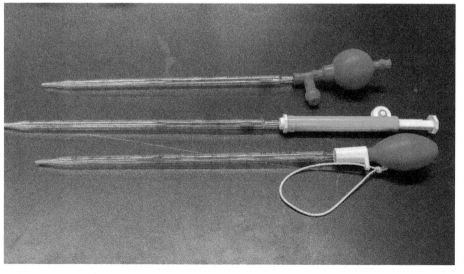

reagent
A pure, chemical substance or solution of that substance that is used to make new products or to measure, detect, or examine other substances in chemical tests.

Figure 9.4   Pipet bulbs and a pipet filler (center)

3. Wipe excess liquid from the tip of the pipet using laboratory tissue paper.
4. Lift finger and allow the liquid to drain to the zero mark.
5. Move the pipet to the receiving container and allow the desired volume to drain from the pipet.

Note: There are some styles of pipet bulbs with valves that can control the flow of liquid from the pipet without removing the bulb.

Check Your Understanding

1. For each type of glassware listed below, describe the item and its use or purpose.
   a. Beakers
   b. Graduated cylinders
   c. Pipets
   d. Burets
2. Why should graduated cylinders never be heated in an open flame?
3. What is a meniscus?
4. What is a reagent?

9.2.3   Chemical Solutions

The strength or concentration of a chemical solution can be expressed in several ways.

9.2.3.1   Mass Concentration

Mass concentration is the mass of a constituent contained in a liter of solution. In water chemistry, the typical units are from the International System of Units or SI System—**milligrams per liter (mg/L)** or micrograms per liter (µg/L). A liter of water has a mass close to 1 kilogram, so a mg/L is approximately one part per million by mass. Similarly, a µg/L is approximately one part per billion by mass.

You should remember that concentrations are expressed in terms of mass per liter of finished solution, not mass added to 1 liter of solution. For instance, suppose the directions for preparing a **standard solution** (or reagent) says to weigh out 7.6992 grams of a chemical and dilute to 1 liter with distilled water. Fill a volumetric flask about 75 percent full. Weigh out the required mass and add it to the water. You will probably have to use a funnel. Use distilled water to wash down any chemical sticking to the funnel or to the neck of the flask. Put a magnetic stir bar into the flask and mix until the chemical is dissolved. Next, remove the stir bar from the solution and top off the flask to the 1-liter mark. **Standardization** is the process of using one solution of known concentration to determine the concentration of another solution. This action often involves **titration**.

It might not be easy to weigh out the exact mass required. An alternative procedure is to weigh out a mass close to the desired value (usually a little less) and mix it into a proportionate volume of water. An example calculation for proportional volume is shown below.

milligrams per liter, mg/L
A measure of the concentration by weight of a substance per unit volume. For practical purposes, 1 mg/L of a substance in water is equal to 1 part per million parts (ppm). A liter of water with a specific gravity of 1.0 weighs 1 million milligrams. If 1 liter of water contains 10 milligrams of calcium, the concentration is 10 milligrams per million milligrams, or 10 milligrams per liter (10 mg/L), or 10 parts of calcium per million parts of water, or 10 parts per million (10 ppm).

standard solution
A solution in which the exact concentration of a chemical or compound is known.

standardize
To compare with a standard. (1) In wet chemistry, to find out the exact strength of a solution by comparing it with a standard of known strength. This information is used to adjust the strength by adding more water or more of the substance to be dissolved. (2) To set up an instrument or device to read a standard. This allows you to adjust the instrument so that it reads accurately, or enables you to apply a correction factor to the readings.

titrate
To titrate a sample, a chemical solution of known strength is added drop by drop until a certain color change, precipitate, or pH change in the sample is observed (endpoint). Titration is the process of adding the chemical reagent in small increments (0.1–1.0 milliliter) until completion of the reaction, as signaled by the endpoint.

Example 6

The directions for preparing a standard reagent indicate that you should weigh out 7.6992 grams and dilute to 1 liter. If you weigh out 7.6761 grams, how much water should be added to produce the desired concentration or normality of the standard reagent?

$$\left(\frac{7.6992\,\text{g}}{1.0\,\text{L}}\right) = \left(\frac{7.6761\,\text{g}}{V}\right)$$

$$V = \frac{7.6761}{7.6992} = 0.997\,\text{L} = 997\,\text{mL}$$

You can measure this out with a graduated cylinder. Alternatively, you can measure out 1,000 mL in advance using a volumetric flask and then remove 3 mL using a pipet. Do this before adding any chemical.

For liquid reagents an "a + b" dilution system may be used. What this means is that "a" volumes of concentrated reagent are to be diluted with "b" volumes of distilled water. For example, "1 + 5 H_2SO_4" means one volume of concentrated sulfuric acid is diluted with five volumes of distilled water.

9.2.3.2 Molar Concentration

In chemical reactions, we are often more interested in the number of molecules than in their mass. For this, we use **molar** concentrations. A **mole** (abbreviated mol) is a unit of measurement for the number (as opposed to the mass) of atoms or molecules. A molar solution (mol/L or *M*) consists of 1 gram molecular weight of a compound dissolved in enough water to make 1 liter of solution. A mass of a substance equal to its **molecular weight** (the sum of the atomic weights of the elements in the compound) in grams is 1 mole, which is 6.02×10^{23} molecules (Avogadro's Number). To convert from grams to moles, use the molecular weight like a unit conversion factor.

Example 7

How many moles are in a liter of water at 15°C, knowing that its mass is 999 g/L? The molecular weight of water (H_2O) = 2 + 16 = 18 g, which means the conversion is 18 g/mol.

$$\frac{999\left[\dfrac{\text{g}}{\text{L}}\right]}{18\left[\dfrac{\text{g}}{\text{mol}}\right]} = 55.5\,\frac{\text{mol}}{\text{L}}$$

M **or molar or mol/L**
A molar solution consists of one gram molecular weight of a compound dissolved in enough water to make one liter of solution. A gram molecular weight is the molecular weight of a compound in grams. For example, the molecular weight of sulfuric acid (H_2SO_4) is 98. A one *M* solution of sulfuric acid would consist of 98 grams of H_2SO_4 dissolved in enough distilled water to make one liter of solution.

Example 8

What is the molar concentration of a solution containing 5 mg/L of sodium (Na^+)? The molecular weight of sodium = 23 g/mol.

$$5\left[\frac{\text{mg}}{\text{L}}\right] \times \frac{\text{g}}{1,000\,\text{mg}} \times \frac{\text{mol}}{23\,\text{g}} = 0.00022\,\frac{\text{mol}}{\text{L}} \text{ or } 2.2 \times 10^{-4}\,M$$

Example 9

If you want to create a 0.025 M solution of sodium, how much table salt (sodium chloride, NaCl) would you have to use per liter of solution?

Molecular weights: Na^+ = 23 g/mol; Cl^- = 35.5 g/mol; NaCl = 23 + 35.5 = 58.5 g/mol.

$$M\left[\frac{\text{mol}}{\text{L}}\right] = \frac{\dfrac{\text{Mass}[g]}{\text{MW}\left[\dfrac{g}{\text{mol}}\right]}}{V[L]} = \frac{\text{Mass}[g]}{\text{MW}\left[\dfrac{g}{\text{mol}}\right] \times V[L]}$$

Where,

 M = Molar Concentration
 Mass = Mass of Chemical
 MW = Molecular Weight of Chemical
 V = Volume of Solution

$$M\left[\frac{\text{mol}}{\text{L}}\right] = 0.025\left[\frac{\text{mol}}{\text{L}}\right] = \frac{\text{Mass}[g]}{58.5\left[\dfrac{g}{\text{mol}}\right] \times 1[L]}$$

$$\text{Mass}[g] = 0.025\left[\frac{\text{mol}}{\text{L}}\right] \times 58.5\left[\frac{g}{\text{mol}}\right] \times 1[L] = 1.46\,g$$

You need to use about 1.5 grams of table salt per liter of solution to make a 0.025 molar solution.

9.2.3.3 Normality

Many laboratory procedures call for concentrations expressed as normality (N). A normal solution contains 1 gram equivalent weight of reactant (compound) per liter of solution. The equivalent weight of an acid is the weight that contains 1 mole of ionizable hydrogen or its chemical equivalent. The equivalent weight is the molar weight (g/mol) divided by the number of hydrogen ions per molecule. For instance, each molecule of H_2SO_4 donates two hydrogen ions. Therefore, to obtain 1 mole of H^+, you need only 0.5 moles of H_2SO_4. Accordingly, its equivalent weight is 49 g (98 g/mol divided by 2). A 1.00 N solution would be made by dissolving 49 g of H_2SO_4 in enough water to make 1 liter of solution. This nomenclature is used for bases as well, except that instead of H^+, we consider OH^- or its chemical equivalent.

Mathematically, we can relate molarity and normality by the following equation:

$$N = M \times z$$

Where,

 M = Molar Concentration (Molarity)
 z = Number of H^+ or OH^- Reaction Sites per Molecule

N (normal)
A normal solution contains 1 gram equivalent weight of reactant (compound) per liter of solution. The equivalent weight of an acid is that weight that contains 1 gram atomic weight of ionizable hydrogen or its chemical equivalent. For example, the equivalent weight of sulfuric acid (H_2SO_4) is 49 (98 divided by 2 because there are 2 replaceable hydrogen ions). A 1.0 N solution of sulfuric acid would consist of 49 grams of H_2SO_4 dissolved in enough water to make 1 liter of solution.

The value for z comes from the molecular formula. In general, the mathematics of normality are the same as for molarity except that equivalent weights are used instead of molar weights.

Example 10

How much sodium carbonate (Na_2CO_3) would you need to make 1.00 L of 0.15 N solution? For sodium carbonate, z is equal to 2 because 2 hydrogens can replace the sodium atoms on the molecule (i.e., these are hydrogen reaction sites). The molarity (M) of the desired solution is:

$$M = \frac{N}{z} = \frac{0.30}{2} = 0.15\frac{mol}{L}$$

The molecular weight (MW) of Na_2CO_3 is:

$$MW = (2 \times 23) + 12 + (3 \times 16) = 106\frac{g}{mol}$$

Therefore, the mass needed for a 0.15 M solution is:

$$0.15\left[\frac{mol}{L}\right] \times 106\left[\frac{g}{mol}\right] = 15.9\frac{g}{L}$$

9.2.4   Data Recording and Recordkeeping

Using a laboratory notebook and worksheets is a must for laboratory analysts and operators for several reasons. Notebooks and worksheets help you record data in an orderly manner. Too often, hours of work are wasted when test results and other data (such as a sample volume) are written down on a scrap of paper that can be misplaced or thrown away by mistake. Likewise, the fill-in-the-blank aspect of standard worksheets helps assure that you collect all the data necessary to calculate analytical results. Forgetting to record an important reading (such as pH) can also result in wasted effort. Notebooks and worksheets provide a valuable record of your work. Such records can be a useful reference in diagnosing problems or identifying trends in plant performance. The routine use of laboratory worksheets and notebooks is the only way an operator or a lab person can be sure that all important information is properly recorded.

There is no standard laboratory form. Most operators usually develop their own worksheets for recording test results and other important data. These sheets should be prepared in a manner that makes it easy for you to record results, review them, and recover these results when it is necessary. Each treatment plant will have different needs for collecting and recording data and may require several different data or worksheets. Figure 9.5 is an example of a worksheet for an alkalinity lab test.

Some large plants use software-based digital notebooks linked to databases. Besides saving paper, this approach records the data in a form that can be easily manipulated such as graphing or calculating statistics. Worksheets can also be constructed with spreadsheet software such as Microsoft Excel™. (Figure 9.5 was done this way.) An advantage to this approach is that formulas can be programmed into the spreadsheet to do the calculations automatically. Also, error checking can be incorporated to catch mistakes such as inputting pH values outside the 0–14 range.

Alkalinity

Analyst: _____
Date: _____

Shaded cells are calculation results

Acid Standardization

Primary standard

Mass of Na_2CO_3	g	
Formula weight of Na_2CO_3		
Volume of distilled water	mL	
Normality of primary standard	meq/L*	

Standardization titration

Volume of primary standard	mL	
Volume of acid titrant	mL	
Normality of acid	meq/L*	

Alkalinity Measurements

		Sample 1	Sample 2	Sample 3	Sample 4
Sample ID					
Sample Volume	mL				
Initial pH					
Titrant volume to pH 8.3	mL				
Total titrant volume to pH 4.5	mL				

Calculated Values

		Sample 1	Sample 2	Sample 3	Sample 4
Phenolphthalein alkalinity	meq/L*				
	mg/L**				
Total alkalinity	meq/L*				
	mg/L**				

*meq/L = milliequivalents per liter
**mg/L as $CaCO_3$

Figure 9.5   Worksheet for an alkalinity lab test

9.2.5   Laboratory Quality Control

Having good equipment and using the correct methods are not enough to ensure correct analytical results. Each operator must be constantly alert to factors in the water treatment process that can lead to poor data quality. Such factors include sloppy laboratory technique, deteriorated reagents and standards, poorly operating instruments, and calculation mistakes. One of the best ways to ensure quality control in your laboratory is to analyze reference-type samples to provide independent checks on your analysis. These reference-type samples are available from the US Environmental Protection Agency (EPA) and from commercial sources. From time to time, it is also a good idea to split a sample with one of your fellow operators or another laboratory and compare analytical results. Frequent self-appraisal and evaluation—from sampling to reporting results—can help you gain full confidence in your results.

Check Your Understanding

1. What is a standard solution?
2. What is the primary purpose of laboratory notebooks and worksheets?
3. List three sources or causes of poor quality of analytical data.

9.3   Laboratory Safety

Safety is just as important in the laboratory as it is outside. State laws and the Occupational Safety and Health Act (OSH Act of 1970) demand that proper safety procedures be followed in the laboratory at all times. OSH Act of 1970 specifically deals with safety in the work place. The act requires that employers have the duty to furnish all employees with employment free from recognized hazards causing, or likely to cause, death or serious physical harm.

Personnel working in the water industry must realize that a number of hazardous materials and conditions can exist. Be alert, careful, and aware of potential dangers at all times. Locate the safety showers and eyewash station (Figure 9.6) before starting to work with dangerous chemicals. Employing safe practices while working around chemicals prevents injuries.

On specific questions of safety, consult your state's General Industry Safety Orders or US Occupational Safety and Health Administration (OSHA) regulations.

9.3.1   Laboratory Hazards

Working with chemicals and other materials in the water treatment plant laboratory can be dangerous. Laboratory hazards include:

- Hazardous materials
- Explosions
- Cuts and bruises
- Electric shock
- Fire
- Burns (heat and chemical)
- Spills and splashes

Hazardous materials include: corrosive, toxic, and explosive or flammable materials.

OSHA

With the Occupational Safety and Health Act of 1970, Congress created the Occupational Safety and Health Administration (OSHA) to assure safe and healthful working conditions for workers by setting and enforcing standards and by providing training, outreach, education and assistance.

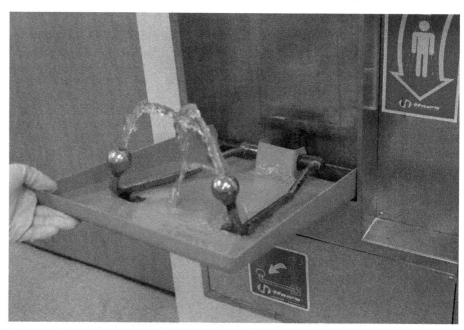

Figure 9.6   Eyewash station

CORROSIVE MATERIALS

Acids

- Examples: Hydrochloric or muriatic (HCl), hydrofluoric (HF), glacial acetic (CH_3COOH), nitric (HNO_3), and sulfuric (H_2SO_4).
- Acids can be extremely corrosive and hazardous to human tissue, metals, clothing, cement, stone, and concrete.
- Commercially available spill cleanup kits should be kept on hand to neutralize the acid in the event of an accidental spill. Baking soda (bicarbonate, not laundry soda) effectively neutralizes acids. Baking soda can be used on lab and human surfaces without worrying about toxicity.

Bases (Caustics)

- Examples: Sodium hydroxide (caustic soda or lye, NaOH), quicklime (CaO), and hydrated lime ($Ca(OH)_2$).
- Bases are extremely corrosive to skin, clothing, and leather. Caustics can quickly and permanently cloud vision if not immediately flushed out of eyes.
- Commercially available spill cleanup materials should be kept on hand for use in the event of an accidental spill. A jug of ordinary vinegar can be kept on hand to neutralize bases and it will not harm your skin.

Miscellaneous Corrosive Chemicals

- Examples: Alum, chlorine, ferric salts (ferric chloride), fluorosilicic acid, and other strong oxidants.

Know where the safety showers and eyewash stations are before using dangerous chemicals.

TOXIC MATERIALS

Examples:
- Solids: Cyanide compounds, chromium, orthotolidine, cadmium, mercury, and other heavy metals.
- Liquids: Chloroform and other organic solvents.
- Gases: Chlorine, ammonia, sulfur dioxide, and chlorine dioxide.

EXPLOSIVE OR FLAMMABLE MATERIALS

Examples:
- Liquids: Acetone, ethers, and gasoline.
- Gases: Propane, hydrogen, and methane (natural gas).

9.3.2   Personal Safety and Hygiene

Laboratory work can be quite dangerous if proper precautions are not taken. Always follow these basic rules:

- Never work alone in the laboratory. Someone should always be available to help in case you should have an accident that blinds you, leaves you unconscious, or starts a fire you cannot handle. If necessary, have someone check on you regularly to be sure you are OK. A good idea is to provide a window in the wall or door so that aid workers outside can see into the lab.
- Wear protective goggles or eyeglasses at all times in the laboratory. Contact lenses may be worn under safety goggles, but contact lenses are not eye protective devices, and wearing them does not reduce the requirement for eye and face protection.

Safety Glasses

Because fumes can seep between the lens and the eyeball and irritate the eye, the National Institute of Occupational Safety and Health (NIOSH) recommends an eye injury hazard evaluation be conducted in accordance with the safety guidelines listed in Current Intelligence Bulletin (CIB) 59 2005-139 (https://www.cdc.gov/niosh/docs/2005-139/pdfs/2005-139.pdf).

- Be sure you know the location and operation of safety showers and eyewash stations before working with dangerous chemicals.
- Never pipet hazardous liquids by your mouth.
- Always wear a lab coat or apron and disposable gloves in the laboratory to protect your skin and clothes.
- Wear insulated gloves when handling hot objects. If there is a danger of hot liquid erupting from a container, wear a face shield, too. Use tongs and clamps as appropriate.
- Do not bring food or drink into the lab or keep food in a refrigerator that is used for chemical or sample storage. Do not eat in the lab.
- Exercise good housekeeping as an effective way to prevent accidents.
- Wear close-toed shoes when in the laboratory.

9.3.3   Preventing Laboratory Accidents

Laboratory accidents are preventable. An operator can prevent accidents through proper chemical storage, correctly transferring chemicals for use, and following proper laboratory techniques. Electric shocks, cuts, burns, toxic fumes, and fires can be avoided by carefully following laboratory safety procedures.

9.3.3.1   Chemical Storage

An adequate chemical storeroom is essential for safety in the water laboratory. The storeroom should be properly ventilated and lighted and laid out to segregate incompatible chemicals. Order and cleanliness must be maintained. Clearly label and date all chemicals and bottles of reagents. Make a first-aid kit available in the laboratory, and there should be a **safety data sheet (SDS)** available for all chemicals used in the laboratory.

Store heavy items on or as near to the floor as possible. **Volatile liquids** that may escape as a gas, such as ether, must be kept away from heat sources, sunlight, and electric switches. Store acids and bases in separate storage cabinets designed for acid and base storage.

Cap and secure cylinders of gas in storage to prevent rolling or tipping. They should also be placed away from any possible sources of heat or open flames.

Provide clamps, raised shelf edges, and proper arrangement to prevent items such as chemicals or glassware in the stockroom from falling off shelves.

Follow usual common-sense rules of storage. Good housekeeping is a most significant contribution toward an active safety campaign.

9.3.3.2   Moving Chemicals

Another area of concern is the transfer of chemicals, apparatus, gases, or other hazardous materials from the storeroom to the laboratory for use. Use cradles or drum tilters to facilitate handling carboys or other large chemical vessels. A carboy is a large (5–15 gallon) protected glass or plastic container with a narrow neck. It is used for corrosive liquids. It may also have a low-level tap and valve.

 Drum Tilter

In transporting cylinders of compressed gases, use a trussed hand truck. Never roll a cylinder by its valve. Immediately after they are positioned for use, cylinders should be clamped securely into place to prevent shifting or toppling.

Carry flammable liquids in safety cans or, in the case of reagent-grade chemicals, protect the bottle with a carrier. Always wear protective gloves, safety shoes, and rubber aprons in case of accidental spilling of chemical containers.

9.3.3.3 Proper Laboratory Techniques

Faulty technique is one of the chief causes of accidents and, because it involves the human element, is one of the most difficult to correct.

Because of their nature and prevalence in the laboratory, acids and other corrosive materials constitute a series of hazards ranging from poisoning, burning, and gassing through explosion. Always flush the outsides of acid bottles with water before opening them. Do not lay the stopper down on the countertop where a person might lay a hand or rest an arm on it. Keep all acids tightly stoppered when not in use and make sure no spilled acid remains on the floor, table, or bottle after use.

When concentrated acids are combined with water, heat is released. If you add acid to water, a dilute solution forms that does not get excessively hot. If you add water to acid, so much heat is created that the water can boil and cause concentrated acid to splash out of the container (and perhaps onto you). To avoid splashing acid, always pour acid into water; do not pour water into acid. A good way to remember this safety advice is that you add chemicals to the swimming pool, you do not add the swimming pool to the chemicals.

Mercury requires special care. Keep all mercury containers tightly stoppered. Even a small amount of spilled mercury can poison the atmosphere in a room. To clean up a small mercury spill (the amount in a thermometer), put on rubber, nitrile, or latex gloves. Use a squeegee or cardboard to gather the mercury beads. Use slow sweeping motions to keep the mercury from becoming uncontrollable. Use a disposable eyedropper to collect or draw up the mercury beads. Slowly and carefully squeeze the mercury onto a damp paper towel. Place the paper towel in a self-sealing bag and secure. Make sure to label the bag as directed by your local health or fire department. Never pour mercury, or any hazardous material, down a drain. It may lodge in the plumbing and cause future problems during plumbing repairs. If discharged, it can pollute the septic tank or wastewater treatment plant. Contact your local health department, municipal waste authority, or fire department for proper disposal in accordance with local, state, and federal laws. Remember to keep the area well ventilated to the outside (windows open and fans in exterior windows running) for at least 24 hours after your successful cleanup. For larger spills, call your local or state health or environmental agency.

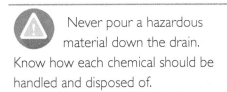

Never pour a hazardous material down the drain. Know how each chemical should be handled and disposed of.

LABELING

It is extremely important that chemical containers be properly labeled to identify their contents and associated hazards. Labeling helps prevent accidental misuse or mixing of incompatible chemicals. Proper labeling also enables quick decision making and action during an emergency. Besides potential safety hazards, improper labeling can lead to poor lab results from using the wrong chemical or the wrong amount of chemical.

Improper labeling can also cost you money. Any chemical that is not labeled must be assumed to be hazardous, and disposal costs for hazardous materials can be quite high.

Some best practices to use in labeling are:

- Use both the chemical name and formula as a double check.
- Clearly indicate the concentration. The same chemical can be harmless at low concentrations but dangerous at high concentrations.
- Coordinate labels with the SDS. If a chemical needs special precautions, such as limits on storage temperatures or the need to use under a fume hood, add those to your label.
- Use labels and ink that are not soluble in water or in the liquid in the container. If the label is accidentally washed clean, replace it immediately.
- For reagents that you formulate for internal lab use, include the name or initials of the person making up the reagent and the date it was made. Many reagents have a limited storage time (see *Standard Methods*).

9.3.3.4 Accident Prevention

Electric Shock. Wherever there are electrical outlets, plugs, and wiring connections, there is a danger of electric shock. The usual dos and don'ts of protection against shock in the home are equally applicable in the laboratory. Do not use worn or frayed wires. Replace connections when there is any sign of thinning insulation. Ground all apparatus using three-pronged plugs. Do not continue to run a motor after liquid has spilled on it. Turn it off immediately and clean and dry the inside thoroughly before attempting to use it again.

Electrical units that are operated in an area exposed to flammable vapors should be explosion-proof. All permanent wiring should be installed by an electrician with proper conduit or BX (armored) cable to eliminate any danger of circuit overloading.

Cuts. Laboratory glassware is fragile so it should only be used for its specific purpose. Proper handling can reduce the risk of injury and accident. Inspect glassware before use for cracks, scratches, and other defects and remove defective glassware from use. Use two hands when carrying glassware and wear gloves if there is a risk of breakage, thermal hazard (handling hot or cold glassware), or chemical contamination. Also, wear safety goggles when working in the laboratory.

Using common sense will prevent many accidents. For example, do not set hot glass on a cold surface, and do not store glassware in fume hoods or drawers without drawer pads. Take precautions when using glassware in vacuum or pressure operations. Use specific glassware designed to withstand vacuum or pressure procedures, and also temperatures. Many glassware accidents occur during cleaning, so protective equipment and good lab techniques should be used.

Some pieces of glass used in the laboratory, such as glass tubing, thermometers, and funnels, must be inserted through rubber stoppers. Never force the glass into place or the glass may break. This is a common source of cuts in the laboratory. Take care in making rubber-to-glass connections. Wear gloves and hold the tubing as close to the end being inserted as possible to prevent bending or breaking. Long glass tubes should be supported while they are being inserted into rubber. The ends of the glass should be **flame polished** and either wet or covered with water, soapy water, glycerin, or a lubricating jelly for ease in joining connections. Never use oil or grease. Also, in disassembling a connection, never try to force glassware off of tubing or out of stoppers. Cut the rubber away from the glass.

Burns. All glassware and porcelain look cold after the red color from heating has disappeared. Hot glass looks like cold glass, so take precautions. The color is gone in seconds but the glass stays hot enough to burn for quite a while. After heating a piece of glass, put it out of the way until cold. Use gloves and tongs to prevent burns from hot glass. Never decide it is too much trouble to put on a pair of gloves or use a pair of tongs to handle a dish or flask that has been heated.

Chemical burns can be caused by spattering from acids, caustic materials, and strong oxidizing solutions. Use gloves, aprons, and goggles to protect yourself. Know the location and operation of emergency deluge showers and eyewash stations. In case of accident, wash the affected area of your body immediately with large quantities of water. Immediately flood your eyes with water from an eyewash station or use a special eyewash solution from a safety kit. Keep vinegar and soda handy to neutralize acids and bases (caustic materials).

Figure 9.7 Fume hood

Toxic Fumes. Use a ventilated laboratory fume hood (Figure 9.7) for volatile reagent preparation. Select a hood that has adequate air displacement and collects harmful vapors and gases at their source. Noxious fumes can be spread by the building's heating and cooling systems, so make sure the hood is properly vented.

Waste Disposal. A good safety program requires constant care in disposal of laboratory waste. Hazardous materials should never be poured down an ordinary sink or drain. Corrosive substances disposed down a drain can corrode away the drain pipe or trap. Corrosive acids should be neutralized and poured down corrosion-resistant sinks and sewers using large quantities of water to dilute and flush the acid.

Never pour hazardous materials down the drain. Corrosives can damage pipes. Flammables can volatilize and in extreme cases the vapors can explode. Wastewater treatment plants do not remove all chemicals, so toxins you pour down the drain can pass through the treatment plant and poison local water bodies or groundwater. This is especially true for septic systems. Always dispose of hazardous materials in accordance with local or state health or environmental regulations.

Provide a dedicated trash receptacle for broken glass and empty it separately. Mixing broken glass with other waste can be a cut hazard for custodial staff.

Fire. The laboratory should be equipped with a fire blanket. The fire blanket is used to smother clothing fires. Small fires that occur in an

flame polished
Melted by a flame to smooth out irregularities. Sharp or broken edges of glass (such as the end of a glass tube) are rotated in a flame until the edge melts slightly and becomes smooth.

evaporating dish or beaker may be put out by covering the container with a glass plate, wet towel, or wet blanket. For larger fires, or ones that may spread rapidly, promptly use a fire extinguisher. Do not use a fire extinguisher on small beaker fires because the force of the spray will knock over the beaker and spread the fire.

Take time to become familiar with the operation and use of your fire extinguishers, and use the proper type of extinguisher for each class of fire. For example, water should not be poured on grease fires, electrical fires, or metal fires because water could increase the hazards, such as splattering of the fire or electric shock.

Fire classifications are important for determining the type of fire extinguisher needed to control the fire. Classifications also aid in recordkeeping. Fires are classified as A, B, C, or D fires based on the type of material being consumed. Fire extinguishers are classified the same as fires to correspond with the class of fire each will extinguish.

Class A fires: Ordinary Combustibles such as wood, paper, cloth, rubber, many plastics, dried grass, hay, and stubble. Use foam, water, soda-acid, carbon dioxide gas, or almost any type of extinguisher.

Class B fires: Flammable and Combustible Liquids such as gasoline, oil, grease, tar, oilbased paint, lacquer, and solvents, and also flammable gases. Use foam, carbon dioxide, or dry chemical extinguishers.

Class C fires: Energized Electrical Equipment such as starters, breakers, and motors. Use carbon dioxide or dry chemical extinguishers to smother the fire; both types are nonconductors of electricity.

Class D fires: Combustible Metals such as magnesium, sodium, zinc, and potassium. Operators rarely encounter this type of fire. Use a Class D extinguisher or use fine dry soda ash, sand, or graphite to smother the fire.

Multipurpose extinguishers are also available, such as a Class BC carbon dioxide extinguisher that can be used to smother Class B and Class C fires. A multipurpose ABC carbon dioxide extinguisher will handle most laboratory fire situations. (When using carbon dioxide extinguishers, remember that the carbon dioxide can displace oxygen—take appropriate precautions.)

There is no single type of fire extinguisher that is effective for all fires so it is important that you understand the class of fire you are trying to control. You must be trained in the use of the different types of extinguishers, and the proper type should be located near the area where that class of fire may occur. Consult with your local fire department about the best methods to use for specific hazards that exist at your facility.

 General Industry, OSHA Safety and Health Standards (CFR Title 29, Pt. 1910). US Government Publishing Office. www.gpo.gov/.

Standard Methods, 22nd Edition, Part 1090, Laboratory Occupational Health and Safety. www.awwa.org. ISBN: 9780-87553-013-0.

Check Your Understanding

1. List seven laboratory hazards.

2. Why should you not work alone in a laboratory?

3. True or False? You may add acid to water, but never water to acid.

4. How would you dispose of a corrosive acid?

9.4   Water Quality Tests

This section contains detailed laboratory methods for several tests that are routinely conducted in water treatment plants. They are divided into three general categories—instrument-based tests (temperature, pH, and turbidity), biological tests (coliform bacteria), and tests for process control (jar test for coagulation and flocculation, and chlorine demand). Generally speaking, these methods mirror those in *Standard Methods for the Examination of Water and Wastewater* (American Public Health Association, American Water Works Association, and Water Environment Federation). They are presented here to educate you on what is involved in these tests. Some details have been left out or summarized with the understanding that on the job you will be following *Standard Methods* procedures or their equivalent (e.g., EPA, your state, or your plant).

9.4.1   Instrument-Based Tests

Instrument-based tests are commonly used in water laboratories. These include tests for temperature, pH, and turbidity.

9.4.1.1   Temperature

Temperature is one of the most frequently taken tests in the water industry. Accurate water temperature readings are important because of its influence on chemical reaction rates, biological growth, dissolved gas concentrations, and water viscosity, not to mention its acceptability by consumers for drinking.

WHAT IS TESTED

Sample	Common Range, °C
Raw and Treated Surface Water	5–25
Well Water	10–20

APPARATUS

One National Institute for Standards and Technology (NIST) thermometer for calibrating the other thermometers.

Thermometer Types
- Liquid-in-glass. This is the classic style in which the temperature is indicated by expansion of a liquid. Mercury and alcohol are the most common liquids. There are two kinds of liquid-in-glass thermometers:
    1. Total immersion. This type of thermometer must be totally immersed when read. Readings with this type of thermometer will change most rapidly when removed from the liquid to be recorded.
    2. Partial immersion. This type thermometer will have a solid line (water-level indicator) around the stem below the point where the scale starts.

- Dial. This type has a dial that is actuated by the differential expansion of two different metal strips joined together. The bi-metal strip bends as one metal expands more than the other.
- Digital. This is a common style both at home and on the job. Many are based on measuring voltage across an element whose resistance varies by temperature. The sensing element is separated from the electronic readout. It might be in a probe attached to the readout or it might be at the end of a wire located far from the readout to allow viewing the temperature at a distance.

REAGENTS

None required.

PROCEDURE

Although temperature is easily measured by placing the thermometer into the water, there are some best practices to follow:

1. Collect as large a volume of sample as is practical. The temperature will have less chance to change in a large volume than in a small container.
2. Immerse the thermometer to the proper depth. Do not touch the bottom or sides of the sample container with the thermometer.
3. Record temperature to the nearest fraction of a degree that can be estimated from the thermometer available.
4. The temperature of water samples collected from a distribution system mainly depends on the soil temperature at the depth of the water main. To get an accurate reading of this temperature, allow the water to continuously overflow a small container (a polystyrene coffee cup is ideal). Place the thermometer in the cup. After there has been no change in the temperature reading for 1 minute, record the temperature. Use this procedure for well water samples, too.

PRECAUTIONS

- To avoid breaking or damaging a glass thermometer, store it in a shielded metal case.
- Periodically, check your thermometer's accuracy against the NIST-certified thermometer by measuring the temperature of a sample with both thermometers simultaneously. Some of the poorer quality thermometers are substantially inaccurate (off as much as 6°F or 3°C).

Example 11

To measure the temperature of treated water, a sample was obtained in a gallon bottle, the thermometer immediately immersed, and a temperature of 15°C recorded after the reading became constant.

Normally, we measure and record temperatures using a thermometer with the proper scale. We could, however, measure a temperature in °C and convert to °F. The following formulas are used to convert temperatures from one scale to the other.

1. If we measure in °F and want °C,

$$\text{Temperature}[°C] = (°F - 32°) \times \frac{5}{9}$$

2. If we measure in °C and want °F,

$$\text{Temperature}[\degree F] = \left(\degree C \times \frac{9}{5}\right) + 32\degree$$

3. The measured treated water temperature was 15°C. What is the temperature in °F?

$$\text{Temperature}[\degree F] = \left(\degree C \times \frac{9}{5}\right) + 32\degree$$

$$= \left(15\degree \times \frac{9}{5}\right) + 32\degree$$

$$= 27\degree + 32\degree$$

$$= 59\degree F$$

Check Your Understanding

1. Why are temperature readings important?

2. Why should a total-immersion thermometer remain immersed in the liquid while being read?

3. Why should thermometers be calibrated against an accurate NIST-certified thermometer?

 Standard Methods, 22nd Edition, Part 2550, Temperature. www.awwa.org. ISBN: 978-0-87553-013-0.

9.4.1.2   pH

The **pH** of a water indicates the intensity of its **acidic** or basic strength. The pH scale runs from 0 to 14. Water having a pH of 7 is at the midpoint of the scale and is considered neutral. Such a water is neither acidic nor basic. A pH of greater than 7 indicates basic water. The stronger the basic intensity, the greater the pH. Conversely, the stronger the acidic intensity of the water, the lower the pH will be.

Mathematically, pH is the logarithm (base 10) of the reciprocal of the hydrogen ion activity, or the negative logarithm of the hydrogen ion activity.

pH SCALE

$0 \leftarrow$ INCREASING ACID $- - - - 7 - - - - -$ INCREASING BASE $\rightarrow 14$

$1 \leftarrow$ $2 \leftarrow$ $3 \leftarrow$ $4 \leftarrow$ $5 \leftarrow$ 6 $8 \rightarrow$ $9 \rightarrow 10 \rightarrow$ $11 \rightarrow 12 \rightarrow 13$

Neutral

$$pH = \text{Log}\frac{1}{\{H^+\}} = -\text{Log}\left(\{H^+\}\right)$$

In freshwater, the hydrogen ion activity is equal to its molar concentration (see earlier discussion on molar concentration). The nomenclature, $[H^+]$, means the concentration of hydrogen ions in mol/L.

Acids have many H^+ ions; bases have few. For example, if a water has a pH of 1 (very acidic), then the hydrogen ion activity $[H^+] = 10^{-1} = 0.1$ mol/L. If the pH is 14, then $[H^+] = 10^{-14} = 0.00000000000001$ mol/L. A change in the pH of one unit represents changing the hydrogen ion level by a factor of 10 (10 times).

In a solution, both hydrogen ions (H^+) and hydroxyl ions (OH^-) are always present. At a pH of 7, the activity of both hydrogen and hydroxyl ions equals 10^{-7} *M*. When the pH is less than 7, the activity (molar concentration) of hydrogen ions is greater than the hydroxyl ions.

pH plays an important role in the water treatment processes such as disinfection, coagulation, softening, and corrosion control. The pH test also indicates changes in raw and finished water quality.

pH (pronounce as separate letters)
pH is an expression of the intensity of the basic or acidic condition of a liquid. Mathematically, this is equivalent to the negative of the base 10 logarithm.
If $\{H^+\} = 10^{-6.5}$, then pH = 6.5. The pH may range from 0 to 14, where 0 is most acidic, 14 most basic, and 7 neutral.

$$pH = \text{Log}\frac{1}{\{H^+\}} \text{ or } -\text{Log}\left(\{H^+\}\right)$$

acidic (uh-SID-ick)
The condition of water or soil that contains a sufficient amount of acid substances to lower the pH below 7.0.

Standard Methods, 22nd Edition, Part 4500-H+, pH Value. www.awwa.org. ISBN: 978-0-87553-013-0.

Approximate pH measurements can be made with chemical indicators. **Indicators** are compounds that change color at certain pH values due to reactions with [H+]. This is called a **colorimetric measurement**. Litmus paper is an example of this.

Accurate lab measurements of pH are done with an electrochemical probe. The tip of the probe is a special glass whose electrical properties change with pH. The meter reads a voltage difference between two internal electrodes.

If you are measuring pH in colored samples or samples with high solids content, or if you are taking measurements that need to be reported to regulatory authorities, you should use a pH electrode and meter instead of a colorimetric measurement or test papers. pH meters are capable of providing ± 0.1 pH accuracy in most applications. In contrast, colorimetric tests provide ± 0.1 pH accuracy only in a limited range. pH papers provide even less accuracy.

WHAT IS TESTED

Sample	Common Range, NTU
Surface Water	6.5–8.5
Finished Water	7.5–9.0
Well Water	6.5–8.0

APPARATUS

- pH meter (see Figure 9.8)
- Glass electrode and reference electrode (inside the probe)
- Magnetic stirrer and stir-bar

REAGENTS

1. Buffer tablets for various pH value solutions (available from laboratory chemical supply houses).
2. Distilled water.

indicator
A substance that gives a visible change, usually of color, at a desired point in a chemical reaction, generally at a specified endpoint.

colorimetric measurement
A means of measuring unknown chemical concentrations in water by measuring a sample's color intensity. The specific color of the sample, developed by addition of chemical reagents, is measured with a photoelectric colorimeter or is compared with color standards using, or corresponding with, known concentrations of the chemical.

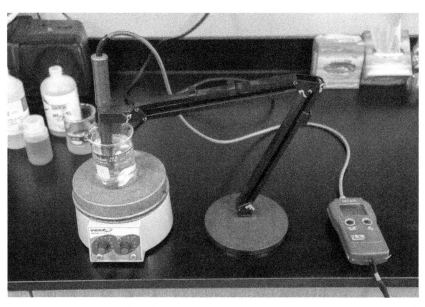

Figure 9.8   A working pH meter (right) setup

PROCEDURE

1. Due to the difference between the various makes and models of pH meters commercially available, specific instructions cannot be provided for the correct operation of all instruments. In each case, follow the manufacturer's instructions for preparing the electrodes and operating the instrument.

2. Standardize the instrument (pH meter) against a **buffer** solution with a pH close to that of the sample, or with buffers over a range as per manufacturer's instructions.

3. Rinse electrodes thoroughly with distilled water after removal from buffer solution.

4. Place electrodes in sample and measure pH.

5. Remove electrodes from sample, rinse thoroughly with distilled water.

6. Immerse electrode ends in beaker of pH 7 buffer storage solution.

7. Turn meter to standby (or OFF).

PRECAUTIONS

- To avoid faulty instrument calibration, prepare fresh buffer solutions as needed, and at least once per week (from commercially available buffer tablets).

- pH meter, buffer solution, and samples should all be near the same temperature because temperature variations will give somewhat erroneous results. Allow a few minutes for the probes to adjust to the buffers before calibrating a pH meter to ensure accurate pH readings.

- Watch for erratic results arising from electrodes, faulty connections, or fouling of electrodes with oily precipitated matter. Films may be removed from electrodes by placing isopropanol on a tissue or a Q-tip and cleaning the probe. (Follow the manufacturer's instructions.)

- The temperature compensator on the pH meter adjusts the meter for changes in electrode response with temperature. However, the pH of water may change with temperature due to certain reactions and the pH meter cannot compensate for this change.

APPLICATIONS OF PH

As described in Chapter 6, "Disinfection," the pH of water has an important influence on the effectiveness of chlorine **disinfection**. Chlorine is a more effective disinfectant at lower pH values. A chlorine residual of 0.2 mg/L at a pH of 7 is just as effective as 1 mg/L at a pH of 10. Five times as much chlorine is required to do the same disinfecting job at pH 10 as it does at pH 7.

As described in Chapter 7, "Corrosion Control," pH is an important parameter in corrosion control. It is used to determine whether or not the finished water is corrosive and needs to be adjusted with lime or caustic soda.

As discussed in Chapter 4, "Sedimentation," chemical coagulants each have a pH range in which they work most effectively.

9.4.1.3 Turbidity

The term turbidity is an expression of the physical cloudiness of water. Turbidity is caused by the presence of suspended matter such as silt, finely divided organic and inorganic matter, and microscopic organisms such as algae. Operators have more control over turbidity than other water quality indicators.

(i) Although they both affect the clarity of a water sample, turbidity and color are not the same thing. Turbidity is caused by particles that scatter (change the direction of) light. Removing these particles is the treatment goal of your plant. If the particles are colored, the sample will have an apparent color when you look at it. True color, in contrast, is caused by dissolved substances preferentially absorbing some wavelengths of light as it passes through the water. The wavelengths that pass through the water (i.e., are not absorbed) are picked up by your eye and interpreted as color. For example, tannins (or tannic acid) are naturally occurring materials that absorb blue and green light. Since they let the yellows and reds pass through water, it has a tan or tea color. Because it is caused by dissolved molecules, color is not removed by coagulation, settling, or sand filtration.

---

buffer
A solution or liquid whose chemical makeup neutralizes acids or bases, thereby maintaining a relatively constant pH.

disinfection (dis-in-FECT-shun)
The process designed to kill or inactivate most microorganisms in water or wastewater, including essentially all pathogenic (disease-causing) bacteria. There are several ways to disinfect, with chlorination being the most frequently used in water and wastewater treatment plants.

The accepted method used to measure turbidity is called the **nephelometric** method. The nephelometric turbidity meter or nephelometer (Figure 9.9) is designed to measure particle-reflected light at an angle of 90 degrees to the source beam. The greater the intensity of scattered light, the higher the turbidity.

Turbidity is measured in turbidity units (TUs) of various kinds. A turbidity unit is an empirical quantity that is based on the amount of light that is scattered by particles in a standard. The usual standard is a suspension of formazin, an insoluble polymer formed by a particular reaction. When formazin is the standard, the units are Formazin Turbidity Units (FTU). The most common unit you will encounter, however, is the Nephelometric Turbidity Unit (NTU) obtained from nephelometric turbidity meters. In practice, NTUs and FTUs are interchangeable because nephelometric turbidity meters are calibrated most commonly using formazin standards*. In some parts of the law, NTUs are referred to as simply turbidity units (TUs).

Secondary turbidity standards are suspensions of various materials formulated to match the primary formazin solutions. These secondary standards are generally used because of their convenience and the instability of dilute formazin primary standard solutions. Examples of these secondary standards include standards that are supplied by the turbidity meter's manufacturer with the instrument. Periodic checks of these secondary standards against the primary formazin standard are recommended to provide assurance of measurement accuracy. Because of their instability, secondary standards should not be used past their expiration date.

The turbidity measurement is one of the most important tests the plant operator performs. The Safe Drinking Water Act stipulates specific monitoring requirements for turbidity. Turbidity of treated surface water must be measured every 4 hours or continuously and the results reported to the appropriate authority. Turbidity testing is the most critical tool in recognizing changes in raw water quality, detecting problems in coagulation and sedimentation, and troubleshooting filtration problems.

(i) *Other turbidity units. In international standards, turbidity is expressed in Formazin Nephelometric Units (FNUs). FNUs are similar to FTUs except FTUs are based on the scatter of white light and FNUs are based on the scatter of infrared light.

WHAT IS TESTED

Sample	Common Range, NTU
Untreated Surface Water	1–300
Filtered Water	0.03–0.50
Well Water	0.05–1.0+

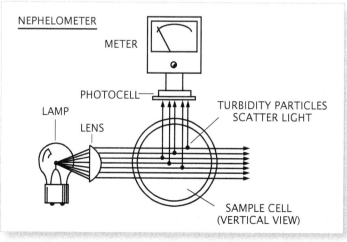

nephelometric (neff-el-o-MET-rick)
A means of measuring turbidity in a sample by using an instrument called a nephelometer. A nephelometer passes light through a sample and the amount of light deflected (usually at a 90-degree angle) is then measured.

Figure 9.9 Nephelometer *(Courtesy of HACH Company)*

APPARATUS

- Turbidity meter: To minimize differences in turbidity measurements, rigorous specifications for turbidity meters are necessary. The turbidity meter should have the following important characteristics:
  - The turbidity meter should consist of a nephelometer with a light source illuminating the sample, and one or more photoelectric detectors to indicate the intensity of scattered light at a 90° angle with a readout device. A typical instrument is shown in Figure 9.10.
  - The light source should be an intense tungsten filament lamp.
  - The total distance traveled by the light through the sample should be less than about 5 centimeters.
  - The instrument should have several measurement ranges. The instrument should be able to measure from 0 to 100 turbidity units, with sufficient sensitivity (on the lowest scale) to detect differences of 0.02 or less in filtered waters having turbidities of less than one unit.
- Sample tubes: These are usually provided with the instrument.

Figure 9.10   Turbidity meter *(Courtesy of HACH Company)*

REAGENTS

1. Turbidity-free water: Pass distilled (or reagent-grade) water through a membrane filter having a pore size small enough to exclude small particles (i.e., no greater than 0.2 microns). Discard the first 200 mL collected.

2. Stock formazin turbidity suspension: Stock secondary standard turbidity suspensions that require no preparation are available from commercial suppliers and approved for use. Instructions for preparing a primary formazin standard are contained in *Standard Methods*. Secondary standards are available commercially. Replace them when their age exceeds their stated shelf life.

3. Dilute turbidity standards: Dilute portions of the standard turbidity suspension with turbidity-free water as required.

PROCEDURE

1. Turbidity meter calibration: Follow the manufacturer's operating instructions. Measure your standard solutions on the turbidity meters covering the range of interest. If the instrument is already calibrated in standard turbidity units, this procedure will check the accuracy of the calibration scales. At least one standard should be run in each instrument range to be used. Some instruments permit adjustments of sensitivity so that scale values will correspond to turbidities. Reliance on an instrument manufacturer's scattering standards for calibrating the instrument is not an acceptable practice unless they are in acceptably close agreement with prepared standards.

2. Measurements:
   a. Sample tubes must be kept scrupulously clean, both inside and out. Discard them when they become scratched or etched. Never handle them where the light strikes them.
   b. Turbidity should be measured in a sample as soon as possible to obtain accurate results. The turbidity of a sample can change after the sample is collected. Shaking the sample will not re-create the original turbidity.

(i) For updated information on monitoring requirements and turbidity requirements of the Surface Water Treatment Rule (SWTR), see the Interim Enhanced Water Treatment Rule and Long Term Enhanced Water Treatment Rules available from the EPA at www.epa.gov/dwreginfo/drinking-water-rule-quick-reference-guide. For more information, check the drinking water regulations chapter in Volume 2 in this series of manuals.

c. Samples should be agitated enough to assure that all the particles are suspended but not so much as to introduce air bubbles. If needed, de-gas samples with vacuum, surfactants, or an ultrasonic bath as per *Standard Methods*.

d. Samples should not be diluted unless absolutely necessary. Besides potential measurement errors, particles in the original sample may dissolve or otherwise change characteristics.

3. Report turbidity results as follows:

Turbidity Reading (NTU)	Record to Nearest (NTU)
0.0–1.0	0.05
1–10	0.1
10–40	1
40–100	5
100–400	10
400–1,000	50
> 1,000	100

Example 12

Dilution Calculation for Standard: You wish to make a 100 NTU calibration standard by diluting a 4,000 NTU commercial standard with turbidity-free water. You measure out 5.00 mL of the 4,000 NTU standard with a volumetric pipet and place it in a beaker. How much dilution water should you add?

$$(NTU \times V)_{\text{before dilution}} = (NTU \times V)_{\text{after dilution}}$$

$$V_{\text{after dilution}} = \frac{(NTU \times V)_{\text{before dilution}}}{NTU_{\text{after dilution}}} = \frac{4,000\,[NTU] \times 5\,[mL]}{100\,[NTU]} = 200\,mL$$

$$V_{\text{after dilution}} = V_{\text{standard}} + V_{\text{dilution water}}$$

$$V_{\text{dilution water}} = V_{\text{after dilution}} - V_{\text{standard}} = 200\,[mL] - 5\,[mL] = 195\,mL$$

Standard Methods, 22nd Edition, Part 2130 B.5, Turbidity. www.awwa.org. ISBN: 978-0-87553-013-0.

Check Your Understanding

1. What is an indicator?

2. What does the pH of water indicate?

3. True or false? The more acidic the water, the lower the pH.

4. What precautions should be exercised when using a pH meter?

5. What are the causes of turbidity in water?

6. How is turbidity measured?

9.4.2 Titration-Based Tests

The following three tests are based on titrations.

9.4.2.1 General Titration Procedure

A titration involves adding a measured amount of a standardized solution, which is usually in a buret, to another solution in a flask or beaker.

The solution in the buret is referred to as the titrant and is added to the other solution until there is a measurable change in the test solution in the flask or beaker. This change is frequently a color change as a result of adding another chemical called an indicator to the solution in the flask before the titration begins. The solution in the buret is added slowly to the flask until the change, which is called the **endpoint**, is reached. The entire process is the titration. Figure 9.11 illustrates the four general steps used during a chemical titration.

In a titration test, the titrant is reacting with a constituent in the sample. The endpoint usually indicates that the reaction has been completed, so the amount of titrant is proportional to the constituent being measured. It is therefore important not to titrate past the endpoint. Doing so will overestimate the parameter being measured.

Check Your Understanding

1. What is an endpoint?

2. In a titration test, what does the titrant react with?

9.4.2.2 Alkalinity

The **alkalinity** of a water sample is a measure of the water's capacity to neutralize acids. In natural and treated waters, alkalinity is caused mainly by bicarbonate (HCO_3^-), carbonate (CO_3^{2-}), and hydroxide (OH^-), which originate from calcium, magnesium, and sodium compounds. Carbonate ions and hydroxide react with hydrogen ions (from acids) as shown below:

$$OH^- + H^+ \rightarrow H_2O$$
$$CO_3^{2-} + H^+ \rightarrow HCO_3^-$$
$$HCO_3^- + H^+ \rightarrow H_2CO_3$$

H_2CO_3 is carbonic acid, which is the result of carbon dioxide reacting with water. (Soda pop is essentially carbonic acid mixed with sugar and flavoring.)

Notice that in the reactions above, the hydrogen ions are consumed rather than contributing to the H^+ concentration of the solution. In turn, this means the pH does not decrease, i.e., the acid is neutralized. Because of equilibrium chemistry, not every hydrogen ion added is removed as implied above. So the pH is not static. It drops slowly as you add acid until all the hydroxide, carbonate, and bicarbonate are consumed. The hydroxide is consumed by about pH 10; the carbonate is consumed by about pH 8.3, and the bicarbonate is consumed by about pH 4.5. In the alkalinity titration, you add acid until you reach these pH endpoints (see Figure 9.12). The amount of acid you add is the amount that the solution can neutralize, that is, it's the alkalinity. Titrating down to pH 4.5 measures the total alkalinity. Titrating down to pH 8.3 measures the phenolphthalein alkalinity, named so because this indicator is often used to show the endpoint. Alkalinity is usually expressed in terms of **calcium carbonate (CaCO$_3$) equivalents**.

Knowing alkalinity is important for several water treatment processes. Many of the chemicals used in water treatment, such as alum, chlorine, or lime, act as acids or bases and can change the pH if the water is

---

i The phenomenon of little pH change with the addition of an acid or a base is called buffering. Buffering is desirable to prevent wide swings in pH. Bicarbonate can neutralize both acids (by reacting with the hydrogen ions) and bases (by giving off a hydrogen ion). Waters with a lot of bicarbonate (and high alkalinity) are well buffered, meaning that relatively large amounts of acid or base can be added without significantly changing the pH. Waters with little bicarbonate (and low alkalinity) are poorly buffered, meaning the pH is sensitive to even relatively small amounts of added acids or bases.

endpoint
The completion of a desired chemical reaction. In the titration of a water or wastewater sample, a chemical is added, drop by drop, to the sample until a certain color change occurs (blue to clear, for example). In addition to a color change, an endpoint may be reached by the formation of a precipitate or by reaching a specified pH. An endpoint may be detected by using an electronic device such as a spectrophotometer or pH meter.

alkaline (AL-kuh-line)
The condition of water or soil that contains a sufficient amount of alkali substances to raise the pH above 7.0.

calcium carbonate (CaCO$_3$) equivalent
An expression of the concentration of specified water constituents in terms of their equivalent value to calcium carbonate. For example, the hardness in water that is caused by calcium, magnesium, and other ions is usually described as calcium carbonate equivalent. Alkalinity test results are usually reported as mg/L CaCO$_3$ equivalents.

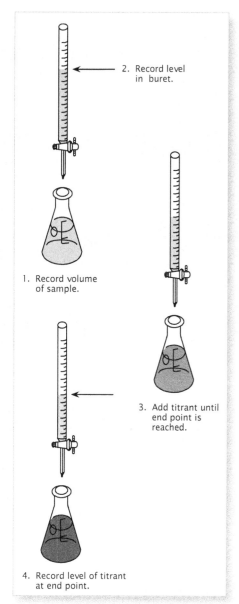

Figure 9.11   Titration steps

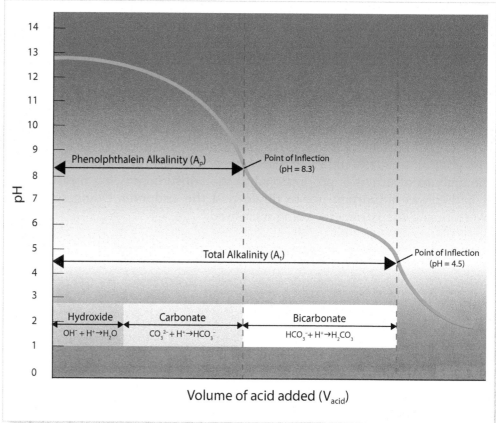

Figure 9.12   Idealized acid titration curve

insufficiently buffered. As noted in Chapter 3, "Coagulation and Flocculation," reactions between the alkalinity minerals and coagulants are what create insoluble flocs. Carbonate is a necessary ingredient in $CaCO_3$ which is important in corrosion protection (Chapter 7, "Corrosion Control") and water softening (Volume II in this series of manuals).

WHAT IS TESTED

Sample	Common Range, mg/L as $CaCO_3$
Raw and Treated Surface Water	20–300
Well Water	80–500

APPARATUS

- pH meter and probe
- Magnetic stirrer and stir bar
- Analytical balance
- Graduated cylinder
- Buret
- Beaker (250 mL)
- Dessicator

REAGENTS

Standardized solutions are commercially available for most reagents. Refer to *Standard Methods* if you wish to prepare your own reagents.

1. Sodium carbonate (Na_2CO_3) solution, approximately 0.05 *N*.
2. Sulfuric acid (H_2SO_4), 0.1 *N* or other.
3. Standard sulfuric acid, 0.02 *N*: Dilute 200 mL 0.10 *N* standard acid to 1 liter with distilled water using a volumetric flask.

PROCEDURE (FIGURE 9.13)

Determine acid titrant normality. (Do not depend on dilution of 0.1 N acid solution to be sufficiently accurate.)

1. Place 15.00 mL of 0.05 N sodium carbonate and about 60 mL of distilled water into a beaker.
2. While stirring, titrate the solution to about pH 5. Remove electrodes.
3. Cover the beaker with a watch glass cover and boil gently for 3–5 minutes.

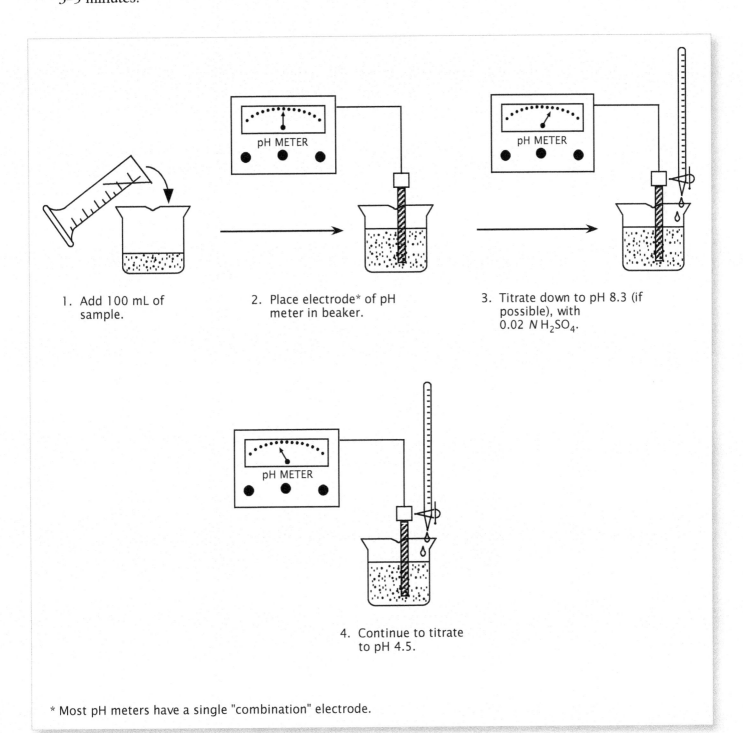

1. Add 100 mL of sample.

2. Place electrode* of pH meter in beaker.

3. Titrate down to pH 8.3 (if possible), with 0.02 N H_2SO_4.

4. Continue to titrate to pH 4.5.

* Most pH meters have a single "combination" electrode.

Figure 9.13 Outline of procedure for alkalinity

4. Cool to room temperature. Rinse the cover glass into the beaker with distilled water. Finish titrating to pH 4.5.

$$N_{acid} = \frac{N_{Na_2CO_3} \times V_{Na_2CO_3}}{V_{acid}}$$

Where,

N_{acid} = Acid Normality

$N_{Na_2CO_2}$ = Normality of Sodium Carbonate Solution

$V_{Na_2CO_3}$ = Volume of Sodium Carbonate Solution

V_{acid} = Total Volume of Acid Used in the Titration

TOTAL AND PHENOLPHTHALEIN ALKALINITY

1. Take a clean beaker and add 100 mL of sample (or other sample volume that will give a titration volume of less than 50 mL of acid titrant).
2. Place the electrodes of pH meter into the beaker containing sample.
3. Stir sample slowly (with a magnetic stirrer if possible).
4. Check pH of sample. If pH is 8.3 or below, then there is no phenolphthalein alkalinity present and you can go to Step 6.
5. If the pH is greater than 8.3, titrate very carefully to a pH of 8.3 with 0.02 N H_2SO_4. Record the amount of acid used to this point.
6. Continue to titrate to pH 4.5 with 0.02 N H_2SO_4. *Standard Methods*, 22nd Edition, Part 2320 B, Titration Method, recommends titrating to a pH of 4.5 for routine or automated analyses—other pH levels are suggested for various levels of alkalinity. Record the total amount of acid used from starting point to finish.
7. Calculate total and phenolphthalein (if present) alkalinities.

   In lieu of using a pH meter, bromcresol green or bromcrosol green-methyl red indicators can be used for pH 4.5 and metacresol purple or phenolphthalein indicators can be used for pH 8.3.

Standard Methods, 22nd Edition, Part 2310 B.2, Titration Method. www.awwa.org. ISBN: 978-0-87553-013-0. *Method 310.1 EPA Methods for Chemical Analysis of Water and Wastes*, US Environmental Protection Agency. 1983. EPA/600/4-79/020. www.epa.gov.

INTERPRETING RESULTS

There are five alkalinity conditions possible in a water sample: bicarbonate alone, bicarbonate and carbonate, carbonate alone, carbonate and hydroxide, and hydroxide alone. These conditions may be distinguished and quantities determined from the results of acid titrations as shown below.

Titration Result	Alkalinity Values by Type, mg/L as $CaCO_3$		
	Bicarbonate	Carbonate	Hydroxide
$A_p = 0$	A_t	0	0
A_p is less than ½ A_t	$A_t - 2A_p$	$2A_p$	0
$A_p = $ ½ A_t	0	$2A_p$	0
A_p is greater than ½ A_t	0	$2A_t - 2A_p$	$2A_p - A_t$
$A_p = A_t$	0	0	A_t

Where,

A_p = Phenolphthalein Alkalinity
        (carbonate alkalinity + hydroxide alkalinity)

A_t = Total Alkalinity (bicarbonate + carbonate + hydroxide)

PRECAUTIONS

- Samples should be analyzed as soon as possible, at least within a few hours after collection.
- The sample should not be agitated, warmed, filtered, diluted, concentrated, or altered in any way.

Example 13

Results from alkalinity titrations on a finished water sample were as follows (see Figure 9.12):

V_{sample} = Sample Size = 100 mL
$V_{8.3}$ = Titrant Used to pH 8.3 = 0.5 mL
$V_{4.5}$ = Total Titrant Used to pH 4.5 = 6.8 mL
N_{acid} = Acid Normality = 0.02 N H_2SO_4

1. Calculate the phenolphthalein alkalinity (A_p) of the sample.*

$$A_p \times V_{sample} = N_{acid} \times V_{8.3}$$

$$A_p = \frac{N_{acid} \times V_{8.3}}{V_{sample}} = \frac{0.02[N] \times 0.5[mL]}{100[mL]} = 10^{-4} N$$

To convert to mg/L as $CaCO_3$, we need to estimate the molecular weight of $CaCO_3$, knowing that its z value is 2 (it takes 2 hydrogen ions to neutralize each carbonate ion).

$$MW_{CaCO_3} = MW_{Ca} + MW_C + 3 \times MW_O = 40 + 12 + 3 \times 16$$

$$= 100 \frac{g}{mol} = 100,000 \frac{mg}{mol}$$

Then, the normality as $CaCO_3$ is 100,000/2 = 50,000 mg/N. The conversion is then:

$$A_p = 10^{-4} N \times 50,000 \frac{mg}{L \cdot N} = 5 \frac{mg}{L}$$

2. Calculate the total alkalinity (A_t) of the sample.

$$A_t \times V_{sample} = N_{acid} \times V_{4.5}$$

$$A_t = \frac{N_{acid} \times V_{4.5}}{V_{sample}} = \frac{0.02[N] \times 6.8[mL]}{100[mL]} = 0.00136 N$$

To convert to mg/L as $CaCO_3$ (as shown above)

$$A_t = 0.00136 N \times 50,000 \frac{mg}{L \cdot N} = 68 \frac{mg}{L}$$

Example 14

Example 13 gave the following results:
Phenolphthalein Alkalinity (A_p) = 5 mg/L
Total Alkalinity (A_t) = 68 mg/L

 Alkalinity test results are usually reported in mg/L as $CaCO_3$. (*Standard Methods* writes mg $CaCO_3$ /L. Either is acceptable.) What this unit expression is saying is that this concentration of dissolved $CaCO_3$ would have the same normality as what was measured in the alkalinity test. There is a simple conversion factor that can be used. The molar weight of $CaCO_3$ is 100 g/mol and the z value is 2 because 2 hydrogen ions can replace the Ca^{2+} (see the discussion on normality earlier in this chapter). Therefore, the equivalent weight is 100/2 = 50 g = 50,000 mg. A 1.0 N solution requires 50,000 mg/L as $CaCO_3$.

Calculate half the total alkalinity.

$$\frac{1}{2}A_t = \frac{1}{2} \times 68 \left[\frac{mg}{L}\right] = 34\frac{mg}{L}$$

Note that $A_p = 5 < 34$ mg/L. According to the table in the "Interpreting Results" section on page 568, the bicarbonate alkalinity is

$$A_t - 2A_p = 68\left[\frac{mg}{L}\right] - 2 \times 5\left[\frac{mg}{L}\right] = 58\frac{mg}{L} \text{ as } CaCO_3$$

and the carbonate alkalinity is

$$2A_p = 2 \times 5\left[\frac{mg}{L}\right] = 10\frac{mg}{L} \text{ as } CaCO_3$$

Check Your Understanding

1. What is alkalinity a measure of?
2. Why is it important to know the alkalinity of a water sample?

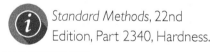 *According to *Standard Methods*, the preferred method for calculating hardness is to determine the calcium (Ca) and magnesium (Mg) concentrations separately and then add them as shown below:

Hardness (mg/L as $CaCO_3$) = 2.497 × Ca concentration (mg/L) + 4.118 × Mg concentration (mg/L)

The titration method is acceptable if you do not have the means of determining calcium and magnesium concentrations in your lab.

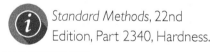 *Standard Methods*, 22nd Edition, Part 2340, Hardness. www.awwa.org. ISBN: 978-0-87553-013-0.

9.4.2.3   Hardness

Hardness* is caused principally by the calcium and magnesium ions commonly present in water. Iron, manganese, aluminum, strontium, and zinc also contribute to hardness if present in significant amounts. Because only calcium and magnesium are present in significant concentrations in most waters, hardness can be defined as the total concentration of calcium and magnesium ions expressed as the calcium carbonate ($CaCO_3$) equivalent.

There are two types or classifications of water hardness: carbonate and noncarbonate. Carbonate hardness is due to calcium/magnesium bicarbonate or carbonate minerals. Hardness due to calcium/magnesium sulfate, chloride, or nitrate minerals is called noncarbonate hardness.

The calcium and magnesium in hard water can precipitate and form solid incrustations (scale) in pipes and on fixtures. Scale is composed mainly of calcium carbonate, magnesium hydroxide, and calcium sulfate minerals. It is a particular problem in hot water systems and associated piping. Hardness-producing substances in water also combine with certain soaps to form insoluble precipitates. The common method of minimizing problems due to hardness is water supply softening. This procedure is discussed in the chapter on softening in Volume II of this series of manuals.

WHAT IS TESTED

**Levels of hardness depend on local conditions.

Sample	Common Range, mg/L as $CaCO_3$
Surface Water	30–500**
Well Water	80–500**

APPARATUS

- Buret (25 mL)
- Graduated cylinder (100 mL)
- Beaker (250 mL)
- Magnetic stirrer and stir bar
- Funnel

REAGENTS

Standardized solutions are available already prepared from laboratory chemical supply companies. Instructions for preparing reagents from scratch are provided in Part 2340 C, *Standard Methods*. The following three reagents are required.

1. Buffer solution (pH 10 with magnesium).
2. Standard EDTA titrant. EDTA is disodium ethylene-diaminetetraacetate dihydrate, also called (ethylenedinitrilo)-tetraacetic acid disodium salt.
3. Indicator solution, Endochrome Black T (EBT) or Calmagite.

In addition, one or more inhibitors or complexing agents may be required to counteract interfering substances in your sample. See *Standard Methods* for details.

Finally, a standard calcium solution is needed to determine the exact strength of the EDTA titrant. *Standard Methods* contains instructions for preparing a standard whose concentration is 1.00 mg $CaCO_3$ per mL.

PROCEDURE (FIGURE 9.14)

The titration procedure is the same for both standardizing the EDTA titrant and measuring samples. For standardizing the EDTA titrant, use the $CaCO_3$ standard solution in place of your water sample.

1. Take a clean beaker and add 25 mL of sample and 25 mL of distilled water.
2. Add 1–2 mL of buffer solution (or as indicated by supplier).
3. Add 1–2 drops indicator solution. The sample will turn red.
4. Titrate with standard EDTA solution until the last reddish tinge disappears from the solution. The solution is pure blue when the endpoint is reached.
5. Calculate total hardness and EDTA strength as shown in the example.

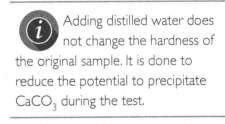

Adding distilled water does not change the hardness of the original sample. It is done to reduce the potential to precipitate $CaCO_3$ during the test.

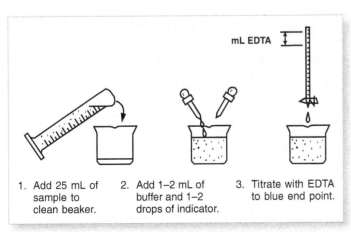

1. Add 25 mL of sample to clean beaker.
2. Add 1–2 mL of buffer and 1–2 drops of indicator.
3. Titrate with EDTA to blue end point.

mL EDTA

Figure 9.14   Outline of procedure for hardness

PRECAUTIONS

- Some metal ions interfere with this procedure by causing fading or indistinct endpoints. In these cases, an inhibitor reagent should be used.
- The titration should be completed within 5 minutes to minimize $CaCO_3$ precipitation.
- A sample volume should be selected that requires less than 15 mL of EDTA titrant to be used.
- For titrations of samples containing low hardness concentrations (less than 150 mg/L as $CaCO_3$) a larger sample volume should be used.

Example 15

Results from a titrant standardization were as follows:

Volume of standard $CaCO_3$ solution (V_{CaCO_3}) = 25.0 mL
Volume of EDTA titrant used (V_{EDTA}) = 24.3 mL
Normality of standard calcium solution (N_{CaCO_3}) = 1.00

$$(N \times V)_{EDTA} = (N \times V)_{CaCO_3}$$

$$N_{EDTA} = \frac{N_{CaCO_3} \times V_{CaCO_3}}{V_{EDTA}} = \frac{1.00\left[\frac{mg}{mL}\right] \times 25.0[mL]}{24.3[mL]}$$

$$= 1.03\frac{mg}{mL} \text{ or } 1.03 \text{ mg of } CaCO_3 \text{ per mL}$$

Results from water hardness testing of a well water sample and a titrant standardization were as follows:

Volume of sample (V_{sample}) = 25.0 mL
Volume EDTA used (V_{EDTA}) = 10.7 mL
Normality of EDTA (N_{EDTA}) from above = 1.03 mg $CaCO_3$/mL

$$(Hardness \times V)_{sample} = (N \times V)_{EDTA}$$

$$Hardness = \frac{N_{EDTA} \times V_{EDTA}}{V_{sample}} = \frac{1.03\left[\frac{mg}{mL}\right] \times 10.7[mL]}{25.0[mL]}$$

$$= 0.441\left[\frac{mg}{mL}\right] \text{ or } 441 \text{ mg/L of } CaCO_3$$

Check Your Understanding

1. What are the principal hardness-causing ions in water?
2. What problems are caused by hardness in water?

9.4.2.4 Chlorine Residual

Chlorine is not only an excellent disinfectant but also reacts with iron, manganese, protein substances, sulfide, and many taste- and odor-producing compounds to help improve the quality of treated water. In addition, chlorine helps to control microorganisms that might interfere with coagulation and flocculation, keeps filter media free of slime growths, and helps bleach out undesirable color.

There are two general types of residual chlorine produced in chlorinated water. They are: (1) free residual chlorine, and (2) combined residual chlorine. Free residual chlorine refers to chlorine gas (Cl_2), hypochlorus acid (HOCl), and the hypochlorite ion (OCl^-). Combined residual chlorine refers to the chlorine-ammonia compounds monochloramine (NH_2Cl), dichloramine ($NHCl_2$), and trichloramine (NCl_3 or nitrogen trichloride). Both types of residuals act as disinfectants, but differ in their capacity to produce a germ-free water supply during the same contact time. Figures 9.15 and 9.16 show a free chlorine sensor and a chlorine residual analyzer.

In addition to all the positive aspects of chlorination, there may be some adverse effects. Potentially **carcinogenic chlororganic** compounds, such as chloroform and other trihalomethanes (THMs), may be formed during the chlorination process. To minimize adverse effects, the operator should be familiar with the concentrations of free and combined residual chlorine produced in a water supply following chlorination. Both residuals are extremely important in producing a **potable water** that is not only safe to drink but is also free of objectionable tastes and odors.

WHAT IS TESTED

Sample	Common Range Residual Chlorine, mg/L	
	Free	Total
Chlorinated Raw Surface Water (Prechlorination)	0.3–3	0.5–5
Chlorinated Finished Surface Water (Post-Chlorination)	0.2–1	0.3–1.5
Well Water	0.2–1	0.2–1

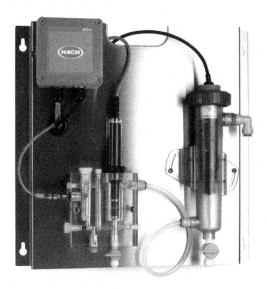

Figure 9.15   Chlorine sensor
(Courtesy of HACH Company)

Figure 9.16   Chlorine residual analyzer
(Courtesy of HACH Company)

carcinogen (kar-SIN-o-jen)
Any substance that tends to produce cancer in an organism.

chlororganic (klor-or-GAN-ick)
Organic compounds combined with chlorine. These often toxic compounds generally originate from chlorinating water containing dissolved organic substances or microbes such as algae.

potable (POE-tuh-bull) water
Water that does not contain objectionable pollution, contamination, minerals, or infective agents and is considered satisfactory for drinking.

METHODS

There are many methods listed for measuring residual chlorine in the 22nd Edition of *Standard Methods*. Selecting the most practical and appropriate procedure in any particular instance generally depends upon the characteristics of the water being examined. The **amperometric titration** method is a standard of comparison for determining free or combined chlorine residual. This method is relatively free of interferences but does require greater operator skill to obtain good results. In addition, the titration instrument is expensive.

The **DPD methods** are simpler to perform than amperometric titration but are subject to interferences due to manganese. Field comparator kits and direct-reading colorimeters are available from several suppliers such as Tintometer Inc., WIKA Instruments, and the HACH Company.

1. Amperometric Titration Method

   In amperometric titration, a reducing agent (phenylarsine oxide, PAO) is added to the sample to reduce residual chlorine to chloride (Cl^-). The titrator measures electrical current through the sample. When all of the chlorine has reacted with the PAO, the current stops, which is the endpoint of the titration. The residual chlorine concentration is proportional to the PAO added.

   a. Apparatus

   See Part 4500-Cl D, Amperometric Titration Method, *Standard Methods*, 22nd Edition, and the amperometric titrator's instruction manual.

   b. Reagents*

   (1) Standard phenylarsine oxide (PAO) solution, 0.00564 *N*. *CAUTION: Toxic— avoid ingestion.*
   (2) Acetate buffer solution, pH 4.
   (3) Phosphate buffer solution, pH 7.
   (4) Potassium iodide solution.

   Store in brown glass, stoppered bottle, preferably in the refrigerator. Discard when solution becomes yellow.

   c. Procedure
   For Free Residual Chlorine

   Because of the specific circumstances associated with each titration instrument, the best advice is to follow the manufacturer's instructions. The general instructions are:

   (1) Place the specified volume of sample in the titrator.
   (2) Start the agitator.
   (3) Add the specified volume of pH 7 buffer.
   (4) Titrate with 0.00564 *N* phenylarsiene oxide (PAO) solution.
   (5) Endpoint is reached when one drop of PAO will cause a deflection on the microammeter and the deflection will remain.
   (6) The PAO volume (in mL) used in titration is equal to mg/L of free chlorine residual.

2. DPD Titrimetric Method (*Standard Methods*, Part 4500-Cl F)

   In this method, N, N-diethyl-p-phenylenediamine (DPD) reacts with free chlorine to produce a red color. Titrating with ferrous ammonium sulfate (FAS) causes the color to disappear (titration

(i) *Prepared reagents may be purchased from laboratory chemical supply houses.

amperometric (am-purr-o-MET-rick)
A method of measurement that records electric current flowing or generated, rather than voltage.

amperometric (am-purr-o-MET-rick) titration
A means of measuring concentrations of certain substances in water (such as strong oxidizers) based on the electric current that flows during a chemical reaction.

DPD method
A method of measuring the chlorine residual in water. The residual may be determined by either titrating or comparing a developed color with color standards. DPD stands for N,N-diethyl-p-phenylenediamine.

endpoint). Adding a small amount of iodide produces color due to monochloromine. Adding excess iodide produces color due to dichloromine. The concentrations of the various compounds are proportional to the FAS titrant used.

a. Apparatus

- Graduated cylinder (100 mL)
- Pipets (1 and 10 mL)
- Flask, Erlenmeyer (250 mL)
- Buret (10 mL)
- Magnetic stirrer
- Balance, analytical

b. Reagents*

(1) Phosphate buffer solution
(2) DPD indicator solution
(3) Standard FAS titrant, 0.00282 N
(4) Potassium iodide (KI) crystals

Other reagents are needed to control for interfering compounds, particularly manganese (Mn). (See *Standard Methods*.)

i *Prepared reagent may be purchased from laboratory chemical supply houses. Instructions for preparing reagents can be found in *Standard Methods*.

c. Procedure (Figure 9.17)

For Free Residual Chlorine

(1) Place 5 mL each of buffer reagent and DPD indicator in a 250 mL flask and mix.
(2) Add 100 mL of sample and mix.
(3) Titrate rapidly with standard FAS titrant until red color disappears.
(4) Record amount of FAS used (Reading A). If combined residual chlorine fractions are desired, continue to next step.

For Combined Residual Chlorine

Monochloramine

(5) Add one very small crystal of KI (about 0.5 mg) and mix. If monochloramine is present, the red color will reappear.
(6) Continue titrating carefully until red color again disappears.
(7) Record the total volume of FAS used to this point (this includes amount used above). This is Reading B.

Dichloramine

(8) Add several crystals KI (about 1 g) and mix until dissolved.
(9) Let stand 2 minutes.
(10) Continue titrating until red color again disappears. Record the cumulative volume of FAS used to this point. This includes amounts used in two previous titrations for free residual chlorine and monochloramine. This is Reading C.

Total Combined Chlorine

If differentiating the monochloramine and dichloramine is not necessary, then skip the monochloramine step above and add several KI crystals to the titrated free chlorine solution.

3. DPD Colorimetric Method

The colorimetric method uses the same chemistry as the DPD Titrimetric Method. The difference is that instead of titrating away the red color, an electronic instrument is used to measure the

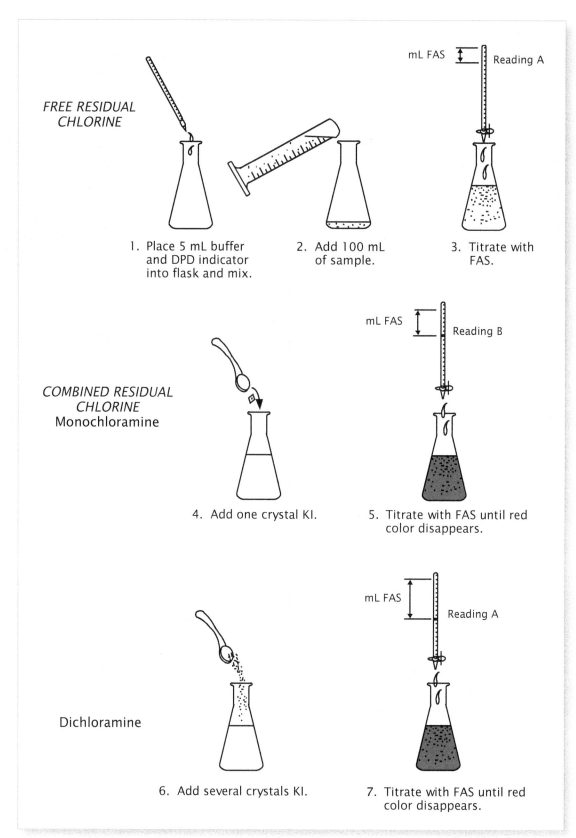

Figure 9.17 Measuring chlorine residual by the DPD method

intensity of the color and convert that to a chlorine concentration using an internal calibration curve. Another option is to visually compare the color developed in the sample with standards corresponding to various chlorine concentrations. For field measurements, the visual option is convenient, although hand-held electronic colorimeters are reasonably affordable today. An example field kit using a portable electronic colorimeter is shown below (Figure 9.18).

a. Apparatus*

- Photometric equipment options
  - Spectrophotometer that can read 515 nm wavelength and has a light path of 1 cm or more.
  - Filter photometer (also called a colorimeter) that can read wavelengths in the 490–530 nm range and has a light path of 1 cm or more.
  - Color comparator kit for visual determination (can be in the form of a wheel or a set of individual standardized cells).
- Miscellaneous glassware for mixing the sample and reagents.

b. Reagents

See the DPD Titrimetric Method. Commercial suppliers of electronic instruments will often supply reagents as well.

c. Procedure

If you are using a commercial kit, follow the manufacturer's instructions. If you are doing the test in your lab and need to prepare a calibration curve, refer to *Standard Methods*. As with the DPD Titrimetric Method, *Standard Methods* contains the procedures for dealing with various interfering substances.

PRECAUTIONS

- For accurate results, careful pH control is essential. The pH of the sample, buffer, and DPD indicator together should be between 6.2 and 6.5.
- Samples should be analyzed as soon as possible after collection.

> (*i*) *Manufacturers of laboratory equipment are continually developing faster and more accurate ways to measure chlorine residual. If you have new equipment, follow the manufacturer's procedures.

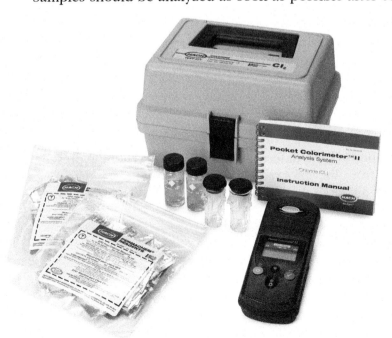

Figure 9.18   Direct-reading colorimeter (free chlorine residual) *(Courtesy of HACH Company)*

- If oxidized manganese, bromine, iodide, chromate, nitrogen trichloride, or chlorine dioxide are present, special procedures are necessary. (See *Standard Methods*.)
- Operators using the DPD Colorimetric Method to test water for a free chlorine residual need to be aware of a potential error that may occur. If the DPD test is run on water containing a combined chlorine residual, a **precipitate** may form during the test. The particles of precipitated material will give the sample a turbid appearance or the appearance of having color. This turbidity can produce a positive test result for free chlorine residual when there is actually no chlorine present. Operators call this error a false positive chlorine residual reading.

Standard Methods, 22nd Edition, Part 4500-Cl F, Chloride. www.awwa.org. ISBN: 978-0-87553-013-0.

INTERPRETING RESULTS

- Any chlorine taste and odor that may result from chlorination would generally be from the dichloramine or nitrogen trichloride nuisance residuals.
- Free residual chlorination produces the best results when the free residual makes up more than 80 percent of the total residual. However, this will not be the case if ammonia is added to the water being treated to form chloramines to prevent the formation of THMs.

Example 16

A sample taken after prechlorination at a filtration plant was tested for residual chlorine using the DPD Titrimetric Method.

mL of sample   = 100 mL
Reading A, mL = 1.4 mL
Reading B, mL = 1.6 mL
Reading C, mL = 2.7 mL

Calculation:

Reading	Chlorine Residual
A =	mg/L free residual chlorine
B – A =	mg/L monochloramine
C – B =	mg/L dichloramine
C – A =	mg/L combined available chlorine
C =	mg/L total residual chlorine

The concentrations of the different types of residual chlorine present can be calculated from the information given above. The standard reagent concentrations are such that 1 mL of FAS is needed to titrate 1 mg/L of chlorine.

Reading	Chlorine Residual
A = 1.4 mL	= 1.4 mg/L free residual chlorine
B – A = 1.6 – 1.4	= 0.2 mg/L monochloramine
C – B = 2.7 – 1.6	= 1.1 mg/L dichloramine
C – A = 2.7 – 1.4	= 1.3 mg/L combined available chlorine
C = 2.7 mL	= 2.7 mg/L total residual chlorine

precipitate (pre-SIP-uh-TATE)
(1) An insoluble, finely divided substance that is a product of a chemical reaction within a liquid. (2) The separation from solution of an insoluble substance.

Check Your Understanding

Check Your Understanding

1. List some of the important benefits of chlorinating water.

2. What is a potential adverse effect from chlorination?

3. What is the DPD method?

9.4.3   Tests for Plant Processes

Tests for plant processes include the jar test for coagulation/flocculation and the chlorine demand test to determine chlorine dosage.

9.4.3.1   Jar Test for Coagulation/Flocculation

Jar tests are tests designed to show the effectiveness of chemical treatment in a water treatment facility. In the broadest sense, the term jar test can refer to any laboratory-scale investigation of a chemical's action. In practice though, the term most often refers to a procedure for testing the effectiveness of coagulants such as alum and polymers. Using the jar test, an operator can approximate the correct coagulant dosage and adjust it for variations in raw water quality, particularly its turbidity. The jar test is also a very useful tool in evaluating new coagulants or polymers being considered for use on a plant scale.

In the procedure presented below, the coagulant being tested is assumed to be alum (aluminum sulfate). In your jar test, you should, of course, substitute the chemicals you want to test. Unlike other tests discussed in this chapter, there is no standard method for jar testing. So do not be afraid to modify parameters, such as mixing speeds and flocculation times and settling times, to create a jar test that faithfully simulates your plant's operation. No simulation will be perfect, of course, but you do want the jar test results to accurately reflect how the plant will respond to the chemicals being tested.

WHAT IS TESTED

The water entering the coagulation process in the plant is tested. Normally, this will be the plant water source. If the treatment train includes processes, such as prechlorination, the jar test sample should be collected downstream from these processes.

APPARATUS

- A stirring machine with six paddles capable of variable speeds from 0 to 100 revolutions per minute (RPM) (see Figure 9.19).
- An illuminator located underneath the stirring mechanism (optional).
- Experimental jars with 1- or 2-liter capacity. Square, 2-liter containers are commercially available from the mixer manufacturers. Standard 1- or 2-liter laboratory beakers can also be used, but they may not be as effective*.
- Pipets (10 mL) and small beakers (50 mL). Plastic syringes can also be used.
- Flask, volumetric (1,000 mL).
- Balance, analytical.

*On first thought, the shape of the jar would not seem to be an issue in evaluating the effectiveness of a chemical. In a jar test, though, the goal is to simulate the coagulation and flocculation processes in the plant. In Chapter 3, "Coagulation and Flocculation," you learned that floc formation depends on bringing the particles together in the slow mix step. In round jars (like beakers) the water tends to form a circular vortex when mixed. In the vortex, the particles swirl around in the same direction and do not contact each other. This limits floc formation. On the other hand, square jars disrupt the formation of circular vortexes, forcing the particles into contact more often, and building larger flocs. Treatment plants generally do not have circular flocculation tanks; so jars intended to simulate them should not be circular either.

If you do not wish to purchase commercial square jars you can construct them by cementing together sheets of clear acrylic plastic. To contain 2 liters, the jars should be 11.5 centimeters square and 21 centimeters deep with a water depth of 15.2 centimeters. Of course, make sure these dimensions are compatible with your mixer before starting construction.

Figure 9.19 Jar test apparatus with mechanical stirrers *(Courtesy of Phipps & Bird, Inc.)*

REAGENTS

Stock Coagulant Solution (for testing alum*)

- Dry alum, $Al_2(SO_4)_3 \cdot 14.3 \, H_2O$. Dissolve 10.0 g dry alum (17 percent) in 600 mL distilled water contained in a 1,000 mL volumetric flask. Fill to mark. This solution contains 10,000 mg/L or 10 mg/mL.

or

- Liquid alum, $Al_2(SO_4)_3 \cdot 49.6 \, H_2O$. The operator should verify the **specific gravity** of the alum with a hydrometer. Liquid alum is usually shipped as 8.0 to 8.5 percent Al_2O_3 and contains about 5.36 pounds of dry aluminum sulfate (48.5 percent dry or 0.485 active alum) per gallon (specific gravity 1.325). This converts to 485,000 mg per L of liquid alum. Use the following volume of liquid alum to create a 10,000 mg/L stock solution:

Mass Alum in Stock Solution = Mass Added in Liquid Alum

$$\left(\frac{10,000 \text{ mg Alum}}{L}\right)(1 \text{ L Water}) = \left(\frac{0.485 \text{ g Alum}}{1 \text{ g Liq Alum}}\right)\left(\frac{1,325 \text{ g Liq Alum}}{1,000 \text{ mL Liq Alum}}\right)$$

$$\left(\frac{1,000 \text{ mg Alum}}{1 \text{ g Alum}}\right) V$$

Solving, V = 15.6 mL of Liquid Alum

To create this stock solution, add 15.6 mL of liquid alum to about 800 mL of water, stir and then add enough water to make 1.0 L of stock solution. Finish the procedure with a final stir. You will add the stock solution to the experimental jar. The goal is to keep the volume of stock solution small, but not so small that it is difficult to measure. Table 9.3 lists the strengths of stock solutions for various dosages. You may use any stock concentration you wish. Formulas used to calculate stock solution volume for any desired test dose are shown in the sidebar. To calculate the volume of liquid alum needed to make different stock solution concentrations, replace the 10,000 mg/L value in the equation above.

(i) *Values of 14.3 H_2O and 49.6 H_2O are typical values. Verify the values used in your laboratory.

(i) Most coagulants react with water (hydrolysis). If your stock solutions are old, the coagulants you are testing are not quite the same molecules as the coagulants in your plant. Ideally, make up your stock solution the same day as your jar test. Be suspicious of stock solutions more than 48 hours old.

specific gravity
(1) Weight of a particle, substance, or chemical solution in relation to the weight of an equal volume of water. Water has a specific gravity of 1.000 at 39°F (4°C). Particulates with specific gravity less than 1.0 float to the surface and particulates with specific gravity greater than 1.0 sink. (2) Weight of a particular gas in relation to the weight of an equal volume of air at the same temperature and pressure (air has a specific gravity of 1.0). Chlorine gas has a specific gravity of 2.5.

Table 9.3   Suggested stock solution concentrations

Desired Concentration in Jar, mg/L	Suggested Stock Concentration, mg/L	Volumes of Stock Solution Required, mL	
		I L Jar	2 L Jar
1–10	1,000	1.0–10.1	2.0–20.2
10–40	2,000	5.0–20.4	10.1–40.8
40–500	10,000	4.0–10.1	8.0–20.2

Polymer and coagulant aid dosages can also be tested during the jar test procedure. In that case, a stock solution must be created for each of those chemicals to simulate the approximate dosages. These solutions can be added during the flash mix or flocculation steps of the jar test (Steps 5 and 6).

PROCEDURE (BASED ON 2-LITER JARS)

1. Collect a 5-gallon (18-liter) sample of the water to be tested.
2. Measure six 2-liter quantities and place into each of six jars.
3. Place all six jars on the stirring apparatus.
4. Set out six small beakers (one for each experimental jar) and fill them with the stock solution volumes needed to achieve the desired coagulant doses.
5. With stirring paddles lowered into the beakers, start stirring apparatus at the highest speed possible (or a lower speed chosen to match the plant's rapid mixer).
6. Add the stock solution from the small beakers as simultaneously as possible and mix rapidly for 30–60 seconds (or a time chosen to match the plant mixer).
7. Reduce the stirring speed to a value that you have determined simulates your plant and stir for 30 minutes. Use stirring speeds and times that are similar to actual conditions in your water treatment plant.
8. Stop stirring and observe the floc quality in each beaker. A hazy sample indicates poor coagulation (see Figure 9.20). Properly coagulated water contains floc particles that are well-formed and

(i) The basic formula for determining the volume of stock solution needed for each jar is:

$$V_s C_s = (V_j + V_s)C_j$$

Where,

V_s = Volume of stock solution

C_s = Concentration of coagulant in stock solution

V_j = Volume of jar

C_j = Concentration desired in jar

After a little algebra, the formula for V_s becomes:

$$V_s = \frac{V_j C_j}{C_s - C_j} \approx \frac{V_j C_j}{C_s}$$

This is the formula used to create the values in Table 9.3. Substitute your own values if your stock solutions are different from those in the table. If the jar concentration (the coagulant dose) is small compared to the stock solution concentration, it can be dropped from the formula without introducing much error.

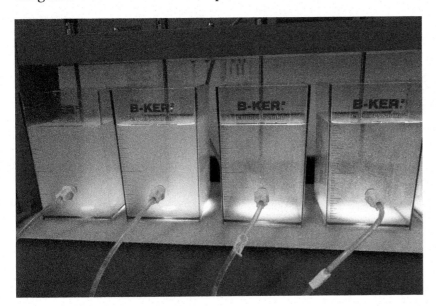

Figure 9.20   Jar test in progress *(Courtesy of Phipps & Bird, Inc.)*

dense, with the liquid between the particles clear. Record results (see Figure 9.21 for a helpful jar test data sheet).

9. Allow samples in beakers to settle for 30 minutes. Observe the floc settling characteristics. Describe results as poor, fair, good, or excellent.

10. Without disturbing the settled floc, carefully remove a sample of the supernatant (clear water on top) using either a pipet or a sampling port built into the experimental jar. Be sure to flush any sampling tubing.

TESTING AND EVALUATION

First of all, do not underestimate the information that can be gained by simple observation. By closely watching the floc form and settle in the jars, you can get a good indication of whether or not you are near the best coagulant dose.

Next, there are a number of tests that can be performed on the sampled water to aid in interpreting the coagulation results. These tests include:

- Turbidity of **supernatant** (before and after)
- pH (before and after)
- Temperature (before and after)
- Alkalinity (before and after)

Alkalinity must be monitored carefully before and after the jar test. For alum and ferrous sulfate, the alkalinity must always be at least half of the coagulant dose. For example, if the optimum coagulant dose is 50 mg/L, the total alkalinity must be at least 25 mg/L. For ferric chloride, the ratio is 1:1. When the alkalinity is too low, the pH drops and flocs may not form. If the natural alkalinity is too low, adjust the total alkalinity up by adding lime or sodium hydroxide (caustic soda). For additional information on how to calculate the amount of lime needed to increase the alkalinity, see Section 9.2.3.3, "Formulas," and page 548, Example 10.

Another test that can be run is a filtered turbidity test. In this case, the sample is run through a Whatman No. 40 filter paper to simulate the action of the plant's filters.

Example 17

A sample of river water was collected for jar test analysis to determine the optimum alum dosage for effective coagulation. Results of the jar testing were recorded as follows:

Beaker No.	Alum Dose, mg/L	Floc Quality	Turbidity, NTU
1	10	poor	25
2	12	fair	15
3	14	good	10
4	16	excellent	5
5	18	excellent	5
6	20	good	10

The above results seem to indicate that a dose of 16 or 18 mg/L would be optimum. The operator should, however, verify this result with visual observation of what is actually happening in the flocculation basin.

supernatant (soo-per-NAY-tent)
The relatively clear water layer between the sludge on the bottom and the scum on the surface of a basin or container. Also called clear zone.

JAR TEST DATA SHEET

JAR NUMBER	SAMPLE SIZE	TEMP START	TEMP FINISH	TURB START	TURB FINISH	pH START	pH FINISH	ALKA START	ALKA FINISH	FILTERED TURBIDITY	COAG DOSE	COAG-AID DOSE	ADJUSTED pH	1ST APPEAR	5 MIN	10 MIN	15 MIN	20 MIN	25 MIN	30 MIN	45 MIN	HARD-NESS START/FINISH	COMMENTS
1																							
2																							
3																							
4																							
5																							
6																							

DESCRIPTION OF FLOC IN JARS

CODE: VS = VERY SMALL; S = SMALL; M = MEDIUM
D = DENSE; VD = VERY DENSE; SP = SPARSE
L = LARGE; VL = VERY LARGE

SOURCE:
COAGULANT:
TIME:
DATE:
COAGULANT AID:
pH ADJUST:
ANALYST:

Figure 9.21   Example jar test data sheet

Additional example results and their interpretation can be found in Chapter 3, Coagulation and Flocculation.

Check Your Understanding

1. What is the purpose of the jar test?

2. What is tested in a jar test?

9.4.3.2 Chlorine Demand

The chlorine demand (Cl_{demand}) of water is the difference between the amount of chlorine applied or dosed (Cl_{dose}) and the amount of free, combined, or total residual chlorine ($Cl_{residual}$) remaining at the end of a specific contact period. Chlorine demand is caused by organics and other **reducing** agents (such as iron, manganese, or sulfide) reacting with the chlorine. Other factors that affect demand include the amount of chlorine applied, length of contact time, pH, and temperature. The chlorine demand test should be conducted with chlorine gas or with granular hypochlorite, depending upon which form you usually use for chlorination.

The purpose of the chlorine demand test is to help you determine the chlorine dosage to achieve specific chlorination objectives. In general:

$$Cl_{dose} = Cl_{demand} + Cl_{residual}$$

The measurement of chlorine demand is performed by treating a series of water samples in question with known but varying amounts of chlorine or hypochlorite. After the desired contact time, measuring residual chlorine in the samples will demonstrate which dosage satisfied the requirements of the chlorine demand in terms of the residual desired.

WHAT IS TESTED

Sample	Common Range Chlorine Demand, mg/L
Surface Water	0.5–5
Well Water	0.1–1.3

APPARATUS

In addition to the apparatus described under one of the methods for chlorine residual, the following items are required:

- Flasks, Erlenmeyer (1,000 mL)
- Pipets (5 and 10 mL)
- Graduated cylinder (500 mL)
- Flask, volumetric (1,000 mL)
- Flask, Erlenmeyer (250 mL)
- Buret (25 mL)

(i) Distilled water is water that has had impurities removed through distillation. Deionized water or DI water is water that has had charged molecules or ions removed through a deionizing process.

---

reduction (re-DUCK-shun)
Reduction is a chemical reaction that results in the addition of hydrogen, removal of oxygen, or the addition of electrons to an element or compound. Under anaerobic conditions (no dissolved oxygen present), sulfur compounds are reduced to odor-producing hydrogen sulfide (H_2S) and other compounds. In the treatment of metal finishing wastewaters, hexavalent chromium (Cr^{6+}) is reduced to the trivalent form (Cr^{3+}). The opposite of *oxidation*.

REAGENTS

In addition to the reagents described under the method you choose to determine chlorine residual, the following items are also needed:

1. Chlorine demand-free water. Prepare chlorine demand-free water from good-quality distilled or deionized water by adding sufficient chlorine to give 5 mg/L free chlorine residual. After standing 2 days, this solution should contain at least 2 mg/L free residual chlorine. If not, discard and obtain better quality water. Remove remaining free chlorine by placing the solution in the direct sunlight. After several hours, measure total chlorine residual. Do not use until last trace of chlorine has been removed. Additional procedural details can be found in *Standard Methods* (Part 4500-Cl.C.3m).

2. Stock chlorine solution. Obtain a suitable solution from a chlorinator solution line in the plant or by purchasing a bottle of household bleach. Store in a dark, cool place to maintain chemical strength. Household bleach products usually contain approximately 5 percent available chlorine, which is about 50,000 mg/L. (Concentrated household bleach may contain 7.86 percent available chlorine, or 78,600 mg/L.)

3. Chlorine dosing solution. If using household bleach, carefully pipet about 10 mL of the bleach into a 1,000 mL volumetric flask. Fill to the mark with chlorine demand-free water and standardize. If the bleach is 50,000 mg/L, this solution will be 500 mg/L. If using a stock chlorine solution obtained from a chlorinator solution line, simply standardize this solution directly.

   Standardization: Measure the concentration of the dosing solution using your desired chlorine residual method. Because of its high concentration, you may need to dilute the dosing solution with distilled or deionized water. A standardization procedure that does not require dilution can be found in *Standard Methods*.

PROCEDURE (FIGURE 9.22)

1. Measure out equal volume samples into each of five to ten 1,000 mL flasks or bottles. The volume does not matter so long as they are all the same. A suggested range is 200–500 mL.

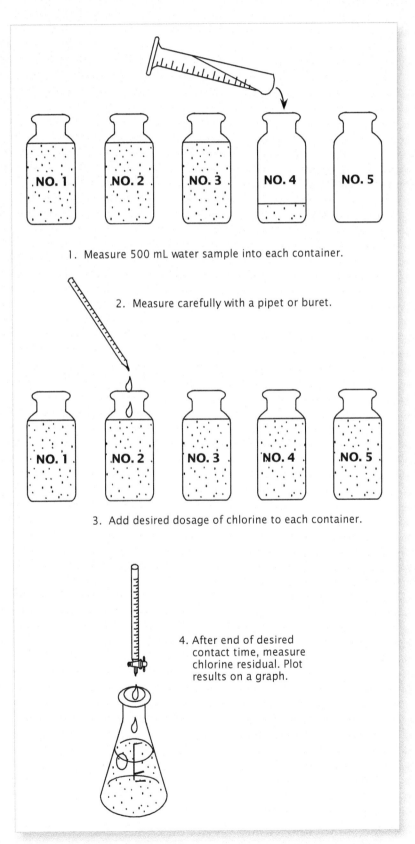

1. Measure 500 mL water sample into each container.

2. Measure carefully with a pipet or buret.

3. Add desired dosage of chlorine to each container.

4. After end of desired contact time, measure chlorine residual. Plot results on a graph.

Figure 9.22 Procedure for chlorine demand

2. To the first flask, add an amount of chlorine that leaves no residual at the end of the contact time. Mix.

3. Add increasing amounts of chlorine to successive flasks and mix. Since the goal is for each flask to have the same contact time, dose sample portions at timed intervals that will leave you enough time between samples to do the chlorine residual testing. Choose concentrations to bracket the expected demand.

4. Measure residual chlorine in each flask after waiting the specific contact time. Record the results.

5. On graph paper or a spreadsheet, plot the residual chlorine (or the amount of chlorine consumed) versus chlorine dose (see example in Figure 9.23).

PRECAUTIONS

- Conduct test over the desired contact time. If test objective is to duplicate your plant contact time, then match plant detention time as closely as possible.
- Keep samples in the dark, protected from sunlight.
- Match the plant temperature and pH as closely as possible.
- Disinfection efficiency can be checked by measuring the bacteria in each bottle. If this is the goal, sterilize the glassware before performing the test.

Example 18

A raw water sample was collected from a river to determine chlorine demand.

Contact Time = 30 minutes (Plant Detention Time)
pH   = 7.6
Temperature = 15°C

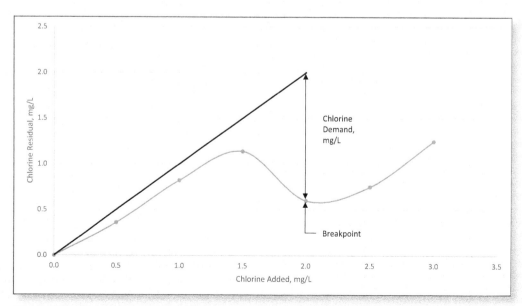

Figure 9.23   Breakpoint chlorination curve

Results from the chlorine demand test were as follows:

Flask No.	Chlorine Added, mg/L	Total Residual Chlorine After 30 min, mg/L
1	0.5	0.36
2	1.0	0.82
3	1.5	1.14
4	2.0	0.60
5	2.5	0.75
6	3.0	1.25

1. Calculate the chlorine demand by using the formula:

$$Cl_{demand} = Cl_{added} - Cl_{residual}$$

Flask No.	Chlorine Added, mg/L	–	Total Residual Chlorine, mg/L	=	Chlorine Demand, mg/L
1	0.5	–	0.36	=	0.14
2	1.0	–	0.82	=	0.18
3	1.5	–	1.14	=	0.36
4	2.0	–	0.60	=	1.40
5	2.5	–	0.75	=	1.75
6	3.0	–	1.25	=	1.75

2. Plot the data from test results. Figure 9.23 is a plot of the chlorine added versus the free chlorine residual after 30 minutes. By drawing a smooth line between the plotting points, a typical breakpoint chlorination curve is produced. The 45° line represents the hypothetical case where there is no demand (i.e., chlorine residual equals chlorine added).

3. Plot the chlorine demand curve. Figure 9.24 shows that the far right end of the curve is flat, which indicates that the chlorine demand has been satisfied. The curve reveals that the chlorine demand will increase as chlorine is added to water until you have gone past the breakpoint.

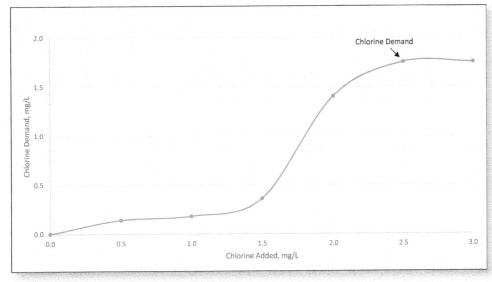

Figure 9.24   Plot of chlorine demand

Standard Methods, 22nd Edition, Part 2350 B, Chlorine Demand/Requirement. www.awwa.org. ISBN: 978-087553-013-0.

> ## Check Your Understanding
>
> 1. What conditions can cause variations in the chlorine demand of water?
> 2. How does the operator use the results from the chlorine demand test?
> 3. What is the formula to calculate the chlorine dosage?

9.4.4   Biological Tests

An improperly treated or unprotected water supply may contain microorganisms that are pathogenic, that is, capable of producing disease. Testing for specific **pathogenic microorganisms** (pathogens that cause diseases such as typhoid, dysentery, cryptosporidiosis, or giardiasis) is very time-consuming and requires special techniques and equipment. So, instead of testing for specific pathogens, water is generally analyzed for the presence of an indicator **organism**, usually the coliform group of bacteria.

The presence or absence of coliform bacteria is a good index of the degree of bacteriologic safety of a water supply. In general, coliform bacteria can be divided into fecal or non-fecal groups. Fecal coliform bacteria occur normally in the intestines of humans and other warm-blooded animals. They are discharged in great numbers in human and animal wastes. Pathogens that cause waterborne diseases are also discharged in wastes. Finding coliform bacteria in a water sample, therefore, indicates the possibility that the water contains pathogens. Coliforms are generally more hardy than true pathogenic bacteria and their absence from water is thus a good indication that the water is bacteriologically safe for human consumption.

The coliform group of bacteria includes all the **facultative** anaerobic gram-negative, nonspore-forming, rod-shaped bacteria that ferment lactose (a sugar) within 48 hours at 35°C (the approximate human body temperature). The fecal coliform can grow at a higher temperature (45°C) than the non-fecal coliform.

The bacteriological quality of water supplies is subject to control by federal, state, and local agencies, all of which are governed by the rules and regulations contained in the Safe Drinking Water Act. This law stipulates the methods to be used, the number of samples required, and the Maximum Contaminant Levels (MCLs) allowed for coliform organisms in drinking water supplies. The number of samples required is generally based on population served by the water system.

The primary standards (MCLs) for coliform bacteria have been established to indicate the likely presence of disease-causing bacteria. The Total Coliform Rule uses a presence–absence approach rather than an estimation of how many coliforms are present to determine compliance with the standards. The Maximum Contaminant Level Goal (MCLG) for coliform is zero.

In general, to be compliant with the Total Coliform Rule, coliform must be absent in at least 95 percent of the samples for larger systems that collect more than 40 samples per month. Smaller systems that collect fewer than 40 samples cannot have coliform-positive results in

pathogenic (path-o-JEN-ick) organisms
Organisms, including bacteria, viruses, protozoa, or internal parasites, capable of causing disease (such as giardiasis, cryptosporidiosis, typhoid fever, cholera, or infectious hepatitis) in a host (such as a person). Nonpathogenic organisms do not cause disease.

organism
Any form of animal or plant life.

facultative (FACK-ul-tay-tive) bacteria
Facultative bacteria can use dissolved oxygen, oxygen obtained from food materials such as sulfate or nitrate ions, or some can respire through GLYCOLYSIS. The bacteria can live under aerobic, anoxic, or anaerobic conditions.

more than one sample per month (check that your local requirements are not different).

Whenever a routine coliform sample is coliform-positive, the regulation calls for determining the presence of fecal coliform (*Escherichia coli* [*E. coli*]) and for repeat sampling. Whenever fecal coliform (*E. coli*) are present in the routine sample and the repeat samples are coliform-positive, there is a violation of the MCL, additional repeat sampling is required, and notification of both the state and the public is required.

Also see *Water Treatment Plant Operation*, Volume 2, for sections on microbial standards and the Total Coliform Rule.

9.4.4.1 Test Methods Overview

In this section, the goal is to become familiar with some of the different approaches used to detect and count bacteria. We will do this by looking at the procedures of different types of tests. Although they may look complicated, the procedures presented here are simplified versions of those in *Standard Methods* and you should not rely on them alone to do these tests in your laboratory. Bacterial testing requires careful execution of specialized procedures. Follow *Standard Methods* or other procedures approved for your plant.

There are a number of approved water testing procedures for total coliform bacteria. This section will provide discussions for the following procedures: (1) the Multiple Tube Fermentation Method (sometimes called the **Most Probable Number or MPN** procedure), (2) the Presence–Absence (P–A) Method, (3) the Enzyme Substrate™ Method, (4) the Membrane Filter (MF) Method, and (5) the Membrane Filter Method—*Escherichia coli* (EPA Method). Other approved testing procedures include the E*Colite™ Test, the m-ColiBlue24® Test, the Readycult® Coliforms 100 Presence–Absence Test, the Chromocult® Coliform Agar Presence–Absence Membrane Filter Test, and the Colitag™ Presence–Absence (P–A) Test. Additional information on these tests is provided at the end of this section. (See Section 9.4.4.5, "Additional Test Methods," page 618.)

Before launching into the details of the various test methods, it will be helpful to consider several broad themes:

- *Bacterial Cultures*. All of the tests are based on growing bacterial cultures in the laboratory. The seeds for these cultures are the microorganisms in the water sample being tested. Because bacterial growth takes time, test results are usually not available for 24 to 48 hours after the test is started. It is possible to have a major outbreak of disease due to contaminated water while waiting for the bacterial tests to finish. Thus, these tests cannot be relied upon to warn you of a potential health problem in advance.
- *Test Goals*. Some tests are designed to simply indicate the presence or absence of coliform bacteria. Others are designed to estimate the number of bacteria present. Tests designed to give counts require more complicated laboratory procedures.
- *Types of Counts*. There are two approaches to counting bacteria. One is the MPN, which is a statistical estimate of the number of bacteria needed to produce a given laboratory response. The second method is to filter bacteria out of the sample, and count the colonies that grow from them during the test.

i The total coliform bacteria group can indicate bacteria that grow in soil (e.g., *Aerobacter*). Therefore, not all coliforms observed in the total coliform test may come from human wastes. To better determine the extent of human wastes in water, additional testing for fecal coliform, or specifically for *Escherichia coli* (*E. coli*) may be required.

MPN
MPN is the Most Probable Number of coliform-group organisms per unit volume of sample water. This is a statistical estimate, expressed as a density or population of organisms per 100 mL of sample water.

- *Identification Strategies.* There are many different kinds of microbes in water. Two different approaches are used to determine the presence of specific kinds of bacteria. One approach is selective culturing. Each step of the culturing (growing) process selects for the bacteria of interest. As shown in Figure 9.25, for instance, many bacteria grow in lauryl tryptose broth, many grow in Brilliant Green Bile Broth, and many grow on LES endo agar, but only coliform grow on all three. Thus, you will see that some tests involve several culturing steps with different media. Another approach is to feed the unknown microbes a specialized chemical nutrient that produces a specific response (a color change, typically) when it is metabolized (eaten) by the specific kind of bacteria we are looking for.

MULTIPLE TUBE FERMENTATION METHOD (MPN PROCEDURE)

The multiple tube coliform test has been a standard method for determining the coliform group of bacteria since 1936. In this procedure, tubes of lauryl tryptose broth are inoculated with a water sample. Lauryl tryptose broth contains lactose, which is the source of carbohydrates (sugar). In metabolizing the sugar, bacteria produce CO_2 gas. The coliform density is then calculated from statistical probability formulas that predict the MPN of coliforms in a 100 mL sample necessary to produce the observed combinations of gas-positive and gas-negative tubes in the series of inoculated tubes. There are three distinct test states for coliform testing using the Multiple Tube Fermentation Method—the presumptive test, the confirmed test, and the completed test. Each test makes the coliform test more valid and specific.

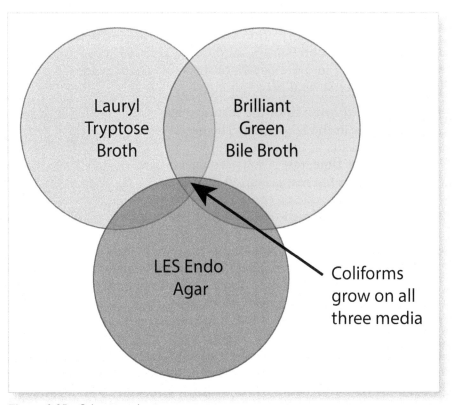

Figure 9.25   Selective culturing strategy

PRESENCE–ABSENCE (P–A) METHOD

In theory, there should be no coliform bacteria in drinking water. Consequently, routine monitoring can be done with a Presence–Absence (P–A) Test. The P–A Method for total coliform is a simple modification of the Multiple Tube (or MPN) Method. Both lactose broth and lauryl tryptose broth are the essential ingredients of the P–A media. In addition, bromocresol purple is added as a pH indicator. A color change indicates the possible presence of coliform.

ENZYME SUBSTRATE COLIFORM TESTS

Enzyme substrate coliform tests (*Standard Methods,* Part 9223) can simultaneously detect total coliform and *Escherichia coli* (*E. coli*) enzymes. Total coliform bacteria produce an enzyme that breaks a molecule in a special organic nutrient (ortho-nitrophenyl-β-galactoside [ONPG]). The complete ONPG molecule is colorless but it turns yellow when broken. *E. coli* bacteria produce an enzyme that does the same thing for another molecule (4-methylumbelliferyl-β-D-galactopyranoside [MUG]). The broken MUG molecule is visible under fluorescent light. Therefore, yellow color and fluorescence indicate the presence of total coliforms and *E. coli*, respectively. These tests are available commercially under the trade names Colilert® and Colisure®. In a manner like the Multiple Tube Fermentation Method, these tests can be used to determine the MPN or in a presence–absence type of test. They are specific enough, however, that a second confirmation test is not needed.

MEMBRANE FILTER (MF) METHOD

This method was introduced as a tentative method in 1955 and has been an approved test for coliform bacteria since 1960. Unlike the multiple tube tests that result in a statistically based probable number, membrane filter results are based on counting colonies. The basic procedure involves filtering a known volume of water through a membrane filter of optimum pore size for full coliform bacteria retention. As the water passes through the microscopic pores, bacteria are entrapped on the upper surface of the filter. The membrane filter is then placed in contact with either a paper pad saturated with liquid medium or directly over a gelatin-like agar medium to provide proper nutrients for bacterial growth. Following incubation under prescribed conditions of time, temperature, and humidity, the cultures are examined for coliform colonies, which are counted directly and recorded as a density of coliforms per 100 mL of water sample.

There are certain important limitations to membrane filter methods. Some types of samples cannot be filtered because of excessive turbidity, high non-coliform bacterial densities, or heavy metal (bactericidal) compounds. In addition, coliform bacteria contained in chlorinated supplies sometimes do not give characteristic reactions on the media and hence special procedures must then be used.

MEMBRANE FILTER METHOD—*ESCHERICHIA COLI* (EPA METHOD)

This method is a membrane filter (MF) procedure for identifying and counting *E. coli*. The method incorporates the use of a selective and differential medium, mTEC agar, followed by incubation with urea substrate medium.

The *E. coli* test is used as a measure of recreational water quality. Epidemiological studies have led to the development of recreational water standards based on established relationships between health effects and water quality. Finding *E. coli* in recreational water samples is significant because of the direct relationship between the density of *E. coli* and the risk of gastrointestinal illness associated with swimming in the water. This test can be applied to fresh, estuarine, and marine waters. Since a wide range of sample volumes or dilutions can be analyzed by the MF technique, a wide range of *E. coli* levels in water can be identified and counted.

9.4.4.2 What is Tested

Sample	Common Range of Total Coliforms per 100 mL
Surface Waters	50–1,000,000
Treated Water Supplies	0
Well Water	0–50

9.4.4.3 General Materials Required for Microbial Testing

For all of the tests described, several common pieces of apparatus and preparation procedures are discussed here.

SAMPLING BOTTLES

Bottles of glass or other material that are watertight, resistant to the solvent action of water, and capable of being **sterilized** may be used for bacteriologic sampling. Plastic bottles made of nontoxic materials are also satisfactory and eliminate the possibility of breakage during transport. Not all plastic bottles can be autoclaved (see below), so check before purchasing. An alternative to consider is presterilized disposable bottles. The bottles should hold a sufficient volume of sample for all tests, permit proper washing, and maintain the samples uncontaminated until examinations are complete.

If you are sampling chlorinated water, you want to stop the disinfection so that you can obtain an accurate bacterial count. To do this, use a reducing agent to neutralize the chlorine (i.e., convert it to a nontoxic chloride ion). Add 0.1 mL 3 percent sodium thiosulfate per 4-ounce bottle (120 mL). This will neutralize a sample containing about 5 mg/L residual chlorine. If the residual chlorine is not neutralized, it would continue to be toxic to the coliform organisms remaining in the sample and give false results. Cap the bottle and then sterilize it.

When filling bottles with sample, do not flush out the sodium thiosulfate. Fill bottles approximately three-quarters full and start the test in the laboratory. Take care to avoid contaminating the bottle, cap, or the sample. If the samples cannot be processed within 1 hour, they should be held for not longer than 6 hours at below 50°F or 10°C.

GLASSWARE

All glassware must be thoroughly cleaned using a suitable detergent and hot water (160°F or 71°C), rinsed with hot water (180°F or 82°C) to remove all traces of residual detergent, and finally rinsed with distilled or deionized water.

WATER

Only distilled water or demineralized water that has been tested and found free from traces of dissolved metals and bactericidal and inhibitory compounds may be used for preparing culture media. Residual chlorine or chloramines can sometimes be found in distilled water made from a

sterilization (STAIR-uh-luh-ZAY-shun) The removal or destruction of all microorganisms, including pathogens and other bacteria, vegetative forms, and spores.

chlorinated water source like tap water. If chlorine is present, neutralize it by adding sodium thiosulfate.

BUFFERED DILUTION WATER

Prepare a stock solution by dissolving 34.0 grams of KH_2PO_4 in 500 mL of distilled water, adjusting the pH to 7.2 with 1 N NaOH and dilute to 1 liter. Prepare dilution water by adding 1.25 mL of the stock solution and 5.0 mL magnesium chloride (81.1 g $MgCl_2 \bullet 6\ H_2O$ per liter) to 1 liter of distilled water. This solution can be dispersed into various sized dilution **blanks**. After sterilization, it can be used as a rinse for the membrane filter test.

MEDIA PREPARATION AND STORAGE

Careful media preparation is necessary for meaningful bacteriological testing. Attention must be given to the quality, mixing, and sterilizing of the ingredients. You want to ensure that if the bacteria being tested for are present in the sample, every opportunity is presented for their development and identification. Bacteriological identification is often done by noting changes in the medium; consequently, the composition of the medium must be standardized. Purchasing dehydrated media (Difco, BBL, or equivalent) from local scientific supply houses not only saves the tedium of preparing the medium, but assures uniformity and quality control. *Standard Methods* recommends never preparing media from basic ingredients if a dehydrated media is commercially available. Follow the manufacturer's instructions for preparing and storing liquid media carefully. Generally speaking, bacterial culture media should not be stored longer than a week, so batch sizes should be chosen accordingly. If only a few tests of a particular type are planned, it may be economical to purchase premixed media in sterile containers.

To sterilize liquid media and associated apparatus on-site, use a steam autoclave. It sterilizes (kills all organisms) at a relatively low temperature of 121°C within 15 minutes using moist heat. Components of the media, particularly sugars such as lactose, may decompose at higher temperatures or longer heating times in an autoclave. For this reason, adhering to time and temperature schedules is vital. The maximum elapsed time for exposure of the media to any heat (from the time the autoclave door is closed to when it is unloaded) is 45 minutes. Preheating the autoclave can reduce total heating time.

Autoclaves operate in a manner similar to a standard kitchen pressure cooker:

1. Water is heated in a boiler to produce steam.
2. The steam is vented to drive out air.
3. The steam vent is closed when the air is gone.
4. Continued heat raises the pressure to 15 lb/in² (103.4 kPa or 1.05 kg/cm²). At this pressure, pure steam has a temperature of 121°C.
5. The heat and pressure are maintained for 15 minutes.
6. The heat is turned off.
7. The steam vent is opened slowly to vent steam until atmospheric pressure is reached. (Fast venting will cause the liquids to boil and overflow tubes.)
8. Sterile material is removed to cool.

blank
A bottle containing only dilution water or distilled water; the sample being tested is not added. Tests are frequently run on a sample and a blank and the differences are compared. The procedure helps to eliminate or reduce errors that could be caused by contaminated dilution water or distilled water.

In autoclaving fermentation tubes, a vacuum is formed in the inverted inner tubes. As the tubes cool, the inner tubes are filled with sterile medium. Capture of gas in the inner tubes from the culture of bacteria is the evidence of fermentation and is recorded as a positive test.

Check Your Understanding

1. Why are drinking waters not tested for specific pathogens?
2. What is the number of samples required for coliform tests generally based on?
3. What are the membrane filter results based on?
4. Why should sodium thiosulfate be added to coliform sample bottles?
5. Steam autoclaves sterilize (kill all organisms) at a temperature of _____ °C during a _____-minute time period (at sea level).

9.4.4.4　Procedures for Testing Total Coliform Bacteria

MULTIPLE TUBE FERMENTATION METHOD

This test for coliform bacteria is used to measure the suitability of a water for human use. It is not only useful in determining the bacterial quality of a finished water, but it can be used by the operator in the treatment plant as a guide to achieving a desired degree of treatment.

In this test, coliform bacteria are detected by placing portions of a water sample in lauryl tryptose broth. Lauryl tryptose broth is a standard bacteriological media containing lactose (milk) sugar in tryptose broth. The coliform bacteria are those that can grow in this media at 35°C temperature and are able to ferment and produce gas from the lactose within 48 hours. To detect these bacteria, the operator inspects the fermentation tubes for gas. A schematic of the coliform test procedure is shown in Figure 9.26.

To meet EPA standards for treated drinking water, the five-tube Multiple Tube Fermentation Method is not appropriate. Instead, the test uses 10 tubes with a 10 mL sample in each tube.

There are three distinct test states for coliform testing using the Multiple Tube Fermentation Method—the presumptive test, the confirmed test, and the completed test. Each test is done with a different media. Coliform and other bacteria metabolize lauryl tryptose. To be conservative in protecting public health, we presume that a positive reading here indicates coliforms. To confirm that the bacteria causing the positive reading are coliforms, we check that they can grow in Brilliant Green Bile Broth, which selects against non-coliform bacteria. A positive reading here is usually enough evidence of the presence of coliform, but to verify this and provide quality control, *Standard Methods* recommends that the completed test be done on 10 percent of the positive confirmed tubes. The completed test involves additional culturing and examination under a microscope. These procedures can be found in *Standard Methods*, 22nd Edition (Part 9221 B).

Not all coliform bacteria are associated with fecal contamination; some live in soil. To determine whether coliform appearing in the presumptive test might be fecal in origin, a separate test based on body temperature incubation is performed (see Figure 9.26).

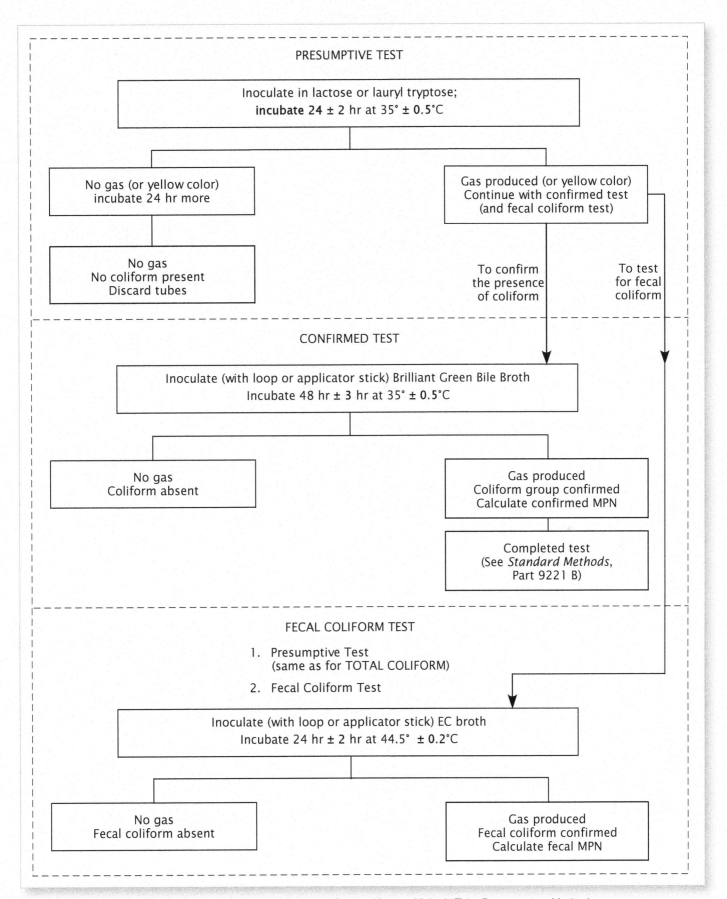

Figure 9.26   Test procedure flowchart for total coliform and fecal coliform—Multiple Tube Fermentation Method

1. Materials Needed

   Media—MPN (Total Coliform)
   a. Lauryl tryptose broth

   For the presumptive coliform test, dissolve the recommended amount of the dehydrated media in distilled water. Dispense solution into fermentation tubes containing an inverted glass vial. For 10 mL water portions from samples, double-strength media is required while all other inoculations require single strength. Directions for preparation are given on the media bottle label.

   b. Brilliant Green Bile (BGB) broth

   For the confirmed coliform test, dissolve 40 grams of the dehydrated media in 1 liter of distilled water. Dispense and sterilize as with lauryl tryptose broth.

   c. EC broth

   Prepare this media following the instructions on the bottle.

   For Untreated Water Samples
   a. Fifteen sterile tubes containing 10 mL of lauryl tryptose broth are needed for each sample. Use five tubes for each dilution.
   b. Dilution tubes or blanks containing 99 mL of sterile buffered distilled water.
   c. A quantity of 1 mL and 10 mL serological pipets. The 1 mL pipets should be graduated in 0.1 mL increments.
   d. Incubator set at $35°C \pm 0.5°C$.

   For Drinking Water Samples
   a. Ten sterile tubes of 10 mL, double-strength lauryl tryptose broth are needed if 10 mL of sample is added to each tube. Ten mL of lauryl tryptose broth is required in all tubes containing 1 mL or less of sample.
   b. Sterile 10 mL pipet for each sample.
   c. Incubator set at $35°C \pm 0.5°C$.
   d. Water bath set at $44.5°C \pm 0.2°C$.

2. Procedure for Presumptive Test

   All inoculations and dilutions of water samples must be accurate and made so that no contaminants from the air, equipment, clothes, or fingers reach the sample, either directly or by way of a contaminated pipet. Clean, sterile pipets must be used for each separate sample.

   For Untreated Water Samples
   a. Shake the sample bottle vigorously 25 times before removing sample volumes.
   b. Pipet 10 mL of sample directly into each of the first five tubes. Each tube must contain 10 mL of lauryl tryptose broth (double strength).

   Note: Use 10 mL pipets for 10 mL sample portions, and 1 mL pipets for portions of 1 mL or less. Handle sterile pipet only near the mouthpiece, and protect the delivery end from external contamination.

   Note: When delivering the sample into the culture medium, deliver sample portions of 1 mL or less down into the culture

---

i If you have too many bacteria in your sample, you will get gas production in all 15 fermentation tubes, which is not a useful result. The first time a sample is analyzed, prepare 25 tubes of lauryl tryptose broth (the 15 tubes described in the procedure plus two extra dilutions as shown in Figure 9.27). To make a 1/100 (0.01) dilution, add 1 mL of well-mixed water sample to 99 mL of sterile buffered dilution water. Mix thoroughly by shaking. Add 1 mL from this bottle directly into each of five more lauryl tryptose broth tubes. To make a 1/1,000 (0.001) dilution, add 0.1 mL from the 1/100 dilution bottle directly into each of five more lauryl tryptose broth tubes. Once you find out what dilutions give usable results for determining the MPN index of your samples, you will need to prepare only 15 tubes to analyze subsequent samples from the same source.

tube near the surface of the medium. Do not deliver small sample volumes at the top of the tube and allow them to run down inside the tube; too much of the sample will fail to reach the culture medium.

c. Pipet 1 mL of water sample into each of the next five lauryl tryptose broth (single strength) tubes.

d. Pipet 1/10 mL (0.1 mL) of water sample into each of the next five lauryl tryptose broth (single strength) tubes. This makes a 0.1 (1 to 10) dilution.

At this point, you have 15 tubes inoculated (Figure 9.27) and can place these three sets of tubes in the incubator.

For Drinking Water Samples

a. Shake the sample bottle vigorously 20 times before removing a sample volume.

b. Pipet 10 mL of sample directly into each of 10 tubes containing 10 mL of double-strength lauryl tryptose broth (Figure 9.28).

3. Incubation (Total Coliform)

a. 24-Hour Lauryl Tryptose (LT) Broth Presumptive Test Place all inoculated LT broth tubes in 35°C ± 0.5°C incubator. After 24 ± 2 hours have elapsed, examine each tube for gas formation in the inverted vial (inner tube). Mark plus (+) on a report form such as the one shown in Figure 9.29 for all tubes that show presence of gas. Mark minus (–) for all tubes showing no gas

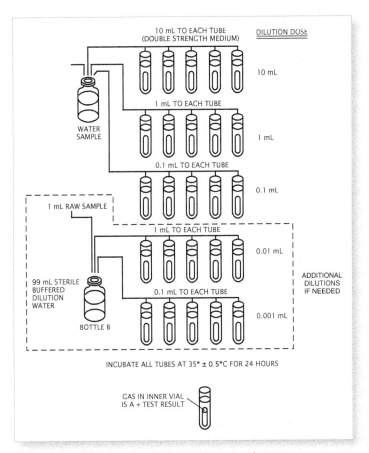

Figure 9.27 Coliform bacteria test—untreated water

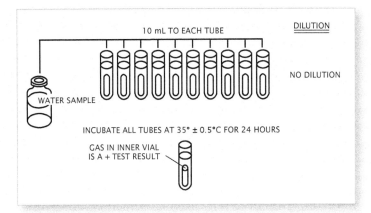

Figure 9.28 Coliform bacteria test—drinking water

formation. Immediately perform confirmation test on all positive (+) tubes (see Paragraph 4.a, "24-Hour Brilliant Green Bile (BGB) Confirmation Test"). The negative (−) tubes must be reincubated for an additional 24 hours.

b. 48-Hour Lauryl Tryptose Broth Presumptive Test
Record both positive and negative tubes at the end of 48 ± 3 hours. Immediately perform total coliform confirmation test on all new positive tubes. For drinking water samples, all positive tubes should also be tested for fecal coliforms (*E. coli*).

4. Confirmed Test
a. 24-Hour Brilliant Green Bile (BGB) Confirmation Test
Confirm all presumptive tubes that show gas at 24 or 48 hours. Transfer, with the aid of a sterile 3 mm platinum wire loop (sterile wood applicator or disposable loops may be used also), one loop-full of the broth from the lauryl tryptose broth tubes showing gas, to a corresponding tube of BGB broth by mixing the loop of broth in the BGB broth. Discard all positive lauryl tryptose broth tubes after transferring is completed.

Always sterilize inoculation loops and needles in flame immediately before transfer of culture. Do not lay loop down or touch it to any nonsterile object before making the transfer. After sterilization in a flame, allow sufficient time for air cooling to prevent the heat of the loop from killing the bacterial cells

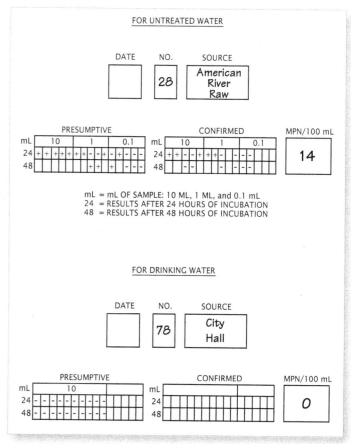

Figure 9.29 Recorded coliform test results

being transferred. Sterile wood applicator sticks also are used to transfer cultures, especially in the field where a flame is not available for sterilization. If using hardwood applicators, sterilize by autoclaving before use and discard after each transfer.

After 24 hours have elapsed, inspect each of the BGB tubes for gas formation. Those with any amount of gas are considered positive and are so recorded on the data sheet. Negative BGB tubes are reincubated for an additional 24 hours.

  b. 48-Hour Brilliant Green Bile Confirmation Test

    (1) Examine tubes for gas at the end of the 48 ± 3-hour period. Record both positive and negative tubes.

    (2) Complete reports by determining MPN index and recording MPN on worksheets.

5. Completed Test

For drinking water, positive confirmed test results are usually enough evidence to spur corrective action. For nonpotable samples, though, a completed test may be desirable to further verify that the bacteria being observed are from the coliform group. The completed test should also be done periodically even on drinking water samples as a quality control step. The multiple steps involved in the completed test are laid out in *Standard Methods,* Part 9221 B.

6. Fecal Coliform (FC) Test (*Standard Methods,* Part 9221 E)

For drinking water samples, all presumptive tubes that show gas at 24 or 48 hours must be tested for fecal coliform. This test more reliably indicates the potential presence of pathogenic organisms than do tests for total coliform. The procedure described is an elevated temperature test for fecal coliform bacteria.

  a. Materials Needed

Equipment required for the tests are the same as those required for the 24-hour lactose broth presumptive test, plus a water bath set at $44.5°C \pm 0.2°C$, EC broth media, and a thermometer certified against a NIST thermometer.

  b. Procedure

    (1) Run lactose broth or lauryl tryptose broth presumptive test.

    (2) After 24 hours, temporarily retain all gas-positive tubes.

    (3) Label a tube of EC broth to correspond with each gas-positive tube of broth from presumptive test.

    (4) Shake or mix positive presumptive tubes by rotating them. Transfer one loop-full of culture from each gas-positive culture in the presumptive test to the correspondingly labeled tube of EC broth.

    (5) Incubate EC broth tubes 24 ± 2 hours at $44.5°C \pm 0.2°C$ in a water bath with water depth sufficient to come up at least as high as the top of the culture medium in the tubes. Place in water bath as soon as possible after inoculation (not more than 30 minutes after inoculation).

    (6) After 24 hours, remove the rack of EC cultures from the water bath, shake gently, and record gas production for each tube. Gas in any quantity is a positive test.

    (7) As soon as results are recorded, discard all tubes. This is a 24-hour test for EC broth inoculations and not a 48-hour test.

(8) Transfer any additional 48-hour gas-positive tubes from the presumptive test to correspondingly labeled tubes of EC broth. Incubate for 24 ± 2 hours at $44.5°C \pm 0.2°C$ and record results on data sheet.

(9) Codify results using the same procedure as for total coliforms and determine MPN of fecal coliforms per 100 mL of sample (Figure 9.29).

7. Method of Calculating the Most Probable Number (MPN)

As noted earlier, the MPN is a probabilistic result (the most likely number of bacteria that produced the observed results). MPN values for different combinations of positive tubes have been tabulated (see Table 9.4). If you have only three dilutions, determining the MPN is straightforward. Look in Table 9.4 for the combinations of positive tubes that reflect your results and read the MPN Index. The MPN and MPN Index are the same when there are no dilutions involved. In the examples that follow, the inoculate volume is the volume of sample put into each tube.

Example 19

Results

Inoculate Volume (mL)	10	I	0.1
Number of Positive Tubes	5	2	I

Combination to read in Table 9.4: 5-2-1
MPN Index from the table = 70
MPN = 70 coliform/100 mL

If you have diluted the sample before running the test, multiply the MPN Index value from Table 9.4 by the dilution factor.

Example 20

Before running the MPN test, the sample was diluted by a factor of 100. Results = 33

Inoculate Volume (mL)	10	I	0.1
Number of Positive Tubes	4	3	I

MPN = 33 × 100 = 3,300 coliform/100 mL

If you have five dilutions, the situation is more complicated because you must choose a three-dilution combination to use Table 9.4. A flowchart showing how to choose the correct combination is shown in Figure 9.30, which is based on *Standard Methods*, Part 9221 C. Follow the lines. For instance, if your five-tube result is 5-5-5-3-2, then the three-tube combination you want to use is 5-3-2 according to Figure 9.30.

Depending on which three dilutions you choose, you may have to multiply the number from the table (called the MPN Index) by a factor of 10 or 100. When to do this is illustrated in Table 9.5. When inoculation volumes of 10, 1, and 0.1 mL are used, the MPN Index (from the table) is the MPN in coliforms/100 mL of sample. If instead, we use 1, 0.1, and 0.01 mL, we are using one-tenth of the volumes and the MPN Index is the MPN per 10 mL not 100 mL. To get the coliforms per 100 mL, multiply the MPN Index by 10. When using the dilutions associated with 0.1, 0.01, and 0.001 mL inoculation volumes, multiply the MPN Index by 100.

Table 9.4 MPN Index for combinations of gas-positive results using five tubes per dilution (10 mL, 1.0 mL, 0.1 mL) (Adapted from the FDA method presented in www.fda.gov/Food/FoodScienceResearch/LaboratoryMethods/ ucm109656.htm.)

Combination of Positives	MPN Index / 100 mL	Combination of Positives	MPN Index / 100 mL	Combination of Positives	MPN Index / 100 mL
0-0-0	<1.8	3-0-0	7.8	5-0-0	23
0-0-1	1.8	3-0-1	11	5-0-1	31
0-1-0	1.8	3-0-2	13	5-0-2	43
0-1-1	3.6	3-1-0	11	5-0-3	58
0-2-0	3.7	3-1-1	14	5-0-4	76
0-2-1	5.5	3-1-2	17	5-1-0	33
0-3-0	5.6	3-2-0	14	5-1-1	46
0-3-1	7.4	3-2-1	17	5-1-2	63
		3-2-2	20	5-1-3	84
1-0-0	2.0	3-3-0	17	5-1-4	110
1-0-1	4.0	3-3-1	21	5-2-0	49
1-0-2	6.0	3-3-2	24	5-2-1	70
1-1-0	4.0	3-4-0	21	5-2-2	94
1-1-1	6.1	3-4-1	24	5-2-3	120
1-1-2	8.1	3-4-2	28	5-2-4	150
1-2-0	6.1	3-5-0	25	5-3-0	79
1-2-1	8.2	3-5-1	29	5-3-1	110
1-2-2	10	3-5-2	32	5-3-2	140
1-3-0	8.3			5-3-3	170
1-3-1	10	4-0-0	13	5-3-4	210
1-3-2	12	4-0-1	17	5-4-0	130
1-4-0	10	4-0-2	21	5-4-1	170
1-4-1	13	4-0-3	25	5-4-2	220
1-4-2	15	4-1-0	17	5-4-3	280
		4-1-1	21	5-4-4	350
2-0-0	4.5	4-1-2	26	5-4-5	430
2-0-1	6.8	4-1-3	31	5-5-0	240
2-0-2	9.1	4-2-0	22	5-5-1	350
2-1-0	6.8	4-2-1	26	5-5-2	540
2-1-1	9.2	4-2-2	32	5-5-3	920
2-1-2	12	4-2-3	38	5-5-4	1,600
2-2-0	9.3	4-3-0	27	5-5-5	>1,600
2-2-1	12	4-3-1	33		
2-2-2	14	4-3-2	39		
2-3-0	12	4-3-3	45		
2-3-1	14	4-4-0	34		
2-3-2	17	4-4-1	40		
2-4-0	15	4-4-2	47		
2-4-1	17	4-4-3	54		
2-4-2	20	4-5-0	41		
2-5-0	17	4-5-1	48		
2-5-1	20	4-5-2	56		
2-5-2	23	4-5-3	64		

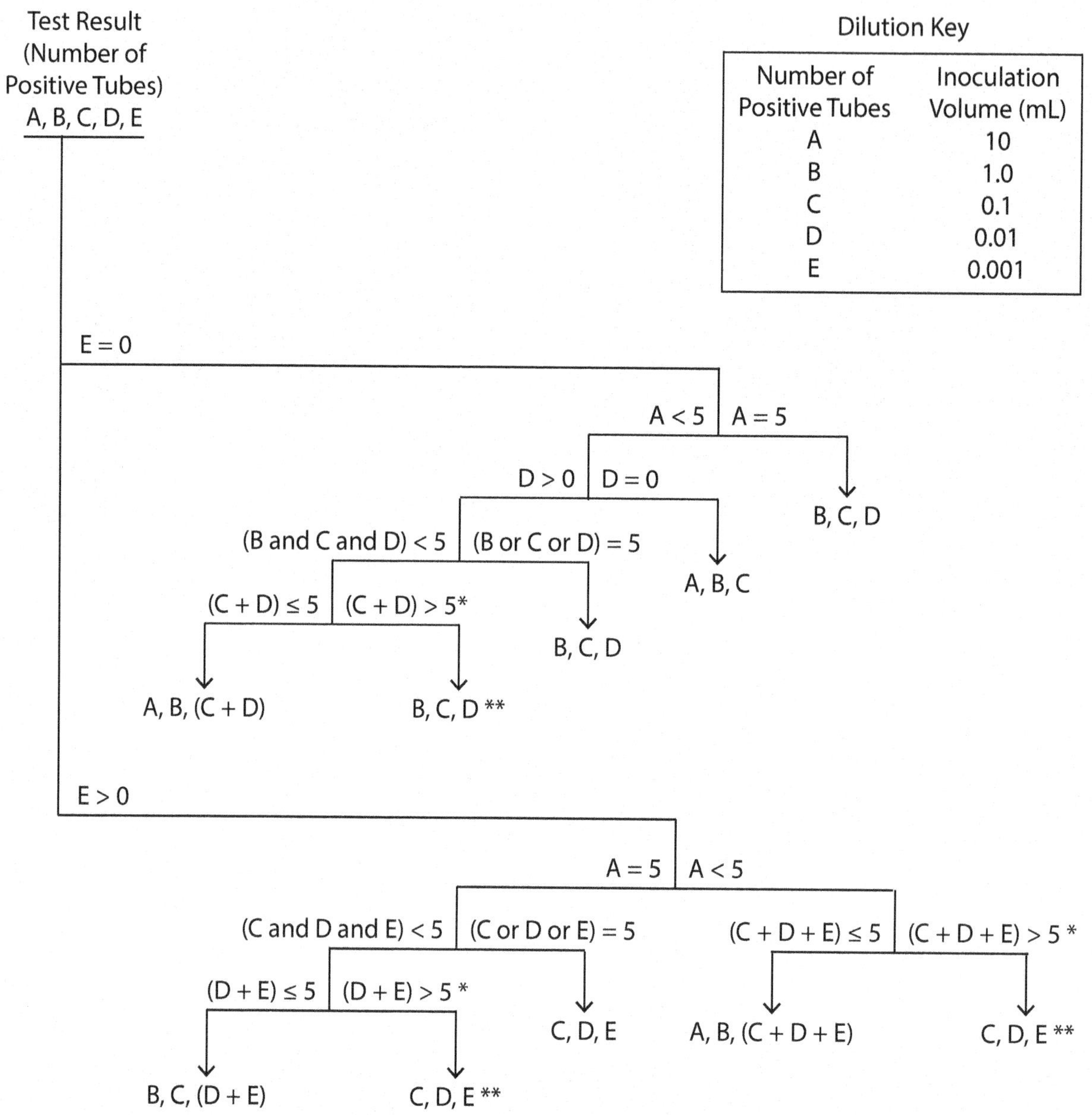

Test Result
(Number of
Positive Tubes)
A, B, C, D, E

Dilution Key

Number of Positive Tubes	Inoculation Volume (mL)
A	10
B	1.0
C	0.1
D	0.01
E	0.001

E = 0

| A < 5 | A = 5 |

| D > 0 | D = 0 |

B, C, D

| (B and C and D) < 5 | (B or C or D) = 5 |

A, B, C

| (C + D) ≤ 5 | (C + D) > 5* |

B, C, D

A, B, (C + D)   B, C, D **

E > 0

| A = 5 | A < 5 |

| (C and D and E) < 5 | (C or D or E) = 5 | | (C + D + E) ≤ 5 | (C + D + E) > 5 * |

| (D + E) ≤ 5 | (D + E) > 5 * | C, D, E   A, B, (C + D + E)   C, D, E **

B, C, (D + E)   C, D, E **

*Something in the serial dilution was unusual. Collecting a new sample is recommended.
**If given set of dilutions is not in Table 9.4, then use the following formula to calculate MPN:

$$MPN = \frac{-230.3}{z_s} \log_{10}\left(1 - \frac{x_s z_s}{\sum\limits_{j=s}^{k} n_j z_j}\right)$$

where   j = a dilution
    s = the highest dilution (lowest inoculation value) with at least one positive tube
    xs = the number of positive tubes in the sth dilution
    nj = the number of tubes in the jth dilution
    zj = the amount of the original sample inoculated into each tube in the jth dilution
    zs = the amount of the original sample inoculated into each tube in the sth dilution
    k = the number of dilutions

Figure 9.30   Flowchart for choosing the best three dilutions in a five-dilution MPN test (Adapted from *Standard Methods*, 22nd Edition, Part 9221 C.2.)

Table 9.5   Calculating MPN from MPN Index for different dilution combinations

Inoculation Volume (mL)				
10	1	0.1	0.01	0.001

MPN = MPN Index × 1

MPN = MPN Index × 10

MPN = MPN Index × 100

Note: The MPN Index is the value read from Table 9.4.

Example 21

This is a five-dilution test. The original sample was not diluted before running the test. See Figure 9.31.

Results

Inoculate Volume (mL)	10	1	0.1	0.01	0.001
Number of Positive Tubes	5	5	2	0	0

Following the procedure in Figure 9.30, ABCDE = 5-5-2-0-0 and the combination of dilutions to use is BCD, or 5-2-0.

Read MPN Index = 49 from Table 9.4. From Table 9.5, multiply the MPN Index by 10 because your combination is from the middle three dilutions of the five-dilution set.

Calculate the MPN as: MPN = 49 × 10 = 490 coliform/100 mL.

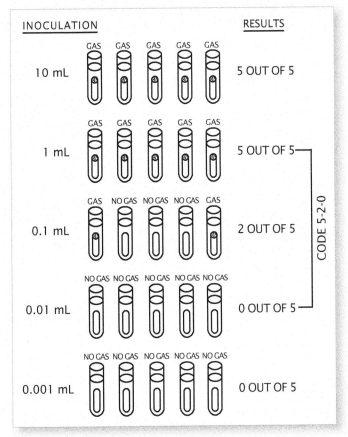

Figure 9.31   Results of coliform test—untreated water

Example 22

This is a five-dilution test. The original sample was not diluted before running the test.

Results

Inoculate Volume (mL)	10	I	0.1	0.01	0.001
Number of Positive Tubes	5	4	5	I	I

Following the procedure in Figure 9.30, ABCDE = 5-4-5-1-1 and the combination of dilutions to use is CDE, or 5-1-1. Note that E > 0, which puts us in the lower part of Figure 9.30.

Read MPN Index = 46 from Table 9.4. From Table 9.5, multiply the MPN Index by 100 because your combination is from the upper three dilutions of the five-dilution set. (Remember, small inoculation volumes mean high dilutions.)

Calculate the MPN as: MPN = 46 × 100 = 4,600 coliform/100 mL.

Example 23

This is a five-dilution test. The original sample was not diluted before running the test.

Results

Inoculate Volume (mL)	10	I	0.1	0.01	0.001
Number of Positive Tubes	4	3	I	I	I

Following the procedure in Figure 9.30, ABCDE = 4-3-1-1-1 and the combination of dilutions to use is AB(C+D+E), which is 4-3-(1+1+1), or 4-3-3.

Read MPN Index = 45 from Table 9.4. From Table 9.5, multiply the MPN Index by 1 because your combination is based on the lower three dilutions (largest inoculation volumes) of the five-dilution set.

Calculate the MPN as: MPN = 45 × 1 = 45 coliform/100 mL.

Interpreting the five-dilution test is a bit tricky, but you should not have to do it often. After you have determined the general range of coliform counts for a water source, in subsequent samples you can dilute the original to get into the range of counts covered by a three-dilution test. For instance, if you dilute the sample in Example 22 by 100 before running the three-dilution test, you would expect to get a 5-1-1 result, which can be read directly from Table 9.4. Multiply the MPN Index by 100 to account for the initial dilution of the sample.

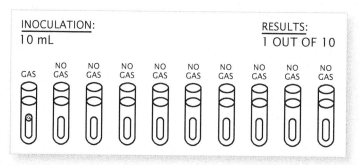

Figure 9.32   Results of coliform test—drinking water

Table 9.6 MPN Index and 95% confidence limits for all combinations of positive and negative results when ten 10 mL portions are used (Source: *Standard Methods* [Part 9221 C])

No. of Tubes Giving Positive Reaction out of 10 (10 mL Each)	MPN Index/ 100 mL	95% Caonfidence Limits (Exact)	
		Lower	Upper
0	<1.1	—	3.4
1	1.1	0.051	5.9
2	2.2	0.37	8.2
3	3.6	0.91	9.7
4	5.1	1.6	13
5	6.9	2.5	15
6	9.2	3.3	19
7	12	4.8	24
8	16	5.8	34
9	23	8.1	53
10	>23	13	—

Reprinted from *Standard Methods [Part 9221 C)*, courtesy of the American Public Health Association.

If you are testing drinking water and using the ten-tube test (see Figure 9.32), the MPN can be read directly from Table 9.6. Only undiluted water samples should be used with the ten-tube test. If you suspect the sample may contain more bacteria than 23 coliform/100 mL, use the five-tube multiple fermentation test described above.

Example 24

A treated drinking water sample is tested with the ten-tube test using 10 mL per tube. After incubation, one tube was found to have bubbles (see Figure 9.32). From Table 9.6, the MPN is 1.1 coliform/100 mL.

Example 25

The results from an MPN test using three dilutions and five fermentation tubes in each dilution were as follows:

Inoculation Volume (mL put into the tube)	10	1	0.1
Number of Positive Tubes (out of 5)	3	1	0
Number of Negative Tubes (= 5 − number of positives)	2	4	5

Calculate the values of the variables in the Thomas equation (see sidebar).

$P = 3 + 1 = 4$

$V_N = (2 \times 10[\text{mL}]) + (4 \times 1[\text{mL}]) + (5 \times 0.1[\text{mL}]) = 24.5 \text{ mL}$

$V_T = (5 \times 10[\text{mL}]) + (5 \times 1[\text{mL}]) + (5 \times 0.1[\text{mL}]) = 55.5 \text{ mL}$

Insert these values into the Thomas equation and calculate the MPN.

$$\frac{\text{MPN}}{100 \text{ mL}} = \frac{4 \times 100}{\sqrt{24.5 \times 55.5}} = 10.8 \text{ / }100 \text{ mL}$$

Usually MPN values are rounded to two digits. Report this result as:

$$\text{MPN} = \frac{11 \text{ Coliforms}}{100 \text{ mL}}$$

Notice that this is the same as the table value.

(*i*) The MPN for combinations not appearing in the table may be estimated by a formula developed by H.A. Thomas, Jr., in 1942:

$$\frac{\text{MPN}}{100 \text{ mL}} = \frac{P \times 100}{\sqrt{V_N \times V_T}}$$

Where,

P = Number of positive tubes in 15-tube series

V_N = Sum of inoculation volumes (mL) in all negative tubes

V_T = Total inoculation volumes (sum of mL in all tubes)

8. Characterizing Bacterial Counts

   When you have numerical results (MPN or plate counts) from multiple tests, how do you choose a representative value for the data set? Numerical data can be summarized by using the arithmetic mean, median, or geometric mean. The arithmetic mean is often called the average.

$$\text{Mean} = \frac{\text{Sum of Data}}{\text{Number of Data Points}}$$

When a data set contains a few very large values, as often happens in the case of bacterial counts, they can bias (distort) the arithmetic mean. Consequently, the median and geometric mean are used more often to summarize bacterial data. For more information, see Appendix A, "Introduction to Basic Math for Operators," Section A.8.3.2, "Numerical Representation of Data."

The median is the middle value in a set or group of data. There are as many values that are larger than the median as there are values smaller than the median. To determine the median, the data should be written in ascending (increasing) or descending (decreasing) order and the middle value identified. If the group has an even number of data points, the median is often reported as the average of the two numbers closest to the middle.

Median = Middle Value in a Set or Group of Data

The geometric mean (X_G) is:

$$\text{Geometric Mean} = \left[\left(X_1 \right)\left(X_2 \right)\left(X_3 \right)\left(X_4 \right) ... \left(X_n \right) \right]^{1/n}$$

If X is a count value and n is the number of measurements, then multiply all the count values together and take the 1/n root of the result. This calculation can be easily performed on many electronic calculators using the power function (y^x). (The geometric mean got its name from the fact that this calculation is the equivalent calculating the mean of the logarithms of the different values and then taking the antilog.)

Example 26

Results from MPN tests during one week were as follows:

Day	S	M	T	W	TH	F	S
MPN/100 mL	2	8	14	6	10	56	4

Estimate the mean, median, and geometric mean of the data in MPN/100 mL.

1. Calculate the mean.

$$\text{Mean}\left[\frac{\text{MPN}}{100 \text{ mL}} \right] = \frac{\left(2 + 8 + 14 + 6 + 10 + 56 + 4 \right)}{7} = \frac{100}{7} = 14.3 \frac{\text{MPN}}{100 \text{ mL}}$$

2. To determine the median, rearrange the data in ascending (increasing) order and select the middle value (three will be smaller and three will be larger in this example).

Order	1	2	3	4	5	6	7
MPN/100 mL	2	4	6	8	10	14	56

$$\text{Median}\left[\frac{\text{MPN}}{100 \text{ mL}}\right] = \text{Middle Value of a Group of Data} = 8\frac{\text{MPN}}{100 \text{ mL}}$$

3. Calculate the geometric mean for the given data.

$$\text{Geometric Mean}\left[\frac{\text{MPN}}{100 \text{ mL}}\right] = (2 \times 8 \times 14 \times 6 \times 10 \times 56 \times 4)^{1/7}$$

$$= (3,010,560)^{0.143}$$

$$= 8.4\frac{\text{MPN}}{100 \text{ mL}}$$

Summary
1. Arithmetic Mean (average) = 14.3 MPN/100 mL
2. Median = 8 MPN/100 mL
3. Geometric Mean = 8.4 MPN/100 mL

As you can see from the summary, the geometric mean describes the central tendency of the data set better than the arithmetic mean. The one large value throws off the arithmetic mean.

Check Your Understanding

1. What is the Multiple Tube Fermentation Method?
2. How do you sterilize inoculation loops and needles before transferring a culture?

Standard Methods, 22nd Edition, Part 9221, Multiple-Tube Fermentation Technique for Members of the Coliform Group. www.awwa.org. ISBN: 978-087553-013-0.

MEMBRANE FILTER METHOD

The general strategy for this test is to capture bacteria on a filter and to place them on a nutrient medium where they will grow colonies that can be counted. The test uses a membrane filter made from cellulose ester, the pore size of which can be manufactured to close tolerances. Not only can the pore size be made to selectively trap bacteria from water filtered through the membrane, but nutrients can be diffused (from an enriched pad) through the membrane to grow these bacteria into colonies. These colonies are recognizable as coliform because the nutrients include fuchsin dye, which peculiarly colors the colony. Knowing the number of colonies and the volume of water filtered, the operator can then compare the water tested with water quality standards.

A two-step pre-enrichment technique for chlorinated samples is included at the end of this section. Chlorinated bacteria that are still living have had their enzyme systems damaged and require a 2-hour enrichment media before contact with the selective M-Endo media.

1. Materials Needed
   a. One sterile membrane filter having a 0.45 μm pore size.
   b. One sterile 47 mm Petri dish with lid.
   c. One sterile funnel and filter holder.
   d. Two sterile absorbent pads (see *Standard Methods* for specifications).

e. One filter flask (1,000 mL).

f. Vacuum pump, trap, suction or vacuum gauge, connection sections of plastic tubing, glass T-hose clamp to adjust pressure bypass.

g. Forceps (round-tipped tweezers), alcohol, Bunsen burner, grease pencil.

h. Sterile, buffered, distilled water for rinsing.

i. M-Endo media (obtain this from a commercial source).

j. Sterile pipets—two 5 mL graduated pipets and one 1 mL pipet for sample or one 10 mL pipet for larger sample. A quantity of 1 mL pipets if dilution of sample is necessary. Also, a quantity of dilution water blanks if dilution of sample is necessary.

k. One moist incubator at 35°C; auxiliary incubator dish with cover.

l. Enrichment media—lauryl tryptose broth (for pre-enrichment technique).

m. A binocular, wide-field, dissecting microscope is recommended for counting. The light source should be a cool white, fluorescent lamp.

2. Selecting Sample Size

The size of the sample or **aliquot** will be governed by the expected bacterial density. An ideal quantity will result in the growth of 20 to 80 coliform colonies, but not more than 200 bacterial colonies of all types. The table below lists suggested sample volumes for MF total coliform testing.

	Quantities Filtered (mL)				
	100	10	1	0.1	0.01
Well Water	X				
Drinking Water	X				
Lakes	X	X	X		
Rivers		X	X	X	X

3. Preparing Petri Dish for Membrane Filter

a. Sterilize forceps by dipping in alcohol and passing quickly through Bunsen burner flame to burn off the alcohol. An alcohol burner may be used also.

b. Place sterile absorbent pad into sterile Petri dish.

c. Add at least 2.0 mL M-Endo medium to the absorbent pad using a sterile pipet. Pour off excess media. An alternative to the absorbent pad is to make a solid gelatinous media in the Petri dish beforehand using LES Endo agar and place the filter directly on the agar. (See *Standard Methods*, Part 9222 B.)

4. Procedure for Filtering Unchlorinated Samples

The procedure presented here is a generalized version of the one contained in *Standard Methods*. It is illustrated in Figure 9.33. Consult *Standard Methods* for details.

In performing this test, avoid letting contaminants from the air, equipment, clothes, or fingers reach the specimen either directly or by way of the contaminated pipet.

a. Secure tubing from the vacuum pump to the filter flask. Place palm of hand on flask opening and start pump. Adjust suction to ¼ atmosphere with hose clamp on pressure bypass. Turn pump switch to OFF.

b. Set sterile filter holder and funnel on receiving flask.

aliquot (AL-uh-kwot)
Representative portion of a sample of any volume; often an equally divided portion of a sample.

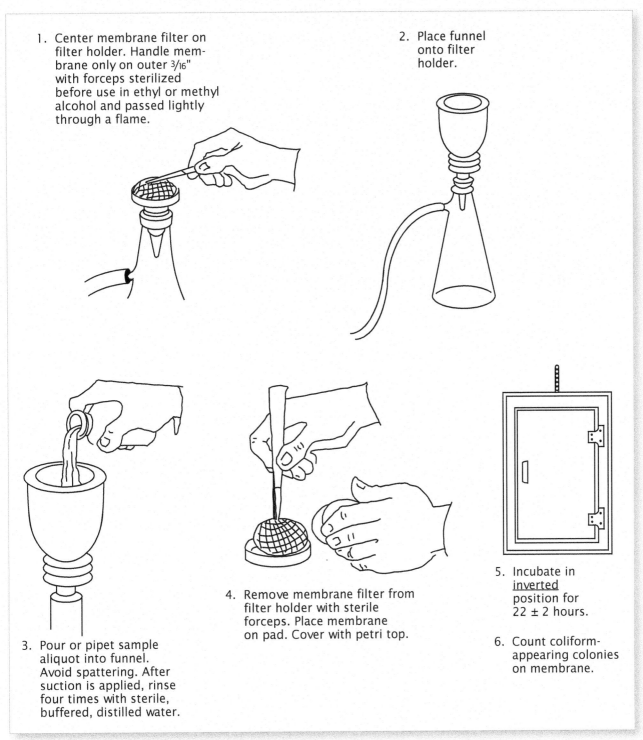

1. Center membrane filter on filter holder. Handle membrane only on outer 3/16" with forceps sterilized before use in ethyl or methyl alcohol and passed lightly through a flame.

2. Place funnel onto filter holder.

3. Pour or pipet sample aliquot into funnel. Avoid spattering. After suction is applied, rinse four times with sterile, buffered, distilled water.

4. Remove membrane filter from filter holder with sterile forceps. Place membrane on pad. Cover with petri top.

5. Incubate in <u>inverted</u> position for 22 ± 2 hours.

6. Count coliform-appearing colonies on membrane.

Figure 9.33 Procedure for inoculation of membrane filter

c. Place Petri dish on the lab bench with lid up. Write identification on lid with grease pencil.

d. Sterilize forceps by dipping in alcohol and passing quickly through Bunsen burner to burn off the alcohol.

e. Center the sterile membrane filter on the filter holder with forceps after lifting funnel. Membrane filter with printed grid should show grid on the upper surface (Figure 9.33, number 1).

f. Replace funnel and lock or clamp in place. Make sure the filter is centered under the funnel and that there are no leaks.

g. Shake sample or diluted sample. Measure proper aliquot with sterile pipet and add to funnel. If less than 10 mL of sample are to be filtered, there is a chance that the bacteria will not be uniformly distributed on the filter. Add the sample to a sterile bottle with 20–30 mL of sterile dilution water, and filter the mixture.

h. Now start vacuum pump.

i. After the sample is filtered, rinse the filter with three 20–30 mL portions of sterile dilution water.

j. When the water has drained through the filter, switch the pump to OFF.

k. Sterilize forceps as before.

l. Remove the membrane filter with forceps after first removing the funnel.

m. Place the membrane filter on the pad containing M-Endo media with a rolling motion to ensure water seal. Inspect membrane to ensure no captured air bubbles are present (Figure 9.33, number 4). Reseat the membrane if necessary to remove bubbles.

n. Place the Petri dish in the incubator inverted (pad on the top) for 22 ± 2 hours. Incubate at 35°C.

5. Procedure for Counting Membrane Filter Colonies

a. Remove Petri dish from incubator.

b. Remove lid from Petri dish.

c. Tilt the Petri dish so that green and yellow-green colonies are most apparent. Do this away from the windows; direct sunlight has too much red to facilitate counting.

d. Count individual colonies using an overhead fluorescent light. Use a low-power (10 to 15 magnifications), binocular, wide-field, dissecting microscope or other similar optical device. The typical colony has a pink to dark red color with a metallic surface sheen. The sheen area may vary from a small pinhead size to complete coverage of the colony surface. Only those showing this sheen should be counted.*

*Connected colonies, mirror reflections from fluorescent tubes, water condensate, and particulate matter can be mistaken for colonies, which could result in counting errors.

e. Report total number of coliform colonies on a worksheet. Results are often reported as colony forming units or CFUs. Use the membranes that show from 20 to 90 colonies and do not have more than 200 colonies of all types (including non-sheen or, in other words, non-coliforms). If the number of colonies exceeds 200, report the result as TNTC (too numerous to count). If the colonies grow together so that individuals cannot be discerned, report the result as confluent growth. If no colonies are observed, report the result as < 1 CFU/100 mL.

For the EPA Total Coliform Rule, any membrane containing a verified total coliform colony must be reported as a positive sample.

Example 27

A total of 42 colonies grew after filtering a 10 mL sample.

$$\frac{\text{Bacteria}}{100 \text{ mL}} = \frac{\text{No. of Colonies Counted}}{\text{Sample Volume Filtered [mL]}} \times 100 = \frac{42}{10 \text{ [mL]}} \times 100 = \frac{4.2}{\text{mL}} \times 100$$

$$= 420 \text{ CFU per 100 mL}$$

6. Verification Procedure

To confirm that the observed colonies are coliform bacteria, especially when colony appearance is ambiguous, use a sterile cotton swab or wire loop to transfer at least five colonies (or pieces from the colonies) to lauryl tryptose broth and Brilliant Green Bile Broth and incubate at 35°C for 24–48 hours (see Multiple Tube Fermentation Method above). Gas produced in both media verifies the colony as a coliform.

MEMBRANE FILTER PROCEDURE MODIFICATIONS

Procedure for Fecal Coliform

For drinking water samples, all 47 positive total coliform colonies must be tested for fecal coliform. When the Membrane Filter Method is used, growth from the positive total coliform colony should be transferred to a tube of EC media to determine the presence of fecal coliform.

1. Transfer, with the aid of a sterile 3 mm platinum wire loop, growth from each positive total coliform colony to a corresponding tube of EC broth.
2. Incubate inoculated EC broth tube in a water bath at 44.5°C ± 0.2°C for 24 hours.
3. Gas production in an EC broth culture within 24 hours or less is considered a positive fecal coliform reaction.

Procedure for Filtering Chlorinated Samples Using Enrichment Technique

Because chlorination damages coliforms, it may be desirable to provide an enrichment step to enable bacterial populations to recover enough to get a reliable count (consult local regulatory requirements). The following steps should be carried out between filtering the sample and counting the colonies.

1. Place a sterile absorbent pad in the upper half of a sterile Petri dish and pipet at least 2.0 mL sterile lauryl tryptose broth. Carefully remove any surplus liquid.
2. **Aseptically** place the membrane filter through which the sample has been passed on the pad.
3. Incubate the filter, without inverting the dish, for 1½ to 2 hours at 35°C in an atmosphere of at least 60 percent humidity (damp paper towels added to a plastic container with a snap-on lid can be used to produce the humidity).
4. Remove the enrichment culture from the incubator. Place a fresh, sterile, absorbent pad in the bottom half of the Petri dish and saturate with at least 2.0 mL M-Endo broth. Pour off excess.
5. Transfer the membrane filter to the new pad. Take care that no air bubbles are trapped between the membrane. The used pad of lauryl tryptose may be discarded.
6. Invert the dish and incubate for 20 to 22 hours at 35°C ± 0.5°C.
7. Count colonies as in previous method.

Alternate Membrane Filter Procedure for Fecal Coliform (Standard Methods, 22nd Edition, Part 9222 D)

The following method is a way to test for fecal coliform directly without first running the total coliform test as a separate procedure.

aseptic (a-SEP-tick)
Free from living microbes that cause disease, fermentation, or putrefaction; sterile.

This membrane filter procedure for fecal coliform uses an enriched lactose medium (M-FC broth) that depends on an incubation temperature of 44.5°C ± 0.2°C for its selectivity. Since the temperature is critical, incubation takes place in a water bath using watertight plastic bags.

1. Materials Needed
   a. M-FC media.
   b. Rosolic acid (17%) and 0.2 N NaOH.
   c. Culture dishes with tight-fitting lids.
   d. Membrane filters.
   e. Watertight plastic bags.
   f. Water bath set at 44.5°C ± 0.2°C.
   g. Use a sample size of 100 mL or whatever size is recommended by your regulatory agency.

2. Preparing Culture Dish
   Rehydrate the media powder as directed by the manufacturer using water and rosolic acid reagent.
   Place a sterile absorbent pad in each culture dish and pipet approximately 2 mL of sterile M-FC medium to saturate the pad. Pour off any surplus liquid.

3. Filtering Sample
   Follow the procedure prescribed for total coliform using membrane filters.

4. Incubation
   Place the prepared culture dishes in waterproof plastic bags and immerse in water bath set at 44.5°C ± 0.2°C for 24 hours. All culture dishes should be placed in the water bath within 30 minutes after filtration.

5. Counting
   Colonies produced by fecal coliform bacteria are blue. The non-fecal coliform colonies are gray to cream colored. Normally, few non-fecal coliform colonies will be observed due to the selective action of the elevated temperature and the addition of the rosolic acid to the M-FC media.
   Examine the cultures under a low-power magnification. Count and calculate fecal coliform density per 100 mL.

$$\frac{\text{Fecal Coliforms}}{100 \text{ mL}} = \frac{\text{Fecal Colonies Counted}}{\text{Sample Volume Filtered}\,[\text{mL}]} \times 100$$

Example 28

What is the fecal coliform count per 100 mL if two colonies grew after filtering a 100 mL sample?

$$\frac{\text{Fecal Coliforms}}{100 \text{ mL}} = \frac{\text{Fecal Colonies Counted}}{\text{Sample Volume Filtered}\,[\text{mL}]} \times 100 = \frac{2}{100\,[\text{mL}]} \times 100$$

$$= \frac{0.02}{\text{mL}} \times 100$$

$$= 2 \text{ per } 100 \text{ mL}$$

Standard Methods, 22nd Edition, Part 9222 D, Membrane Filter Technique for Members of the Coliform Group.
www.awwa.org.
ISBN: 978-0-87553-013-0.

MEMBRANE FILTER METHOD—*ESCHERICHIA COLI*

This method describes a membrane filter (MF) procedure for identifying and counting *E. coli* (EPA Method 1603). Because the bacterium is a natural inhabitant only of the intestinal tract of warm-blooded animals, its presence in water samples is an indication of fecal pollution and the possible presence of enteric (intestinal) pathogens.

This MF method provides a direct count of bacteria in water based on the development of colonies on the surface of the membrane filter. A water sample is filtered through a membrane that retains the bacteria. After filtration, the membrane is placed on mTEC, a specially formulated medium. The membrane is incubated at 35°C for 2 hours to revive injured or stressed bacteria, and then incubated at 44.5°C for 22 hours. *E. coli* appear as red or magenta colonies on the mTEC agar.

1. Materials Needed
   a. Glass lens with magnification of 2 to 5×, or stereoscopic microscope.
   b. Lamp with a cool, white fluorescent tube.
   c. Flasks.
   d. Autoclave for sterilizing media and glassware.
   e. Culture dishes with tight-fitting lids.
   f. Test tubes, screw-cap.
   g. Membrane filters.
   h. Incubator maintained at 35°C ± 0.5°C.
   i. Whirl-Pak® bags (sterile, sealable plastic).
   j. Water bath maintained at 44.5°C ± 0.2°C.
   k. Absorbent pads.
   l. Modified mTEC agar (Difco 0334).
      Preparation: Add 45.3 grams of dehydrated mTEC agar to 1 L of reagent-grade distilled water in a flask and heat to boiling until the ingredients dissolve. Autoclave at 121°C (15 lbs pressure) for 15 minutes and cool in a 50°C water bath. Adjust pH to 7.3 ± 0.2 using HCl or NaOH. Pour the medium into each 9 × 50 mm culture dish to a 4–5 mm depth (approximately 4–6 mL) and allow to solidify. Store in a refrigerator.
   m. Nutrient agar (Difco 0001, BD 4311472).
      Preparation: Add 23 grams of nutrient agar to 1 L of reagent-grade distilled water and mix well. Heat to boiling to dissolve the agar completely. Dispense in screw-cap tubes and autoclave at 121°C (15 lbs pressure) for 15 minutes. Remove the tubes and slant. The final pH should be 6.8 ± 0.2.

2. Procedure
   a. Prepare the modified mTEC agar as directed in (l).
   b. Mark the Petri dish and report form with the sample identification and volume.
   c. Place a sterile membrane filter on the filter holder, grid side up, and attach the funnel to the base. Make sure that the filter is centered under the funnel and that there are no leaks.
   d. Shake the sample bottle vigorously about 25 times to distribute the bacteria uniformly and deliver the desired volume of sample or dilution into the funnel. For surface waters and wastewaters, select sample volumes based on previous knowledge of pollution level to produce 20 to 80 *E. coli* colonies on the mem-

branes. For small sample sizes (< 20 mL), add the sample to 20–30 mL of sterile water in a sterile bottle and filter the mixture. This will allow even distribution of the sample on the filter.

   e. Filter the sample and rinse the sides of the funnel at least twice with 20 to 30 mL of sterile buffered rinse water. Turn off the vacuum and remove the funnel from the filter base.

   f. Use sterile forceps to aseptically remove the membrane filter from the filter base and roll it onto the modified mTEC agar to avoid forming bubbles between the membrane and the agar surface. Reseat the membrane if bubbles occur. Close the dish and invert (agar on the top).

   g. Incubate for 2 hours at 35°C ± 0.5°C. Then transfer the plate to a Whirl-Pak® bag, seal the bag, and submerge the bag with the plate inverted in a 44.5°C ± 0.2°C water bath for 22 ± 2 hours.

   h. After incubation, remove the plate from the water bath.

   i. Count and record the number of red or magenta colonies on the membrane filters.

3. Calculating Results

Select the membrane filter with an acceptable number of yellow, yellow-green, or yellowbrown colonies (20 to 80) on the urea substrate and calculate the number of *E. coli* per 100 mL according to the following general formula:

$$\frac{E.coli}{100 \text{ mL}} = \frac{\text{Number of } E.coli \text{ Colonies Counted}}{\text{mL of Sample}} \times 100$$

4. Reporting Results

Report the results as *E. coli* per 100 mL of sample.

Example 29

A total of 40 *E. coli* colonies were counted after filtering a 50 mL sample. Calculate the *E. coli* per 100 mL of sample.

$$\frac{E. coli}{100 \text{ mL}} = \frac{\text{Number of } E. coli \text{ Colonies Counted}}{\text{mL of Sample}} \times 100$$

$$= \frac{40}{50\,[\text{mL}]} \times 100 = \frac{0.8}{\text{mL}} \times 100$$

$$= \frac{80 \; E. coli}{100 \text{ mL}}$$

 EPA Method 1603: *Escherichia coli (E. coli)* in Water by Membrane Filtration Using Modified Membrane-Thermotolerant *Escherichia coli* Agar (Modified mTEC). EPA-821-R-06-011, July 2006.

Standard Methods, 22nd Edition, Part 9222, Membrane Filter Technique for Members of the Coliform Group.

www.awwa.org.
ISBN: 978-0-87553-013-0.

PRESENCE–ABSENCE METHOD

The Presence–Absence (P–A) Test for the coliform group in drinking water is a simplified version of the multiple tube procedure (described previously) that gives qualitative information on the presence or absence of coliforms. The media used is a mixture of lactose and lauryl tryptose broths with bromocresol purple added to indicate pH. Following incubation, gas or acid (yellow color of media) is produced if coliforms are present. The P–A Test procedure is essentially the same as for the

Multiple Tube Fermentation Test outlined in Figure 9.26 except that only one large test portion (100 mL) is used rather than multiple tubes and dilutions.

1. Materials Needed

    Media

    a. P–A broth. This media is available commercially. Prepare according to the instructions on the media bottle label.
    b. Brilliant Green Bile Broth. Same as for the MPN method.
    c. EC broth. Same as for the MPN method.

    Other Materials

    a. 250 mL screw-cap, milk dilution bottle with 50 mL sterile triple-strength P–A broth.
    b. Autoclave for sterilizing media and glassware.
    c. Incubator set at 35°C ± 0.5°C.
    d. Sterile, 100 mL graduated cylinder.
    e. Water bath or incubator set at 35.5°C ± 0.5°C.

2. Presumptive Test
    a. Vigorously shake sample approximately 25 times.
    b. Inoculate 100 mL of sample into a P–A culture bottle with 50 mL of sterile triple-strength medium and mix thoroughly.
    c. Incubate at 35°C ± 0.5°C and inspect after 24 and 48 hours for acid reactions (yellow color). Record both positive and negative culture bottles.
    d. A distinct yellow color forms in the culture when acid conditions exist. If gas is also being produced, gentle shaking of the bottle will result in a foaming action. Any amount of gas or acid constitutes a positive presumptive test requiring confirmation for total coliform.

3. Confirmation Test
    a. Confirm all presumptive culture bottles that show gas or acid at 24 or 48 hours. Using a sterile platinum wire loop or sterile wood applicator, transfer the liquid from the P–A broth showing gas or acid to a fermentation tube containing a tube of Brilliant Green Bile (BGB) Broth fitted with an inverted vial to collect gas (same as in the MPN method).
    b. After 24 hours have elapsed, inspect the tube for gas formation. Any amount of gas formation is considered positive. Incubate negative BGB tubes for an additional 24 hours. Record both positive and negative results on the data sheet.
    c. Examine tubes for gas at the end of the 48 ± 3-hour period. Record both positive and negative tubes.

        Gas production in the BGB broth culture within 48 hours confirms the presence of coliform bacteria. Report result as Presence–Absence Test positive or negative for total coliforms in 100 mL of sample.

4. Fecal Coliform Test (*E. coli* Test)
    Samples that are positive for total coliform must be tested for fecal coliform or *E. coli*. See procedures presented previously.

Example 30

Given the P–A Test data from several water samples shown below, interpret the results and decide the next step for each sample.

	Sample 1	Sample 2	Sample 3
Presumptive Test			
Appearance after 24 h	Clear (-)	Yellow color, with or without foam-ing/gas (+)	Clear (-)
Appearance after 48 h	Yellow color, with or without foam-ing/gas (+)		Clear (-)
Interpretation and next step	Report a positive P–A presumptive result. Proceed with confirmation test.	Report a positive P–A presumptive result. Proceed with confirmation test.	Report a negative P–A result. No further testing.
Confirmation Test			
Appearance after 24 h	No gas (-)	No gas (-)	
Appearance after 48 h	No gas (-)	Gas (+)	
Interpretation and next step	Report a negative P–A result for total coliforms. No further testing.	Report a positive P–A result for total coliforms. Test for fecal coliform (E. coli).	

ENZYME SUBSTRATE TESTS

Enzyme substrate tests (Colilert® and Colisure®) use a media that contains specific indicator nutrients for total coliform and *E. coli*. As these nutrients are metabolized, yellow color and fluorescence are released confirming the presence of total coliform and *E. coli* respectively. Non-coliform bacteria are chemically suppressed and cannot metabolize the indicator nutrients. Consequently, they do not interfere with identifying the target microbes. Total coliforms and *E. coli* are specifically and simultaneously detected and identified in 24 hours or less. The confirmed and completed tests used in multiple tube fermentation are not needed. Neither is the fecal coliform test needed because the presence of *E. coli* indicates fecal contamination.

Although this method can yield presence–absence results within 24 hours and is less cumbersome to perform than the Membrane Filtration (MF) Method, operators should be aware of the limitations of these tests in evaluating samples for regulatory purposes. The results are not always comparable to MF due to:

• Interferences in the sample that may suppress or mask bacterial growth
• Greater sensitivity of the Colilert® media
• Added stress to organisms related to filtering
• The fact that different media may obtain better growth for some bacteria

Two enzyme substrate tests are commercially available under the names Colilert® and Colisure®, registered trademarks of IDEXX Laboratories

(Westbrook, Maine). Both procedures test for coliform and *E. coli* bacteria simultaneously.

As the media's nutrients are metabolized, yellow color (Colilert®) or magenta (Colisure®) and fluorescence are released, confirming the presence of total coliforms and *E. coli* respectively. Both tests can be used for presence–absence testing or for determining coliform MPN. For an MPN test, either serial dilutions such as those described for the Multiple Tube Fermentation Method, or an IDEXX proprietary method (called a Quanti-tray) can be used.

Both test methods are approved by the EPA for reporting under the Safe Drinking Water Program. When these tests produce conflicting bacteriological results, however, the safest course of action is to increase monitoring and treatment efforts until the results for both tests are negative.

1. Materials Needed
   a. For the Presence–Absence Test: dehydrated Colilert® or Colisure® media and 100 mL jars or flasks with lids.
   b. For the MPN test: 10 culture tubes each containing Colilert® reagent.
   c. Sterile 10 mL pipets.
   d. Incubator at 35°C.
   e. Long wavelength (365 nm) ultraviolet lamp.
   f. Color and fluorescence comparator.

2. Procedures
   General procedures are presented below. Refer to the manufacturer's instructions for details. Also, be sure to check your local regulator's requirements.
   a. Presence–Absence (P–A) Procedure
      (1) Add the contents of one pack of dry media to a 100 mL sample in a sterile, transparent, nonfluorescing container (a jar or flask) with a lid.
      (2) Cap the container and shake until the powdered media has dissolved.
      (3) Incubate at 35 ± 0.5°C for 24 hours.
      (4) Look for color. For Colilert®, compare the color against a manufacturer-provided comparator. For Colisure®, look for a red or magenta color.
      (5) Look for fluorescence with a 6-watt, 365 nm UV light within 5 inches of the sample in a dark environment. Do not look into the UV light; face it toward the sample and away from your eyes. For Colilert®, compare the color against a manufacturer-provided comparator.
      (6) Record results as per Table 9.7.
   b. MPN Test for Drinking Water Samples (using the pre-dispensed MPN tube kit)
      (1) Aseptically fill each of 10 Colilert® tubes with 10 mL of a well-mixed water sample.
      (2) Cap the tubes tightly and mix vigorously to dissolve the reagent by repeated inversion. Some particles may remain undissolved. Dissolution will continue during incubation.
      (3) Incubate at 35±0.5°C for 24 hours.
      (4) Read and record results as described in the Presence–Absence Test.

Table 9.7　Interpretation of observations

Result	Colilert® a	Colisure® b
Negative for coliforms and *E. coli*	Less yellow than comparator	Yellow or gold color
Positive for total coliforms	Yellow color equal to or more intense than comparator	Red or magenta color
Positive for *E. coli*	Yellow color and fluorescence equal to or greater than comparator	Red/magenta and fluorescence

a If the results are ambiguous after 24 hours, incubate up to an additional 4 hours (but not to exceed 28 hours total) to allow the color and/or fluorescence to intensify. Positives for both total coliforms and *E. coli* observed before 24 hours and negatives observed after 28 hours are also valid.
b If the sample is pink or orange, incubate up to an additional 24 hours (48 hours total). A sample is negative if it remains pink or orange at 48 hours. Red/magenta or red/magenta with fluorescence observed before 24 hours is a valid positive result. The lack of red/magenta and/or fluorescence after 24 hours is a negative result.

Look up the MPN Index based on the number of positive tubes in Table 9.6. If the original sample was diluted, be sure to multiply the MPN Index by the dilution factor.

9.4.4.5　Additional Test Methods

For a summary of approved and accepted microbiological methods for EPA water monitoring regulations, see EPA's website at https://www.epa.gov/dwanalyticalmethods/approved-drinking-water-analytical-methods. Information on all of the following methods may be found at www.EPA.gov or by contacting the companies listed below.

- E*Colite™ Test: Information available from Charm Sciences, Inc., www.charm.com/products/ecolite.
- m-ColiBlue24® Test: Information available from the HACH Company, www.hach.com.
- Presence–Absence Broth, EMD Millipore (http://www.emdmillipore.com/US/en/product/Presence–Absence-Broth,MDA_CHEM-100414).
- Coliform Agar ES (Enhanced Selectivity), EMD Millipore (http://www.emdmillipore.com/US/en/product/Coliform-Agar-ES-%28Enhanced-Selectivity%29,MDA_CHEM-100850).
- Colitag™ Presence–Absence (P–A) Test: Information available from CPI International, www.cpiinternational.com or www.colitag.com.

See the Characterizing Bacterial Counts portion of Section 9.4.4.4, "Procedures for Testing Total Coliform Bacteria," for additional information.

Check Your Understanding

1. What is the strategy for the Membrane Filter Method?
2. What is an aliquot?
3. What is the meaning of TNTC?
4. Where does E. coli naturally live and what does its presence in water indicate?
5. What is the Presence–Absence (P–A) Test and what is it used for?

9.5   Sampling

Proper sampling is a vital part of studying the quality of water in a water treatment process, distribution system, or water supply. Most errors in the whole process of obtaining water quality information occur during sampling. Proper sampling procedures are essential to obtain an accurate description of the material or water being tested. This fact cannot be over-emphasized. Water treatment decisions based upon incorrect data may be made if sampling is performed in a careless and thoughtless manner.

In any type of testing program where only small samples (1–2 liters) are withdrawn from perhaps millions of gallons of water under examination, there is uncertainty because of possible sampling errors. Obtaining good test results will depend to a great extent upon the following factors:

- Ensuring that the sample taken is truly representative of the water being tested
- Using proper techniques to obtain the samples
- Protecting and preserving the samples until they are analyzed

(i) The greatest sources of errors in laboratory tests are improper sampling, poor preservation, or lack of enough mixing during testing. The accuracy of your analysis is only as good as the care that was taken in obtaining your sample.

9.5.1   Representative Sampling

Unless a **representative sample** is collected, test results will not reflect actual water conditions.

Obtaining a representative sample of a tank or a lake that is completely mixed is a simple matter because the mixing assures that the water characteristics are the same everywhere in the system. Unfortunately, most bodies of water are not well mixed and obtaining samples that are truly representative of the whole body depends to a great degree upon sampling technique. A sample collected from one location has a significant risk of being unrepresentative of the whole. One way to create a sample that reflects the average conditions in the water body or tank is to take small portions of the water at points distributed over the whole body and mix them together. Such a sample is called a **composite sample**. The more portions taken, the more nearly the sample represents the original. Statistically, the sample error would reach zero when the size of the sample becomes equal to the original volume of material being sampled, but, for obvious reasons, this method of decreasing sample error is not practical.

The size of sample depends on which water quality indicators are being tested and how many. Some tests require only a small volume; others require more.

representative sample
A sample portion of material, water, or wastestream that is as nearly identical in content and consistency as possible to that in the larger body being sampled.

composite (proportional) sample
A composite sample is a collection of individual samples obtained at regular intervals. For larger plants, this is usually every one or two hours during a 24-hour time span. For smaller wastewater treatment plants and lagoons, this sample may be collected three or more times during an eight-hour shift. Each individual sample is combined with the others in proportion to the rate of flow when the sample was collected. Mixtures of equal volume individual samples collected at intervals after a specific volume of flow passes the sampling point, or after equal time intervals, are still referred to as composite samples. The resulting mixture (composite sample) forms a representative sample and is analyzed to determine the average conditions during the sampling period.

9.5.1.1   Source Water Sampling

Rivers. To adequately determine the composition of a flowing stream, each sample (or set of samples taken at the same time) must be representative of the entire flow at the sampling point at that instant. Furthermore, the sampling process must be repeated frequently enough to show significant changes of water quality that may occur over time in the water passing the sampling point.

On small or medium-sized streams, it may be possible to find a single sampling point at which the composition of the water is presumably uniform at all depths and across the stream. Obtaining representative samples in these streams is relatively simple. For larger streams, more than one sample may be required. It may be necessary to sample at several locations across the stream, and/or at several depths and create a composite sample. River edges may contain shallow sub-streams that are loosely connected to the main flow and may not be representative of the main flow. If they are a small fraction of the river flow, they may be omitted from the sampling scheme.

Reservoirs and Lakes. Water stored in reservoirs and lakes is usually poorly mixed. Thermal stratification and associated in water composition with depth (such as dissolved oxygen) are among the most frequently observed effects. Single samples can therefore be assumed to represent only the spot of water from which the sample came. Therefore, a number of samples must be collected at different depths and from different areas of the impoundment to accurately sample reservoirs and lakes.

Groundwater. Most of the physical factors that promote mixing in surface waters are absent or much less effective in groundwater systems. Consequently, there can be considerable spatial variation in groundwater quality. A well sample should be considered to be representative of only that particular location. Vertically, an aquifer can be divided into different zones by impermeable layers (strata). Wells usually draw water from a considerable thickness of saturated rock and often from several different strata. These water components are mixed by the turbulent flow of water in the well before they reach the surface and become available for sampling. Usually, the only means of sampling the water tapped by a well is the collection of a pumped sample. This sample will represent the water only in the layers feeding the well. In Figure 9.34, only Zones 2 and 3 are being sampled. Check the depth(s) of the well screen. Also, remember that well pumps and casings can contribute to sample contamination. If a pump has not run for an extended period of time before sampling, the water collected may not be representative of the normal water quality. Leaking seals and oils are a potential source of contamination.

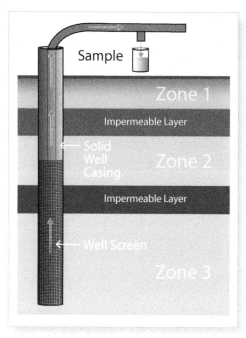

Figure 9.34   Well profile

9.5.1.2   In-Plant Sampling

Collecting representative samples within the water treatment plants is really no different from sample collection in a stream or river. You want to be sure the water sampled is representative of the water passing that sample point. In many water plants, money is wasted installing sample pumps and piping that sample from points that are not representative of the passing water. A sample tap in a dead area of a reservoir or on the floor of a process basin will not help you properly control water quality. Find every sample point and ensure it is located to provide a useful and representative sample. If the sampling point is not properly located, plan to move the piping to a better location.

9.5.1.3   Distribution System Sampling

Representative sampling in the distribution system is a true indication of the water quality being delivered to the customer. Sampling results should show if there are quality changes in the entire system or parts of the system, and may point to the source of a problem (such as tastes and odors). Sampling points should be selected, in part, to trace the course of the finished water from the well or plant, through the transmission mains, and then through the major and minor arteries of the system. A sampling point on a major artery, or on an active main directly connected to it, would be representative of the water quality being furnished to a downstream subdivision of this network. Generally, these primary points are used as official sample points for evaluating prevailing water quality.

Obtaining a representative sample from the distribution system is not as easy as it might seem. One would think almost any faucet would do, but experience has shown otherwise. Local conditions at the tap and in its connection to the main can easily make the point unrepresentative of water being furnished to your consumers. For instance, a tap on a dead-end pipe that conveys little flow will not give a sample that is representative of most of the water flowing through the system.

The truest evaluation of water in a distribution system can be obtained from samples drawn directly from the main. You might think that samples taken from a fire hydrant would prove satisfactory, but this is usually not the case. The problem with fire hydrants as sampling points is that they give erratic (uneven) results due to the way they are constructed and their lack of use. In general, an ideal sample station is one that has a short, direct connection with the main and is made of corrosion-resistant material.

In most smaller water systems, special sample taps are not available. Therefore, customer's faucets must be used to collect samples. The best sample points are front yard faucets on homes supplied by short service lines (homes with short service lines are located on the same side of the street as the water main).

If the customer is home, you should contact the person in the home and obtain permission to collect the sample. Disconnect the hose from the faucet if one is attached and do not forget to reconnect the hose when you are done. Open the faucet to a convenient flow for sampling (usually about half a gallon per minute). Allow the water to flow until the water in the service line has been replaced twice. Since 50 feet (15 m) of three-quarter inch (18 mm) pipe contains over 1 gallon (3.8 liters), 4 or 5 minutes will be required to replace the water in the line twice. Collect the sample. Be sure the sample container does not touch the faucet.

Do not try to save time by turning the faucet handle to wide open to flush the service line. This can disturb sediment and incrustations in the line, which will contaminate your sample. If you accidentally do this, be sure to flush the line thoroughly before collecting your sample.

FORMULAS

To estimate the flow from a faucet, use a gallon jug and a watch (or the timer on your phone). If you want a flow of half a gallon per minute, then the jug should be half full in 1 minute or completely full in 2 minutes.

$$Q\left[\frac{\text{gal}}{\text{min}}\right] = \frac{V[\text{gal}]}{t[\text{min}]}$$

Where,

Q = Flow Rate

V = Volume

t = Time

Pipes can be treated like cylinders. To calculate the volume of a service line, multiply the cross-sectional area of the pipe in square feet times the length of the pipe in feet to obtain cubic feet. The diameter of a pipe is given in inches, so this value must be divided by 12 inches per foot to obtain a volume in cubic feet. Multiply cubic feet by 7.48 gallons per cubic foot to obtain the volume in gallons.

$$V_{pipe}\left[ft^3\right] = A_{pipe}\left[in^2\right] \times \frac{ft^2}{144\ in^2} \times L_{pipe}\left[ft\right]$$

$$V_{pipe}\left[gal\right] = V_{pipe}\left[ft^3\right] \times \frac{7.48\ gal}{ft^3}$$

Where,

V_{pipe} = Pipe Volume
A_{pipe} = Pipe Area
L_{pipe} = Pipe Length

To determine the time needed to flush a service line twice, divide the pipe volume in gallons by the flow rate in gallons per minute. Then, multiply the result by two so the line will be flushed twice.

$$t_{flushing}\left[min\right] = \frac{V_{pipe}\left[gal\right] \times 2}{Q\left[\dfrac{gal}{min}\right]}$$

Where,

$t_{flushing}$ = Flushing Time

Example 31

How long should a three-quarter-inch service line 80 feet long be flushed if the flow is 0.5 gpm?

Known	Unknown
D_{pipe} = Pipe Diameter = ¾ inch	$t_{flushing}$ = Flushing Time, min
= 0.75 inch	
L_{pipe} = Pipe Length = 80 feet	
Q = Flow Rate = 0.5 gpm	

Calculate the pipe volume (V_{pipe}) in cubic feet (ft³) and then in gallons (gal).

$$V_{pipe}\left[ft^3\right] = A_{pipe}\left[in^2\right] \times \frac{ft^2}{144\ in^2} \times L_{pipe}\left[ft\right] = \frac{\pi}{4} \times D^2\left[in^2\right] \times \frac{ft^2}{144\ in^2} \times L\left[ft\right]$$

$$= 0.785 \times 0.75^2\left[in^2\right] \times \frac{ft^2}{144\ in^2} \times 80\left[ft\right] = 0.245\ ft^3$$

$$V_{pipe}\left[gal\right] = V_{pipe}\left[ft^3\right] \times \frac{7.48\ gal}{ft^3} = 0.245\left[ft^3\right] \times \frac{7.48\ gal}{ft^3} = 1.83\ gal$$

Calculate the flushing time ($t_{flushing}$) for the service line in minutes (min).

$$t_{flushing}\left[min\right] = \frac{V_{pipe}\left[gal\right] \times 2}{Q\left[\dfrac{gal}{min}\right]} = \frac{1.83\left[gal\right] \times 2}{0.5\left[\dfrac{gal}{min}\right]} = 7.3\ min$$

or

$$= 7\left[min\right] + 0.3\left[min\right] \times \frac{60\ s}{min} = 7\ min\ and\ 18\ s$$

Check Your Understanding

1. What are frequently the greatest sources of errors in laboratory tests?
2. Why must a representative sample be collected?
3. How are sampling points selected in a distribution system?

9.5.2 Types of Samples

There are generally two types of samples collected by waterworks operators, and either type may be obtained manually or automatically. The two types are grab samples and composite samples.

9.5.2.1 Grab Samples

A **grab sample** is a single water sample collected at a single time. Grab samples will show the water characteristics at the time and place the sample was taken; thus, they should not be used to measure turbidity as a water quality indicator. A grab sample may be preferred over a composite sample when:

- The water to be sampled does not flow continuously.
- The water's characteristics are relatively constant.
- The water is to be analyzed for water quality indicators that may change with time so the test needs to be started quickly. Examples include dissolved gases, coliform **bacteria**, residual chlorine, temperature, and pH.

9.5.2.2 Composite Samples

In many processes, the water quality is changing from moment to moment or hour to hour. A continuous sampler-analyzer would give the most accurate results in these cases. Unfortunately, relatively few parameters (e.g., temperature, pH, turbidity) can be analyzed this way. Composite samples should not be collected for bacteriological examination. Microbial tests are only done on grab samples.

For most parameters, operators themselves would have to be the sampler-analyzer and continuous analysis would leave little time for anything but sampling and testing. Except for tests that cannot wait due to rapid physical, chemical, or biological changes of the sample (such as tests for dissolved oxygen, pH, and temperature), a fair compromise may be reached by taking samples throughout the day at hourly or 2-hour intervals. Each sample should be refrigerated immediately after it is collected. At the end of 24 hours, a composite sample can be made by mixing together portions of each sample. The sizes of the portions should be in direct proportion to the flow at the time the sample was collected and the total size of sample needed for testing. The proportions for a 2-liter total sample are as follows:

Sample	Flow (MGD)	Fraction[a]	Volume of Sample[b]
1	1.2	0.32	0.64
2	1.0	0.27	0.54
3	1.5	0.41	0.82
Sum	3.7		2.00

[a] Fraction = Flow ÷ Total Flow
[b] Volume = Fraction × 2 L

bacteria (back-TEER-e-uh)
Bacteria are living organisms, microscopic in size, that usually consist of a single cell. Most bacteria use organic matter for their food and produce waste products as a result of their life processes.

When the samples are taken, they can either be set aside to be combined later or combined as they are collected. In both cases, they should be stored at a temperature of 39°F (4°C) until they are analyzed.

9.5.3   Sampling Devices

Automatic sampling devices are wonderful time-savers but can be expensive. As with anything automatic, problems do arise and the operator should be on the lookout for potential difficulties.

Manual sampling equipment includes dippers, weighted bottles, hand-operated pumps, and cross-sectional samplers. Dippers consist of wide-mouth, corrosion-resistant containers (such as cans or jars) attached to long handles. A weighted bottle is a collection container that is lowered to a desired depth where a cord or wire removes the bottle stopper so the bottle can be filled (see Figure 9.35).

Some water treatment facilities use sample pumps to collect the sample and transport it to a central location such as the plant lab. The pump and its associated piping should be corrosion-resistant and sized to deliver the sample at a high enough velocity to prevent sedimentation in the sample line.

Many water agencies have designed and installed special sampling stations throughout their distribution systems (see Figure 9.36). These stations provide an excellent location to sample the water in your distribution system.

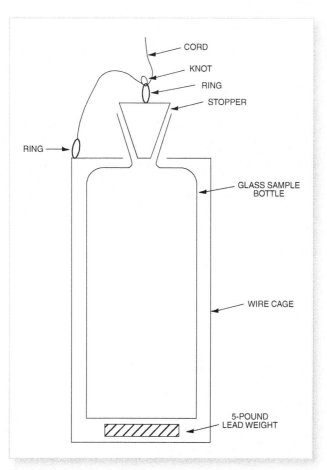

Figure 9.35   Sectional view of homemade depth sampler

Figure 9.36   Distribution system sampling station

9.5.4   Sampling Techniques

A water treatment plant operator must be proficient in collecting samples using several techniques. These techniques include surface sampling, depth sampling, water tap sampling, and first-draw sampling.

9.5.4.1   Surface Sampling

A surface sample is obtained by grasping the sample container at the base with one hand and plunging the bottle mouth down into the water to avoid introducing any material floating on the surface. Position the mouth of the bottle into the current and away from the hand of the collector (see Figure 9.37). If the water is not flowing, then an artificial current can be created by moving the bottle horizontally in the direction it is pointed and away from the sampler. Tip the bottle slightly upward to allow air to exit so the bottle can fill. Tightly stopper and label the bottle.

When sampling from a walkway or other structure above a body of water, place the bottle in a weighted frame that holds the bottle securely. Remove the stopper or lid and lower the device to the water surface. A nylon rope, which does not absorb water and will not rot, is recommended. Face the bottle mouth upstream by swinging the sampling device downstream and then allow it to drop into the water, without slack in the rope. Pull the sample device rapidly upstream and out of the water simulating (imitating) the scooping motion of the hand-sampling described previously. Take care not to dislodge dirt or other debris that might fall into the open sample container from above. Be sure to label the container when sampling is completed.

9.5.4.2   Depth Sampling

Several additional pieces of equipment are needed for collecting depth samples from basins, tanks, lakes, and reservoirs. These depth samplers require lowering the sample device and container to the desired water depth, then opening, filling, and closing the container, and returning the device to the surface. Although depth measurements are best made with a premarked steel

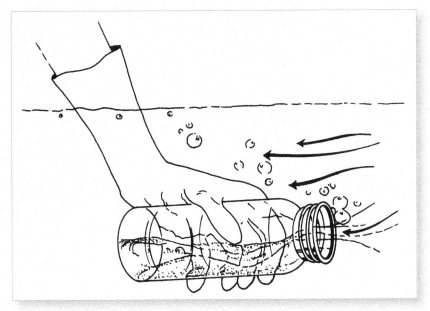

Figure 9.37   Demonstration of technique used in grab sampling of surface waters (*Source: US EPA, Microbiological Methods for Monitoring the Environment*)

cable, the sample depths can be determined by premeasuring and marking a nylon rope at intervals with a non-smearing ink, paint, or fingernail polish. One of the most common commercial devices is called a Kemmerer sampler (see Figure 9.38). This type of depth sampler consists of a cylindrical tube that contains a rubber stopper or valve at each end. The device is lowered into the water in the open position and the water sample is trapped in the cylinder when the valves are closed by the dropped messenger.

Figures 9.35 and 9.39 show typical depth samplers. These samplers are lowered to the desired depth. A jerk on the cord will remove the stopper and allow the bottle in the depth sampler to fill. Good samples can be collected in depths of water up to 40 feet (12 m).

9.5.4.3 Water Tap Sampling

To collect samples from water main connections, first flush the service line for a brief period of time. Samples should not be taken from drinking fountains, restrooms, or taps that have aerators. Aerators can change water quality indicators such as pH and dissolved oxygen, and can harbor bacteria under some conditions. Do not sample from taps surrounded by excessive foliage (leaves, flowers) or from taps that are dirty, corroded, or are leaking. Never collect a sample from a hose or any other attachment fastened to a faucet. Take care that the sample collector does not come in contact with the faucet.

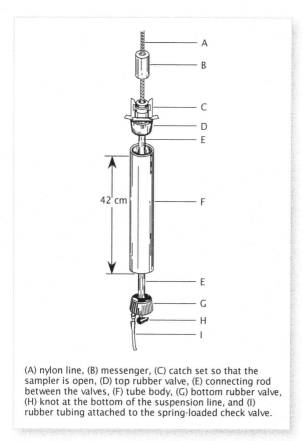

(A) nylon line, (B) messenger, (C) catch set so that the sampler is open, (D) top rubber valve, (E) connecting rod between the valves, (F) tube body, (G) bottom rubber valve, (H) knot at the bottom of the suspension line, and (I) rubber tubing attached to the spring-loaded check valve.

Figure 9.38 Kemmerer depth sampler (*Source: US EPA, Microbiological Methods for Monitoring the Environment*)

Figure 9.39 Depth sampler (*Courtesy of HACH Company*)

9.5.4.4   First-Draw Sampling

The Lead and Copper Rule calls for first-draw or first-flush samples. These are water samples taken at the customer's tap after the water stands motionless in the plumbing pipes for at least 6 hours. This usually means taking a sample early in the day before water is used in the kitchen or bathroom.

9.5.5   Sampling Containers and Preservation of Samples

Some water quality indicators, such as residual chlorine and temperature, require immediate analysis, while others can be preserved and transported to the laboratory. The shorter the time that elapses between collecting the sample and the start of the lab test, the more reliable your results will be. If the samples are not going to be analyzed immediately due to remoteness of the laboratory or workload, they should be preserved according to accepted protocols. Examples of these protocols are shown in Table 9.8. Preserving some sample types is essential to preventing deterioration of the sample.

Whenever you collect a sample for a bacteriological test (coliforms), be sure to use a sterile plastic or glass bottle. If the sample contains any chlorine residual, sufficient sodium thiosulfate should be added to neutralize all of the chlorine residual. Usually, two drops (0.1 mL) of 10 percent sodium thiosulfate for every 100 mL of sample is sufficient, unless you are disinfecting mains or storage tanks. Clearly identify on the container the sample location, date and time of collection, name of collector, and other pertinent information.

Standard Methods, 22nd Edition, Part 1060, Collection and Preservation of Samples. www.awwa.org. ISBN: 978-0-87553-013-0.

Water Quality Sampling Guidelines, Second Edition, January 2005. American Water Works Association. www.awwa.org.

9.5.5.1   Chain-of-Custody Samples

Usually, samples must be collected and transported to a laboratory using chain-of-custody protocol. Normally the laboratory will provide a chain-of-custody form to be filled out by the person collecting the sample.

9.5.6   Reporting

The water system owner (e.g., water utility agency) is responsible for reporting lab results at regular frequencies to the regulatory agency as required by the Safe Drinking Water Act.

Check Your Understanding

1. What are the two general types of samples collected by water treatment personnel?
2. List three water quality indicators that are usually measured with a grab sample.
3. How would you collect a depth sample from a lake?
4. Samples should not be collected from water taps under what conditions?
5. What information should be recorded when a sample is collected?

Table 9.8 Recommendation for sampling and preservation of samples according to measurement ("Required Containers, Preservation Techniques, and Holding Times," *Code of Federal Regulations,* Protection of the Environment, 40, Parts 136–149, 2008)

Measurement	Vol. Req. (mL)	Container[a]	Preservative	Max. Holding Time[b]
Physical Properties				
Color	500	P, G	Cool, 4°C	48 hours
Conductance	500	P, G	Cool, 4°C	28 days
Hardness[c]	100	P, G	HNO_3 to pH <2, H_2SO_4 to pH <2	6 months
Odor	200	G only	Cool, 4°C	6 hours
pH[c]	25	P, G	Determine on site	Immediately
Residue, Filterable	100	P, G	Cool, 4°C	7 days
Temperature	1,000	P, G	Determine on site	Immediately
Turbidity	100	P, G	Cool, 4°C	48 hours
Metals (Fe, Mn)				
Dissolved or Suspended	200	P, G	Filter on site, HNO_3 to pH <2	6 months
Total	100	P, G	Filter on site, HNO_3 to pH <2	6 months
Inorganics, Nonmetallics				
Acidity	100	P, G	Cool, 4°C	14 days
Alkalinity	200	P, G	Cool, 4°C	14 days
Bromide	100	P, G	None required	28 days
Chloride	50	P, G	None required	28 days
Chlorine, Total Residual	500	P, G	Determine on site	Immediately
Cyanide, Total and Amenable to Chlorination	500	P, G	Cool, 4°C, NaOH to pH >12, 0.6 g ascorbic acid[d]	14 days
Fluoride	300	P	None required	28 days
Iodide	100	P, G	Cool, 4°C	24 hours
Nitrogen:				
Ammonia	500	P, G	Cool, 4°C, H_2SO_4 to pH <2	28 days
Kjeldahl and Organic	500	P, G	Cool, 4°C, H_2SO_4 to pH <2	28 days
Nitrate-Nitrite	200	P, G	Cool, 4°C, H_2SO_4 to pH <2	28 days
Nitrate	100	P, G	Cool, 4°C	48 hours
Nitrite	100	P, G	Cool, 4°C	48 hours
Dissolved Oxygen:				
Probe	300	G with top	Determine on site	Immediately
Winkler	300	G with top	Fix on site, store in dark	8 hours
Phosphorus:				
Orthophosphate	50	P, G	Filter on site, cool, 4°C	48 hours
Elemental	50	G	Cool, 4°C	48 hours
Total	50	P, G	Cool, 4°C, H_2SO_4 to pH <2	28 days
Silica	50	P	Cool, 4°C	28 days
Sulfate	100	P, G	Cool, 4°C	28 days
Sulfide	100	P, G	Cool, 4°C, add zinc acetate plus H_2SO_4 to pH >9	7 days
Sulfite	50	P, G	Determine on site	Immediately

[a]Polyethylene (P) or Glass (G). For metals, polyethylene with a polypropylene cap (no liner) is preferred.

[b]Holding times listed above are recommended for properly preserved samples based on currently available data. It is recognized that for some sample types, extension of these times may be possible while for other types, these times may be too long. Where shipping regulations prevent the use of the proper preservation technique or the holding time is exceeded, such as the case of a 24-hour composite, the final reported data for these samples should indicate the specific variance.

[c]Hardness and pH are usually considered chemical properties of water rather than physical properties.

[d]Use ascorbic acid only if residual chlorine is present.

9.6 Math Assignment

Turn to Appendix A, "Introduction to Basic Math for Operators." Read and work through the example problems in sections:

- "Learning Objectives"
- A.9.10, "Laboratory Procedures" (English system)
- A.10.10, "Laboratory Procedures" (Metric system)

Check the math using a calculator. You should be able to get the same answers.

9.7 Additional Resources

Your local library may be a good source for obtaining additional resources related to the field of water treatment.

Manual of Water Utilities Operations (Texas Manual). Texas Water Utilities Association, www.twua.org. ASIN: B000KT9EIU. Chapters on water chemistry and laboratory examinations.

Methods for Chemical Analysis of Water and Wastes. US Environmental Protection Agency. 2013. www.epa.gov. ISBN: 978-1287218173.

Simplified Procedures for Water Examination (M12). American Water Works Association. www.awwa.org. ISBN: 978-1-58321-182-3.

Standard Methods for the Examination of Water and Wastewater, 22nd Edition. 2012. American Public Health Association, American Water Works Association, and Water Environment Federation. www.awwa.org. ISBN: 978-0-87553-013-0.

Handbook for Analytical Quality Control in Water and Wastewater Laboratories. US Environmental Protection Agency. 2013. www.epa.gov. ISBN: 978-1288889006.

Microbiological Methods for Monitoring the Environment—Water and Wastes. US Environmental Protection Agency. 2013. www.epa.gov. ISBN: 978-1492283751.

Water Quality (Principles and Practices of Water Supply Operations), 4th Edition. 2010. American Water Works Association. www.awwa.org. ISBN: 978-1583217801.

Water Analysis Handbook, Fifth Edition. 2008. HACH Company. www.hach.com. ASIN: B007CM3US8.

Chapter Review

Key Term Matching

_____ 1. DPD Method

_____ 2. Indicator

_____ 3. Supernatant

_____ 4. Colorimetric Measurement

_____ 5. Representative Sample

_____ 6. Nephelometric

_____ 7. Amperometric Titration

_____ 8. Organic

_____ 9. Composite (Proportional) Sample

_____ 10. Endpoint

A. A means of measuring concentrations of certain substances in water (such as strong oxidizers like chlorine) based on the electric current that flows during a chemical reaction.

B. A means of measuring unknown chemical concentrations in water by measuring a sample's color intensity.

C. A collection of individual samples obtained at regular intervals. Each individual sample is combined with the others in proportion to the rate of flow when the sample was collected.

D. A method of measuring the chlorine residual in water. The residual may be determined by either titrating or comparing a developed color with color standards.

E. The completion of a desired chemical reaction.

F. A substance that gives a visible change, usually of color, at a desired point in a chemical reaction, generally at a specified end point.

G. A means of measuring turbidity in a sample by using an instrument that passes light through a sample. The amount of light deflected (usually at a 90-degree angle) is then measured.

H. Describes chemical substances that come from animal or plant sources.

I. A sample portion of material, water, or waste stream that is as nearly identical in content and consistency as possible to that in the larger body being sampled.

J. The relatively clear water layer between the sludge on the bottom and the scum on the surface of a basin or container. Also called clear zone.

9.1 Basic Laboratory Concepts, Equipment, and Techniques

11. Convert 62°F to °C. Convert 22°C to °F.
12. Convert 5 grams (g) into milligrams (mg).
    Convert 250 milliliters (mL) to liters (L).
    Convert 50 micrograms (µg) to grams (g).

9.2 Laboratory Equipment and Techniques

13. Which piece of laboratory equipment is the most accurate way to measure a volume of liquid?
    1. Beaker
    2. Flask
    3. Graduated cylinder
    4. Pipet
14. Three types of pipets are _____, _____, and _____.
15. A _____ is the curve of the surface of a liquid in a small tube.
16. A standard solution is a solution in which the exact concentration of a chemical or compound is _____.
17. Which is not a factor in poor quality laboratory data?
    1. Calculation mistakes
    2. Deteriorated reagents
    3. Neatly recorded results in lab notebooks
    4. Sloppy laboratory technique

9.3 Laboratory Safety

18. Which of the following hazards would not be encountered in a laboratory?
    1. Burns and explosions
    2. Drowning
    3. Electric shock
    4. Hazardous materials
19. Do you add water to acid or acid to water?
20. Use _____, _____, or _____ to manage Class B fires.

9.4 Water Quality Tests

21. Which style of thermometer will change most rapidly when removed from the liquid sample?
    1. Dial
    2. Digital
    3. Partial immersion
    4. Total immersion
22. The lower the pH, the more _____ the water is.
23. What is the accepted method used to measure turbidity?
    1. Candle
    2. Jackson
    3. Light
    4. Nephelometric
24. Turbidity should be measured in a sample _____ to obtain accurate results. The turbidity of a sample can change after the sample is collected.
25. Alkalinity is usually expressed in terms of _____.
26. The principal hardness-causing ions in water are _____.

27. What are the general types of residual chlorine produced in chlorinated water?
    1. Chlorine demand plus chlorine dose
    2. Free chlorine plus chlorine dose
    3. Free residual chlorine and combined residual chlorine
    4. Hypochlorous acid plus hydrochloric acid

28. What does DPD stand for?
    1. Diethyl peptide-D
    2. Diphenylpolydiamine
    3. Di-polynucleic-diesterase
    4. N, N-diethyl-p-phenylenediamine

29. After a jar test is completed, what does a hazy water sample indicate?
    1. Poor coagulation
    2. Poor flocculation
    3. Poor mixing
    4. Poor turbidity

30. The presence or absence of _____ is a good index of the degree of bacteriologic safety of a water supply.

31. Which procedure is not an approved water testing procedure for total coliform bacteria?
    1. Enzyme Substrate™ Method
    2. Presence–Absence (P–A) Method
    3. Membrane Filter Method—*Escherichia coli*
    4. Ready-brane Coliforms Presence–Absence Test

32. Results from MPN tests during 1 week were as follows:

Day	S	M	T	W	T	F	S
MPN/100 mL	2	8	14	6	10	56	4

Calculate the arithmetic mean (average), median, and geometric mean.

9.5 Sampling

33. A _____ is a sample portion of material or water that is as nearly identical in content and consistency as possible to that in the larger body of material or water being sampled.

34. Grab samples should not be used to measure which of the following water quality indicators?
    1. Chlorine residual
    2. Dissolved gases
    3. Temperature
    4. Turbidity

35. What is a first-flush sample?
    1. Sample taken at a sampling stations after running the water for a few minutes
    2. Sample taken at a sampling stations after the water stands motionless in the distribution system for at least 6 hours
    3. Sample taken at the customer's tap after running the water for a few minutes
    4. Sample taken at the customer's tap after the water stands motionless in the plumbing for at least 6 hours

Introduction to Basic Math for Operators

APPENDIX OUTLINE

LEARNING OBJECTIVES

1. Perform basic math operations such as addition, subtraction, multiplication, and division.

2. Understand units of measurement.

3. Calculate perimeters, circumferences, areas, and volumes of basic geometric shapes.

4. Analyze and solve math problems.

5. Solve problems related to basic physical concepts such as flow rate, velocity, pressure, and power.

6. Solve unit conversion problems, including metric units.

7. Solve problems related to pumps.

8. Collect and analyze data, including:
   a. Recognize the presence of variation in data.
   b. Read manometers, gauges, and charts.
   c. Analyze and present data using charts, graphs, tables, and numbers.
   d. Calculate arithmetic mean, range, median, mode, geometric mean, moving average, variance, and standard deviation.
   e. Apply prediction equations, trends, and correlations to the analysis of data.

KEY TERMS

density	median	range	specific gravity
lateral surface area	meniscus	representative sample	total dynamic head (TDH)

Introduction

The primary goal of this appendix is to show operators how to solve problems encountered in the field. Some chapters in this training manual include practical math examples to provide you with context for calculations related to the chapter topic. This appendix presents a review of basic math principles and examples of typical math problems. If you can add, subtract, multiply, and divide, you can learn how to solve the math problems presented. This appendix was redesigned to make the sometime challenging subject of math more accessible to operators with varying levels of experience and knowledge. New features include:

- Sidebar containing defined terms, illustrations of key concepts, and other useful information
- Equations presented in a standardized format
- Units and unit conversions shown in a consistent style

The math topics are divided into four sections. The first three are considered foundational for all operators, and include:

1. Basic Concepts (Sections A.1–A.4) focuses on basic calculations and math principles.
2. Intermediate Concepts (Sections A.5–A.6) uses units of measurements as a foundation to present geometry and physical concepts.
3. Advanced Concepts (Section A.7–A.8) discusses pumps and what professional operators need to understand about related math concepts.

In the fourth section (A.9–A.10), we apply the concepts discussed previously by focusing on a particular area of operation: water treatment in this training manual. Operators, such as yourself, can practice solving problems that are encountered during daily plant or facility operation. Water treatment operators need math skills for:

1. Estimating flow rates and water volumes
2. Estimating needed storage volumes
3. Calculating chemical doses and feed rates for water treatment processes such as coagulation/flocculation, sedimentation, and disinfection
4. Estimating loading rates for sedimentation and filtration systems
5. Estimating backwash pumping and loading rates in filter systems
6. Estimating doses and feed rates of chlorine needed for desired chlorine residual
7. Estimating a Langelier Index and related saturation pH for assessing corrosion potential
8. Calculating mean and median values from lab results

Operators use advanced math concepts for tasks such as calibrating and comparing the performance of new equipment with desired operational goals. As you progress in your profession, you will need to use your math skills to understand and communicate plant or facility information for lab or cost analysis and more. These are skills you will also use to communicate with coworkers, managers, the public, and funding authorities.

Verifying Calculations

As a standard practice, it is suggested that when completing calculations in the field, operators should verify their calculations, examine their answers to see if they appear reasonable, and, if possible, have another operator review the calculations before taking any action based on that information.

Basic Concepts (Sections A.1–A.4)

This introductory section outlines the basic math operations of addition, subtraction, multiplication, and division. If you need to review these concepts, continue reading this section to refresh your knowledge before continuing on to the next sections that build on these basic concepts. This section also discusses order of calculations, solving equations (basic algebra), and unit conversion. All these concepts will be applied to solving practical problems as part of the learning process in the intermediate and advanced sections of this appendix. If you need additional resources, math textbooks for all skill levels are available at your local library or bookstore. One title to consider is *Basic Math Concepts for Water and Wastewater Plant Operators,* which provides considerable detail on how to do basic calculations.* Many free math resources are also available online.

Basic Math Concepts for Water and Wastewater Plant Operators and Applied Math for Wastewater Plant Operators by Joanne Kirkpatrick Price. Obtain from CRC Press at www.crcpress.com or phone (800) 272-7737.

A.1   Numbers and Operations

Before covering the basics, we need to understand numbers and how they are represented. Numbers could be whole numbers; examples include 2, 10, and 105. We call these integers. Typically, we use whole numbers to count things such as the number of buckets or containers. Other ways numbers can be represented are as a ratio of two whole numbers; examples include ¼, ⅖, and ⅝. We call these fractions. We also can represent numbers as decimal fractions. For example, 2.4 is a representation of a number.

A.1.1   Addition

2	6.2	16.7	6.12	43
+3	+8.5	+38.9	+38.39	39
5	14.7	55.6	44.51	34
				38
				39
2.12	0.12	63	120	37
+9.80	+2.0	+32	+60	+29
11.92	2.12	95	180	259
				70
4	23	16.2	45.98	50
7	79	43.5	28.09	40
+2	+31	+67.8	+114.00	+80
13	133	127.5	188.07	240

A.1.2   Subtraction

7	12	25	78	83
−5	−3	−5	−30	−69
2	9	20	48	14

61	485	4.3	3.5	123
-37	-296	-0.8	-0.7	-109
24	189	3.5	2.8	14

8.6	11.92	27.32	3.574	75.132
-8.22	-3.70	-12.96	-0.042	-49.876
0.38	8.22	14.36	3.532	25.256

A.1.3   Multiplication

Notation

We indicate multiplication by using

4×2, or

$(4)(2)$

We indicate division by using

$4 \div 2$,

$4/2$, or

$\dfrac{4}{2}$

$(3)(2)$	$= 6$	$(40)(20)(6)$	$= 4{,}800$
$(4)(7)$	$= 28$	$(4{,}800)(7.48)$	$= 35{,}904$
$(10)(5)$	$= 50$	$(1.6)(2.3)(8.34)$	$= 30.6912$
$(10)(1.3)$	$= 13$	$(0.001)(200)(8.34)$	$= 1.668$
$(2)(22.99)$	$= 45.98$	$(0.785)(7.48)(60)$	$= 352.308$
$(6)(19.5)$	$= 117$	$(12{,}000)(500)(60)(24)$	$= 8{,}640{,}000{,}000$
			$= 8.64 \times 10^9$
$(16)(17.1)$	$= 273.6$	$(4)(1{,}000)(1{,}000)(454)$	$= 1{,}816{,}000{,}000$
			$= 1.816 \times 10^9$
$(50)(20{,}000)$	$= 1{,}000{,}000$		
$(40)(2.31)$	$= 92.4$		
$(80)(0.433)$	$= 34.64$		

Exponents for Large Numbers

The term, $\times 10^9$, means that the number is multiplied by 10^9 or $1{,}000{,}000{,}000$. Therefore, $8.64 \times 10^9 = 8.64 \times 1{,}000{,}000{,}000 = 8{,}640{,}000{,}000$.

A.1.4   Division

$\dfrac{6}{3} = 2$	$\dfrac{48}{12} = 4$	$\dfrac{300}{20} = 15$	$\dfrac{20}{7.1} = 2.8$
$\dfrac{11{,}400}{188} = 60.6$	$\dfrac{1{,}000{,}000}{17.5} = 57{,}143$	$\dfrac{861{,}429}{30{,}000} = 28.7$	$\dfrac{4{,}000{,}000}{74{,}880} = 53.4$
$\dfrac{1.67}{8.34} = 0.20$	$\dfrac{80}{2.31} = 34.6$	$\dfrac{62}{454} = 0.137$	$\dfrac{250}{17.1} = 14.6$

A.2   Order of Operations

When you complete this section, you should be able to perform a sequence of calculations and produce the correct answer. To do this successfully, you need to know a few rules.

When adding or subtracting numbers, the sequence in which you do computations does not matter. Let's say we want to perform the following series of calculations:

$$1 + 4 - 2 + 5 = 5 - 2 + 5 = 3 + 5 = 8$$

This calculation proceeded from left to right by first adding 1 to 4 producing 5, then subtracting 2 from 5 to produce 3, and finally adding 3 to produce 8. If we do the intermediate calculations in a different sequence, we get the same answer as shown below.

$$1 + 4 - 2 + 5 = 1 + 2 + 5 = 1 + 7 = 8$$

This time we subtracted 2 from 4 first, then added 5, and finally added 1. Because the sequence does not matter, the arrangement of the numbers on the page also does not matter. In our example,

$$1 + 4 - 2 + 5 = 4 + 5 - 2 + 1 = 5 - 2 + 1 + 4 = -2 + 4 + 1 + 5 = 8$$

The last example above is a reminder that subtracting one number from another is the same as adding the negative of that number.

$$10 - 4 = 10 + (-4) = 6$$

Be careful. $10 - 4$ is not the same as $4 - 10$. You can think of positive numbers as steps we take to the right if the number is positive or left if the number is negative. For this example of $10 - 4$, first you take 10 steps to the right, followed by taking four steps to your left—at the end you will be six steps to the right of your starting point. What happens if you apply the same exercise to do $4 - 10$?

In the same way, the sequence of computations does not matter when multiplying or dividing numbers. Consider this example:

$$10 \times 8/4 = 80/4 = 20$$

First we multiplied 10 by 8, and then divided the intermediate answer (80) by 4. Alternatively, we could first divide 8 by 4, and then multiply the intermediate answer (2) by 10 as shown below.

$$10 \times 8/4 = 10 \times 2 = 20$$

As before, the arrangement of numbers on the page does not matter as long as we are only multiplying or dividing. In our example,

$$10 \times 8/4 = 10/4 \times 8 = (1/4) \times 10 \times 8 = 20$$

The last example is a reminder that dividing one number by another is the same as multiplying the reciprocal of that number. (The reciprocal of a number is 1 divided by that number.) For instance,

$$10/5 = 10 \times (1/5) = 2$$

Be careful. $10/5$ is not the same as $5/10$.

When you have a mix of operations, the sequence of computations does matter and is governed by the Order of Operations. To understand how to complete any calculation, we need to define the concept of a mathematical term. A mathematical term means any group of numbers separated by addition (+ sign) or subtraction (− sign). We will simply call it a term rather a mathematical term. Previously, we did calculations such as this one:

$$1 + 4 - 2 + 5$$

The above calculation has four terms: 1, 4, 2, and 5. Each of the four terms happens to be a single number and, as we learned previously, the

---

Order of Operations

In mixed calculations, perform the different operations in the following order:

1. Parentheses
2. Exponentiation
3. Multiplication or division
4. Addition or subtraction

An aid to remembering the order is PEMDAS or "Please Excuse My Dear Aunt Sally."

Exponents for Large Numbers

Exponents are helpful when we want to express large numbers. For example, for a million (1,000,000), instead of showing six zeros, we can show:

$10^6 = 10 \times 10 \times 10 \times 10 \times 10 \times 10$
$= 1,000,000$

Therefore, when the base is 10, the exponent tells us how many zeroes are in back. In this case we have 6 zeroes.

Practice: Express the number 1,000 in exponent notation and the number 10^4 in regular notation.

Answers: 10^3, 10,000

Exponents for Small Numbers

As an operator, in some cases you need to deal with small numbers. For example, if we want to talk about 1 millionth, we can do the following:

$\dfrac{1}{1,000,000} = \dfrac{1}{10^6} = 0.000001 = 10^{-6}$

So, when we see a negative 6 as the exponent, we mean that we have 1 in the sixth decimal place.

So, as a rule, an exponent in the denominator becomes negative when moved to the numerator. The opposite is true as well.

Practice: Express the number 1 thousandth in exponent notation and the number 10^{-4} in regular notation.

Answers:

$\dfrac{1}{1,000} = \dfrac{1}{10^3} = 0.001 = 10^{-3}$

and

$10^{-4} = \dfrac{1}{10^4} = \dfrac{1}{10,000}$

order of calculation does not matter in this case. But, let's consider the following calculation:

$$1 + 2 \times 10/4 - 3 = ?$$

Let's define the terms first:

Term 1: 1

Term 2: $2 \times 10/4$

Term 3: 3

The second term includes a number of calculations that need to be completed first. Then the calculations can proceed as follows (focus on the second term):

$$1 + 20/4 - 3 = 1 + 5 - 3 = 6 - 3 = 3$$

Here we multiplied 2 by 10 (intermediate answer = 20) then divided by 4 (intermediate answer = 5). As explained above, we could also have first divided 10 by 4 (intermediate answer = 2.5) and then multiplied by 2 (intermediate answer = 5). Either way, once all the terms are single numbers, we can finish the calculation.

To recap, when dealing with mixed calculations, you must separate the terms. For each term follow the order of operations (as shown on the sidebar). Another way of saying this is that parentheses have a higher priority than exponentiation which has a higher priority than multiplication (or division) which has a higher priority than addition (or subtraction).

Above multiplication and division in the order of operations is exponentiation. Exponentiation is a special case of multiplication in which we multiply the same number by itself, such as:

$$3 \times 3 \times 3 \times 3$$

The mathematical way to express this is 3^4. In this term, the 3 is called the base. It is the number we are multiplying. The 4 is called the exponent. It is the number of times the base is to be multiplied by itself. If you are using a calculator, you can do this computation by inputting the number and multiplying it over and over. Your calculator may have a built-in function for this. (See your calculator instruction book for details.) Here's an example that includes exponentiation:

$$2^3 \times 3/4 + 1 = 8 \times 3/4 + 1 = 24/4 + 1 = 6 + 1 = 7$$

First, we have two terms: $2^3 \times 3/4$ and 1. Following the order of operations for the first term, the exponential is done first ($2 \times 2 \times 2 = 8$), followed by multiplying 8 by 3, followed by dividing 24 by 4, resulting in 6. This means the first term is 6. Finally, we add 1 (second term) to get the final answer 7. Alternatively, we could have done the exponentiation first, followed by dividing 3 by 4, followed by multiplying 8 by 0.75, followed by adding 1.

$$2^3 \times 3/4 + 1 = 8 \times 0.75 + 1 = 6 + 1 = 7$$

The key point here is that exponentiation was done first and addition was done last.

Parentheses are the very highest priority in the order of operations. Do the operations inside parentheses first. Consider this calculation:

$$1 + 2 \times 10/(4 - 3) = ?$$

It looks very similar to a calculation we computed earlier, but this one has parentheses. As per the rule, we do the operations inside the parentheses first, then the other operations according to the order of operations. The computation looks like this.

$$1 + 2 \times 10/(4 - 3) = 1 + 2 \times 10/(1) = 1 + 2 \times 10 = 1 + 20 = 21$$

If you follow the logic of identifying terms, we have two terms: 1 and $2 \times 10/(4 - 3)$. The second term has multiple calculations. If you follow the order of operations as outlined previously, you will get the correct answer. That is, do the operation inside the parentheses $(4 - 3)$ first. Then use that intermediate answer in later computations—multiplication or division followed by addition or subtraction. Adding parentheses causes the answer to be very different from our earlier example, even though the numbers and operations are the same.

$$1 + 2 \times 10/4 - 3 = 3$$

$$1 + 2 \times 10/(4 - 3) = 21$$

For the first one, we have three terms: 1, $2 \times 10/4$, and 3. The first and third terms are single numbers. We need to calculate the second term before we can finish the calculation to get 3.

If there is a mix of operations inside the parentheses, follow the regular order, but do all the operations inside each set of parentheses first. Try this:

$$(4 + 2 \times 10)/(6 - 4/2) = (4 + 20)/(6 - 2) = (24)/(4) = 6$$

Inside a set of parentheses, follow the same process of identifying the terms. For the first set of parentheses, we have two terms: 4 and 2×10. First, multiply 2 by 10 and then add 4. Next, do the operations in the second term (the right set of parentheses). First, divide 4 by 2 then subtract this from 6. Finally, divide the first term by the second term. Here's one more example. Follow the order of operations first inside and then outside the parentheses. We'll compute the first term then the second.

$$3 + (4^3/8) \times (3 \times 2 - 1) = 3 + (64/8) \times (3 \times 2 - 1) = 3 + (8) \times (3 \times 2 - 1)$$
$$= 3 + (8) \times (6 - 1)$$
$$= 3 + (8) \times (5) = 3 + 40 = 43$$

Finally, if you have parentheses inside of parentheses, do your computations from the inside out.

$$10 \times \left(4 + \left(6^2/4\right)\right) = 10 \times \left(4 + (36/4)\right) = 10 \times \left(4 + (9)\right) = 10 \times (13) = 130$$

Other Rules for Exponents

What does it mean to have an exponent of zero? Here is the rule: Any number with an exponent of zero is equal to 1.

$10^0 = 1$

$5^0 = 1$

Let's calculate 1,000 multiplied by 1,000. We know the answer is one million. But, let's use exponents with a base of 10:

$10^3 \times 10^3 = 10^6$

So, when the base is the same, all we need to do is add the exponents.

Let's extend this rule. What happens if we divide 10,000 by 100. We know the answer is 100. With exponents:

$$\frac{10,000}{100} = 10,000 \times \frac{1}{100}$$
$$= 10^4 \times 10^{-2} = 10^2 = 100$$

In the second step, we separated the numbers into two parts: 10,000 and 1 hundredth. The intermediate result is reached by subtracting the exponents because we have the same base.

Practice Problem

$$\frac{5^5 \times 2^2}{5^2} = 5^5 \times 5^{-2} \times 2^2$$
$$= 5^3 \times 2^2$$
$$= 125 \times 4$$
$$= 500$$

The exponents for 5 and 2 were not added because the bases are different.

Practice: $\frac{6^3 \times 2^3}{6 \times 2^2}$

Answer: 72

A.2.1   More on Exponents

As we discussed earlier, the exponent indicates the number of times that the base needs to be multiplied by itself. That is,

$$3^4 = 3 \times 3 \times 3 \times 3 = 81$$

When the exponent is less than one, what does that mean? For example,

$$(81)^{\frac{1}{4}} = ?$$

This expression asks what number (base) when multiplied by itself four times equals 81. It is clear from the previous calculations that the answer is 3 because

$$3^4 = 3 \times 3 \times 3 \times 3 = 81$$

Then,

$$(81)^{\frac{1}{4}} = 3$$

We also call this the fourth root of 81 and it can be shown as:

$$\sqrt[4]{81} = 3$$

Most calculators are equipped with a button x^y that calculates any exponent to any base.

Example 1

Calculate the third root of 8.

$$\sqrt[3]{8} = 8^{\frac{1}{3}} = 2$$

That means, if we multiply 2 by itself three times we get 8. Check the calculation:

$$2^3 = 2 \times 2 \times 2 = 8$$

A special root that is used often is the square root, for example:

$$\sqrt[2]{25} = \sqrt{25} = 25^{\frac{1}{2}} = 5$$

We usually do not show the 2 on the root symbol $\sqrt{}$. The square root symbol indicates the number that is multiplied by itself to get the number under the square symbol. For the previous calculation, $5 \times 5 = 25$.

A.3   Basic Algebra (Solving Equations)

Algebra is a set of methods and rules on how to deal with numbers. In algebra, we often use shorthand to name certain things that we measure or calculate. For example, we use Q for flow rate or V for the volume of a tank. So, we can say the volume of a tank is 10,000 gallons or we can use shorthand to say:

V = 10,000 gal

If we want to know how many gallons of water are in a tank if it is half full, then we can say:

$$\frac{V}{2} = \frac{10,000}{2} = 5,000 \text{ gal}$$

The examples in this section are designed to illustrate math concepts. Once the reader is more comfortable with the math, solving practical problems will be much easier.

Example 2

If 3,000 gallons of water are needed to fill a 10,000-gallon tank, how many gallons are already in the tank?

In order to approach this problem, let's list what we know.

V = 10,000 gal

Volume Added = 3,000 gal

We need to find how many gallons are already in the tank. Call this initial unknown volume X.

X + 3,000 = 10,000

This creates an equation. An equation makes a statement and it is a combination of symbols and numbers, including an "equals" sign. (The equals sign makes it an equation.) The equation above says that if we add 3,000 gallons to the initial volume (X) the tank will be full. We know that the full tank is 10,000 gallons. Any equation has a right-hand side (RHS) and a left-hand side (LHS). An equation is like a balance: both sides are the same. So, if you take away or add anything to one side, you need to do the same to the other side. Using this principle, we can solve for X.

X + 3,000 − 3,000 = 10,000 − 3,000

X = 7,000 gal

To solve an equation, put the unknowns on one side and the known numbers on the other. By subtracting 3,000 from both sides, we kept X on the LHS and put the numbers (10,000 and 3,000) on the RHS. As a shortcut for this operation (of adding to or subtracting from both sides of an equation), we just move the number we are adding or subtracting across the equals sign and change its sign. That is, we can just show −3,000 on the RHS without showing the + 3,000 −3,000 on the LHS.

X + ⟨3,000⟩ = 10,000

X = 10,000 ⊖ 3,000 = 7,000 gal

Example 3

If the tank is half full and it contains 7,000 gallons, what is the capacity of the tank?

Unknown

V = Volume of Tank

Setting up the equation:

$$\frac{1}{2}V = 7,000$$

We use ½ to indicate that half of the tank volume (as the problem stated) is equal to 7,000. For example, if the problem statement indicates that a quarter of the tank is full, then we would use ¼.

Another principle using the concept of a balance is if one side of an equation is multiplied or divided by a number, the other side needs to be multiplied or divided by the same number. Multiply both sides by 2.

$$2 \times \frac{1}{2}V = 2 \times 7,000$$

$$V = 14,000 \text{ gal}$$

Example 4

If the tank is two-thirds full and it contains 8,000 gallons, what is the capacity of the tank?

Unknown

V = Volume of Tank

Setting up the equation:

$$\frac{2}{3}V = 8,000$$

Remember, using the concept of a balance, if one side of an equation is multiplied or divided by a number, the other side needs to be multiplied or divided by the same number. Multiply both sides by 3 and divide by 2.

$$\frac{3}{2} \times \frac{2}{3}V = \frac{3}{2} \times 8,000$$

$$V = 12,000 \text{ gal}$$

A shortcut called cross multiplication, shown below, makes these calculations quicker.

$$\frac{2}{3}V = 8,000$$

$$V\left(\frac{2}{3}\right) \diagdown \frac{8,000}{1}$$

$$V = \left(\frac{3}{2}\right)8,000 = 12,000 \text{ gal}$$

In cross multiplication, the numerator (the number on top) of one side of an equation becomes the denominator (the number on the bottom) of the other side and vice versa.

The "cross" in cross multiplication refers to crossing the equals sign. It has nothing to do with the "X" formed by the arrows above. Example 3 also used cross multiplication.

A.4 Percentages

As mentioned earlier, numbers are expressed in many forms, including whole numbers (integers), fractions, decimals, and percentages. In this section, our focus will be on the percentages. However, we need to review the other forms just to make sure that we are clear. An integer is a whole number such as 5, 101, and 15. A fraction is represented as a ratio of two integers; for example, ⅜ of an inch. A decimal number is a shorthand way of expressing an integer plus a fraction (called the decimal fraction because it is a ratio of a power of 10). For example, 10.3 means $10 + (\frac{3}{10})$ where $\frac{3}{10}$ is the decimal fraction.

Expressing a number in a percentage is another, and sometimes simpler, way of writing a fraction or decimal. Percentages can be thought of as parts per 100 parts. That is, the percentage is the numerator of a denominator of 100. For example, 25 percent (represented as 25%) is the same as $\frac{25}{100}$ or 0.25. The % symbol takes the place of $\frac{1}{100}$. The following table compares fractions, decimals, and percents to indicate their relationship to each other.

Decimals

In the number 10.5, the number to the left of the decimal place is a whole number (in this case 10), and the number to the right of the decimal place is a fraction of 10s depending on how many digits there are after the decimal place. In this case because we have one digit, it is 5/10. That is, it is half. Let's try more examples:

10.51 is 10 + 51/100 (the two digits, 51, are divided by 100 because it has two zeroes).

Can you expand this number? 10.515

(Hint: There are three digits after the decimal point, so 515 needs to be divided by 1,000.)

Common Fraction	Decimal Fraction	Percent
$\frac{285}{100}$	2.85	285%
$\frac{100}{100}$	1.0	100%
$\frac{20}{100}$	0.20	20%
$\frac{1}{100}$	0.01	1%
$\frac{1}{1,000}=\frac{0.1}{100}$	0.001	0.1%
$\frac{1}{1,000,000}=\frac{0.0001}{100}$	0.000001	0.0001%

Using this relationship, fractions or decimal fractions can be changed to percentages and percentages can be changed to fractions or decimal fractions using the following procedures.

To change from a fraction to a percent, multiply by 100 and place a percent sign (%) after the number.

Example 5

Express $\dfrac{2}{5}$ and $\dfrac{5}{4}$ as percents.

$$\dfrac{2}{5} \times 100\% = \dfrac{200\%}{5} = 40\%$$

$$\dfrac{5}{4} \times 100\% = \dfrac{500\%}{4} = 125\%$$

Example 6

The input or brake horsepower to a pump is 20 horsepower. Output or water horsepower is 15 hp. What is the efficiency of the pump?

$$\text{Efficiency} = \left(\dfrac{\text{Output}}{\text{Input}}\right) \times 100\% = \left(\dfrac{\text{Water HP}}{\text{Brake HP}}\right) \times 100\%$$

$$= \dfrac{15 \text{ hp}}{20 \text{ hp}} \times 100\% = 75\%$$

To change a percent to a fraction, divide by 100 and remove the % symbol.

Example 7

Convert 15% and 0.4% to fractions and decimal fractions.

$$15\% \times \dfrac{1}{100\%} = \dfrac{15}{100} = 0.15$$

$$0.4\% \times \dfrac{1}{100\%} = \dfrac{0.4}{100} = \dfrac{4}{1,000} = 0.004$$

The two percent signs cancel each other.

To find a percent of a given number, convert the percent to a decimal and multiply that decimal by the given number.

Example 8

Find 7% of 32.

$$\dfrac{7\%}{100\%} \times 32 = 0.07 \times 32 = 2.24$$

Example 9

Find 90% of 5.

$$\dfrac{90\%}{100\%} \times 5 = 0.9 \times 5 = 4.5$$

Example 10

What is the weight of dry solids in a ton (2,000 pounds) of wastewater sludge containing 5% solids and 95% water? *NOTE:* 5% solids means there are 5 pounds of dry solids for every 100 pounds of wet sludge.

$$2{,}000 \text{ lb} \times \frac{5\%}{100\%} = 100 \text{ lb of solids}$$

Similarly, to find a number when a given percent of it is known, divide the percent by 100 and cross multiply.

Example 11

If 5% of a number is 52, what is the number?
If we call the number X, then

$$5\% \text{ of } X = 52$$
$$\left(\frac{5}{100}\right)X = 52$$

Cross multiply.

$$X\left(\frac{5}{100}\right) \begin{array}{c} \nearrow \\ \searrow \end{array} 52$$

$$X = \left(\frac{100}{5}\right)52 = 1{,}040$$

A check calculation may now be performed—what is 5% of 1,040?

$$\left(\frac{5\%}{100\%}\right)(1{,}040) = 52 \, (\text{Check})$$

Example 12

16 is 80% of what amount?
If the full amount is X, then

$$(0.80)X = 16$$
$$X = \frac{16}{0.80} = 20$$

Example 13

An electric motor for a pump is 90% efficient. If the motor must deliver 720 watts, how many watts are required to run the motor?

$$\text{Efficiency} = \left(\frac{\text{Output}}{\text{Input}}\right) \times 100\%$$

$$90\% = \left(\frac{720 \text{ watts Output}}{X \text{ Input}}\right) \times 100\%$$

$$0.9 = \frac{720}{X}$$

$$X = \frac{720}{0.9} = 800$$

Intermediate Concepts (Sections A.5–A.6)

This intermediate section covers units of measure from the basic concepts of distance and length to the advanced concepts of velocity and flow rate. It builds on the concepts presented in the "Basic Concepts" section, and presents additional concepts that will be used in the "Advanced Concepts" section that follows. The following discussion assumes that you are comfortable with concepts outlined in the previous section. Refer to that material as needed to assist you in understanding the concepts presented here.

A.5   Units

In this section, we will introduce the idea of units of measurement. As an operator, it is very important to understand the units of measurement for making decisions that effect the operation of your system based on calculations. The discussion is divided into categories with examples to illustrate the applications.

As background for readers, there are two major unit systems: (1) the US Customary Unit system (most of the units in our training manuals are based on US Customary Units) and (2) the metric system. A special variation of the metric system is the International System of Units (*Le Système International d'Unités*), known internationally by the abbreviation SI. The metric system will be discussed in a separate section. In this section, we will cover conversion in general.

A.5.1   Distance or Length

Let's start with easier units of measurement. When we want to measure distance, we can use inches, feet, yards, or miles. In metric systems, we use centimeters, meters, or kilometers. Therefore, we typically use units that make sense in their context. For example, to measure a football field, we do not use feet or inches; we use yards. When we travel in a car we do not use feet, but rather miles. In most cases, operators are concerned about measurements such as the size of a tank, length of a pipe, or size of a pond. A typical unit of distance is feet (for short we denote feet as ft). Another common distance measurement is centimeters. Using the conversion factors given in the sidebar, convert feet into centimeters.

Example 14

Convert 1 foot into centimeters. How many centimeters in 1 foot? Start with 1 foot.

$$\frac{1 \text{ ft}}{1} \times \frac{12 \text{ in}}{1 \text{ ft}} \times \frac{2.54 \text{ cm}}{1 \text{ in}} = 30.48 \text{ cm}$$

Apples and Oranges

Units are important and you need to be aware of what units you are operating in.

In 1983, an Air Canada airliner ran out of fuel in midflight. Because of a units calculation error involving pounds and kilograms, the plane was loaded with only about half of the fuel it needed. Fortunately, the pilot was an experienced glider pilot and was able to land the plane safely.

Also, do not mix different units (that is, apples and oranges).

A famous and expensive units mismatch occurred with the Mars Climate Orbiter in 1999. One piece of software produced numbers in US units. These numbers were used by a second piece of software, but its coding was based on using the numbers in metric units. Consequently, it miscalculated the rocket thrust needed which caused the orbiter to go off course and crash into the Martian atmosphere.

Distance/Length Conversion Factors

1 ft = 12 in

1 in = 2.54 cm

1 m = 100 cm

Let's break down the above conversion into steps:

1. Start with what we need to convert (in this case 1 foot).
2. Divide by 1. (This does not change the quantity, but it is easier to visualize the conversion this way.)
3. We know that 1 foot is equal to 12 inches.
4. Put the equivalent 12 inches in the numerator (top).
5. Put 1 ft in the denominator (bottom) because it is 12 inches per 1 foot.
6. Cancel out the ft units but not the remaining unit (inches). That is not the final units we want. We need to get to centimeters.
7. Using the conversion of 1 in is equal to 2.54 cm (see the sidebar on the previous page), perform steps 4 and 5 placing inches in the denominator and centimeters in the numerator because we want to convert inches into centimeters.
8. The inches cancel out and we are left with centimeters. When you multiply 12 by 2.54, you find that there are 30.48 cm in 1 foot.

Practice: How many feet in 1 meter? (*Answer:* 3.281)

(Hint: Start with $\dfrac{1\,m}{1}$, and use the conversion information in the "Metric System for Distance/Length" sidebar.)

Metric System for Distance/Length
km = kilometer (1,000 meters)
m = meter (3.281 ft)
cm = centimeter = 1/100 m (2.54 cm = 1 in)
mm = millimeter = 1/10 cm

A.5.2   Area

Distance and length are measurements in one dimension. That is, we travel along a line. When we talk about area, we now are moving in space with two dimensions. For example, we are not just moving north to south, but also east to west. Using area, we can talk about how large a piece of land is or how large the footprint of a tank is.

In this section, we will discuss some basic shapes, including rectangles, triangles, trapezoids, and circles, and how to calculate their areas.

A.5.2.1   Surface Area of a Rectangle

A rectangle is a four-sided shape with square (90°) corners (see the following figure). The area of a rectangle is:

$$A = L \times W$$

Where,

A = Area

L = Length

W = Width

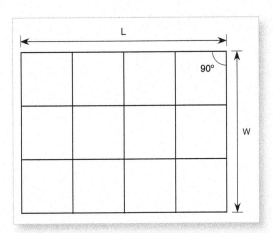

Example 15

Find the area of a rectangle that is 4 ft long and 3 ft wide using the formula. To find the area, calculate how many square feet (in this case) are within the boundary of the rectangle. The unit of measurement in this case is represented as ft × ft = ft². We use the exponentiation notation to indicate that we are measuring area.

A = L × W = 4 ft × 3 ft = 12 ft²

From the previous figure, you can count the number of square feet within that rectangle as shown below.

1	2	3	4
5	6	7	8
9	10	11	12

A square is a special kind of rectangle in which all four sides are the same length.

A.5.2.2   Surface Area of a Triangle

The area (A) of a triangle is equal to ½ the base (B) multiplied by the height (H). The area of any triangle is equal to ½ the area of the rectangle that can be drawn around it as shown in the following figure. The area of the rectangle is B × H. The area of the triangle is $\frac{1}{2}$B × H.

Notation

The expression $\frac{1}{2}$B × H is the same as

$\frac{1}{2}$ × B × H

B × H / 2

$\frac{B \times H}{2}$

$$A = \frac{1}{2}B \times H$$

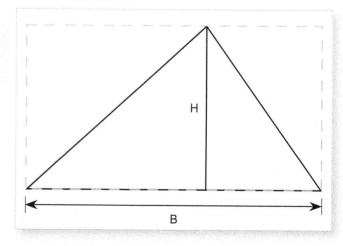

Example 16

Find the area of triangle ABC on the right.

The first step is to make all the units the same. In this case, we convert the height, 48 inches, to feet by dividing by 12 (in/ft) to obtain the desired units in feet.

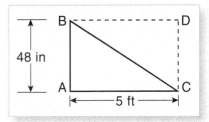

$$H = \frac{48 \text{ in}}{1} \times \frac{\text{ft}}{12 \text{ in}} = 4 \text{ ft}$$

$$A = \frac{1}{2}H \times B = \frac{1}{2} \times 4 \text{ ft} \times 5 \text{ ft} = 2 \text{ ft} \times 5 \text{ ft} = 10 \text{ ft}^2$$

We applied the method that we learned to convert the units.

A.5.2.3 Surface Area of a Trapezoid

A trapezoid is a four-sided shape with two sides that are parallel and corners that are not usually square. You may think about a rectangle as a special kind of trapezoid with square (90°) corners. A trapezoid has two bases (the parallel sides, B_1 and B_2) and a height (H), the vertical (perpendicular) distance between the bases (see the following figure). In operation, open channels (ditches) are usually trapezoids. The area is calculated as:

$$A = \left(\frac{B_1 + B_2}{2}\right) \times H$$

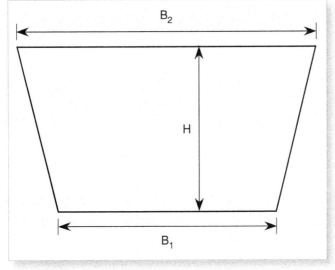

A.5.2.4 Surface Area of a Circle

A square with sides of D can be drawn around a circle with a diameter of D. The area of the square is:

$$A = D \times D = D^2$$

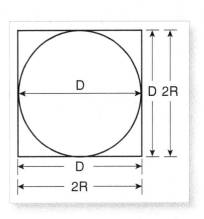

The area of any circle inscribed within a square is smaller than the area of the square. More precisely, the formula for the area of the circle is:

$$A = \frac{\pi}{4}D^2 = 0.785D^2$$

The Greek letter π (pronounced pie and written as "pi") substitutes for its standard or constant value of 3.1416 (usually rounded off to 3.14).

Put another way, the area of a circle is about 78.5% of the area of the square that contains the circle (see the figure in the sidebar). The type of problem and the magnitude (size) of the numbers in a problem will determine which of the two formulas will provide a simpler solution. Either formula will produce the same results.

Another formula for the area of a circle is expressed in terms of the radius (R). That is,

$$A = \pi R^2$$

The diameter is equal to two times the radius (see the figure in the sidebar).

$$D = 2 \times R = 2R$$

Or,

$$R = \frac{1}{2} \times D = \frac{1}{2}D = \frac{D}{2}$$

All these forms mean the radius is half the diameter.

Example 17

What is the area of the top (or bottom) of a gasoline storage tank with a 5-foot radius?

$$A = \pi R^2 = 3.14 \times (5 \text{ ft})^2 = 3.14 \times 25 \text{ ft}^2 = 78.5 \text{ ft}^2$$

We can also use the diameter formula for the area as follows:

$$D = 2R$$

$$A = \frac{\pi}{4}D^2 = \frac{3.14}{4} \times (2 \times 5 \text{ ft})^2 = \frac{3.14}{4} \times (10 \text{ ft})^2 = 0.785 \times 100 \text{ ft}^2 = 78.5 \text{ ft}^2$$

A.5.2.5 Surface Area of a Cylinder

Using the formulas presented thus far, it is a simple matter to find the total wall area of a room that is to be painted. The length of each wall is added together and then multiplied by the height of the walls. This produces the area of the walls. Of course, the areas of any doors and windows are subtracted from the total. If they are rectangular, the area of each can be calculated by multiplying its length by its width.

What if the area to be painted is the total inner area of a circular cylinder instead of a rectangular room? We will work through the problem in two steps. First, we need to find the lateral surface area of the cylinder, starting with the wall's length. To help visualize the problem, suppose we made a vertical cut in the wall and unrolled it. The straightened wall forms a rectangle whose length is the same as the circumference (distance around the circle) of one of the bases.

The circumference of a circle is equal to πD.

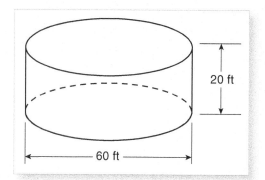

lateral surface area
The area of all the sides but not the top and bottom of a solid object, such as a cube, cylinder, or cone.

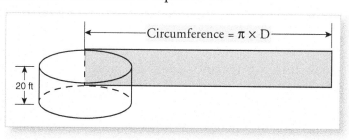

Using this fact, calculate the wall's length.

Length = π × D = 3.14 × 60 ft = 188.4 ft

The area of the wall can now be calculated in the same way any rectangular area is calculated.

Area of Wall = L × H = 188.4 ft × 20 ft = 3,768 ft²

So, the formula for a cylinder's lateral surface area is:

Area = πDH

Second, because we are painting the total inner area of the tank, we need to know the area of the top and bottom of the cylinder (the bases). Use any of the formulas for finding the area of a circle and double your result to account for both bases. Now, find the total area to be painted by adding the area of the bases and the area of the wall.

Total Area to Paint = Area of Bases + Area of Wall = 5,652 ft² + 3,768 ft²
= 9,420 ft²

If you need to paint both the inside and outside surfaces of a container, the total area to paint would be twice the total inner area.

A.5.2.6 Surface Area of a Cone

The lateral surface area of a cone is equal to ½π (Diameter)(Slant Height) or π (Radius)(Slant Height).

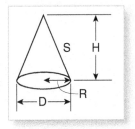

$$A = \frac{1}{2}\pi \times D \times S = \pi \times R \times S$$

Slant height is not always given (see Example 19). We can calculate the slant height by using a property of right-angle triangles.

A right-angle triangle has a square (90°) angle as shown in the sidebar. When we have a right-angle triangle, we can calculate the slanted length (S which is called the hypotenuse) using the following equation.

$$S^2 = H^2 + B^2$$

Or we can say

$$S = \sqrt{H^2 + B^2}$$

The hypotenuse (S) is the square root of the sum of the square of the height and base of the right triangle. This is called Pythagoras' Theorem, after the ancient Greek mathematician Pythagoras.

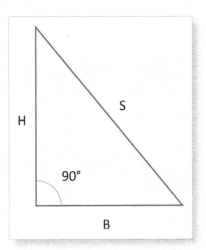

Example 18

Calculate the hypotenuse of a right triangle that has a height of 4 ft and base of 3 ft.

$$S = \sqrt{H^2 + B^2} = \sqrt{4^2 + 3^2} = \sqrt{4 \times 4 + 3 \times 3} = \sqrt{16 + 9} = \sqrt{25} = 5 \text{ ft}$$

When working with a cone, use the perpendicular height of the cone from the base to the top for the height (H) and the radius (R) of the circle for the base. So, to find the slant height of a cone, calculate:

$$S = \sqrt{R^2 + H^2}$$

Example 19

Find the total outside area of a cone (including the base) with a diameter of 30 inches and a vertical height of 20 inches.

$$\text{Slant Height, in} = \sqrt{\left(\text{Radius, in}\right)^2 + \left(\text{Height, in}\right)^2} = \sqrt{\left(15 \text{ in}\right)^2 + \left(20 \text{ in}\right)^2}$$
$$= \sqrt{225 \text{ in}^2 + 400 \text{ in}^2} = \sqrt{625 \text{ in}^2} = 25 \text{ in}$$

Lateral Area of Cone, $\text{in}^2 = \pi$ (Radius, in) (Slant Height, in)
$$= 3.14 \times 15 \text{ in} \times 25 \text{ in} = 1{,}177.5 \text{ in}^2$$

Because the entire area was asked for, the area of the base must be added.

Area of Base, $\text{in}^2 = (0.785)$ (Diameter, in)$^2 = 0.785 \times (30 \text{ in})^2$
$$= 0.785 \times 900 \text{ in}^2 = 706.5 \text{ in}^2$$

Total Area, $\text{in}^2 = $ Lateral Area of a Cone, $\text{in}^2 + $ Area of Base, in^2
$$= 1{,}177.5 \text{ in}^2 + 706.5 \text{ in}^2 = 1{,}884 \text{ in}^2$$

A.5.2.7   Surface Area of a Sphere

The lateral surface area of a sphere or ball is equal to π multiplied by the diameter squared.

$$A = \pi D^2$$

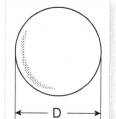

The surface area of a sphere can also be calculated by its radius:

$$A = \pi D^2 = \pi \times 2R \times 2R = 4\pi R^2$$

The surface of a sphere is 4 times its cross-sectional area.

Example 20

What is the surface area of a sphere-shaped water tank 20 feet in diameter?

Area, $\text{ft}^2 = \pi$ (Diameter, ft)$^2 = 3.14 \times (20 \text{ ft})^2 = 3.14 \times 400 \text{ ft}^2 = 1{,}256 \text{ ft}^2$

A.5.3   Volume

The volume of an object is a measure of how large it is in three dimensions. For example, we may need to know how many gallons of water can be contained in a tank. In this section, we will introduce some basic shapes that are typically encountered in the operation of water and wastewater treatment plants, collection systems, and distribution systems.

A.5.3.1 Cube

A cube is the simplest shape. Its depth, height, and width are all the same length. For example, one cubic foot is the volume of a cube whose edges (depth, height, and width) are each one foot. If the edge is more than one unit long (such as 3 feet), then the volume is 3 ft × 3 ft × 3 ft = 27 ft³. The first two numbers represent the area of one face of the cube and then the third counts how many of these cubic feet are stacked on top of each other (see the following figure). The volume of a cube can be written:

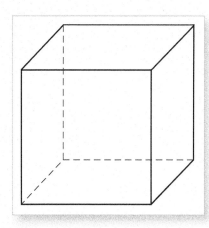

$$V = L^3$$

Where,

L = Length of Edge of the Cube

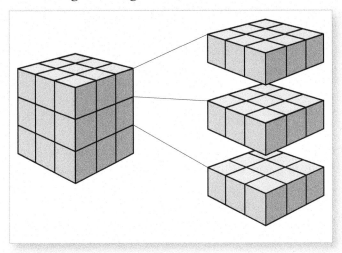

A.5.3.2 Rectangular Prism

A rectangular prism is a shape similar to a cube but the width, length, and height can be different lengths. The following figure illustrates the shape. In operation, some tanks are rectangular prisms. The volume of a prism is calculated as:

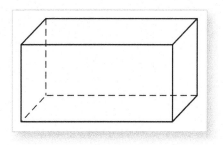

$$V = L \times W \times H$$

We use the same logic as when we calculated the volume of a cube. We calculate the area of the base (L × W) and then we multiply by the height (H) to account for how many units are stacked on top of each other.

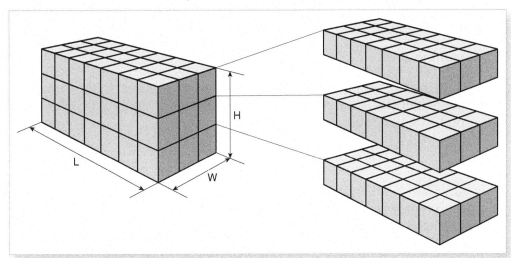

Example 21

The length of a box is 2 feet, the width is 15 inches, and the height is 18 inches. Find its volume in cubic feet.

$$V = L \times W \times H$$

$$= 2 \text{ ft} \times \frac{15 \text{ in}}{1} \times \frac{1 \text{ ft}}{12 \text{ in}} \times \frac{18 \text{ in}}{1} \times \frac{1 \text{ ft}}{12 \text{ in}}$$

$$= 2 \text{ ft} \times 1.25 \text{ ft} \times 1.5 \text{ ft}$$

$$= 3.75 \text{ ft}^3$$

Can you calculate the volume in cubic inches? It is 6,480 in³. From this, we can see that $(12 \text{ in})^3 = 1{,}728 \text{ in}^3$ per ft³.

A.5.3.3   Triangular Prism

The same general rule that applies to the volume of a rectangular prism also applies to a triangular prism.

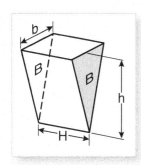

Volume = Area of Base × Height

In this shape, the base is the triangular side. Because the area of a triangle is ½bh (see Example 16), the volume of the prism is also:

$$\text{Volume} = \frac{1}{2}bhH$$

Example 22

Find the volume of a triangular prism with a base area of 10 square feet and a height of 5 feet.

Volume, ft³ = Area of Base, ft² × Height, ft = 10 ft² × 5 ft = 50 ft³

A.5.3.4   Cylinder

The volume of a cylinder is equal to the area of the circular base multiplied by the height.

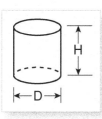

$$V = \pi R^2 \times H = 0.785 D^2 \times H$$

Example 23

A 24-inch sewer pipe is 360 feet long. Find the volume. In the formula, the height (H) is the distance along the length of the cylinder. For this problem, use 360 ft.

$$V = 0.785 D^2 \times H = 0.785 \times (2 \text{ ft})^2 \times 360 \text{ ft} = 1{,}130.4 \text{ ft}^3$$

A.5.3.5   Cone

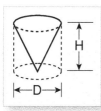

The volume of a cone is equal to ⅓ the volume of a circular cylinder of the same height and diameter.

$$V = \frac{\pi}{3} R^2 \times H$$

Example 24

Calculate the volume of a cone if the height at the center is 4 feet and the diameter is 100 feet (the radius is 50 feet).

$$V = \frac{\pi}{3} R^2 \times H = \frac{\pi}{3} \times \left(50 \text{ ft}\right)^2 \times 4 \text{ ft} = 10,472 \text{ ft}^3 \text{ or about } 10,500 \text{ ft}^3$$

See sidebar for more on rounding.

A.5.3.6 Sphere

The volume of a sphere is equal to $\pi/6$ times the diameter cubed.

$$V = \frac{\pi}{6} \times D^3 = \frac{\pi}{6} D^3$$

Example 25

How much gas can be stored in a sphere with a diameter of 12 feet?

$$V = \frac{\pi}{6} D^3 = \frac{\pi}{6} \times \left(12 \text{ ft}\right)^3 = 904.8 \text{ ft}^3$$

A.5.4   Mass and Weight

In operations, we measure mass or weight of materials used in the treatment systems. For the purposes of these training materials we will define mass as the amount of matter in an object. Weight occurs when a mass is in a gravity field. There is a technical difference between the two terms that we will explore when we discuss forces. For everyday purposes of describing amounts, however, we will sometimes use the terms interchangeably.

We typically measure mass or weight in terms of kilograms (kg) or pounds (lb).

Example 26

If someone weighs 180 lb, what is his mass in kilograms?

$$\frac{180 \text{ lb}}{1} \times \frac{1 \text{ kg}}{2.2 \text{ lb}} = 81.8 \text{ kg}$$

The conversion process to change units of weight to units of mass is the same as shown in Example 14. That is, we start with the known quantity with its unit divided by 1. We find the proper conversion factor. In this case we know 2.2 lb is the same as 1 kg (see sidebar). Then we multiply or divide by the conversion factor as needed. In general, the desired unit is in the numerator and the known unit is in the denominator (the units should cancel out).

Rounding

Operators do not always need to know precise numbers to make accurate calculations. In many cases, approximations will give operators enough information to correctly estimate amounts such as volume, area, and chemical dose. One way to estimate is rounding. Rounding numbers makes them less precise but easier to understand and use. For instance, in Example 24 on calculating the volume of a cone, the precise answer is 10,472 ft^3, but it is also correct to say that the volume is about 10,500 ft^3. In this example, the answer is rounded up to the hundreds place. If we round it to the tenths place, the answer is 10,470 ft^3. Numbers can be rounded to any place value, including decimal places. The number you are rounding to is called the rounding digit. Rounding to different place values will result in slightly different answers, called a rounding error.

The basic rule to remember when rounding is if the number after the rounding digit is 4 or less, the rounding digit stays the same (rounding down) and if the number after the rounding digit is 5 or more, the rounding digit increases by one (rounding up).

Mass versus Weight

Mass does not change. That is, if you weigh yourself and your mass is 180 lb, your mass will not change on the Moon. But your weight will be 1/6th of your weight on Earth. This is because the gravity on the Moon is 1/6th of the gravity on Earth.

Weight = Gravity × Mass

When we talk about mass and weight interchangeably it is because we are using the same gravity.

Unit Conversions

1 kg = 1,000 g

1 kg = 2.205 lb

1 lb = 453.6 g

A.5.5   Density, Specific Weight, and Specific Gravity

Density is the relationship between the mass of a substance and its volume. For example, we know the density of water is 1 g/cm³. Density is influenced by temperature (things expand when they are heated), so 1 g/cm³ is strictly accurate only at 4°C. In the temperature ranges of most treatment operations, the expansion of water is small and we can use 1 g/cm³ in most calculations. It is a good number to remember. Substances that have a higher density than water sink in water; substances that are less dense float.

Specific weight is based on the same idea except using weight instead of mass. In everyday language, people often use specific weight and density interchangeably just as they do weight and mass.

Example 27

What is the specific weight of water (that is, how many pounds of water are in a cubic foot)?

$$1\frac{g}{cm^3} \times \frac{1\ lb}{453.6\ g} \times \frac{(2.54\ cm)^3}{(1\ in)^3} \times \left(\frac{12\ in}{1\ ft}\right)^3 = 62.4\frac{lb}{ft^3}$$

Specific gravity is the ratio of the specific weight of a substance to the specific weight of water.

Example 28

What is the specific gravity of concrete if a cubic foot weighs 150 lb?

From the information given we know that the specific weight of the concrete is 150 lb/ft³. The specific gravity, therefore, is:

$$\frac{150\frac{lb}{ft^3}}{62.4\frac{lb}{ft^3}} = 2.4$$

If a substance has a specific gravity greater than 1 (for example, concrete and stone), it will sink. Substances with specific gravities less than 1 (for example, wood and Styrofoam) will float. If you know a substance's specific gravity, you can calculate its specific weight by multiplying the specific gravity by 62.4 lb/ft³.

A.5.6   Concentration

In operations, we deal with how much mass of a substance is suspended or dissolved in water or wastewater. For example, in water treatment we need to provide a specific concentration of chlorine to assure adequate disinfection. In wastewater treatment, we calculate the mass of waste solids to be disposed of by using the suspended solids concentration and the volume. Concentration is measured in quantities of mass, such as milligrams (mg), of substance in a volume of water in liters (L). Therefore, the concentration (C) of a substance in a water sample is defined as:

$$C = \frac{\text{Mass of Substance}}{\text{Volume of Water}}$$

The typical unit for concentration is mg/L. Another way to measure concentration is in parts per million (ppm). The concentration in ppm is defined as:

$$\text{ppm} = \frac{1 \text{ Part of a Substance}}{1,000,000 \text{ Parts of Another Substance}}$$

The units of measurement of the denominator and numerator are the same units (mass to mass ratio). For example, 1 g of chlorine in 1,000,000 g of water (hence one part in one million) is a proper expression of ppm. When we are dealing with water, which has a density of 1 g/cm³ for all practical purposes, a mg/L is the same as ppm. This is a special case that is only applicable to water solutions and not any other solutions. This point can be illustrated as follows:

$$1 \text{ ppm} = \frac{1.0 \text{ mg Chemical}}{1,000,000 \text{ mg Water}} \times \frac{1,000 \text{ mg}}{1 \text{ g}} \times \frac{1 \text{ g Water}}{\text{cm}^3} \times \frac{1,000 \text{ cm}^3}{L}$$

$$= 1\frac{\text{mg}}{L}$$

We used the density of water of 1 g/cm³ to convert the mass of water to an equivalent volume of water.

Example 29

How many pounds of alum (a chemical often used in drinking water treatment) would we have to add to 1.0 million gallons to make a 30 mg/L solution?

Concentration = mass/volume

If we multiply both sides of the equation by volume, we get:

Mass = concentration × volume = CV

Most of the computational work in this problem is converting units.

$$\begin{matrix}\text{Number}\\\text{of lb}\end{matrix} = \left(\frac{30 \text{ mg}}{L}\right)(1,000,000 \text{ gal})\left(\frac{kg}{1,000,000 \text{ mg}}\right)\left(\frac{2.205 \text{ lb}}{kg}\right)\left(\frac{3.785 \text{ L}}{gal}\right)$$

$$= 250 \text{ lb}$$

A.5.7 Velocity and Flow Rate

Operators deal with how quickly water moves in pipes and channels. Velocity is defined as the speed or rate of movement. Velocity is expressed as:

$$V = \frac{\text{Distance}}{\text{Time}}$$

Typical units for velocity include $\frac{mi}{h}, \frac{m}{s}, \frac{ft}{s}$. The latter is commonly used in water operations in the United States.

Example 30

What is the average velocity of a car that travels 285 miles in 5 hours? The average velocity implies that we are calculating a uniform number that applies for the whole trip. That is, you speed up at some points and slow down at other points, but the average speed is what matters.

$$V = \frac{\text{Distance}}{\text{Time}} = \frac{285 \text{ mi}}{5 \text{ h}} = 57 \frac{mi}{h}$$

Operators use the same process to determine the velocity or speed of water movement in a channel as illustrated in Example 31.

Example 31

What is the average velocity of water in feet per second (fps) if a particle suspended in the flow moves 600 feet in 5 minutes.

$$V = \frac{\text{Distance}}{\text{Time}} = \frac{600 \text{ ft}}{5 \text{ min}} = 120 \frac{ft}{min} \times \frac{1 \text{ min}}{60 \text{ s}} = 2 \frac{ft}{s} \text{ or 2 fps}$$

The conversion of units that occurred in the last step followed the same steps as outlined previously.

Flow rate is the volume of water passing a given point per unit time (see the sidebar for more about rate). The flow rate is represented as Q.

$$Q = \frac{\text{Volume of Water}}{\text{Time}}$$

Typical measurements of flow rate are in cubic feet per second (cfs). Other units commonly used are m^3/s or gal/min (gpm).

Example 32

Calculate the flow rate in gallons per minute (gpm) if the flow rate is 0.5 cfs.

$$0.5 \frac{ft^3}{s} \times \frac{7.48 \text{ gal}}{ft^3} \times \frac{60 \text{ s}}{min} = 224.4 \frac{gal}{min} = 224.4 \text{ gpm}$$

We continue to use the same process in converting units. The first term includes ft^3 in the numerator. Accordingly, our conversion factor has ft^3 in the denominator. Next, we need to find the proper conversion factor for the desired units (in this case, gallons). We used the same process converting the seconds into minutes.

Returning to the concept of velocity of fluid (water) in pipes or channels, let's learn how the velocity and flow rates are related to each other.

If water is flowing in a 1-foot wide channel with a depth of 1 foot, then the cross-sectional area of the channel is 1 ft × 1 ft = 1 ft^2.

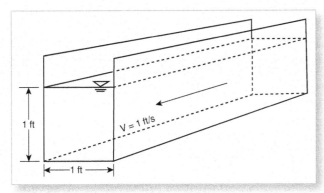

If the velocity in this channel is 1 fps, then each second a volume of water equivalent to 1 ft^2 in area and 1 ft long will pass a given point. The volume of this body of water is 1 cubic foot, and because it passes the point every second, its flow rate is equal to 1 ft^3 per second, or 1 cfs.

Accordingly, we can say that the flow rate Q is expressed as follows:

Q = v × A

Where,

   Q = Flow Rate (Volume of Water per Unit Time, cfs)

   v = Velocity of Water (Distance per Unit Time, fps)

   A = Cross-Sectional Area of the Pipe or Channel (ft^2)

Example 33

A rectangular channel 3 feet wide contains water 2 feet deep, moving at 1.5 feet per second. What is the flow rate in cfs?
   Based on what we learned:

Q = v × A

   We are given the size of the channel and the flow velocity (v). Find the area of the flow.

A = 3 ft × 2 ft = 6 ft^2

   Find the flow rate (Q).

$$Q = 1.5 \frac{ft}{s} \times 6 \ ft^2 = 9 \frac{ft^3}{s} = 9 \ cfs$$

   We are able to combine the units of feet using the exponent property ft × ft^2 = ft^3.

Example 34

Water in a 2.5-foot-wide channel is 1.4 feet deep and the rate of flow is 11.2 cfs. What is the average velocity of the water in the channel in fps?

We are given the dimensions of the channel and the flow rate (Q).

$Q = v \times A$, or

$$v = \frac{Q}{A} = \frac{11.2 \frac{ft^3}{s}}{2.5 \, ft \times 1.4 \, ft} = \frac{11.2 \frac{ft^3}{s}}{3.5 \, ft^2} = 3.2 \frac{ft}{s} = 3.2 \text{ fps}$$

The second equation $\left(v = \dfrac{Q}{A}\right)$ is obtained by the property of cross multiplication (see Example 4 for the cross multiplication shortcut).

Example 35

Flow rate in an 8-inch pipe is 500 gpm. What is the average velocity of the water in the pipe in fps?

The flow rate is given in units of gpm and the answer of velocity needs to be in fps. Our calculations are similar to Example 34, except that we need to convert the area to square feet and the flow rate to cfs. The last step is exactly the same as Example 34. Start with calculating the cross-sectional area of the pipe.

$$A = \frac{\pi}{4} D^2 = 0.785 \times \left(8 \text{ in}\right)^2 \times \frac{ft^2}{\left(12 \text{ in}\right)^2} = 0.35 \, ft^2$$

$$Q = 500 \frac{gal}{min} \times \frac{ft^3}{7.48 \, gal} \times \frac{min}{60 \, s} = 1.114 \frac{ft^3}{s} = 1.114 \text{ cfs}$$

$$v = \frac{Q}{A} = \frac{1.114 \frac{ft^3}{s}}{0.35 \, ft^2} = 3.18 \frac{ft}{s} = 3.18 \text{ fps}$$

A.5.8   Force and Pressure

In general, force is an action of push or pull that is exerted on an object. One of the major forces is gravitational force, which is the force by which Earth's gravity pulls all objects toward its center. Objects tend to fall down toward the Earth's center because of the gravity force the Earth exerts on all objects. We call that force weight. Therefore, we can express weight (W) as a force using the following:

W = m × g

Where,

m = Mass of an Object

g = Gravitational Acceleration

Gravitational acceleration for Earth is 32.2 ft/s² (9.81 m/s²). What matters to us are forces like those that water exerts on the walls of

---

Force (Weight) Units in Metric System

The force unit in the metric system is the Newton, denoted as N:

$$1 \, N = 1 \, kg \times \frac{1 \, m}{s^2} = \frac{kg \times m}{s^2} = 0.225 \, lb$$

tanks or pressure force in pipes. Before we explore that, let's talk about force generated by simple objects. Suppose you stack bricks on your hand (we do not recommend that you do that). For simplicity, assume one brick weighs one pound. Suppose also the brick is 2 inches high. If you continue stacking bricks on your hand, the weight increases. Instead of talking about bricks, we can talk about height. So, we can say that for every two inches the weight increases by one pound.

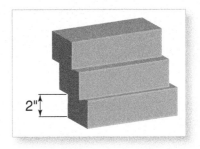

Because water is in a liquid (fluid) form, it cannot keep its shape and it will take the shape of any container. Imagine if we place water in a small container and we break one side of the container; gravity will pull the water out and it will spill. Therefore, the water is pushing the sides of the container because of gravity. The force that the water exerts is due to its weight.

Using the brick analogy, we see that the weight of the water (and therefore the force it exerts) is proportional to the height of the water column. The higher the water column, the more weight it exerts on the bottom of the tank.

Pressure is a concept related to force. Pressure is defined as the force per unit area. Therefore, typical units include pounds per square inch (psi) or N/m² in the metric system. Pressure is expressed as $p = \dfrac{F}{A}$.

Operators typically deal with pressure when they are dealing with tanks containing water (or any liquid). In this situation, when a tank is full of water, there are two types of pressure: pressure on the bottom of the tank and pressure on the side walls of the tank. In this section, we will cover both pressures through examples. The pressure is higher at the bottom of a tank than the average pressure on the wall.

Density of Water

Water weighs 62.4 lb/ft³. In the metric system, water weighs 1 kg/L. The symbol γ is used to denote density in equations.

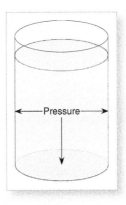

Pressure

Example 36

Calculate the pressure on the bottoms of the two numbered containers shown in the sidebar. Container 1 is a 4-foot-high cylinder with a 2-foot diameter. Container 2 is a rectangular prism with a width of 2 feet and length of 5 feet. The height of the rectangular prism is 4 feet.

To solve this problem, we need to calculate the weight of the water and then calculate the pressure by dividing the weight (force) by the area of the base.

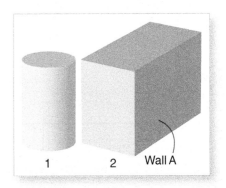

CONTAINER 1

The volume of a cylinder is the volume of the base multiplied by the height of the cylinder.

$$\text{Volume} = \pi R^2 H = \frac{\pi}{4} D^2 H = 0.785 \times (2\text{ ft})^2 \times 4\text{ ft} = 12.56\text{ ft}^3$$

The weight of the water is equal to the volume multiplied by the density of water (see "Density of Water" in the sidebar).

$$\text{Weight} = V \times \gamma = 12.56\text{ ft}^3 \times 62.4\frac{\text{lb}}{\text{ft}^3} = 783.7\text{ lb}$$

Finally, pressure is equal to weight divided by area.

$$\text{Pressure} = \frac{\text{Weight}}{\text{Area}} = \frac{783.7 \text{ lb}}{0.785 \times (2 \text{ ft})^2} = 249.6 \frac{\text{lb}}{\text{ft}^2} \times \frac{(1 \text{ ft})^2}{(12 \text{ in})^2}$$

$$= 1.73 \frac{\text{lb}}{\text{in}^2} = 1.73 \text{ psi}$$

CONTAINER 2

$$\text{Volume} = W \times L \times D = 2 \text{ ft} \times 5 \text{ ft} \times 4 \text{ ft} = 40 \text{ ft}^3$$

$$\text{Weight} = V \times \gamma = 40 \text{ ft}^3 \times 62.4 \frac{\text{lb}}{\text{ft}^3} = 2{,}496 \text{ lb}$$

$$\text{Pressure} = \frac{\text{Weight}}{\text{Area}} = \frac{2{,}496 \text{ lb}}{2 \text{ ft} \times 5 \text{ ft}} = 249.6 \frac{\text{lb}}{\text{ft}^2} \times \frac{(1 \text{ ft})^2}{(12 \text{ in})^2}$$

$$= 1.73 \frac{\text{lb}}{\text{in}^2} = 1.73 \text{ psi}$$

The answer is the same in both containers even though they have different shapes and different areas and volumes because the height of the water is the same in both containers. It turns out that the height is what matters when estimating the pressure of water. Mathematically, we can say:

$$p = \gamma \times H$$

Where,

p = Pressure of Water (in units such as psi, lb/ft², lb/in², or N/m²)

γ = Weight Density of Water (in units such as lb/ft³, lb/in³, or N/m³)

H = Height of Water Column (in units such as ft, in, or meters)

To calculate the pressure on the wall, we need to recognize that the pressure in water at any particular point is the same in all directions (see the figure in the sidebar). So the pressure on any point on a wall is directly related to the depth of that point. In the figure, the pressure on the wall (and everywhere else) at a depth of 1 ft is 0.433 psi. At a depth of 2 ft, it is 0.866 psi. (For practice, see if you can verify these values using the equation $p = \gamma H$.) We will use this information to calculate the pressure and the force on a tank wall in Example 37.

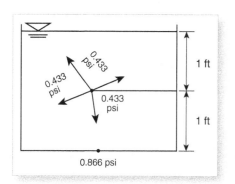

Example 37

Calculate the pressure and then the force exerted by the water on Wall A of the rectangular prism in Example 36.

First, let's estimate the pressure exerted by water for depths 0, 1, 2, 3, and 4 feet from the surface of water in a container (see the following figure) by using the pressure equation.

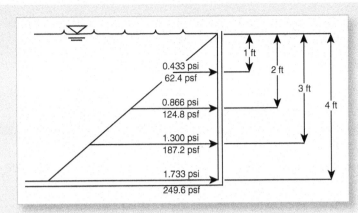

$$p_0 = \gamma \times H = 62.4 \frac{lb}{ft^3} \times 0\,ft = 0\frac{lb}{ft^2} \times \left(\frac{1\,ft}{12\,in}\right)^2 = 0\,psi$$

$$p_1 = \gamma \times H = 62.4 \frac{lb}{ft^3} \times 1\,ft = 62.4\frac{lb}{ft^2} \times \left(\frac{1\,ft}{12\,in}\right)^2 = 0.433\,psi$$

$$p_2 = \gamma \times H = 62.4 \frac{lb}{ft^3} \times 2\,ft = 124.8\frac{lb}{ft^2} \times \left(\frac{1\,ft}{12\,in}\right)^2 = 0.866\,psi$$

$$p_3 = \gamma \times H = 62.4 \frac{lb}{ft^3} \times 3\,ft = 187.2\frac{lb}{ft^2} \times \left(\frac{1\,ft}{12\,in}\right)^2 = 1.300\,psi$$

$$p_4 = \gamma \times H = 62.4 \frac{lb}{ft^3} \times 4\,ft = 249.6\frac{lb}{ft^2} \times \left(\frac{1\,ft}{12\,in}\right)^2 = 1.733\,psi$$

(*NOTE:* p_4 is the same as the pressure on the bottom calculated in Example 36.)

The average pressure of water acting on the tank wall is 1.733 psi/2 = 0.867 psi, or 249.6 psf/2 = 124.8 psf. We divided by two to obtain the average pressure because there is zero pressure at the top and 1.733 psi pressure on the bottom of the wall and in between the pressure increases linearly.

The average pressure in pounds per square foot or pounds per square inch is used to calculate force in pounds. Recall that the wall is 5 feet long. The pressure of the water acts over the entire 20-square-foot (5 ft × 4 ft) area of the wall. Using the average pressure in pounds per square foot, the push or force exerted by water on the surface of the wall is calculated as follows:

$$F = p \times A$$

Where,

F = Force Exerted by Water on the Surface of the Wall

p = Pressure of Water Acting on the Tank Wall

A = Area of the Wall

$$F = p \times A = 124.8 \frac{lb}{ft^2} \times 20\,ft^2 = 2,496\,lb$$

Conversion

$$psi = \frac{lb}{in^2}$$

$$psf = \frac{lb}{ft^2}$$

The Problem with Focusing on Average Pressure

While average pressure is useful for determining the force on the wall of a tank, it is not the pressure at the bottom of the tank. That pressure is much greater than the average pressure.

If the average pressure in pounds per square inch were used, the calculation would be similar, with the area converted to inches:

$$F = p \times A = 0.867 \text{ psi} \times 48 \text{ in} \times 60 \text{ in} = 2{,}497 \text{ lb}*$$

Example 38

Find the force on the wall of a rectangular sedimentation basin that is 100 ft long. The water depth is 12 ft.

We know from the previous discussion that the force is the pressure multiplied by the area.

$$F = p \times A$$

Also, from the previous discussion, the pressure used in this equation is the average pressure, which is one half of the maximum pressure, or the pressure at half of the depth.

$$p_{avg} = \left(\frac{1}{2}\right)\gamma H$$

$$A = LH \qquad \text{where L = the length of the wall}$$

$$F = p \times A = (0.5\gamma H) \times (LH) = 0.5\gamma LH^2$$

Plugging in the numbers produces:

$$F = 0.5\left(\frac{62.4 \text{ lb}}{ft^3}\right)(100 \text{ ft})(12 \text{ ft})^2 = 449{,}280 \text{ lb}$$

Keep in mind this force is acting sideways on the wall.

It is very important from an operation point of view to understand the force that water can exert on tanks. For example, if the groundwater level is high and surrounds the storage tank, operators should not empty the tank because groundwater could push the tank up. Let's explore the upward force generated by groundwater.

Example 39

Find the upward force on the bottom of an empty tank caused by a groundwater depth of 8 feet above the bottom of the tank. The tank is 20 feet wide and 40 feet long, as shown in the following figure.

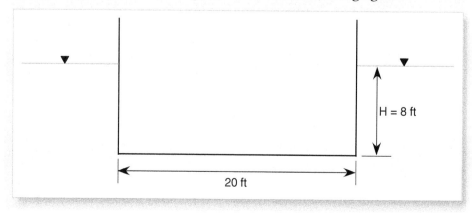

To solve this problem, we need to estimate the pressure at the bottom of the tank. This pressure results from the height of 8 feet of groundwater above the bottom of the tank. The upward force then is estimated by multiplying the pressure by the surface area of the bottom of the tank.

$$F = p \times A = \gamma \times H \times A = \gamma \times H \times W \times L = 62.4 \frac{lb}{ft^3} \times 8\,ft \times 20\,ft \times 40\,ft$$

$$= 399,360\,lb*$$

*For practical purposes, we can round up to 399,400 lb. (See sidebar on page 655 for more on rounding.)

If the tank does not weigh at least this much, the groundwater pressure will cause it to "float" or to raise up off its foundation.

A.5.9   Work, Head, and Power

Work is defined as the amount of force exerted on an object causing it to move a certain distance in the direction of the force. Work is a representation of energy. Based on this definition, work is expressed in units of foot-pounds (ft-lb) or Joule (N·m, Newton-meter). Joule is used in the metric system and usually denoted as J.

$$W = F \times d$$

Where,

W = Work

F = Force

d = Distance in the Direction of the Force

The terms work and energy are interchangeable. For example, if we lift a given volume of water to a certain height (using a pump) we do work because we are moving a mass of water by applying a lifting force against gravity. That volume of water now has stored energy (also called potential energy) that will be released once we allow the water to flow.

Example 40

A 165-pound man runs up a flight of stairs 20 feet high. How much work did he do?

$$W = F \times d = 165\,lb \times 20\,ft = 3,300\,ft\text{-}lb$$

In fluids, we are often interested in the work or energy per unit weight of fluid. We call this quantity head. Here we'll describe two kinds of head.

Static head is the energy or work needed to raise a pound of water a specified height. It can be calculated as:

$$H_s = \frac{work}{weight} = \frac{force \times distance}{weight} = \frac{weight \times height}{weight} = height$$

In the equation above, the force used to lift the water is equal to its weight and the distance traveled is the height of the lift. Often the lift is described as a change in height, denoted by Δz (delta z), so you may see this equation written as:

$$H_s = \Delta z$$

The units on the head are feet (or meters in metric units).

Water under pressure can do work. The following figure shows a tank of water. As discussed above, the water beneath the surface is under pressure, p. If we punch a hole in the side of the tank and a column of water of length L is pushed out, how much work is done?

$$\text{work} = \text{force} \times \text{distance} = p \times A \times L$$

As before, if we define the pressure head as the work per unit weight, we can develop the following equation:

$$H_p = \frac{\text{work}}{\text{weight}} = \frac{\text{work}}{\text{unit weight} \times \text{volume}} = \frac{p \times A \times L}{\gamma \times A \times L} = \frac{p}{\gamma}$$

In the equation above, the volume of water pushed out of the hole is the cross-sectional area times the length. As shown below, the units of pressure head are also feet (or meters).

$$H_p = \frac{p\dfrac{\text{lb}}{\text{ft}^2}}{\gamma\dfrac{\text{lb}}{\text{ft}^3}} = \frac{p}{\gamma}\,\text{ft}$$

In the pressure head equation, the pressure can be caused by the depth of water (hydrostatic pressure) or by a mechanical device such as a pump. When the pressure is hydrostatic there is an interesting trade-off between static head and pressure head, as illustrated in Example 41.

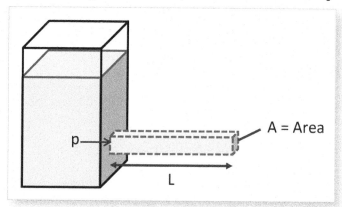

Example 41

Referring to the open water tank in the figure in the sidebar, calculate the static head and pressure head at point A (the water surface) and point B 6 ft below the surface. For convenience we will set the bottom of the tank at z = 0.

For Point A, Δz is the distance above the bottom or 6 ft. The hydrostatic pressure is zero because the depth is zero (no water above point A).

$H_s = \Delta z = 6 \, \text{ft}$

$H_p = \dfrac{p}{\gamma} = 0 \, \text{ft}$

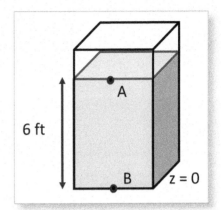

For Point B, Δz is the distance above the bottom or 0 ft. The hydrostatic pressure is equal to the depth times the unit weight of water.

$H_s = \Delta z = 0 \, \text{ft}$

$H_p = \dfrac{p}{\gamma} = \dfrac{\gamma \times (6 \, \text{ft})}{\gamma} = 6 \, \text{ft}$

So, in an open tank, there is a trade-off between static head and pressure head. At any point inside an open tank, the sum of the static head and the pressure head is a constant.

Power is defined as the rate at which work is done. Remember, when we talk about rate, we must include time. Therefore, the units used to denote power would be ft-lb/s. We can express power as:

$$P = \dfrac{W}{t} = \dfrac{F \times d}{t}$$

Where,

P = Power

t = Time

The rest of the terms are the same as we defined them previously.

Example 42

If the man in Example 40 runs up the stairs in 3 seconds, how much power has he exerted?

$$P = \dfrac{W}{t} = \dfrac{3,300 \, \text{ft-lb}}{3 \, \text{s}} = 1,100 \, \dfrac{\text{ft-lb}}{\text{s}}$$

Horsepower is another unit of power measurement and it is defined as 33,000 ft-lb per minute. That is,

$$1 \, \text{hp} = 33,000 \, \dfrac{\text{ft-lb}}{\text{min}}$$

Example 43

Express the power in Example 42 in hp.

$$P = 1,100 \, \dfrac{\text{ft-lb}}{\text{s}} \times \dfrac{60 \, \text{s}}{\text{min}} \times \dfrac{1 \, \text{hp}}{33,000 \, \dfrac{\text{ft-lb}}{\text{min}}} = 2 \, \text{hp}$$

Watt is the unit used to measure power in the metric system. One hp is the same as 746 watts. A watt is defined as:

$$1\,\text{watt} = \frac{\text{m} \times \text{N}}{\text{s}} = \frac{\text{J}}{\text{s}}$$

Where,
m = meters
N = Newtons (units of force)
s = seconds
J = Joules (units of energy)

Newtons are discussed in Section A.5.8, "Force and Pressure."
Electrical equipment is often rated in terms of kilowatts (kW), which are 1,000 watts.
Because power is energy per time, energy is power multiplied by time.

$$E = W = P \times t$$

Consequently, utility bills for energy are commonly expressed in kilowatt-hours (kWh).

A.6   Metric System

Metric System in the United States

In the US, the law was amended to define the metric system to mean the system of units as defined by the International System of Units as established in 1960. The information in this section was obtained from *Guide for the Use of the International System of Units (SI)* (Thompson and Taylor, NIST Special Publication 811, 2008 Edition).

The metric system of measurements is the most commonly used unit system in science, technology, and international commerce. The SI unit system (from the French *Le Système International d'Unités*) is the modern measurement system that is based on the metric system. The National Institute of Standards and Technology (NIST) is the agency within the Commerce Department that works with industry to develop and apply technology, measurements, and standards and is responsible for helping US industry to be competitive in the global marketplace.

A.6.1   SI Base Units

The following table shows the seven base units for the SI system. The last one, luminous intensity, is not typically used.

SI Base Units

Base Quantity	SI Base Unit	
	Name	Symbol
Length	meter	m
Mass	kilogram	kg
Time	second	s
Electric current	ampere	A
Thermodynamic temperature	kelvin	K
Amount of substance	mole	mol
Luminous intensity	candela	cd

These base units form the foundation of other derived units. The following table shows some examples of typical derived units.

SI Derived Units

Derived Quantity	SI Base Unit		
	Name	Symbol	
Area	square meter	m^2	
Volume	cubic meter	m^3	
Speed, velocity	meter per second	m/s	
Density, mass density	kilogram per cubic meter	kg/m^3	
Amount-of-substance concentration, concentration	mole per cubic meter	mol/m^3	

Writing Numbers in the SI System

Which of the following is the correct way to write the number based on the SI system?

12532

12,532

12 532

All of the above

None of the above

Answer: 12 532 is the correct way.

The same rule applies to decimal numbers; therefore, 0.0034501 is written as:

0.003 450 1

Using the SI system, units are expressed as lower case letters. There are two exceptions. The first is when the unit is named after a person; for example, units of electrical current are amperes (or "amps") with the symbol A. The second exception is liter, the measure of volume, which uses an upper case L (in the US) to avoid any confusion with the number 1 and letter l. Liter is a non-SI unit but is allowed by NIST and the International Committee for Weights and Measures (known by its French acronym, CIPM).

To use SI units effectively, the reader needs to be familiar with the terminology used for prefixes to describe multiples or fractions of the basic units. The table below shows a partial list of the prefixes. These prefixes (multiples of 10s) make the SI system attractive and easy to use.

A Partial List of SI Prefixes

Prefix	Symbol	Meaning
giga	G	$10^9 = (10^3)^3 = 1\ 000\ 000\ 000$
mega	M	$10^6 = (10^3)^2 = 1\ 000\ 000$
kilo	k	$10^3 = (10^3)^1 = 1\ 000$
hecto	h	$10^2 = 100$
deka	da	$10^1 = 10$
deci	d	$10^{-1} = 0.1$
centi	c	$10^{-2} = 0.01$
milli	m	$10^{-3} = (10^{-3})^1 = 0.001$
micro	µ	$10^{-6} = (10^{-3})^2 = 0.000\ 001$
nano	n	$10^{-9} = (10^{-3})^3 = 0.000\ 000\ 001$

Pipe-Size Designations

Customary, in	Soft Metric, mm
2	50
4	100
6	150
8	200
10	250
12	300
15	375
18	450
24	600
30	750
36	900
42	1,050
48	1,200
60	1,500
72	1,800
84	2,100

Another feature of the SI system is how numbers are expressed. For numbers with more than four digits, they are separated into groups of three numbers separated with a space; no commas are allowed. This applies to the number to the right of the decimal place as well as illustrated in the table above. See "Writing Numbers in the SI System" in the sidebar for an exercise and an example.

One variation on conversion from inches to millimeters is the agreement of pipe manufacturers to the use of a "soft" metric conversion. That is, they agreed to use 25 mm/in instead the correct conversion of 25.4 mm/in. Accordingly, the pipe-size designation system shown in the table in the sidebar is used.

A.6.2   Measures of Length

The basic measure of length is the meter.

> 1 kilometer (km) = 1,000 meters (m)
> 1 meter (m)   = 100 centimeters (cm)
> 1 centimeter (cm) = 10 millimeters (mm)

Kilometers are usually used in place of miles, meters are used in place of feet and yards, centimeters are used in place of inches, and millimeters are used for inches and fractions of an inch.

Length Equivalents			
1 km	= 0.621 mile	1 mile = 1.61 km	
1 meter	= 3.28 feet	1 foot = 0.305 meter	
1 meter	= 39.37 inches	1 inch = 0.0254 meter	
1 centimeter	= 0.3937 inch	1 inch = 2.54 centimeters	
1 millimeter	= 0.00394 inch	1 inch = 25.4 millimeters	

NOTE: The above equivalents are reciprocals. If one equivalent is given, the reverse can be obtained by division. For instance, if 1 meter equals 3.28 feet, 1 foot equals 1/3.28 meter, or 0.305 meter.

A.6.3   Measures of Capacity or Volume

The basic measure of volume capacity in the SI system is the cubic meter. The liter is used as a measure of volume that is acceptable by the SI system but not part of SI units.

> 1 kiloliter (kL) = 1,000 liters (L) = 1 m³
>
> 1 liter (L) = 1,000 milliliters (mL)

Kiloliters, or cubic meters, are used to measure capacity of large storage tanks or reservoirs in place of cubic feet or gallons. Liters are used in place of gallons or quarts. Milliliters are used in place of quarts, pints, or ounces.

Capacity Equivalents			
1 kiloliter	= 264.2 gallons	1 gallon	= 0.003785 kiloliter
1 liter	= 1.057 quarts	1 quart	= 0.946 liter
1 liter	= 0.2642 gallon	1 gallon	= 3.785 liters
1 milliliter	= 0.0338 ounce	1 fluid ounce	= 29.57 milliliters

A.6.4   Measures of Weight

The basic unit of weight in the metric system is the gram. One cubic centimeter of water at maximum density weighs one gram, and thus there is a direct, simple relation between volume of water and weight in the metric system.

$$1 \text{ kilogram (kg)} = 1,000 \text{ grams (g)}$$
$$1 \text{ gram (g)} = 1,000 \text{ milligrams (mg)}$$
$$1 \text{ milligram (mg)} = 1,000 \text{ micrograms (}\mu\text{g)}$$

Grams are usually used in place of ounces, and kilograms are used in place of pounds.

Weight Equivalents	
I kilogram = 2.205 pounds	I pound = 0.4536 kilogram
I gram = 0.0022 pound	I pound = 453.6 grams
I gram = 0.0353 ounce	I ounce = 28.35 grams
I gram = 15.43 grains	I grain = 0.0648 gram

A.6.5 Temperature

In addition to becoming familiar with the metric system, you should also become familiar with the Celsius (centigrade) scale for measuring temperature. There is nothing magical about the Celsius scale—it is simply a different size than the Fahrenheit scale. For a comparison of the two scales, refer to the figure in the sidebar.

The two scales are related in the following manner:

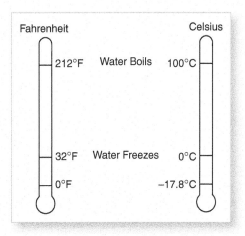

$$\text{Fahrenheit} = \left({}^\circ\text{C} \times \frac{9}{5} \right) + 32^\circ$$

$$\text{Celsius} \quad = ({}^\circ\text{F} - 32^\circ) \times \frac{5}{9}$$

Example 44

Convert 20° Celsius to degrees Fahrenheit.

$${}^\circ\text{F} = \left({}^\circ\text{C} \times \frac{9}{5} \right) + 32^\circ = \left(20^\circ \times \frac{9}{5} \right) + 32^\circ = \frac{180^\circ}{5} + 32^\circ = 36^\circ + 32^\circ = 68^\circ\text{F}$$

Example 45

Convert –10°C to °F.

$${}^\circ\text{F} = \left(-10^\circ \times \frac{9}{5} \right) + 32^\circ = \frac{-90^\circ}{5} + 32^\circ = -18^\circ + 32^\circ = 14^\circ\text{F}$$

Example 46

Convert –13°F to °C.

$${}^\circ\text{C} = ({}^\circ\text{F} - 32^\circ) \times \frac{5}{9} = (-13^\circ - 32^\circ) \times \frac{5}{9} = -45^\circ \times \frac{5}{9} = -5^\circ \times 5 = -25^\circ\text{C}$$

Advanced Concepts (Sections A.7–A.8)

This advanced section covers topics related to pumps (pressure, horsepower, head, and more), and the analysis and presentation of data, building on the basic and intermediate concepts presented in the previous sections. The following discussion assumes that you are comfortable with those concepts. Refer to that material as needed to assist you in understanding the concepts presented here.

A.7 Pumps

In this section, we will discuss the performance of pumps. Operators deal with pumps in many settings. We will discuss pressure, work, power, horsepower, and head in the context of pumps, which will also serve as a review and reinforcement of the basic concepts that were discussed earlier. In addition, pump characteristics and evaluation of pump performance will be covered.

A.7.1 Pressure

Atmospheric pressure at sea level is approximately 14.7 psi. This pressure acts in all directions and on all objects. If a tube is placed upside down in a basin of water and a 1 psi partial vacuum is drawn on the tube, the water in the tube will rise 2.31 feet. We will calculate this rise later.

For now, imagine that initially we invert the tube into a relatively large container of water (refer to the schematic below). Points A and B are at the same level initially and both have the same atmospheric pressure. Now by using a vacuum pump to lower the pressure in the tube by 1 psi, the pressure at point A is higher than the pressure at point B. Therefore, the water in the container is pushed into the tube until the pressures at points A and B are the same (see the schematic below).

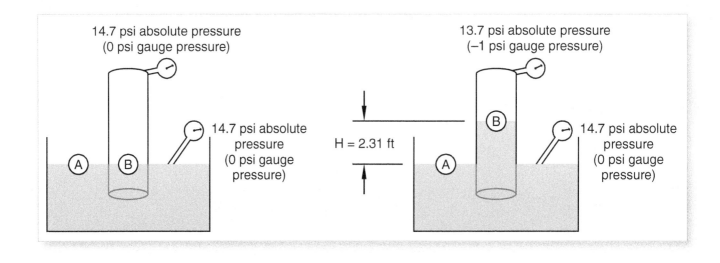

To calculate the height of water in the tube, we need to estimate the pressure that is exerted by the water column of height H. See Section A.5.8, "Force and Pressure."

$$p = \frac{F}{A} = \gamma H$$

Where,

  p = Pressure at the Bottom

  F = Force

  A = Area

  γ = Water Density = 62.4 lb/ft^3

  H = Height of a Water Column

Then, calculate the column height that can exert 1 psi.

$$p = 1\frac{lb}{in^2} \times \left(\frac{12\,in}{ft}\right)^2 = 62.4\frac{lb}{ft^3} \times H$$

$$144\frac{lb}{ft^2} = 62.4\frac{lb}{ft^3} \times H$$

$$H = \frac{144\frac{lb}{ft^2}}{62.4\frac{lb}{ft^3}} = 2.31\,ft$$

The left-hand side of the pressure was in lb/in^2, which needed the conversion to ft^2 to make it consistent with the right-hand side. See the sidebar for an alternative way to calculate the height H.

The action of the partial vacuum is what moves water out of a sump or well and up to a pump. It is not sucked up, but it is pushed up by atmospheric pressure on the water surface in the sump. If a complete vacuum could be drawn, the water would rise 2.31 × 14.7 = 33.96 feet; but this is impossible to achieve. The practical limit of the suction lift of a positive displacement pump is about 22 feet, and that of a centrifugal pump is 15 feet.

A.7.2 Work

Work can be expressed as lifting a weight a certain vertical distance. It is usually defined in terms of foot-pounds (ft-lb).

Example 47

Using a rope, an operator hauls a 5-gallon bucket partially full of water from the bottom of a sump to ground level, a vertical distance of 30 ft. If the bucket and water weigh a total of 40 lb, how much work did he do?

Work, ft-lb = Weight, lb × Height, ft

  = 40 lb × 30 ft

  = 1,200 ft-lb

Calculating H

(Refer to the figure showing the inverted tube.)

If you know that 0.433 psi is equivalent to 1 foot of water height, then 1 psi is equivalent to:

$$H = 1\,psi \times \frac{1\,ft}{0.433\,psi} = 2.31\,ft$$

A.7.3   Power

Power is a rate of doing work and is usually expressed in foot-pounds per minute.

Example 48

If the operator performs the lift described in Example 47 in 8 seconds, how much power did he exert?

$$\text{Power,} \frac{\text{ft-lb}}{\text{s}} = \frac{\text{Work, ft-lb}}{\text{Time, s}}$$

$$= \left(\frac{1,200\,\text{ft-lb}}{8\,\text{s}}\right) \times \left(\frac{60\,\text{s}}{\text{min}}\right)$$

$$= 9,000 \frac{\text{ft-lb}}{\text{min}}$$

A.7.4   Horsepower

Horsepower is a unit of power that is commonly used in the United States. One horsepower is defined as equivalent to 33,000 ft-lb/minute or 746 watts.

Example 49

How much horsepower has the operator exerted in pulling up the bucket? Use the power calculated in Example 48.

$$\text{Power} = P = \left(9,000 \frac{\text{ft-lb}}{\text{min}}\right) \times \left(\frac{1\,\text{hp}}{33,000 \frac{\text{ft-lb}}{\text{min}}}\right) = 0.27\,\text{hp}$$

This is a unit conversion problem. For more details on units and conversions see Section A.5, "Units."

Instead of a bucket, the operator could have lifted water from one level to a higher level using a pump. To estimate the work needed to move a cubic foot of water some distance above where it is, we need to calculate the weight of the water and then multiply it by the height or elevation.

$$W = V\gamma H$$

Where,

$W =$ Work Done, ft-lb

$V =$ Volume of Water, ft^3

$\gamma =$ Density of Water = 62.4 lb/ft^3

$H =$ Height or Elevation, ft

The term $(V\gamma)$ represents the weight of the water and H represents the distance over which the water is lifted (see the following schematic).

Reviewing Work and Power

Work = Force × Distance

$$\text{Power} = \frac{\text{Work}}{\text{Time}}$$

Therefore, the work needed to lift a volume of one ft³ for a distance of one ft can be calculated as:

$$W = 1\,\text{ft}^3 \times 62.4\,\frac{\text{lb}}{\text{ft}^3} \times 1\,\text{ft} = 62.4\,\text{ft-lb}$$

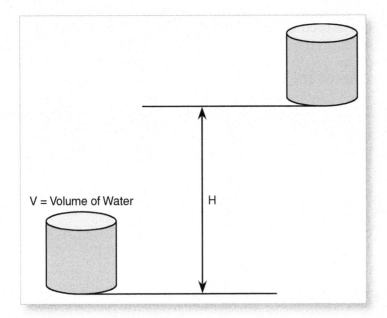

V = Volume of Water   H

The horsepower can be estimated if we have the flow rate. As we discussed, flow rate is volume of flow per unit time. Therefore, if we have a flow rate of Q in gallons per minute (gpm), we can estimate the power needed to move the water.

$$\text{Water HP (WHP)} = \frac{\text{work}}{\text{time}} = \frac{V\gamma H}{\text{time}} = Q\gamma H \text{ because } \frac{V}{\text{time}} \text{ is flow rate}$$

$$\text{WHP} = Q\left(\frac{\text{gal}}{\text{min}}\right) \times H(\text{ft}) \times 62.4\,\frac{\text{lb}}{\text{ft}^3} \times \frac{1\,\text{ft}^3}{7.48\,\text{gal}} \times \frac{1\,\text{hp}}{33,000\,\frac{\text{ft-lb}}{\text{min}}}$$

$$\text{WHP} = Q\left(\frac{\text{gal}}{\text{min}}\right) \times H(\text{ft}) \times \frac{1}{3,960}$$

Water Horsepower

$$\text{WHP} = \frac{Q \times H}{3,960}$$

The flow rate (Q) is in gal/min (gpm) and the lift (H) is feet (ft).

Therefore, water horsepower is calculated using the above equation when the flow rate is in gpm and height (lift) is in feet. Dividing by 3,960 converts the units into horsepower. Also, dividing the water density 62.4 lb/ft³ by 7.48 gal/ft³ converts the water density into the units of pounds per gallon (8.34 lb/gal).

The previous calculation assumes that pumps are 100% efficient. Pumps cannot transmit all the power put into them. The horsepower supplied to the pumps needs to be greater than the water horsepower to account for the less than 100% efficiency. This power is called brake horsepower, which can be calculated as:

$$\text{Brake HP (BHP)} = \frac{Q \times H}{3,960 \times E_p}$$

Brake Horsepower

$$BHP = \frac{Q \times H}{3,960 \times E_p}$$

The flow rate (Q) is in gal/min (gpm) and the lift (H) is in feet (ft). E_p is pump efficiency.

Motor Horsepower

$$MHP = \frac{Q \times H}{3,960 \times E_p \times E_m}$$

The flow rate (Q) is in gal/min (gpm) and the lift (H) is in feet (ft).

Wire-to-Water Efficiency (E_{wtw})

$E_{wtw} = E_p \times E_m \times 100\%$

E_p = Pump Efficiency

E_m = Motor Efficiency

Where,

BHP = Brake Efficiency, hp

E_p = Efficiency of Pump (usually ranging from 50–85%, depending on type and size of pump)

E_p is expressed as a decimal fraction in the equation. That is, if the efficiency is 65%, then E_p is 0.65.

Because the motors that run the pumps are not 100% efficient, the power supplied to the motor needs to be greater than the power that the motor transmits. The motor horsepower can be calculated as:

$$\text{Motor HP} = MHP \frac{Q \times H}{3,960 \times E_p \times E_m}$$

Where,

MHP = Motor Efficiency, hp

E_m = Efficiency of Motor (usually ranging from 80–95%, depending on the type and size of motor)

The formulas above were developed based on the density of water (that is, a specific gravity of 1.0). These formulas work well for water and wastewater, but, if a different liquid is being pumped, the 3,960 factor is incorrect. The power for a different liquid can be calculated by multiplying the formulas above by the specific gravity of the liquid. The specific gravity effect is illustrated in Examples 50 and 51.

Let's illustrate the power loss due to inefficient systems.

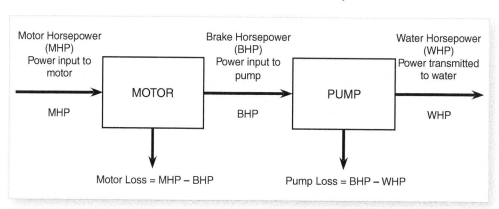

The figure illustrates the concept of wire-to-water efficiency (E_{wtw}). That is, the power delivered to the water relative to the power delivered to the motor. Therefore, the wire-towater efficiency is calculated as:

$$E_{wtw} = \frac{WHP}{MHP} \times 100\% = \frac{\dfrac{Q \times H}{3,960}}{\dfrac{Q \times H}{3,960 \times E_p \times E_m}} \times 100\% = E_p \times E_m \times 100\%$$

Therefore, when we say wire-to-water efficiency, we mean the overall efficiency of the whole system (including motor and pump efficiencies).

Example 50

A flow of 500 gpm of water is to be pumped a height of 100 feet by a pump with an efficiency of 70%. What is the pump horsepower?

$$\text{Brake HP} = \text{BHP} = \frac{Q \times H}{3,960 \times E_p} = \frac{500 \times 100}{3,960 \times 0.7} = 18.0 \, \text{hp}$$

We also could have written

$$\text{BHP} = \frac{\text{WHP}}{E_p}$$

Example 51

Find the horsepower required to pump gasoline (specific gravity = 0.75) instead of water in Example 50. A specific gravity of less than one means it is lighter than water. Therefore, we expect the power to be less.

$$\text{BHP} = \frac{500 \times 100 \times 0.75}{3,960 \times 0.7} = 13.5 \, \text{hp}$$

Example 52

Estimate the motor horsepower for Example 50, if the motor efficiency is 85%.

$$\text{Motor HP} = \text{MHP} = \frac{Q \times H}{3,960 \times E_p \times E_m} = \frac{500 \times 100}{3,960 \times 0.7 \times 0.85} = 21.2 \, \text{hp}$$

We also could have written

$$\text{MHP} = \frac{\text{BHP}}{E_m}$$

The water horsepower is 12.6 hp (verify this for practice). Because of inefficiencies, we need significantly more power to be delivered to the motor in order to extract the water horsepower needed.

Example 53

Estimate the wire-to-water efficiency for Example 52.

$$E_{wtw} = E_p \times E_m \times 100\% = 0.70 \times 0.85 \times 100\% = 59.5\%$$

Another way to calculate this efficiency is by using the WHP and MHP.

$$E_{wtw} = \frac{WHP}{MHP} \times 100\% = \frac{12.6}{21.2} \times 100\% = 59.4\%$$

A rounding error is responsible for the minor difference between the two answers. If you use 12.62 and 21.22, you get the same results.

A.7.5 Head

As illustrated in Example 47, lifting water from one elevation to another requires energy (work). In pump problems, the lifting work required per pound of water is called the static head (H_s).

Two typical conditions for lifting water are shown below.

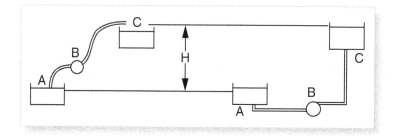

The static head is the vertical distance between water surfaces or between a water surface and the top of a pipe with an open discharge.

In Examples 50–53, if a pump were designed to deliver only head H, the water would never reach the intended point. The reason for this is the water encounters friction in the pipelines. Friction depends on the roughness and length of pipe, the pipe diameter, and the flow velocity. The turbulence caused at the pipe entrance (point A); the pump (point B); the pipe exit (point C); and at each elbow, bend, or transition also adds to these friction losses. Tables and charts are available in Section A.7.9, "Friction or Energy Losses," for calculation of these friction losses. Fiction losses may be added to the static head to obtain the total head. For short runs of pipe that do not have high velocities, friction losses are generally less than 10 percent of the static head. More detailed information on friction losses can be found in Section A.7.9, "Friction or Energy Losses."

Example 54

A pump is to be located 8 feet above a wet well and must lift 1.8 Mgal/d another 50 feet to a storage reservoir. If the pump has an efficiency of 75% and the motor an efficiency of 90%, what is the cost of the power consumed if one kilowatt hour costs 4 cents? Assume that the pump runs continuously and that the friction losses are 10% of the static head.

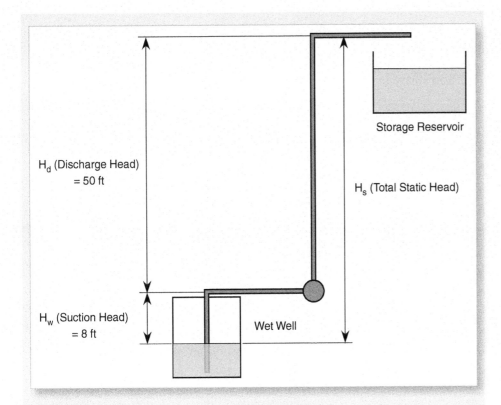

In this example, we are not given any information about the pipe sizes or any additional information about the number of elbows or bends, but we do know that the friction losses are 10% of the static head. We are also given the cost of electricity for one kilowatt hour. That means, we need to be able to convert horsepower to kilowatt hour (see Section A.7.3, "Power").

Estimate motor power needed using the following equation:

$$\text{Motor HP} = \text{MHP} = \frac{Q \times H}{3,960 \times E_p \times E_m}$$

The flow rate (Q) must be in gallons/minute. We also need to estimate the total head (H). Knowing that the pump runs 24 hours a day, we can convert Q to gallons per minute as follows:

$$Q = 1.8\,\frac{\text{Mgal}}{\text{d}} \times \frac{1,000,000\,\text{gal}}{\text{Mgal}} \times \frac{1\,\text{d}}{24\,\text{h}} \times \frac{1\,\text{h}}{60\,\text{min}} = 1,250\,\text{gpm}$$

The total static head is equal to the lift needed from the wet well to the surface plus the lift to the storage reservoir.

$$H_s = H_w + H_d = 8\text{ ft} + 50\text{ ft} = 58\text{ ft}$$

Where,

H_s = Static Head, ft

H_w = Suction Head (from the well to the surface), ft

H_d = Discharge Head (from the surface to the storage reservoir), ft

Finally, we need to add the friction losses, estimated by using the rule of 10%:

Friction Losses = $0.1 \times H_s = 0.1 \times 58$ ft = 5.8 ft

Therefore, the total head is the sum of the static head and the friction losses.

H = 58 ft + 5.8 ft = 63.8 ft

Now, we are ready to calculate the motor horsepower.

$$\text{Motor HP (MHP)} = \frac{Q \times H}{3,960 \times E_p \times E_m} = \frac{1,250 \text{ gpm} \times 63.8 \text{ ft}}{3,960 \times 0.75 \times 0.90}$$

$$= 29.8 \text{ hp, rounded to } 30 \text{ hp}$$

Convert the motor horsepower to kilowatt hours (assuming 24-hour operation). Keep in mind one horsepower is 746 watts or 0.746 kilowatts.

$$\text{Energy Consumed per Day} = E = 30 \text{ hp} \times 24 \frac{h}{d} \times 0.746 \frac{kW}{hp} = 537 \frac{kWh}{d}$$

The cost is then estimated by:

$$\text{Cost} = E \times \frac{\$0.04}{kWh} = 537 \text{ kWh} \times \frac{\$0.04}{kWh} = \$21.48/d$$

A.7.6   Pump Characteristics

Centrifugal pumps are the most common pump type in water and wastewater systems. The discharge of a centrifugal pump varies from zero to a maximum capacity, depending on the speed, head, power, and specific impeller design. The interrelation of capacity, efficiency, head, and power is known as the characteristics of the pump.

The first relation normally considered when selecting a pump is head versus capacity (flow). This relationship reflects the fact that for a given amount of power, a pump can push a lot of water against a low head but only a small flow against a high head. If plotted on a graph, the curve appears as follows:

In these curves, the head is usually labeled total dynamic head (TDH) because it includes static head, friction losses, and other components not addressed here. For our purposes, **total dynamic head** is the sum of the total static head plus friction losses.

Another important characteristic is the pump efficiency. It begins from zero at no discharge, increases to a maximum, and then drops as the capacity is increased. The following graph shows efficiency versus capacity.

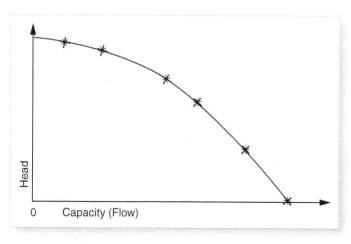

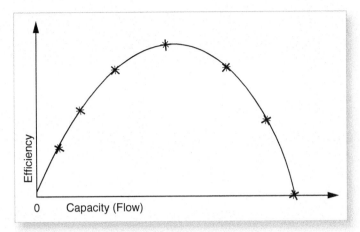

The last important characteristic is the brake horsepower or the power input to the pump. The brake horsepower usually increases with increasing capacity until it reaches a maximum, then it normally reduces slightly.

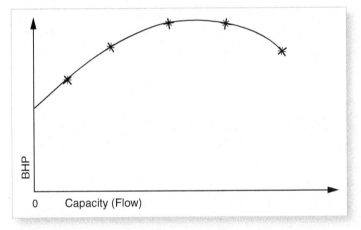

These pump characteristic curves are quite important. Pump sizes are normally picked from these curves rather than calculations. For ease of reading, the three characteristic curves are normally plotted together. A typical graph of pump characteristics is shown below.

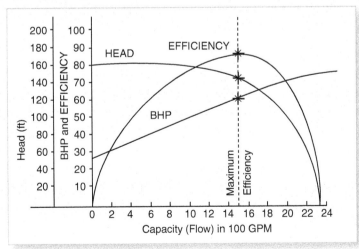

The curves show that the maximum efficiency for the particular pump in question occurs at approximately 1,475 gpm, a head of 132 feet, and a brake horsepower of 58. Operating at this point, the pump has an efficiency of approximately 85 percent. This can be verified by calculating the efficiency from the brake horsepower equation.

$$BHP = \frac{Q \times H}{3,960 \times E_p}$$

After cross multiplication:

$$BHP \times 3,960 \times E_p = Q \times H$$

Solving for E_p,

$$E_p = \frac{Q \times H}{3,960 \times BHP} = \frac{1,475 \, gpm \times 132 \, ft}{3,960 \times 58 \, hp} = 0.85 \, or \, 85\% \, (Check)$$

The preceding is only a brief description of pumps to familiarize the operator with their characteristics. The operator does not normally specify the type and size of pump needed at a plant. If a pump is needed, the operator should be able to supply the information necessary for a pump supplier or engineer to provide the best possible pump for the lowest cost. Some of the information needed includes:

1. Flow range desired
2. Head conditions
   a. Suction head or lift
   b. Pipe and fitting friction head
   c. Discharge head
3. Type of fluid pumped and temperature
4. Pump location

A.7.7   Evaluation of Pump Performance

In this section, we will discuss how operators can determine the capacity of a pump and its efficiency.

A.7.7.1   Capacity

Sometimes it is necessary to determine the capacity of a pump. This can be accomplished by determining the time it takes a pump to fill a tank or empty a portion of a wet well or diversion box when the inflow is blocked off. The best way to illustrate some of these concepts is by showing examples.

Example 55

Calculate the pumping rate or capacity of a pump. To accomplish that, we need to measure the dimensions of the wet well and then record the time it takes for the water level to drop down some given distance. Suppose the following data apply:

Dimensions of the Well (this comes from measurement)

Length = 12 ft

Width = 12 ft

Depth = 6 ft (We are planning to record the time it takes for the water level to go down by 6 feet.)

Suppose after turning the pump on, we record 3 minutes and 30 seconds for the water level to go down by 6 ft. Then we can calculate the pump capacity (Q) in gallons per minute (gpm) by:

$$Q = \frac{\text{Volume Pumped}}{\text{Time}} = \frac{12\,\text{ft} \times 12\,\text{ft} \times 6\,\text{ft}}{3.5\,\text{min}} \times \frac{7.48\,\text{gal}}{\text{ft}^3} = \frac{6{,}463\,\text{gal}}{3.5\,\text{min}}$$

$$= 1{,}846\,\text{gpm}$$

If you know the TDH and have the pump's performance curves (as shown in Section A.7.6, "Pump Characteristics"), you can determine if the pump is delivering at design capacity. After a pump overhaul, the pump's actual performance (flow, head, power, and efficiency) should be compared with the pump manufacturer's performance curves. This method of calculating the rate of filling or emptying a wet well or diversion box can also be used to calibrate flowmeters.

Example 56

For the pump in Example 55, what are its TDH and efficiency at the operating point?

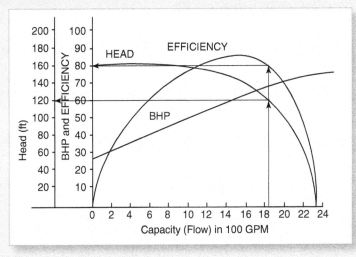

TDH and E_p can be estimated from the pump curve. For a flow of 1,846 gpm, the TDH is about 120 ft and the pump efficiency is 80%.

A.7.7.2   Efficiency

To estimate the efficiency of a pump, the total head and flow must be known. This head may be estimated by measuring the suction and discharge pressures. Assume the pressures and flow were measured as follows:

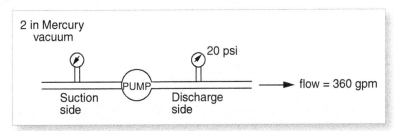

No additional information is necessary if we assume the pressure gages are at the same height and the pipe diameters are the same. Because the pipe is horizontal, there is no elevation change and therefore no change in static head. Because the pipe diameters are equal, the flow velocities entering and leaving the pump are equal. In this situation, the difference in the pressure heads is the energy provided by the pump.

Both pressure readings must be converted to feet. The pressure at the suction side of the pump is reported in inches of mercury. Therefore, we need the specific gravity of mercury, which is 13.56. That is, the pressure for an inch of mercury is the same pressure induced by a column of water 13.56 inches high. We can calculate equivalent head in feet as follows:

$$\text{Suction Head} = H_s = -2\,\text{in} \times 13.56\,\frac{\text{Water, in}}{\text{Mercury, in}} \times \frac{1\,\text{ft}}{12\,\text{in}} = -2.26\,\text{ft}$$

The negative head is because the pressure is a vacuum. The discharge pressure is reported in psi (lb/in²). We calculate the discharge head by using the density of water (see Section A.5.8, "Force and Pressure") and following proper conversions (see sidebar for conversion).

Conversion of psi to feet

Use this conversion factor to calculate feet from psi:

$$\frac{144\,\text{in}^2}{\text{ft}^2} \times \frac{\text{ft}^3}{62.4\,\text{lb}} = 2.31\,\frac{\text{ft}}{\text{psi}}$$

$$\text{Discharge Head} = H_d = 20\,\frac{\text{lb}}{\text{in}^2} \times \left(\frac{144\,\text{in}^2}{\text{ft}^2} \times \frac{\text{ft}^3}{62.4\,\text{lb}} \right) = 46.20\,\text{ft}$$

The total head (H) is the difference between the discharge head (H_d) and the suction lift (H_s):

$$H = H_d - H_s = 46.20\,\text{ft} - (-2.26\,\text{ft}) = 48.46\,\text{ft}$$

You can use the 2.31 ft/psi conversion factor as a quick way to do the calculation. The water horsepower (WHP) is calculated as:

$$\text{WHP} = \frac{Q \times H}{3,960} = \frac{360\,\text{gpm} \times 48.46\,\text{ft}}{3,960} = 4.4\,\text{hp}$$

To estimate the efficiency of the pump, we need to measure the kilowatts drawn by the pump motor. Assume the meter indicates 8,000 watts or 8 kilowatts. The manufacturer claims the electric motor is 80 percent efficient. Recall from our previous discussion that the brake horsepower (BHP) is:

$$\text{BHP} = \text{MHP} \times E_m$$

Where,

MHP = Motor Horsepower, hp

E_m = Motor Efficiency

$$\text{BHP} = 8\,\text{kW} \times 0.80 \times \frac{1\,\text{hp}}{0.746\,\text{kW}} = 8.6\,\text{hp}$$

Now we can calculate the pump efficiency. Recall from Example 50 that the brake horsepower (BHP) is:

$$BHP = \frac{WHP}{E_p}$$

By rearranging the equation above (check cross-multiplication discussed earlier):

$$E_p = \frac{WHP}{BHP} = \frac{4.4\,hp}{8.6\,hp} = 0.51 = 51\%$$

The following diagram summarizes the results:

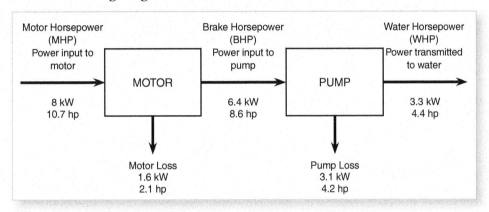

The wire-to-water efficiency is the efficiency of the power input to produce water horsepower, that is, the overall efficiency. Using values on the diagram, we can calculate the wire-to-water efficiency (E_{wtw}):

$$E_{wtw} = \frac{WHP}{Wire\,HP} \times 100\% = \frac{4.4\,hp}{10.7\,hp} \times 100\% = 41\%$$

The wire-to-water efficiency applies to the overall pumping system (pump and electric motor). If the WHP and wire HP are not known, the wire-to-water efficiency can be calculated from the current (amperage) and the voltage. Start with the same wire-to-water efficiency equation.

$$E_{wtw} = \frac{WHP}{Wire\,HP} \times 100\%$$

We know from Section A.7.4, "Horsepower," that the water horsepower (WHP) is:

$$WHP = \frac{Q \times H}{3,960}$$

Q is the flow rate in gallons per minute and H is the overall head in feet. The input horsepower (wire HP) can be estimated based on the current (amperage) and the voltage by using the following equation:

$$Wire\,HP = i \times V \times \frac{1\,hp}{746\,W} = \frac{i \times V}{746}$$

Where,

    i = Electrical current in amperes (A)

    V = Voltage in volts (V)

When the voltage is multiplied by the current, the result is power in watts. Putting the two equations together for the overall wire-to-water efficiency produces:

$$E_{wtw} = \frac{WHP}{Wire\,HP} \times 100\% = \frac{\dfrac{Q \times H}{3,960}}{\dfrac{i \times V}{746}} \times 100\% = \frac{Q \times H}{3,960 \times \dfrac{i \times V}{746}} \times 100\%$$

$$= \frac{Q \times H}{5.308 \times i \times V} \times 100\%$$

We can use the above equation to calculate the wire-to-water efficiency by using 220 volt for voltage and 36 ampere for the current.

$$E_{wtw} = \frac{Q \times H}{5.308 \times i \times V} \times 100\% = \frac{360\,gpm \times 48.46\,ft}{5.308 \times 36\,A \times 220\,V} \times 100\% = 41\%$$

A.7.8   Pump Speed–Performance Relationships

Changing the rotational speed (revolutions per minute) of a centrifugal pump will change its operating characteristics. Changing the speed of a pump changes the flow, head, and power requirements. The operating characteristics of the pump will change with speed approximately as follows:

$$Flow, Q_n = \left(\frac{N_n}{N_r}\right) Q_r$$

$$Head, H_n = \left(\frac{N_n}{N_r}\right)^2 H_r$$

$$Power, P_n = \left(\frac{N_n}{N_r}\right)^3 P_r$$

Where,

    r = Rated (by Manufacturer)

    n = Now (Current Speed Used)

    N = Pump Speed, revolutions per minute (rpm)

Pump efficiency varies somewhat with speed; therefore, these formulas are not quite correct. If speeds do not vary by more than a factor of two (if the speeds are not doubled or cut in half), the results are close enough. Other factors contributing to changes in pump characteristic curves include impeller wear and roughness in pipes.

Example 57

To illustrate these relationships, assume a pump has a rated capacity of 600 gpm at 100 feet of head, and has a power requirement of 15 hp when operating at 1,500 rpm. If the efficiency remains constant, what are the operating characteristics if the speed drops to 1,200 rpm?

Calculate the new flow rate or capacity:

$$\text{Flow}, Q_n = \left(\frac{N_n}{N_r}\right)Q_r = \left(\frac{1,200\,\text{rpm}}{1,500\,\text{rpm}}\right)(600\,\text{gpm}) = (0.8)(600\,\text{gpm})$$

$$= 480\,\text{gpm}$$

Calculate the new head:

$$\text{Head}, H_n = \left(\frac{N_n}{N_r}\right)^2 H_r = \left(\frac{1,200\,\text{rpm}}{1,500\,\text{rpm}}\right)^2 (100\,\text{ft}) = (0.64)(100\,\text{ft}) = 64\,\text{ft}$$

Calculate the new power requirement:

$$\text{Power}, P_n = \left(\frac{N_n}{N_r}\right)^3 P_r = \left(\frac{1,200\,\text{rpm}}{1,500\,\text{rpm}}\right)^3 (15\,\text{hp}) = (0.51)(15\,\text{hp}) = 7.7\,\text{hp}$$

A.7.9   Friction or Energy Losses

Whenever water flows through pipes, valves, and fittings, energy (head) is lost due to pipe friction (resistance) against pipe walls, valves and fittings, and the turbulence resulting from changing the direction of the water. Energy loss inside valves and fittings is a very complicated situation. To simplify head loss calculations, manufacturers and others have published standardized tables and figures that show the equivalent lengths of straight pipe that would produce the same head loss as various valves and fittings (Figures A.1 and A.2).

To estimate the friction or energy losses resulting from water flowing in a pipe system, we need to know water flow rate; pipe size or diameter and length; and number, size, and type of valve fittings. An easy way to estimate friction or energy losses is to follow these steps:

1. Determine the flow rate.
2. Determine the diameter and actual length of pipe.
3. Convert all valves and fittings to equivalent lengths of straight pipe (see Figure A.1).
4. Add equivalent lengths of straight pipe to the to the actual length to get a total equivalent length.
5. Estimate friction or energy losses by using Figure A.2. With the flow in gpm and diameter of pipe, find the friction loss per 100 feet of pipe.
6. Multiply this value by the total equivalent length of straight pipe.

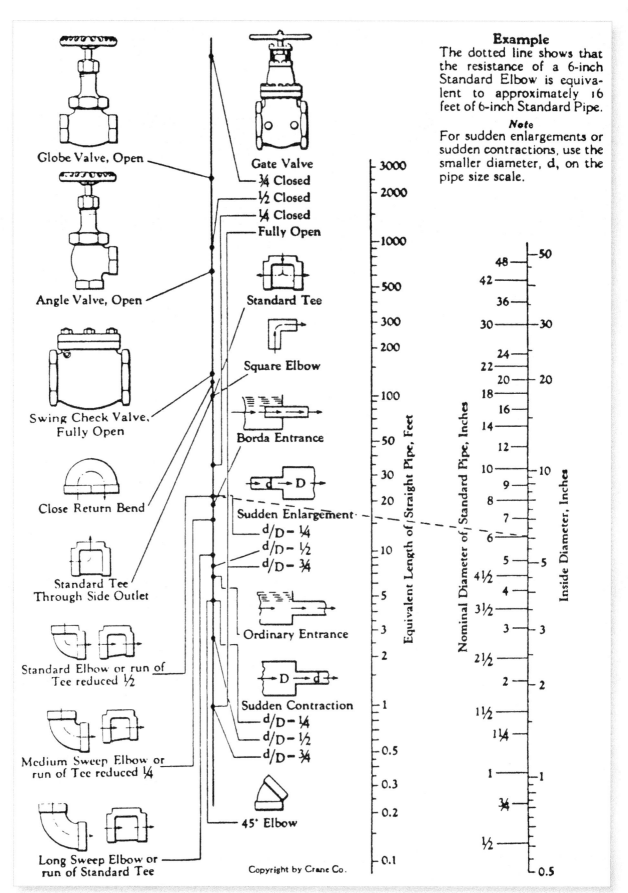

Figure A.1   Resistance of valves and fittings to flow of water *(Courtesy of Crane Company)*

US GPM	0.5 in. Vel.	0.5 in. Frict.	0.75 in. Vel.	0.75 in. Frict.	1 in. Vel.	1 in. Frict.	1.25 in. Vel.	1.25 in. Frict.	1.5 in. Vel.	1.5 in. Frict.	2 in. Vel.	2 in. Frict.	2.5 in. Vel.	2.5 in. Frict.
10	10.56	95.9	6.02	23.0	3.71	6.86	2.15	1.77	1.58	.83	.96	.25	.67	.11
20	....	....	12.0	86.1	7.42	25.1	4.29	6.34	3.15	2.94	1.91	.87	1.34	.36
30	....	....	....	....	11.1	54.6	6.44	13.6	4.73	6.26	2.87	1.82	2.01	.75
40	....	....	....	....	14.8	95.0	8.58	23.5	6.30	10.79	3.82	3.10	2.68	1.28
50	....	....	....	....	....	....	10.7	36.0	7.88	16.4	4.78	4.67	3.35	1.94
60	....	....	....	....	....	....	12.9	51.0	9.46	23.2	5.74	6.59	4.02	2.72
70	....	....	....	....	....	....	15.0	68.8	11.03	31.3	6.69	8.86	4.69	3.63
80	....	....	....	....	....	....	17.2	89.2	12.6	40.5	7.65	11.4	5.36	4.66
90	....	....	....	....	....	....	....	....	14.2	51.0	8.60	14.2	6.03	5.82
100	....	....	....	....	....	....	....	....	15.8	62.2	9.56	17.4	6.70	7.11
120	....	....	....	....	....	....	....	....	18.9	88.3	11.5	24.7	8.04	10.0
140	....	....	....	....	....	....	....	....	....	....	13.4	33.2	9.38	13.5
160	....	....	....	....	....	....	....	....	....	....	15.3	43.0	10.7	17.4
180	....	....	....	....	....	....	....	....	....	....	17.2	54.1	12.1	21.9
200	....	....	....	....	....	....	....	....	....	....	19.1	66.3	13.4	26.7
220	....	....	....	....	....	....	....	....	....	....	21.0	80.0	14.7	32.2
240	....	....	....	....	....	....	....	....	....	....	22.9	95.0	16.1	38.1
260	....	....	....	....	....	....	....	....	....	....	....	....	17.4	44.5
280	....	....	....	....	....	....	....	....	....	....	....	....	18.8	51.3
300	....	....	....	....	....	....	....	....	....	....	....	....	20.1	58.5
350	....	....	....	....	....	....	....	....	....	....	....	....	23.5	79.2

US GPM	3 in. Vel.	3 in. Frict.	4 in. Vel.	4 in. Frict.	5 in. Vel.	5 in. Frict.	6 in. Vel.	6 in. Frict.	8 in. Vel.	8 in. Frict.	10 in. Vel.	10 in. Frict.	12 in. Vel.	12 in. Frict.	14 in. Vel.	14 in. Frict.	16 in. Vel.	16 in. Frict.	18 in. Vel.	18 in. Frict.	20 in. Vel.	20 in. Frict.
20	.91	.15																				
40	1.82	.55	1.02	.13																		
50	2.72	1.17	1.53	.28																		
80	3.63	2.02	2.04	.48	1.28	.14	.91	.06														
100	4.54	3.10	2.55	.73	1.60	.20	1.13	.10														
120	5.45	4.40	3.06	1.03	1.92	.29	1.36	.13														
140	6.35	5.93	3.57	1.38	2.25	.38	1.59	.18														
160	7.26	7.71	4.08	1.78	2.57	.49	1.82	.23														
180	8.17	9.73	4.60	2.24	2.89	.61	2.04	.28														
200	9.08	11.9	5.11	2.74	3.21	.74	2.27	.35														
220	9.98	14.3	5.62	3.28	3.53	.88	2.50	.42	1.40	.10												
240	10.9	17.0	6.13	3.88	3.85	1.04	2.72	.49	1.53	.12												
260	11.8	19.8	6.64	4.54	4.17	1.20	2.95	.57	1.66	.14												
280	12.7	22.8	7.15	5.25	4.49	1.38	3.18	.66	1.79	.16												
300	13.6	26.1	7.66	6.03	4.81	1.58	3.40	.75	1.91	.18												
350	....	....	8.94	8.22	5.61	2.11	3.97	1.01	2.24	.24												
400	....	....	10.20	10.7	6.41	2.72	4.54	1.30	2.55	.30												
450	....	....	11.45	13.4	7.22	3.41	5.11	1.64	2.87	.38	1.84	.12										
500	....	....	12.8	16.6	8.02	4.16	5.67	2.02	3.19	.46	2.04	.15	1.42	.06								
550	....	....	14.0	19.9	8.82	4.98	6.24	2.42	3.51	.56	2.25	.18	1.56	.07								
600	....	....	....	....	9.62	5.88	6.81	2.84	3.83	.66	2.45	.21	1.70	.08	1.25	.04						
700	....	....	....	....	11.2	7.93	7.94	3.87	4.47	.88	2.86	.29	1.99	.12	1.46	.05						
800	....	....	....	....	12.8	10.22	9.08	5.06	5.11	1.14	3.27	.37	2.27	.15	1.67	.07						
900	....	....	....	....	14.4	12.9	10.2	6.34	5.74	1.44	3.68	.46	2.55	.18	1.88	.09						
1000	....	....	....	....	....	....	11.3	7.73	6.38	1.76	4.09	.57	2.84	.22	2.08	.11						
1100	....	....	....	....	....	....	12.5	9.80	7.02	2.14	4.49	.68	3.12	.27	2.29	.13						
1200	....	....	....	....	....	....	13.6	11.2	7.66	2.53	4.90	.81	3.40	.32	2.50	.15	1.91	.08				
1300	....	....	....	....	....	....	14.7	13.0	8.30	2.94	5.31	.95	3.69	.37	2.71	.17	2.07	.09				
1400	....	....	....	....	....	....	....	....	8.93	3.40	5.72	1.09	3.97	.43	2.92	.20	2.23	.10				
1500	....	....	....	....	....	....	....	....	9.57	3.91	6.13	1.25	4.26	.49	3.13	.23	2.34	.12				
1600	....	....	....	....	....	....	....	....	10.2	4.45	6.54	1.42	4.54	.55	3.33	.25	2.55	.13	2.02	.07		
1700	....	....	....	....	....	....	....	....	10.8	5.00	6.94	1.60	4.87	.62	3.54	.29	2.71	.15	2.15	.08		
1800	....	....	....	....	....	....	....	....	11.5	5.58	7.35	1.78	5.11	.70	3.75	.32	2.87	.16	2.27	.09		
1900	....	....	....	....	....	....	....	....	12.1	6.19	7.76	1.97	5.39	.77	3.96	.35	3.03	.18	2.40	.10		
2000	....	....	....	....	....	....	....	....	12.8	6.84	8.17	2.17	5.67	.86	4.17	.39	3.19	.20	2.52	.11		
2500	....	....	....	....	....	....	....	....	....	....	10.2	3.38	7.10	1.33	5.21	.60	3.99	.31	3.15	.17		
3000	....	....	....	....	....	....	....	....	....	....	12.3	4.79	8.51	1.88	6.25	.86	4.79	.44	3.78	.24	3.06	.14
3500	....	....	....	....	....	....	....	....	....	....	14.3	6.55	9.93	2.56	7.29	1.16	5.58	.58	4.41	.32	3.57	.19
4000	....	....	....	....	....	....	....	....	....	....	....	....	11.3	3.31	8.34	1.50	6.38	.75	5.04	.42	4.08	.24
4500	....	....	....	....	....	....	....	....	....	....	....	....	12.8	4.18	9.38	1.88	7.18	.95	5.67	.53	4.59	.31
5000	....	....	....	....	....	....	....	....	....	....	....	....	14.7	5.13	10.4	2.30	7.98	1.17	6.30	.65	5.11	.38
6000	....	....	....	....	....	....	....	....	....	....	....	....	....	....	12.5	3.31	9.57	1.66	7.56	.92	6.13	.53
7000	....	....	....	....	....	....	....	....	....	....	....	....	....	....	14.6	4.50	11.2	2.26	8.83	1.24	7.15	.72
8000	....	....	....	....	....	....	....	....	....	....	....	....	....	....	....	....	12.8	2.96	10.09	1.61	8.17	.94
9000	....	....	....	....	....	....	....	....	....	....	....	....	....	....	....	....	14.4	3.73	11.3	2.02	9.19	1.18
10000	....	....	....	....	....	....	....	....	....	....	....	....	....	....	....	....	....	....	12.6	2.48	10.2	1.45

No allowance has been made for age, differences in diameter, or any other abnormal condition of interior surface. Any Factor of Safety must be estimated from the local conditions and the requirements of each particular installation. For general purposes, 15% is a responsible Factor of Safety.

Figure A.2 Friction loss for water in feet per 100 feet of pipe (*Courtesy of Hydraulic Institute*)

The procedure for using Figure A.1 is very easy. Locate the type of valve or fitting you wish to convert to an equivalent pipe length; find its diameter on the right-hand scale; and draw a straight line between these two points. The equivalent length of straight pipe can be read where your line crosses the scale in the center of the page. For example, Figure A.1 shows that the equivalent pipe length for a sudden enlargement in a 6-inch diameter pipe is a little less than 17 feet. (Be careful about reading the nonlinear scale.)

Example 58

Estimate the friction losses in the piping system of a pump station when the flow is 1,000 gpm. The 8-inch suction line is 10 feet long and contains a 90-degree bend (long sweep elbow) and a gate valve. The 6-inch discharge line is 30 feet long and contains a check valve, a gate valve, and three 90-degree bends (medium sweep elbows). First obtain the equivalent lengths for the fittings from Figure A.1.

Suction Line (8-Inch Diameter)

Item	Equivalent Length, ft
1. Length of pipe	10
2. 90-degree bend (long sweep)	14
3. Gate valve (fully open)	4
4. Ordinary entrance	12
Total equivalent length	40 ft

Friction loss (Figure A.2) = 1.76 ft/100 ft of pipe

Discharge Line (6-Inch Diameter)

Item	Equivalent Length, ft
1. Length of pipe	30
2. Check valve (fully open)	40
3. Gate valve (fully open)	4 (actually 3.5 rounded to 4)
4. Three 90-degree bends (medium sweep)	42 (= 3 × 14 each)
Total equivalent length	116 ft

Friction loss (Figure A.2) = 7.73 ft/100 ft of pipe

Estimate the total friction losses in the pumping system for a flow of 1,000 gpm.

Loss in suction piping   = (1.76 ft/100 ft)(40 ft)   = 0.7 ft

Loss in discharge piping = (7.73 ft/100 ft)(116 ft) = 9.0 ft

Total Friction Losses, ft   = 9.7 ft

A.8 Analysis and Presentation of Data

In the water and wastewater industries, data are collected for decision making related to plant or system performance, informing stakeholders, meeting permit requirements, and other needs. Stakeholders include the public, internal plant staff, system administrators, and regulatory authorities. This section provides you with tools and methods to accurately analyze data. A discussion on presenting data is also included to show you how to make the results of your testing meaningful and easy to interpret and share.

As a professional operator, you may need to explain what is happening at your treatment plant to fix something or to train someone. You may have to request a budgetary increase to support needed operational repairs, upgrades, or improvements that can increase capacity, performance, or cost savings. You can better show what is happening or support your requests and see them approved if you are able to establish and show trends and changes within your system in a way that your intended audience can understand.

Using data effectively generally involves five steps:

Metric Calculations

Metric calculations are not included in this section because the procedures used to analyze and present data are the same for both US and metric systems.

Planning Data Gathering
Different individuals may be involved in each step, so data gathering must be planned and executed properly to ensure the effective application of results.

Collecting Representative Samples
Collecting the right samples to answer a question or to understand a problem also takes advance planning. The samples collected must be **representative**.

Testing Samples in Laboratory
Samples must be tested in a laboratory in a carefully prescribed or standardized manner so the results are accurate and reliable.

Analyzing/Interpreting Data
Lab results (or other data such as flows) must be analyzed and interpreted correctly.

Communicating Results
The results must be communicated effectively so stakeholders understand the meaning of the data and can make use of it.

A.8.1 Causes of Variations in Results

When you collect samples of wastewater or receiving water and measure their characteristics (for instance, temperature, pH, BOD), your results will be affected by several factors. Three principal factors that must be taken into account are:

1. Actual variations in the characteristics of the water or material being examined
2. Sampling procedures
3. Testing or analytical procedures

representative sample
A representative sample is a small portion of material or liquid that has content or consistency that are the same as (or as close as possible to) that of the larger body or volume being sampled.

A.8.1.1   Water or Material Being Examined

The properties or characteristics of the wastewater or receiving water (such as temperature, pH, or BOD) are what you are attempting to measure. These and many other water quality indicators vary continuously depending on what is being discharged into a wastewater collection system, the effectiveness of treatment processes, and the response of the receiving waters based on their changing characteristics (such as temperature or flow). Your objective is to describe the characteristics of the wastewater or receiving water being sampled in terms of average values and to give an indication of variation or the spread of results around the average values.

A.8.1.2   Sampling

Characteristics of a sample can vary even if you always sample at the same location or during the same time of day. Some variations are due to uneven flow in channels and pipes. If you look, you can see differences in characteristics at various depths at different locations (middle versus edge of a pipe or channel), and above or below bends, gates, or valves.

After a sample has been collected in a sampling jar or bottle, it must be treated correctly. Heavy material may settle to the bottom, so the jar must be mixed before the sample is tested. If the sample is not analyzed immediately, its characteristics can undergo chemical or biological changes unless the sample is treated and stored properly following collection. For more information on correct sampling and preservation techniques see "Laboratory Procedures and Chemistry" in *Operation of Wastewater Treatment Plants*, Volume II.

A.8.1.3   Testing

The results from two identical samples can differ depending on the analyzing apparatus and the operator conducting the measurement. Fluctuations in voltage can cause changes in instrument readings, and different individuals may titrate to slightly different end points. Using reagents from different bottles, filter paper from different packages, or different pieces of equipment that were not calibrated identically or were not warmed up during the same time period can cause differences in test results. Variations in test results may be caused by omitting a step in the lab procedure, and interfering substances can cause testing errors.

A.8.2   Controlling Variation

To control variations due to errors, operators follow standard procedures and protocols in sample collections. When reading meters whether they are analog or digital, one has to establish a standard procedure for reading the meters to produce consistent results for whatever is being measured for the water sample. For example, manometers or gauges that are typically used in wastewater treatment plants to measure pressure and pressure differences must be calibrated regularly and reset to zero before they are used. Your objective is to reduce sampling and testing errors as much as possible so you can obtain an accurate description of the water being sampled.

Calibration and zeroing of any instrument means periodically checking the instrument against a known standard to be sure the installed instrument reads properly. Manometers and gauges can be zeroed by making sure the instruments read zero when no pressure is being applied to the manometer or gauge. If the reading is not zero, then the scales should be adjusted according to manufacturer's recommendations.

To read a manometer, note the scale reading opposite the low or high point of the **meniscus**. This reading may have to be converted from inches of mercury to head in feet of water, pressure in psi, or flow in gpm, depending on the use of the manometer.

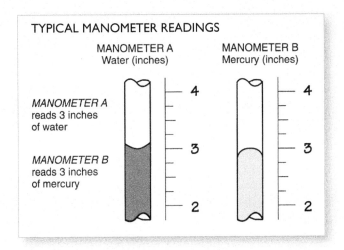

Gauge readings are read directly from a scale behind a gauge pointer and the units must be recorded. Sometimes a gauge will have two scales and care must be taken to be sure the proper scale reading and units are recorded.

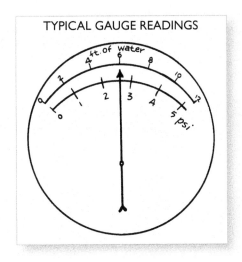

Even digital instruments may not display a steady reading, a standard operating procedure may be needed to record visually the "central" number. Developing standard operating procedures minimizes the difference among the various operators collecting the samples and recording the measurements.

meniscus (meh-NIS-cuss)
The curved surface of a column of liquid (water, oil, mercury) in a small tube. When the liquid wets the sides of the container (as with water), the curve forms a valley. When the confining sides are not wetted (as with mercury), the curve forms a hill or upward bulge.

A.8.2.1   Reading Charts

Another source of variation is reading numbers off of charts and strip recorders. The recorded data are typically transferred from recording devices to a tabular format for further analysis. Depending on how this transfer occurs, it could be a source of variation in the data. For example, if the data are extracted at certain intervals from continuously recorded data, then that transfer of data introduces a source of error and variation. In some case, recorders measure a variable such as depth of flow in a Parshall flume and then the depth is converted to flow rate. This conversion can usually be done using conversion charts provided by manufacturers. Example data is shown first as a chart and then in tabular form.

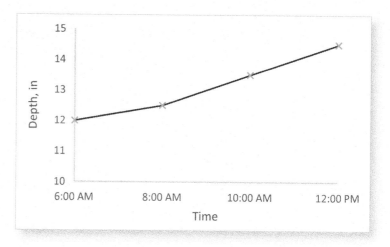

Time	Depth, in[a]	Flow[b], gpd[c]
6:00 am	12.0	0.61
8:00 am	12.5	0.65
10:00 am	13.5	0.74
12:00 pm	14.5	0.84

[a]Multiply the depth in inches by 2.54 to obtain the depth in centimeters.

[b]The flow numbers are obtained from the manufacturer's conversion chart for the flume or flowmeter in use.

[c]Multiply the flow in gpd by 3.785 to obtain the flow in liters per day (L/d).

A.8.3   Describing Data or Results

As mentioned previously, as a professional operator, you will deal with data. In some cases you may have to deal with large data sets. To make sense of the data and to understand what the data tell us, an operator needs to learn techniques to summarize data and communicate the main points of the data. Generally, we have two ways to present summaries of data: (1) graphical representations of data, and (2) computing numerical properties of the data such as mean and median. Before we discuss these two general methods, let's observe the data shown on the sidebar. The data set represents two weeks of BOD measurements. The idea of variation that we discussed earlier is represented in this data set. Meaning, if there are no measurement errors and there is no change in BOD, we will get exactly the same reading of BOD every time we measure BOD at that location. However, as we observe from the measurements that is not the case. Some values are relatively high and some values are low (compared

Two-Week BOD Measurements, mg/L

160
155
160
160
180
165
155
170
160
165
155
150
145
160

to other measurements in the data set). We can see that many of the BOD measurements are around 160. We see some measurements are less than 160 and some are more than 160. These quick observations about the measurements are easier to make when we have a small number of measurements (in this case 14). It gets more challenging to make observations when we have larger numbers of measurements, for instance 100 or more. This motivates operators to look for other ways to present and learn from measurements.

A.8.3.1   Graphs and Charts

Graphs and bar charts make presenting data easier and help users interpret the data and understand its meaning if prepared in a way that converts data into useful information. Professional operators routinely deal with data and need to be able to understand what is needed and what tools are available to prepare visual versions of data. There are many tools and software applications that make preparing graphs and charts easy. Operators are encouraged to become familiar with Microsoft Excel™ (MS Excel™) and other tools if they need to prepare graphs and charts. In this section, we will focus on several kinds of charts and illustrate how to create and use them.

In most cases, it may not be obvious how the measurements that make up the data set are distributed. The kinds of questions we can answer by look at the distribution include: Are all the data tightly clustered around a single value or very scattered? Are the data lumped at the low and high ends of a range (bimodal)? Looking at the two-week BOD measurements shown in the table to the right, it is difficult to visualize the distribution. However, in the graph below, we can see that most of the data are near 160, and that the data range from about 145 to 180. A presentation like this is called a histogram. Histograms, which are usually drawn as bar graphs, show the frequency (number of data points in a range) on the y-axis and the data values (in this case, BOD) on the x-axis. Next, we will discuss how to create histograms.

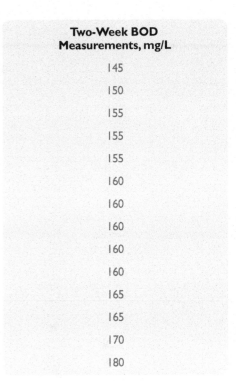

Two-Week BOD Measurements, mg/L
145
150
155
155
155
160
160
160
160
160
165
165
170
180

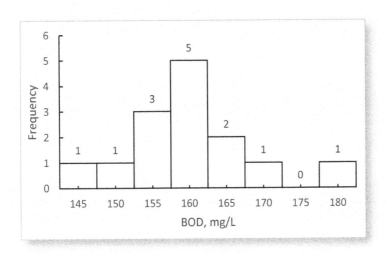

CREATING HISTOGRAMS

As discussed previously, there are software applications that can create histograms. To illustrate the process, however, we will create one manually. Our goal is to create intervals that include all of the data. Typically, the number of intervals (or bins) is between 5 and 10. In the histogram above we have 8 intervals. Here are the steps to recreate it:

1. Arrange the data (measurements) in ascending order (from smallest to largest). Refer to the table in the sidebar.
2. Calculate the **range**. The range is defined as the difference between the largest number and the smallest: $180 - 145 = 35$.
3. Choose a number of intervals (bins) between 5 and 10. We will start with 8. The width of each of these intervals will be the range divided by the number of intervals minus one. In this case, $35/(8 - 1) = 5$. It is best for the bin width to be a whole number. Adjust the number of bins to produce a whole number.
4. Start with the smallest number in the data set (in this case 145). Your first interval will be centered on 145 (we call this midpoint of an interval). The next midpoint is 150 (add five to the previous midpoint). We continue to calculate the midpoints this way until we have 8 midpoints at 145, 150, 155, 160, 165, 170, 175, and 180.
5. With the midpoints calculated, you need to calculate the boundaries of each interval. Start with the first midpoint of 145 and go half of the width below and above of the midpoint. That is, half of the width is $5/2 = 2.5$ (half the bin width). Therefore, the limits of the first midpoint (145) will be:
   Lower Limit $= 145 - 2.5 = 142.5$
   Upper Limit $= 145 + 2.5 = 147.5$

Continue creating all the intervals as shown in the table below:

Lower Limit	Upper Limit	Midpoint
142.5	147.5	145
147.5	152.5	150
152.5	157.5	155
157.5	162.5	160
162.5	167.5	165
167.5	172.5	170
172.5	177.5	175
177.5	182.5	180

NOTE: In MS Excel™ these columns can be created easily.

6. Once all the bin boundaries are created, we want to make sure that the bins include all our measurements. Our bins (intervals) start at 142.5 and end at 182.5. Our data go from 145 to 180, so we can proceed. Now for each interval, count how many data points fall within that interval. This count is called the frequency. After completing the counts, we get the following table. The table indicates that we have one measurement between 142.5 and 147.5, one measurement between 147.5 and 152.5, three measurements between 152.5 and 157.5, and so on.

range
The range of a set of measurements is calculated as the difference between the largest and the smallest values.

Lower Limit	Upper Limit	Midpoint	Frequency
142.5	147.5	145	1
147.5	152.5	150	1
152.5	157.5	155	3
157.5	162.5	160	5
162.5	167.5	165	2
167.5	172.5	170	1
172.5	177.5	175	0
177.5	182.5	180	1

7. We can now create the histogram shown previously by plotting the frequency versus midpoints on a bar graph. If you use MS Excel™, you can plot the data in a bar chart format using the "Column Chart" option under the "Insert" tab on the main menu.

Sometimes the points on a bar graph are connected to form a smooth curve rather than bars. The resulting curve describes the distribution of the data as shown in the following graphs. If the data distribution is a symmetrical bell-shaped curve (often called a "normal" distribution), the center or middle of the data as determined by the mean, median, and mode will be approximately the same (the single line in the first graph). If the distribution is skewed (not symmetrical), the mean, median, and mode values will be different (see the second graph). We discuss mean, median, and mode in detail later in this section.

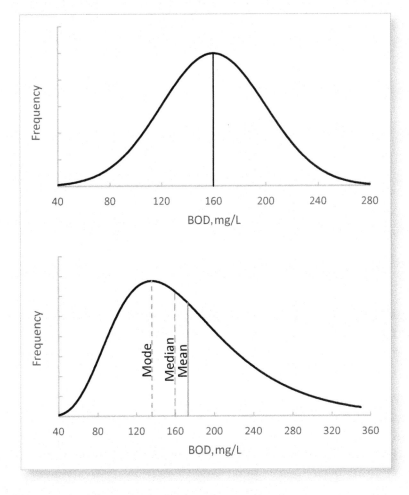

Two-Week BOD Measurements, mg/L
160
155
160
160
180
165
155
170
160
165
155
150
145
160

DOT DIAGRAMS AND STEM-AND-LEAF PLOTS

Dot diagrams are also useful to visualize the distribution of data. For example, referring to the previously used data for BOD, we can plot the data as shown below:

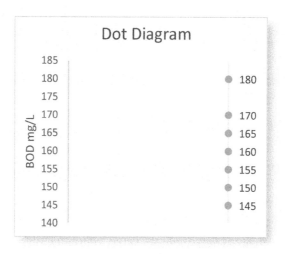

The dot diagram is created by plotting one dot for each unique data value (number). This shows the range or spread of the data. One disadvantage of the dot diagram is that we cannot see how many repeated measurements we have. For example, in the dot diagram we see a dot for 160 mg/L, but the diagram does not tell us that there are five measurements of 160.

As the following figure shows, the stem-and-leaf plot shows the plot similar to the dot diagram, but every measurement get its own dot.

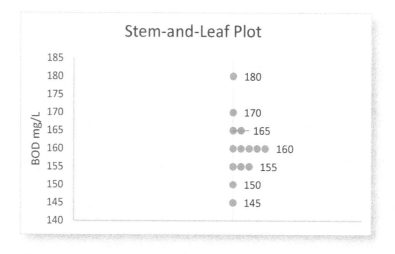

With the stem-and-leaf plot, we know that there are five repeated measurements of 160, three repeated measurements of 155, and two repeated measurements of 165. Note how the shape looks like the bar graph discussed above.

SCATTER PLOTS

Scatter plots (also called xy or x-y plots) are a useful way to show relationships or trends between two measurements or variables. For example, consider the data shown below for the mean daily flow for six years.

Year	2010	2011	2012	2013	2014	2015
Mean Daily Flow (MGD)	1.25	1.38	1.42	1.58	1.65	1.71

The following graph shows the scatter plot of the data.

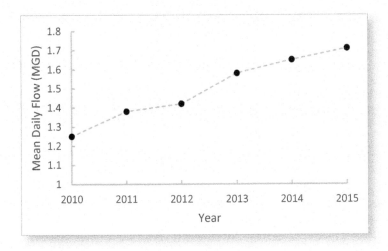

The dashed lines connecting the points have no meaning. The connecting lines are placed there to illustrate the trend or relationship.

The graph shows a continuously increasing mean daily flow. If your plant has a capacity of 2 MGD, the graph would clearly show the need for expansion in the near future if past trends in population growth or industrial expansion continue. You could extend the trend to project future flows and predict when you expect to reach plant capacity (the dark line in the following figure). You can also extend a trend using only the most recent data. The gray line in the following graph shows the projected trend based on only the last three years.

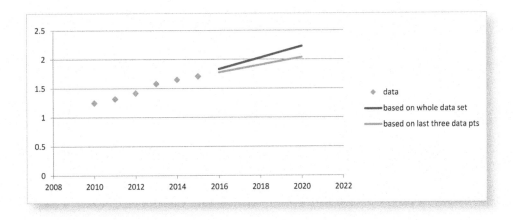

As you can see, a plot like this will be a useful tool to communicate how the plant capacity could be reached by 2020, which helps in planning for expansion.

Plotting data and looking for trends may be helpful to indicate broken pipes and illegal connections or discharges. Make a habit of attempting to identify the cause of a trend. For example, if you plot your wastewater treatment plant data and see a reduction in effluent quality every Friday or Saturday, you might start looking for an industry that cleans up on Friday afternoon and dumps a slug of waste into the collection system that reaches your plant on Friday night.

To recap, scatter plots are good illustrations of trends. To look for or show a trend, plot the value you are interested in (for instance, flow, MGD, or effluent BOD) against time (day, week, month, year). The previous plot shows the trend of flow increasing with time. The moving average can also be useful for establishing trends, especially in xy plots (see Section A.8.4, "Moving Averages"). Such plots are typically created using MS Excel™ spreadsheets.

A.8.3.2   Numerical Representation of Data

In this section we will discuss methods and calculations of numbers that tell us about the data. In general, we can classify these calculations as the "central tendencies" or "dispersion." The central tendency of a set of data represent the middle or center of the data. The most common ways of representing the center of the data are mean (average), median, mode, and geometric mean. We will define these terms in the following sections. The measurement that we call dispersion are numbers that tell us about variation in the data such as how spread out the measurements are or whether the measurements are clustered close to each other or spread very far from each other. Variance, standard deviation, and range are typical numbers that are calculated to represent the spread of data or measurements. We will discuss these terms in this section.

CENTRAL TENDENCY

As mentioned previously, central tendency describes the center or middle of the data. Central tendency is usually represented by the arithmetic mean (also called mean or average), median, mode, or geometric mean.

ARITHMETIC MEAN OR AVERAGE. When you collect representative samples from a plant influent and measure a particular water quality indicator, such as BOD, the results (data) are not always the same. For example, you might measure the BOD of a trickling filter influent to determine the organic loading and find the BOD varying considerably during a six-day period. To calculate an expected daily organic loading, the arithmetic mean (mean or average) of the daily BOD must be calculated.

Example 59

The results of BOD tests on trickling filter influent samples collected at daily intervals during six-day period indicated the BOD to be 150, 200, 250, 200, 100, and 120 mg/L. What is the mean daily BOD?

First organize the data in a table so we can see the data easily (see the table in the sidebar). Add up all the numbers.

$$Sum = 150 + 200 + 250 + 200 + 100 + 120 = 1,020 \frac{mg}{L}$$

A shorthand for representing the sum is:

$$Sum = \sum_{i=1}^{6} BOD_i$$

Day	BOD
1	150
2	200
3	250
4	200
5	100
6	120

The day is indicated by i (it is called the subscript). Therefore, i goes from 1 to 6 because we collected samples for six days. The mean is calculated by dividing the sum by the number of samples, 6 in this case. Therefore, the mean daily BOD is:

$$\text{Mean Daily BOD} = \frac{1,020}{6} = 170\,\frac{mg}{L}$$

To generalize this, we will call the water quality indicator X (in this example, X is BOD). We will call the number of samples n (6 in this example). Then the mean can be represented as $\bar{X}$. That is,

$$\bar{X} = \frac{\sum_{i=1}^{n} X_i}{n} \quad \text{or} \quad \frac{1}{n}\sum_{i=1}^{n} X_i$$

The mean value of any other characteristic (variable) is calculated the same way. For example, if you wanted to calculate the monthly mean of daily flows into a plant, you would add up the daily flows for the month and divide by the number of days in the month. The annual mean of daily flows would be the sum of the daily flows for the whole year divided by 365.

Don't be confused by the terminology. An annual mean is calculated from a year's supply of data. The word in front of "mean" tells you the time interval over which the mean was calculated. The data may also have a time interval, for example daily flows (365), weekly flows (52), or monthly flows (12). When there is a time word on the data, it refers to the time interval of the data points. For example, daily flow is the flow "per day." Daily receipts would be the receipts collected each day. You might be interested in the annual average of daily receipts, or if business is seasonal, the different monthly averages of daily receipts.

MEDIAN AND MODE. Sometimes the mean value is not the best way to describe or analyze data. For example, frequently when running multiple-tube coliform bacteria tests you will obtain some extremely high most probable numbers (MPNs) of coliform group bacteria, especially after a rain, equipment failure, or chlorine dosage mishap.

Example 60

Calculate the average MPN given the following data:

MPN/100 mL: 240 220 240 230 240 7,200 260 250 270 300 250

First, find the arithmetic mean by calculating the sum and then dividing by the number of data points (in this case n = 11). We will refer to the measurement as Y (this is just a label to make it easier to write the equation down as we did previously using X).

$$\text{Median}\,\frac{MPN}{100\,mL} = \bar{Y} = \frac{\sum_{i=1}^{10} Y_i}{11}$$

$$= \frac{240 + 220 + 240 + 230 + 240 + 7,200 + 260 + 250 + 270 + 300 + 250}{11}$$

$$= \frac{9,700}{11}$$

$$= 882\,\frac{MPN}{100\,mL}$$

MPN/100 mL
240
220
240
230
240
7,200
260
250
270
300
250

The mean value, 882 MPN/100 mL, is greater than all the readings except one. In this case, the mean is not a good indicator of the central tendency. An alternative indicator is the median.

The **median** is defined as the middle measurement when the measurements are ranked in order of magnitude. Calculate the median of these measurements. The first step is to rank the data either in ascending order (increasing or from smallest to largest) or in descending order (decreasing or from largest to smallest). Once the measurements are ranked, select the one in the middle. For this example, the solution will be:

MPN/100 mL	220	230	240	240	240	250	250	260	270	300	7,200
Rank	1	2	3	4	5	6	7	8	9	10	11

The middle (shaded cell) or median has exactly five numbers below it and five numbers above it. The median is 250, which indicates the central tendency much better than the mean which was 882. As you can see, most numbers are close to 250. This is the reason we sometimes report the median for multiple-tube coliform results.

What if you had only 10 measurements? Here is a similar but smaller data set shown ranked from largest to smallest:

MPN/100 mL	210	230	240	240	240	250	260	260	280	320
Rank	1	2	3	4	5	6	7	8	9	10

In this case, the middle is between the two shaded numbers. You have five numbers below and five numbers above. The median in this case is the average of the two numbers on each side of the middle which is:

$$\text{Median}\,\frac{\text{MPN}}{100\,\text{mL}} = \frac{240 + 250}{2} = 245\,\frac{\text{MPN}}{100\,\text{mL}}$$

The median is denoted as $\tilde{X}$, so we can generalize the approach for calculating the median as follows. The first step is to arrange the data in ascending or descending order. If n is the number of data points or measurements, then:

When n is odd $\tilde{X} = X_{\frac{n+1}{2}}$

When n is even $\tilde{X} = \dfrac{X_{\frac{n}{2}} + X_{\frac{n}{2}+1}}{2}$

$X_{\frac{n}{2}}$ represents the measurement that is located at the rank of $\left(\dfrac{n}{2}\right)$.

Remember, the data are ranked after they are arranged by magnitude.

The next step is to count the data points starting with the smallest value. For example, X_4 is the fourth data point from the left in the previous table. Use this approach to perform the same calculation as shown earlier in the example.

In the first data set in this example, n = 11, an odd number. So the median will be:

$$\tilde{X} = X_{\frac{n+1}{2}} = X_{\frac{11+1}{2}} = X_{\frac{12}{2}} = X_6 = 250$$

Median Versus Mean

The median is less influenced by a few extreme values in the data than the mean. To illustrate this, there are two ways to calculate the mean and median of the first data set: Case 1—estimate the median and mean for the full data set, and Case 2—estimate the median and mean for the same data set but dropping the largest number (7,200). Our estimates will be:

	Case 1	Case 2
Median	250	245
Mean	882	250

The mean estimate is influenced significantly by including one very large number. However, the median estimates are very close to each other with or without the large number.

Therefore, select the sixth number in the previous table. For the second case when n = 10 (an even number), the median is:

$$\tilde{X} = \frac{X_{\frac{n}{2}} + X_{\frac{n}{2}+1}}{2} = \frac{X_{\frac{10}{2}} + X_{\frac{10}{2}+1}}{2} = \frac{X_5 + X_6}{2}$$

Same as above, you average the fifth and sixth number. The median is 245.

Another measure of central tendency is the mode. The mode is the measurement that occurs most frequently. In our example with 10 data points, the measurement 240 occurs three times, which is more than any other. Therefore: mode MPN/100 mL = 240.

The data sets from earlier in the example show that the median and mode do a better job of describing or predicting the MPN value than the mean. For this reason, these representations are sometimes used to report MPN data.

The mean, median, and mode are calculated to understand how the data are distributed relative to each other. If you refer to the histograms shown on page 705, you will observe that for symmetrical distribution the mean, median, and mode are the same or very close to each other. For asymmetrical data and distributions, the mean, median, and mode will be different.

GEOMETRIC MEAN. The geometric mean is calculated in the following manner (using Example 60):

$$X_G = \left(X_1 \times X_2 \times X_3 \times X_4 \times X_5 \times X_6 \times X_7 \times X_8 \times X_9 \times X_{10} \times X_{11} \right)^{\frac{1}{11}}$$

$$= \left(220 \times 230 \times 240 \times 240 \times 240 \times 250 \times 250 \times 260 \times 270 \times 300 \times 7{,}200 \right)^{\frac{1}{11}}$$

$$= \left(6.6291 \times 10^{27} \right)^{\frac{1}{11}} = 338 \frac{MPN}{100\,mL}$$

The general formula for the geometric mean is:

$$X_G = \left(X_1 \times X_2 \times ... \times X_n \right)^{\frac{1}{n}} = \left(\prod_{i=1}^{n} X_i \right)^{\frac{1}{n}}$$

The symbol $\prod$ is shorthand for multiplying the numbers, while the symbol Σ indicates adding the numbers. See sidebar for using a calculator to solve for the geometric mean.

Example 61

Calculate the geometric mean for the following fecal coliform test data collected using the membrane filter method.

Test Number	1	2	3	4	5	6	7	8
MPN/100 mL	70	225	0	90	TNTC*	<20	325	148

1. Arrange the MPN values you want to use to calculate the geometric mean in increasing order of magnitude (size).

Use of Calculators

To calculate the geometric mean using a calculator, follow these steps:

1. Multiply the numbers
2. Once all the numbers are multiplied, push the key labeled x^y
3. Enter n (number of measurements) and then push the key labeled 1/x
4. Push the "=" button to get your answer

***TNTC**

TNTC means Too Numerous To Count. TNTC cannot be used in calculations. Report this event on your monthly report in the remarks area and give the apparent reason for the high values.

MPN Values 0 < 20 70 90 148 225 325 TNTC

2. Convert any MPN values that are zero to the number 1.
3. Drop all MPN values that are TNTC (see sidebar).
4. If the MPN value is reported as less than (<) or greater than (>) some number, remove the < or > signs and use the values.
5. Multiply all of the values together to get a number.

$$70 \times 225 \times 1 \times 90 \times 20 \times 325 \times 148 = 1.3636 \times 10^{12}$$

6. Determine the value for 1/n where n is the number of MPN values. Divide 1 by 7 and get 0.1429.
7. Calculate the geometric mean. Using your calculator, find the 0.1429 power of 1.3636×10^{12} or

$$\left(1.3636 \times 10^{12}\right)^{\frac{1}{7}} = \left(1.3636 \times 10^{12}\right)^{0.1429} = 54 \, MPN = \frac{54}{100 \, mL}$$

DISPERSION (SPREAD)

As it was mentioned previously, dispersion or spread tells us how scattered the data is from the mean (or center). Two of the measures will be discussed: variance and standard deviation.

VARIANCE AND STANDARD DEVIATION. Variance and standard deviation are terms frequently used in professional journals that report the results of research findings. Knowledge of this material is important in the field of quality control because these terms describe the spread of measurements or results.

Earlier, we discussed variations in data sets and the mean, median, and mode as measures of central tendency; and we introduced histograms as a way to visualize the spread of data. Variance and standard deviation are mathematical descriptions of the spread.

Let's start with an example to illustrate how sometimes the mean and range are not enough to tell us about a data set. The table shows three data sets. Let's calculate the mean and range for these data sets. The mean is calculated by adding all the measurements and then dividing by the number of measurements (in this case we have 15 measurements).

$$\bar{X}_1 = \frac{140 + 140 + \ldots + 160 + 160}{15} = \frac{2,250}{15} = 150 \frac{mg}{L}$$

If you repeat the same calculation for the two other data sets, you will find that the means are the same, that is,

$$\bar{X}_2 = 150 \frac{mg}{L} \quad \bar{X}_3 = 150 \frac{mg}{L}$$

The subscript (the number shown below the X) refers to the set number. It is good practice to see if you can get the same answer.

Because all readings range from 140 to 160, the range will be 20. Recall that the range is defined as the maximum (highest) measurement minus the minimum (lowest) measurement.

Set 1	Set 2	Set 3
140	140	140
140	140	145
145	140	145
145	145	145
145	145	150
150	145	150
150	150	150
150	150	150
150	150	150
150	155	150
155	155	150
155	155	155
155	160	155
160	160	155
160	160	160

		Frequency	
BOD	Set 1	Set 2	Set 3
140	2	3	1
145	3	3	3
150	5	3	7
155	3	3	3
160	2	3	1

So far, the mean and the range do not indicate any differences between the data sets. Another thing we can do is to create a bar graph. Remember that we need to create a frequency table to show the number of measurements that fall within given ranges (bins). The table in the sidebar shows the frequencies of the three data sets. The histograms for these data sets are shown in the following bar graphs. These graphs show definite differences among the data sets. Set 2 is more spread out than Set 1. Set 3 is less spread out, that is, the data are more tightly clustered around the mean. Just looking at the graphs, we expect Set 2 to have a larger variance than Set 1 and Set 3 to have a smaller variance than Set 1. Let's see if this is true.

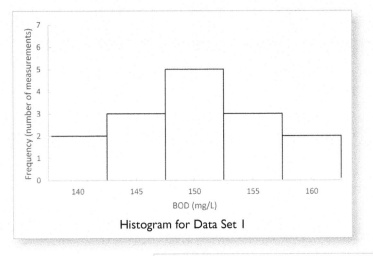

Histogram for Data Set 1

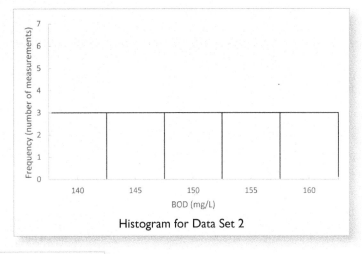

Histogram for Data Set 2

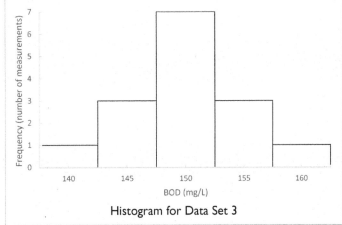

Histogram for Data Set 3

First, we use a notation of S^2 to denote the variance. The standard deviation (S) is the square root of the variance. The best way to describe the variance is to examine the equation that is used to calculate it.

$$S^2 = \frac{\sum_{i=1}^{n} \left(X_i - \bar{X} \right)^2}{n-1}$$

Where,

S^2 = Variance
Σ = Sum
X_i = A Single Measurement
n = Number of Measurements (data points)

Set 1	$(X - \bar{X})$	$(X - \bar{X})^2$
140	−10	100
140	−10	100
145	−5	25
145	−5	25
145	−5	25
150	0	0
150	0	0
150	0	0
150	0	0
150	0	0
155	5	25
155	5	25
155	5	25
160	10	100
160	10	100
Sum 2,250		**550**

$$\bar{X} = \text{Mean} = \frac{\sum\limits_{i=1}^{n} X_i}{n}$$

The summation term in the variance equation says take each measurement and subtract the mean from it, then square that number for each measurement, and add them. Refer to the table on the side that illustrates the calculation for Set 1. The mean of the set is 150 as calculated below:

$$\bar{X} = \frac{\sum\limits_{i=1}^{n} X_i}{n} = \frac{140 + 140 + \ldots + 160 + 160}{15} = \frac{2,250}{15} = 150$$

The variance can then be calculated by using the formula:

$$S^2 = \frac{\sum\limits_{i=1}^{n} (X_i - \bar{X})^2}{n-1}$$

$$= \frac{(-10)^2 + (-10)^2 + (-5)^2 + (-5)^2 + \ldots + (5)^2 + (5)^2 + (10)^2 + (10)^2}{15 - 1}$$

$$= \frac{550}{14} = 39.3$$

We can repeat the same calculation for Sets 2 and 3 to get 53.6, and 25, respectively. (See the tables on sidebar that show the details for the calculations of Sets 2 and 3.)

There is another way to calculate the variance using an equivalent formula:

$$S^2 = \frac{\sum\limits_{i=1}^{n} (X_i - \bar{X})^2}{n-1} = \frac{\sum\limits_{i=1}^{n} X_i^2 - \frac{\left(\sum\limits_{i=1}^{n} X_i\right)^2}{n}}{n-1}$$

The second part of the equation appears very complicated, but it is often easier to do the calculation this way. The first part says add the numbers after you square each one, then add the numbers themselves and square the sum. Finally, divide the last number by n (15 in this case). The calculations are shown below.

$$\sum\limits_{i=1}^{n} X_i^2 = 140^2 + 140^2 + \ldots + 160^2 + 160^2 = 338,050$$

$$\sum\limits_{i=1}^{n} X_i = 140 + 140 + \ldots + 160 + 160 = 2,250$$

$$\left(\sum\limits_{i=1}^{n} X_i\right)^2 = 2,250^2 = 2,250 \times 2,250 = 5,062,500$$

$$S^2 = \dfrac{\sum\limits_{i=1}^{n} X_i^2 - \dfrac{\left(\sum\limits_{i=1}^{n} X_i\right)^2}{n}}{n-1} = \dfrac{338{,}050 - \dfrac{5{,}062{,}500}{15}}{14}$$

$$= \dfrac{338{,}050 - 337{,}500}{14} = \dfrac{550}{14} = 39.3$$

As you see, we get the same answer. For practice, try this method with the other data sets. In the future, use whichever method that makes sense to you.

The following table shows a summary of the variances and standard deviations of the three data sets. This is a good time to review what the variance indicates. In calculating the variance, we calculated how far each data point (or measurement) is from the mean value. We then squared each difference so that positive differences and negative differences do not cancel out. All the squared differences are positive numbers, that is, a difference of −5 (below the mean by 5) or +5 (above the mean by 5) are the same when they are squared (25), so the variance tell us something about how close the measurements are to the mean. Looking at the histogram for Set 1, we observe that most of the readings are around the mean value of 150 (five of them actually are 150) and we have relatively few data points away from 150. Comparing Set 1 to Set 2, notice that we have fewer numbers clustered around the mean and more clustered at the edges (at 140 and 160). Also, the histogram is spread out more and the variance is higher. Comparing Set 3 to Set 1, we observe that more data points are close to the mean and fewer are located on the edges. The data in Set 3 is less spread out and, therefore, the variance is smaller. A large variance indicates a big spread and a small variance indicates a small spread, that is, more clustering around the mean.

The standard deviation is just the square root of the variance (as you can see in the table). Your calculator should have a key for calculating the square root. The key may be labeled with the symbol $\sqrt{\ }$ or the letters SQRT. By entering the variance from Set 1 (39.3) in your calculator and then pressing the square root key, the standard deviation (6.3) should display.

Tools to Calculate Standard Deviation

Some calculators have a standard deviation key. Review your calculator's instructions for using this function. MS Excel™ has built-in functions for calculating variance and standard deviation.

	Set 1	Set 2	Set 3
Variance (S^2)	39.3	53.6	25.0
Standard Deviation (S)	6.3	7.3	5.0

A final note on units. The units for the variance in this case will be $(mg/L)^2$. This is because when we square the numbers, the units are also squared. The units for the standard deviation in this case will be mg/L.

A.8.4   Moving Averages

When a data set contains a lot of fluctuation, it is often hard to see trends over time if you graph the data. For instance, you may want to monitor the long-term treatment effectiveness of a trickling filter, but the daily BOD data may be quite scattered. In these cases, moving averages can help you see underlying trends more clearly. Plots of moving averages tend to smooth out the data, making it easier to see the underlying movement in the data.

7-Day Moving Average

Step			Average of the Shaded Numbers
Step 1	1-Jan	12	
	2-Jan	10	
	3-Jan	5	
	4-Jan	12	
	5-Jan	5	
	6-Jan	15	
	7-Jan	6	9.3
	8-Jan	8	
	9-Jan	20	
	10-Jan	10	
Step 2	1-Jan	12	
	2-Jan	10	
	3-Jan	5	
	4-Jan	12	
	5-Jan	5	
	6-Jan	15	
	7-Jan	6	
	8-Jan	8	8.7
	9-Jan	20	
	10-Jan	10	
Step 3	1-Jan	12	
	2-Jan	10	
	3-Jan	5	
	4-Jan	12	
	5-Jan	5	
	6-Jan	15	
	7-Jan	6	
	8-Jan	8	
	9-Jan	20	10.1
	10-Jan	10	

TRICKLING FILTER PLANT EFFLUENT BOD MOVING AVERAGE

Date	Effluent BOD, mg/L	7-Day Moving Average	Date	Effluent BOD, mg/L	7-Day Moving Average
1-Jan	12		25-Jan	20	13.0
2-Jan	10		26-Jan	13	12.9
3-Jan	5		27-Jan	18	14.9
4-Jan	12		28-Jan	10	15.0
5-Jan	5		29-Jan	25	15.9
6-Jan	15		30-Jan	18	17.0
7-Jan	6	9.3	31-Jan	12	16.6
8-Jan	8	8.7	1-Feb	20	16.6
9-Jan	20	10.1	2-Feb	24	18.1
10-Jan	10	10.9	3-Feb	14	17.6
11-Jan	5	9.9	4-Feb	13	18.0
12-Jan	15	11.3	5-Feb	15	16.6
13-Jan	4	9.7	6-Feb	12	15.7
14-Jan	8	10.0	7-Feb	9	15.3
15-Jan	15	11.0	8-Feb	14	14.4
16-Jan	5	8.9	9-Feb	20	13.9
17-Jan	11	9.0	10-Feb	15	14.0
18-Jan	8	9.4	11-Feb	9	13.4
19-Jan	14	9.3	12-Feb	12	13.0
20-Jan	4	9.3	13-Feb	6	12.1
21-Jan	9	9.4	14-Feb	15	13.0
22-Jan	19	10.0	15-Feb	6	11.9
23-Jan	10	10.7	16-Feb	10	10.4
24-Jan	15	11.3			

Moving averages are commonly seven-day moving averages, which allow each day of the week to be included in the average. The following table contains the raw effluent BOD data and the calculated BOD moving average values for a trickling plant.

To calculate the moving average, simply find the mean for the days being considered. In this example, we are using a seven-day moving average. Therefore, the first average is calculated for the first seven days. That is, using just the first seven days from the measurement table (see sidebar), we calculate the mean by adding the numbers and dividing by 7. So the first average is 65/7 = 9.3. The second step is to move the seven-day period forward by one day. We drop the first number (1-Jan) and add the next one (8-Jan). The average this time is 8.7. The process continues. See Step 3 on the side bar. The mean is 10.1 for the next seven-day interval.

In the Daily Values graph, the individual data are plotted. As you can see, there is considerable scatter in the data set, although the BOD values in late January and early February appear to be somewhat higher than those occurring earlier in the month. The 7-Day Moving Average graph shows the seven-day moving averages plotted on top of the original data. Looking at the seven-day average line, you can see that the average BOD was relatively constant in early January but started to rise about

Daily Values

Data with 7-Day Moving Average

Gaps in the Record

What do you do if the data record has gaps, such as when data are only collected weekdays, not weekends? Don't call the gaps "zero." Jump over them. The example below shows how to calculate a 5-day moving average (Step 1). When the 5-day interval advances, the weekend is skipped and the average is assigned to the following Monday (Step 2). Similarly, in the next step, the gaps are ignored and the interval advances in the usual manner (Step 3).

Step		Average of the Shaded Numbers
Step 1		
Mon	120	
Tues	100	
Wed	95	
Thurs	110	
Fri	95	104
Sat		
Sun		
Mon	85	
Tues	120	
Step 2		
Mon	120	
Tues	100	
Wed	95	
Thurs	110	
Fri	95	
Sat		
Sun		
Mon	85	97
Tues	120	
Step 3		
Mon	120	
Tues	100	
Wed	95	
Thurs	110	
Fri	95	
Sat		
Sun		
Mon	85	
Tues	120	101

January 22, peaked, and then dropped in early February. This shape may suggest that changes are needed in the treatment process.

A.8.5   More Applications of Graphs

Sometimes graphs and charts can be effective computational aids. Figure A.1 (see Section A.7.9, "Friction or Energy Losses") is a sophisticated example. In this section, we will show two other examples.

A.8.5.1   Volume of Sludge in a Digester

Digesters can vary and the depth can be used to calculate the volume of sludge in gallons or in cubic feet. Assume the digester is a cylindrical tank with a top and bottom shaped like a cone (see figure in the sidebar). A quick way to convert depth to volume and to reduce the chance of math errors is to prepare a chart of sludge depth against sludge volume.

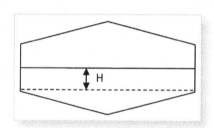

Example 62

Prepare a chart of sludge depth against sludge volume in cubic feet and Mgal for a 100-foot-diameter digester.

Known	Unknown
Measured Sludge Depths	Volume of Digester for Various Sludge Depths, ft³

We need to calculate the sludge volume. In this case, operators will vary the sludge depth from the top of the bottom cone (H = 0) to the bottom of the top cone (H = 20) as shown in the figure in the sidebar. The sludge fills the bottom cone, however, so its volume must be included in the calculation. Therefore, our first task is to estimate the volume of the sludge based on the volume of the cone. The radius of the cone is 50 ft with a height of 12.5 ft (see cross-section above).

$$V_{cone} = \frac{\pi}{3} R^2 H = \frac{\pi}{3} (50\,ft)^2 (12.5\,ft) = 32,725\,ft^3$$

As the digester fills with sludge above the cone-shaped bottom into the cylindricalshaped portion of the tank, we add this volume to the volume of the cone. If the height inside the cylinder is H (as shown in the figure in the sidebar), then the volume of the cylinder is:

$$V_{cylinder} = \pi R^2 H = \pi (50\,ft)^2 H = (3.1416)(2,500)H = 7,854H$$

Then the sludge volume is the sum of the two volumes:

$$V_{sludge} = V_{cone} + V_{cylinder} = 32,725 + 7,854H$$

Using this equation, we can now calculate the volume of sludge in the digester when H = 0, 10, and 20 ft. Calculate the volume when H = 10 ft,

$$V_{sludge} = V_{cone} + V_{cylinder} = 32,725 + (7,854)(10) = 111,265\,ft^3$$

Practice by calculating the volume of sludge in the digester when H = 20 ft. We also need to know the volume of sludge in Mgal. The conversion is calculated as follows:

$$V_{sludge} = 111,265\,ft^3 \times \frac{7.48\,gal}{ft^3} \times \frac{M}{10^6} = 0.832\,Mgal$$

After you calculate the sludge volume for 0, 10, and 20 ft in both cubic feet and Mgal, prepare a table of the values.

Depth of Sludge, ft	Volume of Sludge, ft³	Volume of Sludge, Mgal
0	32,725	0.245
10	111,265	0.832
20	189,805	1.420

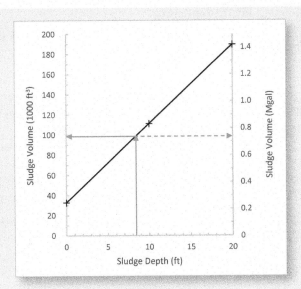

Based on the table, we can create a graph that shows sludge volume as a function of the height of sludge by connecting the points with a straight line. The graph is showing the volume in 1,000 ft³ because the numbers are so large. For example, when you read 80 on the volume scale, you need to multiply that by 1,000. In that case, the volume is 80,000 ft³.

To prepare the Mgal scale (shown on the right side of the graph), we calculate the number of cubic feet corresponding to selected volumes in Mgal. For example, start with a volume of 0.2 Mgal and convert it to the equivalent volume in ft³:

$$0.2\,\text{Mgal} \times \frac{10^6}{M} \times \frac{\text{ft}^3}{7.48\,\text{gal}} = 26,738\,\text{ft}^3$$

Next, select the point that corresponds to 26,738 (shown as 26.7) and draw a line horizontally and mark it as 0.2 Mgal. You can do the same for other points as shown in the following graph. This can be accomplished in MS Excel™ by creating a secondary axis. For more information about how to do this, see MS Excel™ Help or another resource. Now we have a graph that allows us to read (rather than calculate) the sludge volume (in 1,000 ft³ and Mgal) for any depth.

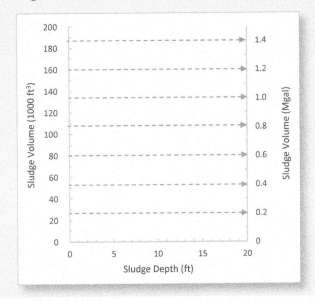

A.8.5.2 Tracking BOD Loading

In this example, we will illustrate how choosing one kind of graph provides more information than another.

Example 63

Construct a graph to track BOD load discharges for permit compliance when the allowable discharge is 80 lb BOD/month.

Treatment plants monitor flows and effluent BOD concentrations on a regular basis. The following table shows a typical data set with flows and BOD concentrations, in this case daily values. Recall that load is obtained by multiplying the flow rate by the concentration as shown in the calculation for Day 1.

$$\text{BOD Load} = Q \times \text{BOD}$$

$$= 19.9 \frac{\text{Mgal}}{\text{d}} \times 9 \frac{\text{mg}}{\text{L}}$$

$$= \left(19.9 \frac{\text{Mgal}}{\text{d}}\right)\left(9 \frac{\text{mg}}{\text{L}}\right)\left(\frac{10^6 \text{ gal}}{\text{Mgal}}\right)\left(\frac{3.785\,\text{L}}{\text{gal}}\right)\left(\frac{\text{g}}{10^3 \text{ mg}}\right)\left(\frac{\text{lb}}{454.5\,\text{g}}\right)$$

$$= 1,492 \frac{\text{lb}}{\text{d}} \text{ or } 1.5 \frac{1,000\,\text{lb}}{\text{d}}$$

TRACKING BOD LOADING

(1) Date	(2) Flow, MGD	(3) BOD, mg/L	(4) BOD Load, 1,000 lb/d	(5) Cumulative BOD Load, 1,000 lb/d[a]
1	19.9	9	1.5	1.5
2	22.6	12	2.3	3.8
3	22.7	10	1.9	5.6
4	23.5	11	2.2	7.8
5	23.8	7	1.4	9.2
6	21.7	20	3.6	12.8
7	22.0	16	2.9	15.7
8	23.3	18	3.5	19.2
9	23.8	16	3.2	22.4
10	23.8	11	2.2	24.6
11	24.0	20	4.0	28.6
12	23.2	15	2.9	31.5
13	22.5	20	3.7	35.2
14	22.1	18	3.3	38.5
15	24.0	27	5.4	43.9
16	25.6	65	13.9	57.8
17	19.5	26	4.2	62.0
18	16.6	8	1.1	63.1
19	17.2	10	1.4	64.5

(1)	(2)	(3)	(4)	(5)
	Flow,	BOD,	BOD Load,	Cumulative BOD Load,
Date	MGD	mg/L	1,000 lb/d	1,000 lb/d[a]
20	17.7	9	1.3	65.9
21	17.1	8	1.1	67.0
22	19.9	10	1.7	68.7
23	20.3	9	1.5	70.2
24	21.0	10	1.7	71.9
25	21.1	10	1.8	73.7
26	21.3	7	1.2	74.9
27	18.0	3	0.4	75.3
28	18.0	5	0.7	76.0
29	21.9	8	1.5	77.5
30	22.4	9	1.7	79.2

TRACKING BOD LOADING (*continued*)

[a]Results calculated using MS Excel™ may be slightly different because this table uses rounded values of daily loads.

The other loads in the table are calculated similarly. Now it is reasonable to track plant efficiency by plotting the effluent loads by day of the month as shown in the following graph.

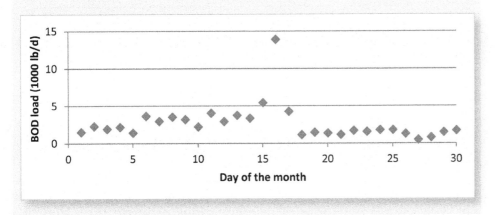

While this graph shows plant day-to-day performance variations, it does not convey much information about whether the plant is meeting its permit requirements. A more informative graph is a plot of cumulative load against day of the month, as shown by the points in the following graph. As before, the dotted line between points is there just to help the eye pick out the trend. Also on this graph, we have plotted a straight line between (0,0) and the monthly BOD limit (80 lb) on day 30. The slope of this line represents the average BOD load (in 1,000 lb/d) that must be achieved to meet the 80-lb limit at the end of the month. Where the data points are below the line, the plant is on track to meet the permit requirement. In the middle of the month, a couple of bad treatment days throw the plant off track. Notice, though, that the plant draws back to the permit requirement through a series of better-than-average treatment days. In the end, the cumulative BOD discharge is 79.2 lb, which meets the requirement. The advantage to a graph in this form as opposed to one like the previous graph is that when things are going well, you can see how much safety buffer is being built up. On the other hand, when things are going poorly you can see how much you have to recover and the time available.

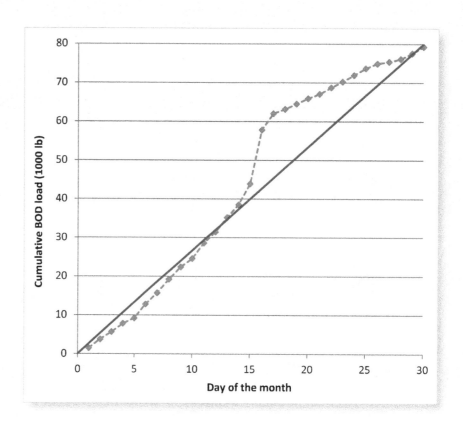

A.8.6   Regression Analysis (Prediction Equations, Trends, and Correlations)

Regression analysis is a powerful math tool to develop prediction (or forecasting) equations. These equations estimate the value of a dependent variable on the basis of a known or measured independent variable. These prediction equations are usually developed because the independent variable is easier to measure than the dependent variable.

For example, the chemical oxygen demand (COD) test is much faster to conduct than a 5-day BOD test, but a treatment plant's permit may be based on the BOD result—so, an operator may wish to predict the BOD value on the basis of a COD measurement. To do this, collect composite effluent samples on a daily basis, split the samples, measure the COD immediately, and measure the BOD five days later. Then use the results to predict BOD levels from COD measurements.

Example 64

Apply regression analysis to develop an equation for estimating BOD values from COD measurements for the effluent data in the following table.

EFFLUENT VALUES FOR
THE FIRST SEVEN DAYS

Date	Day	COD, mg/L	BOD, mg/L
1	Mon	60	9
2	Tue	69	12
3	Wed	62	10
4	Thu	69	11
5	Fri	65	7
6	Sat	131	20
7	Sun	100	16

Plot the effluent COD on the x-axis (horizontal axis) and the effluent BOD on the y-axis (vertical axis) as shown in the following graph. In regression analysis we label the independent variable (in this case COD) as x and the dependent variable (in this BOD) as y. Remember, we want to predict the dependent variable y using independent variable x data. In this case, we are interested in finding the best-fit line through the points that we plotted.

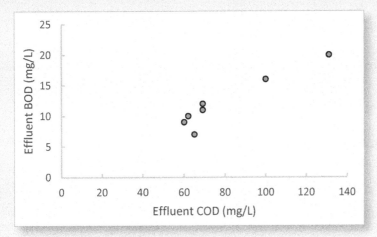

When inspecting the graph, one can imagine drawing a number of straight lines that pass through the points. Selecting the best line and consistently getting the same line independent of the person selecting the line can be accomplished using linear regression.

A straight line is defined by the equation:

$$y = a + bx$$

Where,

a = Intercept
b = Slope of the Line

The following figure shows that the intercept is the value of y when x = 0, and the slope represents the rise in the line when x changes by one unit (if the slope is negative then the line goes down, not up).

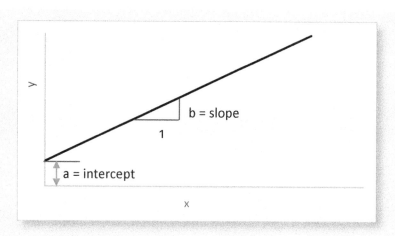

In this example, y is the BOD and x is the COD. Therefore, the equation is:

BOD = a + b(COD)

Now, we need to be able to calculate the intercept (a) and slope of the line (b) from the data set shown in the table. We call (a) and (b) regression parameters. The table below shows the equations needed to calculate the regression parameters. First, we need to calculate some of the numbers from the measurements. Keep in mind anytime you see (x) it refers to the independent variable (in this example, COD). When you see (y), it refers to the dependent variable (in this example, BOD).

Independent Variable (x) Equations	Combined Variable (x and y) Equations	Dependent Variable (y) Equations
n = Number of Measurements	$\left(\sum x\right)\left(\sum y\right)$ = Multiply the Sum of x by the Sum of y	n = Number of Measurements
$\sum x$ = Sum of x	$\dfrac{\left(\sum x\right)\left(\sum y\right)}{n}$ = Divide the Product of the Sum by n	$\sum y$ = Sum of y
$\bar{x} = \dfrac{\sum x}{n}$ = Mean of x		$\bar{y} = \dfrac{\sum y}{n}$ = Mean of y
$\left(\sum x\right)^2$ = Square of the Sum of x (add all x values and then square the sum)	$\sum xy$ = Sum of x Multiplied by y (multiply first and then add)	$\left(\sum y\right)^2$ = Square of the Sum of y (add all y values and then square the sum)
$\sum x^2$ = Sum of Squares of x (this means square x and then add)	$S_{XY} = \sum xy - \dfrac{\left(\sum x\right)\left(\sum y\right)}{n}$ = Corrected Product Sum	$\sum y^2$ = Sum of Squares of y (square first and then add)
$S_{xx} = \sum x^2 - \dfrac{\left(\sum x\right)^2}{n}$ = Corrected Sum of Squares of x		$S_{YY} = \sum y^2 - \dfrac{\left(\sum y\right)^2}{n}$ = Corrected Sum of Squares of y

We can apply the equations to the data from the "Effluent Values" table at the start of the example. Calculate the equations in the side columns first because those values are used in the equations in the center column.

Independent Variable (COD) Equations	Combined Variable (COD and BOD) Equations	Dependent Variable (BOD) Equations
$n = 7$	$(\sum x)(\sum y) = 47{,}260$	$n = 7$
$\sum x = 556$	$\dfrac{(\sum x)(\sum y)}{n} = 6{,}751$	$\sum y = 85$
$\bar{x} = \dfrac{\sum x}{n} = \dfrac{556}{7} = 79.4$	$\sum xy = 7{,}422$	$\bar{y} = \dfrac{\sum y}{n} = \dfrac{85}{7} = 12.1$
$(\sum x)^2 = 309{,}136$	$S_{XY} = \sum xy - \dfrac{(\sum x)(\sum y)}{n} = 671$	$(\sum y)^2 = 7{,}225$
$\sum x^2 = 48{,}352$		$\sum y^2 = 1{,}151$
$S_{xx} = \sum x^2 - \dfrac{(\sum x)^2}{n} = 4{,}190$		$S_{YY} = \sum y^2 - \dfrac{(\sum y)^2}{n} = 119$

Once these numbers are calculated, estimate the regression parameters (the intercept and slope of the line) using the following equations.

$$b = \frac{S_{XY}}{S_{XX}} = \frac{\sum xy - \dfrac{(\sum x)(\sum y)}{n}}{\sum x^2 - \dfrac{(\sum x)^2}{n}}$$

$$a = \bar{y} - b\bar{x}$$

Apply the equations using the values we calculated.

$$b = \frac{7{,}422 - \dfrac{(556)(85)}{7}}{48{,}352 - \dfrac{309{,}136}{7}} = \frac{671}{4{,}190} = 0.16$$

$$a = \bar{y} - b\bar{x} = 12.1 - (0.16)(79.4) = 12.1 - 12.7 = -0.6$$

Therefore, our regression line will be

$$BOD = -0.6 + 0.16(COD)$$

Using this equation we multiply a COD measurement by 0.16 and then subtract 0.6 to obtain an estimate of the BOD level. To visually check our equation, a regression line can be plotted with measurements using MS Excel™ or by picking any two values for COD and connecting these two points with a line. To illustrate this, pick 60 and 135, then plug these numbers into the regression equation.

COD	BOD
60	$BOD = -0.6 + (0.16)\,COD = -0.6 + (0.16)(60) = 9$
135	$BOD = -0.6 + (0.16)\,COD = -0.6 + (0.16)(135) = 21$

If you plot these two points on the same plot and then connect them with a straight line, you will get the following graph.

Performing Regression in MS Excel™

MS Excel™ provides built-in SLOPE and INTERCEPT functions. Another way to do a regression is to graph the data in Excel and display it on a linear trend line. The regression line and equation will be plotted on the same graph as the data, similar to what is displayed in the regression line graph.

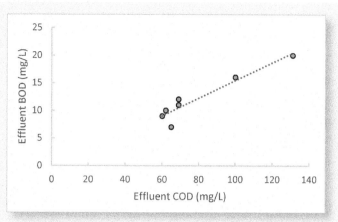

We can also use regression analysis to establish trends. Trends can be either positive (increasing) or negative (decreasing). Example 65 illustrates how to use data and linear regression to establish a trend.

Example 65

Use the BOD data for the first seven days (from the table in Example 64) and determine whether there is a positive or negative trend. First, plot the data with time.

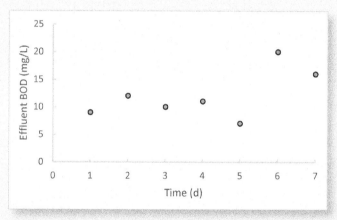

Second, generate all the parameters shown in the summary of linear regression table. In this example, x is time and y is BOD.

Independent Variable (Time) Equations	Combined Variable (Time and BOD) Equations	Dependent Variable (BOD) Equations
$n = 7$	$\left(\sum x\right)\left(\sum y\right) = 2{,}380$	$n = 7$
$\sum x = 28$	$\dfrac{\left(\sum x\right)\left(\sum y\right)}{n} = 340$	$\sum y = 85$
$\bar{x} = \dfrac{\sum x}{n} = \dfrac{28}{7} = 4$		$\bar{y} = \dfrac{\sum y}{n} = \dfrac{85}{7} = 12.1$
	$\sum xy = 374$	
$\left(\sum x\right)^2 = 784$		$\left(\sum y\right)^2 = 7{,}225$
$\sum x^2 = 140$	$S_{XY} = \sum xy - \dfrac{\left(\sum x\right)\left(\sum y\right)}{n} = 34$	$\sum y^2 = 1{,}151$
$S_{XX} = \sum x^2 - \dfrac{\left(\sum x\right)^2}{n} = 28$		$S_{YY} = \sum y^2 - \dfrac{\left(\sum y\right)^2}{n} = 119$

Third, calculate the regression parameters.

$$b = \frac{\sum xy - \dfrac{(\sum x)(\sum y)}{n}}{\sum x^2 - \dfrac{(\sum x)^2}{n}} = \frac{34}{28} = 1.21 \frac{\text{mg BOD}}{\text{L}}\Big/\text{d}$$

$$a = \bar{y} - b\bar{x} = 12.1 - (1.21)(4) = 12.1 - 4.8 = 7.3 \frac{\text{mg BOD}}{\text{L}}$$

The regression equation is BOD = 7.3 + 1.21 (t) and the regression line is shown in the following figure:

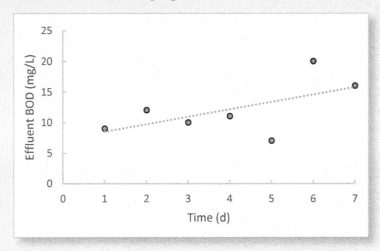

The trend line is increasing by a rate of 1.21 mg BOD/L per day during the 7-day span in which the data were collected. It is important to understand that we do not know what happens outside this data range, so this equation should not be applied to other data sets.

A.8.6.1   Correlations

The correlation coefficient (r) is used as a measure or indication of the "closeness of fit" of the prediction (or forecasting) equation with the actual data. If all the data points lie directly on the regression line, the fit will be perfect and the correlation coefficient will be 1 or −1. A minus (negative) correlation indicates that there is an inverse relationship—meaning the line goes down not up. The closer (r) is to 1, the better the equation fits the data points. If you calculate a number larger than 1 or smaller than −1, you have made an error because (r) is always between −1 and 1. The correlation coefficient is calculated using the following equation.

$$r = \frac{S_{XY}}{\sqrt{S_{XX}S_{YY}}} = \frac{\sum xy - \dfrac{(\sum x)(\sum y)}{n}}{\sqrt{\left(\sum x^2 - \dfrac{(\sum x)^2}{n}\right)\left(\sum y^2 - \dfrac{(\sum y)^2}{n}\right)}}$$

Here, the square root applies to the product of the two terms on the bottom of the fraction (denominator).

Calculating Correlation (r) in MS Excel™

MS Excel™ provides a built-in CORREL function that can calculate the correlation coefficient (r).

Example 66

Referring to Example 65 where x is COD and y is BOD, calculate the correlation coefficient (r).

$$r = \frac{S_{XY}}{\sqrt{S_{XX}S_{YY}}} = \frac{671}{\sqrt{4,190 \times 119}} = 0.95$$

Example 67

Calculate the correlation coefficient (r) for Example 65 concerning BOD change with time.

$$r = \frac{S_{XY}}{\sqrt{S_{XX}S_{YY}}} = \frac{34}{\sqrt{28 \times 119}} = 0.59$$

The results of the last two examples tell us that BOD and COD measurements of the same water samples are much more closely correlated than BOD is to time over a 7-day interval. Considering that BOD and COD are caused by many of the same organic compounds in wastewater, this result makes sense.

A.9 Typical Water Treatment Plant Problems (English System)

A.9.1 Plant Operation

Example 1

Estimate the average use of chlorine in pounds per day (lb/d) based on actual use of chlorine for one week as shown below.

Day of the Week	Sun	Mon	Tue	Wed	Thur	Fri	Sat
Chlorine Used, lb	23	37	35	31	32	36	24

Known	Unknown
Chlorine Used Each Day of Week, lb/d	F_A = Daily Average Use of Chlorine, lb/d

Estimate the average chlorine use in pounds of chlorine per day (lb/d).

$$F_A\left[\frac{lb}{d}\right] = \frac{\text{Sum of Chlorine Used Each Day}}{\text{Total Number of Days}}$$

$$= \frac{23[lb] + 37[lb] + 35[lb] + 31[lb] + 32[lb] + 36[lb] + 24[lb]}{7[d]}$$

$$= \frac{218[lb]}{7[d]} = 31.1\frac{lb}{d}$$

Example 2

A water treatment plant has five 150-pound (lb) chlorine cylinders in storage. The plant uses an average of 28 pounds of chlorine per day (lb/d). How many days' (d) supply of chlorine is in storage?

Known	Unknown
n = Number of Chlorine Cylinders = 5 cyl	Supply of Chlorine, d
M_{cyl} = Cylinder Weight = 150 lb	
F = Average Use = 28 lb/d	

Calculate the available supply of chlorine in storage in days (d).

$$\text{Supply of Chlorine [d]} = \frac{M_{cyl} \times n}{F} = \frac{150\left[\frac{lb}{cyl}\right] \times 5[cyl]}{28\left[\frac{lb}{d}\right]} = 27\ d$$

A.9.2 Flows

Example 3

Convert a flow rate of 800 gallons per minute (gpm) to million gallons per day (Mgal/d).

Known	Unknown
Q = Flow Rate = 800 gpm	Q = Flow Rate, Mgal/d

Convert flow rate from 800 gal/min (gpm) to Mgal/d (MGD).

$$Q\left[\frac{Mgal}{d}\right] = Q\left[\frac{gal}{min}\right] \times \frac{60\ min}{hr} \times \frac{24\ hr}{d} \times \frac{M}{10^6}$$

$$= 800\left[\frac{gal}{min}\right] \times \frac{60\ min}{hr} \times \frac{24\ hr}{d} \times \frac{M}{10^6}$$

$$= 1.15\frac{Mgal}{d} = 1.15\ MGD$$

A.9.3 Chemical Doses

Example 4

Determine the chlorinator setting in pounds per day (lb/d) to treat a flow rate of 2 million gallons per day (Mgal/d) with a chlorine dose of 3.0 mg/L.

Known	Unknown
Q = Flow Rate = 2 Mgal/d	F = Chlorinator Setting, lb/d
Cl_{dose} = Chlorine Dose = 3.0 mg/L	

Determine the chlorinator setting in pounds per day (lb/d).

$$F\left[\frac{lb}{d}\right] = Q\left[\frac{Mgal}{d}\right] \times Cl_{dose}\left[\frac{mg}{L}\right] \times \frac{3.785\ L}{gal} \times \frac{10^6}{M} \times \frac{kg}{10^6\ mg} \times \frac{2.205\ lb}{kg}$$

$$= 2\left[\frac{Mgal}{d}\right] \times 3.0\left[\frac{mg}{L}\right] \times \frac{3.785\ L}{gal} \times \frac{10^6}{M} \times \frac{kg}{10^6\ mg} \times \frac{2.205\ lb}{kg}$$

$$= 50\frac{lb}{d}$$

Example 5

The optimum liquid alum dose from the jar tests is 12 mg/L. Determine the setting on the liquid alum chemical feeder in milliliters per minute (mL/min) when the plant flow rate is 4.7 million gallons per day (Mgal/d). The liquid alum purity delivered to the plant contains 643 milligrams (mg) of alum per milliliter (mL) of liquid solution.

Known	Unknown
D_{alum} = Alum Dose = 12 mg/L	Chemical Feeder Setting, mL/min
Q = Flow Rate = 4.7 Mgal/d	
Liquid Alum Content = 643 mg/mL	

Calculate the liquid alum chemical feeder setting in milliliters per minute (mL/min).

$$\text{Chemical Feeder Setting}\left[\frac{mL}{min}\right]$$

$$= \frac{Q\left[\frac{Mgal}{d}\right] \times D_{alum}\left[\frac{mg}{L}\right] \times \frac{3.785\,L}{gal} \times \frac{10^6}{M} \times \frac{d}{24\,h} \times \frac{h}{60\,min}}{\text{Liquid Alum}\left[\frac{mg}{mL}\right]}$$

$$= \frac{4.7\left[\frac{Mgal}{d}\right] \times 12\left[\frac{mg}{L}\right] \times \frac{3.785\,L}{gal} \times \frac{10^6}{M} \times \frac{d}{24\,h} \times \frac{h}{60\,min}}{643\left[\frac{mg}{mL}\right]}$$

$$= 231\,\frac{mL}{min}$$

Example 6

The optimum liquid alum dose from the jar tests is 12 mg/L. Determine the setting on the liquid alum chemical feeder in gallons per day (gpd) when the flow rate is 4.7 million gallons per day (Mgal/d). The liquid alum delivered to the plant contains 5.36 pounds (lb) of alum per gallon (gal) of liquid solution.

Known	Unknown
D_{alum} = Alum Dose = 12 mg/L	Chemical Feeder Setting, gpd
Q = Flow Rate = 47 Mgal/d	
Liquid Alum Content = 5.36 lb/gal	

Calculate the liquid alum chemical feeder setting in gallons per day (gpd).

$$\text{Chemical Feeder Setting}\left[\frac{gal}{d}\right]$$

$$= \frac{Q\left[\frac{Mgal}{d}\right] \times D_{alum}\left[\frac{mg}{L}\right] \times \frac{3.785\,L}{gal} \times \frac{10^6}{M} \times \frac{kg}{10^6\,mg} \times \frac{2.205\,lb}{kg}}{\text{Liquid Alum}\left[\frac{lb}{gal}\right]}$$

$$= \frac{4.7\left[\frac{Mgal}{d}\right] \times 12\left[\frac{mg}{L}\right] \times \frac{3.785\,L}{gal} \times \frac{10^6}{M} \times \frac{kg}{10^6\,mg} \times \frac{2.205\,lb}{kg}}{5.36\left[\frac{lb}{gal}\right]}$$

$$= 88\,gpd$$

Example 7

An operator is interested in measuring the dry chemical feed rate in pounds per day (lb/d), so he places an empty bucket weighing 0.3 pounds (lb) under the dry chemical feeder. After 30 minutes (min), the bucket weighs 2.1 pounds (lb). Calculate the feed rate based on the information provided.

Known		Unknown
M_{empty} = Weight of Empty Bucket	= 0.3 lb	Dry Chemical
M_{full} = Weight of Bucket and Chemical	= 2.1 lb	Feed Rate, lb/d
t = Feed Time	= 30 min	

1. Determine the chemical fed in pounds (lb).

$$\text{Chemical Fed}[lb] = M_{full}[lb] - M_{empty}[lb]$$
$$= 2.1[lb] - 0.3[lb]$$
$$= 1.8\ lb$$

2. Calculate the dry chemical feed rate in pounds per minute (lb/min).

$$\text{Chemical Feed Rate}\left[\frac{lb}{min}\right] = \frac{\text{Chemical Fed}[lb]}{t[min]}$$
$$= \frac{1.8[lb]}{30[min]}$$
$$= 0.06\frac{lb}{[min]}$$

3. Calculate the dry chemical feed rate in pounds per day (lb/d).

$$\text{Chemical Feed Rate}\left[\frac{lb}{d}\right] = \text{Chemical Feed Rate}\left[\frac{lb}{min}\right] \times \frac{60\ min}{h} \times \frac{24\ h}{d}$$
$$= 0.06\left[\frac{lb}{min}\right] \times \frac{60\ min}{h} \times \frac{24\ h}{d}$$
$$= 86.4\frac{lb}{d}$$

Example 8

A small chemical feed pump lowered the chemical solution in a three-foot (ft) diameter tank one foot (ft) and seven inches (in) during an eight-hour (h) period. Estimate the feed rate delivered by the pump in gallons per minute (gpm) and gallons per day (gpd).

Known		Unknown
D = Tank Diameter = 3 ft		Chemical Used, gal
h = Tank Drop = 1 ft 7 in		Chemical Feed Rate, gpm
t = Time = 8 h		Chemical Feed Rate, gpd

1. Convert the tank drop (h) from one foot (ft) seven inches (in) to feet (ft).

$$h[ft] = 1[ft] + 7[in] = 1[ft] + 7[in] \times \frac{ft}{12\ in} = 1[ft] + 0.58[ft] = 1.58\ ft$$

2. Calculate the volume of chemical used in cubic feet (ft³).

$$\text{Volume}\left[ft^3\right] = \frac{\pi}{4} \times D^2 \left[ft^2\right] \times h\left[ft\right]$$

$$= 0.785 \times 3^2 \left[ft^2\right] \times 1.58\left[ft\right] = 11.16 \; ft^3$$

3. Determine the gallons (gal) of solution pumped.

$$\text{Chemical Used}\left[gal\right] = \text{Volume}\left[ft^3\right] \times \frac{7.48 \; gal}{ft^3}$$

$$= 11.16\left[ft^3\right] \times \frac{7.48 \; gal}{ft^3} = 83.5 \; gal$$

4. Estimate the feed rate delivered by the pump in gallons per minute (gpm) and gallons per day (gpd).

$$\text{Chemical Feed Rate}\left[\frac{gal}{min}\right] = \frac{\text{Chemical Used}\left[gal\right]}{t\left[h\right] \times \dfrac{60 \; min}{h}}$$

$$= \frac{83.5\left[gal\right]}{8\left[h\right] \times \dfrac{60 \; min}{h}} = 0.17 \; gpm$$

or

$$\text{Chemical Feed Rate}\left[\frac{gal}{d}\right] = \frac{\text{Chemical Used}\left[gal\right]}{t\left[h\right] \times \dfrac{d}{24 \; h}}$$

$$= \frac{83.5\left[gal\right]}{8\left[h\right] \times \dfrac{d}{24 \; h}} = 251 \frac{gal}{d}$$

A.9.4   Reservoir Management and Intake Structures

Example 9

A reservoir has a volume of 6.8 acre-feet (ac-ft). What is the reservoir volume in gallons (gal) and million gallons (Mgal)?

Known	Unknown
V = Reservoir Volume = 6.8 ac-ft	V = Reservoir Volume, gal
	V = Reservoir Volume, Mgal

Convert reservoir volume from acre-feet (ac-ft) to gallons (gal) and million gallons (Mgal).

$$V\left[gal\right] = V\left[ac\text{-}ft\right] \times \frac{43,560 \; ft^2}{ac} \times \frac{7.48 \; gal}{ft^3}$$

$$= 6.8\left[ac\text{-}ft\right] \times \frac{43,560 \; ft^2}{ac} \times \frac{7.48 \; gal}{ft^3}$$

$$= 2,215,636 \; gal \; \text{or} \; 2.2 \; Mgal$$

Example 10

A reservoir has a surface area of 51,200 square feet (ft^2) and the desired dose of copper sulfate is 6 pounds per acre (lb/ac). How many pounds (lb) of copper sulfate will be needed?

Known	Unknown
$A_{surface}$ = Surface Area = 51,200 ft^2	$CuSO_4$ = Copper Sulfate, lb
D_{CuSO_4} = Copper Sulfate Dose = 6 lb/ac	

1. Convert the surface area from square feet (ft^2) to acres (ac).

$$A_{surface}[ac] = A_{surface}[ft^2] \times \frac{ac}{43,560\ ft^2} = \frac{51,200}{43,560} = 1.18\ ac$$

2. Calculate the pounds (lb) of copper sulfate needed.

$$CuSO_4[lb] = A_{surface}[ac] \times D_{CuSO_4}\left[\frac{lb}{ac}\right]$$

$$= 1.18[ac] \times 6\left[\frac{lb}{ac}\right] = 7.1\ lb\ Copper\ Sulfate$$

Example 11

The volume of a reservoir is estimated to be 581,000 cubic feet (ft^3). The desired dose of copper is 0.5 mg/L and the copper content of the copper sulfate to be used is 25 percent (%). How many pounds (lb) of copper sulfate will be needed?

Known	Unknown
V = Reservoir Volume = 581,000 ft^3	$CuSO_4$ = Copper Sulfate, lb
D_{Cu} = Copper Dose = 0.5 mg/L	
Copper Content = 25%	

1. Convert the reservoir volume from cubic feet (ft^3) to gallons (gal).

$$V[gal] = V[ft^3] \times \frac{7.48\ gal}{ft^3} = 581,000 \times 7.48 = 4,345,880\ gal$$

2. Calculate the pounds (lb) of copper sulfate needed.

$$CuSO_4[lb] = \frac{V \times D_{Cu} \times 100\%}{Copper[\%]}$$

$$= \frac{4,345,880[gal] \times 0.5\left[\frac{mg}{L}\right] \times \frac{3.785\ L}{gal} \times \frac{kg}{10^6\ mg} \times \frac{2.205\ lb}{kg} \times 100\%}{25\%}$$

$$= 72.5\ lb\ Copper\ Sulfate$$

A.9.5   Coagulation and Flocculation

Example 12

Estimate the actual alum dose in milligrams per liter (mg/L) if a plant treats a raw water with a flow rate of 1.8 MGD. The alum feed rate is 135 pounds per day (lb/d).

Known	Unknown
Q = Flow Rate = 1.8 Mgal/d	D_{alum} = Alum Dose, mg/L
Alum Used = 135 lb/d	

Calculate the alum dose in milligrams per liter (mg/L).

$$D_{alum}\left[\frac{mg}{L}\right] = \frac{Alum\ Used\left[\dfrac{lb}{d}\right]}{Q\left[\dfrac{Mgal}{d}\right] \times \dfrac{3.785\ L}{gal} \times \dfrac{10^6}{M} \times \dfrac{kg}{10^6\ mg} \times \dfrac{2.205\ lb}{kg}}$$

$$= \frac{135\left[\dfrac{lb}{d}\right]}{1.8\left[\dfrac{Mgal}{d}\right] \times \dfrac{3.785\ L}{gal} \times \dfrac{10^6}{M} \times \dfrac{kg}{10^6\ mg} \times \dfrac{2.205\ lb}{kg}}$$

$$= 9.0\frac{mg}{L}$$

Example 13

Determine the strength of a polymer solution as a percent (%) if 80 grams (g) of dry polymer are mixed with four gallons (gal) of water.

Known	Unknown
$M_{polymer}$ = Mass of Dry Polymer = 80 g	Polymer Solution, percent (%)
V_{water} = Volume of Water = 4 gal	

1. Convert the 80 grams (g) of dry polymer to pounds (lb).

$$M_{polymer}[lb] = M_{polymer}[g] \times \frac{lb}{454\ g} = 80[g] \times \frac{lb}{454\ g} = 0.176\ lb$$

2. Convert the four gallons (gal) of water to pounds (lb).

$$Water[lb] = V_{water}[gal] \times \frac{8.34\ lb}{gal} = 4[gal] \times \frac{8.34\ lb}{gal} = 33.36\ lb$$

3. Calculate the polymer solution as a percent (%).

$$Polymer[\%] = \frac{M_{polymer}[lb] \times 100\%}{M_{polymer}[lb] + Water[lb]}$$

$$= \frac{0.176[lb] \times 100\%}{0.176[lb] + 33.36[lb]} = 0.52\%$$

A.9.6   Sedimentation

Example 14

Estimate the detention time in hours (h) for a 30-foot (ft)-diameter circular clarifier when the flow rate is 0.5 Mgal/d. The clarifier is 8 feet (ft) deep.

Known	**Unknown**
D = Diameter = 30 ft	t = Detention Time, h
H = Depth = 8 ft	
Q = Flow Rate = 0.5 Mgal/d	

1. Calculate the clarifier volume (V) in cubic feet (ft³).

$$V\,[\text{ft}^3] = 0.785 \times D^2\,[\text{ft}^2] \times H[\text{ft}] = 0.785 \times 30^2\,[\text{ft}^2] \times 8[\text{ft}] = 5{,}652\ \text{ft}^3$$

2. Estimate the detention time of the clarifier in hours (h).

$$t[\text{h}] = \dfrac{V\left[\text{ft}^3\right] \times \dfrac{7.48\ \text{gal}}{\text{ft}^3}}{Q\left[\dfrac{\text{Mgal}}{\text{d}}\right] \times \dfrac{10^6}{\text{M}} \times \dfrac{\text{d}}{24\ \text{h}}} = \dfrac{5{,}652\left[\text{ft}^3\right] \times \dfrac{7.48\ \text{gal}}{\text{ft}^3}}{0.5\left[\dfrac{\text{Mgal}}{\text{d}}\right] \times \dfrac{10^6}{\text{M}} \times \dfrac{\text{d}}{24\ \text{h}}} = 2.0\ \text{h}$$

Example 15

Estimate the surface loading rate in gallons per minute per square foot (gpm/ft²) for a rectangular sedimentation basin 20 feet (ft) wide and 40 feet (ft) long when the flow rate is 0.5 MGD.

Known	**Unknown**
W = Width = 20 ft	$LR_{surface}$ = Surface Loading, gpm/ft²
L = Length = 40 ft	
Q = Flow Rate = 0.5 MGD	

1. Determine the surface area ($A_{surface}$) of the basin in square feet (ft²).

$$A_{surface}\,[\text{ft}^2] = L \times W = 40[\text{ft}] \times 20[\text{ft}] = 800\ \text{ft}^2$$

2. Convert the flow rate from million gallons per day (MGD) to gallons per minute (gpm).

$$Q\left[\dfrac{\text{gal}}{\text{min}}\right] = Q\left[\dfrac{\text{Mgal}}{\text{d}}\right] \times \dfrac{10^6}{\text{M}} \times \dfrac{\text{d}}{24\ \text{h}} \times \dfrac{\text{h}}{60\ \text{min}}$$

$$= 0.5\left[\dfrac{\text{Mgal}}{\text{d}}\right] \times \dfrac{10^6}{\text{M}} \times \dfrac{\text{d}}{24\ \text{h}} \times \dfrac{\text{h}}{60\ \text{min}}$$

$$= 347\ \text{gpm}$$

3. Estimate the surface loading rate ($LR_{surface}$) in gallons per minute per square foot (gpm/ft²) of surface area.

$$LR_{surface}\left[\dfrac{\text{gal}}{\text{min}\cdot\text{ft}^2}\right] = \dfrac{Q\left[\dfrac{\text{gal}}{\text{min}}\right]}{A_{surface}\left[\text{ft}^2\right]} = \dfrac{347\left[\dfrac{\text{gal}}{\text{min}}\right]}{800\left[\text{ft}^2\right]} = 0.43\ \dfrac{\text{gpm}}{\text{ft}^2}$$

Example 16

Estimate the mean flow velocity in feet per minute (ft/min) through a rectangular sedimentation basin 20 feet (ft) wide and 8 feet (ft) deep when the flow rate is 350 gpm.

Known	**Unknown**
W = Width = 20 ft	v = Mean Flow Velocity, ft/min
D = Depth = 8 ft	
Q = Flow Rate = 350 gpm	

Estimate the mean flow velocity in feet per minute (ft/min).

$$V\left[\frac{ft}{min}\right] = \frac{Q\left[\frac{gal}{min}\right] \times \frac{ft^3}{7.48\ gal}}{A[ft^2]} = \frac{Q\left[\frac{gal}{min}\right] \times \frac{ft^3}{7.48\ gal}}{W[ft] \times D[ft]}$$

$$= \frac{350\left[\frac{gal}{min}\right] \times \frac{ft^3}{7.48\ gal}}{20[ft] \times 8[ft]}$$

$$= 0.29\frac{ft}{min}$$

Example 17

Estimate the weir loading in gallons per minute per foot (gpm/ft) of weir length for a 30-foot (ft)-diameter circular clarifier treating a flow rate of 350 gpm. The weir is located on the water edge of the clarifier.

Known	**Unknown**
D = Weir Diameter = 30 ft	LR_{weir} = Weir Loading, gpm/ft
Q = Flow Rate = 350 gpm	

1. Calculate the weir length in feet (ft).

$$L_{weir}[ft] = \pi \times D = 3.14 \times 30[ft] = 94.2\ ft$$

2. Estimate the weir loading in gallons per minute per foot (gpm/ft) of weir.

$$LR_{weir}\left[\frac{gal}{min \cdot ft}\right] = \frac{Q\left[\frac{gal}{min}\right]}{L_{weir}[ft]} = \frac{350\left[\frac{gal}{min}\right]}{94.2[ft]} = 3.7\frac{gpm}{ft}$$

A.9.7   Filtration

Example 18

A 25-foot (ft)-wide by 30-foot (ft)-long rapid sand filter treats a flow rate of 2,000 gallons per minute (gpm). Calculate the filtration rate in gallons per minute per square foot (gpm/ft²) of filter area.

Known	**Unknown**
W = Width = 25 ft	Filtration Rate, gal/min/ft²
L = Length = 30 ft	
Q = Flow Rate = 2,000 gpm	

1. Calculate the surface area ($A_{surface}$) of the filter in square feet (ft²).

$$A_{surface} \, [ft^2] = L[ft] \times W[ft] = 30[ft] \times 25[ft] = 750 \ ft^2$$

2. Calculate the filtration rate in gallons per minute per square foot (gpm/ft²) of filter surface area.

$$\text{Filtration Rate}\left[\frac{gal}{min \cdot ft^2}\right] = \frac{Q\left[\frac{gal}{min}\right]}{A_{surface}\left[ft^2\right]} = \frac{2{,}000\left[\frac{gal}{min}\right]}{750\left[ft^2\right]} = 2.7\,\frac{gpm}{ft^2}$$

Example 19

With the inflow water to a rapid sand filter shut off, the water is observed to drop 20 inches (in) in nine minutes (min). What is the velocity of the water dropping in feet per minute (ft/min)?

Known	Unknown
h = Water Drop = 20 in	v = Velocity of the Water Dropping, ft/min
t = Time = 9 min	

Calculate the velocity of the water drop in feet per minute (ft/min).

$$v\left[\frac{ft}{min}\right] = \frac{h[ft]}{t[min]} = \frac{20[in] \times \dfrac{ft}{12 \ in}}{9[min]} = 0.185\,\frac{ft}{min}$$

Example 20

Estimate the flow rate through a rapid sand filter in cubic feet per minute (ft³/min) when the velocity of the water dropping is 0.18 feet per minute (ft/min) and the filter is 25 feet (ft) wide and 30 feet (ft) long.

Known	Unknown
v = Velocity of the Water Dropping = 0.18 ft/min	Q = Flow Rate, ft³/min
W = Width = 25 ft	
L = Length = 30 ft	

1. Calculate the surface area ($A_{surface}$) of the filter in square feet (ft²).

$$A_{surface} \, [ft^2] = L[ft] \times W[ft] = 30[ft] \times 25[ft] = 750 \ ft^2$$

2. Estimate the flow rate through the filter in cubic feet per minute (ft³/min).

$$Q\left[\frac{ft^3}{min}\right] = v\left[\frac{ft}{min}\right] \times A_{surface}\left[ft^2\right] = 0.18\left[\frac{ft}{min}\right] \times 750\left[ft^2\right] = 135\,\frac{ft^3}{min}$$

Example 21

Calculate the flow rate through a rapid sand filter in gallons per minute (gpm) when the flow rate is 135 cubic feet per minute (ft³/min).

Known	Unknown
Q = Flow Rate = 135 ft³/min	Q = Flow Rate, gpm

Calculate the flow rate through the filter in gallons per minute (gpm).

$$Q\left[\frac{gal}{min}\right] = Q\left[\frac{ft^3}{min}\right] \times \frac{7.48\ gal}{ft^3} = 135\left[\frac{ft^3}{min}\right] \times \frac{7.48\ gal}{ft^3} = 1,010\ gpm$$

Example 22

Calculate the flow rate through a rapid sand filter in gallons per minute (gpm) when 1.5 million gallons (Mgal) flowed through the filter during a 24-hour (h) filter run.

Known	Unknown
V = Flow Volume = 1.5 Mgal	Q = Flow Rate, gpm
t = Filter Run = 24 h	

Calculate the flow rate through the filter in gallons per minute (gpm).

$$Q\left[\frac{gal}{min}\right] = \frac{V}{t} = \frac{1,500,000\,[gal]}{24\,[h] \times \dfrac{60\ min}{h}} = 1,042\ gpm$$

Example 23

Determine the unit filter run volume (UFRV) in gallons per square foot (gal/ft²) for a filter 20 feet (ft) long and 16 feet (ft) wide if the volume of water filtered between backwash cycles is 2.2 million gallons (Mgal).

Known	Unknown
L = Length = 20 ft	UFRV, gal/ft²
W = Width = 16 ft	
V = Volume Filtered = 2,200,000 gal	

Calculate the UFRV in gallons per square foot (gal/ft²) of filter surface area ($A_{surface}$).

$$UFRV\left[\frac{gal}{ft^2}\right] = \frac{V\,[gal]}{A_{surface}\,[ft^2]} = \frac{V\,[gal]}{L\,[ft] \times W\,[ft]} = \frac{2,200,000\,[gal]}{20\,[ft] \times 16\,[ft]} = 6,875\frac{gal}{ft^2}$$

Example 24

Determine the unit filter run volume (UFRV) in gallons per square foot (gal/ft²) for a filter if the filtration rate was 2.3 gpm/square foot (gpm/ft²) during a 46-hour (h) filter run.

Known	Unknown
Filtration Rate = 2.3 gal/min/ft²	UFRV, gal/ft²
t = Filter Run = 46 h	

Calculate the UFRV in gallons per square foot (gal/ft²) of filter surface area ($A_{surface}$).

$$UFRV\left[\frac{gal}{ft^2}\right] = \text{Filtration Rate}\left[\frac{gal}{min \cdot ft^2}\right] \times t[h] \times \frac{60\ min}{h}$$

$$= 2.3\left[\frac{gal}{min \cdot ft^2}\right] \times 46[h] \times \frac{60\ min}{h}$$

$$= 6,348\,\frac{gal}{ft^2}$$

Example 25

Calculate the backwash pumping rate required in gallons per minute (gpm) to backwash a 25-foot (ft)-wide by 30-foot (ft)-long filter if the desired backwash rate is 20 gallons per minute per square foot (gpm/ft²).

Known	Unknown
W = Width = 25 ft	$Q_{backwash}$ = Backwash Pumping Rate, gal/min
L = Length = 30 ft	
$LR_{backwash}$ = Backwash Loading Rate = 20 gal/min/ft²	

1. Calculate the surface area ($A_{surface}$) of the filter.

$$A_{surface}\ [ft^2] = L[ft] \times W[ft] = 30[ft] \times 25[ft] = 750\ ft^2$$

2. Calculate the backwash pumping rate ($Q_{backwash}$) in gallons per minute (gpm).

$$Q_{backwash}\left[\frac{gal}{min}\right] = A_{surface}\left[ft^2\right] \times LR_{backwash}\left[\frac{gal}{min \cdot ft^2}\right]$$

$$= 750\left[ft^2\right] \times 20\left[\frac{gal}{min \cdot ft^2}\right]$$

$$= 15,000\ gpm$$

Example 26

Convert a filter backwash loading rate ($LR_{backwash}$) of 23 gallons per minute per square foot (gpm/ft²) to inches per minute (in/min) of rise.

Known	Unknown
$LR_{backwash}$ = Backwash Loading Rate = 23 gal/min/ft²	$LR_{backwash}$ = Volume of Backwash Water, in/min

Convert the backwash rate from gpm/ft² to in/min.

$$LR_{backwash}\left[\frac{in}{min}\right] = LR_{backwash}\left[\frac{gal}{min \cdot ft^2}\right] \times \frac{12\ in}{ft} \times \frac{ft^3}{7.48\ gal}$$

$$= 23\left[\frac{gal}{min \cdot ft^2}\right] \times \frac{12\ in}{ft} \times \frac{ft^3}{7.48\ gal}$$

$$= 37\,\frac{in}{min}$$

Example 27

Determine the volume or amount of water in gallons (gal) required to backwash a filter if the backwash pumping rate is 9,500 gpm when the backwash time is seven minutes (min).

Known	Unknown
$Q_{backwash}$ = Backwash Pumping Rate = 9,500 gal/min	$V_{backwash}$ = Volume of Backwash Water, gal
$t_{backwash}$ = Backwash Time = 7 min	

Calculate the volume of backwash water required in gallons (gal).

$$V_{backwash}\left[gal\right] = Q_{backwash}\left[\frac{gal}{min}\right] \times t_{backwash}\left[min\right]$$

$$= 9{,}500\left[\frac{gal}{min}\right] \times 7\left[min\right] = 66{,}500\ gal$$

Example 28

During a filter run, the total volume of water filtered was 13.0 million gallons (Mgal). When the filter was backwashed, 66,500 gallons (gal) of water were used. Calculate the percent (%) of the product or finished water used for backwashing.

Known	Unknown
$V_{filtered}$ = Volume of Water Filtered = 13,000,000 gal	Backwash, %
$V_{backwash}$ = Volume of Backwash Water = 66,500 gal	

Calculate the percent (%) of water used for backwashing.

$$Backwash\left[\%\right] = \frac{V_{backwash}\left[gal\right]}{V_{filtered}\left[gal\right]} \times 100\% = \frac{66{,}500\left[gal\right]}{13{,}000{,}000\left[gal\right]} \times 100\% = 0.5\%$$

A.9.8   Disinfection

Example 29

Calculate the chlorine dose in mg/L when a chlorinator is set to feed 18 pounds (lb) of chlorine in 24 hours (h). The flow rate is 570,000 gallons per day (gpd).

Known	Unknown
Cl_{feed} = Chlorine Feed = 18 lb Cl/24 h (18 lb Cl/d)	Cl_{dose} = Chlorine Dose, mg/L
Q = Flow Rate = 570,000 gpd (0.57 Mgal/d)	

Calculate the chlorine dose in milligrams per liter (mg/L).

$$Cl_{dose}\left[\frac{mg}{L}\right] = 0.120 \times \frac{Cl_{feed}\left[\frac{lb}{d}\right]}{Q\left[\frac{Mgal}{d}\right]} = 0.120 \times \frac{18\left[\frac{lb}{d}\right]}{0.57\left[\frac{Mgal}{d}\right]} = 3.8\frac{mg}{L}$$

Example 30

Estimate the chlorine demand for a water in milligrams per liter (mg/L) if the chlorine dose is 2.9 mg/L and the chlorine residual is 0.6 mg/L.

Known

Cl_{dose} = Chlorine Dose = 2.9 mg/L

$Cl_{residual}$ = Chlorine Residual = 0.6 mg/L

Unknown

Cl_{demand} = Chlorine Demand, mg/L

Estimate the chlorine demand of the water in milligrams per liter (mg/L).

$$Cl_{demand}\left[\frac{mg}{L}\right] = Cl_{dose}\left[\frac{mg}{L}\right] - Cl_{residual}\left[\frac{mg}{L}\right]$$

$$= 2.9\left[\frac{mg}{L}\right] - 0.6\left[\frac{mg}{L}\right] = 2.3\frac{mg}{L}$$

Example 31

Calculate the pounds (lb) of chlorine used to disinfect water if 150 gallons (gal) of hypochlorite as a 2.5 percent (%) chlorine solution was used.

Known

$V_{hypochlorite}$ = Volume of Hypochlorite = 150 gal

Hypochlorite Content = 2.5%

Unknown

Cl = Chlorine Used to Disinfect Water, lb

Calculate the pounds (lb) of chlorine used.

$$Cl[lb] = V_{hypochlorite}[gal] \times \frac{8.34\ lb}{gal} \times \frac{Hypochlorite[\%]}{100\%}$$

$$= 150[gal] \times \frac{8.34\ lb}{gal} \times \frac{2.5\%}{100\%} = 31.3\ lb$$

Example 32

Estimate the desired strength (as a percent chlorine) of a hypochlorite solution that is pumped by a hypochlorinator that delivers 115 gallons per day (gpd). The water being treated requires a chlorine dose of 12 pounds of chlorine per day (lb/d).

Known

$Q_{hypochlorinator}$ = Hypochlorinator Flow Rate = 115 gpd

$Cl_{required}$ = Chlorine Required = 12 lb/d

Unknown

Hypochlorite Strength, %

Estimate the desired hypochlorite strength as a percent (%) chlorine.

$$Hypochlorite\ Strength[\%] = \frac{Cl_{required}\left[\frac{lb}{d}\right] \times 100\%}{Q_{hypochlorinator}\left[\frac{gal}{d}\right] \times \frac{8.34\ lb}{gal}}$$

$$= \frac{12\left[\frac{lb}{d}\right] \times 100\%}{115\left[\frac{gal}{d}\right] \times \frac{8.34\ lb}{gal}} = 1.25\%$$

Example 33

How many gallons (gal) of water must be added to 15 gallons (gal) of 5 percent (%) hypochlorite solution to produce a 1.25 percent (%) hypochlorite solution?

Known	Unknown
$V_{hypochlorite}$ = Volume of Hypochlorite = 15 gal	V_{water} = Volume of Water
Actual Hypochlorite Content = 5%	Added, gal
Desired Hypochlorite Content = 1.25%	

Start with the formula to determine the desired hypochlorite content.

$$\text{Desired Hypochlorite Content}[\%] = \frac{V_{hypochlorite}[gal] \times \text{Actual Hypochlorite Content}[\%]}{V_{hypochlorite}[gal] + V_{water}[gal]}$$

$$1.25[\%] = \frac{15[gal] \times 5[\%]}{15[gal] + V_{water}[gal]}$$

Rearrange by cross multiplying.

$$15[gal] + V_{water} = \frac{15[gal] \times 5[\%]}{1.25[\%]}$$

$$15 + V_{water} = 60$$

$$V_{water} = 60 - 15 = 45 \text{ gal}$$

A.9.9 Corrosion Control

Example 34

Find the pH_S of a water at 10°C having a TDS of 100 mg/L, alkalinity of 40 mg/L, and a calcium hardness of 60 mg/L.

Known	Unknown
T = Water Temp = 10°C	pH_S
TDS = 100 mg/L	
Alkalinity = 40 mg/L as $CaCO_3$	
Ca Hardness = 60 mg/L as $CaCO_3$	

1. Find the formula values from the tables in Chapter 7, "Corrosion Control."

   From Table 7.3 for a water temperature of 10°C,

   A = 2.20

   From Table 7.4 for a TDS of 100 mg/L,

   B = 9.75

   $\log(Ca^{2+})$ = log(60) = 1.78
   $\log(Alkalinity)$ = log(40) = 1.60

2. Calculate pH_S.

$$pH_S = A + B - \log(Ca^{2+}) - \log(\text{Alkalinity})$$
$$= 2.20 + 9.75 - 1.78 - 1.60 = 8.57$$

Example 35

Calculate the Langelier Index for a water with a calculated pH_S value of 8.69 and an actual pH of 8.5.

Known	Unknown
pH_S = 8.69	Langelier Index
pH = 8.5	

Calculate the Langelier Index.

$$\text{Langelier Index} = pH - pH_S = 8.5 - 8.69 = -0.19$$

Since the Langelier Index is negative, the water is corrosive.

A.9.10   Laboratory Procedures

Example 36

Convert the temperature of water from 65° Fahrenheit (°F) to degrees Celsius (°C).

Known	Unknown
T = Temperature = 65° Fahrenheit (°F)	T = Temperature in degrees Celsius (°C)

Change 65 degrees Fahrenheit (°F) to degrees Celsius (°C).

$$T[^\circ C] = (^\circ F - 32^\circ) \times \frac{5}{9} = (65^\circ - 32^\circ) \times \frac{5}{9} = 18.3^\circ C$$

Example 37

Convert a water temperature of 12° Celsius (°C) to degrees Fahrenheit (°F).

Known	Unknown
T = Temperature = 12° Celsius (°C)	T = Temperature in degrees Fahrenheit (°F)

Change 12 degrees Celsius (°C) to degrees Fahrenheit (°F).

$$T[^\circ F] = \left(^\circ C \times \frac{9}{5}\right) + 32^\circ = \left(12^\circ \times \frac{9}{5}\right) + 32^\circ = 21.6^\circ + 32^\circ = 53.6^\circ F$$

Example 38

Results from the MPN tests during one week were as follows:

Day of the Week	Sun	Mon	Tue	Wed	Thu	Fri	Sat
MPN/100 mL	2	4	6	7	9	5	2

Estimate the mean, median, and geometric mean of the data in MPN/100 mL.

1. Calculate the mean.

$$\text{Mean}\left[\frac{\text{MPN}}{100 \text{ mL}}\right] = \frac{(2+4+6+7+9+5+2)}{7} = \frac{35}{7} = 5\frac{\text{MPN}}{100 \text{ mL}}$$

2. To determine the median, rearrange the data in ascending (increasing) order and select the middle value (three will be smaller and three will be larger in this example).

Order	1	2	3	4	5	6	7
MPN/100 mL	2	2	4	5	6	7	9

$$\text{Median}\left[\frac{\text{MPN}}{100 \text{ mL}}\right] = \text{Middle Value of a Group of Data} = 5\frac{\text{MPN}}{100 \text{ mL}}$$

3. Calculate the geometric mean for the given data.

$$\text{Geometric Mean}\left[\frac{\text{MPN}}{100 \text{ mL}}\right] = (2 \times 4 \times 6 \times 7 \times 9 \times 5 \times 2)^{1/7} = (30,240)^{0.143}$$

$$= 4.4\frac{\text{MPN}}{100 \text{ mL}}$$

A.10   Typical Water Treatment Plant Problems (Metric System)

A.10.1   Plant Operation

Example 1

Estimate the average use of chlorine in kilograms per day (kg/d) based on actual use of chlorine for one week as shown below.

Day of the Week	Sun	Mon	Tue	Wed	Thu	Fri	Sat
Chlorine Used, kg	23	37	35	31	32	36	24

Known	Unknown
Chlorine Used Each Day of Week, kg/d	F_A = Daily Average Use of Chlorine, kg/d

Estimate the average chlorine use in kilograms of chlorine per day (kg/d).

$$F_A\left[\frac{\text{kg}}{\text{d}}\right] = \frac{\text{Sum of Chlorine Used Each Day}}{\text{Total Number of Days}}$$

$$= \frac{23[\text{kg}] + 37[\text{kg}] + 35[\text{kg}] + 31[\text{kg}] + 32[\text{kg}] + 36[\text{kg}] + 24[\text{kg}]}{7[\text{d}]}$$

$$= \frac{218[\text{kg}]}{7[\text{d}]}$$

$$= 31.1\frac{\text{kg}}{\text{d}}$$

Example 2

A water treatment plant has five 68-kilogram (kg) chlorine cylinders in storage. The plant uses an average of 12 kilograms of chlorine per day (kg/d). How many days' (d) supply of chlorine is in storage?

Known	Unknown
n = Number of Chlorine Cylinders = 5 cyl	Supply of Chlorine, d
M_{cyl} = Cylinder Weight = 68 kg	
F = Average Use = 12 kg/d	

Calculate the available supply of chlorine in storage in days (d).

$$\text{Supply of Chlorine}\,[d] = \frac{M_{cyl} \times n}{F} = \frac{68\left[\dfrac{kg}{cyl}\right] \times 5\,[cyl]}{12\left[\dfrac{kg}{d}\right]} = 28 \text{ d}$$

A.10.2 Flows

Example 3

Convert a flow rate of 500 gallons per minute (gpm) to liters per second (L/s) and cubic meters per day (m^3/d).

Known	Unknown
Q = Flow Rate = 500 gpm	Q = Flow Rate, L/s
	Q = Flow Rate, m^3/d

1. Convert the flow rate from 500 gpm to liters/seconds (L/s).

$$Q\left[\frac{L}{s}\right] = Q\left[\frac{gal}{min}\right] \times \frac{3.785\ L}{gal} \times \frac{min}{60\ s}$$

$$= 500\left[\frac{gal}{min}\right] \times \frac{3.785\ L}{gal} \times \frac{min}{60\ s}$$

$$= 31.5\frac{L}{s}$$

2. Convert the flow rate from 500 gpm to cubic meters per day (m^3/d).

$$Q\left[\frac{m^3}{d}\right] = Q\left[\frac{gal}{min}\right] \times \frac{3.785\ L}{gal} \times \frac{m^3}{1,000\ L} \times \frac{60\ min}{h} \times \frac{24\ h}{d}$$

$$= 500\left[\frac{gal}{min}\right] \times \frac{3.785\ L}{gal} \times \frac{m^3}{1,000\ L} \times \frac{60\ min}{h} \times \frac{24\ h}{d}$$

$$= 2,725\frac{m^3}{d}$$

A.10.3 Chemical Doses

Example 4

Determine the chlorinator setting in kilograms per day (kg/d) if 4,000 cubic meters of water per day (m^3/d) are to be treated with a desired chlorine dose of 2.5 mg/L.

Known   Unknown

Q   = Flow Rate   = 4,000 m³/d   F = Chlorinator Setting, kg/d
Cl_{dose} = Chlorine Dose = 2.5 mg/L

Determine the chlorinator setting in kilograms per day (kg/d).

$$F\left[\frac{kg}{d}\right] = Q\left[\frac{m^3}{d}\right] \times Cl_{dose}\left[\frac{mg}{L}\right] \times \frac{1,000\ L}{m^3} \times \frac{kg}{10^6\ mg}$$

$$= 4,000\left[\frac{m^3}{d}\right] \times 2.5\left[\frac{mg}{L}\right] \times \frac{1,000\ L}{m^3} \times \frac{kg}{10^6\ mg}$$

$$= 10\frac{kg}{d}$$

Example 5

The optimum liquid alum dose from the jar tests is 12 mg/L. Determine the setting on the liquid alum chemical feeder in milliliters per minute (mL/min) when the plant flow rate is 15 megaliters per day (or million liters per day). The liquid alum delivered to the plant contains 485 milligrams (mg) of alum per milliliter (mL) of liquid solution with a specific gravity of 1.325.

Known   Unknown

D_{alum} = Alum Dose   = 12 mg/L   Chemical Feeder Setting, mL/min
Q   = Flow Rate   = 15 ML/d
SG   = Specific Gravity = 1.325
Liquid Alum Content   = 485 mg/mL

Calculate the liquid alum chemical feeder setting in milliliters per minute (mL/min).

$$\text{Chemical Feeder Setting}\left[\frac{mL}{min}\right]$$

$$= \frac{Q\left[\frac{ML}{d}\right] \times D_{alum}\left[\frac{mg}{L}\right] \times \frac{10^6}{M} \times \frac{d}{24\,h} \times \frac{h}{60\,min}}{SG \times \text{Liquid Alum}\left[\frac{mg}{mL}\right]}$$

$$= \frac{15\left[\frac{ML}{d}\right] \times 12\left[\frac{mg}{L}\right] \times \frac{10^6}{M} \times \frac{d}{24\,h} \times \frac{h}{60\,min}}{1.325 \times 485\left[\frac{mg}{mL}\right]}$$

$$= 195\frac{mL}{min}$$

Example 6

The optimum liquid alum dose from the jar tests is 8 mg/L. Determine the setting on the liquid alum chemical feeder in milliliters per minute (mL/min) when the flow rate is 12 megaliters per day (ML/d). The liquid alum delivered to the plant contains 5.36 pounds (lb) of alum per gallon (gal) of liquid solution.

Known   Unknown

D_{alum} = Alum Dose   = 8 mg/L   Chemical Feeder Setting, mL/min
Q   = Flow Rate   = 12 ML/d
Liquid Alum Content = 5.36 lb/gal

Calculate the liquid alum chemical feeder setting in milliliters per minute (mL/min).

$$\text{Chemical Feeder Setting}\left[\frac{mL}{min}\right]$$

$$= \frac{Q\left[\dfrac{ML}{d}\right] \times D_{alum}\left[\dfrac{mg}{L}\right] \times \dfrac{10^6}{M} \times \dfrac{d}{24\,h} \times \dfrac{h}{60\,min}}{\text{Liquid Alum}\left[\dfrac{lb}{gal}\right] \times \dfrac{gal}{3.785\,L} \times \dfrac{L}{1,000\,mL} \times \dfrac{kg}{2.205\,lb} \times \dfrac{10^6\,mg}{kg}}$$

$$= \frac{12\left[\dfrac{ML}{d}\right] \times 8\left[\dfrac{mg}{L}\right] \times \dfrac{10^6}{M} \times \dfrac{d}{24\,h} \times \dfrac{h}{60\,min}}{5.36\left[\dfrac{lb}{gal}\right] \times \dfrac{gal}{3.785\,L} \times \dfrac{L}{1,000\,mL} \times \dfrac{kg}{2.205\,lb} \times \dfrac{10^6\,mg}{kg}}$$

$$= 104\frac{mL}{min}$$

Example 7

An operator is interested in measuring the dry chemical feed rate in kilograms per day (kg/d), so he places an empty bucket weighing 150 grams (g) under the dry chemical feeder. After 12 minutes (min), the bucket weighs 1,800 grams (g). Calculate the feed rate based on the information provided.

Known		Unknown
M_{empty} = Weight of Empty Bucket	= 150 g	Dry Chemical Feed
M_{full} = Weight of Bucket and Chemical	= 1,800 g	Rate, kg/d
t = Feed Time	= 12 min	

1. Determine the amount of chemical fed in kilograms (kg).

$$\text{Chemical Fed}[kg] = \left(M_{full}[g] - M_{empty}[g]\right) \times \frac{kg}{1,000\,g}$$

$$= \left(1,800[g] - 150[g]\right) \times \frac{kg}{1,000\,g}$$

$$= 1,650[g] \times \frac{kg}{1,000\,g}$$

$$= 1.65\,kg$$

2. Calculate the dry chemical feed rate in kilograms per minute (kg/min).

$$\text{Chemical Feed Rate}\left[\frac{kg}{min}\right] = \frac{\text{Chemical Fed}[kg]}{t[min]}$$

$$= \frac{1.65[kg]}{12[min]}$$

$$= 0.1375\frac{kg}{min}$$

3. Calculate the dry chemical feed rate in kilograms per day (kg/d).

$$\text{Chemical Feed Rate}\left[\frac{\text{kg}}{\text{d}}\right]$$

$$= \text{Chemical Feed Rate}\left[\frac{\text{kg}}{\text{min}}\right] \times \frac{60\,\text{min}}{\text{h}} \times \frac{24\,\text{h}}{\text{d}}$$

$$= 0.1375\left[\frac{\text{kg}}{\text{min}}\right] \times \frac{60\,\text{min}}{\text{h}} \times \frac{24\,\text{h}}{\text{d}}$$

$$= 198\frac{\text{kg}}{\text{d}}$$

Example 8

A small chemical feed pump lowered the chemical solution in an 80-centimeter (cm)-diameter tank 35 centimeters (cm) during an eight-hour (h) period. Estimate the feed rate delivered by the pump in liters per second (L/s).

Known	Unknown
D = Tank Diameter = 80 cm or 0.8 m	Chemical Used, L
h = Tank Drop = 35 cm or 0.35 m	Chemical Feed Rate, L/min
t = Time = 8 h	

1. Calculate the volume of chemical used in cubic meters (m³).

$$\text{Volume}\left[\text{m}^3\right] = \frac{\pi}{4} \times D^2\left[\text{m}^2\right] \times h\left[\text{m}\right]$$

$$= 0.785 \times 0.8^2\left[\text{m}^2\right] \times 0.35\left[\text{m}\right] = 0.176\,\text{m}^3$$

2. Determine the liters (L) of solution pumped.

$$\text{Chemical Used}\left[\text{L}\right] = \text{Volume}\left[\text{m}^3\right] \times \frac{1{,}000\,\text{L}}{\text{m}^3}$$

$$= 0.176\left[\text{m}^3\right] \times \frac{1{,}000\,\text{L}}{\text{m}^3} = 176\,\text{L}$$

3. Estimate the feed rate delivered by the pump in liters per second (L/s).

$$\text{Chemical Feed Rate}\left[\frac{\text{L}}{\text{min}}\right] = \frac{\text{Chemical Used}\left[\text{L}\right]}{t\left[\text{h}\right] \times \dfrac{60\,\text{min}}{\text{h}}}$$

$$= \frac{176\left[\text{L}\right]}{8\left[\text{h}\right] \times \dfrac{60\,\text{min}}{\text{h}}} = 0.37\,\text{L/s}$$

A.10.4 Reservoir Management and Intake Structures

Example 9

A reservoir has a surface area of 0.75 hectares (ha) and an average depth of seven meters (m). Estimate the volume of the reservoir in cubic meters (m³).

	Known		Unknown

$A_{surface}$ = Surface Area = 0.75 ha V = Reservoir Volume, m^3
h = Depth   = 7 m

Estimate the reservoir volume in cubic meters (m^3).

$$V[m^3] = A_{surface}[ha] \times \frac{10,000\,m^2}{ha} \times h[m]$$

$$= 0.75[ha] \times \frac{10,000\,m^2}{ha} \times 7[m] = 52,500\,m^3$$

Example 10

A reservoir has a surface area of 0.48 hectares (ha) and the desired dose of copper sulfate is three kilograms per hectare (kg/ha). How many kilograms (kg) of copper sulfate will be needed?

Known   **Unknown**

$A_{surface}$ = Surface Area   = 0.48 ha $CuSO_4$ = Copper Sulfate, kg
D_{CuSO_4} = Copper Sulfate Dose = 3 kg/ha

Calculate the kilograms (kg) of copper sulfate needed.

$$CuSO_4[kg] = A_{surface}[ha] \times D_{CuSO_4}$$

$$= 0.48[ha] \times 3\left[\frac{kg}{ha}\right] = 1.44\,kg\,Copper\,Sulfate$$

Example 11

The volume of a reservoir is estimated to be 16,000 cubic meters (m^3). The desired dose of copper is 0.5 mg/L and the copper content of the copper sulfate to be used is 25 percent (%). How many kilograms (kg) of copper sulfate will be needed?

Known   **Unknown**

V = Reservoir Volume = 16,000 m^3 $CuSO_4$ = Copper Sulfate, kg
D_{Cu} = Copper Dose   = 0.5 mg/L
Copper Content   = 25%

Calculate the kilograms (kg) of copper sulfate that will be needed.

$$CuSO_4[kg] = \frac{V \times D_{Cu} \times 100\%}{Copper[\%]}$$

$$= \frac{16,000[m^3] \times 0.5\left[\dfrac{mg}{L}\right] \times \dfrac{1,000\,L}{m^3} \times \dfrac{kg}{10^6\,mg} \times 100\%}{25\%}$$

$$= 32\,kg\,Copper\,Sulfate$$

A.10.5　Coagulation and Flocculation

Example 12

Estimate the actual alum dose in milligrams per liter (mg/L) if a plant treats a raw water with a flow rate of 7.0 ML/day (ML/d). The alum feed rate is 60 kg per day (kg/d).

Known	Unknown
Q = Flow Rate = 7.0 ML/d	D_{alum} = Alum Dose, mg/L
Alum Used = 60 kg/d	

Calculate the alum dose in milligrams per liter (mg/L).

$$D_{alum}\left[\frac{mg}{L}\right] = \frac{Alum\ Used\left[\frac{kg}{d}\right] \times \frac{10^6\ mg}{kg}}{Q\left[\frac{ML}{d}\right] \times \frac{10^6}{M}}$$

$$= \frac{60\left[\frac{kg}{d}\right] \times \frac{10^6\ mg}{kg}}{7.0\left[\frac{ML}{d}\right] \times \frac{10^6}{M}} = 8.6\frac{mg}{L}$$

Example 13

Determine the strength of a polymer solution as a percent (%) if 80 grams (g) of dry polymer are mixed with 16 liters (L) of water.

Known	Unknown
$M_{polymer}$ = Mass of Dry Polymer = 80 g	Polymer Solution, percent (%)
V_{water} = Volume of Water = 16 L	

1. Convert the 16 liters (L) of water to grams (g).

$$Water[g] = V_{water} \times \frac{1,000\ g}{L} = 16[L] \times \frac{1,000\ g}{L} = 16,000\ g$$

2. Calculate the polymer solution as a percent (%).

$$Polymer[\%] = \frac{M_{polymer}[g] \times 100\%}{M_{polymer}[g] + Water[g]} = \frac{80[g] \times 100\%}{80[g] + 16,000[g]} = 0.50\%$$

A.10.6　Sedimentation

Example 14

Estimate the detention time in hours (h) for a 10-meter (m)-diameter circular clarifier when the flow rate is 2,000 cubic meters per day (m³/d). The clarifier is 3 meters (m) deep.

Known	Unknown
D = Diameter = 10 m	t = Detention Time, h
H = Depth = 3 m	
Q = Flow Rate = 2,000 m³/d	

1. Calculate the clarifier volume (V) in cubic meters (m³).

$$V[m^3] = 0.785 \times D^2[m^2] \times H[m]$$

$$= 0.785 \times 10^2[m^2] \times 3[m] = 235.5\,m^3$$

2. Estimate the detention time (t) of the clarifier in hours (h).

$$t[h] = \frac{V}{Q} = \frac{235.5[m^3]}{2,000\left[\dfrac{m^3}{d}\right] \times \dfrac{d}{24\,h}} = 2.8\,h$$

Example 15

Estimate the surface loading rate in millimeters per second (mm/s) for a rectangular sedimentation basin 6 meters (m) wide and 12 meters (m) long when the flow rate is 2,000 cubic meters per day (m³/d).

Known	Unknown
W = Width = 6 m	$LR_{surface}$ = Surface Loading, mm/s
L = Length = 12 m	
Q = Flow Rate = 2,000 m³/d	

1. Determine the surface area ($A_{surface}$) of the basin in square meters (m²).

$$A_{surface}[m^2] = L \times W = 12[m] \times 6[m] = 72\,m^2$$

2. Determine the surface loading rate ($LR_{surface}$) for the basin in millimeters per second (mm/s).

$$LR_{surface}\left[\frac{mm}{s}\right] = \frac{Q\left[\dfrac{m^3}{d}\right] \times \dfrac{d}{24\,h} \times \dfrac{h}{60\,min} \times \dfrac{min}{60\,s} \times \dfrac{1,000\,mm}{m}}{A_{surface}}$$

$$= \frac{2,000\left[\dfrac{m^3}{d}\right] \times \dfrac{d}{24\,h} \times \dfrac{h}{60\,min} \times \dfrac{min}{60\,s} \times \dfrac{1,000\,mm}{m}}{72[m^2]} = 0.32\,\frac{mm}{s}$$

Example 16

Estimate the mean flow velocity in meters per minute (m/min) through a rectangular sedimentation basin 6 meters (m) wide and 2.5 meters (m) deep when the flow rate is 20 liters per second (L/s).

Known	Unknown
W = Width = 6 m	v = Mean Flow Velocity, m/min
D = Depth = 2.5 m	
Q = Flow Rate = 20 L/s	

Estimate the mean flow velocity in meters per minute (m/min).

$$v\left[\frac{m}{min}\right] = \frac{Q\left[\frac{L}{s}\right] \times \frac{m^3}{1,000\,L} \times \frac{60\,s}{min}}{A\left[m^2\right]} = \frac{Q\left[\frac{L}{s}\right] \times \frac{m^3}{1,000\,L} \times \frac{60\,s}{min}}{W[m] \times D[m]}$$

$$= \frac{20\left[\frac{L}{s}\right] \times \frac{m^3}{1,000\,L} \times \frac{60\,s}{min}}{6[m] \times 2.5[m]} = 0.08\,\frac{m}{min}$$

Example 17

Estimate the weir loading in cubic meters per day per meter (m³/d/m) of weir length for a ten-meter (m)-diameter circular clarifier treating a flow rate of 20 liters per second (L/s). The weir is located on the outside edge of the clarifier.

Known	Unknown
D = Weir Diameter = 10 m	LR_{weir} = Weir Loading, m³/d/m
Q = Flow Rate = 20 L/s	

1. Calculate the weir length in meters (m).

$$L_{weir}[m] = \pi \times D = 3.14 \times 10\,[m] = 31.4\,m$$

2. Estimate the weir loading in cubic meters per day per meter (m³/d/m).

$$LR_{weir}\left[\frac{m^3}{d \cdot m}\right] = \frac{Q\left[\frac{L}{s}\right] \times \frac{m^3}{1,000\,L} \times \frac{60\,s}{min} \times \frac{60\,min}{h} \times \frac{24\,h}{d}}{L_{weir}\,[m]}$$

$$= \frac{20\left[\frac{L}{s}\right] \times \frac{m^3}{1,000\,L} \times \frac{60\,s}{min} \times \frac{60\,min}{h} \times \frac{24\,h}{d}}{31.4[m]} = 55\frac{m^3}{d \cdot m}$$

A.10.7   Filtration

Example 18

A 7.5-meter (m)-wide by 9-meter (m)-long rapid sand filter treats a flow rate of 125 liters per second (L/s). Calculate the filtration rate in liters per second per square meter (L/s/m²) and also in millimeters per second (mm/s).

Known	Unknown
W = Width = 7.5 m	Filtration Rate, L/s/m²
L = Length = 9 m	Filtration Rate, mm/s
Q = Flow Rate = 125 L/s	

1. Calculate the surface area ($A_{surface}$) of the filter in square meters (m²).

$$A_{surface}[m^2] = L[m] \times W[m] = 9[m] \times 7.5[m] = 67.5\,m^2$$

2. Calculate the filtration rate in liters per second per square meter (L/s/m²) of filter surface area.

$$\text{Filtration Rate}\left[\frac{L}{s \cdot m^2}\right] = \frac{Q\left[\frac{L}{s}\right]}{A_{surface}[m^2]} = \frac{125\left[\frac{L}{s}\right]}{67.5[m^2]} = 1.85\,L/s/m^2$$

3. Calculate the filtration rate in millimeters per second (mm/s).

$$\text{Filtration Rate}\left[\frac{mm}{s}\right] = \frac{Q\left[\frac{L}{s}\right] \times \frac{m^3}{1,000\,L} \times \frac{1,000\,mm}{m}}{A_{surface}[m^2]}$$

$$= \frac{125\left[\frac{L}{s}\right] \times \frac{m^3}{1,000\,L} \times \frac{1,000\,mm}{m}}{67.5[m^2]} = 1.85\frac{mm}{s}$$

Example 19

With the inflow water to a rapid sand filter shut off, the water is observed to drop 50 centimeters (cm) in nine minutes (min). What is the velocity of the water dropping in meters per minute (m/min)?

Known	Unknown
h = Water Drop = 50 cm	v = Velocity of the
t = Time = 9 min	Water Dropping, m/min

Calculate the velocity of the water drop in meters per minute (m/min).

$$v\left[\frac{m}{min}\right] = \frac{h[m]}{t[min]} = \frac{50[cm] \times \frac{m}{100\,cm}}{9[min]} = 0.056\frac{m}{min}$$

Example 20

Estimate the flow rate through a rapid sand filter in cubic meters per minute (m³/min) when the velocity of the water dropping is 0.05 meters per minute (m/min) and the filter is 7.5 meters (m) wide and 9 meters (m) long.

Known	Unknown
v = Velocity of the Water Dropping = 0.05 m/min	Q = Flow Rate,
W = Width = 7.5 m	m³/min
L = Length = 9 m	

1. Calculate the surface area (A_surface) of the filter in square meters (m²).

$$A_{surface}[m^2] = L[m] \times W[m] = 9[m] \times 7.5[m] = 67.5\,m^2$$

2. Estimate the flow rate through the filter in cubic meters per minute (m³/min).

$$Q\left[\frac{m^3}{min}\right] = v\left[\frac{m}{min}\right] \times A_{surface}[m^2]$$

$$= 0.05\left[\frac{m}{min}\right] \times 67.5[m^2] = 3.38\frac{m^3}{min}$$

Example 21

Calculate the flow rate through a rapid sand filter in liters per second (L/s) when the flow rate is 3.5 cubic meters per minute (m^3/min).

Known	Unknown
Q = Flow Rate = 3.5 m^3/min	Q = Flow Rate, L/s

Calculate the flow rate through the filter in liters per second (L/s).

$$Q\left[\frac{L}{s}\right] = Q\left[\frac{m^3}{min}\right] \times \frac{1,000\,L}{m^3} \times \frac{min}{60\,s}$$

$$= 3.5\left[\frac{m^3}{min}\right] \times \frac{1,000\,L}{m^3} \times \frac{min}{60\,s} = 58.3\frac{L}{s}$$

Example 22

Calculate the flow rate through a rapid sand filter in liters per second (L/s) when 6,000 cubic meters (m^3) flowed through the filter during a 30-hour (h) filter run.

Known	Unknown
V = Flow Volume = 6,000 m^3	Q = Flow Rate, L/s
t = Filter Run = 30 h	

Calculate the flow rate through the filter in liters per second (L/s).

$$Q\left[\frac{L}{s}\right] = \frac{V}{t} = \frac{6,000\left[m^3\right] \times \dfrac{1,000\,L}{m^3}}{30\left[h\right] \times \dfrac{60\,min}{h} \times \dfrac{60\,s}{min}} = 55.6\frac{L}{s}$$

Example 23

Determine the unit filter run volume (UFRV) in liters per square meter (L/m^2) for a filter 8 meters (m) long and 5 meters (m) wide if the volume of water filtered between backwash cycles is 8 megaliters (ML).

Known	Unknown
L = Length = 8 m	UFRV, L/m^2
W = Width = 5 m	
V = Volume Filtered = 8 ML	

Calculate the UFRV in liters per square meter (L/m^2) of filter surface area ($A_{surface}$).

$$UFRV\left[\frac{L}{m^2}\right] = \frac{V\left[L\right]}{A_{surface}\left[m^2\right]} = \frac{V\left[L\right]}{L\left[m\right] \times W\left[m\right]}$$

$$= \frac{8,000,000\left[L\right]}{8\left[m\right] \times 5\left[m\right]} = 200,000\frac{L}{m^2} \text{ or } 200\frac{m^3}{m^2}$$

Example 24

Determine the unit filter run volume (UFRV) in liters per square meter (L/m^2) for a filter if the filtration rate was 2 millimeters per second (mm/s) during a 46-hour (h) filter run.

Known	Unknown
Filtration Rate = 2 mm/s	UFRV, L/m^2
t = Filter Run = 46 h	

Calculate the UFRV in liters per square meter of filter surface area (L/m^2).

$$UFRV\left[\frac{L}{m^2}\right]$$

$$= \text{Filtration Rate}\left[\frac{mm}{s}\right] \times t[h] \times \frac{60\,s}{min} \times \frac{60\,min}{h} \times \frac{m}{1,000\,mm} \times \frac{1,000\,L}{m^3}$$

$$= 2\left[\frac{mm}{s}\right] \times 46[h] \times \frac{60\,s}{min} \times \frac{60\,min}{h} \times \frac{m}{1,000\,mm} \times \frac{1,000\,L}{m^3}$$

$$= 331,200\,\frac{L}{m^2}$$

Example 25

Calculate the backwash pumping rate required in cubic meters per day (m^3/d) to backwash a 7.5-meter (m)-wide by 9-meter (m)-long filter if the desired backwash rate is 0.8 cubic meters per day per square meter ($m^3/d/m^2$).

Known			Unknown
W	= Width	= 7.5 m	$Q_{backwash}$ = Backwash Pumping
L	= Length	= 9 m	Rate, m^3/d
$LR_{backwash}$	= Backwash		
	Loading Rate = 0.8 $m^3/d/m^2$		

1. Calculate the surface area ($A_{surface}$) of the filter.

   $$A_{surface}\,[m^2] = L[m] \times W[m] = 9[m] \times 7.5[m] = 67.5\,m^2$$

2. Calculate the backwash pumping rate ($Q_{backwash}$) in cubic meters per day (m^3/d).

   $$Q_{backwash}\left[\frac{m^3}{d}\right] = A_{surface}\left[m^2\right] \times LR_{backwash}\left[\frac{m^3}{d \cdot m^2}\right]$$

   $$= 67.5\left[m^2\right] \times 0.8\left[\frac{m^3}{d \cdot m^2}\right] = 54\,\frac{m^3}{d}$$

Example 26

Convert a filter backwash loading rate ($LR_{backwash}$) of 25 liters per minute per square meter ($L/min/m^2$) to millimeters per minute (mm/min) of rise.

$$\begin{array}{cc} \textbf{Known} & \textbf{Unknown} \\ LR_{backwash} = \text{Backwash Loading} & LR_{backwash} = \text{Volume of Backwash} \\ \text{Rate} = 25 \text{ L/min/m}^2 & \text{Water, mm/min} \end{array}$$

Convert the backwash rate from L/min/m² to mm/min.

$$LR_{backwash}\left[\frac{mm}{min}\right] = LR_{backwash}\left[\frac{L}{min \cdot m^2}\right] \times \frac{m^3}{1,000\,L} \times \frac{1,000\,mm}{m}$$

$$= 25\left[\frac{L}{min \cdot m^2}\right] \times \frac{m^3}{1,000\,L} \times \frac{1,000\,mm}{m}$$

$$= 25\frac{mm}{min}$$

Example 27

Determine the volume or amount of water in liters (L) required to backwash a filter if the backwash pumping rate is 600 liters per second (L/s) when the backwash time is seven minutes (min).

$$\begin{array}{cc} \textbf{Known} & \textbf{Unknown} \\ Q_{backwash} = \text{Backwash Pumping Rate} = 600 \text{ L/s} & V_{backwash} = \text{Volume of} \\ t_{backwash} = \text{Backwash Time} \qquad\qquad = 7 \text{ min} & \text{Backwash} \\ & \text{Water, L} \end{array}$$

Calculate the volume of backwash water required in liters (L).

$$V_{backwash}\left[L\right] = Q_{backwash}\left[\frac{L}{s}\right] \times t_{backwash}\left[min\right] \times \frac{60\,s}{min}$$

$$= 600\left[\frac{L}{s}\right] \times 7\left[min\right] \times \frac{60\,s}{min}$$

$$= 252,000\,L \text{ or } 0.252\,ML$$

Example 28

During a filter run, the total volume of water filtered was 50 megaliters (ML). When the filter was backwashed, 250,000 liters (L) of water were used. Calculate the percent (%) of the product or finished water used for backwashing.

$$\begin{array}{cc} \textbf{Known} & \textbf{Unknown} \\ V_{filtered} = \text{Volume of Water Filtered} \quad = 50 \text{ ML} & \text{Backwash, \%} \\ V_{backwash} = \text{Volume of Backwash Water} = 250,000 \text{ L} & \end{array}$$

Calculate the percent (%) of water used for backwashing.

$$\text{Backwash}\left[\%\right] = \frac{V_{backwash}}{V_{filtered}} \times 100\% = \frac{250,000\left[L\right]}{50\left[ML\right] \times \dfrac{10^6}{M}} \times 100\% = 0.5\%$$

A.10.8   Disinfection

Example 29

Calculate the chlorine dose in mg/L when a chlorinator is set to feed 8 kilograms (kg) of chlorine in 24 hours (h). The flow rate is 2 megaliters per day (ML/d).

Known	Unknown
Cl_{feed} = Chlorine Feed = 8 kg Cl/24 h (8 kg Cl/d)	Cl_{dose} = Chlorine
Q = Flow Rate = 2 ML/d	Dose, mg/L

Calculate the chlorine dose in milligrams per liter (mg/L).

$$Cl_{dose}\left[\frac{mg}{L}\right] = \frac{Cl_{feed}}{Q} = \frac{8\left[\frac{kg}{d}\right] \times \frac{10^6\ mg}{kg}}{2\left[\frac{ML}{d}\right] \times \frac{10^6}{M}} = 4\frac{mg}{L}$$

Example 30

Estimate the chlorine demand for a water in milligrams per liter (mg/L) if the chlorine dose is 2.9 mg/L and the chlorine residual is 0.6 mg/L.

Known	Unknown
Cl_{dose} = Chlorine Dose = 2.9 mg/L	Cl_{demand} = Chlorine Demand,
$Cl_{residual}$ = Chlorine Residual = 0.6 mg/L	mg/L

Estimate the chlorine demand of the water in milligrams per liter (mg/L).

$$Cl_{demand}\left[\frac{mg}{L}\right] = Cl_{dose}\left[\frac{mg}{L}\right] - Cl_{residual}\left[\frac{mg}{L}\right]$$

$$= 2.9\left[\frac{mg}{L}\right] - 0.6\left[\frac{mg}{L}\right] = 2.3\frac{mg}{L}$$

Example 31

Calculate the kilograms (kg) of chlorine used to disinfect water if 600 liters (L) of hypochlorite as a 2.5 percent (%) chlorine solution were used.

Known	Unknown
$V_{hypochlorite}$ = Volume of Hypochlorite = 600 L	Cl = Chlorine Used
Hypochlorite Content = 2.5%	to Disinfect
	Water, kg

Calculate the kilograms (kg) of chlorine used.

$$Cl[kg] = V_{hypochlorite}[L] \times \frac{1\ kg}{L} \times \frac{Hypochlorite[\%]}{100\%}$$

$$= 600[L] \times \frac{1\ kg}{L} \times \frac{2.5\%}{100\%} = 15\ kg$$

Example 32

Estimate the desired strength (as a percent chlorine) of a hypochlorite solution that is pumped by a hypochlorinator that delivers 0.45 cubic meters per day (m^3/d). The water being treated requires a chlorine dose of 6 kilograms of chlorine per day (kg/d).

Known	Unknown
$Q_{hypochlorinator}$ = Hypochlorinator Flow Rate = 0.45 m^3/d	Hypochlorite Strength, %
$Cl_{required}$ = Chlorine Required = 6 kg/d	

Estimate the desired hypochlorite strength as a percent (%) chlorine.

$$\text{Hypochlorite Strength}[\%] = \frac{Cl_{required}\left[\dfrac{kg}{d}\right] \times 100\%}{Q_{hypochlorinator}\left[\dfrac{m^3}{d}\right] \times \dfrac{1,000\,L}{m^3} \times \dfrac{1\,kg}{L}}$$

$$= \frac{6\left[\dfrac{kg}{d}\right] \times 100\%}{0.45\left[\dfrac{m^3}{d}\right] \times \dfrac{1,000\,L}{m^3} \times \dfrac{1\,kg}{L}} = 1.33\%$$

Example 33

How many liters (L) of water must be added to 40 liters (L) of 5 percent (%) hypochlorite solution to produce a 1.33 percent (%) hypochlorite solution?

Known	Unknown
$V_{hypochlorite}$ = Volume of Hypochlorite = 40 L	V_{water} = Volume of Water Added, L
Actual Hypochlorite Content = 5%	
Desired Hypochlorite Content = 1.33%	

Start with the formula to determine the desired hypochlorite content.

$$\text{Desired Hypochlorite Content}[\%] = \frac{V_{hypochlorite}[L] \times \text{Actual Hypochlorite Content}[\%]}{V_{hypochlorite}[L] + V_{water}[L]}$$

$$1.33[\%] = \frac{40[L] \times 5[\%]}{40[L] + V_{water}[L]}$$

Rearrange by cross multiplying.

$$40[L] + V_{water} = \frac{40[L] \times 5[\%]}{1.33[\%]}$$

$$40 + V_{water} = 150$$

$$V_{water} = 150 - 40 = 110\ L$$

A.10.9   Corrosion Control

Example 34

Find the pH_S of a water at 10°C having a TDS of 100 mg/L, alkalinity of 40 mg/L, and a calcium hardness of 60 mg/L.

Known	Unknown
T = Water Temp = 10°C	pH_S
TDS = 100 mg/L	
Alkalinity = 40 mg/L as $CaCO_3$	
Ca Hardness = 60 mg/L as $CaCO_3$	

1. Find the formula values from the tables in Chapter 7, "Corrosion Control."

   From Table 7.3 for a water temperature of 10°C,

   A = 2.20

   From Table 7.4 for a TDS of 100 mg/L,

   B = 9.75

   $\log(Ca^{2+})$ = log(60) = 1.78
   $\log(Alkalinity)$ = log(40) = 1.60

2. Calculate pH_S.

   $pH_s = A + B - \log(Ca^{2+}) - \log(Alkalinity)$
   $= 2.20 + 9.75 - 1.78 - 1.60 = 8.57$

Example 35

Calculate the Langelier Index for a water with a calculated pH_S value of 8.69 and an actual pH of 8.5.

Known	Unknown
pH_S = 8.69	Langelier Index
pH = 8.5	

Calculate the Langelier Index.

Langelier Index = $pH - pH_s$ = 8.5 - 8.69 = -0.19

Since the Langelier Index is negative, the water is corrosive.

A.10.10   Laboratory Procedures

Example 36

Convert the temperature of water from 65° Fahrenheit (°F) to degrees Celsius (°C).

Known	Unknown
T = Temperature = 65° Fahrenheit (°F)	T = Temperature in degrees Celsius (°C)

What If Your Combination Is Not in Table A.1?

The MPN for combinations not appearing in the table may be estimated by a formula developed by H.A. Thomas, Jr. in 1942 (Bacterial Densities from Fermentation Tube Tests. J Amer Water Works Assoc., 34:572):

$$\frac{MPN}{100\ mL} = \frac{P \times 100}{\sqrt{V_N \times V_T}}$$

Where,

P   = number of positive tubes in the 15-tube series

V_N = sum of the inoculation volumes (mL) in all the negative tubes

V_T = total inoculation volumes (sum of mL in all tubes)

Change 65 degrees Fahrenheit (°F) to degrees Celsius (°C).

$$T[°C] = (°F - 32°) \times \frac{5}{9} = (65° - 32°) \times \frac{5}{9} = 18.3°C$$

Example 37

Convert a water temperature of 12° Celsius (°C) to degrees Fahrenheit (°F).

Known	Unknown
T = Temperature = 12° Celsius (°C)	T = Temperature in degrees Fahrenheit (°F)

Change 12 degrees Celsius (°C) to degrees Fahrenheit (°F).

$$T[°F] = \left(°C \times \frac{9}{5}\right) + 32° = \left(12° \times \frac{9}{5}\right) + 32° = 21.6° + 32° = 53.6°F$$

Example 38

The results from an MPN test using three dilutions and five fermentation tubes in each dilution were as follows:

Inoculation volume (mL put into the tube)	10	1	0.1
Number of positive tubes (out of 5)	3	1	0
Number of negative tubes (= 5 − number of positives)	2	4	5

Calculate the values of the variables in the Thomas equation (see sidebar).

P = 3 + 1 = 4

V_N = (2 × 10[mL]) + (4 × 1[mL]) + (5 × 0.1[mL]) = 24.5 mL

V_T = (5 × 10[mL]) + (5 × 1[mL]) + (5 × 0.1[mL]) = 55.5 mL

Insert these values into the Thomas equation and calculate the MPN.

$$\frac{MPN}{100\ mL} = \frac{4 \times 100}{\sqrt{24.5 \times 55.5}} = 10.8/100\ mL$$

Usually, MPN values are rounded to two digits. Report this result as:

$$MPN = \frac{11\ Coliforms}{100\ mL}$$

Notice that this is the same as the table value (see Table A.1).

Table A.1.   MPN Index for combinations of gas-positive results using five tubes per dilution (10 mL, 1.0 mL, 0.1 mL). (Adapted from the FDA method presented in www.fda.gov/Food/FoodScienceResearch/LaboratoryMethods/ucm109656.htm.)

Combination of Positives	MPN Index / 100 mL	Combination of Positives	MPN Index / 100 mL	Combination of Positives	MPN Index / 100 mL
0-0-0	<1.8	3-0-0	7.8	5-0-0	23
0-0-1	1.8	3-0-1	11	5-0-1	31
0-1-0	1.8	3-0-2	13	5-0-2	43
0-1-1	3.6	3-1-0	11	5-0-3	58
0-2-0	3.7	3-1-1	14	5-0-4	76
0-2-1	5.5	3-1-2	17	5-1-0	33
0-3-0	5.6	3-2-0	14	5-1-1	46
0-3-1	7.4	3-2-1	17	5-1-2	63
		3-2-2	20	5-1-3	84
1-0-0	2.0	3-3-0	17	5-1-4	110
1-0-1	4.0	3-3-1	21	5-2-0	49
1-0-2	6.0	3-3-2	24	5-2-1	70
1-1-0	4.0	3-4-0	21	5-2-2	94
1-1-1	6.1	3-4-1	24	5-2-3	120
1-1-2	8.1	3-4-2	28	5-2-4	150
1-2-0	6.1	3-5-0	25	5-3-0	79
1-2-1	8.2	3-5-1	29	5-3-1	110
1-2-2	10	3-5-2	32	5-3-2	140
1-3-0	8.3			5-3-3	170
1-3-1	10	4-0-0	13	5-3-4	210
1-3-2	12	4-0-1	17	5-4-0	130
1-4-0	10	4-0-2	21	5-4-1	170
1-4-1	13	4-0-3	25	5-4-2	220
1-4-2	15	4-1-0	17	5-4-3	280
		4-1-1	21	5-4-4	350
2-0-0	4.5	4-1-2	26	5-4-5	430
2-0-1	6.8	4-1-3	31	5-5-0	240
2-0-2	9.1	4-2-0	22	5-5-1	350
2-1-0	6.8	4-2-1	26	5-5-2	540
2-1-1	9.2	4-2-2	32	5-5-3	920
2-1-2	12	4-2-3	38	5-5-4	1,600
2-2-0	9.3	4-3-0	27	5-5-5	>1,600
2-2-1	12	4-3-1	33		
2-2-2	14	4-3-2	39		
2-3-0	12	4-3-3	45		
2-3-1	14	4-4-0	34		
2-3-2	17	4-4-1	40		
2-4-0	15	4-4-2	47		
2-4-1	17	4-4-3	54		
2-4-2	20	4-5-0	41		
2-5-0	17	4-5-1	48		
2-5-1	20	4-5-2	56		
2-5-2	23	4-5-3	64		

Answer Key

Chapter 1   **INTRODUCTION TO WATER COLLECTION**

1. overpumping fresh water
2. 3
3. 1
4. 2
5. 2
6. 4
7. 3
8. 4
9. 2
10. 4
11. 2
12. 3
13. entrance to the distribution system
14. source water, flocculation effluent, sedimentation effluent, filter effluent
15. pathogenic organisms
16. industrial
17. 2
18. 2
19. 4
20. 4
21. 3
22. 4
23. 3
24. 3
25. 4

Chapter 2   **SOURCE WATER, RESERVOIR MANAGEMENT, AND INTAKE STRUCTURES**

1. B
2. I
3. H
4. E
5. C
6. D
7. G
8. A
9. F
10. J
11. 2
12. 3
13. 2
14. 4
15. 2
16. 3
17. 1
18. 2
19. 1
20. 2
21. 3
22. 4
23. 3
24. 4
25. 1
26. 3
27. 4
28. 3
29. 2
30. 1

Chapter 3   **COAGULATION AND FLOCCULATION**

1. 4
2. 1
3. 2
4. 4
5. 1
6. 1
7. 4
8. 1
9. 1
10. 1
11. 2
12. 2
13. 2
14. 1
15. 4
16. 2
17. 3
18. 3
19. 1
20. 4
21. 3
22. 3
23. 4
24. 3
25. 1

Chapter 4   **SEDIMENTATION**

1. 4
2. 3
3. 4
4. 1
5. 3
6. 2
7. 1
8. 2
9. 4
10. 1
11. solids-contact process units
12. coagulant aid
13. 3
14. 3
15. 2
16. 3
17. sedimentation
18. 4
19. 4
20. 3
21. 2
22. 2
23. 1
24. representative
25. 2

Chapter 5   FILTRATION

1. C
2. G
3. D
4. E
5. J
6. F
7. A

8. B
9. H
10. I
11. 3
12. 1
13. 1
14. 1

15. 4
16. 4
17. 3
18. 2
19. 2
20. 1
21. 2

22. 1
23. 4
24. 1
25. 2

Chapter 6   DISINFECTION

1. A
2. F
3. G
4. I
5. C
6. D
7. E
8. H
9. J
10. B

11. disinfection, sterilization
12. disinfectants, compounds
13. 1
14. 2
15. carcinogenic compounds, ammonia, sulfide compounds, phenolic tastes and odors
16. 4
17. weaken it, start leaking
18. 4

19. 3
20. dry, hydrochloric acid
21. 1
22. 100°, 38°
23. ventilated air, chlorine
24. continuous, water consumers
25. 1
26. 3
27. technological difficulties, high energy costs

28. safety aspects, health effects
29. 1
30. checking UV monitors, cleaning the UV lamps
31. 4
32. 2
33. 2
34. 1
35. 4

Chapter 7   CORROSION CONTROL

1. E
2. A
3. B
4. G
5. D
6. I
7. J
8. C

9. F
10. H
11. 2
12. chemical action
13. algal growths, corrode
14. 1
15. 1
16. 4

17. slowing
18. 3
19. 1
20. change, reduced
21. 3
22. 1
23. 1
24. 2

25. 1
26. reduce, inhibit, reversing
27. ferrous (iron) materials, concrete slab floors
28. 4
29. lead, corrosion-control program
30. 1, 2, 3

Chapter 8   TASTE AND ODOR CONTROL

1. G
2. B
3. H
4. C
5. F
6. I
7. D
8. A
9. J
10. E
11. significant

12. bottled drinking water, alternative water supplies
13. accelerate, worsen
14. night
15. natural runoff
16. 4
17. 3
18. inadequate or incomplete maintenance
19. locate where in the system the problem originates

20. dead ends, low-flow zones, areas subject to wide flow variations or changes in supply source
21. source water
22. standard jar test procedure
23. 1
24. 3
25. 4
26. objectionable byproducts of the reaction do not normally form

27. testing the carbons
28. 3
29. 1
30. 4
31. 1
32. effective communications network
33. recognize that a problem exists, determine the extent of the problem
34. 1
35. 4

Chapter 9   LABORATORY PROCEDURES

1. D
2. F
3. J
4. B
5. I
6. G
7. A
8. H
9. C
10. E

11. 16.7°C, 71.6°F
12. 5,000 mg, 0.250 L, 0.00005 g
13. 4
14. volumetric, graduated or Mohr, and serological
15. meniscus
16. known
17. 3
18. 2

19. acid to water
20. foam, carbon dioxide, or dry chemical extinguishers
21. 4
22. acidic
23. 4
24. as soon as possible
25. calcium carbonate ($CaCO_3$) equivalents
26. calcium and magnesium

27. 3
28. 4
29. 1
30. coliform bacteria
31. 4
32. mean = 14.3, median = 8, geometric mean = 8.4
33. representative sample
34. 4
35. 4

A

absorption (ab-SORP-shun)   The taking in or soaking up of one substance into the body of another by molecular or chemical action (as tree roots absorb dissolved nutrients in the soil).

accountability   When a manager gives power/responsibility to an employee, the employee ensures that the manager is informed of results or events.

accuracy   How closely an instrument measures the true or actual value of the process variable being measured or sensed.

acidic (uh-SID-ick)   The condition of water or soil that contains a sufficient amount of acid substances to lower the pH below 7.0.

acidification (uh-SID-uh-fuh-KAY-shun)   The addition of an acid (usually nitric or sulfuric) to a sample to lower the pH below 2.0. The purpose of acidification is to fix a sample so it will not change until it is analyzed.

acid rain   Precipitation that has been rendered (made) acidic by airborne pollutants.

acre-foot   A volume of water that covers one acre to a depth of one foot, or 43,560 cubic feet (1,233.5 cubic meters).

activated alumina   A charged form of aluminum, used with a synthetic, porous media in an ion exchange adsorption process to remove charged contaminants.

activated carbon   Adsorptive particles or granules of carbon usually obtained by heating carbon (such as wood). These particles or granules have a high capacity to selectively remove certain trace and soluble materials from water.

acute health effect   An adverse effect on a human or animal body, with symptoms developing rapidly.

adsorbate (add-SORE-bait)   The material being removed by the adsorption process.

adsorbent (add-SORE-bent)   The material (activated carbon) that is responsible for removing the undesirable substance in the adsorption process.

adsorption (add-SORP-shun)   The gathering of a gas, liquid, or dissolved substance on the surface or interface zone of another material.

advanced metering infrastructure (AMI)   Refers to a system that measures, collects, and analyzes energy usage while interacting with advanced devices such as water meters through various communication media, either on demand or on predefined schedules. This infrastructure includes hardware, software, communications, consumer energy displays and controllers, customer associated systems, meter data management software, and supplier and network distribution systems.

aeration (air-A-shun)   The process of adding air to water. Air can be added to water by either passing air through water or passing water through air.

aerobic (air-O-bick)   A condition in which atmospheric or dissolved oxygen is present in the aquatic (water) environment.

aesthetic (es-THET-ick)   Attractive or appealing.

age tank   A tank used to store a chemical solution of known concentration for feed to a chemical feeder. An age tank usually stores sufficient chemical solution to properly treat the water being treated for at least one day. Also called a day tank.

air binding   The clogging of a filter, pipe, or pump due to the presence of air released from water. Air entering the filter media is harmful to both the filtration and backwash processes. Air can prevent the passage of water during the filtration process and can cause the loss of filter media during the backwash process.

air gap   An open, vertical drop, or vertical empty space, between a drinking (potable) water supply and the nonpotable point of use. This gap prevents the contamination of drinking water by backsiphonage because it stops wastewater from reaching the drinking water supply. Air gap devices are also used to provide adequate space above the top of a manhole and the end of the hose from the fire hydrant.

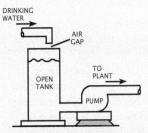

air padding   Pumping dry air (dew point –40°F or –40°C) into a container to assist with the withdrawal of a liquid or to force a liquefied gas such as chlorine or sulfur dioxide out of a container.

air stripping   A physical treatment process used to remove volatile substances from water or wastestreams. The process uses large volumes of air to transfer volatile pollutants from a high concentration in the water or wastestream into a lower concentration in an air stream.

alarm contact   A switch that operates when some preset low, high, or abnormal condition exists.

algae (AL-jee)   Microscopic plants containing chlorophyll that live floating or suspended in water. They also may be attached to structures, rocks, or other submerged surfaces. Excess algal growths can impart tastes and odors to potable water. Algae produce oxygen during sunlight hours and use oxygen during the night hours. Their biological activities appreciably affect the pH, alkalinity, and dissolved oxygen of the water.

algal (AL-gull) **bloom**   Sudden, massive growths of microscopic and macroscopic plant life, such as green or blue-green algae, which can, under the proper conditions, develop in lakes, reservoirs, and lagoons.

algicide (AL-juh-side)   Any substance or chemical specifically formulated to kill or control algae.

aliphatic (al-uh-FAT-ick) **hydroxy acids**   Organic acids with carbon atoms arranged in branched or unbranched open chains rather than in rings.

aliquot (AL-uh-kwot)   Representative portion of a sample of any volume; often an equally divided portion of a sample.

alkali (AL-kuh-lie)   Any of certain soluble salts, principally of sodium, potassium, magnesium, and calcium, that have the property of combining with acids to form neutral salts. They may be used in chemical processes such as water or wastewater treatment.

alkaline (AL-kuh-line)   The condition of water or soil that contains a sufficient amount of alkali substances to raise the pH above 7.0.

alkalinity (AL-kuh-LIN-it-tee)   The capacity of water or wastewater to neutralize acids. This capacity is caused by the water's content of carbonate, bicarbonate, hydroxide, and occasionally borate, silicate, and phosphate. Alkalinity is expressed in milligrams per liter of equivalent calcium carbonate. Alkalinity is not the same as pH because water does not have to be strongly basic (high pH) to have a high alkalinity. Alkalinity is a measure of how much acid must be added to a liquid to lower the pH to 4.5.

alluvial (uh-LOO-vee-ul)   Relating to mud or sand deposited by flowing water. Alluvial deposits may occur after a heavy rainstorm.

alternating current (AC)   An electric current that reverses its direction (positive/negative values) at regular intervals.

altitude valve   A valve that automatically shuts off the flow into an elevated tank when the water level in the tank reaches a predetermined level. The valve automatically opens when the pressure in the distribution system drops below the pressure in the tank.

ambient (AM-bee-ent) **temperature**   Temperature of the surrounding air (or other medium). For example, temperature of the room where a gas chlorinator is installed.

amperage (AM-purr-age)   The strength of an electric current measured in amperes. The amount of electric current flow, similar to the flow of water in gallons per minute.

ampere (AM-peer)   The unit used to measure current strength. The current produced by an electromotive force of one volt acting through a resistance of one ohm.

amperometric (am-purr-o-MET-rick)   A method of measurement that records electric current flowing or generated, rather than voltage.

amperometric (am-purr-o-MET-rick) **titration**   A means of measuring concentrations of certain substances in water (such as strong oxidizers) based on the electric current that flows during a chemical reaction. Also see *titrate*.

amplitude   The maximum strength of an alternating current during its cycle, as distinguished from the mean or effective strength.

anaerobic (AN-air-O-bick)   A condition in which atmospheric or dissolved oxygen (DO) is not present in the aquatic (water) environment.

analog   The continuously variable signal type sent to an analog instrument (for example, 4–20 mA).

analog readout   The readout of an instrument by a pointer (or other indicating means) against a dial or scale. Also see *digital readout*.

analyzer   A device that conducts a periodic or continuous measurement of turbidity or some factor such as chlorine or fluoride concentration. Analyzers operate by any of several methods, including photocells, conductivity, or complex instrumentation.

angstrom (ANG-strem)   A unit of length equal to one-tenth of a nanometer or one-ten billionth of a meter (1 Angstrom = 0.000 000 000 1 meter). One Angstrom is the approximate diameter of an atom.

anion (AN-EYE-en)   A negatively charged ion in an electrolyte solution, attracted to the anode under the influence of a difference in electrical potential. Chloride ion (Cl⁻) is an anion.

anionic (AN-eye-ON-ick) **polymer**   A polymer having negatively charged groups of ions; often used as a filter aid and for dewatering sludges.

annular (AN-yoo-ler) **space**   A ring-shaped space located between two circular objects. For example, the space between the outside of a pipe liner and the inside of a pipe.

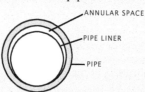

anode (AN-ode)   The positive pole or electrode of an electrolytic system, such as a battery. The anode attracts negatively charged particles or ions (anions).

apparent color   Color of the water that includes not only the color due to substances in the water but suspended matter as well.

appropriative rights   Water rights to or ownership of a water supply that is acquired for the beneficial use of water by following a specific legal procedure.

appurtenance (uh-PURR-ten-nans)   Machinery, appliances, structures, and other parts of the main structure necessary to allow it to operate as intended, but not considered part of the main structure.

aqueous (ACK-wee-us)   Something made up of, similar to, or containing water; watery.

aquifer (ACK-wi-fer)   A natural, underground layer of porous, water-bearing materials (sand, gravel) usually capable of yielding a large amount or supply of water.

arch   (1) The curved top of a sewer pipe or conduit. (2) A bridge or arch of hardened or caked chemical that will prevent the flow of the chemical.

artesian (are-TEE-zhun)   Pertaining to groundwater, a well, or underground basin where the water is under a pressure greater than atmospheric and will rise above the level of its upper confining surface if given an opportunity to do so.

aseptic (a-SEP-tick)   Free from living microbes that cause disease, fermentation, or putrefaction; sterile.

asset management   The process of maintaining the functionality and value of a utility's assets through repair, rehabilitation, and replacement. Examples of utility assets include buildings, tools, equipment, pipes, and machinery used to operate a water or wastewater system. The primary goal of asset management is to provide safe, reliable, and cost-effective service to a community over the useful life of a utility's assets. Another benefit of an effective asset management program is the extension of the life of the asset.

asymmetric (A-sim-MET-rick)   Not similar in size, shape, form, or arrangement of parts on opposite sides of a line, point, or plane.

atm   The abbreviation for atmosphere. One atmosphere is equal to 14.7 psi or 100 kPa.

atom   The smallest unit of a chemical element; composed of protons, neutrons, and electrons.

audit, water   A thorough examination of the accuracy of water agency records or accounts (volumes of water) and system control equipment. Water managers can use audits to determine their water distribution system efficiency. The overall goal is to identify and verify water and revenue losses in a water system.

authority   The power and resources to do a specific job or to get that job done.

automatic meter reading (AMR)   Technology that automatically collects usage, diagnostic, and status data from water metering devices and transfers that data to a central database for billing, troubleshooting, and analyzing.

available chlorine   A measure of the amount of chlorine available in chlorinated lime, hypochlorite compounds, and other materials that are used as a source of chlorine when compared with that of elemental (liquid or gaseous) chlorine.

available expansion   The vertical distance from the sand surface to the underside of a trough in a sand filter. Also called freeboard.

average   A number obtained by adding quantities or measurements and dividing the sum or total by the number of quantities or measurements. Also called the arithmetic mean.

$$\text{Average} = \frac{\text{Sum of Measurements}}{\text{Number of Measurements}}$$

average demand   The total demand for water during a period of time divided by the number of days in that time period. Also called the average daily demand.

axial to impeller   The direction in which material being pumped flows around the impeller or flows parallel to the impeller shaft.

axis of impeller   An imaginary line running along the center of a shaft (such as an impeller shaft).

B

backfill   The material placed over a pipe up to the ground surface. This consists of initial and final backfill. Also see *final backfill* and *initial backfill*.

backflow   A reverse flow condition, created by a difference in water pressures, that causes water to flow back into the distribution pipes of a potable water supply from any source or sources other than an intended source. Also see *backsiphonage*.

back pressure   A pressure that can cause water to backflow into the water supply when a user's water system is at a higher pressure than the public water system.

backsiphonage   A form of backflow caused by a negative or below atmospheric pressure within a water system. Also see *backflow*.

backwashing   The process of reversing the flow of water back through the filter media to remove the entrapped solids.

backwater gate   A gate installed at the end of a drain or outlet pipe to prevent the backward flow of water or wastewater. Generally used on storm sewer outlets into streams to prevent backward flow during times of flood or high tide. Also called a tide gate. Also see *check valve* and *flap gate*.

bacteria (back-TEER-e-uh)   Bacteria are living organisms, microscopic in size, that usually consist of a single cell. Most bacteria use organic matter for their food and produce waste products as a result of their life processes.

baffle   A flat board or plate, deflector, guide, or similar device constructed or placed in flowing water, wastewater, or slurry systems to cause more uniform flow velocities, to absorb energy, and to divert, guide, or agitate liquids (water, chemical solutions, slurry).

bailer (BAY-ler)   A length of pipe equipped with a valve at the lower end used to remove slurry from the bottom or the side of a well as it is being drilled.

ballast   A type of transformer that is used to limit the current to an ultraviolet (UV) lamp.

base-extra capacity method   A cost allocation method used by water utilities to determine water rates for various water user groups. This method considers base costs (O&M expenses and capital costs), extra capacity costs (additional costs for maximum day and maximum hour demands), customer costs (meter maintenance and reading, billing, collection, accounting), and fire protection costs.

base metal   A metal (such as iron) that reacts with dilute hydrochloric acid to form hydrogen. Also see *noble metal*.

batch process   A treatment process in which a tank or reactor is filled, the water is treated or a chemical solution is prepared, and the tank is emptied. The tank may then be filled and the process repeated.

bedding   The soil placed in the bottom of a trench on top of the foundation soil to provide uniform support for a pipe. In addition to bringing the trench bottom to required grade, the bedding levels out any irregularities and ensures uniform support along the length of the pipe.

benchmark   A standard or point of reference used to judge or measure quality or value.

benchmarking   A process an agency uses to gather and compare information about the productivity and performance of other similar agencies with its own information. The purpose of benchmarking is to identify best practices, set improvement targets, and measure progress.

bench-scale analysis (test)   A method of studying different ways or chemical doses for treating water or wastewater and solids on a small scale in a laboratory. Also see *jar test*.

biochemical oxygen demand (BOD)   See *BOD*.

biofiltration   The process of filtering water through a filter medium that has been allowed to develop a microbial biofilm that assists in the removal of fine particulate matter and dissolved organic materials. Also see *biological filtration*.

biological filtration   The process of filtering water through a filter medium that has been allowed to develop a microbial biofilm that assists in the removal of fine particulate matter and dissolved organic materials. Also see *biofiltration*.

biological growth   The activity and growth of any and all living organisms.

blank   A bottle containing only dilution water or distilled water; the sample being tested is not added. Tests are frequently run on a sample and a blank and the differences are compared. The procedure helps to eliminate or reduce errors that could be caused by contaminated dilution water or distilled water.

blockout   The physical prevention of the unexpected movement of machinery, working parts, or other sources of energy within a machine.

blockout device   A device that prevents the unexpected movement of machinery, working parts, or other sources of energy within a machine.

BOD (pronounce as separate letters)   Biochemical Oxygen Demand. A measure of the amount of dissolved oxygen consumed (exertion) by aerobic organisms while stabilizing decomposable organic matter under aerobic conditions. The BOD value is commonly expressed as BOD_5 and refers to the milligrams of oxygen consumed per liter of water during a 5-day incubation period at 68°F (20°C). In decomposition, organic matter serves as food for the organisms and energy results from its oxidation. BOD measurements are used as a measure of the organic strength of wastes in water.

bond (1) A written promise to pay a specified sum of money (called the face value) at a fixed time in the future (called the date of maturity). A bond also carries interest at a fixed rate, payable periodically. The difference between a note and a bond is that a bond usually runs for a longer period of time and requires greater formality. Utility agencies use bonds as a means of obtaining large amounts of money for capital improvements. (2) A warranty by an underwriting organization, such as an insurance company, guaranteeing honesty, performance, or payment by a contractor.

bonnet (BON-it)   The cover on a gate valve.

bowl, pump   The submerged pumping unit in a well, including the shaft, impellers, and housing.

brake horsepower (BHP)   (1) The horsepower required at the top or end of a pump shaft (input to a pump). (2) The energy provided by a motor or other power source.

breakpoint chlorination   Addition of chlorine to water or wastewater until the chlorine demand has been satisfied. At this point, further additions of chlorine will result in a free chlorine residual that is directly proportional to the amount of chlorine added beyond the breakpoint.

breakthrough   A crack or break in a filter bed allowing the passage of floc or particulate matter through a filter. This will cause an increase in filter effluent turbidity. A breakthrough can occur (1) when a filter is first placed in service, (2) when the effluent valve suddenly opens or closes, and (3) during periods of excessive head loss through the filter (including when the filter is exposed to negative heads).

brinelling (bruh-NEL-ing)   Tiny indentations (dents) high on the shoulder of the bearing race or bearing. A type of bearing failure.

buffer   A solution or liquid whose chemical makeup neutralizes acids or bases, thereby maintaining a relatively constant pH.

buffer capacity   A measure of how much a solution or liquid can neutralize acids or bases. For example, this measures the capacity of water or wastewater to resist changes in pH due to the addition of an acid or base.

C

caisson (KAY-sawn)   A structure or chamber that is usually sunk or lowered by digging from the inside. Used to gain access to the bottom of a stream or other body of water.

calcium carbonate ($CaCO_3$) equilibrium   A water is considered stable when it is just saturated with calcium carbonate. In this condition, the water will neither dissolve nor deposit calcium carbonate. Thus, in this water the calcium carbonate is in equilibrium with the hydrogen ion concentration.

calcium carbonate ($CaCO_3$) equivalent   An expression of the concentration of specified water constituents in terms of their equivalent value to calcium carbonate. For example, the hardness in water that is caused by calcium, magnesium, and other ions is usually described as calcium carbonate equivalent. Alkalinity test results are usually reported as mg/L $CaCO_3$ equivalents.

calibration   A procedure that checks or adjusts an instrument's accuracy by comparison with a standard or reference.

call date   First date a bond can be paid off.

capillary (KAP-uh-larry) **action**   The movement of water through very small spaces due to molecular forces.

capillary (KAP-uh-larry) **forces**   The molecular forces that cause the movement of water through very small spaces.

capillary (KAP-uh-larry) **fringe**   The porous material just above the water table that may hold water by capillarity (a property of surface tension that draws water upward) in the smaller void spaces.

capital improvement plan (CIP)   A detailed plan that identifies requirements for the repair, replacement, and rehabilitation of facility infrastructure over an extended period, often 20 years or more. A utility usually updates or prepares this plan annually. The plan is often a part of a water system master plan that combines water demand projections with supply alternatives and facility requirements.

carcinogen (kar-SIN-o-jen)   Any substance that tends to produce cancer in an organism.

catalyst (KAT-uh-list)   A substance that changes the speed or yield of a chemical reaction without being consumed or chemically changed by the chemical reaction.

catalyze (KAT-uh-lize)   To act as a catalyst. Or, to speed up a chemical reaction.

catalyzed (KAT-uh-lized)   To be acted upon by a catalyst.

cathode (KATH-ode)   The negative pole or electrode of an electrolytic cell or system. The cathode attracts positively charged particles or ions (cations).

cathodic (kath-ODD-ick) **protection**   An electrical system for prevention of rust, corrosion, and pitting of metal surfaces that are in contact with water, wastewater, or soil. A low-voltage current is made to flow through a liquid (water) or a soil in contact with the metal in such a manner that the external electromotive force renders the metal structure cathodic. This concentrates corrosion on auxiliary anodic parts, which are deliberately allowed to corrode instead of letting the structure corrode.

cation (KAT-EYE-en)   A positively charged ion in an electrolyte solution, attracted to the cathode under the influence of a difference in electrical potential. Sodium ion (Na^+) is a cation.

cationic (KAT-eye-ON-ic) **polymer**   A polymer having positively charged groups of ions; often used as a coagulant aid.

caution   This word warns against potential hazards or cautions against unsafe practices. Also see *danger*, *notice*, and *warning*.

cavitation (kav-uh-TAY-shun)   The formation and collapse of a gas pocket or bubble on the blade of an impeller or the gate of a valve. The collapse of this gas pocket or bubble drives water into the impeller or gate with a terrific force that can knock metal particles off and cause pitting on the impeller or gate surface. Cavitation is accompanied by loud noises that sound like someone is pounding on the impeller or gate with a hammer.

centrate   The water leaving a centrifuge after most of the solids have been removed.

centrifugal (sen-TRIF-uh-gull) **pump**   A pump consisting of an impeller fixed on a rotating shaft that is enclosed in a casing, and having an inlet and discharge connection. As the rotating impeller whirls the liquid around, centrifugal force builds up enough pressure to force the water through the discharge outlet.

centrifuge   A mechanical device that uses centrifugal or rotational forces to separate solids from liquids.

certification examination   An examination administered by a state agency or professional association that operators take to indicate a level of professional competence. In the United States, certification of operators of water treatment plants, wastewater treatment plants, water distribution systems, and small water supply systems is mandatory. In many states, certification of wastewater collection system operators, industrial wastewater treatment plant operators, pretreatment facility inspectors, and small wastewater system operators is voluntary; however, current trends indicate that more states, provinces, and employers will require these operators to be certified in the future. Operator certification is mandatory in the United States for the Chief Operators of water treatment plants, water distribution systems, and wastewater treatment plants.

certified operator   A person who has the education and experience required to operate a specific class of treatment facility as indicated by possessing a certificate of professional competence given by a state agency or professional association.

C Factor   A factor or value used to indicate the smoothness of the interior of a pipe. The higher the C Factor, the smoother the pipe, the greater the carrying capacity, and the smaller the friction or energy losses from water flowing in the pipe. To calculate the C Factor, measure the flow, pipe diameter, distance between two pressure gauges, and the friction or energy loss of the water between the gauges.

$$C \text{ Factor} = \frac{\text{Flow, GPM}}{193.75 \left(\text{Diameter, ft}\right)^{2.63} \left(\text{Slope}\right)^{0.54}}$$

or

$$C \text{ Factor} = \frac{\text{Flow, m}^3/\text{sec}}{0.278 \left(\text{Diameter, m}\right)^{2.63} \left(\text{Slope}\right)^{0.54}}$$

charge chemistry   A branch of chemistry in which the destabilization and neutralization reactions occur between stable, negatively charged and stable, positively charged particles.

check sampling   Whenever an initial or routine sample analysis indicates that a Maximum Contaminant Level (MCL) has been exceeded, check sampling is required to confirm the routine sampling results. Check sampling is in addition to the routine sampling program.

check valve   A special valve with a hinged disk or flap that opens in the direction of normal flow and is forced shut when flows go in the reverse or opposite direction of normal flows. Also see *flap gate* and *tide gate*.

chelating (KEY-LAY-ting) **agent**   A chemical used to prevent the precipitation of metals (such as copper).

chelation (key-LAY-shun)   A chemical complexing (forming or joining together) of metallic cations (such as copper) with certain organic compounds, such as ethylene diamine tetracetic acid (EDTA). Chelation is used to prevent the precipitation of metals (copper). Also see *sequestration*.

chloramination (KLOR-uh-min-NAY-shun)   The application of chlorine and ammonia to water to form chloramines for the purpose of disinfection.

chloramines (KLOR-uh-means)   Compounds formed by the reaction of hypochlorous acid (or aqueous chlorine) with ammonia.

chlorination (klor-uh-NAY-shun)   The application of chlorine to water, generally for the purpose of disinfection, but frequently for accomplishing other biological or chemical results (aiding coagulation and controlling tastes and odors).

chlorinator (KLOR-uh-nay-ter)   A metering device that is used to add chlorine to water.

chlorine demand   Chlorine demand is the difference between the amount of chlorine added to water or wastewater and the amount of chlorine residual remaining after a given contact time. Chlorine demand may change with dosage, time, temperature, pH, and nature and amount of the impurities in the water.

Chlorine Demand = Chlorine Applied − Chlorine Residual

chlorine requirement   The amount of chlorine that is needed for a particular purpose. Some reasons for adding chlorine are reducing the MPN (Most Probable Number) of coliform bacteria, obtaining a particular chlorine residual, or oxidizing some substance in the water. In each case, a definite dosage of chlorine will be necessary. This dosage is the chlorine requirement.

chlorine residual   The concentration of chlorine present in water after the chlorine demand has been satisfied. The concentration is expressed in terms of the total chlorine residual, which includes both the free and combined or chemically bound chlorine residuals. Also called residual chlorine.

chlorophenolic (KLOR-o-fee-NO-lick)   Chlorophenolic compounds are phenolic compounds (carbolic acid) combined with chlorine.

chlorophenoxy (KLOR-o-fuh-NOX-ee)   A class of herbicides that may be found in domestic water supplies and cause adverse health effects. Two widely used chlorophenoxy herbicides are 2,4-D (2,4-Dichlorophenoxy acetic acid) and 2,4,5-TP (2,4,5-Trichlorophenoxy propionic acid or silvex).

chlorophyll (KLOR-oh-fill)   A group of special green pigments found in plants, algae, and some bacteria that assist in the process of photosynthesis. Also see *photosynthesis*.

chlororganic (klor-or-GAN-ick)   Organic compounds combined with chlorine. These often toxic compounds generally originate from chlorinating water containing dissolved organic substances or microbes such as algae.

chronic health effect   An adverse effect on a human or animal body with symptoms that develop slowly over a long period of time or that recur frequently.

circle of influence   The circular outer edge of a depression produced in the water table by the pumping of water from a well. Also see *cone of influence* and *cone of depression*. [See drawing on page 788.]

circuit   The complete path of an electric current, including the generating apparatus or other source; or, a specific segment or section of the complete path.

circuit breaker   A safety device in an electric circuit that automatically shuts off the circuit when it becomes overloaded. The device can be manually reset.

cistern (SIS-turn)   A small tank (usually covered) or a storage facility used to store water for a home or farm. Often used to store rainwater.

clarifier (KLAIR-uh-fire)   A tank or basin in which water or wastewater is held for a period of time during which the heavier solids settle to the bottom and the lighter materials float to the surface. Also called settling tank. Also see *sedimentation basin*.

class, pipe and fittings   The working pressure rating, including allowances for surges, of a specific pipe for use in water distribution systems. The term is used for cast iron, ductile iron, asbestos cement, and some plastic pipe.

clear well   A reservoir for the storage of filtered water of sufficient capacity to prevent the need to vary the filtration rate with variations in demand. Also used to provide chlorine contact time for disinfection.

clear zone   See *supernatant*.

coagulant (ko-AGG-yoo-lent)   A chemical that causes very fine particles to clump (floc) together into larger particles. This makes it easier to separate the solids from the liquids by settling, skimming, draining, or filtering.

coagulant (ko-AGG-yoo-lent) **aid**   Any chemical or substance used to assist or modify coagulation.

coagulation (ko-agg-yoo-LAY-shun)   The clumping together of very fine particles into larger particles (floc) caused by the use of chemicals (coagulants). The chemicals neutralize the electrical charges of the fine particles, allowing them to come closer and form larger clumps. This clumping together makes it easier to separate the solids from the water by settling, skimming, draining, or filtering.

Code of Federal Regulations (CFR)   A publication of the US government that contains all of the proposed and finalized federal regulations, including safety and environmental regulations.

coliform (KOAL-i-form)   A group of bacteria found in the intestines of warm-blooded animals (including humans) and in plants, soil, air, and water. The presence of coliform bacteria is an indication that the water is polluted and may contain pathogenic (disease-causing) organisms. Fecal coliforms are those coliforms found in the feces of various warm-blooded animals, whereas the term coliform also includes other environmental sources.

colloids (KALL-loids)   Very small, finely divided solids (particles that do not dissolve) that remain dispersed in a liquid for a long time due to their small size and electrical charge. When most of the particles in water have a negative electrical charge, they tend to repel each other. This repulsion prevents the particles from clumping together, becoming heavier, and settling out.

color   The substances in water that impart a yellowish-brown color to the water. These substances are the result of iron and manganese ions, humus and peat materials, plankton, aquatic weeds, and industrial waste present in the water. Also see *true color*.

colorimetric measurement   A means of measuring unknown chemical concentrations in water by measuring a sample's color intensity. The specific color of the sample, developed by addition of chemical reagents, is measured with a photoelectric colorimeter or is compared with color standards using, or corresponding with, known concentrations of the chemical.

combined available chlorine   The total chlorine, present as chloramine or other derivatives, that is present in a water and is still available for disinfection and for oxidation of organic matter. The combined chlorine compounds are more stable than free chlorine forms, but they are somewhat slower in disinfection action.

combined available chlorine residual   The concentration of chlorine residual that is combined with ammonia, organic nitrogen, or both in water as a chloramine (or other chloro derivative) and yet is still available to oxidize organic matter and help kill bacteria.

combined chlorine   The sum of the chlorine species composed of free chlorine and ammonia, including monochloramine, dichloramine, and trichloramine (nitrogen trichloride). Dichloramine is the strongest disinfectant of these chlorine species, but it has less oxidative capacity than free chlorine.

combined residual chlorination   The application of chlorine to water or wastewater to produce a combined available chlorine residual. The residual may consist of chlorine compounds formed by the reaction of chlorine with natural or added ammonia (NH_3) or with certain organic nitrogen compounds.

commodity-demand method   A cost allocation method used by water utilities to determine water rates for the various water user groups. This method considers the commodity costs (water, chemicals, power, amount of water use), demand costs (treatment, storage, distribution), customer costs (meter maintenance and reading, billing, collection, accounting), and fire protection costs.

competent person   A competent person is defined by OSHA as a person capable of identifying existing and predictable hazards in the surroundings, or working conditions that are unsanitary, hazardous, or dangerous to employees, and who has authorization to take prompt corrective measures to eliminate the hazards.

complete treatment   A method of treating water that consists of the addition of coagulant chemicals, flash mixing, coagulation-flocculation, sedimentation, and filtration. Also called conventional filtration. Also see *direct filtration* and *inline filtration*.

composite (proportional) sample   A composite sample is a collection of individual samples obtained at regular intervals. For larger plants, this is usually every one or two hours during a 24-hour time span. For smaller wastewater treatment plants and lagoons, this sample may be collected three or more times during an eight-hour shift. Each individual sample is combined with the others in proportion to the rate of flow when the sample was collected. Mixtures of equal volume individual samples collected at intervals after a specific volume of flow passes the sampling point, or after equal time intervals, are still referred to as composite samples. The resulting mixture (composite sample) forms a representative sample and is analyzed to determine the average conditions during the sampling period.

compound   A pure substance composed of two or more elements whose composition is constant. For example, table salt (sodium chloride, NaCl) is a compound.

concentration polarization   (1) A buildup of retained particles on the membrane surface due to dewatering of the feed closest to the membrane. The thickness of the concentration polarization layer is controlled by the flow velocity across the membrane. (2) Used in corrosion studies to indicate a depletion of ions near an electrode. (3) The basis for chemical analysis by a polarograph.

conditioning   Pretreatment of sludge to facilitate removal of water in subsequent treatment processes.

conductance   A rapid method of estimating the total dissolved solids content of a water supply. The measurement indicates the capacity of a sample of water to carry an electric current, which is related to the concentration of ionized substances in the water. Also called specific conductance.

conductivity   A measure of the ability of a solution (water) to carry an electric current.

conductor   (1) A pipe that carries a liquid load from one point to another point. In a wastewater collection system, a conductor is often a large pipe with no service connections. Also called a conduit. Also see *interceptor (intercepting) sewer* or *interconnector*. (2) In plumbing, a pipe installed to drain water from the roof gutters or roof catchment to the storm drain or other means of disposal. Also called a downspout, roof leader, or roof drain. (3) In electricity, a substance, body, device, or wire that readily conducts or carries electric current. Also called a conduit.

conductor casing   The outer casing of a well. The purpose of this casing is to prevent contaminants from surface waters or shallow groundwaters from entering a well.

conduit   Any artificial or natural duct, either open or closed, for carrying fluids from one point to another. An electrical conduit carries electricity.

cone of depression   The depression, roughly conical in shape, produced in the water table by the pumping of water from a well. Also called the cone of influence. Also see *circle of influence*. [See drawing on page 788.]

cone of influence   See *cone of depression*.

confined space   A space that has the following characteristics: (1) Is large enough and so configured that an employee can bodily enter and perform assigned work; (2) Has limited or restricted means for entry or exit (for example, manholes, tanks, vessels, silos, storage bins, hoppers, vaults, and pits are spaces that may have limited means of entry); (3) Is not designed for continuous employee occupancy. Also see *dangerous air contamination* and *oxygen deficiency*.

confined space, class A A confined space that presents a situation that is immediately dangerous to life or health (IDLH). These include but are not limited to oxygen deficiency, explosive or flammable atmospheres, and concentrations of toxic substances.

confined space, class B A confined space that has the potential for causing injury and illness, if preventive measures are not used, but is not immediately dangerous to life and health.

confined space, class C A confined space in which the potential hazard would not require any special modification of the work procedure.

confined space, non-permit A non-permit confined space is a confined space that does not contain or, with respect to atmospheric hazards, have the potential to contain any hazard capable of causing death or serious physical harm.

confined space, permit-required (permit space) A confined space that has one or more of the following characteristics: (1) Contains or has a potential to contain a hazardous atmosphere; (2) Contains a material that has the potential for engulfing an entrant; (3) Has an internal configuration such that an entrant could be trapped or asphyxiated by inwardly converging walls or by a floor that slopes downward and tapers to a smaller cross section; (4) Contains any other recognized serious safety or health hazard.

confining unit A layer of rock or soil of very low hydraulic conductivity that hampers the movement of groundwater in and out of an aquifer.

consolidated formation A geologic material whose particles are stratified (layered), cemented, or firmly packed together (hard rock); usually occurring at a depth below the ground surface. Also see *unconsolidated formation*.

consumer confidence report (CCR) An annual report prepared by a water utility to communicate with its consumers. The report provides consumers with information on the source and quality of their drinking water and is an opportunity for positive communication with consumers.

contactor An electric switch, usually magnetically operated.

contamination (kun-TAM-uh-NAY-shun) The introduction into water of microorganisms, chemicals, toxic substances, wastes, or wastewater in a concentration that makes the water unfit for its next intended use.

continuous sample A flow of water from a particular place in a plant to the location where samples are collected for testing. This continuous stream may be used to obtain grab or composite samples. Frequently, several taps (faucets) will flow continuously in the laboratory to provide test samples from various places in a water treatment plant.

controller A device that controls the starting, stopping, or operation of a device or piece of equipment.

control loop The combination of one or more interconnected instrumentation devices that are arranged to measure, display, and control a process variable. Also called a loop.

control system An instrumentation system that senses and controls its own operation on a close, continuous basis in what is called proportional (or modulating) control.

conventional filtration A method of treating water that consists of the addition of coagulant chemicals, flash mixing, coagulation-flocculation, sedimentation, and filtration. Also called complete treatment. Also see *direct filtration* and *inline filtration*.

conventional treatment (1) The common wastewater treatment processes such as preliminary treatment, sedimentation, flotation, trickling filter, rotating biological contactor, activated sludge, and

chlorination wastewater treatment processes used by POTWs. (2) The hydroxide precipitation of metals processes used by pretreatment facilities.

coprecipitation A treatment process that occurs when ferrous iron is added to water or metallic wastestreams and subsequently oxidized in an aerator. The oxidized iron, which is insoluble, precipitates along with other metallic contaminants present in the water or wastestream, thereby enhancing metals removal.

corporation stop A water service shutoff valve located at a street water main. This valve cannot be operated from the ground surface because it is buried and there is no valve box. Also called a corporation cock.

corrosion The gradual decomposition or destruction of a material by chemical action, often due to an electrochemical reaction. Corrosion may be caused by (1) stray current electrolysis, (2) galvanic corrosion caused by dissimilar metals, or (3) differential-concentration cells. Corrosion starts at the surface of a material and moves inward.

corrosion inhibitors Substances that slow the rate of corrosion.

corrosive gases Gases that, when dissolved in water, can oxidize materials of construction such as steel and concrete.

corrosivity An indication of the corrosiveness of a water. The corrosiveness of a water is described by the water's pH, alkalinity, hardness, temperature, total dissolved solids, dissolved oxygen concentration, and the Langelier Index.

coulomb (KOO-lahm) A measurement of the amount of electrical charge carried by an electric current of one ampere in one second. One coulomb equals about 6.25×10^{18} electrons (6,250,000,000,000,000,000 electrons).

coupon A steel specimen inserted into water or wastewater to measure corrosiveness. The rate of corrosion is measured as the loss of weight of the coupon or change in its physical characteristics. Measure the weight loss (in milligrams) per surface area (in square decimeters) exposed to the water or wastewater per day. 1 meter = 10 decimeters = 100 centimeters.

coverage ratio The coverage ratio is a measure of the ability of the utility to pay the principal and interest on loans and bonds (this is known as debt service) in addition to any unexpected expenses.

cross-connection (1) A connection between a drinking (potable) water system and an unapproved water supply. (2) A connection between a storm drain system and a sanitary collection system. (3) Less frequently used to mean a connection between two sections of a collection system to handle anticipated overloads of one system.

cross-flow filtration A type of membrane filtration where the water being filtered flows across the surface of the membrane to keep the particle buildup and fouling to a minimum. The flow that is not filtered becomes concentrated and flows out the end of the membrane fiber as a wastestream.

Cryptosporidium (KRIP-toe-spo-RID-ee-um) A waterborne intestinal parasite that causes a disease called cryptosporidiosis (KRIP-toe-spo-rid-ee-O-sis) in infected humans. Symptoms of the disease include diarrhea, cramps, and weight loss. *Cryptosporidium* contamination is found in most surface waters and some groundwaters. Commonly referred to as crypto.

CT value Residual concentration of a given disinfectant in mg/L times the disinfectant's contact time in minutes.

curb stop A water service shutoff valve located in a water service pipe near the curb and between the water main and the building. This valve is usually operated by a wrench or valve key and is used to start or stop flows in the water service line to a building. Also called a curb cock.

Curie (KYOOR-ee) A measure of radioactivity. One Curie of radioactivity is equivalent to 3.7×10^{10} or 37,000,000,000 nuclear disintegrations per second.

current A movement or flow of electricity. Electric current is measured by the number of coulombs per second flowing past a certain point in a conductor. A coulomb is equal to about 6.25×10^{18} electrons (6,250,000,000,000,000,000 electrons). A flow of one coulomb per second is called one ampere, the unit of the rate of flow of current.

cycle A complete alternation of voltage or current in an alternating current (AC) circuit.

D

danger This word is used where an immediate hazard presents a threat of death or serious injury to employees. Also see *caution*, *notice*, and *warning*.

dangerous air contamination An atmosphere presenting a threat of causing death, injury, acute illness, or disablement due to the presence of flammable or explosive, toxic, or otherwise injurious or incapacitating substances. (1) Dangerous air contamination due to the flammability of a gas, vapor, or mist is defined as an atmosphere containing the gas, vapor, or mist at a concentration greater than 10 percent of its lower explosive (lower flammable) limit (LEL). (2) Dangerous air contamination due to a combustible particulate is defined as a concentration that meets or exceeds the particulate's lower explosive limit (LEL). (3) Dangerous air contamination due to the toxicity of a substance is defined as the atmospheric concentration that could result in employee exposure in excess of the substance's permissible exposure limit (PEL). *NOTE:* A dangerous situation also occurs when the oxygen level is less than 19.5 percent by volume (*oxygen deficiency*) or more than 23.5 percent by volume (*oxygen enrichment*).

dateometer (day-TOM-uh-ter) A small calendar disk attached to motors and equipment to indicate the year in which the last maintenance service was performed.

datum line A line from which heights and depths are calculated or measured. Also called a datum plane or a datum level.

day tank A tank used to store a chemical solution of known concentration for feed to a chemical feeder. A day tank usually stores sufficient chemical solution to properly treat the water being treated for at least one day. Also called an age tank.

DBP See *disinfection byproduct (DBP)*.

dead end The end of a water main that is not connected to other parts of the distribution system by means of a connecting loop of pipe.

dead-end filtration A type of membrane filtration where the water being filtered flows through the membrane, but there is no wastestream from the system. All solids accumulate on the membrane during filtration and are removed during backwash.

debt service The amount of money required annually to pay the (1) interest on outstanding debts or (2) funds due on a maturing bonded debt or the redemption of bonds.

decant (de-KANT) To draw off the upper layer of liquid (water) after the heavier material (a solid or another liquid) has settled.

decant (de-KANT) **water** Water that has separated from sludge and is removed from the layer of water above the sludge.

dechlorination (DEE-klor-uh-NAY-shun) The deliberate removal of chlorine from water. The partial or complete reduction of chlorine residual by any chemical or physical process.

decibel (DES-uh-bull) **(dB)** A unit for expressing the relative intensity of sounds on a scale from zero for the average least

perceptible sound to about 130 for the average level at which sound causes pain to humans.

decomposition The conversion of chemically unstable materials to more stable forms by chemical or biological action. Also called decay.

defluoridation (DEE-floor-uh-DAY-shun) The removal of excess fluoride in drinking water to prevent the mottling (brown stains) of teeth.

degasification (DEE-gas-if-uh-KAY-shun) A water treatment process that removes dissolved gases from the water. The gases may be removed by either mechanical or chemical treatment methods or a combination of both.

delegation The act in which power is given to another person in the organization to accomplish a specific job.

demineralization (DEE-min-er-al-uh-ZAY-shun) A treatment process that removes dissolved minerals (salts) from water.

density A measure of how heavy a substance (solid, liquid, or gas) is for its size. Density is expressed in terms of weight per unit volume, that is, grams per cubic centimeter or pounds per cubic foot. The density of water (at 39°F or 4°C) is about 62.4 pounds per cubic foot or 1.0 gram per cubic centimeter.

depolarization The removal or depletion of ions in the thin boundary layer adjacent to a membrane or pipe wall.

depreciation The gradual loss in service value of a facility or piece of equipment due to all the factors causing the ultimate retirement of the facility or equipment. This loss can be caused by sudden physical damage, wearing out due to age, obsolescence, inadequacy, or availability of a newer, more efficient facility or equipment. The value cannot be restored by maintenance.

desalinization (DEE-SAY-lin-uh-ZAY-shun) The removal of dissolved salts (such as sodium chloride, NaCl) from water by natural means (leaching) or by specific water treatment processes.

desiccant (DESS-uh-kant) A drying agent that is capable of removing or absorbing moisture from the atmosphere in a small enclosure.

desiccation (dess-uh-KAY-shun) A process used to thoroughly dry air; to remove virtually all moisture from air.

desiccator (DESS-uh-kay-tor) A closed container into which heated weighing or drying dishes are placed to cool in a dry environment in preparation for weighing. The dishes may be empty or they may contain a sample. Desiccators contain a substance (desiccant), such as anhydrous calcium chloride, that absorbs moisture and keeps the relative humidity near zero so that the dish or sample will not gain weight from absorbed moisture.

destratification (DEE-strat-uh-fuh-KAY-shun) The development of vertical mixing within a lake or reservoir to eliminate (either totally or partially) separate layers of temperature, plant life, or animal life. This vertical mixing can be caused by mechanical means (pumps) or through the use of forced air diffusers that release air into the lower layers of the reservoir.

detention time (1) The theoretical (calculated) time required for a small amount of water to pass through a tank at a given rate of flow. (2) The actual time in hours, minutes, or seconds that a small amount of water is in a settling basin, flocculating basin, or rapid-mix chamber. In storage reservoirs, detention time is the length of time entering water will be held before being drafted for use (several weeks to years, several months being typical).

$$\text{Detention Time, hr} = \frac{(\text{Basin Volume, gal})(24 \text{ hr/day})}{\text{Flow, gal/day}}$$

or

$$\text{Detention Time, hr} = \frac{\left(\text{Basin Volume, m}^3\right)\left(24 \text{ hr/day}\right)}{\text{Flow, m}^3/\text{day}}$$

dewater (1) To remove or separate a portion of the water present in a sludge or slurry. To dry sludge so it can be handled and disposed of. (2) To remove or drain the water from a tank or a trench. A structure may be dewatered so that it can be inspected or repaired.

dew point The temperature to which air with a given quantity of water vapor must be cooled to cause condensation of the vapor in the air.

diatomaceous (DYE-uh-toe-MAY-shus) **earth** A fine, siliceous (made of silica) earth composed mainly of the skeletal remains of diatoms.

diatoms (DYE-uh-toms) Unicellular (single cell), microscopic algae with a rigid, box-like internal structure consisting mainly of silica.

dielectric (DYE-ee-LECK-trick) Does not conduct an electric current. An insulator or nonconducting substance.

digital The encoding of information that uses binary numbers (ones and zeros) for input, processing, transmission, storage, or display, rather than a continuous spectrum of values (an analog system) or non-numeric symbols such as letters or icons.

digital readout The readout of an instrument by a direct, numerical reading of the measured value or variable.

dilute solution A solution that has been made weaker, usually by the addition of water.

dimictic (dye-MICK-tick) Lakes and reservoirs that freeze over and normally go through two stratification and two mixing cycles within a year.

direct current (DC) Electric current flowing in one direction only and essentially free from pulsation. Also see *alternating current*.

direct filtration A method of treating water that consists of the addition of coagulant chemicals, flash mixing, coagulation, minimal flocculation, and filtration. The flocculation facilities may be omitted, but the physical-chemical reactions will occur to some extent. The sedimentation process is omitted. Also see *conventional filtration* and *inline filtration*.

direct runoff Water that flows over the ground surface directly into streams, rivers, or lakes. Also called storm runoff.

discharge head The pressure, in pounds per square inch (psi) or kilopascals (kPa), measured at the centerline of a pump discharge and very close to the discharge flange, converted into feet or meters. The pressure is measured from the centerline of the pump to the hydraulic grade line of the water in the discharge pipe.

Discharge Head, ft = (Discharge Pressure, psi)(2.31 ft/psi)

or

Discharge Head, m = (Discharge Pressure, kPa)(1 m/9.8 kPa)

discrete control ON/OFF control; one of the two output values is equal to zero.

discrete I/O (input/output) A digital signal that senses or sends either ON or OFF signals. For example, a discrete input would sense the position of a switch; a discrete output would turn on a pump or light.

disinfection (dis-in-FECT-shun) The process designed to kill or inactivate most microorganisms in water or wastewater, including essentially all pathogenic (disease-causing) bacteria. There are several ways to disinfect, with chlorination being the most

frequently used in water and wastewater treatment plants. Compare with *sterilization*.

disinfection byproduct (DBP) A contaminant formed by the reaction of disinfection chemicals (such as chlorine) with other substances in the water being disinfected.

dissolved organic carbon (DOC) That portion of the organic carbon in water that passes through a 0.45 μm pore-diameter filter.

dissolved organic matter (DOM) That portion of the organic matter in water that passes through a 0.45 μm pore-diameter filter.

distillate (DIS-tuh-late) In the distillation of a sample, a portion is collected by evaporation and recondensation; the part that is recondensed is the distillate.

distributed control system (DCS) A computer control system having multiple microprocessors to distribute the functions performing process control, thereby distributing the risk from component failure. The distributed components (input/output devices, control devices, and operator interface devices) are all connected by communications links and permit the transmission of control, measurement, and operating information to and from many locations.

divalent (dye-VAY-lent) Having a valence of two, such as the ferrous ion, Fe^{2+}. Also called bivalent.

diversion Use of part of a stream flow as a water supply.

DOC See *dissolved organic carbon (DOC)*.

DOM See *dissolved organic matter (DOM)*.

downspout In plumbing, a pipe installed to drain water from the roof gutters or roof catchment to the storm drain or other means of disposal. Also called a conductor, roof leader, or roof drain.

DPD method A method of measuring the chlorine residual in water. The residual may be determined by either titrating or comparing a developed color with color standards. DPD stands for N,N-diethyl-p-phenylenediamine.

draft (1) The act of drawing or removing water from a tank or reservoir. (2) The water that is drawn or removed from a tank or reservoir.

drawdown (1) The drop in the water table or level of water in the ground when water is being pumped from a well. (2) The amount of water used from a tank or reservoir. (3) The drop in the water level of a tank or reservoir.

drift The difference between the actual value and the desired value (or set point); characteristic of proportional controllers that do not incorporate reset action. Also called offset.

dynamic head When a pump is operating, the vertical distance (in feet or meters) from a point to the energy grade line. Also see *energy grade line (EGL)*, *static head*, and *total dynamic head (TDH)*.

dynamic pressure When a pump is operating, pressure resulting from the dynamic head.

Dynamic Pressure, psi = (Dynamic Head, ft)(0.433 psi/ft)

or

Dynamic Pressure, kPa = (Dynamic Head, m)(9.8 kPa/m)

E

eductor (e-DUCK-ter) A hydraulic device used to create a negative pressure (suction) by forcing a liquid through a restriction, such as a Venturi. An eductor or aspirator (the hydraulic device) may be used in the laboratory in place of a vacuum pump. As an injector, it is used to produce vacuum for chlorinators. Sometimes used instead of a suction pump.

effective range   That portion of the design range (usually from 10 to 90+ percent) in which an instrument has acceptable accuracy. Also see *range* and *span*.

effective size (ES)   The diameter of the particles in a granular sample (filter media) for which 10 percent of the total grains are smaller and 90 percent larger on a weight basis. Effective size is obtained by passing granular material through sieves with varying dimensions of mesh and weighing the material retained by each sieve. The effective size is also approximately the average size of the grains.

effluent (EF-loo-ent)   Water—raw (untreated), partially treated, or completely treated—flowing from a reservoir, basin, treatment process, or treatment plant.

EGL   See *energy grade line (EGL)*.

ejector   A device used to disperse a chemical solution into water being treated.

electrical logging   A procedure used to search for water-bearing formations (aquifers) by determining the porosity (spaces or voids) of geologic materials. Electrical probes are lowered into wells, an electric current is induced at various depths, and the resistance measured indicates the porosity of the formation.

electrochemical reaction   Chemical changes produced by electricity (electrolysis) or the production of electricity by chemical changes (galvanic action). In corrosion, a chemical reaction is accompanied by the flow of electrons through a metallic path. The electron flow may come from an external source and cause the reaction, such as electrolysis caused by a direct current (DC) electric railway, or the electron flow may be caused by a chemical reaction, as in the galvanic action of a flashlight dry cell.

electrochemical series   A list of metals with the standard electrode potentials given in volts. The size and sign of the electrode potential indicates how easily these elements will take on or give up electrons, or corrode. Hydrogen is conventionally assigned a value of zero.

electrolysis (ee-leck-TRAWL-uh-sis)   The decomposition of material by an outside electric current.

electrolyte (ee-LECK-tro-lite)   A substance that dissociates (separates) into two or more ions when it is dissolved in water.

electrolytic (ee-LECK-tro-LIT-ick) **cell**   A device in which the chemical decomposition of material causes an electric current to flow. Also, a device in which a chemical reaction occurs as a result of the flow of electric current. Chlorine and caustic (NaOH) are made from salt (NaCl) in electrolytic cells.

electromotive force (EMF)   The electrical pressure available to cause a flow of current (amperage) when an electric circuit is closed. Also see *voltage*.

electromotive series   A list of metals and alloys presented in the order of their tendency to corrode (or go into solution). Also called the galvanic series. This is a practical application of the theoretical *electrochemical series*.

electron   (1) A very small, negatively charged particle that is practically weightless. According to the electron theory, all electrical and electronic effects are caused either by the movement of electrons from place to place or because there is an excess or lack of electrons at a particular place. (2) The part of an atom that determines its chemical properties.

element   A substance that cannot be separated into its constituent parts and still retain its chemical identity. For example, sodium (Na) is an element.

empty bed contact time (EBCT)   A measure of the time during which a water to be treated is in contact with the treatment

medium in a contact vessel, assuming that all liquid passes through the vessel at the same velocity. EBCT is equal to the volume of the empty bed divided by the flow rate.

enclosed space   See *confined space*.

end bells   Devices used to hold the rotor and stator of a motor in position.

endpoint   The completion of a desired chemical reaction. In the titration of a water or wastewater sample, a chemical is added, drop by drop, to the sample until a certain color change occurs (blue to clear, for example). In addition to a color change, an endpoint may be reached by the formation of a precipitate or by reaching a specified pH. An endpoint may be detected by using an electronic device such as a spectrophotometer or pH meter.

endemic (en-DEM-ick)   Something peculiar to a particular people or locality, such as a disease that is always present in the population.

endocrine (EN-doe-krin) **effects**   An altering of the organs in the human body responsible for secreting hormones into the bloodstream. Endocrine glands include the thyroid gland, the pancreas, and the adrenal glands.

endrin (EN-drin)   A pesticide toxic to freshwater and marine aquatic life that produces adverse health effects in domestic water supplies.

energy grade line (EGL)   A line that represents the elevation of energy head (in feet or meters) of water flowing in a pipe, conduit, or channel. The line is drawn above the hydraulic grade line (gradient) a distance equal to the velocity head ($V^2/2g$) of the water flowing at each section or point along the pipe or channel. Also see *hydraulic grade line (HGL)*. [See drawing on page 789.]

energy-isolating device   A mechanical device that physically prevents the transmission or release of energy, including but not limited to the following: A manually operated electric circuit breaker; a disconnect switch; a manually operated switch by which the conductors of a circuit can be disconnected from all ungrounded supply conductors, and, in addition, no pole can be operated independently; a line valve; a block; and any similar device used to block or isolate energy. Push buttons, selector switches, and other control circuit type devices are not energy-isolating devices.

enteric   Of intestinal origin, especially applied to wastes or bacteria.

entrain   To trap bubbles in water either mechanically through turbulence or chemically through a reaction; or to trap one substance or material by another substance or material.

Environmental Protection Agency   See *EPA*.

enzymes (EN-zimes)   Organic substances (produced by living organisms) that cause or speed up chemical reactions. Organic catalysts or biochemical catalysts.

EPA   United States Environmental Protection Agency. A regulatory agency established by the US Congress to administer the nation's environmental laws. Also called the US EPA.

epidemic (EP-uh-DEM-ick)   A disease that occurs in a large number of people in a locality at the same time and spreads from person to person.

epidemiology (EP-uh-DE-me-ALL-o-jee)   A branch of medicine that studies epidemics (diseases that affect significant numbers of people during the same time period in the same locality). The objective of epidemiology is to determine the factors that cause epidemic diseases and how to prevent them.

epilimnion (EP-uh-LIM-nee-on)   The upper layer of water in a thermally stratified lake or reservoir. This layer consists of the warmest water and has a fairly uniform (constant) temperature. The layer is readily mixed by wind action.

equilibrium, calcium carbonate   A water is considered stable when it is just saturated with calcium carbonate. In this condition, the water will neither dissolve nor deposit calcium carbonate. Thus, in this water the calcium carbonate is in equilibrium with the hydrogen ion concentration.

equity   The value of an investment in a facility.

equivalent weight   That weight that will react with, displace, or is equivalent to one gram of hydrogen.

ester   A compound formed by the reaction between an acid and an alcohol with the elimination of a molecule of water.

eutrophic (yoo-TRO-fick)   Reservoirs and lakes that are rich in nutrients and very productive in terms of aquatic animal and plant life.

eutrophication (YOO-tro-fi-KAY-shun)   The increase in the nutrient levels of a lake or other body of water; this usually causes an increase in the growth of aquatic animal and plant life.

evaporation   The process by which water or other liquid becomes a gas (water vapor or ammonia vapor).

evapotranspiration (ee-VAP-o-TRANS-purr-A-shun)   (1) The process by which water vapor is released to the atmosphere by living plants. This process is similar to people sweating. Also called transpiration. (2) The total water removed from an area by transpiration (plants) and by evaporation from soil, snow, and water surfaces.

F

facultative (FACK-ul-tay-tive) **bacteria**   Facultative bacteria can use dissolved oxygen, oxygen obtained from food materials such as sulfate or nitrate ions, or some can respire through *glycolysis*. The bacteria can live under aerobic, anoxic, or anaerobic conditions.

fail-safe   Design and operation of a process control system whereby failure of the power system or any component does not result in process failure or equipment damage.

feedback   The circulating action between a sensor measuring a process variable and the controller that controls or adjusts the process variable.

feedwater   The water that is fed to a treatment process; the water that is going to be treated.

filtration   The process of passing water through a porous bed of fine granular material to remove suspended matter from the water. The suspended matter is mainly particles of floc, soil, and debris; but it also includes living organisms such as algae, bacteria, viruses, and protozoa.

final backfill   Backfill extending from the top of the initial backfill to the top of the trench. This zone has little influence on pipe performance, but can be important to the integrity of roads and structures. Also see *backfill* and *initial backfill*.

finished water   Water that has passed through a water treatment plant. All the treatment processes are completed or finished. This water is the product from the water treatment plant and is ready to be delivered to consumers. Also called product water.

fixed costs   Costs (rent, insurance, interest) that a utility must cover or pay even if there is no demand for water or no water to sell to customers. Also see *variable costs*.

fixed sample   A sample is fixed in the field by adding chemicals that prevent the water quality indicators of interest in the sample from changing before final measurements are performed later in the laboratory.

flagellates (FLAJ-el-lates)   Microorganisms that move by the action of tail-like projections.

flame polished   Melted by a flame to smooth out irregularities. Sharp or broken edges of glass (such as the end of a glass tube) are rotated in a flame until the edge melts slightly and becomes smooth.

flap gate   A hinged gate that is mounted at the top of a pipe or channel to allow flow in only one direction. Flow in the wrong direction closes the gate. Also see *check valve* and *tide gate*.

float on the system   A method of operating a water storage facility. Daily flow into the facility is approximately equal to the average daily demand for water. When consumer demands for water are low, the storage facility will be filling. During periods of high demand, the facility will be emptying.

floc   Clumps of bacteria and particles or coagulants and impurities that have come together and formed a cluster. Found in flocculation tanks and settling or sedimentation basins.

flocculation (flock-yoo-LAY-shun)   The gathering together of fine particles after coagulation to form larger particles by a process of gentle mixing. This clumping together makes it easier to separate the solids from the water by settling, skimming, draining, or filtering.

flow line   (1) The top of the wetted line, the water surface, or the hydraulic grade line of water flowing in an open channel or partially full conduit. (2) The lowest point of the channel inside a pipe, conduit, canal, or manhole. This term is used by some contractors, however, the preferred term for this usage is invert.

fluidized (FLOO-id-i-zd)   A mass of solid particles that is made to flow like a liquid by injection of water or gas is said to have been fluidized. In water and wastewater treatment, a bed of filter media is fluidized by backwashing water through the filter.

fluoridation (floor-uh-DAY-shun)   The addition of a chemical to increase the concentration of fluoride ions in drinking water to a predetermined optimum limit to reduce the incidence (number) of dental caries (tooth decay) in children. Defluoridation is the removal of excess fluoride in drinking water to prevent the mottling (brown stains) of teeth.

flushing   The removal of deposits of material that have lodged in water distribution lines or sewers because of inadequate velocity of flows. Water is discharged into the lines at such rates that the larger flow and higher velocities are sufficient to remove the material.

flux   A flowing or flow.

foot valve   A special type of check valve located at the bottom end of the suction pipe on a pump. This valve opens when the pump operates to allow water to enter the suction pipe but closes when the pump shuts off to prevent water from flowing out of the suction pipe.

foundation   The material in the bottom of a trench. It may or may not have a layer of bedding soil placed over it. The foundation soil may be (1) undisturbed and remain in place; (2) unsuitable and must be removed and replaced with another material; (3) so wet and soft that it must be displaced by dumping rock into the trench; or (4) removed from the trench, placed back in the trench, and then compacted. Over-excavation and backfill of the foundation is required only when the native trench bottom does not provide a firm working platform for placement of the pipe bedding material.

free available chlorine residual   That portion of the total available chlorine residual composed of dissolved chlorine gas (Cl_2), hypochlorous acid (HOCl), or hypochlorite ion (OCl^-) remaining in water after chlorination at the end of a specified contact period. This does not include chlorine that has combined with ammonia, nitrogen, or other compounds.

freeboard   (1) The vertical distance from the normal water surface to the top of the confining wall. (2) The vertical distance from the sand surface to the underside of a trough in a sand filter. This distance is also called available expansion.

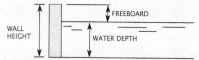

free residual chlorination   The application of chlorine to water to produce a free available chlorine residual equal to at least 80 percent of the total chlorine residual (sum of free and combined available chlorine residual).

friction loss   The head, pressure, or energy (they are the same) lost by water flowing in a pipe or channel as a result of turbulence caused by the velocity of the flowing water and the roughness of the pipe, channel walls, or restrictions caused by fittings. Water flowing in a pipe loses head, pressure, or energy as a result of friction. Also called head loss.

fulvic acid   A complex organic compound that can be derived either from soil or water. Aquatic fulvic acids are major precursors of disinfection byproducts.

fungi (FUN-ji)   Mushrooms, molds, mildews, rusts, and smuts that are small non-chlorophyll-bearing plants lacking roots, stems, and leaves. They occur in natural waters and grow best in the absence of light. Their decomposition may cause objectionable tastes and odors in water.

fuse   A protective device having a strip or wire of fusible metal that, when placed in a circuit, will melt and break the electric circuit if heated too much. High temperatures will develop in the fuse when a current flows through the fuse in excess of that which the circuit will carry safely.

G

galvanic cell   An electrolytic cell capable of producing electric energy by electrochemical action. The decomposition of materials in the cell causes an electric (electron) current to flow from cathode to anode.

galvanic series   A list of metals and alloys presented in the order of their tendency to corrode (or go into solution). Also called the electromotive series. This is a practical application of the theoretical *electrochemical series*.

galvanize   To coat a metal (especially iron or steel) with zinc. Galvanization is the process of coating a metal with zinc.

garnet   A group of hard, reddish, glassy, mineral sands made up of silicates of base metals (calcium, magnesium, iron, and manganese). Garnet has a higher density than sand.

gauge pressure   The pressure within a closed container or pipe as measured with a gauge. In contrast, absolute pressure is the sum of atmospheric pressure (14.7 lb/sq in or 1.0 atm) plus pressure within a vessel (as measured with a gauge). Most pressure gauges read in gauge pressure or psig (pounds per square inch gauge pressure) or kilopascals.

geographic information system (GIS)   An integrated system of computer hardware, software, and trained personnel linking topographic, demographic, utility, facility, images, and other resource data that are geographically referenced. A GIS is used to help operators and maintenance personnel locate utility system features or structures, such as pipes, valves, and manholes, and to assist with the scheduling and performance of maintenance activities.

geological log   A detailed description of all underground features discovered during the drilling of a well (depth, thickness, and type of formations).

geophysical log   A record of the structure and composition of the earth encountered when drilling a well or similar type of test hole or boring.

germicide (JERM-uh-side)   A substance formulated to kill germs or microorganisms. The germicidal properties of chlorine make it an effective disinfectant.

GHS   See *Globally Harmonized System of Classification and Labeling of Chemicals (GHS)*.

Giardia (jee-ARE-dee-ah)   A waterborne intestinal parasite that causes a disease called giardiasis in infected humans. Symptoms of the disease include diarrhea, cramps, and weight loss. *Giardia* contamination is found in most surface waters and some groundwaters.

giardiasis (jee-are-DYE-uh-sis)   Intestinal disease caused by an infestation of *Giardia* flagellates.

GIS   See *geographic information system (GIS)*.

Globally Harmonized System of Classification and Labeling of Chemicals (GHS)   A worldwide initiative to promote standard criteria for classifying chemicals according to their health, physical, and environmental hazards. It uses harmonized pictograms, hazard statements, precautionary statements, and the signal words "Danger" and "Warning" to communicate hazard information on product labels and safety data sheets (SDSs) in a logical and comprehensive way. The primary goals of the GHS are: (1) To enhance the protection of human health and the environment by providing an internationally comprehensible system for hazard communication; (2) To provide a recognized framework for those countries without an existing system; (3) To reduce the need for testing and evaluation of chemicals; (4) To facilitate international trade in chemicals whose hazards have been properly assessed and identified on an international basis. Also see *safety data sheet (SDS)*.

grab sample   A single sample of water collected at a particular time and place that represents the composition of the water only at that time and place.

grade   (1) The elevation of the invert (or bottom) of a pipeline, canal, culvert, sewer, or similar conduit. (2) The inclination or slope of a pipeline, conduit, stream channel, or natural ground surface; usually expressed in terms of the ratio or percentage of number of units of vertical rise or fall per unit of horizontal distance. A 0.5 percent grade would be a drop of one-half foot per hundred feet (one-half meter per hundred meters) of pipe.

gravimetric   A means of measuring unknown concentrations of water quality indicators in a sample by weighing a precipitate or residue of the sample.

gravimetric feeder   A dry chemical feeder that delivers a measured weight of chemical during a specific time period.

greensand   A mineral (glauconite) material that looks like ordinary filter sand except that it is green in color. Greensand is a natural ion exchange material that is capable of softening water. Greensand that has been treated with potassium permanganate ($KMnO_4$) is called manganese greensand; this product is used to remove iron, manganese, and hydrogen sulfide from groundwaters.

ground   An expression representing an electrical connection to earth or a large conductor that is at the earth's potential or neutral voltage.

H

haloacetic (HAL-o-uh-SEE-tick) **acid (HAA)**   A class of disinfection byproducts, formed mainly during the chlorination of water, containing natural organic matter. HAA5 is the sum of the concentrations, in milligrams per liter, of five haloacetic acid compounds.

hardness, water A characteristic of water caused mainly by the salts of calcium and magnesium, such as bicarbonate, carbonate, sulfate, chloride, and nitrate. Excessive hardness in water is undesirable because it causes the formation of soap curds, increased use of soap, deposition of scale in boilers, damage in some industrial processes, and sometimes causes objectionable tastes in drinking water.

hard water Water having a high concentration of calcium and magnesium ions. A water may be considered hard if it has a hardness greater than the typical hardness of water from the region. Some textbooks define hard water as a water with a hardness of more than 100 mg/L as calcium carbonate.

harmful physical agent Any chemical substance, biological agent (bacteria, virus, or fungus), or physical stress (noise, heat, cold, vibration, repetitive motion, ionizing and non-ionizing radiation, hypo- or hyperbaric pressure) that meets any of the following criteria: (1) Is regulated by any state or federal law or rule due to a hazard to health; (2) Is listed in the latest printed edition of the National Institute of Occupational Safety and Health (NIOSH) Registry of Toxic Effects of Chemical Substances (RTECS) (3) Has yielded positive evidence of an acute or chronic health hazard in human, animal, or other biological testing conducted by, or known to, the employer; (4) Is described by a safety data sheet (SDS) available to the employer that indicates that the material may pose a hazard to human health. Also called toxic substance. Also see *acute health effect* and *chronic health effect*.

haunches That portion of pipe from the bottom of the pipe up to the springline or horizontal centerline of the pipe.

haunching The portion of the material placed in an excavation on either side of and under a pipe from the top of the bedding up to the springline or horizontal centerline of the pipe. This backfill layer extends from one trench sidewall to the opposite sidewall. This is the most critical area in providing support for a pipe.

Hazard Communication Standard (HCS) The Occupational Safety and Health Administration (OSHA) mandate, 29 CFR 1910.1200, states that companies producing and using hazardous materials must provide employees with information and training on the proper handling and use of these materials. OSHA's HCS requires the development and dissemination of the following information: (1) Chemical manufacturers and importers are required to evaluate the hazards of the chemicals they produce or import, and prepare labels and safety data sheets to convey the hazard information to their downstream customers. (2) All employers with hazardous chemicals in their workplaces must have labels and safety data sheets for their exposed workers, and train them to handle the chemicals appropriately. The requirements of this standard are intended to be consistent with the provisions of the United Nations Globally Harmonized System of Classification and Labeling of Chemicals (GHS), Revision 3.

hazard statement A statement assigned to a hazard class and category that describes the nature of the hazard(s) of a chemical, including, where appropriate, the degree of hazard.

head The vertical distance, height, or energy of water above a reference point. A head of water may be measured in either height (feet or meters) or pressure (pounds per square inch or kilograms per square centimeter). Also see *discharge head, dynamic head, static head, suction head, suction lift,* and *velocity head*.

header A large pipe to which the ends of a series of smaller pipes are connected. Also see *manifold*.

head loss The head, pressure, or energy (they are the same) lost by water flowing in a pipe or channel as a result of turbulence caused by the velocity of the flowing water and the roughness of the pipe, channel walls, or restrictions caused by

fittings. Water flowing in a pipe loses head, pressure, or energy as a result of friction. The head loss through a filter is due to friction caused by material building up on the surface or by the water flowing through the filter media. Also called friction loss. [See drawing on page 790.]

heat sensor A device that opens and closes a switch in response to changes in the temperature. This device might be a metal contact, or a thermocouple that generates a minute electric current proportional to the difference in heat, or a variable resistor whose value changes in response to changes in temperature. Also called a temperature sensor.

hectare (HECK-ter) A measure of area in the metric system similar to an acre. One hectare is equal to 10,000 square meters and 2.4711 acres.

hepatitis (HEP-uh-TIE-tis) Hepatitis is an inflammation of the liver caused by an acute viral infection. Yellow jaundice is one symptom of hepatitis.

herbicide (HERB-uh-side) A compound, usually a manmade organic chemical, used to kill or control plant growth.

hertz (Hz) The number of complete electromagnetic cycles or waves in one second of an electric or electronic circuit. Also called the frequency of the current.

heterotrophic (HET-er-o-TROF-ick) Describes organisms that use organic matter for energy and growth. Animals, fungi, and most bacteria are heterotrophs.

HGL See *hydraulic grade line (HGL)*.

high-line jumpers Pipes or hoses connected to fire hydrants and laid on top of the ground to provide emergency water service for an isolated portion of a distribution system.

HMI See *human machine interface (HMI)*.

hose bib Faucet. A location in a water line where a hose is connected.

HTH High Test Hypochlorite. Calcium hypochlorite or $Ca(OCl)_2$.

human machine interface (HMI) The device at which the operator interacts with the control system. This may be an individual instrumentation and control device or the graphic screen of a computer control system. Also see *man machine interface (MMI)* and *operator interface*.

humic acid A polymeric constituent of soils, lignite, and peat. Aquatic humic acids are major precursors of disinfection byproducts.

humic substances Natural organic matter resulting from partial decomposition of plant or animal matter and forming the organic portion of soil.

hydrated lime Limestone that has been burned and treated with water under controlled conditions until the calcium oxide portion has been converted to calcium hydroxide $(Ca(OH)_2)$. Hydrated lime is quicklime combined with water. $CaO + H_2O \rightarrow Ca(OH)_2$. Also called slaked lime. Also see *quicklime*.

hydraulic Referring to water flowing through manmade structures such as pipes or channels or natural environments such as rivers.

hydraulic conductivity (K) A coefficient describing the relative ease with which groundwater can move through a permeable layer of rock or soil. Typical units of hydraulic conductivity are feet per day, gallons per day per square foot, or meters per day (depending on the unit chosen for the total discharge and the cross-sectional area).

hydraulic grade line (HGL) The surface or profile of water flowing in an open channel or a pipe flowing partially full. If a pipe is under pressure, the hydraulic grade line is that level water would rise to in a small, vertical tube connected to the pipe. Also see *energy grade line (EGL)*. [See drawing on page 789.]

hydraulic gradient   The slope of the hydraulic grade line. This is the slope of the water surface in an open channel, the slope of the water surface of the groundwater table, or the slope of the water pressure for pipes under pressure.

hydrogeologist (HI-dro-jee-ALL-uh-jist)   A person who studies and works with groundwater.

hydrologic (HI-dro-LOJ-ick) **cycle**   The process of evaporation of water into the air and its return to earth by precipitation (rain or snow). This process also includes transpiration from plants, groundwater movement, and runoff into rivers, streams, and the ocean. Also called the water cycle.

hydrolysis (hi-DROLL-uh-sis)   (1) A chemical reaction in which a compound is converted into another compound by taking up water. (2) Usually a chemical degradation of organic matter.

hydrophilic (hi-dro-FILL-ick)   Having a strong affinity (liking) for water. The opposite of *hydrophobic*.

hydrophobic (hi-dro-FOE-bick)   Having a strong aversion (dislike) for water. The opposite of *hydrophilic*.

hydropneumatic (hi-dro-new-MAT-ick)   A water system, usually small, in which a water pump is automatically controlled (started and stopped) by the air pressure in a compressed-air tank.

hydrostatic (hi-dro-STAT-ick) **pressure**   (1) The pressure at a specific elevation exerted by a body of water at rest. (2) In the case of groundwater, the pressure at a specific elevation due to the weight of water at higher levels in the same zone of saturation.

hygroscopic (hi-grow-SKAWP-ick)   Absorbing or attracting moisture from the air.

hypochlorination (HI-poe-klor-uh-NAY-shun)   The application of hypochlorite compounds to water or wastewater for the purpose of disinfection.

hypochlorinators (HI-poe-KLOR-uh-nay-tors)   Chlorine pumps, chemical feed pumps, or devices used to dispense chlorine solutions made from hypochlorites, such as bleach (sodium hypochlorite) or calcium hypochlorite, into the water being treated.

hypochlorite (HI-poe-KLOR-ite)   Chemical compounds containing available chlorine; used for disinfection. They are available as liquids (bleach) or solids (powder, granules, and pellets) and are packaged and shipped by various methods.

hypolimnion (HI-poe-LIM-nee-on)   The lowest layer in a thermally stratified lake or reservoir. This layer consists of colder and denser water, has a constant temperature, and no mixing occurs.

I

ICR   See *information collection rule (ICR)*.

IDLH   Immediately Dangerous to Life or Health. The atmospheric concentration of any toxic, corrosive, or asphyxiant substance that poses an immediate threat to life or would cause irreversible or delayed adverse health effects or would interfere with an individual's ability to escape from a dangerous atmosphere.

Imhoff cone   A clear, cone-shaped container marked with graduations. The cone is used to measure the volume of settleable solids in a specific volume (usually one liter) of water or wastewater.

impeller   A rotating set of vanes in a pump or compressor designed to pump or move water or air.

impermeable (im-PURR-me-uh-bull)   Not easily penetrated. The property of a material or soil that does not allow, or allows only with great difficulty, the movement or passage of water.

indicator   (1) (Chemical indicator) A substance that gives a visible change, usually of color, at a desired point in a chemical reaction, generally at a specified endpoint. (2) (Instrument indicator) A device that indicates the result of a measurement, usually using either a fixed scale and movable indicator (pointer), such as a pressure gauge, or a moving chart with a movable pen like those used on a circular flow-recording chart. Also called a receiver.

infiltration   The seepage of groundwater into a sewer system, including service connections. Seepage frequently occurs through defective or cracked pipes, pipe joints and connections, interceptor access risers and covers, or manhole walls.

influent   Water—raw (untreated) or partially treated—flowing into a reservoir, basin, treatment process, or treatment plant.

Information Collection Rule (ICR)   The Information Collection Rule (ICR) was promulgated on May 14, 1996, and approved by the Director of the Federal Register on June 18, 1996. It was to remain effective until December 31, 2000. The rule specified requirements for monitoring microbial contaminants and disinfection byproducts (DBPs) by large public water systems (PWSs). It required large PWSs to conduct either bench- or pilot-scale testing of advanced treatment techniques. The data reported under the ICR were used by EPA to learn more about the occurrence of microbial contamination and disinfection byproducts, the health risks posed, appropriate analytical methods, and effective forms of treatment. The ICR data form the scientific basis for EPA's development of the Enhanced Surface Water Treatment Rule and the Disinfectants and Disinfection Byproducts (D/DBP) Rule.

initial backfill   Backfill extending from the springline or horizontal centerline of a pipe to a point above the top of the pipe. This zone provides some pipe support and helps to prevent damage to the pipe during placement of the final backfill. Also see *backfill* and *final backfill*.

initial sampling   The very first sampling conducted under the Safe Drinking Water Act (SDWA) for each of the applicable contaminant categories.

injector water   Service water in which chlorine is added (injected) to form a chlorine solution.

inline filtration   The addition of chemical coagulants directly to the filter inlet pipe. The chemicals are mixed by the flowing water. Flocculation and sedimentation facilities are eliminated. This pretreatment method is commonly used in pressure filter installations. Also see *conventional filtration* and *direct filtration*.

inorganic   Describes materials such as sand, salt, iron, calcium salts, and other minerals. Inorganic materials are chemical substances of mineral origin, whereas organic substances are usually of animal or plant origin. Also see *organic*.

inorganic waste   Waste material such as sand, salt, iron, calcium, and other mineral materials that are only slightly affected by the action of organisms. Inorganic wastes are chemical substances of mineral origin; whereas organic wastes are chemical substances usually of animal or plant origin. Also see *nonvolatile matter*, *organic waste*, and *volatile solids*.

input horsepower   The total power used in operating a pump and motor.

$$\text{Input Horsepower, HP} = \frac{\left(\text{Brake Horsepower, HP}\right)\left(100\%\right)}{\text{Motor Efficiency, }\%}$$

insecticide   Any substance or chemical formulated to kill or control insects.

insoluble (in-SAWL-yoo-bull)   Something that cannot be dissolved.

integrator   A device or meter that continuously measures and sums a process rate variable in cumulative fashion over a given time period. For example, total flows displayed in gallons per minute, million gallons per day, cubic feet per second, or some other unit of volume per time period. Also see *totalizer*.

interceptor (intercepting) sewer   A large sewer that receives flow from a number of sewers and conducts the wastewater to a treatment plant. Often called an interceptor. The term interceptor is sometimes used in small communities to describe a septic tank or other holding tank that serves as a temporary wastewater storage reservoir for a septic tank effluent pump (STEP) system.

interconnector   A sewer installed to connect two separate sewers. If one sewer becomes blocked, wastewater can back up and flow through the interconnector to the other sewer.

interface   The common boundary layer between two substances, such as water and a solid (metal); or between two fluids, such as water and a gas (air); or between a liquid (water) and another liquid (oil).

interlock   A physical device, equipment, or software routine that prevents an operation from beginning or changing function until some condition or set of conditions is fulfilled. An example would be a switch that prevents a piece of equipment from operating when a hazard exists.

internal friction   Friction within a fluid (water) due to cohesive forces.

interstice (in-TUR-stuhz)   A very small open space in a rock or granular material. Also called a pore, void, or void space. Also see *void*.

invert (IN-vert)   The lowest point of the channel inside a pipe, conduit, canal, or manhole. Also called flow line.

ion   An electrically charged atom, radical (such as SO_4^{2-}), or molecule formed by the loss or gain of one or more electrons.

ion exchange   A water or wastewater treatment process involving the reversible interchange (switching) of ions between the water being treated and the solid resin contained within an ion exchange unit. Undesirable ions are exchanged with acceptable ions on the resin or recoverable ions in the water being treated are exchanged with other acceptable ions on the resin.

ion exchange resins   Insoluble polymers, used in water or wastewater treatment, that are capable of exchanging (switching or giving) acceptable cations or anions to the water being treated for less desirable ions or for ions to be recovered.

ionic concentration   The concentration of any ion in solution, usually expressed in moles per liter.

ionization (EYE-on-uh-ZAY-shun)   (1) The splitting or dissociation (separation) of molecules into negatively and positively charged ions. (2) The process of adding electrons to, or removing electrons from, atoms or molecules, thereby creating ions. High temperatures, electrical discharges, and nuclear radiation can cause ionization.

J

jar test   A laboratory procedure that simulates coagulation/flocculation with differing chemical doses. The purpose of the procedure is to estimate the minimum coagulant dose required to achieve certain water quality goals. Samples of water to be treated are placed in six jars. Various amounts of chemicals are added to each jar, stirred, and the settling of solids is observed. The lowest dose of chemicals that provides satisfactory settling is the dose used to treat the water.

jogging   The frequent starting and stopping of an electric motor.

joule (JOOL)   A measure of energy, work, or quantity of heat. One joule is the work done when the point of application of a force of one newton is displaced a distance of one meter in the direction of the force. Approximately equal to 0.7375 ft-lb (0.1022 m-kg).

K

kelly   The square section of a rod that causes the rotation of the drill bit. Torque from a drive table is applied to the square rod to cause the rotary motion. The drive table is chain- or gear-driven by an engine.

kilo   (1) Kilogram. (2) Kilometer. (3) A prefix meaning thousand used in the metric system and other scientific systems of measurement.

kinetic energy   Energy possessed by a moving body of matter, such as water, as a result of its motion.

Kjeldahl (KELL-doll) **nitrogen**   Nitrogen in the form of organic proteins or their decomposition product ammonia, as measured by the Kjeldahl Method.

L

lag time   The time period between the moment a process change is made and the moment such a change is finally sensed by the associated measuring instrument.

Langelier Index (LI)   An index reflecting the equilibrium pH of a water with respect to calcium and alkalinity. This index is used in stabilizing water to control both corrosion and the deposition of scale.

$$\text{Langelier Index} = pH - pH_s$$

where pH = actual pH of the water

pH_s = pH at which water having the same alkalinity and calcium content is just saturated with calcium carbonate

lateral surface area   The area of all the sides but not the top and bottom of a solid object, such as a cube, cylinder, or cone.

laundering weir   Sedimentation basin overflow weir. A plate with V-notches along the top to ensure a uniform flow rate and avoid short-circuiting.

launders   Sedimentation basin and filter discharge channels consisting of overflow weir plates (in sedimentation basins) and conveying troughs.

lead (LEED)   A wire or conductor that can carry electric current.

leathers   O-rings or gaskets used with piston pumps to provide a seal between the piston and the side wall.

LEL   See *lower explosive limit (LEL)*.

level control   A float device (or pressure switch) that senses changes in a measured variable and opens or closes a switch in response to that change. In its simplest form, this control might be a floating ball connected mechanically to a switch or valve, such as is used to stop water flow into a toilet when the tank is full.

lindane (LYNN-dane)   A pesticide that causes adverse health effects in humans and is toxic to freshwater and marine aquatic life.

linearity (lin-ee-AIR-it-ee)   How closely an instrument measures actual values of a variable through its effective range.

littoral (LIT-or-ul) **zone**   (1) That portion of a body of fresh water extending from the shoreline lakeward to the limit of occupancy of rooted plants. (2) The strip of land along the shoreline between the high and low water levels.

lockout   The placement of a lockout device on an energy-isolating device, in accordance with an established procedure, ensuring that the energy-isolating device and the equipment being controlled cannot be operated until the lockout device is removed.

lockout device   A device that uses a positive means such as a lock, either key or combination type, to hold an energy-isolating device in a safe position and prevent the energizing of a machine or equipment. Included are blank flanges and bolted slip blinds.

lockout/tagout   The practices and procedures necessary to disable machinery or equipment, thereby preventing the release of hazardous energy while employees perform servicing and maintenance activities. OSHA's standard, The Control of Hazardous Energy (Lockout/Tagout) (Title 29 CFR Part 1910.147) outlines measures for controlling hazardous energies—electrical, mechanical, hydraulic, pneumatic, chemical, thermal, and other energy sources.

logarithm (LOG-uh-rith-um) **(log)**   The exponent that indicates the power to which a number must be raised to produce a given number. For example, if $B^2 = N$, the 2 is the logarithm of N (to the base B), or $10^2 = 100$ and $\log_{10} 100 = 2$.

lower explosive limit (LEL)   The lowest concentration of a gas or vapor (percentage by volume in air) below which a flame will not spread in the presence of an ignition source (arc, flame, or heat). Concentrations lower than LEL are "too lean" to burn. Also called lower flammable limit (LFL). Also see *upper explosive limit (UEL)*.

lower flammable limit (LFL)   See *lower explosive limit (LEL)*.

M

M **(MOLAR) or mol/L**   A molar solution consists of 1 gram molecular weight of a compound dissolved in enough water to make 1 liter of solution. A gram molecular weight is the molecular weight of a compound in grams. For example, the molecular weight of sulfuric acid (H_2SO_4) is 98. A 1 *M* solution of sulfuric acid would consist of 98 grams of H_2SO_4 dissolved in enough distilled water to make 1 liter of solution.

macroscopic (MACK-row-SKAWP-ick) **organisms**   Organisms big enough to be seen by the eye without the aid of a microscope.

mandrel (MAN-drill)   (1) A special tool used to push bearings in or to pull sleeves out. (2) A testing device used to measure for excessive deflection in a flexible conduit.

manifold   A large pipe to which the ends of a series of smaller pipes are connected. Also called a header.

man machine interface (MMI)   The device at which the operator interacts with the control system. This may be an individual instrumentation and control device or the graphic screen of a computer control system. Also see *human machine interface (HMI)* and *operator interface*.

manometer (man-NAH-mut-ter)   An instrument for measuring pressure. Usually, a manometer is a glass tube filled with a liquid that is used to measure the difference in pressure across a flow measuring device, such as an orifice or a Venturi meter. The instrument used to measure blood pressure is a type of manometer.

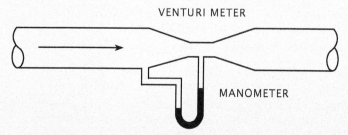

VENTURI METER

MANOMETER

Material Safety Data Sheet (MSDS)   A document that provides pertinent information and a profile of a particular hazardous substance or mixture. An MSDS is normally developed by the manufacturer or formulator of the hazardous substance or mixture. The MSDS is required to be made available to employees and operators or inspectors whenever there is the likelihood of the hazardous substance or mixture being introduced into the workplace. Some manufacturers are preparing MSDSs for products that are not considered to be hazardous to show that the product or substance is not hazardous. Also see *safety data sheet (SDS)*.

maximum contaminant level (MCL)   The largest allowable amount. MCLs for various water quality indicators are specified in the National Primary Drinking Water Regulations (NPDWR).

maximum residual disinfectant level (MRDL)   The highest level of a disinfectant allowed in drinking water without causing an unacceptable possibility of adverse health effects.

MBAS   Methylene Blue Active Substance. Another name for surfactants or surface active agents. The determination of surfactants is accomplished by measuring the color change in a standard solution of methylene blue dye.

MCL   See *maximum contaminant level*.

MCLG   Maximum Contaminant Level Goal. MCLGs are health goals based entirely on health effects. They are a preliminary standard set but not enforced by EPA. MCLs consider health effects, but also take into consideration the feasibility and cost of analysis and treatment of the regulated MCL. Although often less stringent than the corresponding MCLG, the MCL is set to protect health.

measured variable   A factor (flow, temperature) that is sensed and quantified (reduced to a reading of some kind) by a primary element or sensor.

mechanical joint   A flexible device that joins pipes or fittings together by the use of lugs and bolts.

meg   (1) Abbreviation of *megohm*. (2) A procedure used for checking the insulation resistance on motors, feeders, bus bar systems, grounds, and branch circuit wiring. Also see *megger*.

megger (from megohm)   An instrument used for checking the insulation resistance on motors, feeders, bus bar systems, grounds, and branch circuit wiring. A megger reads in millions of ohms. Also see *meg*.

megohm (MEG-ome)   Millions of ohms. Mega- is a prefix meaning one million, so 5 megohms means 5 million ohms.

membrane fouling   The cause of a loss of flow through the membrane as a result of material being retained on the surface of the membrane or within the membrane pores. Membrane fouling may be reversible or irreversible. Loss of flow through the membrane by reversible fouling can be recovered by regular backwashing of the membrane surface. Flow loss by irreversible fouling cannot be recovered.

meniscus (meh-NIS-cuss)   The curved surface of a column of liquid (such as water, oil, or mercury) in a small tube. When the liquid wets the sides of the container (as with water), the curve forms a valley. When the confining sides are not wetted (as with mercury), the curve forms a hill or upward bulge.

WATER
(READ BOTTOM)

MERCURY
(READ TOP)

mesh   One of the openings or spaces in a screen or woven fabric. The value of the mesh is usually given as the number of openings per inch. This value does not consider the diameter of the wire or fabric; therefore, the mesh number does not always have a definite relationship to the size of the hole.

mesotrophic (MESS-o-TRO-fick)   Reservoirs and lakes that contain moderate quantities of nutrients and are moderately productive in terms of aquatic animal and plant life.

metabolism   All of the processes or chemical changes in an organism or a single cell by which food is built up (anabolism) into living protoplasm and by which protoplasm is broken down (catabolism) into simpler compounds with the exchange of energy.

metalimnion (met-uh-LIM-nee-on)   The middle layer in a thermally stratified lake or reservoir. In this layer there is a rapid decrease in temperature with depth. Also called thermocline.

methoxychlor (meth-OX-e-klor)   A pesticide that causes adverse health effects in domestic water supplies and is toxic to freshwater and marine aquatic life. The chemical name for methoxychlor is 2,2-bis(p-methoxyphenol)-1,1,1-trichloroethane.

methyl orange alkalinity   A measure of the total alkalinity in a water sample. The alkalinity is measured by the amount of standard sulfuric acid required to lower the pH of the water to a pH level of 4.5, as indicated by the change in color of methyl orange from orange to pink. Methyl orange alkalinity is expressed as milligrams per liter equivalent calcium carbonate.

MF   See *microfiltration (MF)*.

mg/L   See *milligrams per liter, mg/l*.

microbial (my-KRO-bee-ul) **growth**   The activity and growth of microorganisms, such as bacteria, algae, diatoms, plankton, and fungi.

microfiltration (MF)   A pressure-driven membrane filtration process that separates particles down to approximately 0.1 μm diameter from influent water using a sieving process.

micron (MY-kron)   μm, micrometer, or micron. A unit of length. One millionth of a meter or one thousandth of a millimeter. One micron equals 0.00004 of an inch. A typical bacterium cell diameter ranges from 1 to 10 μm.

microorganisms (MY-crow-OR-gan-is-ums)   Living organisms that can be seen individually only with the aid of a microscope.

mil   A unit of length equal to 0.001 of an inch. The diameter of wires and tubing is measured in mils, as is the thickness of plastic sheeting.

milligrams per liter, mg/L   A measure of the concentration by weight of a substance per unit volume. For practical purposes, 1 mg/L of a substance in water is equal to 1 part per million parts (ppm). A liter of water with a specific gravity of 1.0 weighs 1 million milligrams. If 1 liter of water contains 10 milligrams of calcium, the concentration is 10 milligrams per million milligrams, or 10 milligrams per liter (10 mg/L), or 10 parts of calcium per million parts of water, or 10 parts per million (10 ppm).

millimicron (MILL-uh-MY-kron)   A unit of length equal to 10^{-3}μ (one thousandth of a micron), 10^{-6} millimeters, or 10^{-9} meters; correctly called a nanometer, nm.

molar   See *M (molar)* or *mol/L*.

molarity   A measure of concentration defined as the number of moles of solute per liter of solution. Also see *M (molar)* or *mol/L*.

mole   A unit of measurement for the number (as opposed to the mass) of atoms or molecules. A mass of a substance equal to its molecular weight in grams is 1 mole, which is 6.02×10^{23} molecules (Avogadro's Number). Also see *M (molar)* or *mol/L* and *molecular weight*.

molecular weight   The molecular weight of a compound in grams per mole is the sum of the atomic weights of the elements in the compound. The molecular weight of sulfuric acid (H_2SO_4) in grams is 98.

Element	Atomic Weight	Number of Atoms	Molecular Weight
H	1	2	2
S	32	1	32
O	16	4	64
			98

molecule   The smallest division of a compound that still retains or exhibits all the properties of the substance.

monomer (MON-o-mer)   A molecule of low molecular weight capable of reacting with identical or different monomers to form polymers.

monomictic (mah-no-MICK-tick)   Lakes and reservoirs that are relatively deep, do not freeze over during the winter months, and undergo a single stratification and mixing cycle during the year. These lakes and reservoirs usually become destratified during the mixing cycle, usually in the fall of the year.

monovalent   Having a valence of one, such as the cuprous (copper) ion, Cu^+.

most probable number (MPN)   See *MPN*.

motile (MO-till)   Capable of self-propelled movement. A term that is sometimes used to distinguish between certain types of organisms found in water.

motor efficiency   The ratio of energy delivered by a motor to the energy supplied to it during a fixed period or cycle. Motor efficiency ratings will vary depending on motor manufacturer and usually will be near 90.0 percent.

MPN   MPN is the Most Probable Number of coliform-group organisms per unit volume of sample water. This is a statistical estimate, expressed as a density or population of organisms per 100 mL of sample water.

MRDL   Maximum Residual Disinfectant Level. The highest level of a disinfectant allowed in drinking water without causing an unacceptable possibility of adverse health effects.

MSDS   See *Material Safety Data Sheet (MSDS)*.

mudballs   Material, approximately round in shape, that forms in filters and gradually increases in size when not removed by the backwashing process. Mudballs vary from pea-sized up to golf-ball-sized or larger.

multistage pump   A pump that has more than one impeller. A single-stage pump has one impeller.

N

N **(normal)**   A normal solution contains 1 gram equivalent weight of reactant (compound) per liter of solution. The equivalent weight of an acid is that weight that contains 1 gram atomic weight of ionizable hydrogen or its chemical equivalent. For example, the equivalent weight of sulfuric acid (H_2SO_4) is 49 (98 divided by 2 because there are 2 replaceable hydrogen ions). A 1.0 *N* solution of sulfuric acid would consist of 49 grams of H_2SO_4 dissolved in enough water to make 1 liter of solution.

nameplate   A durable, metal plate found on equipment that lists critical installation and operating conditions for the equipment.

nanofiltration (NF)   A pressure-driven membrane filtration process that separates particles down to approximately 0.002 to 0.005 μm diameter from influent water using a sieving process.

natural organic matter (NOM)   Humic substances composed of humic acid and fulvic acid that come from decayed vegetation. Also see *humic substances, humic acid,* and *fulvic acid.*

nephelometric (neff-el-o-MET-rick)   A means of measuring turbidity in a sample by using an instrument called a nephelometer. A nephelometer passes light through a sample and the amount of light deflected (usually at a 90-degree angle) is then measured.

newton   A force that, when applied to a body having a mass of one kilogram, gives it an acceleration of one meter per second per second.

NF   See *nanofiltration (NF).*

nitrification (NYE-truh-fuh-KAY-shun)   An aerobic process in which bacteria oxidize the ammonia and organic nitrogen in water into nitrite and then nitrate.

nitrogenous (nye-TRAH-jen-us)   A term used to describe chemical compounds (usually organic) containing nitrogen in combined forms. Proteins and nitrate are nitrogenous compounds.

noble metal   A chemically inactive metal (such as gold). A metal that does not corrode easily and is much scarcer (and more valuable) than the so-called useful or base metals. Also see *base metal.*

NOM   See *natural organic matter (NOM).*

nominal diameter   An approximate measurement of the diameter of a pipe. Although the nominal diameter is used to describe the size or diameter of a pipe, it is usually not the exact inside diameter of the pipe.

nonionic (NON-eye-ON-ick) **polymer**   A polymer that has no net electrical charge.

non-permit confined space   See *confined space, non-permit.*

nonpoint source   A runoff or discharge from a field or similar source, in contrast to a point source, which refers to a discharge that comes out the end of a pipe or other clearly identifiable conveyance. Also see *point source.*

nonpotable (non-POE-tuh-bull)   Water that may contain objectionable pollution, contamination, minerals, or infective agents and is considered unsafe or unpalatable for drinking.

nonvolatile matter   Material such as sand, salt, iron, calcium, and other mineral materials that are only slightly affected by the actions of organisms and are not lost on ignition of the dry solids at 1,022°F (550°C). Volatile materials are chemical substances usually of animal or plant origin. Also see *inorganic waste* and *volatile solids.*

normal   See *N (normal).*

normality   The number of gram-equivalent weights of solute in 1 liter of solution. The equivalent weight of any material is the weight that would react with or be produced by the reaction of 8.0 grams of oxygen or 1.0 gram of hydrogen. Normality is used for certain calculations of quantitative analysis. Also see *N (normal).*

notice   This word calls attention to information that is especially significant in understanding and operating equipment or processes safely. Also see *caution, danger,* and *warning.*

NPDES permit   A National Pollutant Discharge Elimination System permit is the regulatory agency document issued by either a federal or state agency that is designed to control all discharges of potential pollutants from all point sources of pollution and stormwater runoff into US waterways. NPDES permits regulate discharges from industries, municipal wastewater treatment plants, sanitary landfills, large animal feedlots, and return irrigation flows, among others.

NPDWR   National Primary Drinking Water Regulations.

NSDWR   National Secondary Drinking Water Regulations.

NTU   Nephelometric Turbidity Units. See *turbidity units (TU).*

nutrient   Any substance that is assimilated (taken in) by organisms and promotes growth. Nitrogen and phosphorus are nutrients that promote the growth of algae. There are other essential and trace elements that are also considered nutrients. Also see *nutrient cycle.*

nutrient cycle   The transformation or change of a nutrient from one form to another until the nutrient has returned to the original form, thus completing the cycle. The cycle may take place under either aerobic or anaerobic conditions.

Occupational Safety and Health Act of 1970 (OSHA)   See *OSHA.*

odor threshold   The minimum odor of a gas or water sample that can just be detected after successive dilutions with odorless gas or water. Also called threshold odor.

offset   (1) The difference between the actual value and the desired value (or set point); characteristic of proportional controllers that do not incorporate reset action. Also called drift. (2) A pipe fitting in the approximate form of a reverse curve or other combination of elbows or bends that brings one section of a line of pipe out of line with, but into a line parallel with, another section. (3) A pipe joint that has lost its bedding support, causing one of the pipe sections to drop or slip, thus creating a condition where the pipes no longer line up properly.

OHM   The unit of electrical resistance. The resistance of a conductor in which one volt produces a current of one ampere.

olfactory (all-FAK-tore-ee) **fatigue**   A condition in which a person's nose, after exposure to certain odors, is no longer able to detect the odor.

oligotrophic (ah-lig-o-TRO-fick)   Reservoirs and lakes that are nutrient poor and contain little aquatic plant or animal life.

operating pressure differential   The operating pressure range for a hydropneumatic system. For example, when the pressure drops below 40 psi in a system designed to operate between 40 psi and 60 psi, the pump will come on and stay on until the pressure builds up to 60 psi. When the pressure reaches 60 psi the pump will shut off. The operating pressure differential in this example is 20 psi.

operating ratio   The operating ratio is a measure of the total revenues divided by the total operating expenses.

operator interface   The device at which the operator interacts with the control system. This may be an individual instrumentation and control device or the graphic screen of a computer control system. Also see *human machine interface (HMI)* and *man machine interface (MMI).*

organic   Describes chemical substances that come from animal or plant sources. Organic substances always contain carbon though not every compound that contains carbon is organic. Carbon dioxide (CO_2), bicarbonate (HCO_3), and carbonate (CO_3^{-2}) are considered to be inorganic. Also see *inorganic.*

organics   (1) A term used to refer to chemical compounds made from carbon molecules. These compounds may be natural materials (such as animal or plant sources) or manmade materials (such as synthetic organics). Also see *organic*. (2) Any form of animal or plant life. Also see *bacteria*.

organic waste   Waste material that may come from animal or plant sources. Natural organic wastes generally can be consumed by bacteria and other small organisms. Manufactured or synthetic organic wastes from metal finishing, chemical manufacturing, and petroleum industries may not normally be consumed by bacteria and other organisms. Also see *inorganic waste* and *volatile solids*.

organism   Any form of animal or plant life. Also see *bacteria*.

organizing   Deciding who does what work and delegating authority to the appropriate persons.

orifice (OR-uh-fiss)   An opening (hole) in a plate, wall, or partition. An orifice flange or plate placed in a pipe consists of a slot or a calibrated circular hole smaller than the pipe diameter. The difference in pressure in the pipe above and at the orifice may be used to determine the flow in the pipe. In a trickling filter distributor, the wastewater passes through an orifice to the surface of the filter media.

ORP (pronounce as separate letters)   See *oxidation-reduction potential (ORP)*.

orthotolidine (or-tho-TOL-uh-dine)   Orthotolidine is a colorimetric indicator of chlorine residual. If chlorine is present, a yellow-colored compound is produced. This reagent is no longer approved for chemical analysis to determine chlorine residual.

OSHA (O-shuh)   The Williams-Steiger Occupational Safety and Health Act of 1970 (OSHA) is a federal law designed to protect the health and safety of industrial workers and the operators of water supply systems and treatment plants. The act regulates the design, construction, operation, and maintenance of water supply systems and water treatment plants. OSHA also refers to the federal and state agencies that administer the OSHA regulations.

osmosis (oz-MOE-sis)   The passage of a liquid from a weak solution to a more concentrated solution across a semipermeable membrane. The membrane allows the passage of the water (solvent) but not the dissolved solids (solutes). This process tends to equalize the conditions on either side of the membrane.

OUCH principle   This principle says that as a manager when you delegate job tasks you must be **O**bjective, **U**niform in your treatment of employees, and the tasks must be **C**onsistent with utility policies, and **H**ave job relatedness.

overall efficiency, pump   The combined efficiency of a pump and motor together. Also called the wire-to-water efficiency.

overdraft   The pumping of water from a groundwater basin or aquifer in excess of the supply flowing into the basin. This pumping results in a depletion or mining of the groundwater in the basin.

overflow rate   One of the guidelines for the design of settling tanks and clarifiers in treatment plants. Used by operators to determine if tanks and clarifiers are hydraulically (flow) over- or underloaded. Also called surface loading or surface settling rate.

$$\text{Overflow Rate} = \frac{\text{Flow}}{\text{Surface Area}}$$

overhead   Indirect costs necessary for a water utility to function properly. These costs are not related to the actual treatment and delivery of water to consumers, but include the costs of rent, lights, office supplies, management, and administration.

overturn   The almost spontaneous mixing of all layers of water in a reservoir or lake when the water temperature becomes similar from top to bottom. This may occur in the fall/winter when the surface waters cool to the same temperature as the bottom waters and in the spring when the surface waters warm after the ice melts. This is also called turnover.

oxidation   Oxidation is a chemical reaction that results in the addition of oxygen, removal of hydrogen, or the removal of electrons from an element or compound; in the environment and in wastewater treatment processes, organic matter is oxidized to more stable substances. The opposite of *reduction*.

oxidation-reduction potential (ORP)   The electrical potential required to transfer electrons from one compound or element (the oxidant) to another compound or element (the reductant); used as a qualitative measure of the state of oxidation in water and wastewater treatment systems. ORP is measured in millivolts, with negative values indicating a tendency to reduce compounds or elements and positive values indicating a tendency to oxidize compounds or elements.

oxidizing agent   Any substance, such as oxygen (O_2) or chlorine (Cl_2), that will readily add (take on) electrons. When oxygen or chlorine is added to w-ater or wastewater, organic substances are oxidized. These oxidized organic substances are more stable and less likely to give off odors or to contain disease-causing bacteria. The opposite is a *reducing agent*.

oxygen deficiency   An atmosphere containing oxygen at a concentration of less than 19.5 percent by volume.

oxygen enrichment   An atmosphere containing oxygen at a concentration of more than 23.5 percent by volume.

ozonation (O-zoe-NAY-shun)   The application of ozone to water for disinfection or for taste and odor control.

P

packer assembly   An inflatable device used to seal the tremie pipe inside the well casing to prevent the grout from entering the inside of the conductor casing.

palatable (PAL-uh-tuh-bull)   Water at a desirable temperature that is free from objectionable tastes, odors, colors, and turbidity. Pleasing to the senses.

parshall flume (PAR-shul FLOOM)   A device used to measure the flow in an open channel. The flume narrows to a throat of fixed dimensions and then expands again. The rate of flow can be calculated by measuring the difference in head (pressure) before and at the throat of the flume.

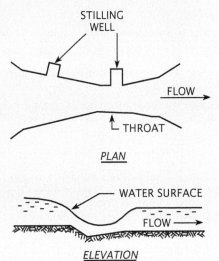

particle count   The results of a microscopic examination of treated water with a special particle counter, which classifies suspended particles by number and size.

particle counter   A device that counts and measures the size of individual particles in water.

particle counting A procedure for counting and measuring the size of individual particles in water. Particles are divided into size ranges and the number of particles is counted in each of these ranges. The results are reported in terms of the number of particles in different particle diameter size ranges per milliliter of water sampled.

particulate (par-TICK-yoo-let) A very small solid suspended in water that can vary widely in size, shape, density, and electrical charge. Colloidal and dispersed particulates are artificially gathered together by the processes of coagulation and flocculation.

parts per million (PPM) Parts per million parts, a ratio, is a measurement of concentration on a weight or volume basis. Parts per million by mass is equivalent to milligrams per liter (mg/L). Parts per million by volume is a common way to measure concentrations of gases.

pascal (Pa) The pressure or stress of one newton per square meter.

1 psi = 6,895 Pa = 6.895 kN/m² = 0.0703 kg/cm²

pathogenic (path-o-JEN-ick) **organisms** Organisms, including bacteria, viruses, protozoa, or internal parasites, capable of causing disease (such as giardiasis, cryptosporidiosis, typhoid fever, cholera, or infectious hepatitis) in a host (such as a person). Nonpathogenic organisms do not cause disease.

pathogens (PATH-o-jens) See *pathogenic organisms*.

PCBs Polychlorinated biphenyls. A class of organic compounds that cause adverse health effects in domestic water supplies.

pCi/L picoCurie per liter. A picoCurie is a measure of radioactivity. One picoCurie of radioactivity is equivalent to 0.037 nuclear disintegrations per second.

pcu (platinum cobalt units) A measure of color using platinum cobalt standards by visual comparison.

peak demand The maximum momentary load placed on a water treatment plant, pumping station, or distribution system. This demand is usually the maximum average load in one hour or less, but may be specified as the instantaneous load or the load during some other short time period.

percent saturation The amount of a substance that is dissolved in a solution compared with the amount dissolved in the solution at saturation, expressed as a percent.

$$\text{Percent Saturation}, \% = \frac{\text{Amount of Substance That Is Dissolved} \times 100\%}{\text{Amount Dissolved in Solution at Saturation}}$$

percolating (PURR-ko-lay-ting) **water** Water that passes through soil or rocks under the force of gravity.

percolation (purr-ko-LAY-shun) The slow passage of water through a filter medium; or, the gradual penetration of soil and rocks by water.

performance indicator A measurable goal used to determine system performance and level of service provided. Examples of performance indicators include the number of stoppages per 100 miles of sewer per year and the number of lost time accidents per year—measurements of how well a utility is doing rather than how much a utility is doing. Also called a production indicator.

periphyton (pair-e-FI-tawn) Microscopic plants and animals that are firmly attached to solid surfaces under water, such as rocks, logs, pilings, and other structures.

permeability (PURR-me-uh-BILL-uh-tee) The property of a material or soil that permits considerable movement of water through it when it is saturated.

permeate (PURR-me-ate) (1) To penetrate and pass through, as water penetrates and passes through soil and other porous materials. (2) The liquid (demineralized water) produced from the reverse osmosis process that contains a low concentration of dissolved solids.

permit-required confined space (permit space) See *confined space, permit-required (permit space)*.

pesticide Any substance or chemical designed or formulated to kill or control animal pests. Also see *insecticide* and *rodenticide*.

pet cock A small valve or faucet used to drain a cylinder or fitting.

pH (pronounce as separate letters) pH is an expression of the intensity of the basic or acidic condition of a liquid. Mathematically, this is equivalent to the negative of the base 10 logarithm.
If {H⁺} = 10⁻⁶·⁵, then pH = 6.5. The pH may range from 0 to 14, where 0 is most acidic, 14 most basic, and 7 neutral.

$$pH = \text{Log}\frac{1}{\{H^+\}} \text{ or } -\text{Log}\left(\{H^+\}\right)$$

phenolic (fee-NO-lick) **compounds** Organic compounds that are derivatives of benzene. Also called phenols (FEE-nolls).

phenolphthalein (FEE-nol-THAY-leen) **alkalinity** The alkalinity in a water sample measured by the amount of standard acid required to lower the pH to a level of 8.3, as indicated by the change in color of phenolphthalein from pink to clear. Phenolphthalein alkalinity is expressed as milligrams per liter of equivalent calcium carbonate.

photosynthesis (foe-toe-SIN-thuh-sis) A process in which organisms, with the aid of chlorophyll, convert carbon dioxide and inorganic substances into oxygen and additional plant material, using sunlight for energy. All green plants grow by this process. Also see *chlorophyll*.

phytoplankton (FIE-tow-plank-ton) Small, usually microscopic plants (such as algae), found in lakes, reservoirs, and other bodies of water.

pico- (PEE-ko) A prefix used in the metric system and other scientific systems of measurement which means 10⁻¹² or 0.000 000 000 001.

picoCurie (PEE-ko-KYOOR-ee) A measure of radioactivity. One picoCurie (pCi) of radioactivity is equivalent to 0.037 nuclear disintegrations per second.

pictogram A graphical composition that may include a symbol plus other graphic elements, such as a border, background pattern, or color, that is intended to convey specific information about the hazards of a chemical. There are nine pictograms under the Globally Harmonized System (GHS) to convey the health, physical, and environmental hazards.

piezometer (pie-ZOM-uh-ter) An instrument used to measure the pressure head in a pipe, tank, or soil. For pipes and tanks, it usually consists of a small pipe or tube. One end is connected or tapped through the wall of the pipe or tank being measured. The other end is most often open to the air, in which case the height of the water in the tube above the connection is the pressure head in inches or feet. For high pressures the tube may be connected to a mercury column in which case the pressure head is expressed in inches of mercury. For soil applications, the tube is connected to a permeable tip which is then buried in or driven into the soil. In this case, the piezometer measures the pore pressure of water in the soil. Also see *manometer*.

pilot-scale study A method of studying different ways of treating water or wastewater and solids or to obtain design criteria on a small scale in the field.

pinpoint floc Very small floc (the size of a pin point) that does not settle out of the water in a sedimentation basin or clarifier. Also see *floc*.

pipe embedment   The material placed around a pipe that supports the pipe.

pipe gauge   A number that defines the thickness of the sheet used to make steel pipe. The larger the number, the thinner the pipe wall.

pipe schedule   A sizing system of numbers that specifies the inside diameter (ID) and outside diameter (OD) for each diameter pipe. The schedule number is the ratio of internal pressure in psi divided by the allowable fiber stress multiplied by 1,000. Typical schedules of iron and steel pipe are Schedules 40, 80, and 160. Other forms of piping are divided into various classes with their own schedule schemes.

pitless adapter   A fitting that allows the well casing to be extended above ground while having a discharge connection located below the frost line. Advantages of using a pitless adapter include the elimination of the need for a pit or pump house and it is a watertight design, which helps maintain a sanitary water supply.

pitot (PEA-toe) **tube**   An instrument used to measure fluid (liquid or air) velocity by means of the differential pressure between the tip (dynamic) and side (static) openings.

plan   A drawing or photo showing the top view of sewers, manholes, streets, or structures.

plankton   (1) Small, usually microscopic, plants (phytoplankton) and animals (zooplankton) in aquatic systems. (2) All of the smaller floating, suspended, or self-propelled organisms in a body of water.

planning   Management of utilities to build the resources and financial capability to provide for future needs.

PLC   Programmable logic controller. A microcomputer-based control device containing programmable software; used to control process variables.

plug flow   A type of flow that occurs in tanks, basins, or reactors when a slug of water or wastewater moves through a tank without ever dispersing or mixing with the rest of the water or wastewater flowing through the tank.

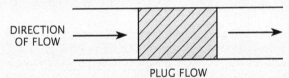

PLUG FLOW

PMCL   Primary Maximum Contaminant Level. Primary MCLs for various water quality indicators are established to protect public health.

point-of-use/point-of-entry (POU/POE)   Point-of-use applications refer to a water treatment device that treats water at the location of the end user. Point-of-entry application refers to a water treatment device that is located at the inlet to an entire building or facility.

point source   A discharge that comes out the end of a pipe or other clearly identifiable conveyance. Examples of point source conveyances from which pollutants may be discharged include ditches, channels, tunnels, conduits, wells, containers, rolling stock, concentrated animal feeding operations, landfill leachate collection systems, vessels, or other floating craft. A *nonpoint source* refers to runoff or a discharge from a field or similar source.

polarization   The concentration of ions in the thin, boundary layer adjacent to a membrane or pipe wall.

pole shader   A copper bar circling the laminated iron core inside the coil of a magnetic starter.

pollution   The impairment (reduction) of water quality by agricultural, domestic, or industrial wastes (including thermal and radioactive wastes) to a degree that the natural water quality is changed to hinder any beneficial use of the water or render it offensive to the senses of sight, taste, or smell or when sufficient amounts of wastes create or pose a potential threat to human health or the environment.

polyanionic (poly-AN-eye-ON-ick)   Characterized by many active negative charges.

polychlorinated biphenyls (PCBs) (POLY-KLOR-uh-nate-ed BI-FEEN-alls)   A class of organic compounds that cause adverse health effects in domestic water supplies.

polyelectrolyte (POLY-ee-LECK-tro-lite)   A high-molecular-weight substance, having points of positive or negative electrical charges, that is formed by either natural or synthetic processes. Natural polyelectrolytes may be of biological origin or obtained from starch products or cellulose derivatives. Synthetic polyelectrolytes consist of simple substances that have been made into complex, high-molecular-weight substances. Used with other chemical coagulants to aid in binding small suspended particles to larger chemical flocs for their removal from water. Often called a polymer.

polymer (POLY-mer)   A long-chain molecule formed by the union of many monomers (molecules of lower molecular weight). Polymers are used with other chemical coagulants to aid in binding small suspended particles to larger chemical flocs for their removal from water. Also see *polyelectrolyte*.

pore   A very small open space in a rock or granular material. Also called an interstice, void, or void space. Also see *void*.

porosity   (1) A measure of the spaces or voids in a material or aquifer. (2) The ratio of the volume of spaces in a rock or soil to the total volume. This ratio is usually expressed as a percentage.

$$\text{Porosity}, \% = \frac{\left(\text{Volume of Spaces}\right)(100\%)}{\text{Total Volume}}$$

positive bacteriological sample   A water sample in which gas is produced by coliform organisms during incubation in the multiple tube fermentation test.

positive displacement pump   A type of piston, diaphragm, gear, or screw pump that delivers a constant volume with each stroke. Positive displacement pumps are used as chemical solution feeders.

postchlorination   The addition of chlorine to the plant discharge or effluent, following plant treatment, for disinfection purposes.

potable (POE-tuh-bull) **water**   Water that does not contain objectionable pollution, contamination, minerals, or infective agents and is considered satisfactory for drinking.

POU/POE   See *point-of-use/point-of-entry (POU/POE)*.

power factor   The ratio of the true power passing through an electric circuit to the product of the voltage and amperage in the circuit. This is a measure of the lag or lead of the current with respect to the voltage. In alternating current, the voltage and amperes are not always in phase; therefore, the true power may be slightly less than that determined by the direct product.

PPM   See *parts per million (PPM)*.

precautionary statement   A phrase that describes recommended measures to be taken to minimize or prevent adverse effects resulting from exposure to a hazardous chemical, or improper storage or handling of a hazardous chemical.

prechlorination   The addition of chlorine at the headworks of the plant before other treatment processes mainly for disinfection and control of tastes, odors, and aquatic growths. Also applied to aid in coagulation and settling.

precipitate (pre-SIP-uh-TATE)   (1) An insoluble, finely divided substance that is a product of a chemical reaction within a liquid. (2) The separation from solution of an insoluble substance.

precipitation   (1) The total measurable supply of water received directly from clouds as rain, snow, hail, or sleet; usually expressed as depth in a day, month, or year, and designated as daily, monthly, or annual precipitation. (2) The process by which atmospheric moisture is discharged onto a land or water surface. (3) The chemical transformation of a substance in solution into an insoluble form (precipitate).

precision   The ability of an instrument to measure a process variable and repeatedly obtain the same result. The ability of an instrument to reproduce the same results.

precursor (PRE-curse-or), **THM**   Natural, organic compounds found in all surface and groundwaters, which may react with halogens (such as chlorine) to form trihalomethanes (THMs); they must be present in order for THMs to form.

prescriptive (pre-SKRIP-tive) **rights**   Water rights that are acquired by diverting water and putting it to use in accordance with specified procedures. These procedures include filing a request (with a state agency) to use unused water in a stream, river, or lake.

present worth   The value of a long-term project expressed in today's dollars. Present worth is calculated by converting (discounting) all future benefits and costs over the life of the project to a single economic value at the start of the project. Calculating the present worth of alternative projects makes it possible to compare them and select the one with the largest positive (beneficial) present worth or minimum present cost.

pressure control   A switch that operates on changes in pressure. Usually this is a diaphragm pressing against a spring. When the force on the diaphragm overcomes the spring pressure, the switch is activated.

pressure head   The vertical distance (in feet or meters) equal to the pressure (in psi or kPa) at a specific point. The pressure head is equal to the pressure in psi (or kPa) times 2.31 ft/psi (or 1.0 m/9.81 kPa).

prestressed   A prestressed pipe has been reinforced with wire strands (which are under tension) to give the pipe an active resistance to loads or pressures on it.

preventive maintenance   Regularly scheduled servicing of machinery or other equipment using appropriate tools, tests, and lubricants. This type of maintenance can prolong the useful life of equipment and machinery and increase its efficiency by detecting and correcting problems before they cause a breakdown of the equipment.

preventive maintenance crews   Crews assigned the task of cleaning sewers (for example, high-velocity cleaning crews) to prevent stoppages and odor complaints. Also see *preventive maintenance*.

primacy   Under the Safe Drinking Water Act (SDWA), primacy is the responsibility for ensuring that a law is implemented, and the authority to enforce a law and related regulations (40 CFR 142.2). A primacy agency has the primary responsibility for administrating and enforcing regulations.

primary element   (1) A device that measures (senses) a physical condition or variable of interest. Floats and thermocouples are examples of primary elements. Also called a sensor. (2) The hydraulic structure used to measure flows. In open channels, weirs and flumes are primary elements or devices. Venturi meters and orifice plates are the primary elements in pipes or pressure conduits.

prime   The action of filling a pump casing with water to remove the air. Most pumps must be primed before startup or they will not pump any water.

process variable   A physical or chemical quantity that is usually measured and controlled in the operation of a water, wastewater, or industrial treatment plant. Common process variables are flow, level, pressure, temperature, turbidity, chlorine, and oxygen levels.

product water   Water that has passed through a water treatment plant. All the treatment processes are completed or finished. This water is the product from the water treatment plant and is ready to be delivered to consumers. Also called finished water.

profile   A drawing showing elevation plotted against distance, such as the vertical section or side view of sewers, manholes, or pipelines.

programmable logic controller (PLC)   A microcomputer-based control device containing programmable software; used to control process variables.

proportional weir (WEER)   A specially shaped weir in which the flow through the weir is directly proportional to the head.

Prussian blue   A blue paste or liquid (often on a paper like carbon paper) used to show a contact area. Used to determine if gate valve seats fit properly.

psig   Pounds per square inch gauge pressure. The pressure within a closed container or pipe measured with a gauge in pounds per square inch. Also see *gauge pressure*.

pump bowl   The submerged pumping unit in a well, including the shaft, impellers, and housing.

pumping water level   The vertical distance from the centerline of the pump discharge to the level of the free pool while water is being drawn from the pool.

purveyor (purr-VAY-or), **water**   An agency or person that supplies water (usually potable water).

putrefaction (PYOO-truh-FACK-shun)   Biological decomposition of organic matter, with the production of foul-smelling and -tasting products, associated with anaerobic (no oxygen present) conditions.

Q

quicklime   A material that is mostly calcium oxide (CaO) or calcium oxide in natural association with a lesser amount of magnesium oxide. Quicklime is capable of combining with water, that is, becoming slaked. Also see *hydrated lime*.

R

radial to impeller   Perpendicular to the impeller shaft. Material being pumped flows at a right angle to the impeller.

radical   A group of atoms that is capable of remaining unchanged during a series of chemical reactions. Such combinations (radicals) exist in the molecules of many organic compounds; sulfate (SO_4^{2-}) is an inorganic radical.

range   The spread from minimum to maximum values that an instrument is designed to measure. Also see *effective range* and *span*.

ranney collector   This water collector is constructed as a dug well from 12 to 16 feet (3.5 to 5 m) in diameter that has been sunk as a caisson near the bank of a river or lake. Screens are driven radially and approximately horizontally from this well into the sand and gravel deposits underlying the river. [See drawing on page 791.]

rate of return   A value that indicates the return of funds received on the basis of the total equity capital used to finance physical facilities. Similar to the interest rate on savings accounts or loans.

raw water   (1) Water in its natural state, before any treatment. (2) Water entering the first treatment process of a water treatment plant.

reactive maintenance   Maintenance activities that are performed in response to problems and emergencies after they occur.

readout   The reading of the value of a process variable from an indicator or recorder or on a computer screen.

reaeration (RE-air-A-shun)   The introduction of air through forced air diffusers into the lower layers of the reservoir. As the air bubbles form and rise through the water, oxygen from the air dissolves into the water and replenishes the dissolved oxygen. The rising bubbles also cause the lower waters to rise to the surface where oxygen from the atmosphere is transferred to the water. This is sometimes called surface reaeration.

reagent (re-A-gent)   A pure, chemical substance or solution of that substance that is used to make new products or to measure, detect, or examine other substances in chemical tests.

recarbonation (re-kar-bun-NAY-shun)   A process in which carbon dioxide is bubbled into the water being treated to lower the pH. The pH may also be lowered by the addition of acid. Recarbonation is the final stage in the lime–soda ash softening process. This process converts carbonate ions to bicarbonate ions and stabilizes the solution against the precipitation of carbonate compounds.

receiver   A device that indicates the result of a measurement, usually using either a fixed scale and movable indicator (pointer), such as a pressure gauge, or a moving chart with a movable pen like those used on a circular flow-recording chart. Also called an indicator.

recorder   A device that creates a permanent record, on a paper chart, magnetic tape, or in a computer, of the changes in a measured variable.

reducing agent   Any substance, such as base metal (iron) or the sulfide ion (S^{2-}), that will readily donate (give up) electrons. The opposite is an *oxidizing agent*.

reduction (re-DUCK-shun)   Reduction is a chemical reaction that results in the addition of hydrogen, removal of oxygen, or the addition of electrons to an element or compound. Under anaerobic conditions (no dissolved oxygen present), sulfur compounds are reduced to odor-producing hydrogen sulfide (H_2S) and other compounds. In the treatment of metal finishing wastewaters, hexavalent chromium (Cr^{6+}) is reduced to the trivalent form (Cr^{3+}). The opposite of *oxidation*.

reference   A physical or chemical quantity whose value is known exactly, and thus is used to calibrate instruments or standardize measurements. Also called a standard.

regulatory negotiation   A process whereby the US Environmental Protection Agency acts on an equal basis with outside parties to reach consensus on the content of a proposed rule. If the group reaches consensus, the US EPA commits to propose the rule with the agreed upon content.

reliquefaction (re-lick-we-FACK-shun)   The return of a gas to the liquid state; for example, a condensation of chlorine gas to return it to its liquid form by cooling.

representative sample   A sample portion of material, water, or wastestream that is as nearly identical in content and consistency as possible to that in the larger body being sampled.

residual chlorine   The concentration of chlorine present in water after the chlorine demand has been satisfied. The concentration is expressed in terms of the total chlorine residual, which includes both the free and combined or chemically bound chlorine residuals. Also called chlorine residual.

residue   The dry solids remaining after the evaporation of a sample of water or sludge. Also see *total dissolved solids*.

resins   See *ion exchange resins*.

resistance   That property of a conductor or wire that opposes the passage of a current, thus causing electric energy to be transformed into heat.

respiration   The process in which an organism takes in oxygen for its life processes and gives off carbon dioxide.

responsibility   Answering to those above in the chain of command to explain how and why you have used your authority.

reverse osmosis (oz-MOE-sis) **(RO)**   The application of pressure to a concentrated solution, which causes the passage of a liquid from the concentrated solution to a weaker solution across a semipermeable membrane. The membrane allows the passage of the water (solvent) but not the dissolved solids (solutes). In the reverse osmosis process, two liquids are produced: (1) the reject (containing high concentrations of dissolved solids) and (2) the permeate (containing low concentrations). The clean water (permeate) is not always considered to be demineralized. Also see *osmosis*.

riparian (ri-PAIR-ee-an) **rights**   Water rights that are acquired together with title to the land bordering a source of surface water. The right to put to beneficial use surface water adjacent to your land.

RO   See *reverse osmosis (RO)*.

rodenticide (row-DENT-uh-side)   Any substance or chemical formulated to kill or control rodents.

roof leader   In plumbing, a pipe installed to drain water from the roof gutters or roof catchment to the storm drain or other means of disposal. Also called a conductor, downspout, or roof drain.

rotameter (ROTE-uh-ME-ter)   A device used to measure the flow rate of gases and liquids. The gas or liquid being measured flows vertically up a tapered, calibrated tube. Inside the tube is a small ball or bullet-shaped float (it may rotate) that rises or falls depending on the flow rate. The flow rate may be read on a scale behind or on the tube by looking at the middle of the ball or at the widest part or top of the float.

rotor   The rotating part of a machine. The rotor is surrounded by the stationary (nonmoving) parts (stator) of the machine.

routine sampling   Sampling repeated on a regular basis.

S

sacrificial anode   An easily corroded material deliberately installed in a pipe or tank. The intent of such an installation is to give up (sacrifice) this anode to corrosion while the water supply facilities remain relatively corrosion free.

Safe Drinking Water Act (SDWA)   An act passed by the US Congress in 1974. The act establishes a cooperative program among local, state, and federal agencies to ensure safe drinking water for consumers. The act has been amended several times, including the 1980, 1986, and 1996 amendments.

safety data sheet (SDS)   Safety data sheets (SDSs) are an essential component of the Globally Harmonized System of Classification and Labeling of Chemicals (GHS) and are intended to provide comprehensive information about a substance or mixture for use in workplace chemical management. They are used as a source of information about hazards, including environmental hazards, and to obtain advice on safety precautions. In the GHS, they serve the same function that the material safety data sheet (MSDS) does in OSHA's Hazard Communication Standard. The SDS is normally product related and not specific to the workplace; nevertheless, the information on an SDS enables the employer to develop an active program of worker protection measures, including training, which is specific to the workplace, and to consider measures necessary to protect the environment.

safe water   Water that does not contain harmful bacteria, or toxic materials or chemicals. Water may have taste and odor problems, color, and certain mineral problems and still be considered safe for drinking.

safe yield   The annual quantity of water that can be taken from a source of supply over a period of years without depleting the source permanently (beyond its ability to be replenished naturally in wet years).

salinity   (1) The relative concentration of dissolved salts, usually sodium chloride, in a given water. (2) A measure of the concentration of dissolved mineral substances in water.

sanitary survey   A detailed evaluation or inspection of a source of water supply and all conveyances, storage, treatment, and distribution facilities to ensure protection of the water supply from all pollution sources.

saprophytes (SAP-row-fights)   Organisms living on dead or decaying organic matter. They help natural decomposition of organic matter in water or wastewater.

saturation   The condition of a liquid (water) when it has taken into solution the maximum possible quantity of a given substance at a given temperature and pressure.

saturator (SAT-yoo-ray-tor)   A device that produces a fluoride solution for the fluoridation process. The device is usually a cylindrical container with granular sodium fluoride on the bottom. Water flows either upward or downward through the sodium fluoride to produce the fluoride solution.

SCADA (SKAY-dah) **system**   Supervisory Control And Data Acquisition system. A computer-monitored alarm, response, control, and data acquisition system used to monitor and adjust treatment processes and facilities.

scale   (1) A combination of mineral salts and bacterial accumulation that sticks to the inside of a collection pipe under certain conditions. Scale, in extreme growth circumstances, creates additional friction loss to the flow of water. Scale may also accumulate on surfaces other than pipes. (2) The marked plate against which an indicator or recorder reads, usually the same as the range of the measuring system. Also see *range*.

SCFM   Standard Cubic Feet per Minute. Cubic feet of air per minute at standard conditions of temperature, pressure, and humidity (0°C, 14.7 psia, and 50 percent relative humidity).

schedule, pipe   See *pipe schedule*.

schmutzdecke (shmoots-DECK-ee)   A layer of trapped matter at the surface of a slow sand filter in which a dense population of microorganisms develops. These microorganisms within the film or mat feed on and break down incoming organic material trapped in the mat. In doing so, the microorganisms both remove organic matter and add mass to the mat, further developing the mat and increasing the physical straining action of the mat.

SDS   See *safety data sheet (SDS)*.

SDWA   See *Safe Drinking Water Act (SDWA)*.

Secchi (SECK-key) **disc**   A flat disc approximately 8 inches (20 cm) in diameter with alternating black and white quarters that is lowered into the water by a rope until it is just barely visible. At this point, the depth of the disc from the water surface is the recorded Secchi disc transparency.

sedimentation (SED-uh-men-TAY-shun)   The process of settling and depositing of suspended matter carried by water or wastewater. Sedimentation usually occurs by gravity when the velocity of the liquid is reduced below the point at which it can transport the suspended material.

sedimentation (SED-uh-men-TAY-shun) **basin**   A tank or basin in which water or wastewater is held for a period of time during which the heavier solids settle to the bottom and the lighter materials float to the surface. Also called settling tank. Also see *clarifier*.

seizing or seize up   Seizing occurs when an engine overheats and a part expands to the point where the engine will not run. Also called freezing.

sensitivity   The smallest change in a process variable that an instrument can sense.

sensitivity (particle counters)   The smallest particle a particle counter will measure and count reliably.

sensor   A device that measures (senses) a physical condition or variable of interest. Floats and thermocouples are examples of sensors. Also called a primary element.

septic (SEP-tick)   A condition produced by bacteria when all oxygen supplies are depleted. If severe, the bottom deposits produce hydrogen sulfide, the deposits and water turn black, give off foul odors, and the water has a greatly increased chlorine demand.

sequestration (SEE-kwes-TRAY-shun)   A chemical complexing (forming or joining together) of metallic cations (such as iron) with certain inorganic compounds, such as phosphate. Sequestration prevents the precipitation of the metals (iron). Also see *chelation*.

service pipe   The pipeline extending from the water main to the building served or to the consumer's system.

set point   The position at which the control or controller is set. This is the same as the desired value of the process variable. For example, a thermostat is set to maintain a desired temperature.

sewage   The used household water and water-carried solids that flow in sewers to a wastewater treatment plant. The preferred term is *wastewater*.

sheave   V-belt drive pulley, which is commonly made of cast iron or steel.

shim   Thin metal sheets that are inserted between two surfaces to align or space the surfaces correctly. Shims can be used anywhere a spacer is needed. Usually shims are 0.001 to 0.020 inch (0.025 to 0.50 mm) thick.

shock load   The arrival at a water treatment plant of raw water containing unusual amounts of algae, colloidal matter, color, suspended solids, turbidity, or other pollutants.

short-circuiting   A condition that occurs in tanks or basins when some of the flowing water entering a tank or basin flows along a nearly direct pathway from the inlet to the outlet. This is usually undesirable because it may result in shorter contact, reaction, or settling times in comparison with the theoretical (calculated) or presumed detention times.

signal word   A single word used to indicate the relative level of severity of a chemical hazard and to alert the reader to a potential hazard on the label. The signal words are "Danger" (used for the more severe hazards) and "Warning" (used for less severe hazards).

simulate   To reproduce the action of some process, usually on a smaller scale.

single-stage pump A pump that has only one impeller. A multistage pump has more than one impeller.

slake To mix with water so that a true chemical combination (hydration) takes place, such as in the slaking of lime.

slaked lime See *hydrated lime*.

slope The slope or inclination of a trench bottom or a trench side wall is the ratio of the vertical distance to the horizontal distance or rise over run. Also see *grade* (2).

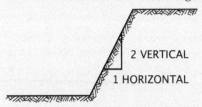

2:1 SLOPE (OR 2 IN 1 SLOPE)

sludge (SLUJ) (1) The settleable solids separated from liquids during processing. (2) The deposits of foreign materials on the bottoms of streams or other bodies of water or on the bottoms and edges of wastewater collection lines and appurtenances. Also referred to as biosolids. However, biosolids typically refers to treated waste.

slurry A watery mixture or suspension of insoluble (not dissolved) matter; a thin, watery mud or any substance resembling it (such as a grit slurry or a lime slurry).

SMCL Secondary Maximum Contaminant Level. Secondary MCLs for various water quality indicators are established to protect public welfare.

SNARL Suggested No Adverse Response Level. The concentration of a chemical in water that is expected not to cause an adverse health effect.

software program Computer program; the list of instructions that tells a computer how to perform a given task or tasks. Some software programs are designed and written to monitor and control treatment processes.

soft water Water having a low concentration of calcium and magnesium ions. According to US Geological Survey guidelines, soft water is water having a hardness of 60 milligrams per liter or less.

solenoid (SO-luh-noid) A magnetically operated mechanical device (electric coil). Solenoids can operate small valves or electric switches.

solution A liquid mixture of dissolved substances. In a solution it is impossible to see all the separate parts.

sounding tube A pipe or tube used for measuring the depths of water.

span The scale or range of values an instrument is designed to measure. Also see *range*.

specific capacity A measurement of well yield per unit depth of drawdown after a specific time has passed, usually 24 hours. Typically expressed as gallons per minute per foot or cubic meters per day per meter (gpm/ft or m/day/m).

specific capacity test A testing method used to determine the adequacy of an aquifer or well by measuring the specific capacity.

specific conductance A rapid method of estimating the total dissolved solids content of a water supply. The measurement indicates the capacity of a sample of water to carry an electric current, which is related to the concentration of ionized substances in the water. Also called conductance.

specific gravity (1) Weight of a particle, substance, or chemical solution in relation to the weight of an equal volume of water.

Water has a specific gravity of 1.000 at 39°F (4°C). Particulates with specific gravity less than 1.0 float to the surface and particulates with specific gravity greater than 1.0 sink. (2) Weight of a particular gas in relation to the weight of an equal volume of air at the same temperature and pressure (air has a specific gravity of 1.0). Chlorine gas has a specific gravity of 2.5.

specific yield The quantity of water that a unit volume of saturated permeable rock or soil will yield when drained by gravity. Specific yield may be expressed as a ratio or as a percentage by volume.

split sample A single grab sample that is separated into at least two parts such that each part is representative of the original sample. Often used to compare test results between field kits and laboratories or between two laboratories.

spoil Excavated material, such as soil, from the trench of a water main or sewer.

spore The reproductive body of certain organisms, which is capable of giving rise to a new organism either directly or indirectly. A viable (able to live and grow) body regarded as the resting stage of an organism. A spore is usually more resistant to disinfectants and heat than most organisms. Gangrene and tetanus bacteria are common spore-forming organisms.

springline The horizontal centerline of a pipe.

stale water Water that has not flowed recently and may have picked up tastes and odors from distribution lines or storage facilities.

standard A physical or chemical quantity whose value is known exactly, and thus is used to calibrate instruments or standardize measurements. Also called a reference.

standard deviation (S) A measure of the spread or dispersion of data or measurements. See math appendix for the procedure to calculate the standard deviation.

standardize To compare with a standard. (1) In wet chemistry, to find out the exact strength of a solution by comparing it with a standard of known strength. This information is used to adjust the strength by adding more water or more of the substance to be dissolved. (2) To set up an instrument or device to read a standard. This allows you to adjust the instrument so that it reads accurately, or enables you to apply a correction factor to the readings.

standard solution A solution in which the exact concentration of a chemical or compound is known.

starters (motor) Devices used to start up large motors gradually to avoid severe mechanical shock to a driven machine and to prevent disturbance to the electrical lines (causing dimming and flickering of lights).

static head When water is not moving, the vertical distance (in feet or meters) from a reference point to the water surface is the static head. Also see *dynamic head*, *dynamic pressure*, and *static pressure*.

static pressure The static pressure in psi (or kPa) is the static head in feet times 0.433 psi/ft (or meters × 9.81 kPa/m). Also see *dynamic head*, *dynamic pressure*, and *static head*.

static water depth The vertical distance in feet (or meters) from the centerline of the pump discharge down to the surface level of the free pool while no water is being drawn from the pool or water table.

static water level (1) The elevation or level of the water table in a well when the pump is not operating. (2) The level or elevation to which water would rise in a tube connected to an artesian aquifer, basin, or conduit under pressure.

stator   That portion of a machine that contains the stationary (nonmoving) parts that surround the moving parts (rotor).

sterilization (STAIR-uh-luh-ZAY-shun)   The removal or destruction of all microorganisms, including pathogens and other bacteria, vegetative forms, and spores. Also see *disinfection*.

stethoscope   An instrument used to magnify sounds and carry them to the ear.

storativity (S)   The volume of groundwater an aquifer releases from or takes into storage per unit surface area of the aquifer per unit change in head. Also called the storage coefficient.

storm runoff   Water that flows over the ground surface directly into streams, rivers, or lakes. Also called direct runoff.

stormwater   The excess water running off from the surface of a drainage area during and immediately after a period of rain. Also see *storm runoff*.

stratification (STRAT-uh-fuh-KAY-shun)   The formation of separate layers (of temperature, plant life, or animal life) in a lake or reservoir. Characteristics within each layer are similar; for instance, all water in the same layer has the same temperature. Also see *thermal stratification*.

stray current corrosion   A corrosion activity resulting from stray electric current originating from some source outside the plumbing system such as DC grounding on phone systems.

submergence   The distance between the water surface and the media surface in a filter.

subsidence (sub-SIDE-ence)   The dropping or lowering of the ground surface as a result of removing excess water (overdraft or overpumping) from an aquifer. After excess water has been removed, the soil will settle, become compacted, and the ground surface will drop, which can cause the settling of underground utilities.

suction head   The positive pressure in feet (meters) of water or pounds per square inch (kilograms per square centimeter) of mercury vacuum on the suction side of a pump. The pressure can be measured from the centerline of the pump up to the elevation of the hydraulic grade line on the suction side of the pump.

suction lift   The negative pressure in feet (meters) of water or inches (centimeters) of mercury vacuum on the suction side of a pump. The pressure can be measured from the centerline of the pump down to (lift) the elevation of the hydraulic grade line on the suction side of the pump.

superchlorination (SOO-per-KLOR-uh-NAY-shun)   Chlorination with doses that are deliberately selected to produce free or combined residuals so large as to require dechlorination.

supernatant (soo-per-NAY-tent)   The relatively clear water layer between the sludge on the bottom and the scum on the surface of a basin or container. Also called clear zone.

supersaturated   An unstable condition of a solution (water) in which the solution contains a substance at a concentration greater than the saturation concentration for the substance.

surface loading rate   One of the guidelines for the design of settling tanks and clarifiers in treatment plants. Used by operators to determine if tanks and clarifiers are hydraulically (flow) over- or underloaded. Also called overflow rate or surface settling rate.

$$\text{Surface Loading, Rate} = \frac{\text{Flow}}{\text{Surface Area}}$$

surfactant (sir-FAC-tent)   Abbreviation for surface-active agent. The active agent in detergents that possesses a high cleaning ability.

surge chamber   A chamber or tank connected to a pipe and located at or near a valve that may quickly open or close or a pump that may suddenly start or stop. When the flow of water in a pipe starts or stops quickly, the surge chamber allows water to flow into or out of the pipe and minimize any sudden positive or negative pressure waves or surges in the pipe. [See drawing on page 792.]

suspended solids   (1) Solids that either float on the surface or are suspended in water or wastewater, and that are largely removable by laboratory filtering. (2) The quantity of material removed from water or wastewater in a laboratory test, as prescribed in *Standard Methods for the Examination of Water and Wastewater*, and referred to as Total Suspended Solids Dried at 103–105°C.

T

tagout   The placement of a tagout device on an energy-isolating device, in accordance with an established procedure, to indicate that the energy-isolating device and the equipment being controlled may not be operated until the tagout device is removed.

tagout device   A prominent warning device, such as a tag and a means of attachment, that can be securely fastened to an energy-isolating device in accordance with an established procedure to indicate that the energy-isolating device and the equipment being controlled may not be operated until the tagout device is removed.

tailgate safety meeting   Brief (10–20 minutes) safety meetings held every 7–10 working days. The term comes from the safety meetings regularly held by the construction industry around the tailgate of a truck.

TCE   See *trichloroethane (TCE)*.

TDS   See *total dissolved solids (TDS)*.

telemetry (tel-LEM-uh-tree)   The electrical link between a field transmitter and the receiver. Telephone lines are commonly used to serve as the electrical line.

temperature sensor   A device that opens and closes a switch in response to changes in the temperature. This device might be a metal contact, or a thermocouple that generates a minute electric current proportional to the difference in heat, or a variable resistor whose value changes in response to changes in temperature. Also called a heat sensor.

thermal stratification (strat-uh-fuh-KAY-shun)   The formation of layers of different temperatures in a lake or reservoir. Also see *stratification*.

thermocline (THUR-moe-kline)   The middle layer in a thermally stratified lake or reservoir. In this layer there is a rapid decrease in temperature with depth. Also called the metalimnion.

thermocouple   A heat-sensing device made of two conductors of different metals joined together. An electric current is produced when there is a difference in temperature between the ends.

thickening   Treatment to remove water from the sludge mass to reduce the volume that must be handled.

THM   See *trihalomethanes (THMs)*.

THM precursor   See *precursor, THM*.

threshold odor   The minimum odor of a gas or water sample that can just be detected after successive dilutions with odorless gas or water. Also called odor threshold.

threshold odor number (TON)   The greatest dilution of a sample with odor-free water that still yields a just-detectable odor.

thrust block   A mass of concrete or similar material appropriately placed around a pipe to prevent movement when the pipe is carrying water. Usually placed at bends and valve structures.

tide gate   A gate installed at the end of a drain or outlet pipe to prevent the backward flow of water or wastewater. Generally used on storm sewer outlets into streams to prevent backward flow during times of flood or high tide. Also called a backwater gate. Also see *check valve* and *flap gate*.

time lag   The time required for processes and control systems to respond to a signal or to reach a desired level.

timer   A device for automatically starting or stopping a machine or other device at a given time.

titrate (TIE-trate)   To titrate a sample, a chemical solution of known strength is added drop by drop until a certain color change, precipitate, or pH change in the sample is observed (endpoint). Titration is the process of adding the chemical reagent in small increments (0.1–1.0 milliliter) until completion of the reaction, as signaled by the endpoint.

topography (toe-PAH-gruh-fee)   The arrangement of hills and valleys in a geographic area.

total chlorine   The total concentration of chlorine in water, including the combined chlorine (such as inorganic and organic chloramines) and the free available chlorine.

total chlorine residual   The total amount of chlorine residual (including both free chlorine and chemically bound chlorine) present in a water sample after a given contact time.

total dissolved solids (TDS)   All of the dissolved solids in a water. TDS is measured on a sample of water that has passed through a very fine mesh filter to remove suspended solids. The water passing through the filter is evaporated and the residue represents the total dissolved solids. Also see *specific conductance*.

total dynamic head (TDH)   When a pump is lifting or pumping water, the vertical distance (in feet or meters) from the elevation of the energy grade line on the suction side of the pump to the elevation of the energy grade line on the discharge side of the pump. The total dynamic head is the static head plus pipe friction losses.

totalizer   A device or meter that continuously measures and sums a process rate variable in cumulative fashion over a given time period. For example, total flows displayed in gallons per minute, million gallons per day, cubic feet per second, or some other unit of volume per time period. Also see *integrator*.

total organic carbon (TOC)   A measure of the amount of organic carbon in water.

toxaphene (TOX-uh-feen)   A chemical that causes adverse health effects in domestic water supplies and is toxic to freshwater and marine aquatic life.

toxic   A substance that is poisonous to a living organism. Toxic substances may be classified in terms of their physiological action, such as irritants, asphyxiants, systemic poisons, and anesthetics and narcotics. Irritants are corrosive substances that attack the mucous membrane surfaces of the body. Asphyxiants interfere with breathing. Systemic poisons are hazardous substances that injure or destroy internal organs of the body. Anesthetics and narcotics are hazardous substances that depress the central nervous system and lead to unconsciousness.

toxic substance   See *harmful physical agent*.

transducer (trans-DUE-sir)   A device that senses some varying condition measured by a primary sensor and converts it to an electrical or other signal for transmission to some other device (a receiver) for processing or decision making.

transmission lines   Pipelines that transport raw water from its source to a water treatment plant. After treatment, water is usually pumped into pipelines (transmission lines) that are connected to a distribution grid system.

transmissivity (TRANS-miss-SIV-it-tee)   A measure of the ability to transmit (as in the ability of an aquifer to transmit water).

transpiration (TRAN-spur-RAY-shun)   The process by which water vapor is released to the atmosphere by living plants. This process is similar to people sweating. Also called evapotranspiration.

tremie (TREH-me)   A device used to place concrete or grout under water.

trichloroethane (TCE) (try-KLOR-o-ETH-hane)   An organic chemical used as a cleaning solvent that causes adverse health effects in domestic water supplies.

trihalomethanes (THMs) (tri-HAL-o-METH-hanes)   Derivatives of methane, CH_4, in which three halogen atoms (chlorine or bromine) are substituted for three of the hydrogen atoms. Often formed during chlorination by reactions with natural organic materials in the water. The resulting compounds (THMs) are suspected of causing cancer.

true color   Color of the water from which turbidity has been removed.

TU   See *turbidity units (TU)*.

tubercle (TOO-burr-kull)   A crust of corrosion products (rust) that builds up over a pit caused by the loss of metal due to corrosion.

tuberculation (too-BURR-kyoo-LAY-shun)   The development or formation of small mounds of corrosion products (rust) on the inside of iron pipe. These mounds (tubercles) increase the roughness of the inside of the pipe thus increasing resistance to water flow (decreases the C Factor).

tube settler   A device that uses bundles of small-bore (2 to 3 inches or 50 to 75 mm) tubes installed on an incline as an aid to sedimentation. The tubes may come in a variety of shapes including circular and rectangular. As water rises within the tubes, settling solids fall to the tube surface, and as the resulting sludge gains weight, it moves down the tubes and settles to the bottom of the basin for removal by conventional sludge collection means. Also called high-rate settlers.

turbid   Having a cloudy or muddy appearance.

turbidimeter   See *turbidity meter*.

turbidity (ter-BID-it-tee)   The cloudy appearance of water caused by the presence of suspended and colloidal matter. In the waterworks field, a turbidity measurement is used to indicate the clarity of water. Technically, turbidity is an optical property of the water based on the amount of light reflected by suspended particles. Turbidity cannot be directly equated to suspended solids because white particles reflect more light than dark-colored particles and many small particles will reflect more light than an equivalent large particle.

turbidity (ter-BID-it-tee) **meter**   An instrument for measuring and comparing the turbidity of liquids by passing light through them and determining how much light is reflected by the particles in the liquid. The normal measuring range is 0 to 100 and is expressed as nephelometric turbidity units (NTUs). Also called a turbidimeter.

turbidity (ter-BID-it-tee) **units (TU)**   Turbidity units are a measure of the cloudiness of water. If measured by a nephelometric (deflected light) instrumental procedure, turbidity units are expressed in nephelometric turbidity units (NTU) or simply TU.

turn-down ratio   The ratio of the design range to the range of acceptable accuracy or precision of an instrument. Also see *effective range*.

U

UEL   See *upper explosive limit (UEL)*.

UF   See *ultrafiltration (UF)*.

ultrafiltration (UF)   A pressure-driven membrane filtration process that separates particles down to approximately 0.01 μm diameter from influent water using a sieving process.

ultraviolet (UV)   Pertaining to a band of electromagnetic radiation just beyond the visible light spectrum. Ultraviolet radiation is used in water treatment to disinfect the water. When ultraviolet radiation is absorbed by the cells of microorganisms, it damages the genetic material in such a way that the organisms are no longer able to grow or reproduce, thus ultimately killing them.

unconsolidated formation   A sediment that is loosely arranged or unstratified (not in layers) or whose particles are not cemented together (soft rock); occurring either at the ground surface or at a depth below the surface. Also see *consolidated formation*.

uniformity coefficient (UC)   The ratio of (1) the diameter of a grain (particle) of a size that is barely too large to pass through a sieve that allows 60 percent of the material (by weight) to pass through to (2) the diameter of a grain (particle) of a size that is barely too large to pass through a sieve that allows 10 percent of the material (by weight) to pass through. The resulting ratio is a measure of the degree of uniformity in a granular material, such as filter media.

$$\text{Uniformity Coefficient} = \frac{\text{Particle Diameter}_{60\%}}{\text{Particle Diameter}_{10\%}}$$

upper explosive limit (UEL)   The highest concentration of a gas or vapor (percentage by volume in air) above which a flame will not spread in the presence of an ignition source (arc, flame, or heat). Concentrations higher than UEL are "too rich" to burn. Also called upper flammable limit (UFL). Also see *lower explosive limit (LEL)*.

upper flammable limit (UFL)   See *upper explosive limit (UEL)*.

US EPA   United States Environmental Protection Agency. A regulatory agency established by the US Congress to administer the nation's environmental laws.

UV   See *ultraviolet (UV)*.

V

variable, measured   A factor (flow, temperature) that is sensed and quantified (reduced to a reading of some kind) by a primary element or sensor.

variable, process   A physical or chemical quantity that is usually measured and controlled in the operation of a water, wastewater, or industrial treatment plant.

variable costs   Costs that a utility must cover or pay that are associated with the actual treatment and delivery of water. These costs vary or fluctuate on the basis of the volume of water treated and delivered to customers (water production). Also see *fixed costs*.

variable frequency drive   A control system that allows the frequency of the current applied to a motor to be varied. The motor is connected to a low-frequency source while standing still; the frequency is then increased gradually until the motor and pump (or other driven machine) are operating at the desired speed.

velocity head   The energy in flowing water as determined by a vertical height (in feet or meters) equal to the square of the velocity of flowing water divided by twice the acceleration due to gravity ($V^2/2g$).

venturi (ven-TOOR-ee) **meter**   A flow measuring device placed in a pipe. The device consists of a tube whose diameter gradually decreases to a throat and then gradually expands to the diameter of the pipe. The flow is determined on the basis of the difference in pressure (caused by different velocity heads) between the entrance and throat of the Venturi meter. *NOTE:* Most Venturi meters have pressure sensing taps rather than a manometer to measure the pressure difference. The upstream tap is the high pressure tap or side of the manometer.

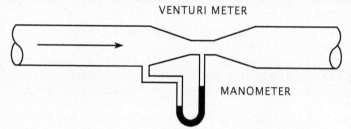

VENTURI METER

MANOMETER

viscosity (vis-KOSS-uh-tee)   A property of water, or any other fluid, that resists efforts to change its shape or flow. Syrup is more viscous (has a higher viscosity) than water. The viscosity of water increases significantly as temperatures decrease. Motor oil is rated by how thick (viscous) it is; 20-weight oil is considered relatively thin while 50-weight oil is relatively thick or viscous.

void   A pore or open space in rock, soil, or other granular material, not occupied by solid matter. The pore or open space may be occupied by air, water, or other gaseous or liquid material. Also called an interstice, pore, or void space.

volatile (VOL-uh-tull)   (1) A volatile substance is one that is capable of being evaporated or changed to a vapor at relatively low temperatures. Volatile substances can be partially removed from water or wastewater by the air stripping process. (2) In terms of solids analysis, volatile refers to materials lost (including most organic matter) upon ignition in a muffle furnace for 60 minutes at 1,022°F (550°C). Natural volatile materials are chemical substances usually of animal or plant origin. Manufactured or synthetic volatile materials, such as plastics, ether, acetone, and carbon tetrachloride, are highly volatile and not of plant or animal origin. Also see *nonvolatile matter*.

volatile acids   Fatty acids produced during digestion that are soluble in water and can be steam-distilled at atmospheric pressure. Also called organic acids. Volatile acids are commonly reported as equivalent to acetic acid. Important volatile acids include acetic, propionic, and butyric, which are all important in the biological removal of phosphorus.

volatile liquids   Liquids that easily vaporize or evaporate at room temperature.

volatile solids   Those solids in water, wastewater, or other liquids that are lost on ignition of the dry solids at 1,022°F (550°C). Also called organic solids and volatile matter.

voltage   The electrical pressure available to cause a flow of current (amperage) when an electric circuit is closed. Also see *electromotive force (EMF)*.

volumetric   A measurement based on the volume of some factor. Volumetric titration is a means of measuring unknown concentrations of water quality indicators in a sample by determining the volume of titrant or liquid reagent needed to complete particular reactions.

volumetric feeder   A dry chemical feeder that delivers a measured volume of chemical during a specific time period.

vortex   A revolving mass of water that forms a whirlpool. This whirlpool is caused by water flowing out of a small opening in the bottom of a basin or reservoir. A funnel-shaped opening is created downward from the water surface.

W

warning   This word is used to indicate a hazard level between "Caution" and "Danger." Also see *caution, danger,* and *notice.*

wastewater   A community's used water and water-carried solids (including used water from industrial processes) that flow to a treatment plant. Stormwater, surface water, and groundwater infiltration also may be included in the wastewater that enters a wastewater treatment and reuse plant. Also called sewage.

water audit   A thorough examination of the accuracy of water agency records or accounts (volumes of water) and system control equipment. Water managers can use audits to determine their water distribution system efficiency. The overall goal is to identify and verify water and revenue losses in a water system.

water cycle   The process of evaporation of water into the air and its return to earth by precipitation (rain or snow). This process also includes transpiration from plants, groundwater movement, and runoff into rivers, streams, and the ocean. Also called the hydrologic cycle.

water hammer   The sound like someone hammering on a pipe that occurs when a valve is opened or closed very rapidly. When a valve position is changed quickly, the water pressure in a pipe will increase and decrease back and forth very quickly. This rise and fall in pressures can cause serious damage to the system.

water purveyor   (purr-VAY-or)   An agency or person that supplies water (usually potable water).

watershed   The region or land area that contributes to the drainage or catchment area above a specific point on a stream or river.

water table   The upper surface of the zone of saturation of groundwater in an unconfined aquifer.

water table head   See *pressure head.*

watt   A unit of power equal to one joule per second. The power of a current of one ampere flowing across a potential difference of one volt.

weir   (WEER)   (1) A wall or plate placed in an open channel and used to measure the flow of water. The depth of the flow over the weir can be used to calculate the flow rate, or a chart or conversion table may be used to convert depth to flow. Also see *proportional weir.* (2) A wall or obstruction used to control flow (from settling tanks and clarifiers) to ensure a uniform flow rate and avoid short-circuiting.

weir   (WEER)   **diameter**   Many circular clarifiers have a circular weir within the outside edge of the clarifier. All the water leaving the clarifier flows over this weir. The diameter of the weir is the length of a line from one edge of a weir to the opposite edge and passing through the center of the circle formed by the weir.

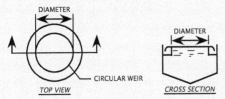

weir   (WEER)   **loading rate**   A guideline used to determine the length of weir needed on settling tanks and clarifiers in treatment plants. Used by operators to determine if weirs are hydraulically (flow) overloaded.

$$\text{Weir Loading Rate} = \frac{\text{Flow}}{\text{Length of Weir}}$$

wellhead protection area (WHPA)   The surface and subsurface area surrounding a public water system water well or well field, through which contaminants are reasonably likely to move toward and reach such water well or well field. Also see *well isolation zone.*

well isolation zone   A surface or zone with restricted land uses surrounding a public water system water well or well field. The zone is established to prevent contaminants from a nonpermitted land use to move toward and reach the water well or well field. Also see *wellhead protection area.*

well log   A record of the thickness and characteristics of the soil, rock, and water-bearing formations encountered during the drilling (sinking) of a well.

wet chemistry   Laboratory procedures used to analyze a sample of water using liquid chemical solutions (wet) instead of, or in addition to, laboratory instruments.

wholesome water   Water that is safe and palatable for human consumption.

wire-to-water efficiency   The combined efficiency of a pump and motor together. Also called the overall efficiency.

wye strainer   A screen shaped like the letter Y. Water flows through the upper parts of the Y and the debris is trapped by the screen at the fork.

Y

yield   The quantity of water (expressed as a rate of flow—gpm, gph, gpd, m/day, ML/day, or total quantity per year) that can be collected for a given use from surface or groundwater sources. The yield may vary with the use proposed, with the plan of development, and with economic considerations. Also see *safe yield.*

Z

zeolite   A type of ion exchange material used to soften water. Natural zeolites are siliceous compounds (made of silica) that remove calcium and magnesium from hard water and replace them with sodium. Synthetic or organic zeolites are ion exchange materials that remove calcium or magnesium and replace them with either sodium or hydrogen. Manganese zeolites are used to remove iron and manganese from water.

zeta potential   In coagulation and flocculation procedures, the difference in the electrical charge between the dense layer of ions surrounding the particle and the charge of the bulk of the suspended fluid surrounding this particle. The zeta potential is usually measured in millivolts.

zone of aeration   The comparatively dry soil or rock located between the ground surface and the top of the water table.

zone of saturation   (1) The soil or rock located below the top of the groundwater table. By definition, the zone of saturation is saturated with water. Also see *water table.* (2) Where raw wastewater is exfiltrating from a sewer pipe, the area of soil that is moistened around the leak point is often called the zone of saturation.

zooplankton   (ZOE-uh-PLANK-ton)   Small, usually microscopic animals (such as protozoans), found in lakes and reservoirs.

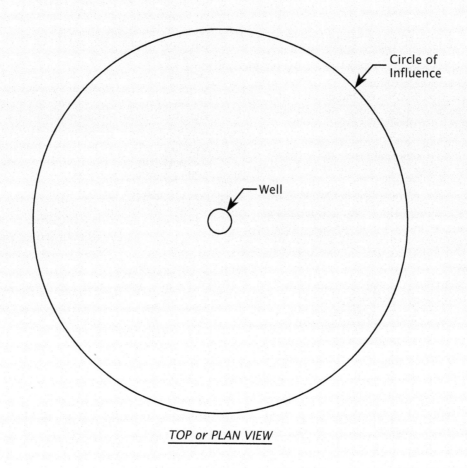

TOP or PLAN VIEW

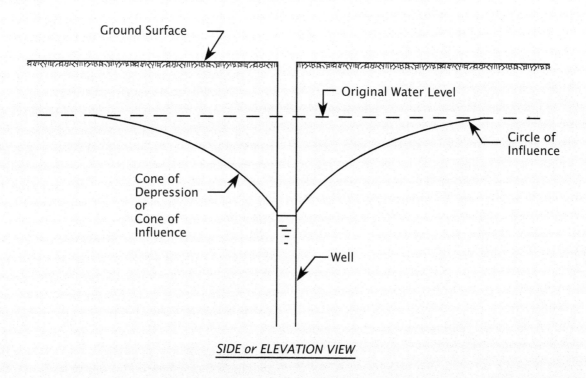

SIDE or ELEVATION VIEW

Circle of influence and cone of depression/cone of influence

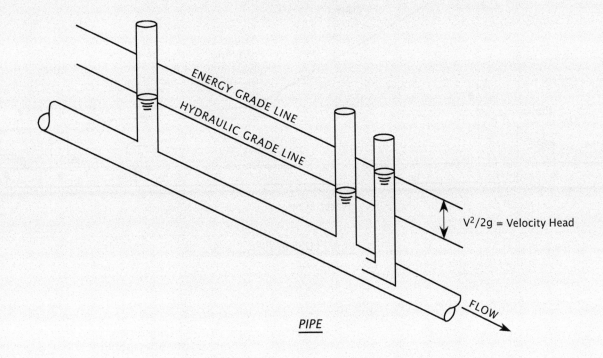

$V^2/2g$ = Velocity Head

FLOW

PIPE

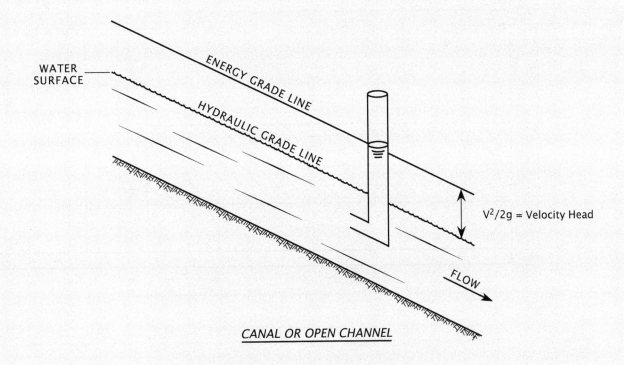

WATER
SURFACE

ENERGY GRADE LINE

HYDRAULIC GRADE LINE

$V^2/2g$ = Velocity Head

FLOW

CANAL OR OPEN CHANNEL

Energy grade line and hydraulic grade line

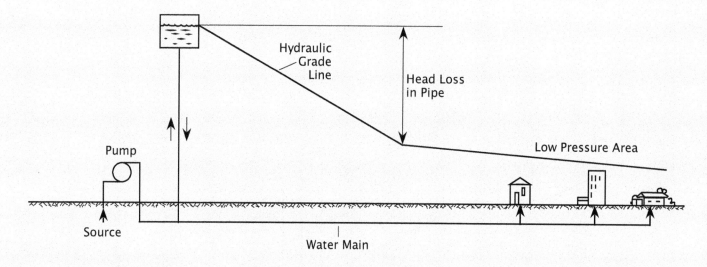

HEAD LOSS IN PIPE

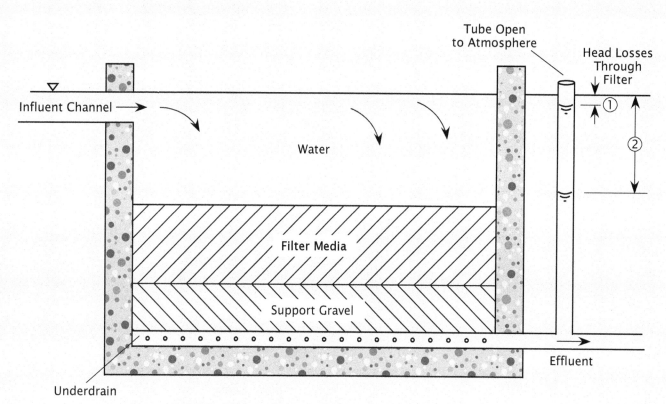

Note: If a tube, open to the atmosphere, were installed in
the filter effluent, then:
① = Head loss through filter at start of run
② = Head loss through filter before start of
backwash cycle

HEAD LOSS THROUGH FILTER

Head loss

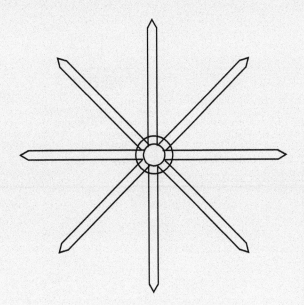

PLAN VIEW OF COLLECTOR PIPES

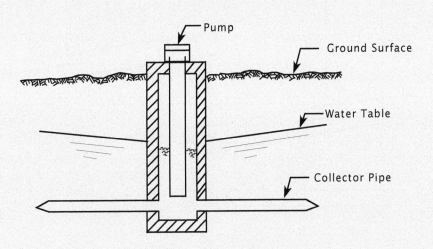

ELEVATION VIEW

Ranney collector

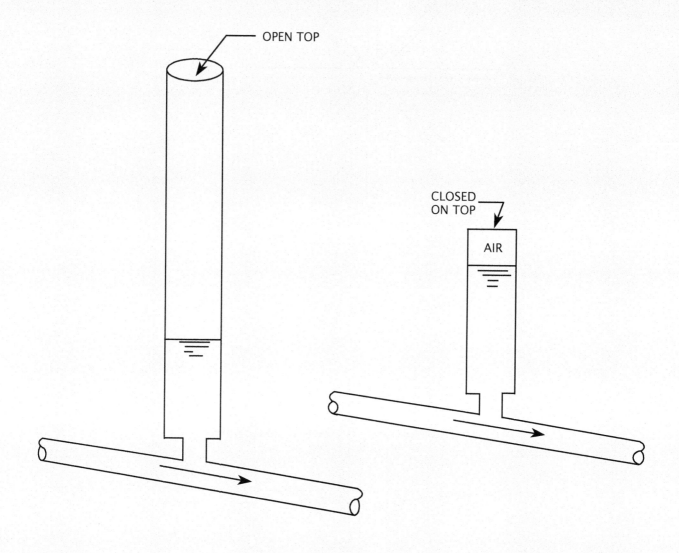

Types of surge chambers

Logarithm, 457–459
Logging, 73
Long Term 1 Enhanced Surface Water
    Treatment Rule (LT1ESWTR), 307
Long Term 2 Enhanced Surface Water
    Treatment Rule (LT2ESWTR), 307

M

Mains, disinfection, 351
Maintenance
    ballasted flocculation, 196
    budget, 38
    chlorination, 368, 385, 402
    chlorine dioxide, 396
    coagulation, 170
    costs, 38, 44
    dams, 108
    emergency repairs, 38, 44
    equipment, 42
    filtration, 284, 294, 304
    flocculation, 170
    inlet structures, 116
    inspection, 42
    instructions, 38
    intake screens, 106–107, 112–117
    intake structures, 119–120
    management, records, 43
    odor control, 497, 503, 518
    operator's duties, 38
    ozone, 420–422
    particle counter, 316–320
    planning, 38
    preventive, 38
    program, 38
    records, 36
    reservoirs, 108
    routine, 42
    scheduling, 38
    sedimentation, 252–253
    solids-contact, 253
    spare parts management, 43–44
    taste and odor control, 497, 503, 518
    tools, 42
    training, 44–45
    ultraviolet (UV) systems, 414–417
    work orders, 44
Management
    energy conservation, 46
    maintenance, 38
    records, 43
Manganese, 7
Manifold, 359, 390
Manual sampling, 624
Manual sludge removal, 240
Marble Test, 448, 456, 463
Mass concentration, 545
Material examination, corrosion, 453
Math assignment
    basic math overview, 53
    chlorination, 431

coagulation, 199
corrosion control, 482
disinfection, 431
filtration, 320
flocculation, 199
laboratory procedures, 629
sedimentation, 255
source water, reservoir management,
    and intake structures, 121
taste and odor control, 528
Maximum contaminant level (MCL)
    (primary standards), 5, 6, 327
    organic chemicals, 306
MBAS (foaming agents), 7
m-ColiBlue24® Test, 589, 618
Mean flow velocity, 225
Mechanical equipment, safety, 50
Mechanical mixing, 105, 132
Mechanical rakes, 240
Mechanical signal transmission, 29
Mechanical sludge removal, 240,
    241, 247
Media
    expansion, 284
    filtration, 8, 260, 262, 265, 268, 283,
        284, 294
    loss, 280, 283, 284, 291, 294, 297, 298
    preparation and storage, 593–594
Membrane filter
    counting colonies, 612
    inoculation, 608
    need, 591
Membrane filter (MF) method, 589,
    607–611
Membrane filter method–E. coli (EPA
    Method), 589, 591, 613–614
Membrane filter procedures modifications,
    611–612
Meniscus, 542
Mercury, 5, 408, 417, 553
Mesotrophic, 78
Metalimnion, 75, 76, 83, 85
Metallic salts, 128
Metering, chlorine, 356, 361
Methoxychlor, 6
Methylisoborneol (MIB), 81, 489, 527
Methyl orange alkalinity, 94
Metric system, 534–536
MIB. See Methylisoborneol (MIB)
Microbes, 590
Microbial populations, 491, 492
Microbial standards, 7, 327
Microbial testing, materials, 592
Micron, 129
Microorganisms, 330
Microprocessors, 31
Microscopic organisms controlling,
    101
Mining, 88, 90
Mixed media, 263, 268
Mixed-oxidant disinfection,
    423

Mixing
    baffles, 131, 133, 135
    blenders, pumped, 132
    chemicals, 130, 131
    chlorine, 345
    coagulation, 130, 132
    diffusers, 131
    energy consumption, 131
    flash, 126, 131, 132, 139
    flocculation, 130–133, 139
    grid systems, 132
    hydraulic, 105, 131
    mechanical, 105, 131
    paddles, 132, 133
    propellers, 132, 133
    pumped blenders, 131, 132
    reservoirs, 102
    throttling valves, 131
    turbines, 132, 133
    valves, 131
Mole, 546
Molecular weight, 129
Molecule, 440
Monitoring
    algae in reservoirs, 101
    chlorination, 326
    coagulation, 138, 140
    control systems, 534
    disinfection, 328
    drinking water, 25, 27, 28
    filtration, 278, 282, 284, 294, 298, 303
    flocculation, 138, 140
    instrumentation, 28
    reservoirs, 101, 109, 117
    rivers, 497
    sedimentation, 244, 247
    taste and odor control, 497, 502
Monochloramine, 339, 573
Monomers, 129
Monomictic, 75, 83
Mosquitos, 109
Most Probable Number (MPN) procedure,
    589, 594–607
Motors, electric, 46
Mouth-to-mouth resuscitation, 405
MPN Index, 601, 605
Mudballs, 270, 272, 280, 283, 294, 297,
    298, 304
Multilevel intake-outlet structures, 79, 114
Multi-media filter, 263, 268, 269
Multiple Tube Fermentation Method (MPN
    Procedure), 589, 590, 594–607
Municipal wastes, 495
Musty tastes and odors, 81, 494, 497, 503, 527

N

National Pollutant Discharge Elimination
    System (NPDES) Permit, 27, 480,
    495
Natural disasters, 35–36

Index